2002

LAND DEVELOPMENT FOR CIVIL ENGINEERS

LAND DEVELOPMENT FOR CIVIL ENGINEERS

SECOND EDITION

T. R. Dion, P.E. & L.S., LLC

Thomas R. Dion, Member

Professor of Civil Engineering, The Citadel

JOHN WILEY & SONS, INC.

This publication is designed to provide accurate and authoritative information in regard to the subject matter covered. It is sold with the understanding that the publisher is not engaged in rendering professional services. If professional advice or other expert assistance is required, the services of a competent professional person should be sought.

Library of Congress Cataloging-in-Publication Data

Dion, Thomas R., 1946–
    Land development for civil engineers / by Thomas R. Dion.—2nd ed.
       p.   cm.
    ISBN 0-471-43500-7 (cloth : alk. paper)
       1. Civil engineering—Handbooks, manuals, etc.   2. Building sites—Handbooks, manuals, etc.   I. Title.
    TA151 .D56 2002
    624—dc21   2001026918

Printed in the United States of America.
10 9 8 7 6 5 4 3

To my wife, Susan, my sons, Tommy and Richard, and my family

CONTENTS

PREFACE TO THE SECOND EDITION

During the seven-year period following the publication of the premier edition of *Land Development for Civil Engineers*, many changes affecting the development of land have taken place. The areas where most of these changes occurred are in the fields of technology, regulation, and land planning. In most cases, these changes have been brought about where better utilization of land, resources, and effort (time and money) result. The focus of this second edition is to place emphasis on defining some of these changes that are most significant and then to provide the reader an opportunity to learn how they apply to land development.

In the technology area many new tools previously under development or locked in the domain of the government are now available to the general public and are commonly used today. For example, the explosive growth in use and reliance on the electronic "information highway" offered by the World Wide Web (WEB) has outstripped the capabilities of the early 1990s vintage computers where both hardware and software improvements have gone through several cycles. Computer networking has expanded user access and capability where electronic data today commonly flows in venues such as e-mail, drawing or document files, electronic spreadsheet files, and even files having multimedia formats including those supporting graphics, sound, and animation.

Another area where technology has driven change is in computer software offerings. With increased storage and computing capability being made available in desktop PC computers, software vendors have increased their offerings since the early 1990s in land development–related fields such as geographic information systems (GISs), surveying and design packages, and improved graphical data-handling capabilities such as the emergence of the triangulated irregular network (TIN).

Because information sharing saves time, eliminates confusion, and provides ready access, there is a trend being followed by software vendors to produce and offer integrated packages or "suites" of software modules (e.g., mapping, surveying, and civil design) that share common databases and conventions. This approach allows multiple users on networked systems to work as a multidisciplinary team, and even allows connected access through the WEB with some enterprise package offerings.

Other technological changes have occurred in instrumentation and application techniques pertaining to land surveying, photogrammetry, and mapping. The biggest boon has occurred in the giant increase of global positioning system (GPS) usage. Improvements in receiver technology have made GPS affordable to most surveyors. To this end, the federal government is in the process of upgrading its geodetic control network and has placed into operation a series of continuously operating reference stations (CORSs) that allow users the opportunity to have access to positional data at no cost just by downloading it over the Internet.

Changes in regulatory requirements have also been seen throughout the 1990s. Some of these changes have resulted from evolving technology changes. For example, the emergence of GIS applications has in some cases precipitated the requirement for design professionals to submit electronic drawing files as part of the review process. These files are then imported into the GIS database to update the system. Other changes that have evolved are the increased usage of GIS and on-line format applications for regulatory construction permit reviews, review status, and issuance.

One area of regulatory growth has occurred in the formation of stormwater utilities. Traditionally, most drainage systems were operated, maintained, and controlled by the municipal public works department until the early 1990s. Since that time, municipalities have formed separate stormwater utilities that raise operation and maintenance revenues from monthly fees specifically targeting storm runoff from properties.

Changes have also occurred in the area of land planning. In many parts of the United States buildable land is scarce and renewed interest in improving the use of the land that is available is an effort defined as the "new urbanism." This groundswell of activity is focused on limiting urban sprawl which resulted from land planning approaches that were practiced in the 1960s, 1970s, and 1980s. Neotraditional towns are being promoted as ways to make better use of available property, limit encroaching on areas more suited as conservation and natural areas, reduce vehicular dependence, promote pedestrian and bicycle activities, create a sense of neighborhood by providing a "central" open area, similar to a town park, and improve the streetscape by making use of back-lot access alleys.

What I have mentioned here are but a few of the many changes affecting the development of land that have evolved since I wrote the first edition. The text has been updated to reflect the most important changes that have occurred. Because this book is used in some colleges and universities as a course text, it is an important undertaking to prepare this second edition where students have access to current engineering practices and technologies. Likewise, it is anticipated that practicing engineers who use this book as a reference will find this refreshed presentation of value.

February 2002
Summerville, S.C.   Thomas R. Dion

PREFACE TO THE FIRST EDITION

I have had the distinct privilege over the past decade and a half of teaching a course on land development to various senior civil engineering students. This course, which builds on most basic civil engineering courses in the nonstructural area, includes the integrated design of roads, storm drainage systems, potable water systems, sanitary sewer systems, site layout, and regulatory agency conformance. The value of this classroom experience is that it allows students to be exposed to the design process prior to graduation and thus illustrates to them practical application of learned theory in an academic setting. Students interested in structural engineering also find this course helpful, because structures are located on improved sites with utilities, parking, and other types of infrastructure.

One major problem associated with teaching a course on land development has been the lack of an adequate textbook. As a result, I have attempted to set forth in a single volume information pertaining to land development that will not only introduce the prospective graduating engineer to practical integrated design applications but also enlighten the student that there is much to be learned after graduation.

It is emphasized that this book is not an end unto itself, and certainly it is not intended to be a substitute for the individual courses from which a foundation for the use of this book must be acquired. Therefore, application of the principles presented in this book can be applied only after adequate related course work has been completed in graphic science, land surveying, photogrammetry, transportation engineering, highway engineering, fluid mechanics, geotechnical engineering, environmental engineering, and engineering administration. Also, throughout the text numerous standards and procedures have been referenced. A prudent engineer should thoroughly review the complete source before making a determination of its applicability for a particular design application. The examples included in this book are provided only to illustrate a way to approach solving various problems. In many cases, other approaches can be used satisfactorily.

I feel that this book will be a valuable reference for practicing engineers because integrated design and analysis procedures are included. In addition, other persons involved professionally with various aspects of land development should find this book a useful reference.

June 1993
Summerville, S.C.

T. R. Dion

ACKNOWLEDGMENTS
TO THE SECOND EDITION

Because the development of land includes many different topics, the author is fortunate to have been offered suggestions and advice from a number of diverse people who reviewed portions of the new edition. To each, I am indebted, especially my colleagues at The Citadel, who again helped me in a variety of ways. They include Thomas J. Anessi, Kenneth P. Brannan, William J. Davis, Dennis J. Fallon, John A. Murden, Russell H. Stout, Sam Tongtoe, and Michael H. Woo.

I am also thankful for the following persons who have reviewed various sections of this book and have offered many valuable comments and suggestions: John F. Adams, U.S. Army Corps of Engineers; Thomas O. Barnwell, U.S. Environmental Protection Agency; Ed Baruth, American Water Works Association; William B. Beauchene, East Coast Development Corporation; Ronald E. Benson, Hole Montes & Associates; John E. Bonkoski, U.S. Department of Housing and Urban Development; Benjamin Bunting, Summerville Fire and Rescue Department; James E. Chellis, Chellis & Frampton; Marc E. Cherry, Berenyi; Christopher M. Childs, S.C. Department of Health and Environmental Control; Joseph Christie, Town of Summerville, S.C.; Charles Crosby, U.S. Army Corps of Engineers; Robert P. Dean, Building Systems Design, Inc.; Lori A. Duncan, U.S. Fish and Wildlife Service; George R. Foster, USDA Agricultural Research Service; John G. Frampton, Chellis & Frampton; Matthew J. Halter, Town of Summerville, S.C.; David G. Hansen, Dalhousie University; Lloyd Herman, Ohio University; Robert D. Horan, The Asphalt Institute; Stephen F. Hutchinson, East Coast Development Corporation; Mark A. Isaak, Post Buckley Schuh and Jernigan; Danny Johnson, S.C. Department of Natural Resources; John E. Keen, Land Surveyor's Workshops; John Kwist, Newton Builders; John A. Laflin, Purdue University; Lewis A. Lapine, South Carolina Geodetic Survey; Dennis A. Leverett, Palmetto Environmental & Site Consultants; Susan McClendon, Building Systems Design; J. P. Mohsen, University of Louisville; Joseph R. Molinaro, National Association of Home Builders; Kenneth Myers, Federal Highway Administration; Fazil T. Najafi, University of Florida; J. Frank Newham, Town of Mount Pleasant, S.C.; Jon C. Noble, CBCL Limited; Donald F. Nye, Olde Towne Mortgage; James P. Reilly, New Mexico State University; Kenneth G. Renard, USDA Agricultural Research Service; Joseph L. Rucker, S.C. Department of Health and Environmental Control; Antony M. Sease, Duany Plater-Zyberk & Company; Stephen W. Skelton, S.C. Department of Archives and History; Edward F. Straw, Insurance Services Office; Brent A. Straka, Surveyors Module International; James H. Suddeth, Keenan Insurance Co.; C. A. Tiltrum, South Dakota State University; Mark A. Vieira, Federal Emergency Management Agency; Richard C. Walker, ROK; H. Fred Waller, The

Asphalt Institute; Glenn A. Weesies, USDA Natural Resources Conservation Service (NRCS); and James Wojcik, Environmental Systems Research Institute.

An expression of gratitude is also in order to William L. Walker, Jr., for his assistance in keeping the issues focused and Joel E. Stein for his encouragement and assistance as editor.

ACKNOWLEDGMENTS TO THE FIRST EDITION

The author is indebted to many persons who have assisted in the preparation of this book. Unfortunately, space limitations preclude all names from being included here; however, thanks are given to my colleagues at The Citadel whose guidance has helped to shape the scope of this book. They include Thomas J. Anessi, Kenneth P. Brannan, Dennis J. Fallon, Loring K. Himelright, John A. Murden, Charles Lindbergh, Arnold B. Strauch, Russell H. Stout, and Michael S. Woo.

Special thanks are given to the following persons who have reviewed various portions of the book and have supplied the author with valuable comments and suggestions: William B. Beauchene, East Coast Development Corporation; Ronald E. Benson, Jr., Hole, Montes & Associates, Neale E. Bird, S. C. Coastal Council; James A. Broody, First Trident Savings and Loan Company; Robert D. Carr, U.S. Department of Housing and Urban Development; J. Edwin Clark, Clemson University; George L. Craft, American Water Works Association; Stephen P. Dix, National Small Flows Clearing House; Jack M. Ellis, C.R.S. Sirrine; Craig K. Haney, Introspec; Gene E. Hardee, U.S. Department of Agriculture—SCS; Stephen F. Hutchinson, East Coast Development Corporation; Mark A. Isaak, Post, Buckley, Schuh & Jernigan; John Kwist, Newton Builders; Edward J. McKay, National Geodetic Survey; D. Sherwood Miler, III, Pine Forest Properties; Sidney C. Miller, S.C. Geodetic Survey; Joseph R. Molinaro, National Association of Home Builders; J. Herbert Moore, Virginia Polytechnic Institute; James K. Nelson, Clemson University; Joseph L. Rucker, S.C. Department of Health and Environmental Control; Harry C. Saxe, Norwich University; Edward F. Straw, ISO Commercial Risk Services; H. Fred Waller, Jr., the Asphalt Institute; Glenn A. Weesies, National Soil Erosion Laboratory; Donald E. Woodward, U.S. Department of Agriculture—SCS; and David B. Zilkoski, National Geodetic Survey.

Thanks are also given to Thomas S. Mc Canless for preparing most of the graphical displays used in this book and to William L. Walker, Jr., for his encouragement and guidance. Finally, appreciation is also given to Daniel R. Sayre for his continued support as editor throughout the preparation of this book.

PART I
Overview of Activities

1 Introduction

Land development is a complex, highly coordinated, planned undertaking that requires for success a profit motive on the part of the developer as well as an appreciation of human and municipal needs. It depends on the availability of land that conforms with areawide plans, codes, and regulations. Finding buildable property is becoming increasingly difficult in a number of locations in the United States because of population growth, a trend projected by the U.S. Census Bureau for all states that is expected to continue through the year 2025 (Campbell, 1997). This trend coupled with increased income trends (U.S. Bureau of Census, 1999) will drive new opportunities for those persons engaged in the development of land.

The development of property usually requires the services of a team of professionals. Development activities relating to the planning and design of a project can be provided by civil engineers, land planners, landscape architects, and/or architects working together. Other team members can include soil scientists, biologists, and archaeologists to inventory and analyze the site, while financial analysts, lenders, and marketing specialists provide economic analysis and support. The success of a project may lie in the ability of its team members to deal with political and public opinion issues. This is especially true if a desired site is not zoned to support the intended type of development.

1.1 DEFINING LAND DEVELOPMENT

Land development normally embodies the conversion of unimproved property into a site that possesses features capable of supporting a desired activity. The three general categories of land development that describe intended usage are commercial, industrial, and residential.

1.1.1 Commercial Development

Commerce, which can include the buying, selling, or trading of products, materials, or services, is supported by commercial development. This support includes the conversion of strategically located property that minimizes consumer transportation requirements. Commercial developments can provide for office, retail, wholesale, and financial activities and also provide facilities for general services and traveler needs. The success of a commercial development is usually dependent upon the purchasing power of the supporting population within the anticipated trade areas. Commercial developments can take place within the downtown central business district (CBD) of the trade area or along thoroughfares that lead to the CBD. Other types of commercial developments include outlying shopping districts and neighborhood

3

facilities. Planned shopping centers can accommodate neighborhood, community, and regional needs.

1.1.2 Industrial Development

In contrast, industrial development, which can vary from heavy to light, focuses on providing suitable property for activities associated with processing of materials and production of consumer goods. Land that is improved for industrial use can be located within an industrial park or urban area or can exist as a stand-alone facility. These developments are predicated on favorable social and political factors, affordable property, and the availability of material and labor. An attractive industrial site also is dependent upon climate, topography, transportation access, the availability of sewer and water, future expansion potential, and environmental compatibility. Today, most industrial developments place emphasis on aesthetics and the preservation of natural open space while minimizing noise, smoke, odor, dust, noxious gases, heat, glare, fire hazards, explosions, and vibrations. (Claire, 1969; So et al., 1979).

The number of industrial developments being planned and built on previously abandoned sites labeled as "contaminate with hazardous materials—or brownfields" is on the increase and referred to as "redevelopment" projects. These projects include cleaning up the site environment as part of their scope of activities where productive use of the land can be regained. This redevelopment creates a positive asset (usually in an area where buildable land is scarce) that can take advantage of existing labor and material markets.

1.1.3 Residential Development

The conversion of property for production of safe, serviceable, and affordable residences along with associated facilities is within the scope of residential land development. Although commercial and industrial developments have individualized and specialized needs, many aspects are common to all three types, including transportation, utility, and drainage considerations. For example, both commercial and industrial developments often require related off-site improvements (e.g., new highway turning lanes and signals or rail spurs) be completed for use during construction well in advance of the actual site improvements. Because the largest percentage of land area is developed for residential use, the main focus of this book is on that activity, beginning with conformance to areawide planning activities.

Typical types of residential development include single-family, multifamily, and manufactured housing developments. Single-family developments entail a single dwelling unit (either detached or attached) on each building lot, whereas multifamily includes more than one. Figure 1.1 illustrates a typical single-family suburban setting (Figure 1.2 shows a single-family neotraditional setting), where each dwelling unit has direct access to a street. Multifamily developments might encompass townhouses, apartments, or other configurations and are typically represented as in Figure 1.3. Manufactured housing (or factory-built housing—comprising one-quarter of all housing starts in 1996) developments, similar to that shown in Figure 1.4, include completed dwelling units, called manufactured homes, which are factory assembled and towed on their own chassis to the site, rather than constructed on site. Another type of manufactured housing includes factory-built modular units that are transported to the site where they are erected on a site-prepared foundation. This type of manufactured housing is usually classified with site-constructed dwelling.

Some jurisdictions allow for mixed-use or integrated-use developments, integrating residential areas with commercial activities that harmonize with the development plan. Examples of this integrated approach are given in Section 2.5.3.

1.2   AREAWIDE DEVELOPMENT PLANNING

To establish an overall process that considers an area's present and future growth while fostering a livable environment that safeguards property values, areawide planning must be undertaken. Regulations have evolved as tools to guide the developer in attaining satisfactory project completion for the good of the public. This can include the setting aside of property subjected to site hazards where the development of land is prohibited. The resulting project serves the populace through benefits reaped by many public service activities involved with varying aspects of society. To the tax official, adequate land title records are made available for public record. For the municipal engineer, properly designed and adequately constructed road and drainage systems are provided to meet the public's needs. For health department officials, clean water supplies and adequate sewage/solid waste disposal are available to the public. For the fire chief, standard hose fittings, with proper plugs and water pressures, are available to provide effective fire protection to the citizens. The educator and recreation official can preserve and secure land to serve residents. To the residents, it means having an investment that will maintain its value and be adequate for everyday needs. To the land planner, the result is a coordinated system of public activities that otherwise would be unrelated in regard to such items as thoroughfares, major utility systems, and school/recreational areas. To the developer, regulations protect projects from unscrupulous operators who might competitively undertake substandard development activities resulting in a degradation of project marketability and use. To the local governing body, regulations are a deterrent to making premature and additional taxpayer expenditures as they require minimum land improvements that will provide a reasonable service life to the public. Finally, regulations allow for discriminatory acquisition of land made available for public use by dedication.

1.2.1   Planning Objectives

One major objective of areawide planning can be to improve the quality of life, which can result in lower rates of crime, suicide, and disorders associated with urbanization. This process takes into consideration the needs of the individual, family, neighborhood, and community. Different age groups have different interests, which can be addressed through proper planning, resulting in increased enjoyment of life.

A second major planning objective can be to provide safety to the inhabitants— safety in travel, safety from fire, safety from crime, safety from disease, and safety from peril. Safety from disease is associated with sanitation, which is a primary objective in planning a healthy environment.

FIG. 1.1 Example of a single-family suburban development application. Courtesy of R. H. Stout.

FIG. 1.2 Example of a single-family neotraditional development application. Courtesy of R. H. Stout.

FIG. 1.3   Example of a multifamily development in a neotraditional development setting. Courtesy of R. H. Stout.

FIG. 1.4   Example of a manufactured housing development.

A third major objective can be to provide a reasonably quiet environment. Noise associated with airports, manufacturing, railroads, and truck and bus traffic may require costly abatement programs if proper planning is not undertaken.

Many other planning objectives can also be incorporated into areawide planning. These other objectives can include affordable housing, economic development, historic preservation, environmental protection, aesthetics, urban design issues, and so forth. The scope of the objectives is established based on local needs.

1.2.2 Purpose of Land Planning

Land planning addresses ways to provide services to the area in an efficient manner. Public funds must be expended, necessitating that a budget be formulated by the jurisdictional authority and followed, not only for day-to-day activities but also for long-range capital improvements. If economic responsibility is practiced in the planning process, improved employment, a larger tax base, and improved services will result.

1.2.3 Land Planning Types

Local planning is normally undertaken through state enabling legislation empowering governing bodies to investigate needs and promulgate plans to shape the area's growth. Planning can be categorized as metropolitan planning, community planning, and neighborhood planning. Metropolitan planning normally includes metropolitan areas with populations of 50,000 or more and areas defined by general land use, population distributions, and economic activities. In addition to areawide transportation planning, such as that stipulated in the Urban Mass Transportation Act of 1964, as amended, drainage, water, sewer, and air purity planning may be undertaken. Other planning activities might include areawide water quality management planning (208 Planning) as set forth in the Clean Water Act [Public Law (PL) 92-500], as amended, to control pollution from point and nonpoint source discharges that might degrade the environment.

Community planning focuses on subareas within the metropolitan planning area or possibly on small cities and municipalities. Community planning is similar to metropolitan planning but is carried out on a more localized basis. It also addresses local land use regulation. Neighborhood planning includes activities of segmental areas within communities, addressing the coordination of community needs and capital expenditures while protecting the integrity of planning units, as illustrated in Figure 1.5.

1.2.4 Forming a Plan

The formulation of a general plan includes the establishment of citizen and governmental area goals over a period of 20–25 years. When formulating an area plan, one must consider the community and surrounding area, especially in cases where historic and natural resources are present and require protection. For example, resort communities have different needs than do metropolitan areas. Consideration must be given to the amount of economic activity in the area because gainful employment is necessary for the quality of life of the residents. Population trends must be established to assess future impacts on housing, transportation, utilities, and other urban systems. Adequate community facilities and services in police, fire, and health protection must

Source: 1990 US Census

FIG. 1.5   Example of census tracts that can be used as planning units. Reprinted from *Summerville Tomorrow—1999 Comprehensive Plan* with permission of Summerville Town Council, 1999.

be anticipated. Educational and socioeconomic needs must be addressed. With urban planning, projected area needs can be translated into a capital improvements program that will allow staged construction of municipal improvements to accommodate anticipated growth, usually in five-year increments. Figure 1.6 illustrates the relational aspects of various planning activities.

An inventory of existing land uses must be undertaken for the planning area. General types of uses are single-family residential, multifamily residential, apartments, commercial, light industrial, heavy industrial, floodway, public semiprivate (e.g., churches), streets, and vacant. In the past hardcopy (paper printed) aerial photographs, Sanborn Fire-Insurance Maps produced by The Sanborn Map Company in Pelham, New York, and tax maps have been used to compile this initial data, which then was verified in the field. Today this material is becoming increasingly available in the form of digital orthophotos (Section 4.14.7). Other digital products are becoming more accessible using the World Wide Web (Internet), such as the Federal Emergency Management Agency's (FEMA's) Digital Flood Insurance Rate Map (DFIRM) and even by-products* obtained from a geographical information system (GIS) discussed in Section 4.1, if one exists that includes the area of study.

A map of existing land uses can then be made similar to the one illustrated in Figure 1.7. Various uses can be indicated with different colors on the map. Other factors to

---

*for example, see www.geodataconsult.com.

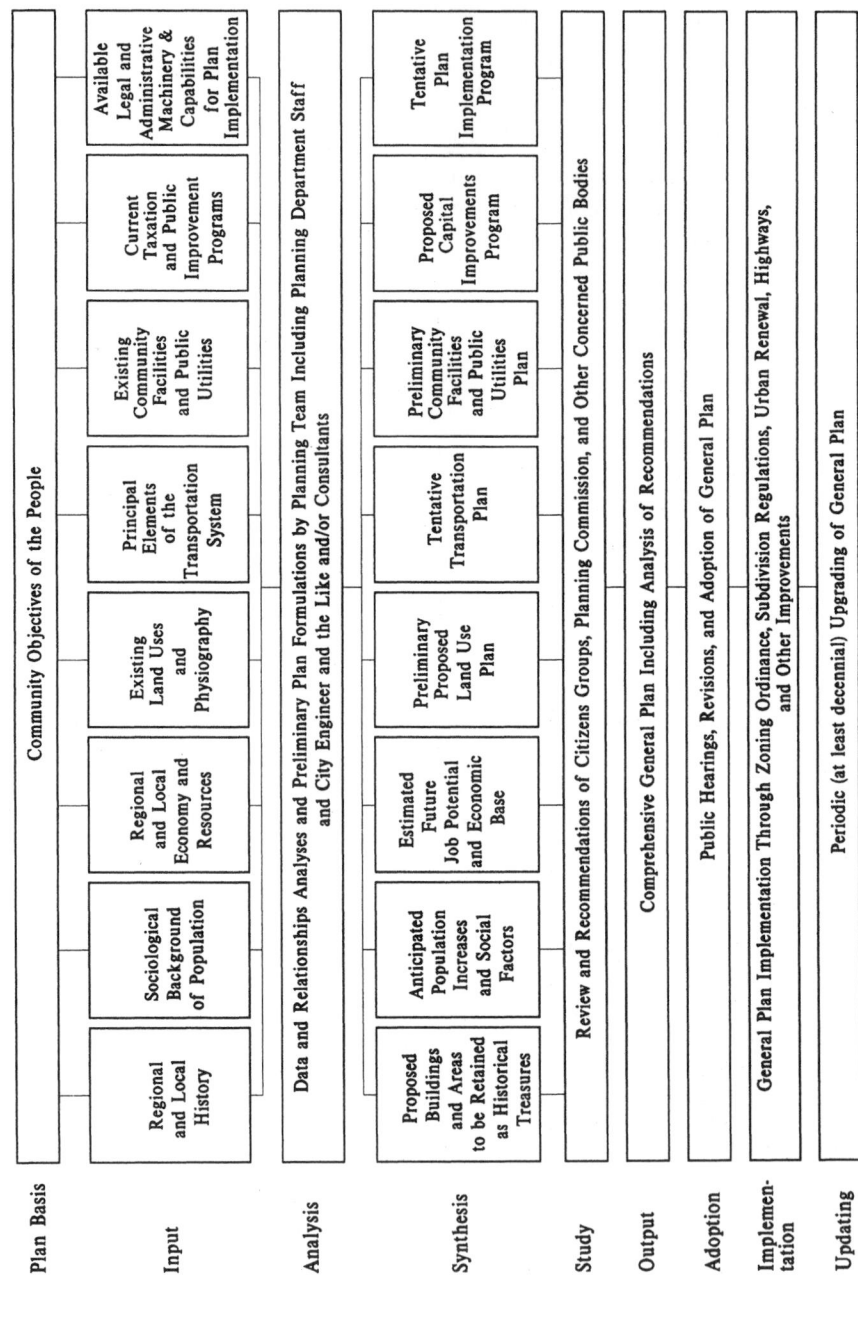

FIG. 1.6 Simplified diagrammatic outline of general planning process. Source: William H. Claire, ed. *Urban Planning Guide—ASCE Manual No. 49*. New York: ASCE, 1969. Reprinted by permission of the American Society of Civil Engineers.

be considered include topography and ridge line locations for drainage basin boundary delineation, flood areas, and a soils inventory.

An important aspect of urban planning includes transportation activities. During the 1920s, increased automobile ownership allowed greater mobility. This increased the movements of persons, which allowed the day-to-day range of daily activities to expand greatly. Movement to the suburbs was experienced, and residential development followed. As vehicle travel expanded, new types of streets allowing higher capacities were needed. Rail, truck, water, air, and mass transit all have been impacted by growth.

Due to this migration away from cities, CBDs, which are the heart of the city, began to deteriorate. Inadequate parking, traffic congestion, crime, blight, outdated utilities, and depressed markets were some of the problems facing the declining CBDs. As a result of urban planning, renewal efforts have helped bring about revitalization of some CBDs, including new opportunities for development associated with in-fill and brownfield reclamation projects.

1.2.5 Plan Implementation

Upon completion of the economic, transportation, population, housing inventory, community facilities, and environmental studies, along with the preparation of various base maps showing existing land use, the planning agency should integrate into its strategy not only the goals and objectives of the planning area but also those requirements set forth by governmental bodies having broader and more encompassing authority. Future requirements are formulated by forecasting and projecting trends. Needs associated with land use, community facilities, transportation, and capital improvements can be established that will guide future growth in an orderly manner. Plan implementation utilizes such devices as generalized land use maps (Figure 1.8), zoning maps, zoning ordinances, subdivision regulations, urban renewal, and the adoption of codes for building, electrical, and plumbing activities.

The general land use plan or master plan is not a zoning map, as it is used only as a guide for controlling future growth patterns. A zoning map, on the other hand, is an integral part of the zoning ordinance. It is site specific and cannot be altered without legal process.

1.2.6 Plan Maintenance

Maintaining the currency of these devices is vital to ensure that the plan remains responsive throughout the entire planning period. Periodic jurisdictional modifications may be necessary to update the plan. As the future unfolds, various "planned" elements based on previous uncertainties give rise to a variable regulatory condition that requires the successful developer (and in turn the development team) to be well informed.

1.2.7 Coping with Evolving Problems

As a result of fragmented low-density growth beginning in the 1920s which reached out from urban centers that were poorly planned or managed and automobile dependent, urban sprawl evolved and is today one of the major issues focused on by many communities across the United States. Urban sprawl has been identified by many as the source of traffic congestion, inadequate infrastructure, loss of open space, and higher taxes.

FIG. 1.7 Existing land use map for Town of Summerville, SC, 1999. Reprinted from *Summerville—Tomorrow—1999 Comprehensive Plan* with permission of Summerville Town Council, 1999.

FIG. 1.8 Future land use plan for Town of Summerville, SC. Reprinted from *Summerville—Tomorrow—1999 Comprehensive Plan* with permission of Summerville Town Council, 1999.

The term "smart growth" has evolved in connection with programs that are designed to help control some of these problems. In the past, sprawl-related issues have for the most part been addressed at the local and state levels; however, the federal government has now proposed new ideas to help curb it. Some of them include considering ways for communities to help acquire open space, protect water supplies, and also reclaim productive use of now idle brownfield sites.

Other local measures consider the allocation of capital improvement project funds based on designated high-priority areas where controlled growth is being encouraged. Funding for low-priority areas would then be restricted. Some communities have even gone so far as to try to restrict growth related to urban sprawl only to find, for example, that the cost of existing housing within their jurisdiction escalated to a point where sales sky-rocketed. There are no simple answers; however, it is clear that governments at all levels are interested in pursuing avenues to reduce sprawl. Along these lines, the federal government undertook a General Assembly Office (GAO) study and report to Congress entitled, *Community Development: Extent of Federal Influence on "Urban Sprawl" Is Unclear* (April 1999). Although this study examines how the federal government may have contributed to sprawl, it does not fully examine the extend of sprawl. It is, however, a step that demonstrates the government's interest in evaluating and seeking ways to reduce sprawl-related problems.

The State of Florida undertook a Department of Community Affairs study in 1996 called *Best Development Practices—Doing the Right Thing and Making Money at the Same Time* which examines ways to arrest sprawl statewide and serves as a model for other states to consider. There are four categories of "best practices" that are offered by the study team, including land use, transportation, the environment, and housing. Figure 1.9 shows each category and the corresponding practice elements associated with that category; however, the reader is invited to review the entire document where each element is fully explored with ample illustrations.

It is evident from these studies that there are many problems yet to be resolved, and various approaches will be used in an effort to improve the quality of life of the citizenry. This requires the developer of a project to work within a constantly changing system of rules, regulations, and demands if success is to be achieved.

1.3 THE DEVELOPER'S ROLE

The person or organization that is the catalyst for land development is called a developer, or owner. This may be an individual, partnership, corporation, or trustee. The developer is usually credited with the conception of the project. The preliminary market potential must be assessed by the developer so that an initial risk assessment can be made. If the developer believes the project has potential and has reason to believe that project funding can be obtained, several preliminary activities must be undertaken. First, an attorney may be retained to handle legal matters, including title searching and agreement drafting, which are required when developing land. Second, the developer may schedule a formal market analysis to obtain an indication of potential success while investigating and soliciting property availability for consideration as the potential site.

A. Best Land Use Practices
    1. Keep vehicle miles of travel (VMT) below the area average.
    2. Contribute to the area's jobs-housing balance.
    3. Mix land uses at the finest grain the market will bear and include civic uses in the matrix.
    4. Develop in clusters and keep the clusters small.
    5. Place higher density housing near commercial centers, transit lines, and parks.
    6. Phase convenience shopping and recreational opportunities to keep pace with housing.
    7. Make subdivisions into neighborhoods with well-defined centers and edges.
    8. Reserve school sites and donate them if necessary to attract new schools.
    9. Concentrate commercial development in compact centers or districts.
    10. Make shopping centers and business parks into all-purpose activity centers.
    11. Tame auto-oriented land uses, or at least separate them from pedestrian-oriented uses.

B. Best Transportation Practices
    1. Design the street network with multiple connections and relatively direct routes.
    2. Space through-streets no more than a half mile apart.
        or the equivalent route density in a curvilinear network.
    3. Use traffic calming measures liberally.
    4. Keep speeds on local streets down to 20 mph.
    5. Keep speeds on arterials and collectors down to 35 mph (at least inside communities).
    6. Keep all streets as narrow as possible, and never more than four travel lanes wide.
    7. Align streets to give buildings energy-efficient orientations.
    8. Avoid using traffic signals wherever possible and always space them for good traffic progression.
    9. Provide networks for pedestrians and bicyclists as good as the network for motorists.
    10. Provide pedestrians and bicyclists with shortcuts and alternatives to travel along
        high-volume streets.
    11. Incorporate transit-oriented design features.
    12. Establish TDM programs at employment centers.

C. Best Environmental Practices
    1. Use a systems approach to environmetal planning.
    2. Channel development into areas that are already disturbed.
    3. Preserve patches of high-quality habitat, as large and circular as possible,
        feaathered at the edges, and connected by wildlife corridors.
    4. Design around significant wetlands.
    5. Establish upland buffers around all retained wetlands and natural water bodies.
    6. Preserve significant uplands, too.
    7. Restore and enhance environmetal functions damaged by prior site activities.
    8. Minimize runoff by clustering development on the least porous soils,
        using infiltration devices and permeable pavements.
    9. Detain runoff with open, natural drainage systems.
    10. Design man-made lakes and stormwater ponds for maximum habitat value.
    11. Use reclaimed water and integrated pest management on large landscaped areas.
    12. Use and requre the use of Xeriscape landscaping.

D. Best Housing Practices
    1. Offer "life cycle" housing.
    2. Achieve an average net residential density of six to seven units per acre
        (without the appearance of crowding).
    3. Use cost-effective site development and construction practices.
    4. Design in energy-saving features.
    5. Supply affordable single-family housing for moderate-income households.
    6. Supply affordable multifamily and accessory housing for low-income households.
    7. Tap government housing programs to broaden and deepen the housing/income mix.
    8. Mix housing to the extent the market will bear.

FIG. 1.9   Best development practices for (a) land use, (b) transportation, (c) the environment, and (d) housing. Reprinted from *Best Development Practices* by Reid Ewing; copyrighted by the State of Florida, published by the American Planning Association (Planners Press), 1996.

1.3.1   Design and Other Professionals

The developer would then solicit the services of design professionals to investigate, plan, design, and complete the desired project. Design professionals normally associated with land developments include registered civil engineers, land surveyors, architects, land planners, and/or landscape architects. A number of other professionals can also be involved, including biologists, archeologists, archivists, and botanists. These latter professionals play a vital role in identifying areas for preservation and conservation. The design professionals work around any identified areas and shape the project while influencing its cost. Combined with the cost of land, the cost of site improvements and structure cannot be so high as to make the final product unmarketable. Figure 1.10 illustrates the cost components of a new single-family home in a residential land development.

It is the responsibility of the developer to establish the project parameters and goals that are used to guide the design professional. If the proposed development is to include commercial areas, the developer usually contacts key regulatory personnel to inspect and approve the completed project. Involved in this process are zoning, building, electrical, fire, and plumbing inspectors, along with local police agencies and jurisdictional highway, health, and utility organizations.

During the design process, the design professional will produce data, studies, alternatives, and recommendations that must be analyzed by the developer for decision-making activities. These decisions form the basis for final project design and include preliminary studies, cost estimates, and a reanalysis of the project's risk by the developer.

1.3.2 Project Approval Activities

The developer can then seek preliminary project approval from regulatory agencies through the assistance of the design professional and at the same time solicit tentative financial commitments to fund the project. Once notified of preliminary plan approval and having received a tentative commitment for project funding, the developer will engage the design professional to complete work on final construction plans, specifications, and cost estimates. Review of these documents allows the developer to evaluate the project's pro forma (estimate of project costs and income to establish the rate of return) and to make yet another final risk analysis. Assuming the developer is

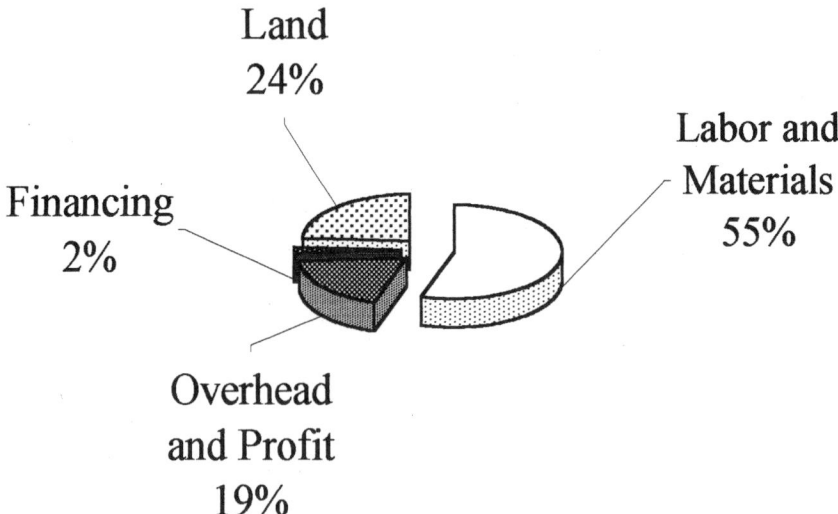

FIG. 1.10 Cost components of a new single-family home. Source: National Association of Home Builders (NAHB). *The Truth About Regulatory Barriers to Housing Affordability*, 1998. Reprinted by permission of the National Association of Home Builders.

still comfortable with the project's scope and potential, design professionals will seek approval for the developer from regulatory authorities for permission to construct the project. The financial commitment for property acquisition and construction improvements is formalized at this time by the developer. Any temporary utility service hookups required during construction can be arranged by the developer; however, some contractors may have this burden placed upon them in the construction specifications.

1.3.3 Activities During Construction

At this stage, a construction contractor can be engaged through the bid process or though direct negotiations. The design professional's assistance is normally utilized by the developer in this process.

If premarketing activities are to be undertaken by the developer prior to the satisfactory completion of all required improvements, many jurisdictional authorities will accept a bond or other posted security to ensure the project's completion.

Safety is another factor that the developer must ensure is a priority on a project site. In 1998 the American Society of Civil Engineers (ASCE) revised its policy on safety relating to construction sites. The new position places a more active role on the part of owners (developers) during the contractor selection process to increase emphasis on safety performance when considering a bidder's qualifications and also to contractually assign safety responsibilities to an individual or organization for continuous site monitoring (including final shop drawing/detail approval). Reasonable care must also be exercised to protect those who pass by or trespass on the development. Those persons, especially children, must not be exposed to unnecessary risk or harm. Prudent developers must take precautions to protect the public during construction and also require the contractor to assume a similar role (Sweet, 1985). (See references at end of chapter.)

Progress by the contractor will be monitored by the design professional or construction manager and the developer. Upon completion, the project will be inspected by the design professional and all appropriate regulatory review agencies having jurisdiction over the development to ascertain whether the finished site improvements meet the scope and intent of the approved plans and specifications. Certificates for occupancy or system operational approval will be issued when construction work is substantially completed and satisfactory. Final marketing arrangements which may include sales or even lease agreements can be undertaken once all constraints have been met by the developer. Compliance with all rules and regulations, such as the Federal Interstate Land Sales Full Disclosure Act (15 USC 1701), can allow the developer access to the funds necessary to undertake and complete the project.

1.4 LAND DEVELOPMENT MEASUREMENT CONVENTIONS

As a practical matter, the development of land in the United States today is still heavily dependent on the use of measurements based on systems other than the *Système International* (SI). Property conveyances based on acres, distances based on Gunter's chains or feet, culvert diameters based on inches, and precast sanitary sewer manhole diameters based on feet are some examples of measurement usage seen today. Because land development is localized and for the most part locally produced materials are used in its construction, there is little consumer pressure to convert to SI. Any changes in measurement conventions therefore are predicted to occur slowly in this field.

This is not to say that land development projects are not being designed and built based on the SI system. The Metric Conversion Act of 1975 sought conversion through volunteerism, while later government actions now require metric use on all federally funded (even housing) projects. Geodetic survey references and now design criteria for streets and roads are based on the SI system. Other examples pertaining to the development of land can be cited to demonstrate that the process of switching measurement systems is proceeding, and as a result, the use of more than one system exists today. This book follows this trend, where the reader will find references to more than one measurement system. The SI system, internationally adopted in 1960, makes use of the conversions shown in Appendix A.

REFERENCES

Campbell, P. R. *Population Projections: States, 1995–2025—Current Population Reports P 25-1131*. Washington: U.S. Bureau of Census, May 1997.

Claire, W. H. (ed.). *Urban Planning Guide—ASCE Manual No. 49*. New York: American Society of Civil Engineers, 1969, pp. 27, 82–83, 149. Referenced material is reproduced by permission of The American Society of Civil Engineers.

Davies, V. J., and K. Tomasin. *Construction Safety Handbook*, 2nd ed. London: Telford, 1996.

Ewing, Reid with Marybeth DeAnna, Christine C. Heflin, and Douglas R. Porter in cooperation with the Urban Land Institute and the Florida Department of Community Affairs. Best Development Practices: Doing the Right Thing and Making Money at the Same Time. Chicago: American Planning Association, 1996, pp. 17, 53, 95, 133.

Heberle, D. *Construction Safety Manual*. New York: McGraw-Hill, 1998.

Hislop, R. D. *Construction Site Safety—A Guide for Managing Contractors*. New York: Lewis, 1999.

Scanlon, B. *Marketing of Engineering Services*. London: Telford, 1988.

So, F. S., I. Stollman, F. Beal, and D. S. Arnold, eds. *The Practice of Local Government Planning*. Washington: International City Management Association, 1979, pp. 262–272, 389–415.

Sweet, J. *Legal Aspects of Architecture, Engineering, and the Construction Process*, 3rd ed. St. Paul, MN: West, 1985, pp. 129–134.

United States Bureau of Census, Income Statistics Branch. *Median Income for 4-Person Families, by State*. Washington: U.S. Bureau of Census, 1999.

United States General Accounting Office. *Community Development: Extent of Federal Influence on "Urban Sprawl" Is Unclear, GAO/RCED-99-87*. Washington: U.S. Government Printing Office, 1999.

Wandmacher, C., and A. I. Johnson. *Metric Units in Engineering—Going SI*. Reston: American Society of Civil Engineers, 1995.

2 Developer's Constraints

A major problem affecting a developer is the time required to complete a project, during which interest payments accumulate and consumer markets, along with construction costs, fluctuate. The engagement level a developer seeks in a project directly relates to the demand for committed personal time. There are several approaches a developer might use to undertake a project, as illustrated in Figure 2.1. If the project is small and all work will be accomplished in a single phase, the project would reflect a consummate approach. If the project is large and will be undertaken in incremental sections through phased construction, a staged approach would be reflected. Within each of these approaches the developer must further establish a role of providing only improved land activities or total development activities. Improved land activities normally include the procurement and transfer of raw property into improved building sites. These improved sites, having transportation access, utilities, and storm drainage facilities, could be sold to others, including potential individual homeowners or professional builders, who would then undertake the completion of all remaining site improvements. Possible site improvements include erection of structures, landscaping, driveways, electrical, water, and sanitary sewer service ties. This type of sale generally contains restrictions on the minimum cost of the proposed dwelling unit, along with its type of construction, size, and possibly other elements. Total development activities include those undertaken in improved land developments and also all other activities required to produce a complete project. A professional builder could be engaged in a total development situation to build on an improved site owned by the developer. The finished product might then be jointly sold. Another approach might be for the developer to engage a single firm to both design and build the project.

2.1 MARKET ANALYSIS

An evaluation of the project area is necessary to determine whether

(a) there exists a need for development,

(b) the resultant product will possess utility when completed,

(c) there is a scarcity of residential units due to a lack of inventory or created by filtration,

(d) there is a capability of a buyer to purchase the product,

(e) there is reasonable expectation that the project will be absorbed by the target market and that the project is economically feasible, and

(f) a profit potential exists.

(a)

(b)

FIG. 2.1 Examples of two approaches used to develop property: (a) consummate approach; (b) staged approach.

These essential conditions, which are normally investigated by a marketing research consulting firm for the developer, are evaluated using a procedure known as market analysis (Ashbury, 1970; Casey, 1968; FHA, 1970a). The results of the analysis are used to determine the highest and best use of a particular site. The analysis must be subject to certain boundary conditions within which the market area exists.

2.1.1 Market Area

It is not an easy task to select a market area. Generally, the approximate realm of influence exerted by a development can only be guessed. This may appear easy, but when one begins to draw lines of delineation on a map, many problems arise. Questions confronting the analyst include whether to follow natural features (ridge lines or drainage basins), or political subdivisions (county, city, or state boundaries), or census tracts, or municipal districts (school, water, fire, or sewer), or even road systems for defining the market area. Information from various sources, which was gathered according to specific originator needs, may be required to aid in the conduct of an analysis. Initial study immediately shows that overlapping or incomplete information will probably be present, since resource dividing lines do not always coincide. The best approach is to pick a combination of sources (e.g., census information and utility company records) that produces the most complete area coverage, along with estimated partial values obtained from other self-serving sources that have data gathered within unique differing confines. An alternate, more accurate, method is to disregard existing sources and conduct a first-hand study of the area for market analysis information, although time and expense usually preclude this approach.

Once a market area has been determined, a two-part analysis can commence. The macroanalysis will yield a telescopic view of the area. The microanalysis will zoom in on specific entities. Together they should indicate development marketability.

2.1.2 Macroanalysis

By placing the market area in broad perspective, the macroanalysis will give an indication of prosperity, viability, and durability of existing conditions. Some of the topics to be addressed are demographic analysis of population characteristics and trends, income distribution, employment trends and levels, existing housing stock, potential and firm housing starts, military facilities (if applicable), and retiree population. Short-term population trends calculated using linear regression or other curve-fitting methods will normally yield reasonable projections from which family numbers can be estimated, based on a locally determined factor that represents the number of persons per household. Future income distributions can also be projected, enabling the developer to determine the project's target price range.

Projecting age distributions allows the analyst to determine future numbers of people within certain age brackets; these projections are required to establish bedroom demands.

Most families spend about 35 percent of their gross income on housing. This knowledge enables one to determine the magnitude of housing indebtedness that can be supported by funds, from which future mortgage sizes can be estimated for economic levels within the population.

Combining this information yields the housing demand, which includes the number of dwelling units, bedrooms, and price range at a given future time. Existing housing numbers, reduced by a factor that reflects dilapidation, can be combined with firm and potential housing starts in the market area (determined by interviews, building permit issues, or governmental sources) to produce an estimate of the housing supply. The numeric difference between the housing supply and the housing demand yields the housing need. If factors such as new industry and low unemployment simultaneously indicate economic stability and growth, the projected annual housing need is considered reasonably valid.

2.1.3   Microanalysis

A microanalysis of possible development sites can be undertaken once the housing need has been established. The fair market value of the land and the effective drawing power of the proposed facility in relation to existing competition are worthy of consideration. A professional market analyst should be consulted for a more thorough assessment of the problem.

Sometimes area market analyses are prepared by governmental bodies such as the U.S. Department of Housing and Urban Development (HUD) and local planning agencies. Inspection of these analyses may reveal that an area is nearly in equilibrium; however, some private sector studies might indicate that the same market reflects differing conditions. To prevent market saturation, a more conservative approach may have been taken by the government as compared to some private sector analyses. If equilibrium were the actual situation, one might wonder why vacancy rates do not rise appreciably. The reason might be that new industry seeks geographic areas where housing is already available; consequently the supply may precede the actual need.

If the developer plans to include some commercial area as part of the project's scope, the market analysis must include the influence, magnitude, location, and distribution of competitive facilities as well as the income, social groups, family size, ages, and lifestyles of people in the market area.

2.2   SITE LOCATION

Having selected a handful of possible sites within the market area as a result of the microanalysis, one must choose which is the most suitable for the particular development—be it a manufactured home park, manufactured home subdivision, single-family development, multifamily development, or apartments. Consideration should be given to any effect the potential development might have on the adjoining properties, and vice versa, relating to light, air, view access, and traffic generation.

2.2.1   Municipal Services

A residential site should provide the occupants with municipal services necessary for everyday family needs. These services include the availability of schools, religious facilities, places of employment, shopping centers, and medical facilities. Police and fire protection availability will make the development a safer and more appealing place to live. To mobilize the area, transportation links are required, both for private conveyance and for public modes. Arrangements must be made for solid waste disposal. Finally, if the development is to attract a particular consumer market, appropriate accommodating facilities should be included as part of the development plan. Figure 2.2 graphically illustrates several of these desired relationships (APHA, 1960; DeChiara and Koppelman, 1969).

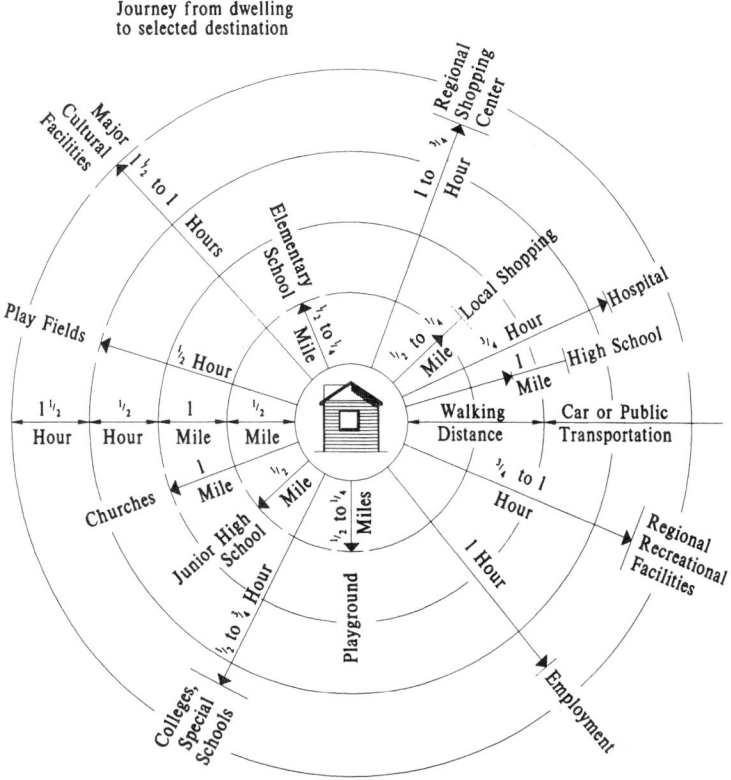

FIG. 2.2 Maximum distance for community facilities. Source: Joseph DeChiara and Lee Koppelman. *Planning Design Criteria.* Van Nostrand Reinhold, 1969, p. 190. Reprinted with permission.

Of significance to the developer is the relationship of the site to existing municipal utility lines. If a site is chosen that is not close to existing sewer and water service, additional development costs will accrue in providing these necessities. Additionally, depending on the size of the development, the governing body may require certain areas to be reserved for future locations of some of the above-listed municipal services.

2.3 SITE GEOMETRY

Property parcels comprised of boundary lines that avoid acute convergence angles do not contribute to spurious land slivers or extremely disjointed shapes and hence offer the greatest ease to the developer in providing site improvements. Figure 2.3*a* illustrates an attractive site, while Figure 2.3*b* represents an undesirable one.

FIG. 2.3   Boundary configurations affecting development planning: (a) site conducive to promoting economical development; (b) site having an irregular shape where development costs are increased.

2.4   SITE CONDITIONS

When a site is being considered for a certain proposed type of development, several physical features must be investigated to determine its suitability. The developer normally utilizes the services of design professionals who strive to minimize construction and improvement costs while providing a safe and workable project that, with normal maintenance, will endure.

Some potential project locations require additional considerations. Site habitation by endangered species, marine mammals, or migratory birds can appear to hamper the activities of an aggressive developer; however, if preservation and conservation are practiced in concert with established regulations, natural features supporting such habitats can be used as beneficial assets to enhance the project's attractiveness.

2.4.1   Endangered Species

The Endangered Species Act (ESA) of 1973, as amended, protects both plants and animals (and their critical habitats, if designated) that are either endangered or threatened with extinction. This protection extends across all federal activities as well as the private sector. The ESA is administered jointly by the U.S. Department of Commerce through the National Marine Fisheries Service (NMFS) and the U.S. Department of Interior through the Fish and Wildlife Service (FWS). The FWS and NMFS are charged with maintaining lists of endangered and threatened species along with critical habitat areas, if designated, for each species listed. According to the ESA, critical habitat is published in the Federal Register using a map and written description; however, a number of areas remain unlisted for a variety of reasons, including bringing public attention to locations where observers/collectors might upset an already delicate situation. Each listed species is protected from actions that modify its habitat, commercial exploration, and hunting. In addition, all federal agencies must avoid actions (including the actions resulting from permits issued and projects funded by or authorized by that agency) that might jeopardize the species survival and recovery or adversely modify critical habitat (50 CFR 402). The FWS/NMFS must also create a recovery plan to help each species.

The ESA term "take" as applied to animals includes to kill, capture, hunt, collect, harass, shoot, harm, pursue, trap, or wound a listed animal. The terms harm and harass are further defined by FWS regulations (50 CFR 17.3) as follows:

> Harass means an intentional negligent act or omission which creates the likelihood of injury to wildlife by annoying it to such an extent as to significantly disrupt normal behavior patterns which include, but are not limited to, breeding, feeding, or sheltering.

> Harm means an act which actually kills or injures wildlife. Such acts may include significant habitat modification or degradation when it actually kills or injures wildlife by significantly impairing essential behavioral patterns including breeding, feeding or sheltering.

If a land development is located within an area designated as a critical habitat and requires a federal permit authorization or is receiving federal funding (federal links), the linked agency must conduct a Section 7 consultation with the FWS/NMFS. Thus, critical habitat only comes into play if a Section 7 link exists. Plant protection is different in that endangered plants cannot be transported across state lines or used for trade. Also, plants located on federal land cannot be reduced to possession, while endangered plant species on any land may be protected by state law. Incidental take may be authorized by FWS/NMFS through its mandatory Section 7(a)(2) consultation process for federal agencies and voluntary Section 10(a)(1) permit program for private land owners. The public is given an opportunity under Section 10 to comment prior to permit approval. Section 10 permits are predicated where the applicant is obligated to minimize and mitigate the impacts of the taking [through a submitted habitat conservation plan (HCP)] if such taking does not appreciably reduce the chances of survival and recovery of the species.

A prudent developer contemplating a particular project should be proactive and determine beforehand whether or not the ESA has an effect on the project's outcome. Having early knowledge that the site is free from federally protected species will allow the developer to concentrate on other issues; thus early agency contact is recommended. If federally listed species or their critical habitats are on-site, the developer should work with the FWS/NMFS to determine if there are ways to avoid and/or minimize the impacts. Local FWS/NMFS offices have available species lists and other pertinent information (such as the *Endangered Species Consultation Handbook: Procedures for Conducting Consultation and Conference Activities Under Section 7 of the Endangered Species Act* (Section 7 Consultation Handbook) and the *Endangered Species Habitat Conservation Planning Handbook* (Section 10 HCP Handbook). The FWS offers the public information access through its Internet web page (http://endangered.fws.gov). For a more complete source on this subject, the reader is invited to peruse the National Association of Home Builder's (1996) *Developer's Guide to Endangered Species Regulation*, the Endangered Species Act, the *Section 7 Consultation Handbook*, and the *Section 10 HCP Handbook*.

2.4.2 Wetlands

Potential development sites must be investigated for the presence of wetlands. A wetland is an area that is inundated or saturated for a frequency or duration to support, and under normal conditions do support, hydrophitic vegetation. There are three parameters that make up a wetland: (a) hydrophitic plants, plants that are adapted to wet environments; (b) hydric soils, soils that are formed under wet conditions; and (c) hydrology, soil that is saturated to the surface for at least 5 percent of the growing season. According to the Conservation Foundation (Train, 1988), wetlands "provide habitat indispensable to a great and varied array of aquatic, avian, and terrestrial wildlife." Salt and freshwater wetlands can be categorized into five different system types: marine, estuarine, riverine, lacustrine, and palustrine systems (Cowardin et al., 1979). Figure 2.4 depicts these classifications of wetlands. Although the marine system is not shown, Figure 2.5 provides important details on the estuarine and riverine systems, while Figure 2.6 illustrates the lacustrine and palustrine systems. Should wetlands occur on site, an investigation and evaluation must be undertaken to determine whether the wetlands fall under the jurisdictional control of regulatory agencies. Regulated wetlands come under the jurisdiction of the U.S. Army Corps of Engineers (COE) and the Environmental Protection Agency (EPA), in concert with state and local units of government (Dunkle et al., 1989). These organizations are responsible for maintaining the chemical, biological, and physical well-being of the country's waters. Potential discharge hazards resulting from dredging or filling activities must be minimized and therefore cannot be undertaken without permit approval as provided under Section 404 of the Federal Clean Water Act (PL 92-500), as amended. The U.S FWS and the NMFS play important advisory roles in this permitting process (DeVereaux et al., 1982; U.S. Army Corps of Engineers, 1991).

Today wetland areas are often incorporated as part of a stormwater management system to help control runoff and remove pollutants. Wetlands are also being used to help treat wastewater. When a project adversely impacts jurisdictional wetlands, jurisdictional authorities may allow mitigation, or the creation of new wetlands, in exchange for approving wetland alterations within critical project areas. Without mitigation an otherwise satisfactory and beneficial project might be prohibited. On-site mitigation requires permits and time after construction to demonstrate that the wetland is functioning properly. Mitigation can be achieved off-site through the use of a wetlands mitigation bank if (a) the project demonstrates little or no practicable on-site mitigation opportunities, (b) the nature of the project is minor (such as an individual lot not connected with a land development), or (c) the project is linear (pipelines, highways) where numerous minor impacts occur.

Mitigation banks are formed having an organizational structure (a government agency, corporation, or nonprofit organization) that is sanctioned by federal (COE/EPA) and state agencies having jurisdictional control over wetlands. Often banks are created using previously drained or degraded forested wetlands that can be restored or enhanced. The bank organization procures the property, and through a series of planned improvements, gradually re-creates (or even creates and preserves) wetland areas for the purpose of selling credits to developers or organizations that meet the criteria listed above. The concept of wetland banking is relatively new; however, in 1996 there were 50 public and 20 private banks nationwide. Today there are many more. Prior to granting any credits, the bank must be functioning in a successful manner

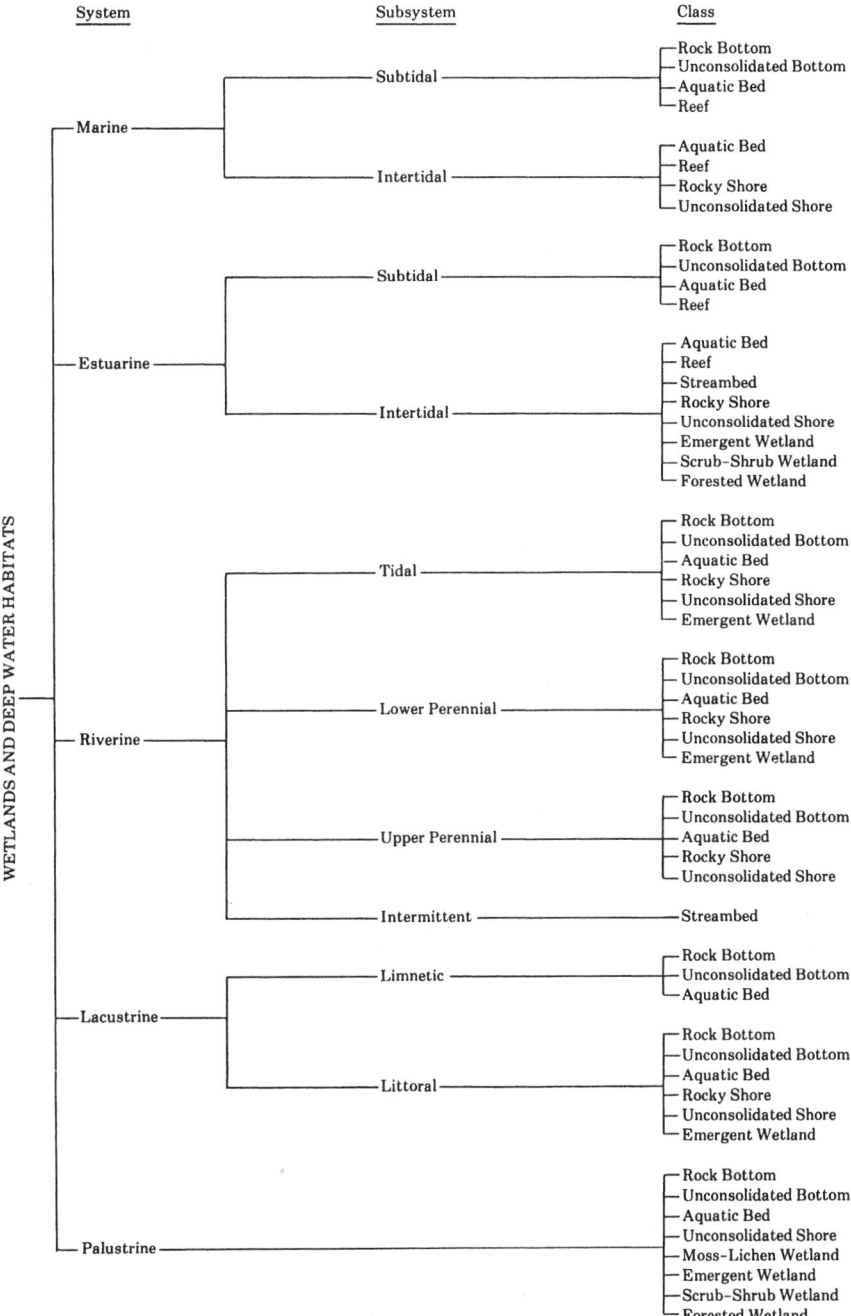

FIG. 2.4 Classification hierarchy of wetlands and deep-water habitats, showing systems, subsystems, and classes. Palustrine system does not include deep-water habitats. Source: Cowardin et al., 1979.

FIG. 2.5 Distinguishing features and examples of habitats: (*a*) in the estuarine system; (*b*) in the riverine system (EHWS = extreme high water of spring tides; ELWS = extreme low water of spring tides). Source: Cowardin et al., 1979.

FIG. 2.5 *continued*

(b)

29

FIG. 2.6 Distinguishing features and examples of habitats: (*a*) in the lacustrine system; (*b*) in the palustrine system.
Source: Cowardin et al., 1979.

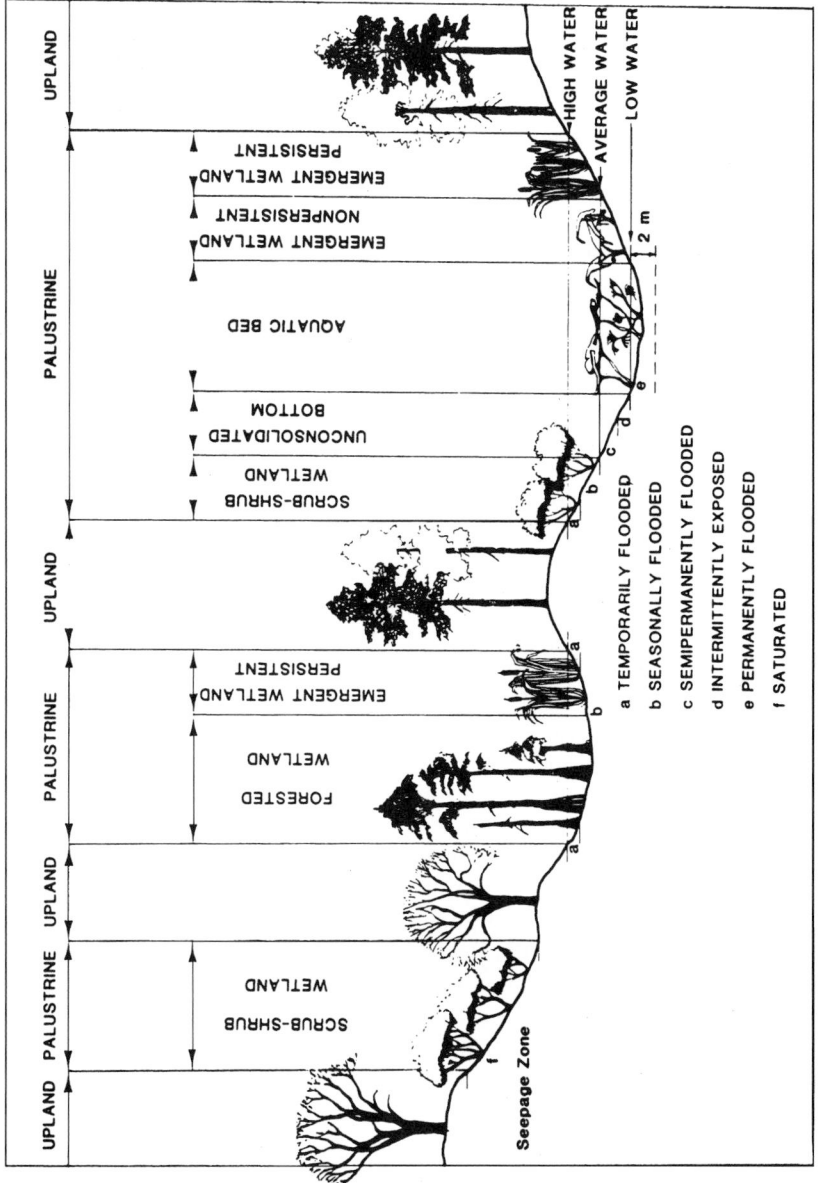

FIG. 2.6 *continued*

(b)

31

as defined by its specific performance criteria. Wetlands of different classifications, to the maximum extent possible, require their mitigation with in-kind bank credits representing similar site conditions.

For a developer dealing with mitigation-related problems, wetland banking can reduce uncertainty and delays in obtaining regulatory approvals and at the same time promote the utilization of a single large wetland for increased ecological benefits, rather than individual smaller fragmented wetlands. Figure 2.7 illustrates an example of a reclamation project by the city of Orlando, Florida.

2.4.3 Environmental Assessment

Another condition that might constrain a developer's activity is the receipt of an adverse environmental report on a potential site. Under the Comprehensive Environmental Response, Compensation, and Liability Act of 1980 (CERCLA), owners of property contaminated with hazardous substances are strictly, jointly, and severally liable for conduction or financing cleanup of such substances under provisions contained under Section 107. In 1986 the Superfund Amendment and Reauthorization Act (SARA) allowed "innocent purchases," a defense to this liability if, prior to property acquisition, "all appropriate inquiry into the previous ownership and uses of the property" along with an "appropriate inspection" of the property was undertaken. A developer who acquires property today with even the slightest indication of possible hazardous substances being present might not enjoy the benefits embodied in SARA and could be liable for all future cleanup costs. In an effort to avoid potential cleanup liability resulting from foreclosure action, where remediation costs may exceed the property's market value, financial institutions and prudent developers are undertaking "due diligence" investigations to protect themselves and the end user. Should contamination be discovered later, developers will bear the burden of proof to demonstrate that due diligence was in fact exercised in order to acquire innocent purchaser status (Gustin and Neal, 1990; Leifer, 1989; Moskowitz, 1989).

Nationally between 130,000 and 400,000 sites have been identified as contaminated, or thought to be so. A number of these most heavily polluted sites have been placed on a Superfund national priorities list for clean-up; however, each of these listed sites takes years and many dollars to remediate. The other remaining sites have varying amounts of contaminate levels that would result in the property being classified as a "brownfield" (property that falls between being a significantly contaminated listed site and one without contamination—a greenfield). Because liability and associated litigation have caused redevelopment gridlock, a large number of these properties (many of which are located in urban areas) sit and degenerate while new development takes place in suburban greenfields, thus promoting and adding to urban sprawl problems (Section 1.2.7). As a result, many state legislatures have passed laws (currently favored by the EPA) supporting redevelopment activities of brownfields, with the idea of attracting new industry and development back where abandoned sites can be remediated and returned to productive use. These sites can take advantage of existing infrastructure, including the availability of water and sewer, transportation, and an accessible work force. State-passed remediation incentives can include tax relief, relaxed remediation requirements, greater flexibility in cleaning the site, and liability relief if the site is remediated according to the approved plan. Although the project developer may be prohibited from using a remediated site to locate dwelling units, it is possible that it

The wetlands are comprised of three vegetative communities: a deep marsh, a mixed marsh, and a hardwood swamp.

Prior to its transformation into a wetland area, the site was an active cattle pasture.

Community organizations like the the Boy Scout troop pictured above utilize the wetlands to learn about nature.

The Orlando Easterly Wetlands was built on a 1,640-acre site in east Orange County near the St. Johns River.

Orlando Easterly Wetlands Reclamation Project Site Schematic

Discharge

Influent

■ Sample station

Permit compliance

FIG. 2.7 Example wetlands mitigation project—The Orlando Easterly Wetlands—City of Orlando, Florida. Shown by permission Post Buckley Schuh & Jernigan, Inc. and the City of Orlando Florida.

33

could be used for recreation, open space, or conservation purposes. This is an option the developer might seriously consider because improved technologies for cleaning up different contaminates are becoming increasingly available, such as air stripping, air sparging (soil vapor extraction, carbon adsorption, biological treatments), pumping and treating, and use of patented granular iron media to degrade chlorinated organic compounds, to name a few examples.

Another type of environmental investigation, precipitated by the National Environmental Policy Act (NEPA) of 1969, may be required if the federal government is involved in the development. HUD requires local environmental reviews for community block grants, rental rehabilitation, and housing development grant programs under the format of an environmental assessment. This process will indicate the project's compliance with applicable rules and regulations while determining whether an environmental impact statement (EIS) or a finding of no significant impact (FONSI) is appropriate (U.S. HUD, undated).

2.4.4 Coastal Zone Regulations

If the potential project is located in an area that comes under state jurisdiction, being regulated through the Federal Coastal Zone Management Act (PL 92-583), the developer may be required to limit development by conforming to established critical zones, setback lines, and pollution control activities that conform to the regulatory coastal zone requirements (South Carolina Coastal Council, 1979, 1988). These regulations help protect coastal land and waterways while providing a coordinated approach to minimizing the effects of erosion-related property losses and flood damage. Structural improvements, such as the installation of docks, bulkheads, or revetments within the coastal zone, may require COE approval under Section 10 of the Federal Rivers and Harbors Act of 1899.

2.4.5 Floodplain Management Regulations

Floodplains can offer lucrative opportunities for developers to take advantage of the waterfront's appeal and natural beauty. As a result, many people have moved into coastal counties, near lakes and rivers, and in other potentially hazardous areas such as seismic zones. The National Flood Insurance Program (NFIP), created in 1968 and later modified through various amendments, focuses on flood hazard mitigation by reducing the amount of property exposed to flooding. This program is administered by the Federal Insurance Administration (FIA) of the Federal Emergency Management Agency (FEMA). Participating communities located in flood-prone areas are offered federal government-backed flood insurance; however, the community must adopt a floodplain management ordinance and enforce it. This community must regulate new construction and development in designated flood hazard areas, established at the 100-year flood level (1 percent chance flood). Flood insurance studies for the community are performed and result in the production of a flood insurance rate map (FIRM) and prior to January 1985 a flood boundary floodway map (FBFM) that are incorporated into the management plan (see Figure 4.8). For coastal communities the FIRMs depict the 100-year coastal floodplain and the elevation of the 100-year flood. The 100-year coastal floodplain is divided into two adjacent zones: the "V zone" and the "A zone." The V zone is the area that is expected to be inundated by tidal surges with velocity wave action and is generally taken to the limit where a 3-foot breaking wave intersects the 100-year 4-foot stillwater flood depth. A zones are not subject to wave action and are also used for riverine conditions. FIRMs also show the 500-year flood boundary, and the area in between is designated a moderate flood hazard area. All remaining areas are above the 500-year flood level.

Communities use regulations such as zoning, subdivision regulations, and building codes to help minimize flood damage in high-risk areas. In addition, low-lying land can be purchased by the community where structures are then relocated and the flood-prone property is used for public recreation and open space. Flood control levees have also been constructed; however, many of them do not reach the required elevation to restrain a 100-year flood event.

The terrain controls much of the development's layout and makes certain parcels unbuildable due to their location within a jurisdictional wetland, floodplain, or drainage basin. Floodplain management practices require the developer to provide minimum grades on roads, establish minimum finished floor elevations on dwellings, and add structural integrity to dwellings also located within floodplain "V zones", or velocity zones. As most developers approach the economics of a project from the standpoint of minimizing costs while maximizing profits, the cost of improvements to accommodate certain topographic features such as ravines, natural bottoms, and sloughs may preclude development of a particular site.

2.4.6 Cultural Resources

Sometimes developers choose a site that either contains or is located near a building, structure, or object that is included (or eligible for inclusion) in the National Register of Historic Places. According to Section 106 of the National Historic Preservation Act of 1966 (36 CFR P 800, as amended), registered sites possess a "quality of significance in American history, architecture, archeology, engineering, and objects." The act also addresses historic properties where prehistoric or historic archaeological resources exist. The goal of this legislation is to further the national policy of historic preservation where federal agencies must take into account the effects of their activities relative to historic properties. Properties identified by the act must be assessed relative to any potential adverse impacts, and alternative ways to avoid, minimize, or mitigate the identified impacts must be explored. Under recent changes to the law that became effective June 1999, early compliance with Section 106 is now encouraged where ample opportunity to explore alternatives and allow enough time to properly plan a project can be found. In addition, federal agencies can now use NEPA (see Section 2.4.3) compliance activities as an alternative means to address these issues.

A land development project that is federally assisted, federally licensed, or federally permitted is subject to the provisions just outlined. In other projects where federal involvement is not incurred, state laws may require an approach following the federal procedures to be used where the state historic preservation officer (SHPO) coordinates the process.

2.4.7 Existing Site Terrain

Attributes of a well-developed project often preexist with the developmental plans, merely magnifying them. For example, if a knoll is present with a deep ravine running through the property, the knoll should be considered a good location to place structures due to view, drainage, and maybe even exposure to prevailing favorable winds. The ravine, a natural drainage feature, should be maintained if possible.

When considering drainage, a gradual and uniformly sloped parcel that has a natural drainage basin into which stormwater can be diverted is the most desirable since this represents site conditions that would require minimal improvement costs. Should the site be level, higher development costs will be incurred because of the increased excavation costs required for gravity drainage line installations. Land that possesses little topographic relief normally suffers from a high groundwater table and, as a result, requires additional earthwork activities to provide positive site drainage features. If the site is located where the closest direct access to a natural drainage basin exists across another's property, then an easement for a drainage outfall may be required to provide proper discharge. Should this easement be granted, potential floodwaters can be properly channeled, and better drainage throughout the area will result. Where natural drainage basin access is unavailable, provisions must be made for holding stormwater runoff until it can evaporate, percolate, or slowly be piped away. This holding process is known as ponding and can be in the form of either retention (runoff is retained without discharge) or detention (runoff is held for a period of time while being slowly discharged).

It would be foolish for a developer to confine all interests in topography strictly to the proposed site just because it has boundaries that limit ownership and development interests. Topographic boundaries (e.g., ridge lines, valleys, and edges of plateaus) usually are not confined to the site boundaries, as the site may be located somewhere between a ridge and a valley. If this is the case, any alteration to the area within the natural drainage boundaries may affect the physical characteristics of interior areas. Should a natural drainage system be realigned to provide better road access, the altered course may result in flooding of adjacent land if there is no place for the water to be channeled to allow continued flow downstream. Developing land also causes runoff to increase considerably, which results from several factors. One factor that accounts for more runoff volume is the increase in impervious areas. The coefficient of runoff associated with development-related impervious surfaces, such as asphalt pavements, concrete sidewalks, driveways, shingle roofs, grass lawns, and others, is much higher than that of forest land, with its prolific foliage and ground cover. This additional runoff must be controlled and channeled through existing outfall drainage facilities. An investigation will determine whether there will be an overload placed on these facilities, resulting in flooding.

Road systems also have a relationship to the existing terrain. Most developers are aware that earthwork is expensive and that it is desirable to minimize cutting and filling operations as far as it is consistent with overall improvement requirements. The better a road adapts to a site's topography, the more construction costs are reduced. Figure 2.8 illustrates three situations reflecting placement of development streets as they relate to site topography.

Terrain directly affects utility improvement costs. Depths of embedment necessary for proper covering and for attaining the required installation slopes are dependent on hilly ground and variation in slope.

Topography also plays an important role in the development of the sanitary sewer system plan. If municipal service is not available adjacent to the site, then an on-site centralized treatment facility might be considered. The project site layout may dictate

SECTION A·A

A. Design Road to Diagonally
Cross Steep Terrain

SECTION B·B

B. Avoid Road Placement
Parallel to Contours

SECTION C·C

C. Streets Placed Perpendicular
to Contours Can Have Excessive
Grades and Necessitate the Need
for Retaining Walls.

FIG. 2.8 Placement of development streets. Source: U.S. Department of Housing and Urban Development.

building the plant in an area that conflicts with the unifying theme of the development, thus causing relational problems. If municipal services are available some distance from the site, the topography of the site and that of the land separating the site from the municipal line are determining factors when contemplating utility line extensions to join existing off-site facilities. These extensions may require the installation of a sanitary lift or pumping station with an emergency generator when gravity flow is not feasible for sanitary sewer lines.

2.4.8   Subsurface Site Conditions

An important concern is the investigation of the underlying soil at the site. Results of this investigation have an effect on mucking and excavation costs, fill material needs, pavement thickness requirements, and foundation requirements. Soil permeability is a parameter that relates to the soil's percolation capability necessary for drain field sewage effluent dissipation associated with septic tank applications.

An attractive site would have fairly stable soils free from rock outcrops and organic content; thus a soil exploration program is necessary to ensure satisfactory results. If the proposed site contains unsuitable soil types, the soil exploration program should yield useful information for correcting the problem. For example, if an area drains poorly as a result of underlying impervious material, subsurface drains may be used to improve these conditions. Data on the soil's permeability would be helpful in this analysis, despite the fact there was no initial requirement to determine the soil's permeability for purposes of evaluating drain field effluent dissipation needs because municipal sewage facilities were available.

Certain basic properties of the soils should be determined for all projects. These are usually confined to the liquid limit, plastic limit, grain size distribution, and possibly the unconfined compression tests, along with the optimum moisture content (OMC) determination for soil located in subgrade and foundation areas. If a multifamily development has sizable structures, additional soil testing may be required.

Geology is another consideration of subsurface investigation. Investigation of a site may reveal fault lines or the potential for landslides or mudslides, which would deter locating the project there. Sinkholes in limestone areas also could pose a problem and therefore should be investigated (U.S. Department of Interior and U.S. HUD, 1970). In addition, the presence of expansive soils on-site can pose a potential problem to the integrity of completed site improvements and structures (Jones and Jones, 1987).

Although groundwater table elevations fluctuate, their range is important when considering a community potable water system. Removal of groundwater from a particular unconfined aquifer may result in lowering the stabilized groundwater table. This results in a cone of depression, called drawdown, which may be of significance. If groundwater is used as the supply, this information must be available to determine a suitable design. On sites having rolling terrain it is necessary to investigate all potential locations where groundwater might seep to the surface through hillsides, resulting in piping, which is a phenomenon that can occur in an earth dam. Should piping exist, it would be hazardous to build in that location, as the ground is saturated and generally unstable. Incorporation of a subsurface filter blanket may minimize this problem.

2.4.9   Climate

Climatic activities, including temperature, humidity, wind, and precipitation, can affect a potential development site and thus give rise to various conditions the developer must address. The area's temperature range should be considered as some design parameters are temperature dependent. Subgrade depths must be of sufficient magnitude to withstand frost action and heave. Also, building foundations and depth of utility line burials are dependent on frost penetration depth. All of these items reflect costs to the developer to improve the land. The range of temperature also plays an important part in the amenities to be provided in the development. For example, if the area is subject to warm summers, a community swimming pool may be included as part of the recreational amenities afforded the occupants. Air conditioning and heating demands are directly related to the ambient air temperature and relative humidity, resulting in costs to the developer in providing these services. Use of active and passive solar heating and cooling systems has proven efficient in providing comfort, with consideration being given to properly positioning structures to provide comfort control and energy conservation.

Precipitation is another factor that affects a developer's final product. Areas subject to snowfall require increased structural strength and abatement provisions for removal. Rainfall affects the scope and magnitude of stormwater collection, delivery, and containment facilities and the frequency and magnitude of flooding. Precipitation also affects the availability of water for consumption by recharging ground aquifers or by contributing to the resupply of surface water sources.

2.4.10 Blight

Unappealing visual, environmental, or auditory conditions can affect the developer's assessment of a potential project's location. Scenes of unpleasant neighboring land use activities may render the site unappealing to the passerby, who might be a potential renter or buyer. Adverse environmental conditions such as hazardous wastes or contaminated groundwater on or close to the site pose serious and significant development potential liability. Loud noises from heavy traffic, manufacturing, aircraft operations, or industrial related activities are considered auditory blight and pose potential problems that can undermine the peace and tranquility of the development.

2.5 SITE LAYOUT

Once the proposed location is chosen, a development plan must be drawn up so that economic analysis can be undertaken to ensure a feasible and workable project. These include a preliminary cost estimate, preliminary financing negotiations, and preliminary negotiations with regulatory officials. The site layout on the preliminary plat is usually one of many that have been generated by a design professional during the project's planning process. Many site plans can be fitted to a particular land plot; however, there is usually one plan that suits best by fulfilling requirements of workability, safety, economy, personal edification on the part of the developer, and finally, compliance with developmental regulatory guidelines.

2.5.1 Development Theme

Site configuration is largely influenced by the theme of the development. If the developer's goal is strictly to provide a neighborhood atmosphere, similar to that proposed by Perry (1929), it would include all necessary services, facilities, and activities for a coordinated and connected setting that would accommodate the everyday needs of the occupants.

Should the developer attempt to attract a certain clientele using a particular marketing theme for appeal, amenities such as golf courses, marinas, and airports have potential for project success.

On a large project, the developer may wish to include some flexibility in his or her marketing plan by sequentially phasing the development. This employs the staged approach in combination with flexible or multiple land use zoning (if available) to compensate for future market fluctuations and trends.

2.5.2   Allocation of Nonresidential Areas

Depending on the size and location of the development, the developer may have to consider allocating some areas of the site to nonresidential uses such as commercial, recreational, educational, and open-space areas (FHA, 1970a).

Commercial facilities such as shopping centers have estimated drawing radii depending on their classification. In addition, there is an estimated minimum population required to support such facilities (Casey, 1968). A neighborhood shopping center of about 50,000 ft^2 of floor space and about 150,000 ft^2 of parking can be supported by about 1000 families. If the development's scope and magnitude warrant consideration of including commercial areas, it would be advantageous to do so, as high rates of return may be obtained if the commercial area prospers. Normally, the developer undertakes the development of a commercial area as a total development activity where all improvements including buildings, parking, and appurtenant work are installed. Figure 2.9 illustrates a typical neighborhood shopping facility.

Theme enhancement can be promoted with the inclusion of recreational areas that involve active or passive participation. Examples of active areas include swimming pools, tennis courts, basketball courts, and tot lots. Passive areas encompass neighborhood and community parks. User safety is paramount; therefore, high visibility from motorists must be provided or adequate spatial separation included. Figure 2.10 shows various types of recreational facilities and their nomenclature.

Nature and conservation areas, which would include fresh- and saltwater wetlands, create passive recreation areas and provide natural beauty while also protecting habitat and timber. Barrier protection and flood-routing benefits also can be attained. Other uses of open space and green belt areas include protecting archeological sites and sites of historic or scenic value. One example is Carter's Grove Country Road from Carter's Grove Plantation to Williamsburg in Virginia. It is a private, one-way drive extending approximately 8 miles in length. When the Country Road was created in the 1970s, it was done simultaneously with the establishment of a private, planned community. This allowed for the road to be protected by a consistent right-of-way that is owned by the Colonial Williamsburg Foundation and by scenic easements in some portions. By following various setback requirements and providing green belts along each side of the corridor, the views along the Country Road remain very colonial and are preserved for posterity.

The developer of a large project may be confronted with issues that center on allocation of areas for educational facilities. Figure 2.11 exhibits one set of guidelines that could vary based on jurisdictional requirements.

2.5.3   Configurations for Residential Developments

A development can take form in many ways; however, there are five basic configurations. The first utilizes a series of orthogonal streets, named gridiron. This configuration is appealing because it is usually the most economical to design and lay out. Also,

FIG. 2.9 Example of a neighborhood shopping center.

41

FACILITY TYPE	SIZE	ORGANIZATIONAL LEVEL (POPULATION)	TARGET AGE GROUP	STANDARD[1]	SERVICE AREA	RECOMMENDED ACTIVITIES AND REMARKS
tot lot	2000-2500 sq. ft.	dwelling group (40)	5 years & under	20f/1000	300 ft. radius	defined areas for individual creative play and areas which encourage social order (stand in line, take turns, share, etc.) remark: this facility primarily intended for use in urban residential areas; may not be feasible in suburban and rural residential areas.
sand lot	½-1 a	small neighborhood (250)	6 - 12 years	4f/1000	¼-½ mile radius	multiuse open area (sports field, social activities, etc.), some play equipment, plus areas like those in tot lot above, but more advanced. remark: this facility may not prove suitable in rural residential areas.
neighborhood park	3-6 a	neighborhood (1500)	all ages	3a/1000	½-1 mile radius	multiuse fields, paved sports facilities (tennis, basketball, etc.), shelters, picnic facilities and areas designed to stimulate individual and social growth.
community park	10-30 a	community (9000)	all ages	2a/1000	3-5 mile radius	recreation center building (possibly with gymnasium), natural and/or nature study area, possibly a swimming facility, plus the same areas which are found in the neighborhood park.
county park	100-300 a	county (54,000)	all ages	3a/1000	10-20 m radius	organized and individual camping facilities (possibly cabins), possibly fishing and boating (especially applicable in Pickens County), plus the same areas found in the community park, natural environmental areas, recreation building (including hardcourt sports facilities) multiuse fields, daycamp facilities, paved sports facilities, could also include golf and large scale spectator sports facilities.
regional park	500-2000a	region (300,000)	all ages	5a/1000	1-1½ hr. drive	wide range of active and passive recreation including day use and overnight use facilities, should provide much the same facilities and services as the county park but with more natural areas and more extensive hiking, camping, nature trail and overnight facilities.

[1]The following abbreviations are used: f-facility; a-acre; m-mile. STANDARDS are expressed in facilities (f) or acres (a) per 1000 population.

FIG. 2.10 Recreational facility standards. Source: E. L. Falk, Planner in Charge. *Pickens County Recreation and Open Space Plan.* Clemson University, 1975, p. 27. Reprinted by permission of Clemson University.

Type	Area Required
Elementary school	10 acres + 1 acre per 100 students
Middle and junior high school	20 acres + 1 acre per 100 students
Senior high school	30 acres + 1 acre per 100 students

FIG. 2.11 Educational facility standards. Source: S.C. Department of Education. *South Carolina Facilities Planning and Construction Guide,* 1983.

construction costs are lower when considering form work for curbing and sidewalks. Certain multifamily developments are conducive to this type of configuration because cost reductions may be possible. In the past gridiron configurations were applicable to communities that experienced little growth and did not want to alter street patterns. Today, however, traditional neighborhood developments (see Section 2.5.5) rely on the use of an orthogonal street system to help promote streets that are interconnected.

Gridiron development, illustrated in Figure 2.12, can produce hazards that are minimal in other configuration types. Long, straight streets can induce high vehicular speeds that should be lowered within the confines of a residential area. One way to achieve this is by using tee intersections, a configuration which exists typically as well-functioning intersections in many older areas of cities all across the United States. Through traffic can intrude into the area because of the accessibility provided. One may find the gridiron arrangement presents a monotonous appearance on paper; however, when viewed in the field, where local topographical features and site improvements interface with the streets, an exhilarating effect can be achieved. Drainage and construction costs may be appreciably higher in this type of layout because it may not conform naturally with the land when executed in the purely abstract sense. However, design of a grid with subtle variations according to the site topography can be very

FIG. 2.12 Example of the gridiron development configuration. Courtesy of Post, Buckley, Schuh, and Jernigan, Inc., Tampa, FL, 1992.

efficient in terms of costs. It produces more efficient building footprints and more efficient use of available land. If the gridiron design is forced, then a curvilinear design might suit more pleasingly and efficiently the terrain and boundary configuration. The curvilinear layout, illustrated in Figure 2.13, consists of roads that follow the terrain in an unrestricted manner while adhering to regulations. The curvilinear method allows greater freedom for the designer to work with the land. This results in winding road patterns that extend through the project and help to reduce traffic speeds. The length of curvilinear roads, compared to an equivalent gridiron design, is usually less, resulting in reduced improvement costs; however, less connectivity is often achieved, resulting in less dispersion of traffic where curvilinear streets need more capacity and channelization. The curvilinear concept provides little protection to residents from off-site traffic. Vacant land and open space are not usually accommodated when either of the aforementioned patterns of land developments are undertaken; however, they do allow for preservation of some natural terrain features, which may result in theme improvement while reducing construction costs attributed to drainage, earthwork, and accessibility requirements. A single-family design technique that essentially utilizes a curvilinear street network can be found in "coved" developments. A coved subdivision is one where larger than usual lots are created so that the front foot distance of each lot is reduced. Because the street system meanders through the site and the dwellings are set a uniform distance back from the right-of-way, the feeling of open space is gained by one riding along the road. The position of each dwelling is such that only the front of the structure is seen from the road. Because this approach sets the dwelling well

FIG. 2.13 Example of a curvilinear development configuration. Courtesy of Post, Buckley, Schuh, and Jernigan, Inc. Tampa, FL, 1992.

away from the right-of-way, the setback area can be used for utilities and walkways. Streets are designed where traffic speed is reduced and cul-de-sacs are often reduced.

Another type of development, known as the density-zoned development, at times called a cluster development, is shown in Figure 2.14. This approach allows even more flexibility in preserving natural topographical features and vegetation. This type of development is controlled by zoning that restricts the number of dwelling units per gross acre. Some regulations require the developer to allocate a minimum percentage to open space and public facilities while maintaining a maximum allowable percentage for commercial and residential areas. This configuration results in groups of dwelling units being surrounded by greenbelts and open-space corridors.

Density-zoned developments provide many things missing from rectilinear and curvilinear developments. They allow the developer to create a project of any size, incorporating any dwelling type or size, provided it conforms with the intent of the zoning laws. Unbuildable areas, such as wetlands, are utilized as open space, while more suitable areas are used for dwelling unit locations. Improvement costs are thus reduced. Increased access and serviceability are obtained through the use of density-zoned developments. Amenities such as variations in parking, interesting courtyard layouts, and sculptured features can be provided by development. Lot sizes are small, providing reduced maintenance obligations to the occupant. Density-zoned developments also provide for seclusion from nonessential traffic, thus affording greater safety for residents. For the developer, this type of development allows for reduced improvement costs and often eliminates the need for curb and gutter and sidewalks. This concept accommodates amenities such as golf courses, picnic areas, playgrounds, conservation, and historic preservation because adequate land is available with no sacrifice in density.

There are, however, additional problems associated with density-zoned developments. One is maintaining and controlling open and vacant land available for use by residents. There are questions about whether the common land should be dedicated to the public, maintained by the dwelling unit owners through a homeowners' association, or maintained by the developer. Another problem might be that local ordinances do not permit density-zoned developments. Legal action may be required to amend the local ordinances.

A common misconception about a density-zoned development is that inadequate recreational areas are better than none. This may seem true until a golf ball shatters the picture window of a dwelling fronting on a cramped golf course. Amenities and basic essentials must be adequate for their intended usage.

Figure 2.15 depicts a cluster development, which is a particular kind of density-zoned development. What distinguishes the cluster-type development is that dwelling unit sites are located in clusters centering on cul-de-sacs or common courts and each of these clusters is separated from the other by open space. In density-zoned developments the dwelling unit sites are not necessarily located in scattered groups or clusters but may all be centrally positioned. Both, however, allow for buildings to be placed for maximum effectiveness while minimizing construction improvement costs.

Planned unit developments (PUDs), which evolved during the 1960s and 1970s, are similar to density-zoned developments. Planned unit developments provide for a more flexible approach to development where overall sites are planned and approved, rather than using a lot-by-lot approach. They contain both residential areas and recreational open space and may include commercial sites that can be centrally and conveniently located. A PUD is designed to self-regulate common areas using

VILLAGE 20

75' x 110'

247 ac.
80 units

FIG. 2.14 Example of a density-zoned configuration. Courtesy of Post, Buckley, Schuh, and Jernigan, Inc., Tampa, FL, 1992.

FIG. 2.15 Example of a cluster development configuration. SFD1 and SFD2 denote zoning classifications. Courtesy of Post, Buckley, Schuh, and Jernigan, Inc., Tampa, FL, 1992.

47

a homeowners' association that is legally organized and chartered. Such an organization, which can be formed for other types of developments as well, is composed of members of the planned unit. The homeowners' association governs, through authority embodied in its charter, such regulatory items as new development conformance, restrictive covenant enforcement, grounds maintenance, operation, and general character of the area. Most PUDs have a central theme such as a golf course, swimming pool, marina, beach, or air facility. Other amenities might include parks, boat docks, jogging trails, club houses, tennis courts, and bicycle facilities (Tomioka and Tomioka, 1984). Owners of dwelling units within the PUD have membership privileges to any or all of the facilities that are provided. The homeowners' association usually assesses each dwelling unit owner a flat fee, depending on the situation, to support and maintain all common facilities. A developer should recognize that a sizable amount of legal work is necessary to set up and protect such a development. In addition, the Federal Housing Administration (FHA) has ruled that any development that has mandatory homeowner association fees (even if it is only one dollar) qualifies as a PUD; therefore, if potential purchasers are to have the opportunity of utilizing FHA-insured financing, the developer must submit specific information to the FHA for approval.

Condominium developments, which can include high-rise, townhouse, and detached units, follow the general pattern of the PUD, except that the individual units may have a specific amount of land sold with each unit designated for yard space.

2.5.4 Types of Dwelling Units

There are many types and styles of residential dwelling units, ranging from single-family detached to multifamily units. A single-family detached unit represents a stand-alone unit situated on a single lot (Figure 1.1). Most often these units are constructed on-site; however, they could be manufactured elsewhere and then transported to the site. The single-family detached unit is used in low-density developments (Figures 2.12 and 2.13); however, it can be used in density-zoned and cluster developments (Figures 2.14 and 2.15). Sometimes these units are designed using clustering where a common driveway (and possibly a courtyard) is shared by several homes. (Density ranges are shown later in Figure 2.30.)

Another type of dwelling unit is the detached patio home. Patio, or zero-lot-line (ZLL), homes are situated along (or close to) one of the lot's side lot lines where the contiguous wall of the unit has no access or windows. As a result, all access and views are oriented to the opposite side of the unit. The usable portion of the lot provides a private yard space and views not shared by neighbors. These dwelling units must be properly designed to take advantage of any side and rear orientations often using large expanses of glass. A modification to the zero-lot-line dwelling is the detached Z (or angled Z) lot configuration. This design eliminates having one long windowless wall on one side of the dwelling because windows are provided on all four sides while maintaining privacy. (Density ranges for ZLL units can be found later in Figure 2.30.) Another similar unit is the wide-shallow lot where the street frontage is much greater than the side lot-line distance. The resulting small backyard requires a privacy fence or extensive landscaping to retain an area free from outside intrusion. Small-lot developments require a more integrated use of the indoor and outdoor areas while providing adequate parking and privacy.

When two single-family units are attached to each other, they are called a duplex unit (three units—triplex). When additional numbers of single-family units (usually six to eight) are attached in rows, they are called town houses or row houses and their net density can range between 10 and 12 units per acre.

Multifamily units include structures where many dwellings are located. Often these units are used as part of apartment complexes and condominium projects. The services of an architect are required to achieve optimum spatial utilization and preservation of open and free space in this type of development.

2.5.5   New Approaches to Development Planning

As a result of urban sprawl (Section 1.2.7), development-related problems, and a reduction in sites suitable within areas of expanding growth, land planners and architects during the 1980s formulated an innovative approach to urban designs initially called "traditional new towns" but now referred to as traditional neighborhood developments (TNDs). They can vary from several hundred acres in size to tens of acres and are usually not incorporated. The goal is to provide the connectivity and mixed uses which are characteristic of a small town or a more dense urban fabric. The movement is termed "new urbanism." According to Lewis (1990), Miami architects Andres Duany and Elizabeth Plater-Zyberk have pioneered the use of

> Traditional townscape imagery, streetscapes, and civic spaces, street grid patterns, mixed uses and higher densities, new approaches to zoning, multi-modal transportation and "pedestrianization."

Traditional new towns have a town center and identifiable civic spaces with spatial identity, a system of tree-lined narrow streets that integrate with site conditions to produce rectilinear or curvilinear grids that range from alleyways to avenues and pedestrian/bicycle facilities. Since the use of cul-de-sacs are discouraged, the short blocks promote traffic interconnectivity and dispersion and at the same time allow for interfacing with mass transit facilities. Parking is provided on nearly all streets and within blocks, at building sides or to the rear. Pedestrian paths/bikeways are networked with regional contiguous greenways. Zoning and mandatory design codes define building typologies which help shape the character of neighborhood elements, including streets, alleys, pocket parks, neighborhood greens, town centers, and regional green spaces.

Civic buildings are located to provide identifiable landmarks. As terminated vistas, they form part of the streetscape system, along with a matrix of various dwelling types and mixed uses. The TND evolved as successor to the PUD. Figure 2.16 shows a neo-traditional development.

According to Spielberg (1989), the TND has regulated maximum and minimum proportions of land allocated to various uses (housing, retail, commercial, services, and light industry). In addition, other requirements, such as pedestrian accommodations, lot sizes, layout, and street frontages have been determined by Duany Plater-Zyberk & Company to be within a "transect" of conditions ranging from "rural to urban, or center-general-edge"[*]

---

[*] Duany Plater-Zyberk & Company, 1999.

FIG. 2.16   Illustration of I'On Neighborhood in Mount Pleasant, SC. Rendering by Grenfell Architecture, courtesy of the I'On Company, Mt. Pleasant, SC.

The TND is contrasted in Figure 2.17 with typical development patterns that have prevailed since the 1950s, and is predicated on historical dwelling patterns from ancient civilizations all the way to the early 1900s. In the mid-1900s, development became increasingly predicated solely on the needs of the automobile, rather than jointly on the needs of the human. For the last thirty years municipalities have increasingly become burdened by sprawling development, straining public infrastructure and their ability to extend municipal services. As a result, many traffic and land-use-related problems which have evolved over this time period are addressed through TNDs. The planning and design of a TND project must be approached where all TND principles are wholly integrated by an interdisciplinary team in order to achieve the

Suburban Sprawl

Traditional Neighborhood

FIG. 2.17   Example of a community developed as a traditional neighborhood development (TND) (bottom) versus that developed under suburban development guidelines (top). Taken from *ITE Journal*, September 1989, p. 17, "The Traditional Neighborhood Development: How Will Traffic Engineers Respond?" by Frank Spielberg. Copyright © 1989 by the Institute of Transportation Engineers. Reprinted with permission.

desired results that promote good urbanism. Thus, the application of one or two TND standards (e.g., reducing street widths) will not necessarily create a successful TND. The reader is referred elsewhere for more complete coverage of this important topic (Spielberg, 1999).

Changes in local development regulations may be required to accommodate a TND because streets are scaled to slow traffic and promote the accommodation of pedestrian movements while street intersections are more prevalent to promote the dispersion of vehicular traffic. Some communities, however, have promoted "affordable housing" developments, where development regulations have been modified to accommodate small-lot developments. These modifications include some provisions similar to those used by a TND such as a reduction in lot areas, frontage, and setbacks, although smaller lots and increased densities alone fail to provide the connective, fine-grained fabric needed for authentic urbanism.

2.5.6 Aesthetics

To promote a project's beauty and harmony, the developer usually relies on architects and landscape architects. As part of the professional team, they work to refine spatial relationships that yield pleasant views and vistas, resulting in an aesthetically pleasing experience (Rubenstein, 1987). Development planting schemes can enrich aesthetic response by including trees that are suitable to the area's climate, exposure, and soil conditions. Subdivision and roadway design regulations used for suburban neighbor-hoods can permit trees and plantings of specified sizes using strict controls where sight distances are not compromised for driver safety. In addition, the location of new plant-ings is often far enough away from the curb or shoulder to reduce serious accidents if the driver loses temporary control of the vehicle. Suburban neighborhood design often uses front setbacks of 15 feet or more, which results in lots having large depths.

Neotraditional neighborhoods rely on the presence of trees to give continuity to the streetscape. In addition to providing a visual transition between street and structures, trees offer relief by reducing temperatures and creating shade. More importantly, continuous street trees help provide spatial definition, an essential part of creating a comfortable street environment. Because dwellings in neotraditional developments are placed close to the road right-of-way, trees play an important role in promoting the usefulness of front porches touted by TND principles.

Regardless of the design approach used, most developers today are interested in preserving open areas and green space because home buyers like these features. The use of conservation easements that afford interconnectivity between adjacent sites links the community together and provides the public synergistic benefits not gained through indiscriminate creation of unrelated green-belt areas.

Some communities have tree ordinances that help to protect mature on-site trees and provide for plant renewal as part of the construction activities. Deciduous and evergreen shrubs, hedges, and ground cover can control erosion and also provide beauty. Natural open spaces and curvaceous lake perimeters can help promote aesthetic continuity. Planting associated with parking lots helps provide visual screens and sight continuity.

The Historical Preservation Act of 1966 as amended (Section 2.5.2) may affect the aesthetics of a development if federal participation is involved. If the site includes areas of historic significance similar to those in the National Register of Historic Places or if it has natural resource features that warrant preservation, the developer should consider the inclusion of conservation easements to protect these valuable assets.

2.5.7 Transportation Considerations

Transportation systems within a development are major concerns of the developer. Facilities to allow the safe and efficient movement of people and goods are essential and should be made within the framework of the existing urban transportation plan. Vehicular driver requirements must be coordinated with pedestrian and bicycle activities to promote the development's transportation network (ASCE et al., 1974).

Initial concerns to the developer are the proximity of adequate highway and street accesses to allow service to the site. Off-site studies of adjacent land can provide data that will be useful in determining the adequacy of existing access facilities. Streets within the development can augment and reinforce the jurisdictional area's transportation plan.

Generally, road system requirements are prescribed in local regulations that attempt to create networks that promote ease in vehicular accessibility and require minimal maintenance. These regulations exist in the expectation that sound engineering practice and principles, as well as quality construction, will yield a workable, efficient, and safe development that, during the life of the improvement, will minimize cost to the taxpayer by minimizing maintenance activities.

Public residential streets are peculiar in that they produce no tax revenue but require tax dollars for maintenance. In addition, the developer receives no direct monetary compensation for the sale of streets but is obligated to pay for their construction. To minimize construction costs for the developer and maintenance costs for the public body assuming custody, the roads should be as narrow as possible per dwelling unit while still yielding a safe and workable system. Utility costs generally are closely related to the length of road per dwelling unit in a project; thus, a reduction in road length usually results in reducing utility construction costs.

When at least two entrances into a site are included, emergency access can still be provided even if one of the entrances becomes blocked. Another benefit of providing multiple entrances is to reduce the concentration of traffic by allowing the dispersion of vehicular movements.

Figure 2.18 shows the general types of publicly dedicated streets found in suburban neighborhood designs based on their use and function. Arterial streets handle large traffic volumes, resulting in heavy usage. Driveway entrances should be avoided along these streets for safety purposes. Collector streets act as feeder elements to convey traffic from local to arterial streets. The main entrance to a residential development can act as a collector street if the development is large enough. Local streets are usually designed to prohibit through traffic and to provide vehicular access to collector streets from interior property locations. A local street that allows frontage property protection from heavy arterial traffic is called a frontage street. Figure 2.19 illustrates two methods of protecting property contiguous to arterial streets by using a frontage street access approach or reverse frontage lots with protective plant barriers, fences, or walls that are located in a nonaccess easement to prevent double-frontage access. Lots with access to more than one street, other than those located at street intersections called corner lots, have double frontage, which is undesirable unless precautions are taken similar to the use of a nonaccess easement (U.S. HUD, 1973b).

FIG. 2.18 Types of streets found in developments. Taken from A *Policy on Geometric Design of Highways and Streets.* Copyright © 1994 by the American Association of State Highway and Transportation Officials, Washington, DC. Used by permission.

FIG. 2.19 Arterial street lot access: (*a*) frontage streets; (*b*) reverse frontage with protective screen easement. Source: U.S. Department of Housing and Urban Development.

A cul-de-sac is another suburban neighborhood street type that permits flexibility in utilizing various difficult areas within the site. Normally cul-de-sacs are limited to a maximum length by local regulations. Cul-de-sacs assume different configurations that allow flexible applications, including circular, "T," and "Y" shapes. Figure 2.20 illustrates several circular applications.

FIG. 2.20 Examples of cul-de-sac configurations. Source: U.S. Department of Housing and Urban Development.

Neotraditional neighborhood street design relies on a network of interconnected streets and sidewalks. The function of this network is to promote safe pedestrian and bicycle activities and at the same time disperse motor vehicular traffic throughout. Local streets therefore are assumed capable of handling not only local residential trips but also some through traffic. Furthermore, with mixed uses and higher densities, fewer trips are generated. Shared parking facilities also become feasible and result in more efficient use of space relative to conventional parking facilities. On-street parallel parking serves the dual purpose of efficient use of streets and acts as a protective barrier between pedestrian and travel lanes. The use of cul-de-sacs is discouraged because dead ends cannot promote networking. Long street segments can be routed where open spaces, such as a park, are used to interrupt the direction of travel so that motor vehicles must alter direction, which helps to discourage speeding.

Narrow streets, called alleys, usually have a minimum width of 20 feet, which allows on and off loading activities in commercial and industrial areas. Alleys are often not allowed in suburban neighborhoods; however, they can be used in neotraditional neighborhoods and small-lot residential developments to improve the streetscape or eliminate curb cuts for driveways. Alleys can also be used to locate underground utilities, provide access for garbage collection, and allow emergency vehicle access. By providing off-street access at the rear of each lot, more on-street parking can be achieved.

Another transportation-related consideration involves reserve strips that can cause problems for the developer. Reserve strips occur at the end of a terminal road between a road right-of-way and the property boundary for the purpose of limiting road access and thus prohibiting adjacent landowners' use of the road. If off-site street access is needed to service an interior portion of the site and a public road exists close to the boundary but is separated by a very small strip of land called a reserve strip, the chances are very good that access though the strip will be granted by the jurisdictional authorities. In the case of a private road, a reserve strip is legal; however, a knowledgeable landowner who seeks safety and orderly traffic patterns may allow access to promote good planning if the imposed traffic does not cause problems. An exception occurs when a development site is landlocked. Then the matter is usually decided in court if the adjacent landowner does not concede an access. Should the developer require access to a residential development through adjacent off-site property, an ingress-egress easement, if granted by the adjoining landowner, would accommodate this access.

Finally, the name of streets is usually a requirement for all developers to undertake as part of the regulatory process. Figure 2.21 illustrates a method used by the Montgomery County (Ohio) Planning Commission to preserve continuity; however, care must be exercised to ensure that existing names are not duplicated, resulting in confusion.

The developer must consider the right location within the project for signs that provide both information and traffic control. Motorists expect traffic signs and pavement markings to conform to the standard established in the *Manual on Uniform Traffic Control Devices* (MUTCD) (U.S. Department of Transportation, 1988), and the developer should insist that these standards be maintained if traffic control devices are required. Should a traffic accident occur as a result of a nonstandard traffic control device being installed, the developer may be liable as a contributory party.

Street intersection design is an integral element of the road system. No more than a maximum of four streets should be allowed to intersect at any one point. Three-way intersections reduce the number of conflicts and are preferred if centerline offsets

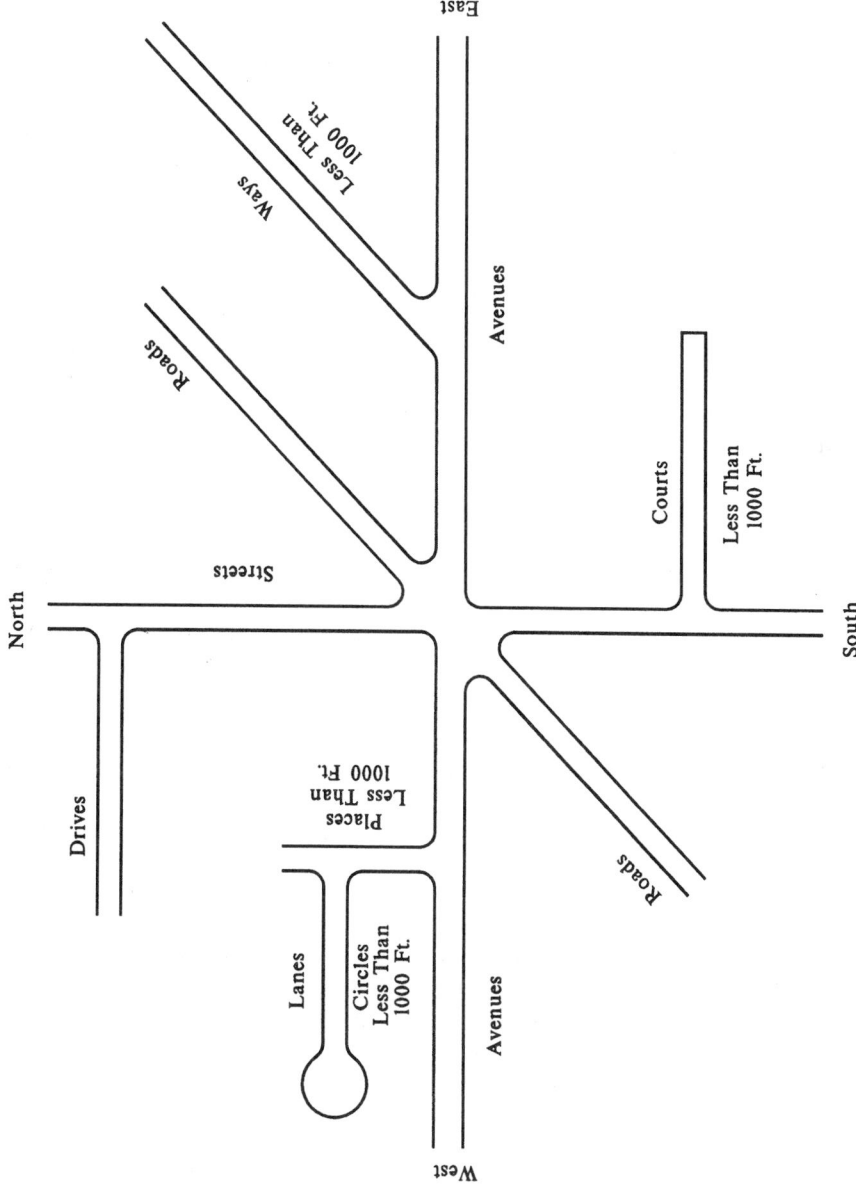

FIG. 2.21 Method for naming streets. Source: Montgomery County Planning Commission, Dayton, OH. Reproduced with permission.

of 125 feet or more are available to provide adequate sight distances, as shown in Figure 2.22.

The angles used in intersections should be as close to right angles as possible, as indicated in Figure 2.23. If large volumes of traffic pass through an intersection, consideration should be give to the inclusion of turning lanes.

Parking requirements vary depending on the type of development and zoning requirements. Most single-family developments utilize off-street parking; therefore, no special provisions are usually required. For multifamily and commercial developments, adequate parking facilities must be available to provide convenience and safety to the user. Parking areas must be proportioned for ease in vehicle maneuvering and proper drainage.

2.5.8   Pedestrian Considerations

In large-scale developments there is an acute need for included pedestrian facilities to augment and complement circulation patterns formulated by local authorities. Large developments may contain open space or recreational areas where safe pedestrian access can enhance the overall project cohesiveness. This will simultaneously allow for such activities as lingering and browsing or an opportunity for a user to socialize with neighbors or perhaps enjoy the scenery. Other pedestrian activities might include trips related to work, shopping, school, and a means to interconnect with public transportation facilities.

Pedestrian facilities should be fairly level. Obstacles, obstructions, lack of maneuvering room to avoid others, and congestion resulting in an inability to maintain reasonable walking rates are potential problems. System continuity that incorporates a minimum number of potentially conflicting situations with vehicles and other pedestrian movements enhances user efficiency and comfort.

User safety entails not only a consideration of vehicle and pedestrian crossings but also protection from pollution associated with noise, refuse, and odor. Amenities such as street furniture, fountains, and other aesthetic improvements are beneficial and can help to convey the underlying project theme. A developer's use of adequately designed ramps, steps, guards, and safety barriers can provide pedestrian safety, as can the use of night or security lighting to enhance visual safety.

2.5.9   Handicap Access

Physically handicapped access should be of importance to a developer. The Americans with Disabilities Act (ADA; PL 101-336), provides that all private establishments/facilities considered "public," such as those in some commercial developments, must provide "reasonable accommodations" to disabled persons. Also, the 1988 Fair Housing Act amendments address handicap access in multifamily developments. The Architectural Barriers Act (PL 90-480) requires all developments that contain public facilities, such as retail stores, or that rely on federal grants, loans, or utilization to provide ample accessibility to all pertinent building and site facilities, including those for trash disposal and mail pickup. This encompasses not only federal-assisted rental and home ownership units but also federally owned properties and qualified assembly areas.

Initially, the American National Standards Institute (ANSI) approved the first national standard for access design in 1961, which was designated as ANSI A117.1. In

FIG. 2.22 Street intersection types and resulting traffic patterns. Source: U.S. Department of Housing and Urban Development.

59

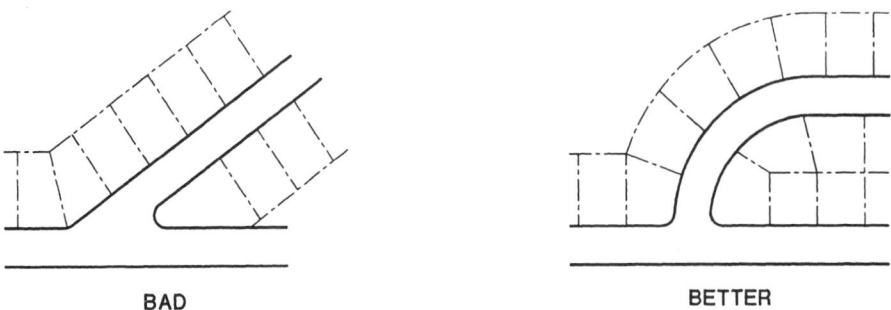

BAD   BETTER

FIG. 2.23 Examples of desirable and undesirable street intersection configurations in suburban neighborhood developments. Source: U.S. Department of Housing and Urban Development.

1980 modifications were made and incorporated as part of the *Minimum Guidelines and Requirements for Accessible Design* (MGRAD) that was mandated by Congress. In 1984 the General Services Administration, the Department of Defense, the Postal Service, and the Department of Housing and Urban Development promulgated the Uniform Federal Accessibility Standards (UFAS), which were consistent with MGRAD.

According to the UFAS, a number of aspects must be addressed, including provisions for adequate passage and maneuvering space to accommodate handicap transit from public streets, sidewalks, and passenger loading zones to the closest serviceable building access. Handicap parking spaces bearing the recognized symbol and loading zones must be contiguous to a 5-foot free-access zone, as illustrated in Figure 2.24. The minimum number of designated spaces should be in accordance with Figure 2.25. Provisions promoting handicap access are illustrated in Figure 2.26. If the access grade is greater than 20 : 1, then it is treated as a ramp, which may require landings and handrails, and cannot exceed a 1 : 12 gradient (General Services Administration et al., 1988).

2.5.10   Bicycle Considerations

Some projects allow developers to provide special facilities for bicyclists, called bikeways. Bikeways should be smooth riding surfaces free from construction joints, holes, and differential settlement. Special attention must be given to drop inlet grates, manhole covers, rumble strips, rails, and bridge expansion joints where bicycle tires are afforded safe support that promotes operator control while crossing (AASHTO, 1999).[*]

Bikeways often share usage with other transportation modes. Typically these facilities are designated as shared roadway with no bikeway designation, shared roadway with designated bike route signs, bicycle lane, and shared-use path. Figure 2.27 illustrates these different bicycle facilities. The designer should incorporate adequate horizontal and vertical design controls to allow proper maintenance of sight distances. Bike/vehicular/pedestrian crossings should be kept to a minimum for improved safety.

---

[*] From *Guide for the Development of Bicycle Facilities*, Copyright © 1999 by The American Association of State Highway and Transportation Officials, Washington, D.C. Used by permission.

FIG. 2.24 Handicap parking details. Source: General Services Administration, et. al. 1988.

Total Parking in Lot	Required Minimum Number of Accessible Spaces
1 to 25	1
26 to 50	2
51 to 75	3
76 to 100	4
101 to 150	5
151 to 200	6
201 to 300	7
301 to 400	8
401 to 500	9

FIG. 2.25 Handicap parking requirements. Source: General Services Administration, et. al. 1988.

2.5.11 Storm Water Management

In addition to local regulations, storm drainage facilities installed by the developer are subject to provisions of an areawide water quality plan implemented as a result of provisions contained within Section 208 of the Clean Water Act, as amended. This plan coordinates the overall surface water quality within the area. Point sources, such as culvert discharges, and nonpoint sources, such as overland sheet flows, are included in the analysis. These waters contain settled air hydrocarbons, metals, eroded soils, salts,

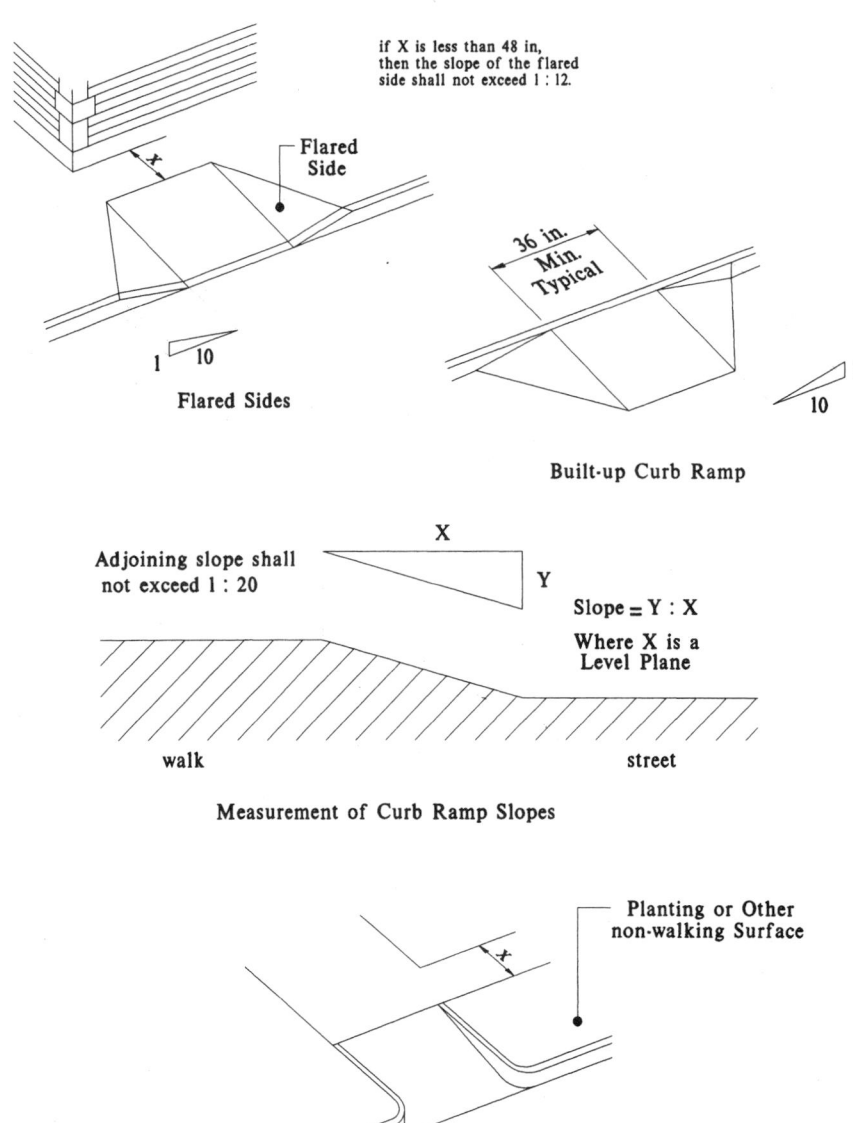

if X is less than 48 in,
then the slope of the flared
side shall not exceed 1 : 12.

Flared
Side

1 ⎓ 10

Flared Sides

36 in.
Min.
Typical

1
10

Built-up Curb Ramp

X

Adjoining slope shall
not exceed 1 : 20

Y

Slope = Y : X

Where X is a
Level Plane

walk   street

Measurement of Curb Ramp Slopes

Planting or Other
non-walking Surface

X

Returned Curb

FIG. 2.26 Handicap access requirements. Source: General Services Administration, et. al. 1988.

chemicals, litter, and animal refuse, with the highest concentration for pollution being contained within the first 1 inch of runoff. Best management practices (BMPs) are encouraged to reduce adverse clean water conditions. Wetlands can be used to remove nutrients, soils, and other pollutants, as can retention/detention ponds with filtration. Local authorities may not issue building or grading permits prior to the developer's compliance with the overall management plan.

Under the EPA's General Storm Water Permits Program, which became effective October 1, 1992, development sites comprised of one or more acres (as of March 1999)

B. Bike Lanes--On Street Parking

C. Two-way Shared Use Path

FIG. 2.27 Types of bicycle facilities: (a) shared roadway; (b) bike lanes incorporated with on-street parking; (c) two-way shared use path. Taken from *Guide for the Development of Bicycle Facilities*, Copyright © 1999 by The American Association of State Highway and Transportation Officials, Washington, DC. Used by permission.

of disturbed land are affected and require a National Pollution Discharge Elimination System (NPDES) general permit. To qualify under the general permit, a "notice of intent" (NOI) must be filed at least 48 hours prior to construction ground breaking, and a stormwater pollution prevention plan must be available on-site for review by regulatory authorities if requested.

This plan must include a description of the development site and a map showing areas that will be disturbed during construction. The plan must also show the location of stabilization activities and control measures. The location of surface waters at discharge points as well as the location of slope and drainage patterns after development must be included. This plan must also contain the sequencing of construction activities (supplied by the contractor). Other requirements of the plan include

(a) an estimate of the coefficient of runoff from the site after construction,

(b) data on the quality of the stormwater draining from the site, and

(c) an inventory of the types of soils located on-site.

Provisions for sediment basins are also required for development sites where 10 or more acres of land are to be disturbed. These basins must provide at least 3600 ft^3 of storage for each acre drained. In addition, the plan must include controls for vehicle soil tracking and disposal of construction solid wastes while providing for site inspections every seven days, or within 24 hours after a rainfall of 0.5 inch or more has occurred.

Should the development be located within a designated urban area or within the jurisdiction of a medium or large municipality where the population exceeds 100,000, point source discharges will be subject to provisions of an NPDES permit, as provided under Section 402 of the Clean Water Act, as amended. Discharge data collection, sampling, and analysis are required with remedial treatment action being prescribed if necessary (U.S. EPA, 1990).

Diffused surface stormwater has been considered a "nuisance" (Rice and White, 1987) as each landowner can dispose of it as long as it does not affect another property owner in a detrimental manner. Installed drainage facilities therefore should be placed to protect other site improvements from flooding while minimizing erosion and excessive ponding in streets, which would hamper vehicle travel.

Operation, maintenance, control, and ownership of the completed facility is normally undertaken by the jurisdictional governing body as public works activities. However, with increased emphasis being placed on managing stormwater facilities such as detention ponds, the developer may find some municipalities have formed separate storm and surface water utilities. These storm and surface water utilities have vested authority to generate operational revenues by assessing monthly charges, similar to those for water and sewer, from landowners based on individual runoff contributions (Bissonnette, 1985). Typically these storm drainage utilities access the adequacy of the drainage system under their jurisdiction while taking into account existing and new developments. Proposed improvements to accommodate developing watersheds are evaluated so that an estimate of the cost of improvements can be ascertained. A plan is then developed where improvements are funded based on priorities to meet new system demands.

2.5.12 Wastewater Management

Proper wastewater disposal is an essential element of a successful project. The developer may be fortunate enough to find an existing wastewater main which was installed in concert with an areawide wastewater plan, provided by Section 201 of the Clean Water Act, as amended, that is close enough to the project that feasible tie-in arrangements can be made. Improvements would include collection lines, with appurtenances, and, if required, a pumping facility with force main to allow a means to transport wastewater to the connection point.

Should the development be distantly located from an existing system, an alternate means of handling the wastewater might include on-site ground disposal through the use of properly designed septic tank and drain field systems that will neither contaminate groundwater nor leach into waterways. Another means might be the installation of a central wastewater treatment plant facility.

In compliance with requirements of the Clean Water Act, as amended, a central facility would have to play an integral role within the areawide 208 plan. The developer could obtain a Section 401 water quality certificate by utilizing spray irrigation on suitable land, such as a golf course, to dispose of the treated wastewater. Should a point discharge be proposed, the developer would need an NPDES permit for treated effluent release based on dissolved oxygen sag curve depletion analysis and other parameters.

2.5.13 Potable Water Considerations

Every effort should be made by the developer to provide potable water from an established municipal system. Should this not be available, then a suitable supply from a reliable groundwater aquifer or surface impoundment must be located. The U.S. Bureau of Reclamation and the Natural Resources Conservation Service (NRCS)— formerly the U.S. Soil Conservation Service (SCS)—are federal organizations involved with water supply activities. Treatment and storage facilities must be provided so that safe drinking water can comply with the National Primary Drinking Water Regulations and the Safety of Public Water Systems Act (P.L. 93-523). Should conditions warrant the withdrawal of water from one basin with ultimate discharge being made into another, a regulatory interbasin transfer permit likely will be required.

Availability of water for a development's supply from groundwater aquifers and surface water impounds involves water rights doctrines that vary throughout the country. Surface waters can be classified as following riparian or appropriation doctrines. The riparian doctrine allows a contiguous landowner the right to "continual natural flow of the stream" (Rice and White, 1987) with reasonable use. The appropriation doctrine, which prevails where water is scarce, promotes the taking of water for beneficial use based on "first in time is first in right" (Rice and White, 1987). Control of groundwater falls under several doctrines. One is the absolute privilege doctrine, which basically provides overlying rights, without taking into account groundwater movement. Another is the reasonable use doctrine, which controls extracted groundwater strictly to use on the overlying property. Correlative rights define shared usage based on overlying ownership, whereas the prior appropriation doctrine examines percolating groundwater as underground streams. Permitting encompasses the remaining control system, which is normally handled on a state and regional basis, along with Federal Reserved Water Rights. Water lawyers and engineers proficient in water rights allocations will be required by the developer if a water supply is to be perfected for a development. For further details, consult specialized references on this important topic (Rice and White, 1987).

2.5.14 Electrical Considerations

Most power company policies vary with respect to developer responsibilities (South Carolina Electric and Gas, 1967). To qualify for underground service, the developer may be required to have at least 24 contiguous subscribers situated in a configuration acceptable to the power company. A minimum of 15 feet dispersion may be required between the dwelling unit and the sewage system (either centrally located system and appurtenances or individual septic systems). A manufactured housing development (mobile home type), which can take the form of a subdivision or park, must have a minimum lot width of 40 feet. The terrain must not be excessively rocky or entail unusually nonuniform slopes. Many companies prefer rear-lot electric distribution configurations (a must for manufactured housing developments), as most water and sewer services are along the dwelling unit front. In the case of the manufactured housing (mobile home) park, a five-year contract between the power company and the developer could allow the power company to amortize its installation investment. The preliminary plat generally contains the minimum information needed by the power company for initial design purposes. In addition, the developer must indicate the planned project construction time and phasing, which must be consistent and logical. To assist the power company in the system design, an estimated electrical load for each dwelling unit must be provided by the developer. Once these requirements have been satisfied, a wiring diagram for the project can be formulated by the power company. The developer is then able to review the proposed wiring diagram and grant utility easements necessary for plan implementation. If the developer expresses dissatisfaction with the plan, which is most likely to occur in street lighting schemes and transformer pad locations, the power company will normally make revisions with the understanding that any additional construction costs will be paid by the developer.

Electrical lights are necessary along streets, parking lots, walkways, and areas requiring nighttime illumination. In suburban neighborhood designs lights are usually provided using high poles spaced at relatively large distances. In neotraditional designs street lighting is scaled back where less intense luminaries are used on lower pole mounts. The intent is to provide illumination to the streetscape where the nighttime sky can also be seen. Electrical lights are available in several styles, with the developer usually paying for nonstandard types, if selected. A contractual agreement that stipulates how revenue will be raised by the power company to support lighting services centers either on the developer or dwelling unit occupant and is normally undertaken as part of the project planning process.

Wiring configurations vary according to development type and are based on where the power company terminates its facilities and where the developer must make connection. In the case of apartments and mobile homes, the power company may supply all material and labor through the installation of meter boxes. The service entrance must be provided by the developer, and if materials of a certain established quality are provided by the developer, power company reimbursement may be available. In most single-family developments the developer must include materials and labor from the secondary connection unit to the meter box and then within the dwelling unit. If materials meet certain standards, the developer may receive reimbursement for some incurred costs.

The power company would prefer a firm commitment from regulatory agencies prior to commencement of distribution system design. Once an agreement has been reached on the configuration, the developer must grant easements, usually 10 feet wide, for installation and maintenance purposes. The developer normally is required to clear the easements and remove existing stumps and debris. Often the telephone company will utilize one side of the easement while the power company uses the other.

Construction of the power distribution system will not start until all lots are staked and the road system has been established in the field. Care must be taken to ensure close coordination between the power, telephone, gas, and construction companies installing the sewer, storm drainage, and water distribution systems, as carelessness and lack of coordination often result in cut power lines and broken utilities.

2.5.15 Gas Distribution Considerations

Major gas companies have policies that establish relationships between developers and the gas company. Gas service is normally provided by the company to the dwelling unit meter. Piping from the meter through the dwelling unit is the dwelling unit owner's responsibility.

As in the case of the electrical distribution system, a gas line configuration usually will be provided by the gas company for review by the developer. Necessary easements must be granted, similar to those required for electrical distribution. If gas service cannot be provided by a public utility company, the developer then must decide whether to provide a centrally located master tank and distribution system or individual liquid propane tanks.

2.5.16 Telephone Considerations

Telephone companies have requirements that should be satisfied to simplify service availability. Developers often do not consult initially with the phone company because installation and services are usually provided at no cost. The need for on-site electric power occurs early because construction tools must be energized; however, telephone service need is not usually felt until the dwelling unit occupants arrive. Below are several reasons why the telephone company should be consulted punctually:

(a) It alerts the telephone company that installation cables must be ordered from the supplier as materials generally are not locally stocked.

(b) It allows time to study the relationship of present and future demands to trunk line capacities to determine whether adequate capacities exist.

(c) It allows the telephone company to avoid piecemeal planning by addressing areawide service needs.

(d) It allows time for the telephone company to obtain any additional easements or encroachment permits (e.g., rail, highway, and others) necessary to locate service cables within controlled rights-of-way.

The telephone company will provide the developer a proposed distribution plan for review. Conflicting interests can be discussed and the plan modified if necessary. Ten-foot minimum easements must be provided. Construction procedures are much the same as for electric and gas companies, except that during the interior construction of a structure the telephone company should have access to install or inspect interior wiring prior to wall finishing. The telephone company, as in the case of the other utility companies, experiences cut cables during the construction phases. Maximum coordination and respect for one another's services must command prime consideration.

2.5.17 Cable Television Considerations

Cables for the distribution of television broadcasts should be installed much the same as telephones. The utility that has the franchise to service a development area should be contacted early to provide efficient service.

2.5.18   Adjacent Utility Usage

On occasion utility systems installed by a developer are replaced by larger capacity lines, even though the original lines would accommodate the development's needs. This may be necessary because of new demands that have been placed on the system by the development of adjacent land. A prudent developer who knows adjacent property will be developed in the future should make arrangements initially with regulatory officials to install facilities compatible with all anticipated future loadings. This will reduce future public expenses, assist in areawide planning, and prevent unnecessary future burdens on the completed development. Incurred expenses over and above that required for the developer's installation are usually reimbursable.

2.5.19   Solid Waste Collection

Solid waste collection services must be available to occupants of the development. The service may be provided by a public body, such as the city or county, or by a private firm acting on an agreement with the developer or the homeowners' association. Depending on the development, certain types of pickups are available. For single-family developments generally house-to-house service is standard. Manufactured housing communities and multifamily developments utilize centrally located dumpsters for point pickups. The developer must provide easy access to these collection points to ensure adequate receptacle capacity. High subgrade compaction and sufficient pavement depth are necessary to accommodate the applied concentrated tire loads that are nearly static and the additional loads of the container and waste.

If the project includes total development activities, provisions for trash compactors can be included to help reduce solid waste volumes and the separation of waste materials (plastics, newspapers, metal, glass) for recycling. This will help to prolong the life of the receiving sanitary landfill.

2.5.20   Municipal Services

Many projects are regulated to incorporate various user service facilities within the development plans. For example, postal delivery facilities must be available and must meet requirements of the U.S. Postal Service. If public transportation or schools have routes that will provide service to the development, then access stops must be investigated. In addition, fire alarms and police call facilities may be required to enhance neighborhood protection and provide increased security.

2.5.21   Property Ownership

A developer may dispose or depart from property located within the development site by deed or gift conveyance, forced sale for back taxes, foreclosure sale, bankruptcy, adverse possession, eminent domain, or dedication. Planned activities would focus on the conveyance of improved property by sale; however, land dedication may be considered appropriate for certain parcels. During the planning process, the developer should avoid creating random residual parcels that could be left over as a result of a boundary that has an irregular configuration. If such parcels are created, the developer may be unable to market them, resulting in continued ownership, liability, tax obligations, and maintenance requirements.

2.5.22 Dedication of Land

Many times a site plan has provisions for pedestrian easements, natural conservation areas or buffer zones, tot lots, storm drainage retention or detention ponds, and recreational areas. In the case of the PUD the homeowners' association usually assumes maintenance and control of such areas. When such an organization does not exist, a decision must be made by the developer to either maintain these areas or turn them over to a public body (So et al., 1979). The latter is known as dedication. For land to be legally dedicated, there must be

(a) an offer by the developer to turn the land over to the public body, and

(b) an acceptance by the public body once the offer has been formally made, usually done by the adoption process constituting legal acceptance of the offer.

If either element is not present, the act of dedication has not transpired, and the public body may rightfully assume no responsibility. Therefore, intent of dedication by a developer is not a valid reason for a public body to assume responsibility.

The petitioned public body customarily investigates to determine whether the proposed land offered for dedication will serve the public or only a chosen few, whether the area can be easily accessed by service vehicles, and whether any sizable expenses for maintenance are foreseeable. Such factors may result in a "white elephant" gift.

2.6 REGULATORY REQUIREMENTS

A number of previously discussed issues confronting a developer are related to federal involvement in protecting the country's environment. Figure 2.28 illustrates a typical regulatory network that includes some interrelationships that could affect project permitting activities. Sizable developments having the potential to create impacts over large areas relative to infrastructural capacities, environmental quality, or economic activities may require special regional review.

2.6.1 Regulations

Developments are controlled by local jurisdictional requirements, which include but are not limited to subdivision regulations, zoning, land use maps, environmental regulations, and requirements of the municipal engineer.

The guidelines referred to most often by a developer are the subdivision regulations adopted by the unit of government having regulatory power over the area. The local planning organization, or board, oversees the enforcement of these regulations through project reviews. Extraterritorial police power may be included in the organization's mandate, which would allow the regulation of areas that could be annexed in the future into the jurisdictional area. A subdivision is usually defined as any division of a tract or parcel of land into two or more lots, building sites, or other divisions for purposes of sale, legacy, or building. Included are divisions of land involving new or altered streets and parcel redivisions when the following occur:

PROPOSED RESIDENTIAL DEVELOPMENT SITE

FIG. 2.28 Regulatory permitting and commenting agencies that influence land development.

ANY INVOLVEMENT OF PROPERTY THAT IS OWNED, OPERATED, MAINTAINED, AND/OR CONTROLLED BY FEDERAL, STATE, OR LOCAL GOVERNING BODIES REQUIRES DIRECT APPROVAL OF THAT BODY.

(a)   the combination or recombination of portions of previously platted lots where the total number of lots is not increased and the resultant lots are equal to the standards of the governing authority and

(b)   the subdivision of land into parcels of 5 acres or more where no new street is involved.

This definition may vary from regulation to regulation, depending on the scope and wording; however, the basic idea should remain constant. A typical review procedure for developments is shown in Figure 2.29. The purpose of subdivision regulations are (Charleston County Planning Board, 1955)

> To provide for the harmonious development of the [area] . . . secure a coordinated street and road layout with relation to major streets and adjoining subdivisions and adequate provisions for traffic, to protect residential areas from through traffic and other traffic hazards, to insure proper street intersection design, to achieve individual property lots of maximum utility, to secure adequate provisions for light, air, water supply, drainage, and sanitary sewer facilities and other public health requirements, to secure adequate provisions for transportation, recreational areas, school sites, and other public services and facilities, to provide accurate land records for the convenience and protection of the public and for adequate identification and permanent location of real estate boundaries, and to ensure the recording of necessary survey data prior to the selling of land.

Because subdivision regulations vary from locale to locale, no attempt will be made to discuss detailed information contained within a set of regulations.

2.6.2 Zoning

A study of the legally adopted zoning ordinance for the market area, if one exists, must be undertaken. This legal device is supplemented by the zoning map, which physically ties the zoning ordinance to the land it polices. Zoning provides compatibility of land use and ensures availability of adequate space for a particular land use. By controlling the use of land in areas of future growth, the local governing body maintains a reasonable expectation of providing adequate streets, schools, refuse collection, police and fire protection, utility systems, and recreational areas. Additionally, zoning protects existing property by causing developments to conform to certain established standards that allow for adequate air, light, and privacy for persons who live and work in the area. Land use, setbacks, and parking requirements are some of the elements regulated by zoning (So et al., 1979). Figure 2.30 contains various relationships that are often regulated through zoning.

Zoning is an exercise of police power on future situations, rather than past, and therefore cannot be made retroactive to correct undesirable existing conditions other than through natural attrition. A physical investigation of a possible development site may uncover blighted areas, public nuisances, or nonconforming land uses. These undesirable land uses may be undergoing elimination by such actions as urban renewal, health department actions, or building code enforcement. Nonconformances may be the result of a variance from the zoning ordinance granted by a zoning board of appeals. These items must be investigated carefully as the land use of surrounding areas will influence the development.

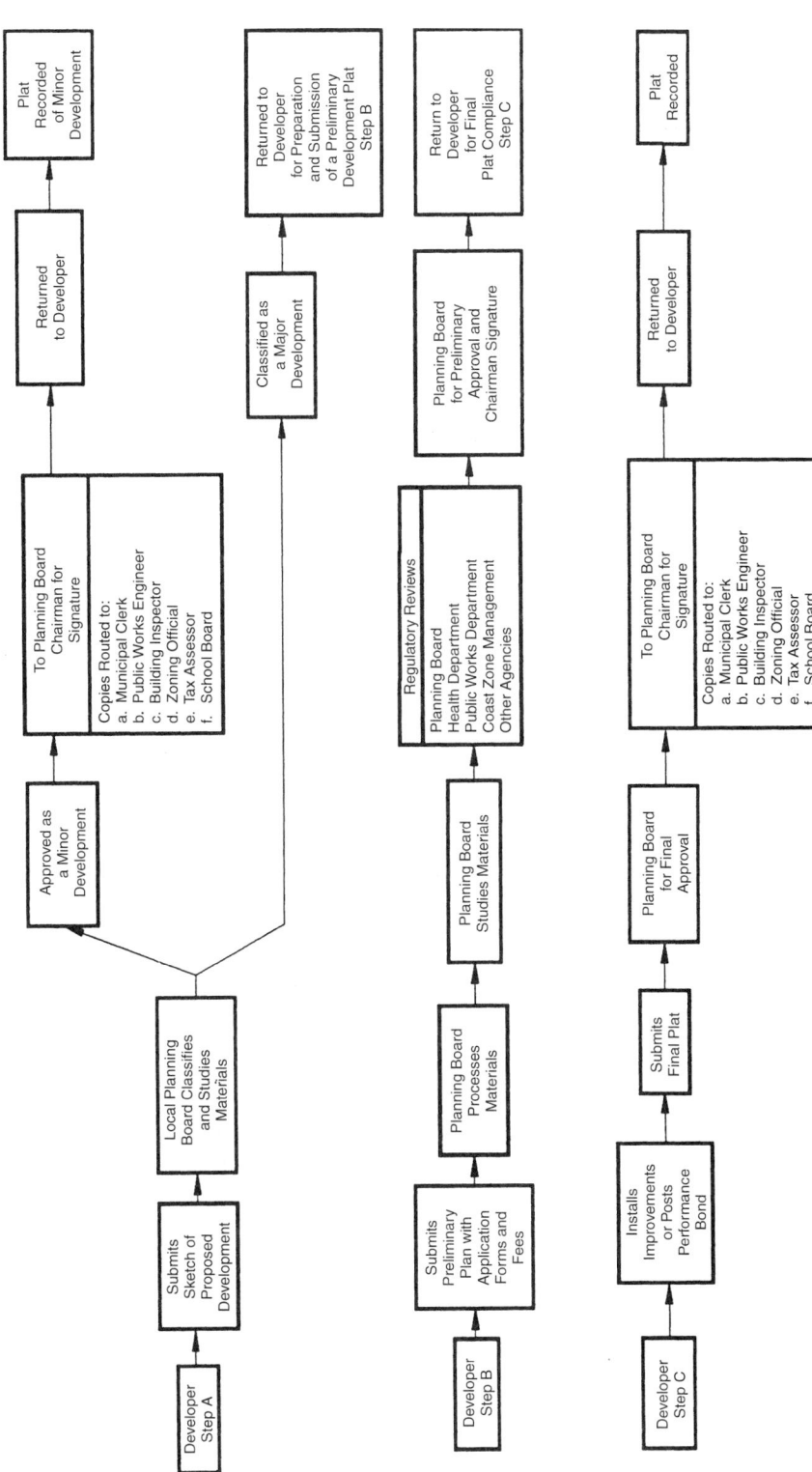

FIG. 2.29 Typical procedure for local regulatory approval of land developments.

73

INTENSITY STANDARDS FOR RESIDENTIAL DEVELOPMENT

DWELLING UNIT TYPE	DWELLING UNITS PER ACRE	COMMON OPEN SPACE AS PERCENTAGE OF TOTAL SITE	PARKING PER UNIT (1)	TREES PER ACRE OF TOTAL SITE AREA (2)	PRIVATE OPEN SPACE (3)
Single family estate		Usually not provided	3+	15+	Depends on lot size
Single family	3 to 5 depending on lot size	Usually not provided	2.5+	15+	Depends on lot size
Duplex	5 to 10 depending on lot size	Usually not provided	2+/(4+ for each structure)	15+	Depends on lot size
0-lot line (4)	4 to 7 units per acre	Usually not provided	2.25+	15+	Depends on lot size
Single family cluster (5)	4 to 7	25 to 50%	2.25+	15 to 20± for first 3 acres, 10+ for remaining acreage	Depends on lot size
Atrium (6)	4 to 8 units per acre	Usually not provided	2.25+	15+	25% of total square footage
Suburban townhouse (7)	6 to 9	25% to 40%	2.25+	15 to 20± for first 3 acres, 10± for remaining acreage	500 to 700 sq ft
Urban townhouse (8)	8 to 16	15% to 25% if provided	2.0+	15 to 20± for first 3 acres, 10± for remaining acreage	400 to 600 sq ft
Walk-up apartments	10 to 25	25% to 40%	2.0± depending on number of bedrooms	15 to 20± for first 3 acres, 10± for remaining acreage	Usually not provided except on ground floor apartments
Midrise apartment up to 6 stories	15 to 35	30% to 40%	2.0± depending on number of bedrooms	15 to 20± for first 3 acres, 10± for remaining acreage	Usually not provided except on ground floor apartments
Highrise apartments	30 to 75	35% to 60%	2.0± depending on number of bedrooms	15 to 20± for first 3 acres, 10± for remaining acreage	Usually not provided except on ground floor apartments
Planned unit development (9)	10 to 25 depending on type of planned unit development	25% to 60%	2.0± depending on unit type	15 to 20± for first 3 acres, 10± for remaining acreage	Depends on individual unit type

NOTES

1. PARKING PER UNIT: In determining the parking per unit, the site design should consider the influence of public transportation and location of the development in relation to employment centers and supporting facilities. An excessive amount of parking spaces to accommodate infrequent special activities could result in excess pavement; depending on soil conditions, it is sometimes better to accommodate infrequent overflow parking on grassed areas or, where conditions permit, on commercial parking areas when not in use or at community centers.

2. TREES: Trees provide one of the major unifying design elements and act as climate modifiers; where a substantial number of trees do not already exist, extensive tree planting should be undertaken as part of most planning programs. The determination of the number of trees per acre should be based more on the particular locale than on a uniform standard.

3. PRIVATE OPEN SPACE: Private open space should be private to the unit concerned and may be provided in the form of courtyards, entrance courts, and rear, side, and front yards. Both visual and aural privacy is important in the design of these spaces.

4. 0-LOT LINE: A single family unit located directly on one side property line and possibly on the rear property line; the emphasis is on eliminating small side yards for larger, more usable open spaces.

5. SINGLE FAMILY CLUSTER: Single family units grouped in clusters in order to maximize open space; units may or may not be attached.

6. ATRIUM: A single family unit related to the early Greek and Roman prototypes which incorporate interior living spaces fronting on an interior court.

7. SUBURBAN TOWNHOUSE: A single family unit attached to other single family units with a common party wall; open space is usually a major element in the suburban townhouse development.

8. URBAN TOWNHOUSE: Early prototypes appear in the urban areas of many older cities; similar to suburban townhouse, although densities are usually higher and common open space is usually provided in the form of public open space.

9. PLANNED UNIT DEVELOPMENT: A mixture of housing types with emphasis on total community design; all the housing types included in the chart, plus associated retail support facilities and community amenities, may be found in a PUD development.

Gary Greenan, AIP, ASLA; Miami, Florida

FIG. 2.30 Intensity standards for residential developments. Source: John Hoke, Ed. *Architectural Graphic Standards*, 8th ed. Copyright © 1988 by John Wiley & Sons, Inc. Reprinted by permission of John Wiley & Sons, Inc.

Zoning is undoubtedly the most commonly used legal device for implementing land use plans. Zoning can regulate the size of buildings, parcel areas, density, and use. The landmark court case that upheld, the constitutionality of zoning was *Euclid v. Ambler Realty Co.* in 1926 (Yokley, 1978).

What happens if a site is chosen in an unzoned area and (a) the surrounding area is undeveloped and (b) the surrounding area has undesirable uses present? Two courses of action the developer might follow entail either an effort to enact a zoning ordinance that would support the development or an effort to purchase the problem land. The latter method is an expensive but effective undertaking and, by the use of restrictive deed covenants, the property could be resold as an asset to the development.

Zoning also can regulate the location and size of signs. If the development includes a neighborhood shopping center, these regulatory requirements may be of concern.

Additionally, certain provisions are sometimes utilized by municipalities to control activity outside their jurisdiction. Such display of extraterritorial regulation of unincorporated land adjacent to a municipality's boundary ensures compatibility and coordination with the existing municipality plan should future site annexation be instituted.

2.6.3 Restrictive Covenants

Another land use restraint is a restrictive deed covenant, which may designate such items as allowable land uses, minimum lot size if the land is subdividable, regulations regarding construction of substructures, minimum house size or relative value, and numerous other possibilities. Covenants are placed on the land by the developer. They remain in force no matter who owns the property and run with the land. During formulation of the municipality's land use plan, the restrictive deed covenants of all properties under the jurisdiction of the municipality should be investigated and incorporated provided they are reasonable and compatible to the zoning ordinance. Sometimes a conflict exists between zoning and the restrictive deed covenant for a particular parcel. In practice, when the restrictive deed covenant is more stringent than the zoning ordinance, then the zoning ordinance must be satisfied along with the more stringent requirements set forth in the covenant. Should covenants exist, they could further limit the usability of the land under investigation.

2.6.4 Political Considerations

An otherwise perfect development site might only need its zoning classification changed or a variance granted to be a complete success. Public hearings during the regulatory review process are common because citizen input is important. The role of the developer during this process is critical because the life of the project is usually predicated by the outcome of the hearing. The developer must assemble the development team of professionals to prepare for the hearing presentation as well as for answering questions.

One of the best ways to prepare is to find out what concerns the regulatory agencies have as well as those of the public. If the development team has been in contact with the regulatory agencies during the preliminary stages of the development, the team probably has a good feel for what regulatory concerns exist relative to the project's elements. Finding out the concerns of the public can entail sampling the public through interviews, telephone calls, or questionnaires. Once the concerns are identified, the team needs to explore all possible avenues and scenarios to address the issues and build a presentation that takes all of these into consideration.

The team then needs to determine who will present what parts of the presentation. Several rehearsals should be scheduled where the remaining team members critique and ask questions. This helps to stimulate the exchange of group ideas and promotes brainstorming. Full audiovisual props should be used when practicing to work any glitches out. A well-rehearsed presentation given at a public hearing goes a long way toward gaining approval for the project.

2.6.5   Permit Approvals

Some state and local permits that may be required by the developer are:

(a)   to construct and operate potable water systems;

(b)   to construct and operate wastewater systems;

(c)   to construct and operate storm drainage systems;

(d)   to construct across navigable streams;

(e)   to encroach on public highways, railroads, and public utility rights-of-way;

(f)   to construct within the coastal zone;

(g)   to withdraw surface and subsurface water;

(h)   for the interbasin transfer of water;

(i)   to construct public facilities such as swimming pools;

(j)   for building construction;

(k)   to grade sites and construct streets and roads;

(l)   for landscaping and tree removal; and

(m)   to dredge and fill wetlands.

In some locations the developer may find that local jurisdictions have planned programs for extending and increasing the capacity of various infrastructure systems. Using this approach their capital improvements plan can proceed in an orderly manner. If the issuance of any development-related permits is predicated on the completion of various capital improvements, the developer may find the project's start delayed. This could affect the outcome of the project so the developer needs to find out early.

The developer also must adhere to various rules and regulations pertaining to employment conditions if federal money is used under various laws such as the Davis Bacon Act and the Civil Rights Act of 1964. Federal and state labor laws, tax regulations, business licenses, unemployment insurance requirements, and workman's compensation are examples of other constraints.

2.7   ECONOMIC CONSIDERATIONS

A potential project must be economically feasible for a developer and financial institution to desire undertaking its completion. This feasibility is based not only on the market's supply and demand but also on the anticipated absorption rate of the new dwelling units that will be produced. Developers undertaking total development activities may establish a two-year product inventory as a maximum that will govern the rate of project phasing.

2.7.1   Initial Costs

The developer is faced early with development-related expenditures resulting from initial studies and evaluations conducted to establish a basis for financial lending and project potential. Some of the initial expenditures are costs associated with preliminary engineering studies, land surveying activities, architectural/landscape architectural services, environmental assessment, archeological studies, legal costs associated with title and abstracts, interim interest, permit application review fees, and other miscellaneous costs.

2.7.2   Impact and Linkage

Impact fees and linkage are two variable items of great economic concern to a developer as they can amount to appreciable expenditures. Impact fees are used to create jurisdictional capital funds based on units, structures, or development density for water, sewer, roads, fire protection, or recreational improvements necessary to respond to new demands created by the development. Linkage or inclusionary zoning is used to create some housing units that will be sold at below market values in an effort to increase supplies of affordable housing in exchange for permitting the developer to increase the usual density allocations. Linkage also can be accomplished through incentive zoning, where the developer receives increased density allocations in exchange for paying mitigating fees or providing public amenities to offset development-related strains being placed on urban systems (Nelson, 1988).

Other regulatory fees might include posting a performance bond or negotiable instrument with the appropriate jurisdictional body. This will provide project completion security, allowing the developer to premarket the project without having to wait until all improvements have been installed and approved by regulatory authorities.

2.7.3   Project Cost Estimate

The estimated cost to install development-related improvements is provided by the design professional. This estimate includes the costs for constructing all transportation, drainage, wastewater, and potable water systems if it is an improved land development. Care must be taken by the developer when undertaking total development activities to include not only the costs to erect structures but also the costs associated with service entrances to the building, driveways, parking, and other improvements. This is true for most single-family and multifamily developments. Manufactured housing parks and subdivisions usually require the developer to include the costs of utility service entrances through the concrete pad. If there are any additional costs involved with the development, such as costs to mitigate significant cultural or historical resources, they also must be included.

A complete cost estimate includes all anticipated nonconstruction developmental costs. These costs usually include applicable engineering, architectural, surveying, interest, taxes, and legal fees. Also, water supply and treatment facility surety bond costs, mortgage loan insurance, filing fees, and mortgage application fees are included and vary according to the company to which applications are submitted.

Architectural fees vary according to the type of development. For a multifamily development 12 percent of the construction costs are generally allowed for the architectural firm. This should include building plans and any engineering fees involved with the structure. Single-family-unit architectural fees can be as high as 15 or 20 percent of the value of the house.

Engineering services for the design of the sewer system, water distribution system, storm drainage system, and road network are negotiable and generally range from 6 to 10 percent of the cost of construction. Costs associated with soils investigations can vary depending on the project size, terrain, equipment accessibility, and prevailing subsurface conditions. In addition, surveying costs for site boundary work are variable, based on the property complexity, the number of contiguous landowners, and their properties' relationships along with topographic conditions. Topographic survey costs vary greatly depending on

(a) amount of site vegetation,
(b) ground surface uniformity,
(c) amount of detail required,
(d) maximum allowable distance between rod shots,
(e) desired contour interval, and
(f) datum requirements.

For a single-family development lot staking generally is negotiated on a per-lot basis, whereas multifamily developments normally use a per-unit basis. Unforeseeable problems in the field could increase surveying costs.

Legal fees are negotiable for title research and deed transactions, with the developer usually purchasing land title insurance for protection. Once the initial land transaction has taken place, legal fees are charged for such things as drafting partnership or corporate declarations, negotiations with governmental agencies, obtaining rights-of-way for connection to existing services if required, drafting protective deed covenants, and drafting individual lot deeds. Planned unit developments require much legal work, and as in the case of surveying, many uncertainties exist. Provisions are normally made to provide for compensation for professional services arising from unforeseeable problems. Although there are no hard and fast rules for such fee determination, one method used is to base additional professional fees on the additional time expended.

Mortgage loan insurance is available to mortgage lenders with the fee being paid by the developer upon loan closing. Most institutions limit their maximum loan to 75 percent of the raw land fair market value, or 75 percent of the raw land plus improvement value, regardless of the insurance provisions. Mortgage application fees generally range from 1 to 2 percent of the mortgage value. Some lending institutions account for this fee in their first payment disbursement.

2.7.4   Recurring Costs

A developer must determine, for feasibility analysis, costs of a recurring nature that can be expected during project development. Recurring costs might include mortgage retirement payments, interest on the mortgage, taxes, insurance, promotion, advertising, real estate agent fees, management fees, utilities, depreciation, solid waste collection, fire protection, power, and maintenance/repair costs.

The life of a mortgage on a land improvement loan can range from three to five years. Retirement payments are flexible and can be arranged to be paid weekly, bi-weekly, monthly, quarterly, semiannually, or annually. A mortgage life of 25 years may be available on a manufactured home park. Mortgage interests are variable and can range from 9 to 15 percent, depending on the market.

The valuation of the land before and after the development varies, as does the corresponding *ad valorem* tax liability imposed on the developer. Taxes must be paid as long as the developer owns the property and should be included in the analysis.

Promotion, advertising, and real estate agent fees vary according to the type of development. For an apartment complex or a manufactured home park, continuous advertising may be necessary throughout the life of the development. Contrarily, a single-family or multifamily development may require a short-term but intense advertising campaign because the need for advertising decreases as dwelling units are sold. Real estate agents normally charge commissions of 6 percent of the sale price of a dwelling unit. Agent's fees can approach 10 percent of a month's rent in apartment complexes or manufactured home parks. Fees may vary and must be negotiated for a particular situation.

Development management costs vary with the type of development and the amount of management required. Management costs of a mobile home park or an apartment complex would be appreciably more than those of a multifamily development. These costs must be estimated on an individual basis.

Utility costs are variable according to development type. Electric power costs should be included for the construction period, while street light expenditures will be absorbed by the developer, the renter, or the dwelling unit purchaser.

Development depreciation must be investigated based on the remaining economic life of the facility. An economic life can be assigned to buildings, privately owned utility systems, and amenities. As the development depreciates, it may allow the developer to decrease tax payments, because of decreased valuation, and allow incremental savings to replace the facility in the future.

Where solid waste collection services are contracted, a variety of costs can result, depending on the service provider. Single-family and some multifamily developments utilize individualized pick-up contractual operations, whereas apartment facilities may require point pick-up contracts.

When municipal fire coverage is not available, negotiations with the nearest municipality may be necessary. Such charges for fire protection are generally on an annual basis, in the form of either a bill or a bond payment.

Maintenance and repair incurred by the developer vary with the type of development and construction. Weather conditions play an important part in the amount of maintenance required, as does the unit's quality of workmanship and materials. Estimation of such costs must be made on an individual basis.

2.7.5 Revenues

Revenue is money earned. Direct revenue results from consideration paid by the renter or buyer during the course of renting or purchasing a lot and dwelling unit. Indirect revenue results from money generated by vending machines, telephones, laundry facilities, and other such operations that are provided in conjunction with the development. Their combination represents the earning potential of the development. The amount of anticipated revenue can be estimated from the preliminary site plan and market analysis. The number of dwelling units to be sold or rented within a certain price range determines the gross direct revenue. This should be adjusted downward to indicate something less than 100 percent occupancy or immediate sellout. When the adjusted value is added to the indirect revenue, the result yields the gross adjusted income. Once the gross adjusted income is known, the feasibility of the preliminary site plan can be determined.

2.7.6 Feasibility

The feasibility of a project, as indicated in a pro forma analysis, is simply an indicator that the project will be a success or a potential failure. The difference between the gross adjusted income, the developmental costs, and operating expenses represents the developer's profit. This is usually expressed as a percentage of the developer's equity in the project. A feasible development could produce from 10 to several hundred percent return on an initial investment. However, some developers expect to produce gross sales that approach 200 percent of the acquisition and development costs from which 100 percent of the acquisition and development costs can be satisfied, leaving the developer approximately a 40 percent profit. The other 60 percent is accounted for by legal fees, interest, real estate agent fees, and other costs of doing business. Once the developer has determined that the project will be feasible, arrangements for a mortgage loan must be made.

2.7.7 Financing

Agencies that finance land development loans are banks, savings and loan associations, mortgage investment companies, insurance companies, and others. All have policies and directives that limit their amount of project financing. This protects their investment. The developer strives to obtain as much financial leverage as possible to accomplish the development. The developer uses the least amount of personal capital possible and secures the maximum loan available to accomplish project financing. There are two basic reasons for this. First, it allows the developer to maintain diversity in investing capital that is not committed to one development. Second, the developer can obtain a larger rate of investment return, which means additional earning power.

The lending institution is very interested in who is borrowing the money for a development, be it an individual, partnership, or corporation. Certified financial statements concerning the development organization and persons involved are required by the lending agencies to determine financial stability. The developer's past credit history provides the lender with an idea of the willingness of the borrower to repay any incurred debt. Past experience in developing and competent members in the development organization are also definite assets in obtaining a loan. Lenders are interested in the ability of a borrower to repay the loan so they review and verify assets available to the developer should the project not perform as expected. Financial institutions also require the submission of a Master Appraiser Institute (MAI) certified commercial appraisal report on the project. The institution then undertakes an independent assessment of the report and those same items the developer investigated during site selection and site planning to ensure

(a) that the proposed development is competitive in costs and amenities,

(b) that harmony is maintained throughout the interior and surrounding areas of the development,

(c) that there are no local hazards adjacent to the development and the development does not create any hazards, and

(d) that the contractors engaged for the work are competent.

Financial institutions require a written appraisal for all real-estate-related financial transactions normally handled by savings institutions, in accordance with the Financial Institutions Reform, Recovery, and Enforcement Act of 1989 (FIRREA). This would encumber all sales, of leases, of purchases, of interest in, or the exchange of real property, or property financing.

The appraisal must be independently and impartially prepared by a licensed or certified appraiser. The appraiser must set forth an opinion of value, which is described as of a particular date. All materials used as a basis for value, including market data, must be included, along with a legal description of the appraised property that exactly mirrors conveyance documents contents.

Upon establishment of a favorable developer profile, completion of a formal supporting appraisal, and submission of a feasible plan, the lending institution will process the mortgage request, pending regulatory compliance certification. If the applicant is accepable to the lending institution's system of checks and balances, the loan is granted.

2.7.8 Mortgages

The mortgage loan is normally issued to the developer in increments, called disbursements, for work completed. For example, if 20 percent of the water distribution system were completed and the developer received a bill from the contractor covering this phase of the construction, the mortgage company would disburse to the developer the percentage of the loan allocated to cover that percent of the water system costs. The developer would in turn pay the contractor. The total amount of money disbursed on a percentage basis never exceeds the percent financing for the entire project.

Mortgage retirement, or repayment, is accomplished after all improvements have been installed and approved, by releasing land according to a predetermined amount, which might be 120 percent of the improved property value. In addition to retiring the mortgage, the property release provides a way to offer clear title on a lot or dwelling unit to a potential buyer, even though the lending institution still holds the development mortgage.

In summary, most developers, while being confronted by numerous constraints, concentrate their efforts on seeking project success and completion. A high degree of dependence therefore is placed on the various professionals engaged by the developer to provide safe, economical, and acceptable solutions to the many problems that arise out of land development. The civil engineer plays a key role in the total process.

REFERENCES

American Association of State Highway and Transportation Officials (AASHTO). *A Policy on Geometric Design of Highways and Streets.* Washington: AASHTO, 1994, p. 8.

American Association of State Highway and Transportation Officials (AASHTO). *Guide for the Development of Bicycle Facilities.* Washington: AASHTO, 1999, pp. 7–9, 24, 35.

American Public Health Association (APHA), Committee on the Hygiene of Housing. *Planning the Neighborhood.* Chicago: Public Administration Services, 1960.

American Society of Civil Engineers (ASCE), National Association of Home Builders, and Urban Land Institute. *Residential Streets.* New York: ASCE, 1974.

Ashbury, N. G. "A Formula for Financing Mobile Home Housing Developments." An unpublished report. Chicago: Mobile Home Manufacturers Association, 1970.

Bissonnette, P. "Bellevue Experiences with Urban Runoff Quality Control Strategies." In *Perspectives on Nonpoint Source Pollution. Proceedings of a National Conference.* Washington: U.S. Environmental Protection Agency, 1985, pp. 279–280.

Casey, W. J. *Real Estate Investments and How To Make Them,* 3rd ed. New York: Institute for Business Planning, 1968.

Charleston County Planning Board (CCPB). "Subdivision Regulations: Charleston County." An unpublished regulation. Charleston: Charleston County Government, 1955.

Cowardin, L.M., V. Carter, F. C. Golet, and E. T. La Roe. *Classification of Wetlands and Deepwater Habitats of the United States.* Washington: U.S. Department of the Interior, 1979, pp. 3–14.

De Chiara, J., and L. Koppelman. *Planning Design Criteria.* New York: Van Nostrand Reinhold, 1969, pp. 180–190.

DeVereaux, A. B., V. J. Tschinkel, E. Gissendanner, and State of Florida. "Joint Application for Permit Dredge–Fill–Structures." An unpublished report. Jacksonville, FL: U.S. Army Corps of Engineers, 1982.

Duany Plater-Zyberk & Co. *The Lexicon of the New Urbanism.* Miami: Duany Plater-Zyberk & Co, 1999, p. C1.

Dunkle, F., R. Hanmer, R. W. Page, and W. Scaling. *Federal Manual for Identifying and Delineating Jurisdictional Wetlands.* Washington: U.S. Government Printing Office, 1989.

Federal Housing Administration (FHA). *FHA Techniques of Housing Market Analysis.* Washington: U.S. Department of Housing and Urban Development, 1970a.

General Services Administration, U.S. Department of Defense, U.S. Department of Housing and Urban Development, and U.S. Postal Service. *Uniform Federal Accessibility Standard.* Washington: U.S. Government Printing Office, 1988, pp. 5, 23–25.

Gustin, J. D., and L. A. Neal. "Site Assessments." In *Civil Engineering.* New York: American Society of Civil Engineers, 1990, pp. 53–55. Referenced material is reproduced by permission of the American Society of Civil Engineers.

Jones, D. E. Jr, and K. A. Jones. "Treating Expansive Soils." In *Civil Engineering.* New York: American Society of Civil Engineers, 1987, pp. 62–65. Referenced material is reproduced by permission of the American Society of Civil Engineers.

Leifer, S. L. *EPA's Innocent Landowner Policy: A Practical Approach to Liability under Superfund.* Washington: Bureau of National Affairs, 1989, pp. 646–649. Reprinted with permission from *Environmental Reporter*, Vol. 20, No. 14, pp. 646–649 (August 4, 1989). Copyright 1989 by The Bureau of National Affairs, Inc. (800-372-1033). <http://www.bna.com>

Lewis, R. K. "Shaping the City: Designers Are Going Back to Basics with the "Traditional New Towns." *Washington Post,* June 13, 1990. Referenced material is reproduced by permission of Roger K. Lewis, 5034 ½ Dana Place, N.W., Washington, DC 20016.

Moskowitz, J. S. *Environmental Liability and Real Property Transactions: Law and Practice.* New York: Wiley, 1989.

National Association of Home Builders. *Developer's Guide to Endangered Species Regulation* Washington: Home Builder Press, 1996.

Nelson, A. C. (ed.). *Development Impact Fees: Policy Rationale, Practice, Theory, and Issues.* Chicago: Planners Press, American Planning Association, 1988, pp. 219–277.

Perry, C. A. *The Neighborhood Unit, a Scheme of Arrangement for the Family Life Community.* New York: Russell Sage Foundation, 1929.

Rice, L., and M. D. White. *Engineering Aspects of Water Law.* New York: Wiley, 1987, pp. 13–19.

Rubenstein, H. M. *A Guide to Site and Environmental Planning,* 3rd ed. New York: Wiley, 1987.

So, F. S., I. Stollman, F. Beal, and D. S. Arnold, (Eds.). *The Practice of Local Government Planning.* Washington: International City Managers' Association, 1979, pp. 389–415.

South Carolina Coastal Council (SCCC). "Storm Water Management Guidelines." An unpublished report. Charleston: SCCC, 1988.

South Carolina Coastal Council. *Guidelines and Policies of the South Carolina Coastal Management Program.* Washington: U.S. Department of Commerce, 1979.

South Carolina Electric and Gas Co. (SCE & G). "Electrical Underground Distribution." An unpublished manual. Columbia: SCE&G, 1967.

Spielberg, F. "The Traditional Neighborhood Development: How Will Traffic Engineers Respond?" *ITE Journal,* September 1989, pp. 17–18.

Spielberg, F. *Traditional Neighborhood Development Street Design Guidelines.* Washington: Institute of Transportation Engineers, 1999.

Tomioka, S., and E. M. Tomioka. *Planned Unit Developments—Design and Regional Impact.* New York: Wiley, 1984.

Train, R. E., Chairman. *Protecting American's Wetlands: An Action Agenda. The Final Report of the National Wetlands Policy Forum.* Washington: Conservation Foundation, 1988, p. 1.

U.S. Army Corps of Engineers and the South Carolina Coastal Council. "Developer's Handbook for Freshwater Wetlands." An unpublished handbook. Charleston: S.C. Coastal Council, 1991.

U.S. Department of Housing and Urban Development (HUD). *Land Planning Principles for Home Mortgage Insurance, A HUD Handbook, 4140.1.* Washington: HUD, 1973b, pp. 5-9 through 5-18.

U.S. Department of Housing and Urban Development (HUD). *Environmental Reviews at the Community Level.* Washington: undated.

U.S. Department of the Interior and U.S. Department of Housing and Urban Development. *Environmental Planning and Geology.* Washington: U.S. Government Printing Office, 1970.

U.S. Department of Transportation. *Manual on Uniform Traffic Control Devices for Streets and Highways.* Washington: U.S. Government Printing Office, 1988.

U.S. Environmental Protection Agency (EPA). "Release of Final Storm Water Regulations." An unpublished memorandum. Washington: EPA, 1990.

U.S. Fish and Wildlife Service. *Endangered Species Consultation Handbook: Procedures for Conducting Consultation and Conference Activities Under Section 7 of the Endangered Species Act.* Washington: U.S. Government Printing Office, 1998.

U.S. Fish and Wildlife Service and National Marine Fisheries Service. *Endangered Species Habitat Conservation Planning Handbook.* Washington: U.S. Government Printing Office, 1996.

Yokley, E.C. *Zoning Law and Practice,* 4th ed. Charlottesville: Michie Company, 1978, pp. 10–11.

SUGGESTED READINGS

American Society for Testing and Materials (ASTM). *ASTM Standards on Environmental Site Assessments for Commercial Real Estate, E 1527-97 and E 1528-96,* 3rd ed. Philadelphia: ASTM, 1997.

American Society of Civil Engineers (ASCE). *Environmental Site Investigation Guidance Manual—ASCE Manuals and Reports on Engineering Practice No. 83.* Reston: ASCE, 1996.

American Society of Civil Engineers (ASCE), National Association of Home Builders, and Urban Land Institute. *Residential Streets,* 2nd ed. New York: ASCE, 1990.

Arendt, R. G. *Conservation Design for Subdivisions.* Island Press; Covelo, CA: 1996.

Asante-Duah, D. K. *Managing Contaminated Sites: Problems, Diagnosis, and Development of Site Restoration.* New York: Wiley, 1996.

Bookout, L. W., M. D. Beyard, and S. W. Fader. *Value by Design: Landscape, Site Planning, and Amenities.* Washington: Urban Land Institute, 1994.

Federal Housing Administration (FHA). *Mobile Home Park Development Standard G 4200.7a.* Washington: U.S. Department of Housing and Urban Development, 1970b.

Flug, M., and F. A. Klancnik. *Mariners, Parks, and Recreational Developments: Proceedings from the International Conference.* Reston: American Society of Civil Engineers, 1994.

Goodman, W. I., and E. C. Freund (Eds.). *Principles and Practice of Urban Planning.* Washington: International City Managers' Association, 1968, pp. 441–484.

Handy, R. L. *The Day the House Fell.* Reston: American Society of Civil Engineers, 1995.

Hoke, J. (Ed.). *Architectural Graphic Standards,* 8th ed. New York: Wiley, 1988, p. 148.

Home Builder Press. *Developer's Guide to Endangered Species Regulation.* Washington: Home Builder Press, 1996.

Jarvis, F. D. *Site Planning and Community Design for Great Neighborhoods.* Washington: Home Builder Press, 1993.

Johnson, D. E. *Residential Land Development Practices—A Textbook on Developing Land into Finished Lots.* Reston: American Society of Civil Engineers, 1997.

Jury, M. A., and S. C. Geniesse. *Residential Cluster Development.* Chicago: Council of Planning Librarians, 1994.

Kone, L. D. *Land Development,* 8th ed. Washington: Home Builder Press, 1994.

Lovell, D. D., and R. S. Martin. *Subdivision Analysis.* Chicago: Appraisal Institute, 1993.

Marriott, B. B. *Practical Guide to Environmental Impact Assessment.* New York: McGraw-Hill, 1997.

McNamara, R. *Impact Fee Handbook.* Washington: National Association of Home Builders, 1997.

Montgomery County Planning Commission (MCPC). "A Method of Naming Streets." Dayton, OH: MCPC, 1992.

Morris, M. *Subdivision Design in Flood Hazard Areas.* Chicago: American Planning Association, 1997.

Payne, S. M. *Strategies for Accelerating Cleanup at Toxic Waste Sites—Tracking Environmental Actions and Decision Making.* New York: Lewis, 1998.

Petit, J., D. L. Basseert, and C. Kollin. *Building Greener Neighborhoods—Trees as Part of the Plan.* Washington: American Forests and the Home Builder Press, 1995.

Salvesen, D. *Wetlands—Mitigating and Regulating Development Impacts.* Washington: Urban Land Institute, 1990.

Sanders, W. *Manufactured Housing: Regulations, Design Innovations, and Development Options—PAS #478.* Chicago: American Planning Association, 1998.

Sanders, W., J. Getzels, D. Mosena, and J. A. Butler. *Affordable Single-Family Housing—Review of Development Standards—Planning Advisory Service Report No. 385.* Chicago: American Planning Association, 1984.

Sellers, K. *Fundamentals of Hazardous Waste Site Remediation.* New York: Lewis, 1999.

South Florida Water Management District (SFWMD). "Guidance for Preparing an Application for Surface Water Management Permit." An unpublished report. West Palm Beach: SFWMD, 1990.

Sowers, G. F. *Building on Sinkholes—Design and Construction of Foundations in Karst Terrain.* Reston: American Society of Civil Engineers, 1996.

Spielberg, F. *Traffic Engineering for Neo-Traditional Neighborhood Design.* Washington: Institute of Transportation Engineers, 1994.

State of Maine. *Report of the Patterns of Development Task Force—Maine Environmental Priorities Project—A Response to Sprawl: Designing Communities to Protect Wildlife Habitat and Accommodate Development.* Augusta: State of Maine, 1997.

U.S. Department of Housing and Urban Development (HUD) and the NAHB Research Foundation, INC. *The Affordable Housing Demonstration—A Case Study: Phoenix, Arizona HUD-PDR0-761-1.* Washington: HUD, 1984.

U.S. Department of Housing and Urban Development (HUD) and the NAHB Research Foundation. *The Affordable Housing Demonstration—Two Case Studies: Charlotte, N.C. and Greensboro, N.C. HUD-1073-PDR.* Washington: HUD, 1987a.

U.S. Department of Housing and Urban Development (HUD) and the NAHB Research Foundation. *The Affordable Residential Land Development HUD-1128-PDR (vi).* Washington: HUD, 1987b.

U.S. Department of Housing and Urban Development (HUD). *Land Planning Data Sheet Handbook, Circular 4140.3.* Washington: HUD, 1973a, Data Sheet 6.

U.S. Department of Housing and Urban Development (HUD). *Manual of Acceptable Practices, 1973 ed., 4930.1 Vol. IV.* Washington: HUD, 1973c.

U.S. Government. *Protection of Historic Properties 36 CRF Part 800.* Washington: U.S. Government Printing Office, 1999.

U.S. League of Savings Institutions (LSI). *Special Management Bulletin.* Chicago: LSI, 1990.

Ward, C. H., J. A. Cherry, and M. R. Scalf. *Subsurface Restoration.* Chelsea: Ann Arbor Press, 1997.

Wray, W. K. (Ed.). *So Your Home Is Built on Expansive Soils.* Reston: American Society of Civil Engineers, 1995.

3 Engineering Administration

Civil engineering focuses on solving problems that pertain to the health, safety, welfare, and economic well-being of the populace. This involves the principle that the engineer is an independent thinker who makes decisions based on factual information and not as a result of outside influence. Classical undergraduate civil engineering programs provide students with special knowledge in mathematical, physical, and engineering sciences along with studies in land surveying, transportation, structures, geotechnical, and environmental engineering. As a result of their education, training, and experience, these students are generally well suited for participating in land development activities. Civil engineers who have met various criteria set forth by state law are qualified to offer professional services to the public as registered professional engineers.

These activities center on problem solving within the constraints of various regulations, economics, and project goals set forth by the developer. The design professional is therefore not involved in the valuation, potential marketability, or financeability of the project and would not normally be in a position to provide the developer with any financial recommendations other than those pertaining to costs and operation of improvements.

3.1 ORGANIZATION

A civil engineer who provides consulting engineering services can offer professional services as a sole proprietor, as a partnership or professional association member, or even under the umbrella of a corporation. Traditionally, sole proprietorships and partnerships are the means most utilized by design professionals. As a sole proprietor, the professional engineer is personally responsible for all actions that arise from the conduct of business. As a result, the engineer reaps all profits from produced work while at the same time being responsible for any financial losses.

A partnership allows for two or more professional engineers to conduct business activities as co-owners, with each partner acting as an agent for the organization. Each partner therefore shares in the management and operations of the business, generally using the premise of majority rule unless otherwise agreed. Unanimous consent is needed for a change in business activities. Obligations made by individual partners during the course of ordinary partnership business transactions are binding on all partners, just as tort liability (Section 3.4) created by one of the partners is answerable by all. Likewise, partnership assets are subject to co-ownership status while each partner shares in any profits and losses. It is advisable to form a partnership by written agreement to establish a clear understanding of the rights and obligations of the members. The Uniform Partnership Act can provide an easy form to follow.

A corporation for profit may allow design professionals to organize effectively, especially if large numbers of engineers are involved. The corporation acts as a legal entity with the shareholders or stockholders possessing undivided ownership of assets. Stockholders are only liable to pay for stock shares purchased in the firm, and not for organizational indebtedness or business liability. Normally the articles of incorporation set forth authorized activities of engagement, shareholder rights, and the organizational structure and governance. Should the corporation enter into a contract that falls outside the purview of its authorized activity, the contract is unenforceable at law and termed *ultra vires* (Abbett, 1963). Corporate profits can be distributed by the board of directors in accordance with the articles of incorporation or reinvested into the organization. Many design professionals have adopted the professional corporation form of organization to conduct business, which allows various tax advantages.

Other forms of organizations include associations and joint ventures. Regardless of the organizational forum within which the design professional operates, all engineering work is considered personal service performed by a professional. Therefore, each professional is individually responsible for the work's adequacy, including work performed in a design-build situation.

3.2 MARKETABILITY OF PROFESSIONAL SERVICES

Many land development consulting firms provide "full-service" operations, where not only engineering is offered, but also planning, architectural, and even surveying/mapping services. Specialized firms offering just engineering usually work closely with other planning, landscape architecture/architecture, and land surveying firms which form a combined development consulting team. Often developers prefer the single-shop approach where the selected firm offers more than one service. For example, many engineering firms also have the capability of providing surveying and mapping services. Multifaceted firms allow the developer to obtain services that are coordinated in-house, which helps to reduce confusion, improve communication, and streamline the flow of work. This is especially important when developers consider the regulatory review process. Consultants having knowledge of contemporary issues (e.g., urban sprawl), local regulatory procedures, markets, and the history of an area often find favor with developers. In addition, consulting firms that have the ability to promote the development to the public are crucial to the developer, especially if the project requires public hearings as part of the regulatory review process.

The workload incurred by consulting firms engaged in providing land development services fluctuates over time because of changing interest rates, regulations, local initiatives, and land availability. The last drop in new housing starts occurred at the end of the 1980s and early 1990s. Because of its cyclic nature, many firms are providing a diversity of services where not only residential projects are involved, but also commercial and industrial jobs. Other approaches call for opening new service market areas in geographically dispersed locations. These approaches are used to take advantage of the breadth of public sector markets to hopefully ensure workload stability when the housing market fluctuates.

3.3 CONDITIONS OF EMPLOYMENT

The developer can engage a design professional to perform various services using one of several procedures, which include the customary approach, the customary approach with project management, and services rendered in conjunction with a professional construction manager (Fisk, 1988). On occasion, the developer may use a design-build firm where the project is not only designed but also constructed by the same firm.

3.3.1 Customary Approach

The most common situation, which historically has been the method of choice, is for the developer to solicit the design services from the engineer with an arrangement for intermittent site visitations being performed by the engineer during construction. Site visitations are for the purposes of monitoring the general compliance of material and accomplished work with the construction documents. Intermittent observation made by the design professional in an attempt to prevent project deficiencies generally does not imply a guarantee of the contractor's work on the part of the engineer, nor does it relieve the contractor of contractual responsibility. The engineer therefore would act as the developer's agent in all activities associated with the project except when performing design activities that require the design professional to act as an independent contractor. Other duties would entail the review of shop drawings, approving pay requests for the contractor, activating change orders, and final job inspection. Figure 3.1 illustrates this relationship.

3.3.2 Customary Approach with Project Management

Should the developer request a full-time job inspector or should a jurisdictional regulatory agency require a full-time inspector, the design professional could provide, in addition to those customary services previously described, various management services as shown in Figure 3.2. Again, the presence of a full-time inspector does not relieve the contractor of any obligation to properly complete all contracted work, nor does it place any conditions on the design professional to assume responsibility for consequences arising from methods, procedures, or construction sequencing utilized by the contractor, nor is there any obligation for the engineer to guarantee any of the completed work other than for adequacy of design.

3.3.3 Professional Construction Manager Approach

Should the project be large or complex, requiring either the services of several design professional firms or the use of several prime construction contractors, a professional construction manager may be procured by the developer for overall project coordination. This manager would in turn prosecute for the owner the engagement of the design professionals and contractors required to complete the project, resulting in the developer entering into individual contracts with the professional construction manager, the design professional(s), and the construction contractor(s). The construction manager would provide guidance and make recommendations relating to design reviews, bid packaging, scheduling, coordination, budgeting, quality control, progress analysis, construction methods, and claims management, as illustrated in Figure 3.3.

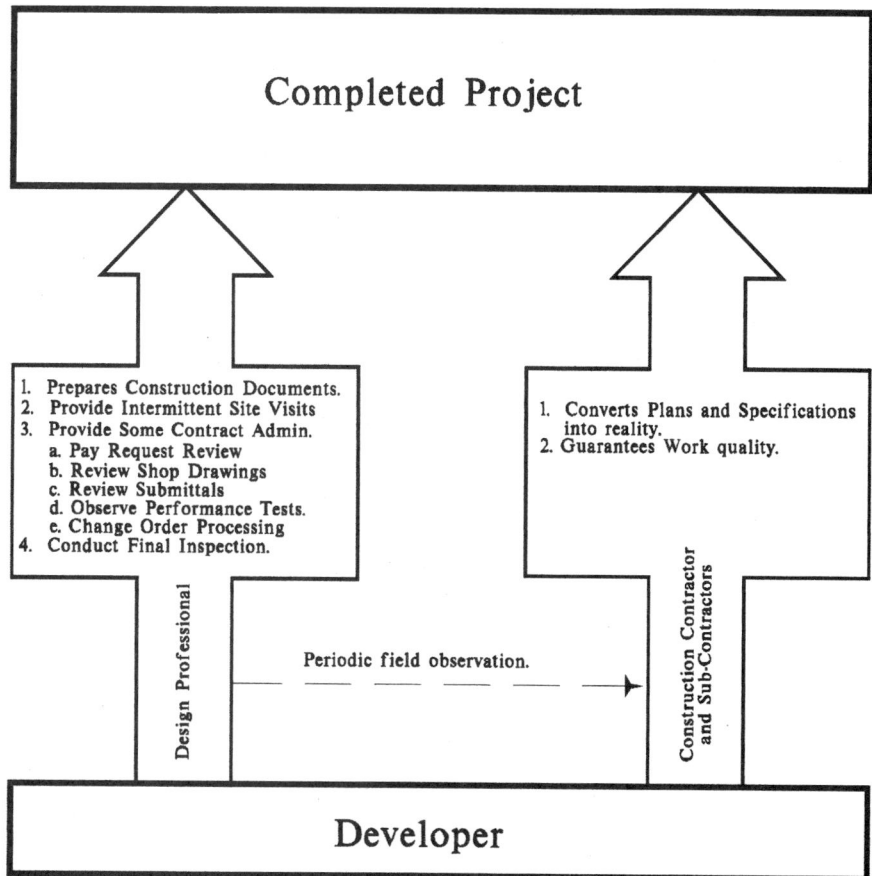

FIG. 3.1 Customary approach for land developments.

3.3.4 Design-Build Approach

There is an increased utilization of the design-build method for creating new projects in the United States. This single-entity approach improves communication and scheduling while eliminating many redundancies. If a developer selects this approach, the contractual agreement with the design-build firm would include provisions for designing the project and its construction.

Many public agencies in the past have been unable to engage in design-build activities because of procurement regulations. With the passing of the Federal Acquisition Reform Act of 1996 (FARA), the agencies of the federal government have new initiatives to consider the design-build option. Section 4105 sets forth criteria to determine if a project is suitable for the design-build approach. A two-step selection process can be used when at least three offers are anticipated to be tendered from firms seeking the work, or the scope of the project is such that some design must be undertaken (or substantial expenses incurred) before a cost can be developed. Other considerations include the capability of potential firms to deliver the final product within certain constraints.

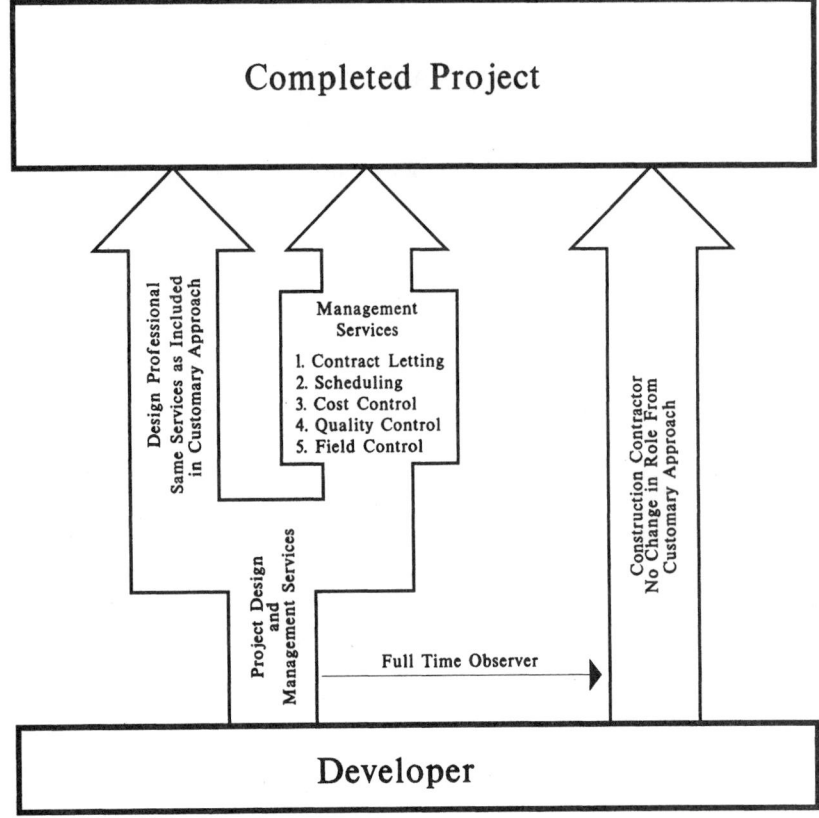

FIG. 3.2 Customary approach for land developments that incorporate project management.

The design-build selection process includes a request for proposals (RFP) similar to that used for selecting design professionals (see Section 3.7.1). In this first-step request are listed the criteria that will be used to evaluate each proposal (e.g., experience, complexities of past projects, performance record). Interested parties also are asked to include in their response the approach they will use to complete the project; however, no cost factors are to be included or considered at this stage.

The second step requires the public agency to review and rank order the proposals. From this ordered listing, invitations are issued to the top candidates (usually no more than five) to prepare and submit competitive proposals. The firm issuing the proposal that is considered the best value is then awarded the contract.

Some states are following suit and are changing their laws to better accommodate the use of the design-build concept. Brownfield remediation sites, for example, can take advantage of the design-build approach where site evaluation and testing become an integral part of the remediation efforts. Developers could therefore be involved with design-build projects if they are considering use of a remediated site for a recreational area.

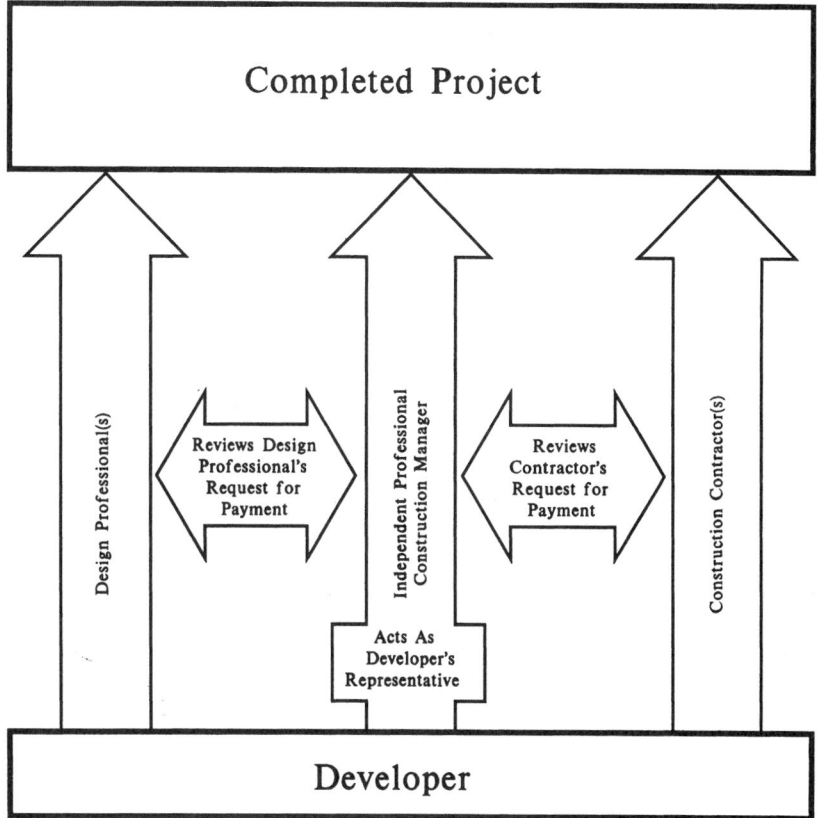

FIG. 3.3 Professional construction manager approach for land developments.

3.3.5 Registration Law Compliance

On occasion, developers may seek engineering-type services from nonlicensed sources, as a result of personal friendship or otherwise, which would in effect bypass state registration laws that are there to protect the public. This practice should be discouraged; however, the nonlicensed source can only persist as a project participant by working under the supervision of a licensed engineer. The registered supervising party must be completely responsible as the primary design consultant to prevent circumventing the law.

3.3.6 Engineering Services: Private and Public

Work associated with consulting engineering can deal with either public or private interests. Public interest work involves public funding and is associated with the needs of or in conjunction with a governmental agency. Private work is conducted using private business relationships and private funds. This distinction is important because competitive activities are required when public involvement exists.

3.4 CONDUCT OF AFFAIRS

The engineer who undertakes a project must possess the developer's good faith and loyalty in order to foster the mutual trust and confidence required for successful project completion. The engineer's position is one of great responsibility; therefore, any engineer who accepts a job implies that adequate competence is possessed to undertake such a job, and that diligence will be exerted to complete the project professionally and satisfactorily.

3.4.1 Engineer's Liability

The engineer's responsibility has been recognized since the early Code of Hammurabi. Under this code, if the owner of a dwelling was killed due to structural failure, the builder would suffer death (Nischwitz, 1984). Professional services normally are offered with the understanding that the engineer will exercise ordinary professional skill and diligence while conforming to acceptable standards of practice. These standards include building and design codes, recognized textbook procedures, and standard practices in the industry. Possible sources of liability could center on defective plans and specifications where the measurement of damage may range from repair costs for rectification of the resulting deficiency to the resulting loss in value of the finished product. Other areas of possible liability stem from specifying inferior materials, delaying the project, or even significantly underestimating the project's cost. The engineer can be held liable, as defined by applicable state law, if the resulting plans are grossly defective or for failure to observe obvious construction activity deficiencies during site visits.

When the customary approach is used to contract for engineering services (see Figure 3.1), the general conditions of the construction contract should unequivocally state that the owner and engineer are not responsible for the procedures or techniques used by the contractor or for enforcement of site safety. The reason for this is the contractor directs the work schedule, controls the job site, and has authority to hire, fire, schedule, purchase, or rent. The general conditions should assign contractual responsibility for site safety to an individual or organization (safety supervisor) for continuous site monitoring to include safety reviews of submittals and shop drawings. This individual could be under the employ of the contractor; however, this individual must be professionally licensed to review shop drawings. The fact that a safety supervisor is contractually designated and responsible for site safety will not excuse a consulting design engineer from being negligent if on-site observations uncover impending accidents or accidents in progress and no action is taken (see Section 18.7.1).

Because engineers are assuming multiple roles and responsibilities, such as participating as a project manager or part of a design-build team, their responsibilities for site safety are also changing. When engineers, as part of their duties, control or supervise schedules or workers, influence the means and methods of construction, perform safety inspections, or control safety procedures, they have assumed some responsibility (and liability) for site safety over and above that associated with the customary approach.

3.4.2 Governing Laws

Two forms of law that regulate a consulting engineer's activities are statutory law and common law. Statutory law consists of rules of conduct enacted by a duly authorized legislative authority; it includes federal, state, and local statutes. Common law consists of various doctrines and maxims that originated in court decisions and generally are not found in the statutes.

Torts are civil wrongs caused either by intentional actions or through negligence. Common torts are assault and battery; defamation of character, which is verbal slander, or libel, which results from disreputable written or pictorial pronouncements; infringements of copyrights, trademarks, or patents; trespassing; conversion, or seizing of other's property for personal use; creating a nuisance; and riparian rights violations.

3.4.3 Negligence

If an engineer is found negligent as a result of some less than desirable condition being associated with the project, several conditions must be met to show that the engineer had a duty to conform to various standards of conduct to protect against unreasonable danger and that the engineer did not conform to those standards. In addition, a "legally protected interest" must have been infringed upon as a result of a "reasonably close causal connection" that existed (Sweet, 1985).

Many states are making efforts to alter tort-related liability by abolishing or limiting the use of joint and several liability, which provides for damages to be assessed on ability to make restitution rather than on the degree of contribution when several parties are at fault. The engineer may find relief from a negligence action if the damaged party in fact contributed to the negligent condition or risk in use was assumed by the damaged party.

3.4.4 Fraud

In addition to confronting liability resulting from the nonperformance of contractual duties, the engineer can also be subject to liability resulting from fraudulent activity—intentionally misrepresenting facts that were reasonably relied upon by others who, as a result, suffered damage. An example would be an engineer who had knowledge of design errors on a set of plans but took no action to correct them prior to job bidding, causing the bids to come in too low.

3.4.5 Equity

Good practice dictates that a practicing engineer should use no short cuts and should heed standards that could be applied to scrutinize professional performance. One set of standards used in settling matters concerning questions in equity are the maxims of equity. Some of these maxims are included in Figure 3.4.

3.4.6 Insurance

Professional liability insurance can be purchased to offset risks that endanger the practicing engineer during the performance of professional services. The cost of coverage may be 5 percent of the engineering firm's gross income.

3.4.7 Ethics

The American Society of Civil Engineers (ASCE) instills professionalism within its membership by upholding a professional code of ethics (ASCE, 1999). Figure 3.5 includes the fundamental principles and canons that are included in the code.

He who seeks equity must do equity.
He who comes into equity must come with clean hands.
Equity aids the vigilant, not those who slumber on their rights.
Equity acts specifically.
Equity follows the law.
Equity delights to do justice and not by halves.
Equity will not suffer a wrong to be without a remedy.
Equity regards that as done which ought to be done.
Equity regards substance rather than form.
Equity imputes an intention to fulfill an obligation.
Equity is equality.
Between equal equities the first in order of time shall prevail.

FIG. 3.4   Maxims of equity. Source: Robert W. Abbett. *Engineering Contracts and Specifications,* 4th ed. Copyright © 1963 by John Wiley & Sons, Inc. Reprinted by permission of John Wiley & Sons, Inc.

CODE OF ETHICS
FUNDAMENTAL PRINCIPLES

Engineers uphold and advance the integrity, honor and dignity of the engineering profession by:

1. *Using their knowledge and skill for the enhancement of human welfare and the environment.*
2. *Being honest and impartial and serving with fidelity the public, their employers and clients;*
3. *Striving to increase the competence and prestige of the engineering profession; and*
4. *Supporting the professional and technical societies of their disciplines.*

ASCE GUIDELINES TO PRACTICE UNDER THE FUNDAMENTAL CANNONS OF ETHICS

Canon 1.   Engineers shall hold paramount the safety, health and welfare of the public and shall strive to comply with the principles of sustainable development in the performance of their professional duties.

Canon 2.   Engineers shall perform services only in areas of their competence.

Canon 3.   Engineers shall issue public statements only in an objective and truthful manner.

Canon 4.   Engineers shall act in professional matters for each employer or client as faithful agents or trustees, and shall avoid conflicts of interest.

Canon 5.   Engineers shall build their professional reputation on the merit of their services and shall not compete unfairly with others.

Canon 6.   Engineers shall act in such a manner as to uphold and enhance the honor, integrity, and dignity of the engineering profession.

Canon 7.   Engineers shall continue their professional development throughout their careers, and shall provide opportunities for the professional development of those engineers under their supervision.

FIG. 3.5   A civil engineer's code of ethics. Source: The American Society of Civil Engineers (ASCE). *ASCE Official Register—1999.* Reprinted by permission of the American Society of Civil Engineers.

3.5 ENGINEERING COMMUNICATION

As professionals, civil engineers must be able to conduct effectively all duties required for the successful execution of the project, which includes possessing adequate skills to communicate through verbal, written, and graphical means. Public hearings and regulatory review proceedings often require engineers to make verbal presentations where effective public speaking ability is essential (see Section 2.6.4). Similarly, report preparation and correspondence require the engineer to clearly and adequately express intended issues. Finally, well-organized drawings that contain adequate information will minimize confusion, and it is therefore essential that graphical communication skills are mastered by the design professional.

Advances in communication technology have produced innovative devices that allow the design professional to utilize time more efficiently. Global electronic personal pagers, telephone answering machines, and electronic mailboxes, along with global cellular and fax telephone services, allow avenues of communication for engineers to conduct everyday business. In addition, various improvements to digital computers, including laptop, desktop, and mainframe systems, provide the engineer a means to send or receive data, including reports, calculations, and even files of drawings. A communication tool in use today that was not generally available to the public when the first edition of this book was published is access to the World Wide Web, or Internet. This tool has become an integral part of everyday activities, from firms advertising and promoting consulting activities/services to professionals collaborating with geographically separated colleagues as part of a development team (see Section 3.10).

3.6 TYPES OF ENGINEERING SERVICES

Most civil engineering services are categorized into two general areas. The first involves either performing investigations or making consultations, with the final product being the formation of a report. The second involves the provision of services associated with the construction of a project.

The engineer requires knowledge of the existing site conditions to form a basis for design; this requires a certain amount of investigative work. A detailed discussion is included in Chapter 4 concerning most required activities associated with land development; however, the services requested may include not only feasibility, life-cycle, planning, and rate studies but also specialty services such as materials evaluation, pilot and equipment evaluations, site evaluations, engineering-related surveying services, and others.

Professional services that pertain to land development construction projects are discussed in detail throughout the remainder of this book and include the following:

(a) background investigation,
(b) preliminary design,
(c) final project design,
(d) job-bidding activities,
(e) construction-related activities, and
(f) project completion.

3.7   SELECTION OF A DESIGN PROFESSIONAL

The developer should have a procedure established for selecting a consulting engineer prior to soliciting proposals from consulting firms interested in performing services connected with a projected job. Several things that will probably be evaluated by the developer are the consulting firm's reputation, the breadth and depth of talent within the organization, information pertaining to completed projects, and evidence of financial stability. Typical information that might be requested is shown in Figure 3.6.

CONSULTING ENGINEERING FIRM'S VITAE

Firm Name
Year established
Former firm names
Business address
Telephone number and name of person to contact
Type of service particularly qualified to perform
Names of principals of firm and where registered
Names of key personnel, with experience of each and years of membership in the firm, and identified under special headings such as:

Civil Engineers	Architects
Structural Engineers	Transportation Engineers
Mechanical Engineers	Environmental Engineers
HVAC Engineers	Geotechnical Engineers
Electrical Engineers	Surveying & Mapping Engineers
Industrial Engineers	Computer Specialists
Planners—site, city, community, other	

Number of staff at present time
Approximate square feet of office space
Outside consultants and associates usually employed
Completed similar work on which the firm was the Engineer of Record
Present activities:
   Number and identity of projects
   Estimated total construction cost
Completed similar work on which the firm was associated with others
Present activities on which the firm is associated with others
Estimated annual work capacity (revenue) in dollars
Average annual volume of work (revenue) in last five years in dollars
Largest project in last five years (construction cost)
Largest current project (construction cost)
Financial capability
Banking references
Date of submission

FIG. 3.6   Consulting engineering firm's vitae. Source: the American Society of Civil Engineers (ASCE) Committee on Standards of Practice. *ASCE Manuals and Reports on Engineering Practice No. 45.—Consulting Engineering—A Guide for the Engagement of Engineering Services.* 1988. Reprinted by permission of the American Society of Civil Engineers.

3.7.1 Selection Process

The developer should extend notice to potential consultants by making available details that depict the nature of the solicited work and should request interested parties to respond by providing submitted prerequisite information and a statement of interest.

The developer would select possibly no more than five of the most promising firms, based on review of the requested material, to continue further negotiations and evaluation. The preferred firms would be given a more complete description of the anticipated scope of work, and the developer would request these firms to prepare a proposal that would outline their anticipated conduct and scope of proposed services for completing the associated work if awarded the job.

Upon receipt of these proposals, the developer normally conducts individual interviews with each candidate firm where all shared information would be discussed. After completing inquiries into past performance by contacting supplied references, the developer would be able to rank order in decreasing preference the remaining firms. Following the order of preference, each consultant is afforded an opportunity to negotiate compensation and other matters for contractual agreement. If agreement cannot be reached with the highest ranked firm, that firm is eliminated and the process is repeated.

3.7.2 Bidding for Professional Service

Good business procedures would dictate that a developer follow the process just outlined in order to procure the best possible service for each particular job; however, many developers operating with private funds will engage a design professional based on favorable past work experiences using an informal selection process. In addition, developers who rely too heavily on the bidding process to solicit professional services often receive unsatisfactory results because professional abilities vary among firms, as does the quality of work. In much the same manner, a client creates the potential of being ill served if solicitation for professional services is made on a contingency basis. According to *ASCE Manual No. 45* (ASCE, 1988):

> Such practice, which requires a favorable recommendation for the proposed project as a condition of receiving subsequent compensation, could project an undesirable appearance of tending to influence the engineer's objectivity.[*]

3.8 ENGINEERING SERVICES CONTRACT

A contract is an agreement to exchange things of value—an agreement that can be enforced between two or more parties bound by specific terms contained therein. Engineering contracts normally provide for documents resulting from services rendered to remain the design professional's property unless otherwise specified. The developer can obtain reproduced copies of drawings and job-related documents at cost. Contracts for auxiliary professional services are made with either the primary design professional

---

[*] Taken from ASCE (1988) by permission.

or the developer. The ensuing discussion is focused on the primary design professinal's contract.

3.8.1   Elements of a Contract

If it is to be enforceable, the contract must include a clear description of an agreement to which all parties concur. Should the agreement include a mistake in the terms or conduct of the contract, the contract may void utilizing a doctrine called unreality of consent. This basis can also be used to void a contract if fraud, duress, or undue influence was used (Abbett, 1963). Contracts can be expressed or implied. Express contracts involve a definite offer by one party and an unqualified acceptance by the other. Most contracts may be assigned to a third party and an unqualified acceptance by the other. Most contracts may be assigned to a third party unless the terms of the contract specifically prohibit such activity. Normally the assignor is not relieved of obligations unless all original parties agree to such action. Contracts also can be changed, modified, or terminated by mutual agreement.

An enforceable contract must contain material that is lawful, and there must be an obligation or something must be accomplished. Should the contractual matter violate federal, state, or local law or should it be contradictory to acceptable public processes (if, e.g., bid rigging were involved), the contract would be unenforceable. In addition, should the contract not include real and valuable consideration as part of the contractual elements, that too would cause the contract to be unenforceable.

All parties to a contract, acting in good faith, must be competent to produce a binding contract. An incompetent person, such as a minor, an insane person, or a substance abuse victim, a corporation operating outside of its articles of incoporation, or a governmental agency operating outside of its relegated authority cannot produce a binding contract.

Finally, certain contracts by law must be reduced to writing before they can become enforceable. Included are conveyances of interests in real property, agreements that extend more than a year, contracts for sale of goods over a minimum value, and suretyship and guarantees. Forms of agreement typically used by civil engineers for various professional services have been prepared by the Engineers Joint Contract Documents Committee (EJCDC), the National Society of Professional Engineers, the American Consulting Engineers Council, the American Society of Civil Engineers, and the Construction Specification Institute, as indicated in Figure 3.7.

3.8.2   Contract Termination

Contract termination can be attained by several means. The most common is the satisfactory completion of all contractual provisions in the form of specific performance, while substantial performance indicates substantial completion of the contractual elements excepting some minor conditions that by uncontrollable conditions were unattainable. When a party refuses or fails to perform contractual obligations, a termination by breach occurs, with the resulting damages being assessed for nonperformance, while the damaged party is relieved of contractual liability of obligations. Finally, if unforeseeable conditions should make it impossible or radically more expensive to carry out the terms of the contract, such as, in the case of personal services, the death of one of the parties, violation of a recently passed law, or destruction of

The Engineers Joint Contract Documents Committee
Typical Standard Forms

(a)   *Standard Form of Agreement Between Owner and Engineer For Study and Report Professional Services (EJCDC No. 1910-19).*

(b)   *Standard Form of Agreement Between Owner and Geotechnical Engineer For Professional Services (EJCDC No. 1910-27A).*

(c)   *Standard Form of Agreement Between Engineer and Geotechnical Engineer for Professional Services (EJCDC No. 1910-27B).*

(d)   *Standard Form of Agreement Between Owner and Engineer For Professional Services (EJCDC No. 1910-1).*

(e)   *Standard Form of Letter Agreement Between Owner and Engineer For Professional Services (EJCDC No. 1910-2).*

(f)   *Limitation of Liability in Design Professional Contracts (EJCDC No. 1910-9-E).*

(g)   *Standard Form of Agreement Between Engineer and Architect For Professional Services (EJCDC No. 1910-10).*

(h)   *Standard Form of Agreement Between Engineer and Consultant For Professional Services (EJCDC No. 1910-14).*

(i)   *Standard General Conditions of the Contract Between Owner and Designer/Builder. (EJCDC No. 1910-40).*

(j)   *Standard Form of Agreement Between Owner and Design/Builder on the Basis of Stipulated Price (EJCDC No. 1910-40-A).*

(k)   *Standard Form of Subagreement Between Design/Builder and Engineer for Design Professional Services (EJCDC No. 1910-41).*

(l)   *Standard Form of Agreement Between Owner & Owner's Consultant for Design Professional Services in Design/Build Projects (EJCDC No. 1910-43).*

(m)   *Standard General Conditions of the Construction Subagreement Between Design/Builder and Subcontractor (EJCDC No. 1910-48).*

(n)   *Standard Form of Construction Subagreement between Design/Builder & Subcontractor on Basis of Stipulated Price (EJCDC No. 1910-48-A).* Reston: ASCE, 1995.

(o)   *Standard Form of Construction Subagreement Between Design/Builder and Subcontractor on Basis of Cost Plus (EJCDC No. 1910-48-B).* Reston: ASCE, 1995.

(p)   *Standard Form of Agreement Between Owner and Design/Builder on the Basis of Cost Plus (EJCDC No. 1910-40B).* Reston: ASCE, 1995.

FIG. 3.7 Typical standard forms of agreement for various professional services. Source: The Engineers Joint Contracts Documents Committee (EJCDC): National Society of Professional Engineers, American Consulting Engineers Council, American Society of Civil Engineers, The Construction Specifications Institute.

the subject matter of the contract, the contract can be terminated by impossibility or impracticability of performance.

3.9   COMPUTING FACILITIES

Computer systems in use today are sometimes referred to as parts of the "information highway." If viewed from this perspective, these electronic systems do have components that parallel the nomenclature of streets and roads depicted in Figure 2.18 as follows (Dion, 1995):

1. Arterial component—providing mobility to users. This can include regional or national/international networks where distant operators are contacted or distant resources are utilized on a time-shared basis.
2. Collector component—providing a way to access local and regional components through the use of local-area networks. Data and program software can be stored on a host computer or server, as an example, to support collector component activities.
3. Local component—providing local access by utilizing data acquisition equipment with local storage.*

Figure 3.8 shows the overall "information highway" system where local computing facilities, collector networks, and arterial components are illustrated. The World Wide Web, or Internet, provides the arterial routes. Intranets, or private local networks (secured behind a so-called fire wall) for internal communication among employees and management, act as a collector and provide controlled access to company records, policies, specifications, and costs. With recent improvements in operating systems and security (including the digital signature), more reliance has been placed on the utilization of computer networks for e-mail, scheduling, conferencing, and even archiving information. Recently the term "Extranet" has been applied to dedicated web sites serving a select audience reached through the Internet using secure means. These web sites have evolved as a result of the collaboration of architects, engineers, and contractors (A/E/C) working together to improve project communication and management. The information contained in a project extranet site is accessed through security controls so that only those who have a need to know can enter. Authorized users can access drawings, specifications, submittals, shop drawing logs, and other documents pertaining to the job. Normally users do not need copies of parent software programs to access the material because embedded "viewers" are provided. This is important to promote accessibility to multiple users such as financial consultants being able to view the contract drawings. By providing instant access 24 hours a day, confusion is reduced along with expenses associated with copying, couriers, and conferences. Project logs can be kept up to date where managers in distant offices do not have to wait a day or so for progress-related information.

3.9.1 Field-gathered Information

The information highway is available and being used by a growing number of land development professionals. As a result, it affects the handling of information that is field gathered, office generated, and imported from or exported to distant users. Today's technology makes use of data that are often field recorded and stored in a hand-held electronic data collector (see Section 4.9.3). These collectors can be used

---

*Taken from Dion, Thomas R. "The Information Highway's Present and Potential Impact On Teaching Undergraduate Civil Engineering Students" Proceedings of the Southeastern Section ASEE 1995 Conference. Washington: American Society for Engineering Education, 1995. Reprinted by permission of ASEE.

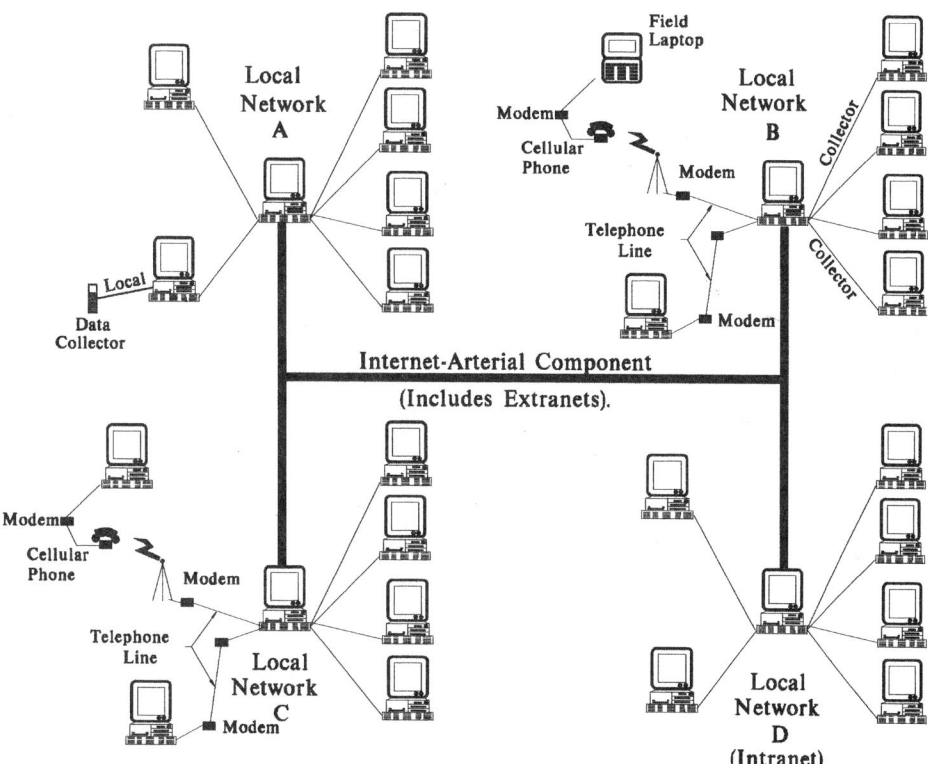

FIG. 3.8 Information highway nomenclature, showing local, collector, and arterial computer network components. Source: Thomas R. Dion. *Proceedings of the Southeastern Section ASEE 1995 Conference*, Washington, 1995. Reprinted with permission.

in conjunction with total stations (Section 4.9.2), global positioning systems (GPSs, Section 4.10), and other geographic information systems (GISs, Section 4.1) to locally access the information highway, as indicated in local network A shown in Figure 3.8. If collected data are downloaded directly to an office-based computer, then office-generated information procedures can begin. However, if the computer receiving the downloaded data is a field laptop, the field-collected information must first be conveyed to the office using a mobile cellular telephone link and modem. This is illustrated as local network B in Figure 3.8. Depending on where the sender is relative to the receiving office, the transmitted information may travel along collector and arterial components of the system.

Field observations made where meteorological, environmental, or transportation factors are being monitored in real time can give land development professionals additional valuable information. Through improved sensors and data-handling facilities linked to the information highway, field conditions are being better defined where increased reliability can be placed on their value. Data reflecting historical trends and long-term effects can now be remotely collected and routed to the design professional where more efficient and reliable engineered solutions evolve that improve the life-cycle performance.

3.9.2   Office-generated Information

Field-gathered information conveyed to office equipment allows one to analyze the data in an environment that is conducive for this type of work. Error analysis, adjustment activities, and data transformation can be completed efficiently in the office where digital graphical representations are often created in the form of drawing files or graphs. Some types of engineering data are used to investigate future needs. One illustration would be the use of real-time traffic data obtained through remote sensors to measure peak-hour volumes and growth trends. System capacity analysis using these data can be used to determine levels of service for streets and intersections. From this office-generated traffic information future capital improvement projects can be derived.

In many cases additional information (over and above that collected in the field) is required by office personnel to complete an engineering design or the analysis of a problem. Often this new material is searched for and obtained using the Internet.

3.9.3   Imported Information

Numerous sources exist today on the Internet where on-line data can be accessed using either a modem or network connection. On-line electronic bulletin boards usually offer a selection of choices that can be downloaded for local use. One example would be downloading continuously operating reference station (CORS) data files (see Section 4.10.1) provided 24 hours a day by the National Geodetic Survey (NGS) for the purpose of differential GPS corrections. Another example would be acquisition of digital maps or other forms of mapping data discussed in Section 4.1.3. Still another source of imported information is that associated with GISs, as discussed in Section 4.1. Other forms of imported information useful to the engineer center on jurisdictional review commenting, redlining drawings, permitting status, and revision requests. Some municipalities and local governments now offer on-line services to expedite the permitting process (Section 4.1.1).

3.9.4   Exported Information

Consulting engineers can produce a variety of documents based on the terms of their employment (Section 3.6) that can be placed on the information highway and conveyed to end users. Often engineering drawings are delivered to the client in soft- (electronic drawing files) and hard-copy formats. Drawing files sent to distant locations over the information highway can be uploaded into distant computers or workstations and used there. Files containing text and/or spreadsheet information routinely are conveyed to end users over the information highway as attachments to e-mail. If a final hard copy must be conveyed, then a facsimile (fax) can be sent.

3.9.5   Computer Utilization Standards

Most organizations have established computer standards and protocols that must be followed. These standards can range from observing copyright and licensing requirements to the utilization of established file structures and formats. Often the system administrator promulgates a set of guidelines that instruct the user how to access the company system, what etiquette practices are expected, and what precautions must be taken such as password protection.

Company directives could address the Digital Millennium Copyright Act enacted in 1998 that protects copyrighted digital works. The law makes the circumvention of encryption codes used to block access to these protected works illegal. This includes the creation or selling of any device or technology capable of decryption. Another directive element might address the absolute requirement for every computer to have a designated license in effect for each software program installed. By following this procedure, the company will avoid being paid a visit by the Business Software Alliance (BSA), an industry-sponsored organization based in Washington, D.C., that seeks out intellectual property offenders in an effort to combat software piracy.

Another policy element might address standards for managing drawing files (e.g., DWG, the file format for Autodesk's AutoCAD®-based products, including AutoCAD®, AutoCAD LT®, LDDT, and AutoCAD Map®, or DGN™ file format for Bentley's MicroStation®-based products,* including MicroStation®, Triforma®, MicroStation Modeler®, MicroStation PowerDraft®, and MicroStation ReproGraphics®.) Items that might be defined include file naming, feature layering, line types, fonts, and standard sheets. The company might also designate a designer notation or comment layer to record information pertaining to the drawing or the design procedures used to create the drawing. Until recently there was no universally recognized system for organizing computer-aided design (CAD) drawings. The Facility Information Council the National Institute of Building Sciences (NIBS) formed the National CAD Standards Project Committee. This committee developed the voluntary *National CAD Standard*, which consists of the Committee's *Introduction and Amendments to Industry Publications* and the following individual documents: (a) The American Institute of Architects' CAD Layer Guidelines, (b) The Construction Specifications Institute's *Uniform Drawing System, Modules 1–3*, and (c) the U.S. Department of Defense Tri-Service CADD/GIS Technology Center's *Plotting Guidelines of the United States Coast Guard*. Using a system such as this would provide for consistent classification of project data, ease of information exchange between design team members, and reduced time for translations and reformatting.

Word-processed, spreadsheet, and database files can also have stipulated company formats where confusion among users is reduced and ease in access is promoted, especially when increased productivity can be achieved through cutting and pasting operations.

Because multiple users access company-generated files, the ability to translate these files from one format to another is critical when different software packages are involved. This is especially true when dealing with drawing file exchanges and multimedia formats. Multimedia presentation products can include word-processed, graphical, audio, and video media. Video clips, digital still camera images, and flatbed scanned raster images are being used today more and more to record site conditions, construction progress, and other job-related information. Digital images can even be incorporated into GISs as well as being draped behind conventional engineering drawings (Section 4.1.2). Company-produced multimedia CD-ROM electronic bid sets can save time, money, effort, and confusion, especially if a large number of participants are

---

* See http://www.bentley.com.

involved with project bidding or construction. As in the case with Extranets (Section 3.9), viewers are provided where the user can read the included files without having to purchase the specialized software used for its creation.

3.9.6   New Software Suites for Enterprise Applications

New computer software packages are available that extend across the enterprise of land development. What is offered is multitasking capabilities with cut-and-paste features and document linking. Multiple users can also access drawings and extract layers or entities from other drawings or database files, which supports collaborative efforts and the team approach. Digital imaging and GIS mapping applications can be included as well as add-on packages for surveying and site design. For example, one such program, called *AutoCAD® Land Development Desktop* (LDDT) Suite, is produced by Autodesk®, based in San Rafael, California. LDDT makes use of AutoCAD® where the only menus and options provided are those referring to civil engineering, digital terrain modeling, coordinate geometry, and parcel/alignment tools. Mapping, surveying, and civil design are included parts of the integrated suite family. LDDT-generated drawings can be posted on project web sites for Internet dissemination. Figure 3.9 shows one of the suite members, AutoCAD Map®, being used to perform GIS functions.

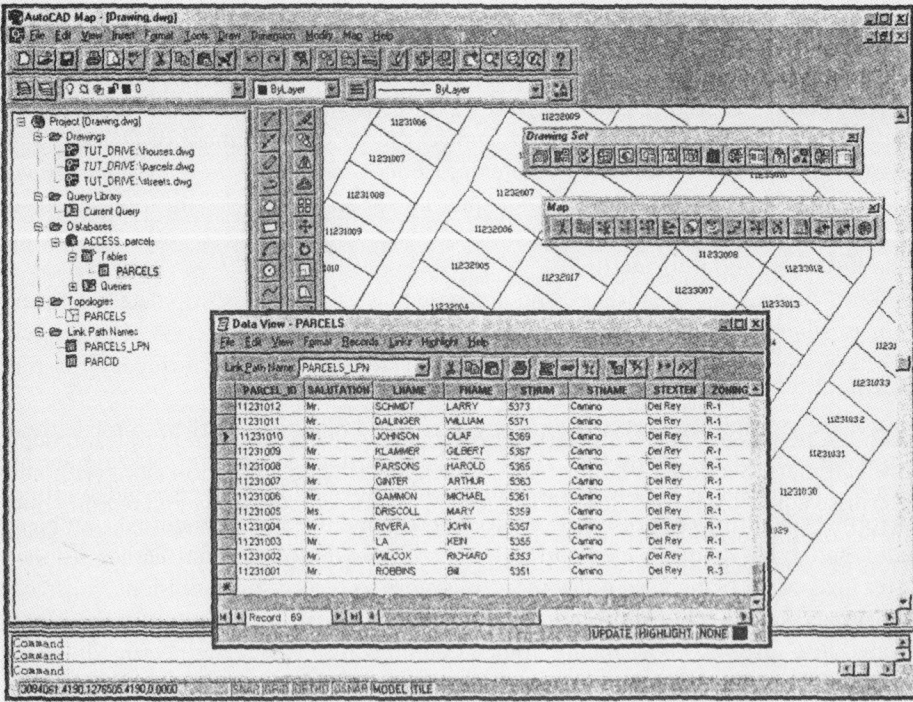

FIG. 3.9   Example of an enterprise-based development software package. Shown is a view of AutoCAD Map® (one of the LDDT suite members) in use with an external database application. Source: © Autodesk, Inc. Reproduced by permission of Autodesk, Inc., San Rafael, CA.

3.10   MOBILIZATION PLAN

The design professionals on which the remainder of this book focuses are consulting civil engineers who have a primary responsibility to their clients to undertake and see projects through to completion. Details of responsibilities outlined in the engineering services contract relegate activities that the consulting engineer must undertake. Work assignments based on available staffing with proper planning will provide systematic efforts to ensure timely completion of the project. Good practice includes the assignment of a permanent project coordinator who will provide continuity of services. Various memoranda of understanding are helpful, especially when coordinating the efforts associated with the procurement of auxiliary services, such as subsurface site investigations. Of utmost importance is the necessity for all parties involved to maintain clear and factual records that will minimize confusion and hopefully prevent the occurrence of major problems.

3.10.1   Management of the Project Design Team

Once the design team members have been selected and engaged, the team must be organized and then mobilized. The lead design professional would assemble the entire team along with the developer. Two of the leader's roles are to build trust and relationships among other team members. Positive attitudes and motivation to prosecute the work diligently are necessary requisites for success where the project is completed on time and within the budget.

At this first meeting (which may in the form of a one- or two-day workshop) the developer may assist in the drafting of a mission statement along with project goals and objectives which could include some form of partnering agreement. The roles of the various members are also defined, including those identified as filling key team positions. An assignment list tasking various team members with various project responsibilities can also be adopted. A work schedule that includes estimated times and costs to produce the preliminary and final construction documents is often developed, along with an agreement to use standard formats for drawing files, documents, and filing systems.

Once the project design is underway and the team is functioning, periodic meetings can be scheduled where members receive/make status reports, discuss problems, and share ideas. These meetings should address action items left over from previous meetings, new items, and budgetary/scheduling problems. As a benefit to team members, extranets (Section 3.9) offer real-time engineering collaboration along with new enterprise edition software suites. Production, review, commenting, and revising documents on-line provide instantaneous feedback to all participants. When the project enters the construction phase, the duties of the team need to focus on job completion. These duties can be found in Part IV.

3.11   REMUNERATION FOR ENGINEERING SERVICES

The consulting engineer can receive compensation for services rendered in any number of ways. Most land developments entail the design consultant being paid a percentage of the costs of construction. Figure 3.10 illustrates possible relationships between engineering services costs and construction costs for projects similar to land developments. Although this graph is not included in the current edition of *ASCE Manual No. 45*, it is useful to illustrate general cost trends. Negotiated engineering percentage costs associated with an improved land development could range from 6 to 10 percent of the cost of construction, with provisions of on-site observation ranging from 3 to 5 percent. Other services, such as engineering surveying for construction control, may range between 2 and 4 percent of the construction cost.

Some projects warrant payments to the consulting engineer on a lump-sum basis, especially if the job is clearly defined and small. Should this method be inappropriate, a cost-plus-fixed-fee approach could be employed, where the engineer would collect reimbursements for all costs associated with salaries, overhead costs, auxiliary expenses, and a fixed fee for profit and intangible costs. A final method of compensation uses a cost multiplier approach where all costs associated with salaries are increased by a factor that may range from 2 to 3 to provide income for profit, contingencies, and overhead costs. The cost multiplier approach usually provides that incidental costs associated with the necessary conduct of work be reimbursed, including an administrative charge of up to 15 percent being assessed. Other methods of payment include retainage and *per-diem* fees.

Although various methods of allocating partial payments to the consulting engineer can be used, one effective method is the use of monthly billings. These may be based

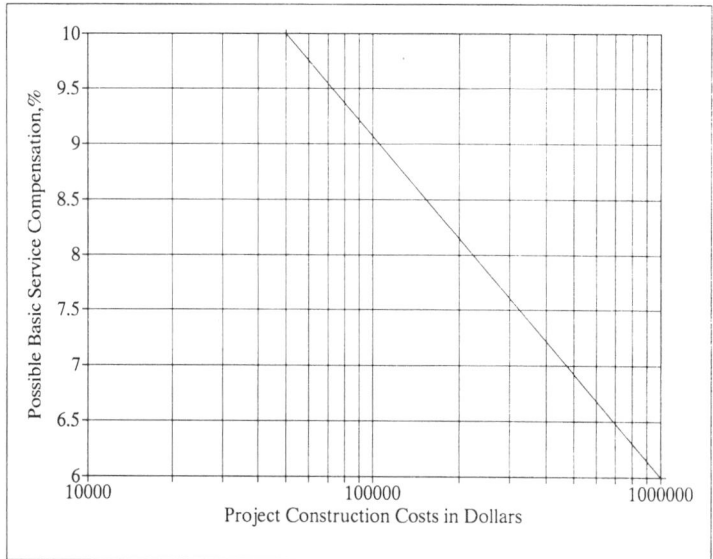

FIG. 3.10   Possible costs of engineering services for land developments. Source: The American Society of Civil Engineers (ASCE) Committee on Standards of Practice. *ASCE Manuals and Reports on Engineering Practice No. 45—Consulting Engineering—A Guide for the Engagement of Engineering Services,* 1972. Reproduced by permission of the American Society of Civil Engineers.

either on percentage of work completed, if the method of payment is related to a percent of construction costs, or on the number of actual accrued hours and expenses. Another schedule of payment might be:

Completion through preliminary phase: 40 percent.

Completion through final design phase: 90 percent.

Completion of the construction phase: 100 percent.

Prior to entering into an agreement with a developer, it is important for the engineer to determine that the project is financially sound. This will help to ensure that the engineer will receive payment for professional services rendered.

When there is no express agreement as to the compensation for services rendered, the law implies that an engineer is worthy of the services provided. The doctrine of *quantum meruit* provides that a party who accepts professional services is obligated to pay reasonable and commensurate remuneration in return.

3.12 ENGINEER'S ROLE

A civil engineer who has been engaged to perform professional services must function within the terms of the contractual agreement. Simultaneously the engineer must assist the developer in addressing the various constraints associated with the development. An initial step in this process requires the engineer to begin with an investigation into existing conditions that affect the project. This step is normally associated with what is called preliminary engineering activities and is further described in Part II.

REFERENCES

Abbett, R. W. *Engineering Contracts and Specifications.* New York: Wiley, 1963, pp. 23–24, 85–88, 112–113.

American Society of Civil Engineers (ASCE). *ASCE Manuals and Reports on Engineering Practice—No. 45—Consulting Engineering: A Guide for the Engagement of Engineering Services.* Reston: ASCE, 1972, p. 29, and 1988, pp. 4, 42–43. Referenced material is reproduced by permission of The American Society of Civil Engineers.

American Society of Civil Engineers (ASCE). *ASCE Official Register 1999.* Reston: ASCE, 1999, pp. 10–11. Referenced material is reproduced by permission of the American Society of Civil Engineers.

Dion, T. R. "The Information Highway's Present and Potential Impact on Teaching Undergraduate Civil Engineering Students." In *Proceedings of the Southeastern Section ASEE 1995 Conference.* Washington: American Society for Engineering Education (ASEE), 1995, p. 48.

Fisk, E. R. *Construction Project Administration,* 3rd. ed. New York: Wiley, 1988.

Nischwitz, J. L. "The Crumbling Tower of Architectural Immunity Evolution and Expansion of the Liability to Third Parties." *Ohio State Law Journal,* Vol. 45, p. 217, 1984.

Sweet, J. *Legal Aspects of Architecture, Engineering, and the Construction Process,* 3rd. ed. New York: West, 1985, p. 105.

SUGGESTED READINGS

American Society of Civil Engineers (ASCE). *Consulting Engineering: A Guide for the Engagement of Engineering Services, Rev. Ed.—ASCE Manuals and Reports on Engineering Practice No. 45.* Reston: ASCE, 1996.

American Society of Civil Engineers (ASCE). *Selling Effectively to the Civil Engineering Market.* Reston: ASCE, 1991.

Autodesk. *AutoCAD® Land Development Desktop.* San Rafael: Autodesk, 1998a.

Autodesk. *AutoCAD Map® Release3 Users Guide.* San Rafael: Autodesk, 1998b, p. 15.

Construction Specifications Institute (CSI). *The Uniform Drawing System (UDS)—Modules 1,2, and 3.* Alexandria: CSI, 1997.

Cox, A., and I. Thompson. *Contracting for Business Success.* London: Telford, 1998.

Davies, V. J., and K. Tomasin. *Construction Safety Handbook,* 2nd ed. London: Telford, 1996.

Heberle, D. *Construction Safety Manual.* New York: McGraw-Hill, 1998.

Hislop, R. D. *Construction Site Safety—A Guide for Managing Contractors.* New York: Lewis, 1999.

Levin, P. *Construction Contract Claims, Changes, and Dispute Resolution,* 2nd. ed. Reston: American Society of Civil Engineers, 1998.

National Institution of Building Sciences (NIBS). *U.S. National CAD Standard, Version 1.0.* Washington: NIBS.

National Society of Professional Engineers, American Consulting Engineers Council, and American Society of Civil Engineers (ASCE). *Standard Form of Letter Agreement between Owner and Engineer for Professional Services (EJCDC No. 1910-2).* Reston: ASCE, 1985.

National Society of Professional Engineers, American Consulting Engineers Council, and American Society of Civil Engineers (ASCE). *Limitation of Liability in Design Professional Contracts (EJCDC No. 1910-9-E).* Reston: ASCE, 1986.

National Society of Professional Engineers, American Consulting Engineers Council, American Society of Civil Engineers (ASCE), and Construction Specifications Institute. *Standard Form of Agreement between Owner and Geotechnical Engineer for Professional Services (EJCDC No. 1910-27A).* Reston: ASCE, 1989.

National Society of Professional Engineers, American Consulting Engineers Council, and American Society of Civil Engineers (ASCE). *Standard General Conditions of the Contract between Owner and Designer/Builder (EJCDC No. 1910-40).* Reston: ASCE, 1995a.

National Society of Professional Engineers, American Consulting Engineers Council, and American Society of Civil Engineers (ASCE). *Standard Form of Agreement between Owner and Design/Builder on the Basis of Stipulated Price (EJCDC No. 1910-40-A).* Reston: ASCE, 1995b.

National Society of Professional Engineers, American Consulting Engineers Council, and American Society of Civil Engineers (ASCE). *Standard Form of Subagreement between Design/Builder and Engineer for Professional Services (EJCDC No. 1910-41).* Reston: ASCE, 1995c.

National Society of Professional Engineers, American Consulting Engineers Council, and American Society of Civil Engineers (ASCE). *Standard Form of Agreement between Owner & Owner's Consultant for Design Professional Services in Design/Build Projects (EJCDC No. 1910-43).* Reston: ASCE, 1995d.

National Society of Professional Engineers, American Consulting Engineers Council, and American Society of Civil Engineers (ASCE). *Standard General Conditions of the Construction Subagreement between Design/Builder and Subcontractor (EJCDC No. 1910-48).* Reston: ASCE, 1995e.

National Society of Professional Engineers, American Consulting Engineers Council, and

American Society of Civil Engineers (ASCE). *Standard Form of Construction Subagreement between Design/Builder & Subcontractor on Basis of Stipulated Price (EJCDC No. 1910-48-A)*. Reston: ASCE, 1995f.

National Society of Professional Engineers, American Consulting Engineers Council, and American Society of Civil Engineers (ASCE). *Standard Form of Construction Subagreement between Design/Builder and Subcontractor on Basis of Cost Plus (EJCDC No. 1910-48-B)*. Reston: ASCE, 1995g.

National Society of Professional Engineers, American Consulting Engineers Council, and American Society of Civil Engineers (ASCE). *Standard Form of Agreement between Owner and Design/Builder on the Basis of Cost Plus (EJCDC No. 1910-40B)*. Reston: ASCE, 1995h.

National Society of Professional Engineers, American Consulting Engineers Council, and American Society of Civil Engineers (ASCE). *Standard Form of Agreement between Owner and Engineer for Study and Report Professional Services (EJCDC No. 1910-19)*. Reston: ASCE, 1996a.

National Society of Professional Engineers, American Consulting Engineers Council, and American Society of Civil Engineers (ASCE). *Standard Form of Agreement between Engineer and Geotechnical Engineer for Professional Services (EJCDC No. 1910-27B)*. Reston: ASCE, 1996b.

National Society of Professional Engineers, American Consulting Engineers Council, and American Society of Civil Engineers (ASCE). *Standard Form of Agreement between Owner and Engineer for Professional Services (EJCDC No. 1910-1)*. Reston: ASCE, 1996c.

National Society of Professional Engineers, American Consulting Engineers Council, and American Society of Civil Engineers (ASCE). *Standard Form of Agreement between Engineer and Architect for Professional Services (EJCDC No. 1910-10)*. Reston: ASCE, 1997a.

National Society of Professional Engineers, American Consulting Engineers Council, and American Society of Civil Engineers (ASCE). *Standard Form of Agreement between Engineer and Consultant for Professional Services (EJCDC No. 1910-14)*. Reston: ASCE, 1997b.

O'Reilly, M. *Civil Engineering Construction Contracts*. London: Telford, 1996.

Scanlon, B. *Marketing of Engineering Services*. London: Telford, 1988.

Warne, T. R. *Partnering for Success*. Reston: ASCE, 1994.

PART II
Preliminary Engineering Activities

4 Inventory, Analysis, and Reporting

A consulting engineer's initial activity would be to hold informal detailed discussions with the developer to review pertinent factors pertaining to the project. Topics of discussion might include potential phasing requirements and development theme guidelines that are necessary for project design. Economic and technical aspects also must be discussed. The developer may be considering several sites for the project, requiring the consulting engineer to investigate the conditions of each site. The investigation would include analysis of how well the site potentially supports the development's theme as well as the impact on surrounding areas. The design professional can ascertain during the initial discussions whether services of other professionals are required and mutually agree with the developer on the scope of these auxiliary services. Of keen interest to the developer is the potential cost of the development, which the design professional will have to estimate.

It is imperative that the consulting engineer make initial contact with jurisdictional regulatory authories on local, state, and federal levels (Figure 2.28) to establish communications that will aid in the progression of subsequent review activities. Hopefully, the consulting engineer will glean through these early contacts important guidance that otherwise might not surface until later, during the review process. The engineer can take proper action while planning the project to prepare for the project's review. If regulatory concerns have been satisfactorily addressed in the preliminary development plan, the project will receive approval. The consultant's role is to seek regulatory approval on the developer's behalf while incorporating into the project plans and specifications provisions that address concerns raised by regulatory agencies. The developer, however, is obligated to perfect and obtain regulatory approval (or the approval of others in the case of an easement) since many approvals require negotiation and fee payments.

4.1 GEOGRAPHIC INFORMATION SYSTEMS

A geographic information system (GIS) is a computer-based tool that utilizes not only geographic data but also alphanumeric attribute data to analyze situations and display the results where informed decisions based on complex issues can be made. Spatial relationships can be organized into two groups. The first group is projection dependent (cartographic) and includes area, shape direction, and distance. The second is projection independent (topologic) and includes connectivity, adjacency, and containment (within). In a GIS geographical data are stored as features called points (X and Y), arcs (at least two points connected by a straight line), polylines (one or more arcs that have the same attributes), polygons (connected arcs that define an area), and text. GIS users are able to define spatial relationships called "topology" among data elements, which is more than just defining the location and geometry of a particular feature. Analytical procedures profit through the use of topologic relationships because the outcome is the same, regardless of the coordinate system used. A GIS user can make and print copies of maps, model networks, store and edit attribute data, and perform spatial analysis.

A GIS is therefore much more than a computer-aided mapping (CAM) system, where only maps are produced. CAM systems organize data using layers for map features and themes or for different line types where elements are referenced to a defined coordinate system. CAM systems can have functions similar to those which comprise computer aided drafting (CAD) systems such as San Rafael–based Autodesk® AutoCAD®, INC's products. Another system which is similar to the CAM system is an automated mapping/facilities management (AM/FM) system often used to manage utility system data. This system also relies on the use of layers by type to associate map features, although the mapping capability associated with this system is not usually as accurate or detailed as the CAM system. An AM/FM system allows the user to link attribute data to graphic features. A network is created by defining how the geographic features are linked. The network is then used to identify components that are connected and how their connections are made either spatially or through their attributes. For example, no two similar utility lines can cross in an AM/FM system, except at an intersection, which sets AM/FM apart from CAM systems. These data connections are usually stored in a separate file from the graphic data file. Attribute data can describe system component characteristics such as lengths, capacities, materials, installation dates, and so forth. By linking attribute data and spatial features to network data, the system's operation can be modeled and analyzed. [It must be noted, however, that recently some companies, such as Environmental Systems Research Institute (ESRI™) have developed AM/FM tools that are based upon current GIS technology].

A land information system (LIS) is similar to AM/FM in that it also makes use of spatial data to define parcel geometry and location and nongraphic attribute data linked usually to parcel numbers. But like CAM systems, LIS data are not linked where spatial relationships can be developed.

Many governmental entities either have or are in the process of creating a GIS of their domain because of its usefulness in providing an integrated approach to cataloging, analyzing (through topology), and maintaining jurisdictional data pertaining to their needs. Likewise, many consulting engineers and land surveyors have found it useful to create a GIS that is tailored to their particular area of interest and practice. The existence of a GIS for use during the inventory, analysis, and reporting phase of a land development project allows the engineer, land planner, surveyor, and other professionals on the development team to have ready access to information that could affect project planning and design.

4.1.1   Growth In GIS Utilization

Many cities, counties, and even states are placing increased reliance on GIS for planning and management activities. This increase can be attributed in part to improvements being made in desktop computers, networking, operating systems, software, and relational databases where many users can have access to remote data and even to geospatial information stored in one location. Today it is possible for a city to save large sums of taxpayer's money by creating, operating, and maintaining an integrated and comprehensive GIS database where custom viewing applications are made available to city employees and taxpayers through on-line access to GIS maps, orthophotos, and land records. Often the heaviest users are employees associated with the city engineer's office, public works, utilities, and planning departments. Other government users could include the fire department and voter registration office. The Internet provides a new medium for taxpayers to view and query parcel information. One site in particular, Forsyth County, NC, at http://maps.co.forsyth.nc.us, shows users geographically the particular parcel of interest and all of its associated data.

The GIS often includes layers for features such as land parcels, easements, street intersections, addresses, zoning districts, political or police districts, school districts, and even easements. Activities or events which can be defined in space as points, lines, or areas also qualify for GIS inclusion. The base map is often created from digital orthophotos (see Section 4.14.7) where land parcels formed by closed polygon topology can be added. (It is noted, however, that the use of three-dimensional GIS applications is growing.) Users can navigate through these data using a viewing engine where selected criteria could include street address, parcel number, property owner, or subdivision lot and block designation. A system such as this would also be capable of producing scaled hard-copy maps and generating mailing labels through object linking. Likewise, such a system would allow city employees to utilize the GIS for functional tasks such as looking up and printing information pertaining to the general land use map, defining the limits of a prescribed notification impact area set for seeking public input to zoning variance requests, and even maintaining emergency response 911 data, accident/crime analysis, code enforcement, building permitting/inspections, and areawide master planning.

The land development professional should make inquiries during the preliminary stages of planning a land development as to whether a GIS is locally available as well as determining how to gain access. A growing number of these same professionals are beginning to maintain their professional records in a company GIS that is easily accessible for reference and planning new work. Citywide or countywide GIS data might even be available for the professional to use as part of a company database, and often these same professionals assist the city or county in updating their database by preparing and submitting electronic files of plats and maps as part of the local review process (see Section 4.8.4 as an example).

4.1.2   GIS Components

The data components of a GIS can be organized into two groups: (1) where and (2) what. The position and graphical representation as a point, line, or area constitutes "where," and the characteristics, or attributes, at the position constitute "what." Access to each of these "sides" of the data is enforced while relational joins permit GIS functions are bidirectional.

Many times the base layer of a GIS consists of digital orthophotographic panels that have been registered to a selected datum (see Sections 4.8.1 and 4.8.2) upon which all other geographic data are referenced, as indicated in Figure 4.1. (In many cases geographic coordinates in the form of latitude and longitude are used.) Digital orthophoto panels are comprised of raster graphic files where pixels of different color shades illuminate the computer monitor to varying intensities to create a graphical image. Vector graphics, on the other hand, use coordinate values and equations to create lines, circles, and points, but cannot be used to represent subtle light and shadow variations that exist in the real world. Raster images are therefore useful to represent land information and can be enhanced using computer programs. Over time, however, each program developed its own raster file format. Today there are many different raster file formats, and as a consequence copious software conversion programs exist to access each file type. One raster format widely used today called GIF® was developed by CompuServe® (a registered trademark of America Online) to transfer files via modem from one computer architecture to another. Adobe Systems Incorporated maintains and owns the Tagged Image File Format (TIFF) originally developed by Aldus Corporation, which is widely used on multiple platforms for a broad range of image storage needs. PCX format raster files evolved from PC Paintbrush, created by Z-Soft in Marietta, Georgia, while probably the most widely known raster file format is the Microsoft® BitMaP image, or BMP. BMP files tend to be larger than other formats because image data are not compressed. Most GIS applications require BMP data to be translated into another file format before use.

GIS map topology defines spatial features shown in Figure 4.2 using points, lines, nodes, and areas, each of which is uniquely numbered. Point features are depicted as single, independent x,y coordinates. Lines are defined by an ordered set of x,y coordinates. Where lines are straight, only two coordinate pairs are needed. Curved lines require any number of additional coordinates to define cartographic shape. At the beginning and ending of each line and at line intersections are nodes. Lines that begin or end with the same position as other lines will share the same node. Nodes, therefore, provide the means to define topologic connectivity. A polygon is an ordered set of lines, each of which also contains the location of its centroid. Where polygons

FIG. 4.1 Example of a geographic information system (GIS) where six themes are shown as layers. Source: Graphic Mapping Group, Inc. 2001 Asheville Highway, Hendersonville, NC. Reprinted with permission.

have also been defined, each line on the polygon's boundary also identifies by number the polygon to its left and right, relative to the direction in which the coordinates are stored. This is the property of topologic adjacency. Figure 4.3 illustrates a view of a "study area" where five themes are activated: blocks, drawing annotation, buildings, trees, and parcels.

Mapped feature attributes can reflect a level of measurement as defined by the following four categories:

(a) A label or class assigned to a feature is classified under the nominal information category.

(b) The ordinal information category assigns a rank order to the feature such as a highway classification (e.g., arterial, collector, or local).

(c) Relative scale measurements such as elevation fall under the interval category.

(d) The ratio category relies on measurements based on an absolute scale.

Mapped feature attributes also consider when the data have a continuous distribution (e.g., contour maps) or a discontinuous distribution, where data are classed and often displayed by areas colored using varying intensities to account for their individual class ranking. Geographic features can exhibit various properties based on individual feature characteristics. These properties can include the use of different symbolic shapes, sizes, patterns, distributions, scales, and orientations to visually depict data. Other categories used include depicting data located within established neighborhoods or the depiction of contiguous geographical areas that exhibit some type of relationship.

Spatially distributed activities stored in a GIS database often reflect distributions and patterns that pertain to the general population in the form of census maps, traffic accident maps, or maps showing crime or disease incidents. Spatially distributed events stored as attributes place value on the importance of time. For example, the highway location of an overturned tractor-trailer marking a toxic spill incident is an example of a point event, while the flow of water within a distribution network segment could

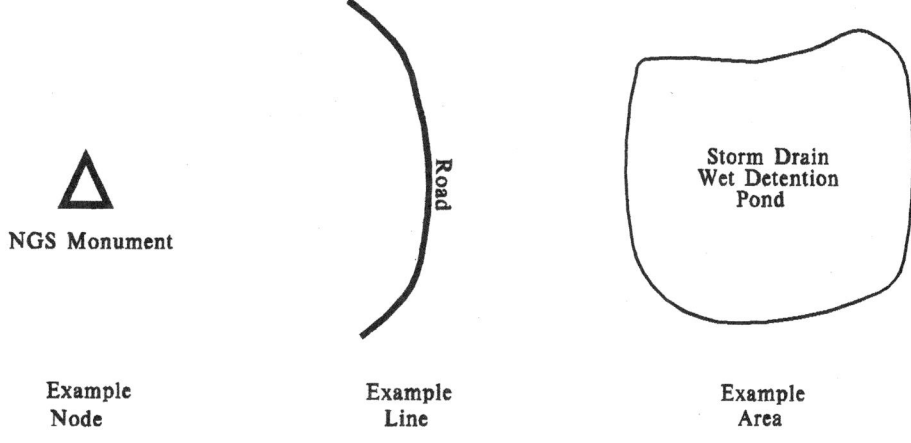

FIG. 4.2   GIS mapping features.

FIG. 4.3 Example of GIS graphical user interface. The ArcView GIS® GUI is the intellectual property of ESRI and is used herein with permission. Copyright © 1999 ESRI. All rights reserved. ArcView® is a registered trademark of ESRI.

represent a line event. An area event could be a reservoir's threshold low water surface area when approaching drought conditions.

Attribute data consisting of descriptive information linked to geometric and spatial data defined by nodes, lines, and areas by some unique feature identification allow GIS users to explore both spatial and attributed data simultaneously. Stand-alone relational databases such as Oracle® or Microsoft® SQL Server™ can be used to establish a link between the geometric and spatial data and the attribute data. A user can query the database and display the results geographically, and vice versa. AutoCAD Map® is capable of these functions (see Section 3.9.6). Because AutoCAD Map® is integrated into LDDT, all job information can be shared by other suite segments for access, design, and analysis. Other software platforms capable of accessing relational databases include ESRI's™ ArcView ® GIS and ARC/INFO®.

GIS data handling varies according to the software capabilities. One common system utilizes client/server technology, where one main server stores and distributes all data (geographic and tabular). Users that have access to the server can manipulate the data via client software that resides on the user's computer.

4.1.3 GIS Data Sources

Traditionally, GIS geospatial data have been obtained from conventional surveying techniques described in Section 4.9. Although these data sets are usually very accurate (often more accurate than necessary for GIS applications), they often require a large amount of time and money to produce. Some GIS data associated with property lines, rights-of-way, and the location of underground utilities require high accuracy; however, there many other data sets are only accurate to the nearest foot or even 10 feet.

Aerial photography (Section 4.14) is used to create many GIS data sets, including street centerlines and building footprints. However, resolution (i.e., the area one pixel of an image encompasses) on aerial photographs is generally greater than 1 foot. Therefore, the accuracy of the GIS data based on imagery cannot be less than 1 foot or objects that are less than 1 ft² are undetectable.

GIS usage has grown significantly in recent years, which has placed demands on gathering field information, most of which is hard (or even impossible) to discern using aerial photographs. For example, water main valves and boxes are small and usually not seen on aerial photographs. Their location and the location of other small spatial objects are better determined using global positioning system (GPS) methods described in Section 4.10.3. These data are electronically gathered (often with attributes) and stored in the field using a receiver memory bank or a data collector. Once collected, these data can be electronically downloaded to a computer and differentially corrected, translated to a desired datum, and then graphically plotted for later import as part of the GIS database.

Rather than procuring geospatial data first hand from efforts associated with conventional, GPS, or photogrammetric surveying, one might consider using archived images that are readily available for use. For example, Microsoft has compressed browsable images on the Internet of the entire U.S. Geological Survey (USGS) digital orthophoto quadrangle library. Other USGS mapping products include digital line graphs (DLGs), digital elevation models (DEMs), land use and land cover digital data (GIRAS), digital cartographic text—Geographic Names Information System (GNIS), and digital raster graphics (DRGs). The National Geophysical Data Center of the National Oceanographic and Atmospheric Administration (NOAA) has available many different types of digital map sets, including oceanic bathymetry, land surface topography, and geodetic/magnetic data. Another federal agency, the U.S. Census Bureau, has topologically integrated geographic encoding and referencing (TIGER) digital files available that are built on street-level geography where blocks, census tracts, and enumeration districts are delineated. Commercially available satellite imagery from EOSAT and others has also been used as a source of geospatial data for GISs encompassing large land areas.

Another way geospatial data can be procured for GIS use is by electronically digitizing or scanning existing paper products, such as sanitary sewer line plans. Scanning produces bit-mapped raster images previously discussed in Section 4.1.2. If a digitizer is used and is calibrated to the coordinate system of the GIS, the paper product can be directly added to the GIS. However, scanned imagery must be registered and rectified or the image coordinates are manually matched to known ground coordinate values. When vector-based geography created from CAD programs are imported as part of a GIS application, these images must be translated into GIS node, line, and area topology format. Most of the time the CAD data will need to be translated to a coordinate system. To minimize translation problems, the following should be considered:

(a)   The lines at the edges of adjoining digitized material must have identical coordinates for endpoints, common layer assignments, and visual alignment (edge matching).

(b)   Common boundaries shared by graphic features must be digitally represented the same, regardless of the layer, and within a map theme all features must represent the boundary by a single graphic element. Different boundary types are distinguished by associated attribute data.

(c) Lines representing the same type of data in a network must not cross except at an intersection. In addition, network linear element lines should not have gaps unless the gaps reflect visual or attribute code characteristics, or unless the database forces the break because of limitations.

(d) Each polygon must have the same beginning and ending point coordinate values. Lines or line strings are allowed to cross a polygon of the same data type, unless it joins the polygon at an intersection. Each polygon must have a unique interior point to which attribute data can be associated. When single data layers exist, polygons must cover the area completely and be exclusive within that area.

(e) All lines must not cross except at an intersection. Straight lines are defined by the beginning and ending nodes, while curvilinear features are represented by a minimum number of points where a smooth curve can be fitted when plotted. Unnecessary points along curves should be eliminated to reduce processing time and storage requirements.

(f) All graphical features that visually connect must also digitally connect where intersection points are exact at the same point where the lines also terminate. In cases where a line connects to a polygon, the line endpoint must also be a point on the polygon's boundary.

Although some GIS software is capable of directly importing and viewing CAD-generated files, translator programs are available to make the conversion. Direct translators efficiently make use of binary source files to create binary translated files. Indirect file translators make use of ASCII file formats that include DLGs—(U.S. Geological Survey, Reston, VA), Drawing Exchange Format (DXF, Autodesk®,INC.-San Rafael, CA); Initial Graphic Exchange Specification (IGES; National Institute of Standards and Technology, Boulder, CO), and Standard Interchange Format (SIF; Intergraph® Corporation, Huntsville, AL), to name a few.

GIS software packages also allow the user to determine topology from the imported CAD data. Associated attribute data can be manually added using the software or relinked to external bases.

4.1.4   GIS Data Relationships

All GISs rely on the use of digital data, represented by numbers. Information pertaining to a mapped object described in Section 4.1.2 is stored in a tabular file format where each row represents a record and each column an attribute, or what is usually called a field. Figure 4.4 illustrates a typical file data structure for a polygon consisting of more than one arc (or related line segments). Records created sequentially and stored in a file can be searched line by line for a particular entry or sorted using built-in database options. Usually a report generator is included so that a hard copy of the database records can be produced if needed.

Triangulated irregular networks (TINs) were developed to efficiently depict geographic surfaces in three dimensions (e.g., topography). Two types of TIN file formats exist: (1) the information pertaining to each arc connecting the points is stored in a file

FIG. 4.4 GIS data storage techniques and topologic structures. Adapted from Environmental Systems Research Institute, Inc. *Understanding GIS— The ARC/INFO® Method*, Redlands, CA, 1997. Adapted diagrams provided courtesy of ESRI and are used herein with permission. Copyright © 1990–1997 Environmental Systems Research Institute, Inc. All rights reserved.

and (2) all data pertaining to each triangle are stored together. Many GIS applications that work with CAD use TIN data structures. See Section 4.12 for more detail.

4.1.5   GIS Analysis

Analysis in a GIS is usually performed by searching the attributes for particular value and searching for projection-dependent and projection-independent spatial relationships. A GIS query usually includes making some sort of database record retrieval either by searching or browsing through stored records until the required information is found. Records can also be sought by restricting selection to attribute values with set limits. If data exist in other database tables it can be added to the current database table being queried by using a join function to preserve relational integrity. Attribute queries are usually built of standardized syntax, called Structured Query Language (SQL) first developed at the IBM® San Jose Research Laboratory in the early 1970s.[*] Databases such as Oracle® or Microsoft® Access, for example, can be queried using SQL procedures and syntax to return specific attribute data.

In addition, queries can include numeric operations performed on attributes such as area greater than 100,000 ft^2 (SQL: Area > 100000). Geospatial searches of the map database records can include extracting coordinate values, arc lengths, and areas of polygons. Most GIS software packages offer to the user the ability to conduct queries using wizardlike interfaces. To make GIS more user friendly, many vendors have provided a means for the user to write code that allows the software to perform repetitive tasks automatically. ESRI's™ ArcView® Avenue is one such language, as is MapInfo Corporation's MapBasic® language.

One of the benefits offered by GIS through topology is thematic mapping, or the creation of maps that depict relational outcomes resulting from attribute and geographic queries. Thematic maps utilize altered colors, shapes, and/or appearances to show analysis outcomes. To create maps of this nature, database queries must be made where spatial relationships have been defined. Thematic maps can be valuable when planning a land development project. Developers, market analysis, surveyors, planners, and site engineers want to find out as much about a potential site as possible and as soon as possible. For example, GIS can provide these types of engineering-related analyses depending on the database contents:

(a)   transportation-related studies for selecting sites conveniently located where trip lengths are minimized from the site to various trip destination points (e.g., schools), as gauged by Figure 2.2;

(b)   traffic accident studies occurring at intersections close to alternative development sites;

(c)   noise pollution studies when evaluating a potential site's proximity to a nearby airport using noise contour plots or field-gathered decibel readings;

(d)   water supply studies when evaluating the effects of a hydrostatic drawdown curve associated with a well water supply source on surrounding properties;

(e)   wastewater disposal studies when evaluating the effects of seasonal groundwater levels and soil types when considering on-site disposal methods; and

[*] Printed with permission from International Business Machines Corporation.

(f) storm drainage analysis of a potential site when considering drainage basins, wetlands, floodway limits, and bodies of water.

4.2 DESKTOP STUDY AND SITE INSPECTION

Initially the engineer must make every reasonable effort to obtain existing information pertaining to the potential site that might reflect material useful to the evaluation process.

4.2.1 Reference Sources

In addition to GIS-produced information, an excellent source of material is a USGS map on a scale of 1 : 24,000, similar to that shown in Figure 4.5. This map shows general locations, transportation systems, and topography. (These maps are now available in digital format and can be downloaded using the Internet.) Information pertaining to waterways can be obtained from the U.S. Department of Commerce hydrographic charts. Local tax maps show general property boundaries and streets, including serialized identification of individual parcels for specific property information such as current landowner names. Figure 4.6 shows details associated with a typical tax map.

Local authorities have available zoning and/or land use maps, similar to Figure 4.7, which are useful in determining how the proposed site fits within an area's overall land use plan. If the proposed development site would create an incongruent situation, the developer may have to consider petitioning local authorities to rezone the property.

Flood insurance rate maps (FIRMs) (Section 2.4.5) contain drainage basin information that could affect a potential development by delineating areas of potential inundation. Figure 4.8 shows a portion of a typical flood boundary floodway map. Because many of the map panels in inventory are aged and a number of inaccurate map features have been discovered over the years, the Federal Emergency Management Agency (FEMA) now reviews each map on a five-year cycle. New maps are to be made available in digital format with the intent of Internet distribution. The reader is referred to FEMA's (1995) *Guide to Flood Maps—How to Use a Flood Map to Determine Flood Risk for a Property* for proper utilization of FEMA-produced flood maps.

Another important reference is a U.S. Department of the Interior national wetlands inventory (NWI) map, which indicates possible jurisdictional wetlands being present. Figure 4.9 shows one of these maps.

Other sources for potential information include documents contained within local public record offices such as recorded deeds, plats, easements, and other legal instruments. Local inquiries can also include interviews with residents who are familiar with the area. Transportation maps, water resources management maps, soils inventory maps, stream gage data, historical and archaeological maps, aerial photographs, satellite imagery, and utility system maps similar to Figure 4.10 can also provide information that documents prevailing conditions.

4.2.2 Base Map Construction

After compiling and assembling existing site boundary data, the design professional can either sketch a draft copy of a base map by hand or create it as a CAD product, similar

FIG. 4.5 Example of a USGS 1 : 24,000 quad map—Summerville, SC. Source: U.S. Geological Survey, 1990.

to the one illustrated in Figure 4.11. The map would be referenced to a chosen datum if registered data are used. This base map would include preliminary information that pertains to on-site and significant off-site areas, including transportation networks, adjoining properties and owners, existing utilities and storm drainage facilities, easements, rights-of-way, and critical geologic and topographic features such as wetlands.

FIG. 4.6   Sample section of a tax map—town of Summerville, Dorchester County, SC. Source: Dorchester County Tax Assessor's Office. 1980. Reprinted with permission.

4.2.3   Site Inspection

The base map is useful when conducting an on-site inspection since it provides a background where observed field conditions can be recorded for later office use. Photographs, sketches, and significant findings should be included in the field investigation notations. Digital images from still cameras or camcorder clips are especially valuable because they can be stored in a computer and used in later reports. In addition,

FIG. 4.7   Portion of the Official Zoning Map for the Town of Summerville, SC, showing the following zones: R-1 = single-family residential, R-2 = single-family residential, R-3 = single-family residential (attached buildings), R-4 = planned residential development, R-5 = mixed residential, R-6 = multifamily residential, B-1 = residential business, B-2 = neighborhood business centers, B-3 = general business, I-1 = limited industrial, I-2 = industrial, AC = agricultural preservation, and PL = public lands. Not shown on this figure is R-7—mobile home and trailer parks. Source: Town of Summerville, SC, *Official Zoning Map,* 1999. Reprinted with permission.

hand-held GPS receivers such as Sunnyvale, CA-based Trimble Navigation, Ltd's *GeoExplorer® 3** six-channel receiver, can be used to log point, line, and area features during this inspection for later incorporation into the registered base map CAD file.

Apparent property corner monuments found during this inspection should be noted and flagged using plastic surveying ribbon; these corners would subsequently be verified during the boundary survey. Obvious encumbrances, such as contaminated waste materials or property line encroachments, should also be noted for further investigation, although appropriate auxiliary professionals who conduct subsequent evaluative work should investigate them. The resulting site sketch, similar to the one illustrated in Figure 4.12, would include base map information and field notations. On completion of the site inspection, the design team may begin planning conceptual alternatives from which the developer could later choose a preference. These alternatives are subject to revision since a number of formal studies must be undertaken prior to forming final concepts. Some of these studies follow.

4.3   FLOODPLAIN ANALYSIS

Established floodplain boundaries delineated on FEMA community FIRM panels must be recognized by communities participating in the National Flood Insurance Program

---

*A registered trademark of Trimble Navigation Limited.

FIG. 4.8 Example of a national flood insurance program floodway and flood boundary map, Town of Summerville, SC. Source: Federal Emergency Management Agency. *Flood Boundary Floodway Map 450073—Town of Summerville, South Carolina*, 1985.

(NFIP; see Section 2.4.5). These boundaries are determined based on selected return period storms (see Chapter 7) and their affect on a given drainage basin. For those developments that include land situated in the floodplain, additional floodplain studies may be required as part of the state and local regulatory review process to determine how the proposed development might impact the floodway area. If a study is required, the best point of beginning is to obtain any existing information used for previous studies, if it exists. If other developments are being planned for the basin, their impact should also be considered. Additional sources of background information necessary for the study include the drainage basin topography, anticipated rates of flow determined from hydrology, stream cross sections and roughness coefficients, and the limits of the area being evaluated. From these data the stream flow characteristics can be estimated using the resulting water surface profile in conjunction with the various cross sections to establish floodwater levels. The final report would include a plan and profile of the drainage features studied (including representative cross sections) and a summary of

FIG. 4.9 Portion of a national wetlands inventory (NWI) map, Raccoon Island, Charleston, County, SC. Source: U.S. Department of Interior. National Wetlands Inventory Map (NWI), Bennetts Point, SC, 1989.

all procedures, assumptions, calculations, results, and interpretations used in arriving at the final conclusions.

Any new residential structures built in the floodplain must be positioned above the base flood elevation (BFE), while nonresidential structures must be elevated to the BFE, or flood proofed to that level, unless local regulations require additional height. New structures and site improvements placed within the floodplain could constrict flow where the predevelopment BFE is increased. New structures proposed for beyond the floodway location are often regulated to ensure that the postdevelopment water

FIG. 4.10 Inventory of a portion of the existing water and sewer mains in the Town of Summerville, SC, using the following nomenclature: V = water main valve and box, H = fire hydrant, and M = sanitary sewer manhole. Source: Commissioners of Public Works (CPW), Town of Summerville, SC, Water and Sewer Utility Distribution System, Sheet 29, 1999. Reprinted by permission of the Commissioners of Public Works for Town of Summerville.

surface profile is not elevated above the allowable surcharge limits. Figure 4.13 illustrates how floodway fringe encroachments can affect the water surface.

According to 44 CFR 60.3(b)(3), participating communities must require that all new subdivision proposals and other proposed development (including proposals for manufactured home parks and subdivisions) greater than 50 lots or 5 acres, whichever is the lesser, include such BFE data. If BFE data are required, a detailed study might be required if an approximate A zone exists on-site. This study is to be submitted to FEMA within six months as a letter of map revision (LOMR) request. A LOMR-F change is one based on new fill in the floodplain, whereas other LOMRs can be based on bridges, culverts, and relocation of a channel. Another way maps can be changed is through a letter of map amendment (LOMA). The purpose of a LOMA is to remove property from a FIRM flood zone when it in fact has always been above the 100-year flood level. Details for submitting these requests and other forms/certificates can be obtained from FEMA over the Internet at www.fema.gov. Normally local permits for construction will not be issued by NIFP participating communities until all other federal and state permits have been obtained.

4.4   WETLANDS DETERMINATION STUDY

Under Section 404 of the Clean Water Act, the U.S. Corps of Engineers (COE), in conjunction with the Environmental Protection Agency (EPA), issues permits for dredge and fill activity that occurs within jurisdictional wetlands. In addition, the COE has, under the Rivers and Harbors Act, similar authority, while state and local authorities may have additional control over wetlands, especially in coastal zone areas. The

FIG. 4.11 Example of a base map for a land development.

130

FIG. 4.12   Example of a site sketch showing existing features.

Line A-B is the flood elevation before encroachment.
Line C-D is the flood elevation after encroachment.

*Surcharge not to exceed 1.0 foot (FIA requrement) or lesser amount if specified by state.

FIG. 4.13   Floodway schematic. Adapted from FEMA, 1985.

presence of hydrophytic vegetation, hydric soils, and wetland hydrology constitutes a jurisdictional wetland. Private consultants are usually employed to perform extensive on-site studies to identify and delineate jurisdictional saltwater and freshwater wetlands (Dunkle et al., 1989; U.S. Army COE and South Carolina Coastal Council, 1991).

4.4.1 Defining Wetlands

> Hydrophytic vegetation is defined as macrophytic plant life growing in water, soil, or on a substrate that is at least periodically deficient in oxygen, as a result of excessive water content. (Dunkle et al., 1989, p. 5)

As part of an on-site inventory, dominant plant genus and species must be inventoried to determine the existence of significant occurrences in obligate, or various facultative plants, which would establish one criterion for jurisdictional wetland declaration. Significant occurrences could include having obligate and/or facultative species comprising more than 50% of the dominant types or a prevalence index of less than 3 based on specific criteria.

Hydric soils have been defined for jurisdictional wetland identification purposes as "soils that are saturated, flooded, or ponded long enough during the growing season to develop anaerobic conditions in the upper part" (Dunkle et al., 1989). This would include soils that are poorly drained, inundated for long durations, or frequently flooded during the growing season. They are normally categorized into organic muck, organic peat, or a combination thereof. The Natural Resources Conservation Service (NRCS) and the National Technical Committee for Hydric Soils (NTCHS) have compiled a listing of hydric soils. Local soil surveys published by the NRCS can be used to determine initially whether prevalent soil types in the vicinity of the site fall within hydric soil classifications. If so, soil samples will be required for further investigation.

Wetland hydrology has been defined for jurisdictional wetland determination purposes as the "permanent or periodic inundation, or soil saturation to the surface, at least seasonally, [which] are the driving forces behind wetland formation" (Dunkle et al., 1989). Water from precipitation, runoff, groundwater, or tidal activity may be contributory sources. The local offices of the COE, USGS, NRCS, NOAA, along with state and local units of government, can provide information relating to site hydrology. Aerial photographs and field inspections provide amplifying data that can reveal hydrologic activities affecting a site.

The definition of jurisdictional wetlands has undergone an evolutionary change in the past and is still being debated by both governmental and private sectors. As a result of these debates, modifications to current regulations defining wetlands could be forthcoming.

4.4.2 Jurisdictional Wetland Determination

Professional consultants can choose from four recognized methods a means to conduct a wetlands inventory. Figure 4.14 indicates a method of conducting an off-site analysis that might be useful during the initial study of a site; however, a prudent developer should always provide for an on-site investigation prior to completing the preliminary development plan. On-site physical investigations consist of routine, intermediate, or comprehensive assessments. Routine procedures are used primarily on areas of 5 acres

(1) Locate and delineate the development site on a USGS 1:24,000 map and note the topographic features that are encompassed by the site that connote wetlands (e.g., lakes, marshes, and swamps).

(2) Study the appropriate National Wetlands Inventory map to determine the potential wetland areas on site.

(3) Study the soil survey map by SCS to determine if hydric soils exist on site.

(4) Study any available aerial photographs to seek wetland indications.

(5) Review any amplifying information on the area to include other wetland studies.

(6) Make a determination if wetlands exist based on the reviewed information. If an indication of wetlands is present, an on-site inspection would establish areas directly on the ground that could be surveyed and platted.

FIG. 4.14   Off-site wetlands inventory procedure.

or less, while intermediate and comprehensive methods can apply to any site size, including small ones, with intensive wetland activities.

Figure 4.15 illustrates a process for undertaking an on-site jurisdictional wetland inventory that a professional consultant would follow, with routine results being recorded in a style similar to Figure 4.16.

Although in the past authorization for dredging or filling of nontidal or isolated jurisdictional wetlands comprising 10 acres or less was authorized under a nationwide permit, No. 26, the evolving policy indicates that in the near future some type of review of all wetlands, regardless of location and size, will be required by the COE. Other regulations included in most coastal zone management programs may require strict permitting provisions for any size jurisdictional wetland. Jurisdictional authorities may allow mitigation, or the creation of new wetlands, in exchange for approving wetland alterations within critical project areas. Without mitigation an otherwise satisfactory and beneficial project might be prohibited.

4.5   ENDANGERED SPECIES EVALUATION

The developer should include the services of a qualified biologist either as part of the design team or as an independent consultant to investigate at the earliest possible date any site being considered for development for the presence of endangered or protected species (see Section 2.4.1). As part of these services, the biologist should seek a list of presence of federally listed, proposed, and candidate species. The presence of federally listed species on-site may require a Fish and Wildlife Service/National Marine Fisheries Service (FWS/NMFS) permit to complete the survey with the stipulation that the results are forwarded to the FWS/NMFS. If the survey produces evidence that federally protected species are present, the developer is in a position to abandon the site or explore mitigation and conservation alternatives up front as part of the design process. Should the presence of candidate species be confirmed on or near the site, the developer can help create an administrative record by information pertaining to the identified species to expand the body of knowledge associated with that species. This additional information might influence the FWS in its final determination in listing the species and, if listed, the extent of protection offered. Foreknowledge of the presence of candidate species would also allow a developer to seek alternative conservation options through the FWS/NMFS, known as a candidate conservation agreement, that could

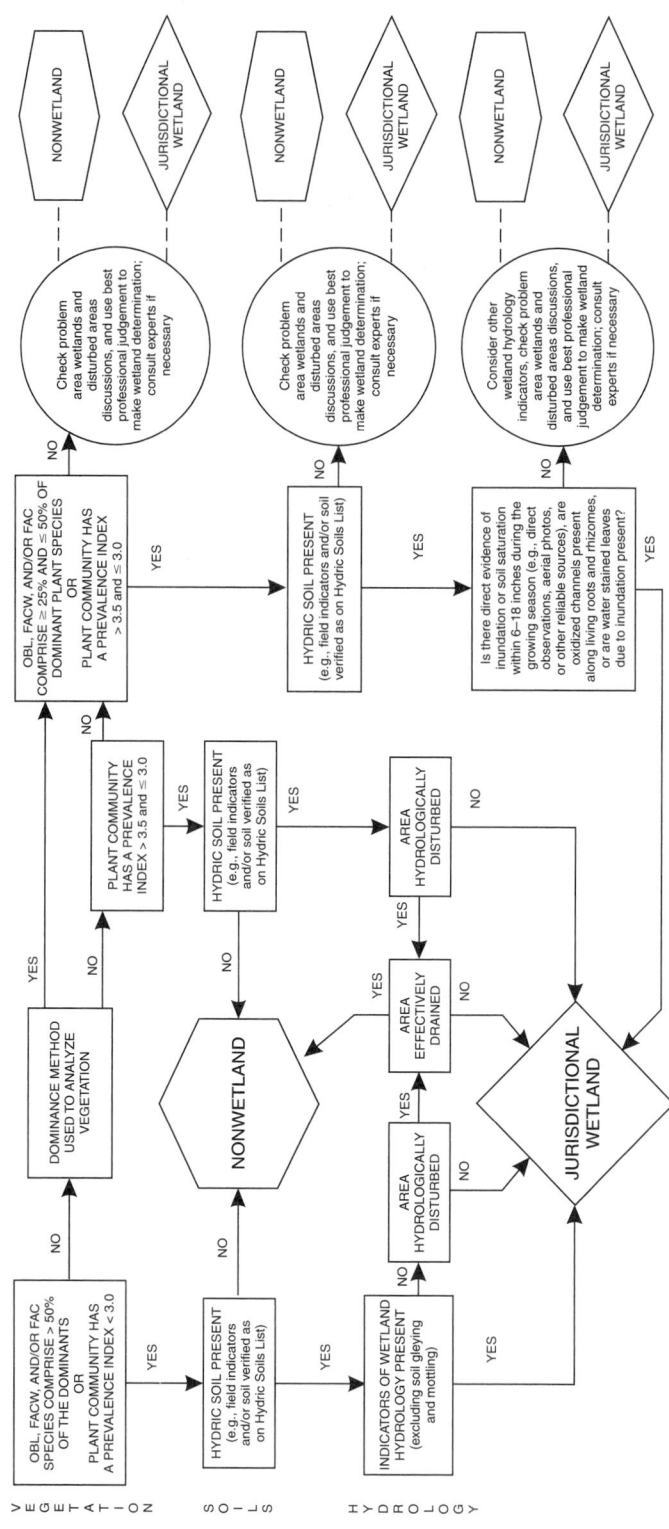

FIG. 4.15 On-site jurisdictional wetland determination showing conceptual (not decision making) approaches: OBL = obligate wetland plants, or plants that almost always occur in wetlands; FACW = faculative wetland plants, or plants that usually occur in wetlands; FAC = faculative plants, or plants that are equally likely to occur in wetlands or nonwetlands. Source: Dunkle et al., 1989.

134

DATA FORM
ROUTINE ONSITE DETERMINATION METHOD

Field Investigator(s): _____   Date: _____

Project/Site:_____ State: _____ County: _____

Applicant/Owner: _____ Plant Community #/Name: _____

Note: If a more detailed site description is necessary, use the back of data form or a field notebook.
- -

Do normal environmental conditions exist at the plant community?

Yes _____ No _____ (If no, explain on back)

Has the vegetation, soils, and/or hydrology been significantly disturbed?

Yes _____ No _____ (If yes, explain on back)
- -

VEGETATION

Dominant Plant Species	Indicator Status	Stratum	Dominant Plant Species	Indicator Status	Stratum
1. _____	____	____	11. _____	____	____
2. _____	____	____	12. _____	____	____
3. _____	____	____	13. _____	____	____
4. _____	____	____	14. _____	____	____
5. _____	____	____	15. _____	____	____
6. _____	____	____	16. _____	____	____
7. _____	____	____	17. _____	____	____
8. _____	____	____	18. _____	____	____
9. _____	____	____	19. _____	____	____
10. _____	____	____	20. _____	____	____

Percent of dominant species that are OBL, FACW, and/or FAC _____

Is the hydrophytic vegetation criterion met?   Yes _____ No _____

Rationale: _____

SOILS

Series/phase: _____ Subgroup: _____

Is the soil on the hydric soils list?   Yes _____ No _____ Undetermined _____

Is the soil a Histosol? Yes _____ No _____ Histic epipedon present? Yes _____ No _____

Is the soil: Mottled? Yes _____ No _____ Gleyed? Yes _____ No _____

Matrix Color: _____ Mottle Colors: _____

Other hydric soil indicators: _____

Is the hydric soil criterion met?   Yes _____ No _____

Rationale: _____

HYDROLOGY

Is the ground surface inundated?   Yes _____ No _____ Surface water depth: _____

Is the soil saturated?   Yes _____ No _____

Depth to free-standing water in pit/soil probe hole: _____

List other field evidence of surface inundation or soil saturation.
_____

Is the wetland hydrology criterion met?   Yes _____ No _____

Rationale: _____

JURISDICTIONAL DETERMINATION AND RATIONALE

Is the plant community a wetland?   Yes _____ No _____

Rationale for jurisdictional decision: _____

FIG. 4.16   Routine wetlands analysis report: OBL = obligate wetland plants, or plants that almost always occur in wetlands; FACW = faculative wetlands plants, or plants that usually occur in wetlands; FAC = faculative plants, or plants that are equally likely to occur in wetlands or nonwetlands. Source: Dunkle et al., 1989.

be included in the project's design so that if the species is later listed, the agreed-to procedures would not halt construction.

Section 7 of the Endangered Species Act (ESA; see Section 2.4.1) applies to private landowners only if there is a federal link to the action. The FWS/NMFS can authorize incidental take through the Section 7 incidental take statement to the applicable federal agency. Reasonable and prudent measures and terms of conditions to minimize the amount of take are required.

Section 10(a)(1) of the ESA (see Section 2.4.1) authorizes FWS/NMFS authority to consider the issuance of a permit for incidental taking if the developer submits a habitat conservation plan (HCP) that describes the impact of the take and what procedures will be followed to minimize and mitigate the effects of the take. In addition, the developer must indicate what alternatives were considered up to the point where the proposed incidental take was chosen, and why the alternatives were discarded. Permit applicants might be asked to evaluate other measures as part of the process.

4.6   ENVIRONMENTAL EVALUATION

Provisions contained within federal regulations such as the Comprehensive Environmental, Compensation, and Liability Act (CERCLA) and the Superfund Amendment and Reauthorization Act (SARA) or other parallel state requirements such as New Jersey's Environmental Cleanup Responsibility Act (ECRA) place a high priority on the investigation for hazardous wastes.

4.6.1   Innocent Landowners

Relying on the provisions for an innocent landowner embodied in SARA, a developer requires professional services of an environmental consultant to undertake and perfect "due diligence" activities. For legal and ethical purposes the intent of this action would be to establish that the developer did not know contamination existed and that there was not reason to know of such occurrence. Considerable burden is placed on the developer to obtain adequate information to ensure the production of safe and usable dwelling units. Environmental consultants, in conjunction with the developer and legal council, may determine the need for investigative activities that will satisfy the developer that due diligence has been met. Professional guidance can be obtained from *The American Society of Civil Engineer's ASCE Manual No. 83—Environmental Site Investigation Guidance Manual* (1996). The American Society for Testing and Materials (ASTM, 1997a) has also developed a standard on "due diligence" performance under the title *E1527-97 Standard Practice for Environmental Site Assessments: Phase 1 Environmental Site Assessment Process.*

4.6.2   Phase I Environmental Audit

Due diligence activities are used to determine whether hazardous substances are present or threaten a site. A Phase I environmental audit includes the following courses of action:

(a)   A check of the ownership title chain to include wills, deeds, easements, leases, restrictions, and covenants for at least 50 years to determine whether any previous owners appear capable of potentially contaminating the site.

(b)   An inspection of any available aerial photography or topographic maps that might reflect prior property usage.

(c)   An investigation of the existence of any recorded environmental clean-up liens placed against the property as a result of federal, state, or local action within 3 miles of the site.

(d) A check of all federal, state, and local governmental regulatory records that would indicate a release of hazardous substances, including

(1) landfills and other disposal records;

(2) underground storage tank records;

(3) hazardous waste handling and generation records; and

(4) accidents and spill reporting records, including crash and rescue, police, fire fighting records and any local planning board records.

(e) A visual inspection of the property that includes looking for distressed vegetation, above and underground storage tanks, trash, rubbish, fertilizers, old electrical transformers, junk cars, batteries, and tires. In addition, this inspection should include investigating past potential on-site chemical use, storage, or treatment and disposal practices. Surface water wells present should be inventoried to determine whether driller's logs are available. Also, information from adjoining landowners about the site's past usage can be helpful, along with information pertaining to the use of their contiguous property (Weldon et al., 1989).

A number of naturally occurring environmental hazards might be present at the site. These could detrimentally affect the completed project's quality and include any release of subsurface methane or radon gas. Special testing requirements to identify the presence of hazards similar to these must be discussed by the consultant and developer.

When information and site reconnaissance results indicate contamination is unlikely, no further activity may be required. Although the environmental consultant has the obligation to investigate contamination potential within the bounds of SARA and the scope of engagement services, the decision to conclude the investigation is solely the developer's. Some may wish to undertake some type of subsurface investigation even if a favorable Phase I environmental audit is received, especially if the site was previously used.

Prior to commencing a Phase I audit, most environmental consultants require the current landowner or developer to complete a questionnaire pertaining to the site. Figure 4.17 shows typical questions that might be included.

4.6.3 Phase II and Phase III Activities

A Phase II activity reflects a discovery stage that includes some type of site exploration and testing. This action would be necessary if the Phase I assessment indicates a possibility of hazardous wastes on site. Should no contamination be found during the Phase II activities, due diligence requirements most likely would have been met. However, if contamination is confirmed during Phase II, then a Phase III site clean-up program must be undertaken by the current landowner, unless the developer elects to pursue the property's purchase. The developer would assume at purchase the obligation for full site remediation.

Installing groundwater monitoring wells and collecting subsurface soil samples for chemical testing can be expensive in terms of time and money; nevertheless, adequate coverage is necessary. When a Phase I audit does not clearly indicate the site is contamination free or when land use activities on adjacent property are questionable, a preliminary exploration program should be undertaken to obtain initial data. The environmental consultant would have to make a determination of methods and proce-

I.   Existing and Previous Site Conditions and Use
   (a)   Primary use including agricultural, commercial, or industrial.
   (b)   Do above- or underground storage tanks exist?
   (c)   Are buildings on-site that contain asbestos?
   (d)   Is this site an EPA identified waste generating site?
   (e)   Has this site been identified as a hazardous waste clean-up site?
   (f)   What permits have been issued pertaining to this site?
   (g)   What zoning classification has been assigned to this parcel in the past?
   (h)   Are there any wells present on site?
   (i)   Has any metal plating, chemical, or petroleum processing been conducted on site?
   (j)   Has electrical equipment been manufactured, repaired, or stored on site?

II.   Future Site Conditions
   (a)   Will this parcel be used as a residential area?
   (b)   Will child care facilities be located on-site?
   (c)   Will commercial facilities be located on-site?

III.   Adjacent Property Use Both Past and Present

IV.   Past and Present Site Owners

V.   Special Sampling Requirements
   (a)   Owner specified for radon, methane, or other hazards.

VI.   Special Requirements
   (a)   Confidentiality of Report
   (b)   Was the site ever a location for an accident?
   (c)   Does the report require special certifications?
   (d)   Do you have access to reference plats, soil borings, or site drawings?

FIG. 4.17   Preaudit Phase I questionnaire for conducting an environmental site assessment.

dures to use, which could range from surface soil sampling to subsurface investigations. Care must be exercised to account for migratory patterns associated with certain types of contaminants, such as polychlorinated biphenyls (PCBs) or metallic compounds. These substances should be sampled close to the suspected source.

When a Phase I assessment indicates the likelihood of site contamination, it is essential to conduct a comprehensive exploratory program that includes groundwater sampling. Test borings, soil gas analysis, groundwater monitoring wells, and test pits are possible ways to investigate a site. Other methods could include ground-penetrating radar or magnetometer, seismic, gravimetric, or electromagnetic conductivity surveys. Sample testing might utilize a variety of means, including mass spectrometry, atomic absorption, or gas chromatography.

Environmental consultants that perform Phase I audits and Phase II site evaluation activities should consider potential conflicts of interest, as results of the investigation could affect many personal contacts. These consultants also must maintain knowledge of rapidly changing events that occur in this rapidly evolving field (Moskowitz, 1989).

4.6.4   Water and Wastewater Service

Other environmental issues that the design professional must consider include the investigation of a potable water supply source and an assessment of wastewater disposal availabilities. This would necessitate checking with local public service utilities to determine whether service is available, as these organizations will be required to approve and accept the proposed system once installed and placed into service for operation, maintenance, control, and ownership. If potable water service is available, the engineer will have to gather existing flow data to determine whether excess system capacity is sufficient to accommodate the new demands. Should capacity not be available, the developer may be required to provide various capital improvements in the project's development, such as the inclusion of a booster pumping station to maintain adequate operating pressure. Similarly, the availability of wastewater treatment and disposal must be investigated to determine if the capacity on hand is sufficient to accomodate new development loads. The system's gravity and pressure mains and the wastewater treatment plant must be capable of accommodating all loads. Various improvements might be required to accommodate the new project, such as the installation of an improved sanitary sewer pumping station for increased wastewater flows. The construction costs associated with these types of improvements are usually shared on a *pro rata* basis with the developer's financial obligation being linked to the amount of utilization. Remaining costs may be borne by the public service utility until other users request service.

If utility service is not available, an alternate source such as a surface water impoundment or a subsurface water well for water supply must then be investigated. Alternate wastewater disposal methods include on-site subsurface disposal facilities or a wastewater treatment plant.

4.6.5 Other Environmental Concerns

Noise is comprised of unwanted sounds resulting from rapid changes in air pressure below and above prevailing atmospheric pressure. These sounds travel as pressure waves outward from the vibrating source; therefore, they will not travel through a vacuum. When sounds travel through various media, they lose energy as a result of geometric spreading and heat energy loss. Noise is measured using a sound-level meter which measures sound pressure levels. When noise affects the general health, welfare, and well-being of the public, it becomes noise pollution. The physical intensity or pressure levels exerted by the sound is quantified in a dimensionless unit referred to as the decibel (dB). The term *noise level* refers to the weighted sound pressure level expressed as dBA, where the A-weighted scale approximates human frequency responses. Zero dBA is barely audible to young people with excellent hearing, while 120 dBA represents a pain threshold. Figure 4.18 illustrates typical ranges of common sounds.

Control of noise affecting residential property can be accomplished in many ways. One is for local governments to situate high-noise-producing land use zones (e.g., major roadways, air operations, industrial activities) in areas set apart from residences. Other measures can include the use of highway green-belt areas, improving the quality of riding surfaces, and vehicle noise suppression enforcement (mufflers). Also highway-related noise reduction procedures can include enclosing (or elevating/depressing) the highway, the use of wider rights-of-way that incorporate buffer areas, the use of noise-absorbing or noise-deflecting walls/barriers (earth berms), lowering speed limits/minimizing stop-and-go traffic, and rerouting truck traffic. Figure 4.19 illustrates the application of some of these highway mitigating alternatives and their effects on adjoining property. Although Figure 4.19 includes the use of vegetation as a noise control technique, some states do not place high value in the ability of trees and shrubs to control noise, especially from aircraft sources or when protecting high-rise structures.

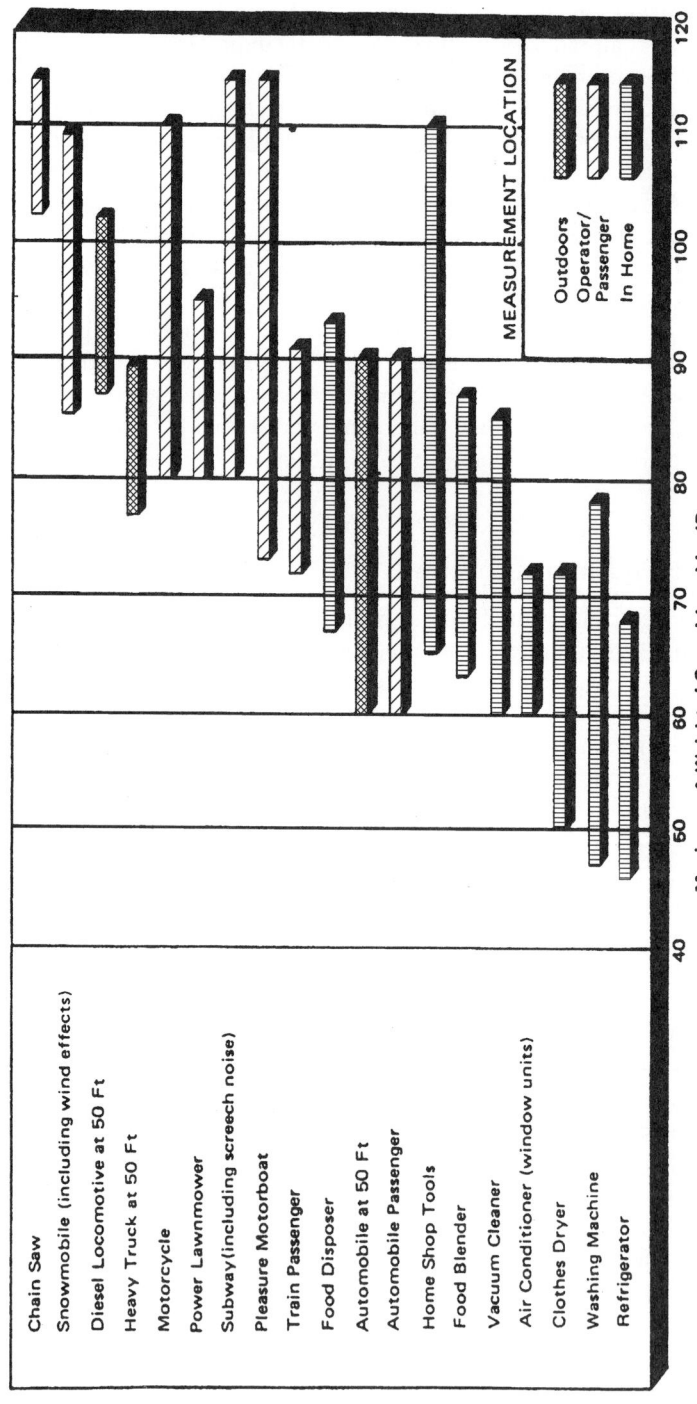

FIG. 4.18 Typical range of common sounds. Source: U.S. EPA, 1978.

Technique	Description	Approximate Reduction in Noise
Noise Barriers	Solid obstruction built between highway and dwelling, high enough to block road view. Ends extend four times as far in each direction as the distance from receiver to barrier.	10-15 decibels reduction which reduces the loudness of traffic by half.
Vegetation	Dense vegetation extending at least 61 meters wide between source and receiver that is high enough and cannot be seen through. (Note: No credit given for sparsely planted vegetation)	10 decibels reduction which reduces the loudness of traffic by half.
Traffic Management	Prohibation of trucks; traffic signals synchronozed to smooth the flow of traffic; and reducing the speed of traffic.	A 33 Km/h reduction in speed is required to achieve a noticeable reduction in noise.
Building Insulation	Windows sealed; cracks and other openings filled; and noise absorbing materials placed on walls.	All buildings-windows opened- 10 dB reduction; light frame, ordinary sash closed--20 dB reduction; light frame, storm windows closed--25 dB reduction; masonry, single glazed closed--25 dB reduction; and masonry, double glazed closed--35 dB reduction.
Buffer Zones	Undeveloped open spaces which border a highway. This alternative is often not possible due to the tremendous amount of land required.	---

FIG. 4.19 Highway traffic noise abatement techniques. Adapted from U.S. Department of Transportation, 1995.

Figure 4.20 shows Federal Highway Administration (FHWA) noise abatement criteria for various land uses. Should traffic noise approach or exceed these criteria or substantially exceed existing noise levels, noise abatement measures can be included with other highway project improvements. The impact of in-fill or redevelopment projects undertaken by a developer can affect the outcome of a noise abatement analysis made for future highway improvements.

Noise generated by aircraft operations will also have a negative impact on the surrounding area. The Federal Aviation Administration's (FAA's) Federal Aviation Regulation (FAR Part 150) includes recommended standards for the creation of airport noise exposure (contours or lines showing various noise levels) maps. Where L_{dn} exposure values are less than 65 dB, no site or structural noise protections are usually required (where L_{dn} represents average day-night sound levels). When L_{dn} values range between 65 and 75 dB, the affected area is where restrictions are placed on educational and residential faculties. Over 75 dB, all residential use (even with structural abatement) is deemed unacceptable according to the U.S. Department of Housing and Urban Development (HUD), where these areas are often reserved for open space or industrial land use. Land development sites slated for residential use that are located near an air facility should be checked with the facility manager or the FAA to determine future facility plans and traffic projection in order to consider future potential adverse impacts when planning the development. Other noise-producing sources include railroads, industry, and noise generated from numerous background sources (e.g., race tracks, quarries, 24-hour commercial operations). HUD assesses noise impacts relative to new residential construction according to the standards in Figure 4.21.

One other environmental concern is associated with quality of air surrounding a potential development site. Air pollution limits are controlled through the Federal Air Quality Act (PL 90–148), as amended, using the National Ambient Air Quality Standards. The amount of air pollution affecting a proposed development site is a function of seasonal pollutant loads and prevailing atmospheric conditions. A site air quality study should include not only the long-term pollution effects but also any potential "quick-release" effects attributed to emergencies or accidents. Other factors that could be included in this report is the handling of any on-site asbestos during construction, potential emissions during the construction phase, containment of lead-paint particulate matter if demolition is included as part of the project, and considerations for air-handling equipment (to promote clean indoor air) installed in on-site structures. This study is part of the overall environmental assessment of the site. Consideration should be given by the development team to provide as much green areas as possible to help counter the effects of air pollution.

4.7   HISTORICAL AND NATURAL RESOURCE SURVEY

Some sites proposed for land developments might be subject to federally mandated site surveys (Section 2.4.6), while other sites might be subject to state or local survey requirements, especially if marked or unmarked graves are present. If so, a suitable inventory must be made. This cultural resource survey should be conducted prior to undertaking any physical land survey so that potential "finding" locations can accurately be tied to the property boundary.

A natural resource survey is another survey where findings must be related to the property boundary. This survey is focused on trees and vegetation. Because trees help

Activity Category	Description of Activity Category	Noise Level L_{eq}(h)	Noise Level L_{10}(h)
A	Lands on which serenity and quiet are of extraordinary significance and serve an important public need and where the preservation of those qualities is essential if the area is to continue to serve its intended purpose.	57-Exterior	60-Exterior
B	Picnic areas, recreation areas, playgrounds, active sports areas, parks, residences, motels, hotels, schools, churches, libraries, and hospitals.	67-Exterior	70-Exterior
C	Developed lands, properties, or activities not included in Categories A or B above.	72-Exterior	75-Exterior
D	Undeveloped lands.	*	*
E	Residences, motels, hotels, public meeting rooms, schools, churches, libraries, hospitals, and auditoriums.	52-Interior	55-Interior

Note: Either L_{eq}(h) or L_{10}(h) (but not both) may be used on a project. These sound levels are only to be used to determine impact. These are not the absolute levels where abatement must be considered. Noise abatement should be designed to achieve a substantial noise reduction—not the noise abatement criteria.

FIG. 4.20 FHWA noise abatement criteria for various land uses where the following abbreviations are used: $L_{eq}(h)$ = equivalent sound level (or A-weighted average measure) over a given period of time; L_{10} = probability where the sound levels are exceeded 10 percent of the time; * = noise abatement criteria not established. They may be treated as developed lands if the probability for development is high. Source: U.S. Department of Transportation, 1995.

Category	External Exposures*	Special Approvals and Requirements
Acceptable	Not exceeding 65 dB(1)	None
Normally Unacceptable	Above 65 dB but not exceeding 75 dB	Special approvals (2); environmental review(3); attenuation(4).
Unacceptable	Above 75 dB	Special approvals(2); environmental review(3); attenuation(5).

Notes:

(1) Acceptable threshold may be shifted to 70 dB in special circumstances.

(2) Special environmental clearance is required for a proposed porject located in the normally unacceptable zone.

(3) Environmental clearances shall be conducted pursuant to the requirements of HUD's Departmental Policies, Responsibilites and Procedures for Protection and Enhancement of Environmental Quality (38 FR 19182 as amended) or other environmental regulations which may be issued by the Department.

(4) 5 dB additional attenuation requred for sites above 65 dB but not exceeding 70 dB and 10 dB additional attenuation required for sites above 70 dB but not exceeding 75 dB.

(5) Attenuation measures to be submitted to the Assistant Secretary for approval on a case-by-case basis.

* The magnitude of the external noise environment at a site is determined by the value of the day-night average sound level produced as the result of the accumulation of noise from all sources contributing to the external noise environment at the site. Day-night average sound level, abbreviated as DNL and symbolized as Ldn, is the 24-hour average sound level, in decibels, obtained after addition of 10 decibels to sound levels in the night from 10 p.m. to 7 a.m.

Source: U.S. Department of Housing and Urban Development—Office of Community Planning and Development. *The Noise Guidebook, HUD-953-CPD(1).* Washington, 1991.

FIG. 4.21 HUD site acceptability standards. Source: U.S. Department of Housing and Urban Development, 1991.

us to control stormwater, conserve energy, improve the quality of air, and provide areas for wildlife habitat/recreational settings, the results of the natural resource survey is invaluable when considering conceptual designs for the project.

4.7.1 Historical Resource Survey Procedure

Historical resource surveys are conducted by certified professional archaeologists, who normally use a two-step approach. The first step is to check all pertinent records including historical maps, state and local archives, and the register of potential historical locations to obtain direct and corollary information. Upon completion, a site inspection is conducted to determine visually whether any artifacts or features of cultural importance are present. (These features can include historic structures, landscape features, roads, or a field of artifacts.) This process usually entails a visual inspection of the property in conjunction with spot shovel testing. Should the first step indicate the possibility that noteworthy features exist, additional and more intense field inspection procedures may be required in concentrated locations to ascertain whether the findings are significant. Investigators conclude the inspection process by preparing a written report that justifies the site's potential eligibility for inclusion in the National Register.

Should significant findings occur on private property, a memorandum of agreement is normally drafted. Either a protective easement is instituted to ensure the area's preservation or the area is set aside to be sold to a preservation-type organization. A last resort would be to approve the site for excavation so that information can be recorded prior to the removal of important features which will result from the development construction.

4.7.2 Natural Resource Survey Procedure

The services of a professional urban forester are required for conducting an inventory of the trees located on a potential development site. Typically the site is cruised and prevalent stands of trees are identified and marked. Grand trees (trees of significant size and age) are located and vital information given identifying the tree type and characteristics (including condition and canopy spread). The identified stands and grand trees are located either by the forester or land surveyor when the property is surveyed. The forester would include in the natural resource survey report general recommendations on preserving, transplanting, and even new plantings that might enhance the site. (The landscape architect, if one is included on the design team, would work closely with the forester to refine these options once the design was underway.) Regulatory requirements triggered by the ESA and local tree or landscape ordinances would also be addressed in the report. For example, some local governments require a tree-cutting permit for any tree over 6 inches in diameter measured 3 feet above the ground. The report should also contain construction specification recommendations for preserving and protecting grand trees during construction where their limbs are not broken, their root systems are not disturbed, and the bark on their trunks is not scarred.

Part of the natural resource survey should address the proximity of other natural systems contiguous to the site under investigation. Connecting on-site areas with off-site areas will promote an integrated and interconnected network of conserved areas where the overall effect is better than a plan that promotes scattered pockets of conserved areas. The report should also address the potential use of protective easements for the conserved areas as well as how the areas might be maintained.

4.8   LAND SURVEY CONTROLS

The development of a piece of property must begin with a knowledge of the metes and bounds (bearings and distances) of its boundary. This information precisely defines the shape, area, and physical geographic orientation necessary for placement of certain recreational facilities and for proper residential unit orientation. In addition, the site's position relative to other geographic features is required for mapping integrity and for coordinating the design of off-site improvements necessary for the development's success. This necessitates the use of reference datums.

4.8.1   Reference Datums

Some regulatory agencies require a project's horizontal and vertical control to be based on adopted systems of reference. Within the United States most past vertical control has been linked to the National Geodetic Vertical Datum of 1929 (NGVD 29), which is the result of a national vertical adjustment performed by constraining the heights of 26 tidal stations in the United States and Canada. One geographic area, relying on geopotential surfaces for hydraulic and engineering projects, encountered problems using NGVD 29. As a result, it developed the International Great Lakes Datum of 1955 (IGLD 55), based on mean sea level at Father's Point, Quebec, Canada. While NGVD 29 has served the engineering, surveying, and mapping community well, it contains distortions not understood when the adjustment was made. This, coupled with the dynamic nature of the earth's crust in certain parts of the country, resulted in problems with the overall integrity of NGVD 29. As a consequence, the National Geodetic Survey (NGS), a component of the Coast and Geodetic Survey, in conjunction with international parties, redefined the vertical reference surface to better service user needs. This resulted in the North American Vertical Datum of 1988 (NAVD 88), publication of which began in 1991. This datum holds the local mean sea-level value (1970–1988 tidal epoch) for the tidal benchmark at Father's Point/Rimouski fixed, thus making IGLD 85 and NAVD 88 "one and the same," according to Zilkoski (1992). Differences in elevation between NGVD 29 and NAVD 88 will vary up to approximately 1.5 meters, as indicated in Figure 4.22 (where the longitude values are shown in degrees east). NAVD 88 is not designed to coincide with local mean sea level at any point in the United States. Local mean sea level and NAVD 88 zero geodetic elevation values will differ by large enough amounts to be of much concern, especially on the west coast.

NAVD 88 has replaced NGVD 29. Appropriate NAVD 88 elevations pertaining to NGVD 29 marks not yet published can be obtained from the NGS or converted for mapping purposes using NGS-prepared computer software under the name of VERTCON.

Likewise, the reference surface for horizontal control used for plane coordinate systems has undergone recent revisions through the redefinition and readjustment of the North American Datum of 1927 (NAD 27). NAD 27 was established using the previous coordinates of horizontal control station Meades Ranch in Kansas as the origin and the Clarke Spheroid of 1866 to define the earth's shape. To remove distortions that developed in the NAD 27 coordinates as new surveys were forced to fit the existing horizontal network and to provide for a system that is compatible with satellite geodesy, the ellipsoid was changed to the Geodetic Reference System of 1980 (GRS 80). [*Note:* A geodetic datum or reference frame such as GRS 80 is defined internationally by seven

FIG. 4.22 Contour map depicting height differences between NAVD 88 and NGVD 29 (units = mm). Source: David B. Zilkoski, 1992. Reprinted with permission.

146

parameters. Three parameters define the origin of a three-dimensional rectangular coordinate system. Three parameters define the orientation of the axes (generally orthometric). The last parameter represents scale (see Section 4.10.1).] The origin was moved to the earth's center, and the entire U.S. horizontal control network of approximately 250,000 control points was simultaneously recomputed or "readjusted," resulting in a more accurate set of coordinates. The result was the North American Datum of 1983 (NAD 83/86).

NAD 83 is used not only for navigation and mapping but also in support of transportation, emergency 911, GIS, and even electric power distribution. Maintaining its accuracy is most important because the earth's coordinate system is dynamic as a result of crustal movement, glacial rebound, and polar motion. Global positioning system technology has helped to promote an increase in its accuracy. Early in the 1990s the NGS began relying on GPS technology for field data acquisition. Although an analysis of NAD 83 had shown that the first-order network was better than expected with a relative accuracy of about 1 : 250,000, it did not match the accuracy of 1 : 1,000,000 often achieved with GPS. This benefit offered by the new technology coupled with the fact that many network triangulation marks were located based on extended line-of-sight parameters and not optimum GPS operations forced the NGS to upgrade NAD 83 using a cooperative state/regional/local approach.

In 1987 the NGS decided to establish a system of national GPS fiducial points to support the High Accuracy Reference Network (HARN) project composed of two classes of stations as defined by the Federal Geodetic Control Subcommittee (see Section 4.8.5). A-order stations, also known as federal base network stations, have a positional accuracy of 1 cm and are spaced at least 100 km apart. B-order stations, also known as cooperative base network stations, also have a positional accuracy of 1 cm but can be spaced as close as 25 km. Each A-order station maintains a proportional accuracy of 1 : 10,000,000 relative to its nearest A-order station. Each B-order station maintains a proportional accuracy of 1 / 1,000,000 relative to its closest B-order or A-order station. Faced with downsizing and financial constraints in 1990, the NGS began in Tennessee the cooperative 7100-point GPS survey, which was later completed in Indiana during 1997 (see Figure 4.23).

The HARN project was only one effort of several being pursued by the NGS during this period of time. In 1994 a strategic plan consisting of 15 goals was adopted including movement towards the National Spatial Reference System (NSRS), network densification, an improved geoid definition, better data access, and continuously operating reference stations (CORSs) as described in Section 4.10.1. Around the year 2002 the NGS will be releasing new HARN coordinates that will update NAD 83/86. These new coordinates will be expressed in latitude, longitude, ellipsoid height, and state plane coordinates (SPCs).

4.8.2   State Plane Coordinate Mapping Systems

Figure 4.24 shows current state plane coordinate system (SPCS) 83 zones for which unique zoned projection constants apply to define either Lambert or transverse Mercator state plane coordinate projection surfaces similar to those shown in Figure 4.25 (NOAA, 1990). One exception, however, is the state of Alaska, which uses an oblique Mercator projection surface. Each SPCS has been redefined by the states (in conjunction with NGS) to conform to NAD 83. Judicial authorities are placing increased em-

FIG. 4.23 Map showing the distribution of HARN stations throughout the United States. Source: National Geodetic Survey, Rockville, MD.

FIG. 4.24   Index of SPC zone codes. Source: NOAA, 1990.

phasis on the use of the SPCS to control land development data uniformity necessary for compatibility between varying GIS-based activities.

When the project's boundary is tied to a SPCS, all measured horizontal distances must be reduced to the ellipsoid to obtain the equivalent geodetic length using an elevation factor which takes into account the geoid (sea level) and ellipsoid separation called the geoid height. This value must further be reduced to its equivalent grid length by using a grid scale factor. Grid scale factors increase in magnitude as the distance of a point increases from the central parallel in a Lambert projection (scale varies with latitude), while in a transverse Mercator projection, scale increases as the distance of a point increases from the central meridian. Grid scale factors can be different at each end of a surveyed line. The greatest difference occurs for a north-south line in a Lambert projection and an east-west line in a transverse Mercator projection. Several

FIG. 4.24   (continued)

methods to obtain a grid scale factor, depending on the accuracy desired according to the *NOAA Manual NOS NGS 5* (1990) include the following:

(a)   For high accuracy the grid scale factor can be obtained for each line in the survey using

$$K_{12} = \tfrac{1}{6}(K_1 + 4K_m + K_2) \tag{4.1}$$

where K_1 and K_2 are the grid scale factors at each end of the line and K_m is the grid factor at the midpoint of the line.

(b)   A less accurate method includes using the grid scale factor of the line's midpoint or the mean of the line's endpoint grid scale factors.

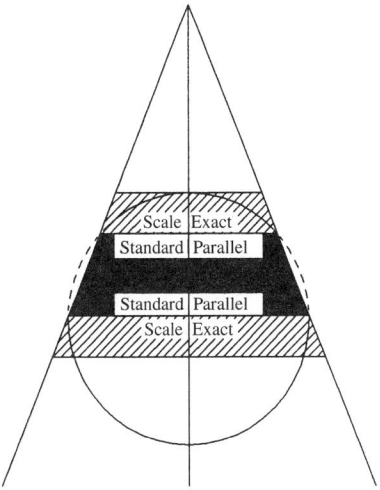

Lambert Projection - Cone Secant to Sphere
Defined by Two Standard Parallels and the Origin

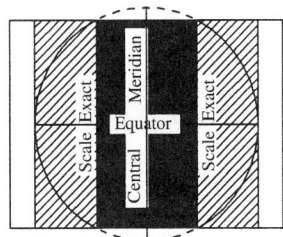

Transverse Mercator Projector -
Cylinder Secant to Sphere
Defined by Central Meridian and Its Scale
Factor, and the Origin

FIG. 4.25 Projections for state plane systems. Source: NOAA, 1990.

(c) For lower accuracy, the grid scale factor for a project can be obtained by using the grid scale factor that corresponds to the project's center point.

Figure 4.26 graphically depicts the relationship of various distances as they apply to SPC calculations. Elevation factor relationships are shown in Figures 4.26a, c, and d, while grid scale factor relationships are depicted in Figure 4.26b. The product of these two factors is known as the combined factor and its application is demonstrated in Section 4.11.4.

FIG. 4.26 State plane system distance relationships: (a) geodetic-ellipsoid surface relationships; (b) geodetic versus grid distance; (c) reduction to the ellipsoid; (d) reduction to the ellipsoid (shown with a negative geoid height). Source: NOAA, 1990.

Some states have legislation that defines one of the two available standards for converting meters to feet, which are:

(a) International foot: 1 inch = 2.54 cm (exactly).
(b) U.S. Survey foot: 1 meter = 39.37 inches (exactly).

It is noted that the difference in these two conversion factors is only 1 part in 500,000. While the difference is insignificant for a survey measurement, it is very important when converting large numbers, such as SPCs, which are usually in the millions. For instance, using the wrong conversion on a SPC of 2,000,000 would result in a 4-foot error. This is especially important because the NGS only publishes SPCs in meters.

Also, observed azimuths require reduction to grid equivalents. This is done by first applying a Laplace correction to the astronomic observations, which yields an equivalent geodetic value. Laplace corrections can be obtained from the NGS or by using NGS-prepared computer software titled DEFLECT 96. (Laplace corrections may be small in most parts of the United States, while the largest corrections can occur in the Rocky Mountains.) Obtaining a reduced grid equivalent for this geodetic value requires corrections for mapping angle convergence and arc-to-chord corrections $(t - T)$ to obtain a reduced grid equivalent. Mapping angle convergence is the relationship between the projected geodetic azimuth and grid azimuth.

The convergence angle (or mapping angle) depends on the position of the azimuth's origin in relation to the projection's central meridian and therefore is a function of longitude. In the Lambert system the convergence angle can be obtained using

$$\text{Convergence angle } = (L_0 - L_p) \sin(B_0) \tag{4.2}$$

where L_0 is the central meridian (longitude of the true and grid origin), L_p the meridian of geodetic longitude (positive west), and B_0 the central parallel (the latitude of the true projection origin).

For example, at a point where the longitude is 82°01'05" W, the convergence angle can be determined using Equation 4.2. If the central meridian for this example is taken as 81° W and B_0 as 33.6693534716, then

$$\begin{aligned} \text{Convergence angle } &= (81°00'00" - 82°01'05")(0.554399350) \\ &= (-3665")(0.554399350) \\ &= -33'52" \end{aligned}$$

NOAA Manual NOS NGS 5 (1990) contains a series of equations for determining convergence angles in transverse Mercator projections, while NGS-prepared computer software titled SPCS 83 can provide both the convergence angle and scale factor at a point in either projection (in addition to converting latitude and longitude to SPC northing and easting values, and vice versa).

The $(t - T)$ correction is usually small and not needed for ordinary surveys associated with land developments except for extreme cases or where high precision is required. These angular relationships are illustrated in Figure 4.27.

Corrections for $(t - T)$ can be determined using the following equation:

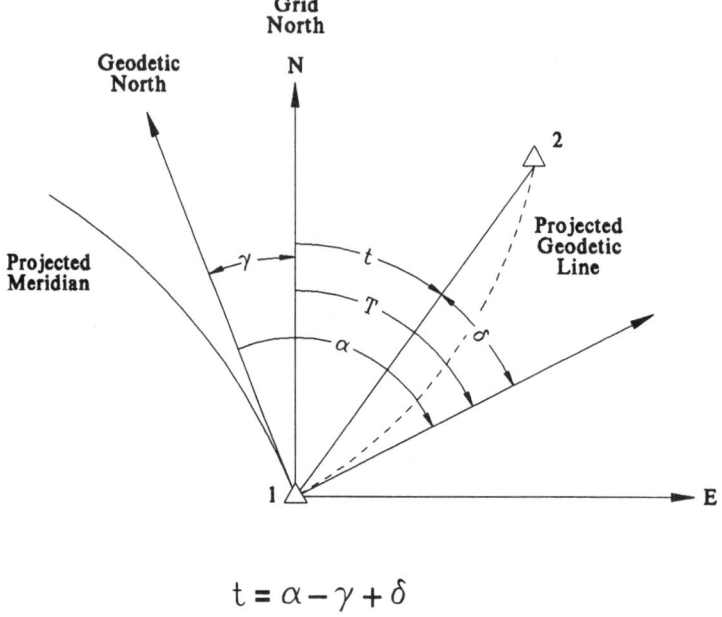

$$t = \alpha - \gamma + \delta$$

α = Geodetic Azimuth Reckoned From North
T = Projected Geodetic Azimuth
t = Grid Azimuth Reckoned From North
γ = Mapping Angle = Convergence Angle
δ = t − T = Second Term Correction = Arc-to-Chord Correction

FIG. 4.27   State plane system azimuth relationships. Source: NOAA, 1990.

$$(t - T) = 2.36(\Delta N)(\Delta E) \times 10^{-10} \tag{4.3}$$

where $(t - T)$ is the correction in seconds of arc and N and E are in feet.

For a Lambert projection, ΔE is the departure of a line and ΔN is the difference between the mean north coordinate value of the line and the projection's central axis. In a Transverse Mercator projection, ΔE is the difference between the east value of the projection's central meridian and the mean east coordinate value of the line. As shown in Figure 4.28, ΔN is the line's latitude. Section 4.11.4 includes a procedure for determining a line's latitude and departure.

Other horizontal control systems that the development might reference are the public land survey system or local urban control networks resulting from city surveys.

4.8.3   Public Land Surveys

According to the U.S. Department of the Interior's Bureau of Land Management (1988) *Manual of Instructions for the Survey of the Public Lands of the United States*, public lands of 29 continental states and Alaska in the United States were surveyed to create as many 1-mile-square sections as possible. Thirty-six of these sections comprised each 6 × 6 mile township—the primary subdivision unit. Sections within each township

FIG. 4.28   State plane system geodetic and grid angle relationships: (*a*) geoid-ellipsoid relationships, deflection of the vertical; (*b*) projected geodetic versus grid angles, arc to chord ($t - T$) relationships; (*c*) Lambert and Mercator convergence angles. Source: NOAA, 1990.

were numbered consecutively from 1 to 36. Section 1 is always the most northeastern section in the township (the one situated north of section 12), while section 6 is the most northwestern. Section 7 is south of 6, and 8 is east of 7. Section 36 is the most southeastern section in the township. This numbering system is shown in Figure 4.29. An exception, however, is in Ohio, the first state that used this system.

Survey controls used for dividing public lands originate at the intersection of a principal meridian (true north-south line) and baseline (true east-west line) called the initial point. Thirty-seven initial points were established in the continental United States. Five more were established in Alaska. All initial points, principal meridians,

FIG. 4.29 System for numbering sections in the United States in which large numbers indicate section designation and small numbers indicate order in which the sections were established on the ground. Adapted from U.S. Department of Interior, Bureau of Land Management, 1988, p. 76.

and/or baselines are referenced in legal descriptions involving aliquot descriptions in the public land survey states. Some initial points are the "starting points" for the public land survey work in many states, and some states have many initial points (Ohio has seven along with some metes-and-bounds survey work). Figure 4.30 shows how various townships are positioned relative to various standard parallels (true east-west lines) and guide meridians (true north-south lines 24 miles apart) used to orient and position included townships within quadrangles. Each quadrangle was slightly less than 24 miles in length along its northern boundary because of meridian convergence. Because the baseline was a true parallel (curved line), special care had to be taken to correctly lay it out on the ground, usually employing the secant, solar, or tangent methods more fully described in most surveying texts and the U.S. Department of Interior's Bureau of Land Management (1988) *Manual of Instructions for the Survey of the Public Lands of the United States*. The standard unit of measurement was the Gunter chain, 66 feet in length. Each chain was divided into 100 links (each 7.92 inches) and 10 square chains equaled 1 acre. Sections were 80 × 80 chains, hence 6400 square chains (640 acres).

Each township is identified by its position north or south of the baseline and east or west of the principal meridian. All townships contiguous to but north of the baseline

FIG. 4.30 Public land surveys nomenclature showing: initial point, baseline, principal meridian, standard parallels, guide meridians, townships (TWP), and ranges (R). Adapted from U.S. Department of Interior, Bureau of Land Management, 1988, p. 62.

occupy the tier (or township) 1 north position (T1N). The townships with northerly boundaries located 18 miles south of the baseline occupy the row known as T4S. The number of columns of townships east or west of the principal meridian determine its range. For example, a township east of (but contiguous to) the principal meridian occupies what is known as the range 1 east (R1E) position. A R4W designation locates townships that are west of the principal meridian and separated from the meridian by three intervening townships.

Figure 4.29 shows in small numbers the order that the 80-chain section lines were laid out after the township boundary lines had been previously established on the ground, beginning with the western boundary of section 36 and followed by its northern boundary. Since the work progressed north and west, discrepancies resulting from curvature and other error sources were pushed to the northern most and westernmost sections in the township. Sections 1–7, 18, 19, 30, and 31 can ordinarily be expected to have irregular measurements usually in the last half mile (40 chains ±). Within each of these sections fractional parts, referred to as "government lots," were created along the northern boundary of sections 1–6 and along the western boundary of sections 6, 7, 18, 19, 30, and 31.

In public land survey states the original boundaries of many land developments follow the original lines established by the aliquot description used when subdividing the section (e.g., section lines, quarter line, quarter-quarter line). For a new land development it is necessary for the surveyor to resurvey the section and subdivide it according to the applicable descriptions. The early public land surveyors set section corners and quarter corners, thus subdividing the section into quarters (160 acres ±). The quarter section was often patented as a homestead in some areas during early settlement of the West. Today's retracement surveyor may have hours and days of research and field work time needed to reestablish these lines and then further subdivide the land into smaller subdivisions for development.

The reader is invited to consult more complete instructions on public land surveys nomenclature and procedures to be used to resurvey and restore lost or obliterated corners. These instructions are available in the U.S. Department of Interior's Bureau of Land Management (1988) *Manual of Instructions for the Survey of the Public Lands of the United States—1973.*

4.8.4   Local Coordinate Systems

Many units of government have established local coordinate systems that are intended to provide horizontal and vertical control for surveys conducted within their jurisdiction. The purpose is to provide a common reference for basing measurements to minimize property/right-of-way conflicts and position uncertainty. Local coordinate systems established years ago often have been superceded or augmented by growing use of the state plane coordinate system. Some municipalities are just recognizing the need to improve local mapping controls.

One example is the Town of Summerville, South Carolina. Summerville was one of the first railroad towns established in the United States in 1832 by the South Carolina Canal and Railroad Company. C. E. Detmold, a civil engineer located in Washington, DC, prepared a gridiron street layout, but there is no record of this plan being monumented on the ground. City blocks and individual parcels were surveyed over the years using "lines of occupation," sometimes called "the island concept." Between 1892 and 1981 there were more than 200 occasions where the town council took corrective action to amend street rights-of-way or alignment. With the advent of improved surveying technology and the desire for the town to create a GIS, the town council appointed a committee to study the problem, make recommendations for implementing improved survey controls, and create a best fit "Official Map," shown in Figure 4.31. After an 18-month study using both conventional and GPS equipment, the committee completed its work, which the town later officially adopted. As part of this effort, the South Carolina Geodetic Survey installed 19 new NGS monuments (in pairs) within the study area. Now all surveys must be tied to one of these monuments.

To reduce confusion, the committee offered in its recommendations the option of surveyors working with "project" or local coordinates by assigning an X value of 200,000.00 feet and a Y value of 100,000.00 feet to the NGS monument in the center of town, point DOTY. Corresponding project (local) coordinate values for the remaining 18 monuments were calculated using a combined factor (elevation and scale) applicable

Detail--N.T.S.

1997 Detmold Baseline
Readjustment Map

Prepared by the Detmold Base Line Committee
for the Town of Summerville, located in Dorchester
and Berkeley Counties, State of South Carolina
30 June 1998

Committee Membership:

T.R. Dion, P.E. & L.S.--Chairman
T.W. Bailey, P.E. & L.S. Town Council Liaison
R.D. Branton, P.E. & L.S.
J.B. Priestner, P.E. & L.S.
J.E. Chellis, J.D.
S.B. Jones III
D. S. Miler, Sesquicentennial Committee Liaison
R. Still, Ex-Officio
S. Seegers, Ex-Officio
B. Dukes, Ex-Officio

FIG. 4.31 Example of a local "Official Map": The 1997 Detmold Baseline Readjustment Map. Source: Town of Summerville, SC, *Official Zoning Map*, 1999. Reprinted with permission.

to DOTY. The benefit to this is that local surveyors can either base their work on the state plane coordinate system or conduct their surveys as usual, relying on horizontal ground distances. Since both approaches are oriented to the state plane grid north, traverse (boundary) rotation is eliminated when going from one system to the other. Graphically, it is a simple matter to obtain the corresponding state plane coordinates if the survey is accurately plotted in CAD using local coordinates. All that needs to be done is for the traverse (or boundary) to be graphically scaled using the combined factor about point DOTY, and then the entire image, including DOTY, must be moved, or translated to DOTY's published state plane coordinate values. All translated image points are now graphically positioned in the state plane system where grid distances can be determined by measuring the distance between desired points depicted on the computer monitor or the state plane coordinate of a point can be found by performing a position inquiry instead. This system allows user flexibility while preserving mapping integrity necessary for GIS base map applications.

4.8.5 Survey Specification Requirements

Surveying field observations are subject to personal, natural, and instrument errors that directly affect the quality of any data gathered. To define the minimum accuracy data must meet for a particular application, surveying standards have been established. Some standards exist at the federal level, while others have evolved at the state and/or local level. For example, at the federal level, horizontal and vertical control standards along with those for relative GPS surveys (Section 4.10) have been promulgated by the Federal Geodetic Control Committee (FGCC) in their publication *Standards and Specifications for Geodetic Control Networks* (1984). (*Note:* In Canada control survey specifications are set by Surveys and Mapping Branch, Department of Energy, Mines, and Resources, Ottawa, Ontario.)

Specifications define the field operations that must be followed to meet a desired survey standard. Specifications can also include the specified precision and tolerance allowances for observed data, geometric form constraints of the surveyed points, monumentation requirements, and point descriptions.

Referring to *Standards and Specifications for Geodetic Control Networks*, (1984) most survey control for the development of land conforms to third-order traverse constraints (second order for vertical control in urban areas), unless state, local, or lender requirements relegate a higher standard. For example, most states have legally adopted standards for the practice of land surveying that must be followed. If a municipality has in place survey requirements, these stipulations are as stringent (or more so) than state regulations and are usually enacted to promote improved mapping and recordkeeping which are prerequisites for reliable GIS applications (see Section 4.8.4).

If part of the intended use of the survey is to provide control for photogrammetric topographic mapping (Section 4.14.2), then supplemental control might be considered necessary to provide enough control points to provide distributed coverage on-site. The accuracy associated with this additional control survey is a function of the intended topographic map scale and desired contour interval. *National Map Accuracy Standards* are standards developed to guide preparation of U.S. Geological Survey quadrangle maps and are also often used with photogrammetric mapping. These standards contain horizontal and vertical constraints that govern accuracy of data used to construct maps.

Many financial institutions require, as part of a loan package, a survey conforming to ALTA/ACSM Land Title Survey Standards of the property that is proposed to be mortgaged. The American Land Title Association (ALTA) and the American Congress on Surveying and Mapping (ACSM) jointly developed in 1962 this standard, which was updated in 1979, 1986, 1988, 1992, 1997, and most recently in 1999, where two documents are now involved: (a) *Minimum Standard Detail Requirements for ALTA-ACSM Land Title Surveys* (1999b) and (b) *Accuracy Standards for ALTA/ACSM Land Title Surveys* (1999a). These current standards eliminate the previously used four survey classes defined by land use where all surveys are treated the same. An ALTA/ACSM land title survey can meet accuracy standards by any one of three approaches stipulated. The first approach is by meeting "positional uncertainty" tolerances. An allowable 95%

confidence level threshold for point location uncertainty resulting from random errors is required. The second approach is to meet each and every angle, distance, and closure specification requirement. The third way is for the surveyor to use proper field procedures and instrumentation and adequate survey personnel to achieve results comparable with the first two approaches. Figure 4.32 outlines some of the provisions contained in these accuracy standards. For a more complete discussion on ALTA/ACSM land title survey requirements, the reader is referred elsewhere (Keen, 1999).

4.9 CONVENTIONAL SURVEYING

Conventional surveying refers to surveying techniques associated with the use of equipment and procedures employed for the purpose of making terrestrial measurements without the benefit of satellite technology (see Section 4.10). Conventional surveying instruments can include transits, levels, tapes, and theodolites. More recently electronic distance-measuring (EDM) devices, total stations, levels, and data collectors have been added to the mix.

4.9.1 Levels

Conventional vertical control can be established using levels and procedures similar to those depicted in Figure 4.33. Automatic levels (self-leveling) with internal compensators are often used; however, newer electronic models record the rod reading on a display where the operator does not have to personally view the rod. The instrument interrogates a specially marked rod using a laser device. An example is shown in Figure 4.34.

4.9.2 Theodolites/EDMs/Total Stations

Most often today conventional horizontal control is established in the field using a theodolite (instrument used for observing horizontal and vertical angles) in conjunction with an EDM instrument or a total station (a single instrument that combines the features of the theodolite and EDM unit). Examples of two total stations are shown in Figure 4.35. Total stations and EDM devices that use infrared-emitting sources require positioning a reflector prism (or prisms) at a far point where emitted energy can be reflected back to the emitting source for distance determination. Three-dimensional coordinates (X, Y, and Z) of distant points can be obtained when the instrument is set over a point of known location and referenced to another known point—the backsight point. The height of the instrument above datum must be known, as well as the height of the prism rod. The horizontal angle must be observed relative to the backsight point as well as the zenith angle and slope distance. Once these measurements are corrected for field conditions, reduced to the horizontal, and scaled to the datum used, the distant coordinates can be obtained. This procedure is often used today when running a topographic survey (see Section 4.12) with a total station and data collector in lieu of using a level.

4.9.3 Data Collectors

The collection of field data is commonly achieved today with the use of an electronic data collector. This collector can be in the form of an electronic field book or an

Accuracy Standards for ALTA-ACSM Land Title Surveys

I.   These Accuracy Standards address positional uncertainty and minimum angle, distance, and closure requirements for ALTA-ACSM Land Title Surveys . In order to meet these standards, the surveyor must assure that the positional uncertainties resulting from the survey measurements made on the survey do not exceed the allowable positional tolerance defined by:

$$0.07 \text{ feet (or 20 mm)} + 50 \text{ ppm}$$

II.   If the size or configuration of the property to be surveyed or the relief, vegetation, or improvements on the property will result in survey measurements for which the positional uncertainty will exceed the allowable positional tolerance determine above, the surveyor must alternatively apply the following table of "Minimum angle, Distance, and Closure Requirements for Survey Measurements Which control Land boundaries for ALTA-ACSM Land Title Surveys" to the measurements made on the survey.

Minimum angle, Distance, and Closure Requirements for Survey Measurements Which control Land boundaries for ALTA-ACSM Land Title Surveys (Note 1)

Dir. Reading of Instrument (Note 2)	Instrument Reading Estimate (Note 3)	Number of Observations Per Station (Note 4)	Spread From Mean of D&R Not To Exceed (Note 5)	Angle Closure Where N=No of Stations Not to Exceed	Linear Closure (Note 6)	Distance Measurement (Note 7)	Minimum Length of Measurements (Notes 8,9,10)
20"<1'>[10"]	5"<1'>N.A.	2 D&R	5"<0.1'>[5"]	10"√N	1:15,000	EDM or Doubletape with steel tape	(8) 81m, (9) 153m, (10) 20m.

Notes:

(1)   All requirements of each class must be satisfied in order to qualify for that particular class of survey. The use of a more precise instrument does not change the other requirements, such as number of angles turned, etc.

(2)   Instrument must have a direct reading of at least the amount specified (not an estimated reading), i.e.: 20" = Micrometer reading theodolite, <1'> = Scale reading theodolite [10"] = Electronic reading theodolite.

(3)   Instrument must have the capability of allowing an estimated reading below the direct reading to the specified reading.

(4)   D & R means the Direct and Reverse positions of the instrument telescope. i.e. Urban surveys require that two angles in the direct and two angles in the reverse position be measured and meaned.

(5)   Any angle measured that exceeds the specified amount from the mean must be rejected and the set of angles re-measured.

(6)   Ratio of closure after angles are balanced and closure calculated.

(7)   all distance measurements must be made with a a properly calibrated EDM or steel tape, applying atmospheric, temperature, sag, tension, slope, scale factor and sea level corrections as necessary.

(8)   EDM having an error of 5 mm, independent of distance measured (Manufacturer's specifications).

(9)   EDM having an error of 10 mm, independent of distance measured (Manufacturer's specifications).

(10)   Calibrated steel tape.

III.   If the survey data is still out of conformance, then the surveyor must employ, in his or her judgement, proper field procedures, instrumentation and adequate survey personnel in order to achieve comparable results to sections I or II above.

FIG. 4.32 ALTA/ACSM Standards. Source: American Land Title Association and the American Congress on Surveying and Mapping. *Accuracy Standards for ALTA/ACSM Land Title Surveys,* 1999. Reprinted by permission of ACSM.

automatic recorder. In the case of the electronic field book entries are manually keyed into the collector, whereas with the automatic recorder observational values can be sent from the instrument through a cable directly to the collector for annotation and storage. Some collectors are constructed where all included components are specifically made as an integrated product. Other collectors are designed to work in conjunction with programmable calculators. Surveyors Module International (SMI) offers one such

FIG. 4.33 Examples of leveling procedures: (a) differential leveling; (b) turning point locations; (c) temporary bench mark (TBM) check points; (d) three-wire leveling.

system where a read-only memory (ROM) card is installed in a Hewlett-Packard[*] HP48GX, as shown in Figure 4.36. This system offers the user the option of automatic record keeping when an electronically compatible total station is connected by cable to the collector or the user can at any time use the collector as an electronic notebook.

[*] Hewlett Packard, 3000 Hanover Street, Palo Alto, CA 94304-1185 (www.hp.com).

FIG. 4.34   Example of a modern electronic level. Courtesy of SOKKIA Corporation, Overland Park, KS, 2000.

A prudent surveyor will still carry a conventional surveyor's paper notebook to record various field conditions, including sketches that will show graphically any features that would clarify and record the survey's progress and sequencing of events. In addition, as part of the surveyor's permanent records, if the collector is the intended place of storage for all numeric field measurements, then the collector should be set where a raw data file is created for a permanent record. This file can be downloaded in ASCII file format and printed for the purpose of serving as a permanent hard-copy job record.

Data collectors can also be used for field stake-out during construction by storing known point coordinate values and then in the field using the process of inversing (see Section 4.11.4) to determine the field angles and distance to orient the total station (or EDM instrument and Theodolite) where the reflector prism is either at the desired location or the offset location controlling a feature's ground position.

4.9.4   Data Analysis/Adjustment/Plotting

There are many commercial vendors that offer PC-based software packages capable of downloading field data stored in collectors and then performing integrity (or blunder)

FIG. 4.35 Example of two types of modern electronic total stations: (a) represents a conventional instrument and (b) represents a robotic system capable of one-man operation. Copyright © TOPCON America Corporation, Paramus, NJ, 2001. Reproduced by permission of TOPCON America Corporation.

FIG. 4.36 Example of an electronic data collector using a Hewlett Packard* HP 48 GX Graphic Calculator. Surveyors Module International™, www.smi.com.

*Hewlett Packard 3000 Hanover Street, Palo Alto, CA, 94304-1185 (www.hp.com).

analysis and then adjusting the data using procedures such as the compass rule for traverses or least squares for survey networks. Statistical analysis of the collected data is often provided along with error ellipses showing 95% confidence limits of adjusted values.

For example, one vendor offers a software plotting package called SMI Plot where the operator is able to create "a complete plat in 10 minutes," starting with downloading adjusted data from a collector to plotting the final product on a laser printer or drum plotter. Figure 4.37 shows the main work area window. This package is also capable of creating digital terrain models (DTMs), contour plots (as depicted in the figure), summaries of earthwork volumes, and three-dimensional views and exporting∗.DWG format files for use by other vendor software.

Another vendor, Autodesk, offers a survey package that resides in its LDDT suite that has more comprehensive capabilities than SMI Plot. For example, it includes the use of an extensive point database, support of a wide range of data collectors and survey field equipment, and a choice of data adjustment procedures. Because Autodesk® Survey is integrated into the LDDT package, all job information can be shared by other suite segments for access, design, and analysis.

4.10 GLOBAL POSITIONING SYSTEMS

The first GPS Block I satellite, NAVSTAR 1, was launched on February 22, 1978, for use by the U.S. Department of Defense (DoD). It was not until five years later that the first truck-mounted GPS receiver was introduced at a cost of $150,000. In 1989, using Block II satellites, GPS surveying techniques allowed one to fix geodetic baselines up to lengths of 1000 km. It was also during this year that improved accuracy dual-frequency receivers were first introduced commercially. GPS equipment made available during 1992–1993 allowed surveyors to perform topographic mapping, construction stakeouts, and data acquisition for GISs to centimeter-level accuracy.

The last of the Block I satellites were taken out of service in 1994 when the full 24 NAVSTAR Block II satellite constellation was in place. Previous to the GPS becoming fully operational, the DoD warned users that the system was still in the developmental stages, and they (the users) assumed any resulting risk. However, on December 8, 1993, when the satellite constellation was declared fully operational, the government assumed certain obligations, the violation of which may result in its liability. Full operation also allowed the DoD to intentionally degrade signal accuracy, referred to as Selective Availability (S/A) and Anti-Spoofing (A/S). In 1996, after much national debate, the government set in motion a decade plan to phase out S/A. The phase-out, in actuality, was completed May 1, 2000. It was also during this year that the first dual receiver was introduced capable of utilizing either GPS or GLONASS (former USSR) signals. Currently, the U.S. government is committed to continue providing basic GPS services, at no cost to direct users, for the purposes of peaceful civil, commercial, and scientific worldwide utilization.

4.10.1 GPS System Components

The original GPS constellation consisted of 21 satellites and 3 spares. Each satellite orbited the globe twice a day at an inclination of 55 degrees to the equator in one of

FIG. 4.37 SMI Plot work area as displayed on a computer monitor. Printed by permission of Survey Module International LLC, www.smi.com.

six orbital planes. This configuration allowed at least six satellites to be viewed from the earth at all times while it orbits at an altitude of approximately 10,900 miles. Each satellite can be viewed from the earth in the same part of the sky each day, but 4 minutes earlier for each succession. Figure 4.38 shows the full 24-satellite constellation. As of December 21, 2000, there are 28 satellites in orbit.

Each satellite has the ability to continuously orient its 12 navigational antennae toward the earth and also orient its solar power arrays toward the sun. Within each satellite are contained two cesium and two rubidium atomic clocks, each having an accuracy of one-billionth of a second timing error.

Broadcast signals made by each satellite are transmitted continuously on two L-band frequencies: L1 = 1575.42 MHz using a wavelength of 19 cm; L2 = 1227.60 MHz with a 24-cm wavelength. (*Note:* A third civilian frequency is programmed to be added to the system by the year 2005.) The L1 signal is modulated with a Precise code (P code) and also a Coarse Acquisition (C/A) code. The C/A code is available worldwide to any user, whereas the P code is restricted to military use for high precision and can be encrypted as the Y code. Superimposed on the C/A- and P-code signals is a binary data stream (data message) which contains information on, for example, satellite clock status, ephemeris data, and satellite health.

The DoD places each satellite in an orbit according to a master plan. The orbits are known in advance; however, to ensure system integrity, each satellite is monitored twice each day by ground-tracking stations. Each satellite's position, speed, and altitude are measured. Differences between the projected and observed orbit parameters, often caused by gravitational pull from the sun and moon or solar radiation pressure, are called ephemeris errors, which are then transmitted back to the satellite for further relay to GPS receivers as part of the data message mentioned previously. GPS receiver units use these corrections to update their internal ephemeris/almanacs, giving them up-to-date precise positions of each satellite. Ground control stations also monitor and correct for any timing errors in satellite clocks.

The reference system used by the GPS satellite constellation was first developed in

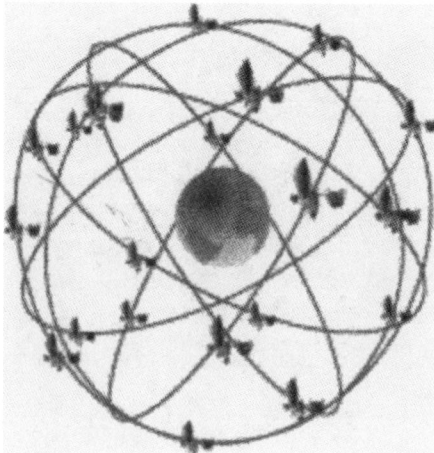

FIG. 4.38 GPS satellite constellation configuration. Source: National Geodetic Survey, Rockville, MD.

1960 and refined over the next two decades by the DoD. Called the World Geodetic System (WGS), the current parameters define what is known as WGS 84. The WGS 84 coordinate system origin and axes are the geometric center of the earth—earth centered, earth fixed (ECEF) and the X, Y, and Z axes of the WGS 84 ellipsoid as defined by the location of the zero meridian (the Greenwich meridian) and the poles at 1984. All coordinates are referenced to WGS 84 when only GPS is used; however, astronomic observations, for example, must be rotated to account for several conditions before an equivalent WGS 84 referenced direction can be determined. Figure 4.39 more clearly illustrates this reference system.

GPS receivers used for surveying and mapping purposes similar to that shown in Figure 4.40 are designed to receive satellite signals and also process the time and position information to obtain position, velocity, and timing measurements. Typically, a receiver consists of an antenna, clock, code-tracking loop, phase-lock loop, navigation processor, control/display components, and a power supply. The antenna gathers L1 or L1/L2 carrier waves from viewable GPS satellites.

Some GPS receivers are capable of observing the following five quantities: (1) C/A code on L1, (2) P code on L1, (3) P code on L2, (4) phase of the L1 carrier, and (5) phase of the L2 carrier wave. Low-cost hand-held GPS receivers not used for surveying and mapping are capable of determining positions in real time (known as autonomous) by observing only the C/A code. With four or more satellites present, this receiver can sequentially determine the range to each satellite and then compute a horizontal position (sometimes called absolute positioning) accurate to ± 10 meters with S/A turned off 95% of the time.

FIG. 4.39 The WGS 84 coordinate system. Adapted from DMA Technical Report 8350.2, Department of Defense "World Geodetic System 1984, Its Definitions and Relationships with Local Geodetic Systems," 30 September 1987. Size of projection surface is defined by the semimajor axis equal to 6,378,165 meters.

FIG. 4.40   Trimble's latest surveying system, the GPS Total Station 5700. Portions © 2001 Trimble Navigation Limited, Sunnyvale, CA.

An improved receiver includes the addition of an on-board data storage memory bank which will allow differential correction of the observed data (DGPS). Concurrent GPS data gathered locally (unknown position) and at a distant point of known location (base station) can be postprocessed to give a ±1- to ±2-meter local position accuracy. Postprocessing data in this context refer to comparing base station positions determined using satellite data with its known (or previously published) position. Any discrepancies are attributed to errors that are assumed to also be present in the roving unit's data. The rover's position is corrected using these discrepancies as a basis for correction.

Receivers giving submeter accuracy are similar to those supporting DGPS described above, except they also observe a number of phase measurements of the L1 carrier in addition to the C/A code. Some receivers are improved a step further to accomplish real-time DGPS with the incorporation of a radio which receives corrections broadcast from the base station.

Another similar GPS receiver that can be used for relative positioning (not DGPS) is likewise capable of observing the L1 C/A codes and carrier phase. Code and phase measurements must be recorded at every epoch using at least two receivers. Postprocessing techniques known as "double differencing" solve for the differences in position between the two receivers at an accuracy of ±1 cm +2 ppm, where baseline lengths are generally restricted to less than 10 km, and these are more fully described in Section 4.10.2.

Yet another type of receiver observes both L1 and L2 and is known as a dual-frequency receiver. These receivers measure the C/A code and the phase of L1 and L2. For work requiring very long lines (>30 km) or short lines using fast static methods, a dual-frequency receiver is needed that is also capable of observing the P codes and the carrier phase of L1 and L2. This type of receiver produces what is often called an ionospheric free solution because errors resulting from GPS signals traveling through the earth's ionosphere (an electrically charged layer 80–120 miles above the earth's surface) can be estimated with this type receiver. Different radio frequencies react differently to the electrically charged layer of the ionosphere. This reaction causes the frequency to change slightly. Each frequency undergoes its own unique change where the difference in the changes is a direct function of ion refraction and used as corrections.

To take advantage of the presence of additional satellite vehicles orbiting the earth for improved real-time positioning, some manufacturers have produced receivers capable of measuring ranges to both GPS and former Soviet Block GLONASS satellites. After full implementation, GLONASS will consist of 24 satellites in a three-orbital-plane constellation with evenly distributed satellites that maintain a height approximately 1060 km lower than that used by GPS. In combination the systems will provide visibility to at least 12 satellites at any time. Both GPS and GLONASS are one-way ranging systems that exhibit similar radio signal structures but utilize different transmission frequencies. GLONASS clock and frequency offsets are related to Coordinated Universal Time (UTC) kept in Russia whereas GPS clocks relate to Coordinated Universal Time as kept by the U.S. Naval Observatory. Ephemeris data are unique for each system: GLONASS broadcasts every half hour each satellite's position, velocity, and acceleration (user interpolates for intervening data to determine ECEF coordinates based on the Soviet Geocentric System 1985) while GPS updates its ephemeris data every hour as related to a linearly varying ellipse with small corrections to account for orbit irregularities where WGS 84 coordinates can be determined.

The CORS network, shown in Figure 4.41, started being assembled in 1994 by NGS as a multiagency cooperative GPS network of base stations supporting carrier and code range data for postprocessing activities. CORS stations (expected to exceed 200 in number) must meet rigorous standards for instrumentation and operation where continuous monitoring has amassed sufficient data for NGS to declare their positions as absolute with centimeter accuracy coast to coast. Each CORS consists of a GPS receiver, a computer, and a telecommunication support system to convey position and elevation corrections to real-time pseudorange observations where submeter accuracy to 200 km away from the station can be achieved 24 hours a day. In addition, carrier-phase observables are also provided by NGS using the Receiver Independent Exchange (RINEX) format via the Internet (http://www.ngs.noaa.gov/CORS/cors-data.html) for cost-free postprocessing data associated with geodetic, land surveying, and the like. CORS reduces the requirements for additional GPS receivers when positioning using differential procedures, and corrections derived from more than one CORS, if available, will ensure the integrity of the results without the addition of extra field expenses for equipment and personnel occupying existing control stations.

Although CORS was originally envisioned as a U.S. navigation system that made use of U.S. Coast Guard facilities, additional efforts made by the U.S. Department of Transportation (USDOT) in creating a nationwide Ground Wave Emergency Network (GWEN)—consisting of 22 sites—resulted in the creation of a combined National Differential GPS Network (NDGPS). Some states are aggressively pursuing the establishment of additional CORS stations. For example, the State of South Carolina currently has two CORS stations—one in Charleston that is operated by the U.S.

FIG. 4.41   CORS Coverage, December 2000: NOS = National Ocean Survey; NGS = National Geodetic Survey; TXDOT = Texas Department of Transportation; FSL = Forecast System Lab (NOAA); NDGPS = National Differential GPS Network; USCG = United States Coast Guard; PANGA = Pacific Northwest Geodetic Array; USACE = United States Army Corps of Engineers; IGS = International GPS Service; COOP = Cooperative CORS. Source: NOAA, NGS, Rockville, MD, 2001.

Coast Guard and another in Columbia that is operated by the South Carolina Geodetic Survey. A third station in Florence is scheduled to be operational in the spring of 2001 by the South Carolina Geodetic Survey and a local technical college.

The goal is to create and operate enough CORS facilities where dual coverage is available anywhere in the United States within the next five years. This will allow precise GPS surveys to be conducted where only a single roving GPS receiver will be required by the surveyor (time and equipment savings) since base station requirements will be met by CORS facilities when conducting relative or differential positioning as described below.

4.10.2   Operation of GPS

Each GPS satellite has a unique pseudorandom noise PRN code consisting of a binary data string of 0's and 1's which it continuously transmits. GPS receivers have in their memory a replica of every satellite PRN code. Receivers detecting satellite signals can identify the satellite and immediately recall the code replica from memory. The code received from the satellite is compared to the replica generated in the receiver. The time differential the receiver must allow in order for the two codes to coincide is the delay caused by the signal travel time, which is Δt, as shown in Figure 4.42a. The transmission time Δt is the only observation made by a code-phased GPS receiver; therefore, the distance between the satellite and receiver (or pseudorange) can be determined as $\Delta t \times$ speed of light, which is 186,000 miles per second.

If four or more satellites are visible at the same time, as shown in Figure 4.42b, the receiver can calculate its three-dimensional position (point positioning) referenced to the WGS 84 datum.

Relative positioning requires two receivers to observe the same satellites simultaneously. With one receiver positioned on a point of known location, the other at an unknown position, it is possible to determine the difference in ΔX, ΔY, and ΔZ between the two receivers within an accuracy of a few millimeters. Using Figure 4.43a as a guide, each receiver observes a particular visible satellite at some moment in time (epoch) where pseudoranges to the satellite from each receiver (S_1–A and S_1–B) can be determined. Distance A–B is unknown but is part of a plane triangle S_1–A–B–S_1. However, as shown in Figure 4.42b, when each receiver simultaneously observes four or more satellites, the position of each receiver can be determined. Using this as a guide where Figure 4.43a now depicts point A at a known position and point B at another known position, S_1–A–B–S_1 forms a vector diagram where the distance and direction of each side is now known. Since the vector diagram must form a closed figure, vector$_{AB}$ = vector$_{AS1}$ − vector$_{BS1}$.

Thus, for each satellite observed by the two receivers, a set of observations to determine the vector from A to B are available. This describes the concept used for relative positioning, which is the only way that a GPS can be utilized for accurate surveying once phase measurements and differencing are employed. The carrier phase is determined by comparing the phase of the received satellite carrier wave with the one generated internally by the GPS receiver. Problems in determining the number of full carrier wavelengths involved in the measurement, called ambiguity, arise when determining the true ranges from carrier-phase measurements. [*Note:* This is similar to an instrumentman's inability to read the whole foot value (usually colored in red) of a rod shot taken during leveling when the sight distance to the rod is short, forcing the instrumentman to say "raise for red." The rodman then raises the rod vertically for the instrumentman to see the number of whole feet involved.] To eliminate this problem as it relates to GPS, pseudoranges are represented by observation equations. A procedure called triple differencing is utilized to obtain the first approximation for the ambiguity. By using differencing techniques, errors can be eliminated or vastly reduced.

First, a single difference must be defined. There are two types of single differences— "between receivers" and "between satellites." Using Figure 4.43b to illustrate single differencing between receivers, side $S_1 A$ represents the range from receiver A to the satellite S_1 and $S_1 B$ represents the range from receiver B to the same satellite. Then $S_1 B - S_1 A$ yields a single difference where any error (naturally occurring or imposed S/A) in the satellite's clock predicted position of the satellite as determined by the DoD is eliminated. In addition, between satellites, single-difference receiver clock errors can be eliminated when two satellites are received simultaneously by the receiver using, again, single differencing.

To remove all of the above-mentioned errors, a double-difference procedure can be employed. Using again Figure 4.43b with $S_1 A$ representing the range from receiver A to satellite S_1, $S_1 B$ representing the range from receiver B to satellite S_1, $S_2 A$ representing the range from receiver A to satellite S_2, and $S_2 B$ representing the range from receiver B to satellite S_2, a double difference between receivers and satellites can be illustrated as $S_1 A - S_2 A - S_1 B + S_2 B$. This same equation can be obtained by subtracting two "between-receiver" or two "between-satellite" single differences. Any

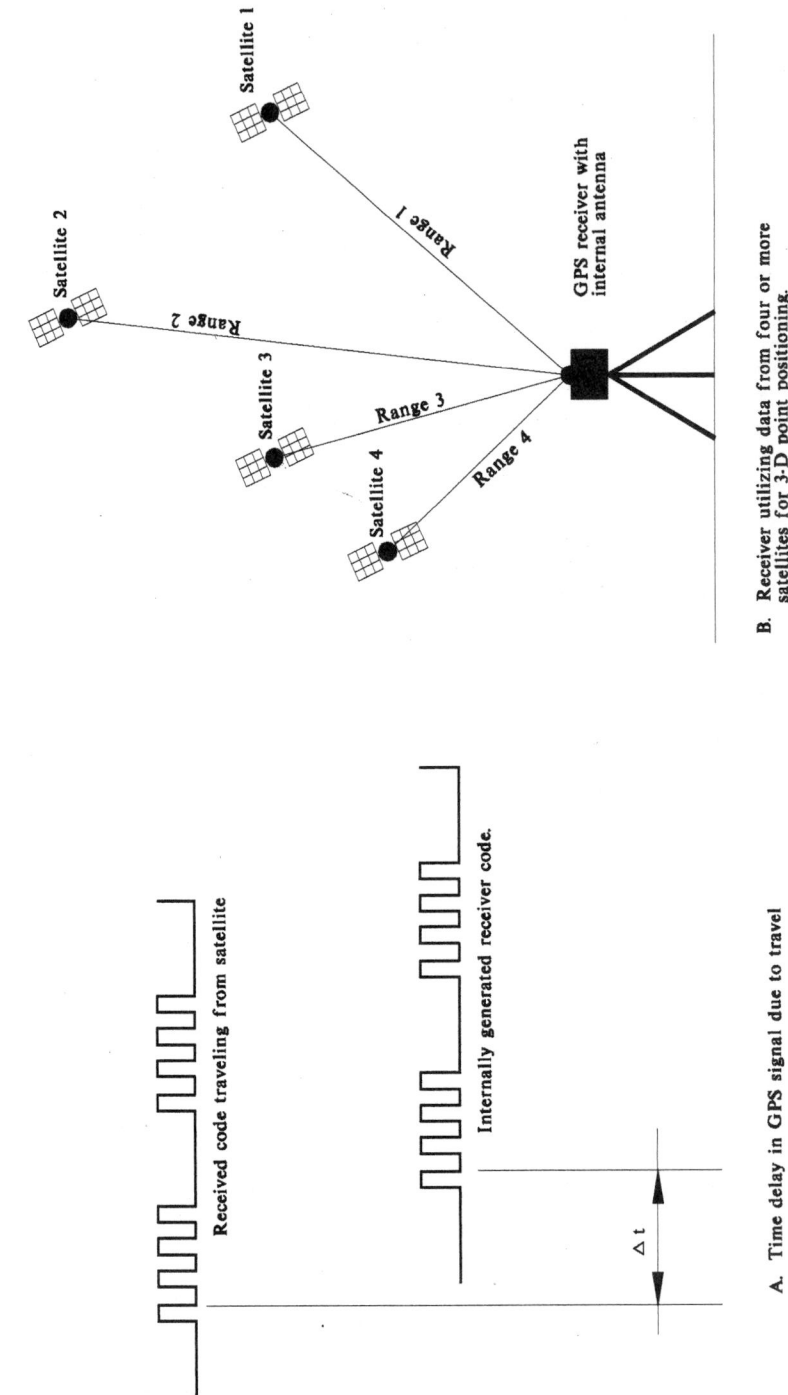

A. Time delay in GPS signal due to travel

B. Receiver utilizing data from four or more satellites for 3-D point positioning.

FIG. 4.42 Examples of (a) pseudocode measuring and (b) three-dimensional point positioning.

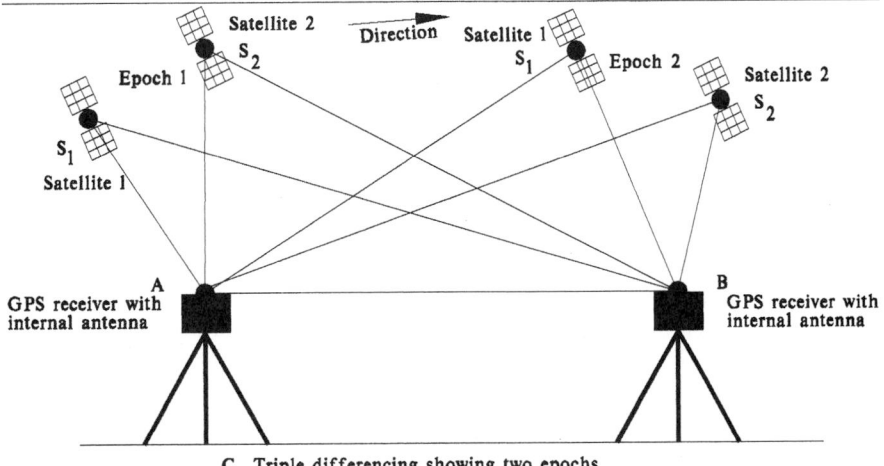

A. Vector determination between two GPS antennae.

B. Single and double differencing techniques derived from a single epoch.

C. Triple differencing showing two epochs.

FIG. 4.43   Vector determination and differencing techniques.

receiver/satellite clock errors and other system errors are now eliminated (Wolf and Ghilani, 1997).[*]

Using a similar approach between various epochs, triple differencing eliminates phase ambiguity, as can be seen in Figure 4.43c. The double differences obtained from two different epochs are subtracted from each other to yield the triple-difference. This solution may not be accurate if the receivers are less than 30 km apart. Typically the triple difference solution is used to determine phase ambiguity; however, the correct phase measurement is obtained using a double-difference solution. An accurate distance from the receiver to the satellite can be obtained by multiplying the phase ambiguity by the carrier wavelength, and the product is added to the carrier-phase measurement.

Code-phase receivers used with differential techniques can determine positions within submeter accuracy, but not to the same accuracy gained from ranges from carrier-phase measurements. Relative positioning receivers using carrier phase can reach $\pm$ 1 cm + 2 ppm. The reader is invited to consult various excellent texts which address these topics in a more complete manner.

One remaining topic important to surveyors involved with land development projects pertains to GPS-derived elevations. The ellipsoid height difference, Δh, between two points is referenced to the ECEF ellipsoid defined by WGS 84 and is defined by a transformation of ΔX, ΔY, ΔZ into $\Delta\varphi$, $\Delta\lambda$, and Δh (where φ is latitude and λ is longitude). Since there are very few monuments described using WGS 84 coordinates in the United States, the preferred reference datum is NAD 83. The difference in WGS 84 and NAD 83 is a matter of where the coordinate origin (center of mass of the earth) is physically located within the earth. Both coordinate systems use the same defining datum parameters, which refer to the center of mass of the earth. In deriving that location from terrestrial observations using GPS and Doppler satellite systems, however, the "noise" in the derivation is about 2 meters. Depending on where on the earth a position is determined, the orientation of the two different determinations of the center of mass of the earth location to that location will range between 0 and 2 meters. To this end, the National Imagery Mapping Agency (NIMA), formerly the Defense Mapping Agency (DMA), has been slowly changing the WGS 84 datum parameters and tracking station coordinates to conform exactly with the International Terrestrial Reference Frame (ITRF 94) where NAD 83 can be transformed to this reference with very little error. In the United States for relative positioning, the known station coordinates are in NAD 83/86 and all new station coordinates will therefore also be in NAD 83/86, despite the fact that the satellite orbits are in WGS 84 coordinates. This is because accurate heights are obtained from double-difference solutions based on the starting coordinates of the base station. Accurate heights can only be obtained once the geoid model is sufficiently accurate to support the translation of ellipsoid height to orthometric height (i.e., better than 2 cm).

The ellipsoid height differences are not the same as elevation (or orthometric height) differences. Orthometric height differences are determined using the following relationship (Hothem et al.,):

$$\Delta H = \Delta h + \Delta N \tag{4.4}$$

where ΔH is the orthometric height difference, Δh the ellipsoid height difference, and ΔN represents the geoid height difference.

GPS-derived elevations rely on using differences in ellipsoid heights to represent Δh by using relative positioning techniques where the known station coordinates are based on NAD 83. The NGS has recently pioneered new standards and specifications for determining ellipsoid heights accurate to 2 cm. This represents an order of magnitude better than ever before. As a result, it is possible to establish accurate benchmark elevations by GPS, which is significantly more cost effective than standard leveling techniques. Since January 2000, the South Carolina Geodetic Survey, for example, has used this technology for establishing all new control in the station, where every station has position and elevation. [See NOAA (1997) for complete details].

The geoid height difference, ΔN, is obtained by determining the geoid height of each terminal point based on its latitude and longitude using a geoid model, such as NGS's GEOID99, and then differencing. The geoid model is based on gravity measurements that defines an equipotential surface best representing the earth's surface and is deliberately biased to allow users to directly convert between NAD 83 GPS and NAVD 88. GEOID99 is a grid model where geoid height values are spaced apart using a 2×2-minute arc increment. Software is available to extract and interpolate values from the grid under the name GEOID99.

To illustrate how GPS-derived elevations are obtained, if point A has an orthometric height of 100.00 feet and the GPS receiver's antenna is 5.00 feet vertically above this mark, the orthometric height (or elevation) of the antenna is 105.00 feet. Assuming that the ellipsoid height difference between the receiver at A and another point B is -2.00 feet, and the geoid difference between the two points is $+1.00$ feet, the elevation of the receiver's antenna at point B is found to have a value of 104.00 feet. The elevation of point B is the difference between this value and the height of the antenna at point B.

4.10.3 GPS Observation Methods

Depending on the receiver's capabilities, GPSs can be used for observations either in a stand-alone absolute (autonomous) positioning, differential positioning, or relative positioning mode. Absolute positioning requires only one receiver while differential and relative positioning utilize more than one. Absolute positioning can be performed where the receiver is positioned over a fixed point for some period of time (static positioning) or is subjected to movement (kinematic). Receivers used for absolute positioning typically rely only on the C/A code as described in Section 4.10.1. The following technique is only used for real-time corrections for navigation and some mapping grade applications and should not be confused with static techniques which many consider as "differential" but use double differencing to remove clock, orbit, and atmospheric errors.

When differential positioning techniques (DGPS) are utilized, at least one receiver called a base station must be positioned over a point of known location (X, Y, Z), while at least one other receiver called a rover receiver is used to occupy a point or points of unknown position. (*Note:* This base/rover set-up is a form of relative positioning because corrections are based on one receiver being located over a known point and corrections are provided to a receiver over an unknown point.) Base station receiver(s) and rover(s) must lock on and track simultaneously at least four common satellites. The base station GPS measures timing errors by determining what the GPS signal travel times should be at the known base location and then compares them to the travel times that are based on actual data received by the rover. These timing errors then can be used by the rover to correct its measurements. Because the base station has no way of determining which satellites the rover is using for position determination, timing errors for all visible satellites are determined and then transmitted to the rover for real-time positioning or used later in post processing. There are several ways that DGPS can be conducted. If the roving unit is stationary during observation, it is considered a static DGPS procedure. If the roving unit is moving during observation, then the procedure is considered kinematic. Other procedures utilized rely on some variance of the static/kinematic modes. DGPS and the previously mentioned autonomous mode are generally used for navigation.

The most common form of relative positioning used to achieve high accuracy described in Section 4.10.2 utilizes the static procedure where both the base station unit and the rover's unit remain stationary during observations. Typically all receivers must record both pseudorange and carrier-phase data for at least 30 minutes (maybe up to 2 hours). All data are downloaded and postprocessed by computer software where corrected X, Y, Z values are determined for the rover's position and the vector relationship between the two receivers. An accuracy along the order of 10 mm ± 1 or 2 ppm can regularly be achieved for baseline measurements between the receivers. Two drawbacks of the static procedure are the inability to perform in the field any staking and data checking.

Rapid static utilizes the previous procedure with slight modification where the rover occupation time is much reduced. This method relies on improved ambiguity resolution by utilizing redundant carrier-phase measurements from multichannel receivers. In addition, improved receiver technology reduces the noise level of the signal, and with the availability of more satellites coupled with the removal of S/A, point data can be observed from 30 seconds to 5 minutes when spaced 1–2 km apart.

Another procedure, "stop-and-go" positioning, requires both the base and rover units to initially lock on at least four common satellites where any ambiguity can be determined. The roving unit then moves to points of unknown position that are fairly close together (not forming very long baselines) where observations for at least $1\frac{1}{2}$ minutes are taken at each location. For redundancy, it is recommended that each point be visited at least two or three times. All receiver data are postprocessed to determine the differential positions of the distant points relative to the base station. If satellite lock is lost, however, this procedure must be restarted.

Kinematic positioning essentially is the same as stop and go except that the roving unit remains in constant motion (e.g., aircraft, boat, vehicle). All measurements are again postprocessed as in stop and go.

An improved variation of stop-and-go positioning that allows temporary loss of satellite lock is called pseudokinematic positioning. The base station remains in continuous operation throughout the procedure. The roving unit occupies each remote position in sequence collecting data for at least 5 minutes. Approximately 1 hour later each point is again occupied for 5 minutes in the same order while receiving data from the same satellites previously utilized. All collected data are postprocessed as before. To successfully utilize this procedure, (1) the remote points must be fairly close together and (2) session planning must be accurate to ensure that the same four (or more) satellites will be visible for both sets of observations.

Finally, real-time kinematic (RTK) positioning techniques offer accuracy similar to that obtained with static methods (±2 cm) but can be used where instantaneous position of the roving unit is determined in the field, unlike the previous methods, where postprocessing give positions in the office after the fact. RTK procedures require the base station be connected to a radio transmitter to relay satellite carrier-phase information to a rover-connected radio receiver. Precise positions (centimeter-level accuracy) are computed using the double-difference procedure explained in Section 4.10.2 by the rover using observed rover-collected satellite data and data provided via radio link from the base station. Data collectors are often included as part of a GPS RTK package where positions and point attribute data are stored.

RTK positioning requires the presence of at least four satellites for continuous tracking. If the roving receiver loses satellite lock, it can be reinitialized by reoccupying the last known point prior to loss of lock. Loss of radio lock with the base station is different, in that carrier-phase differential positioning can be regained when radio contact is reestablished. The time of rover occupation using this procedure is only 5–30 seconds, which is much shorter than requirements for other kinematic procedures. RTK positioning is being improved with the use of a new technology called "on-the-fly" satellite acquisition. This development allows roving units to reacquire satellites during RTK surveys without having to return to a control point to reinitialize the equipment.

RTK systems represent the rover's antenna position as WGS 84 coordinate differences from those of the base station. Some manufacturers (e.g., SOKKIA) offer equipment which allows the user to transform WGS 84 coordinates to other planar reference systems by taking into account curvature and performing both horizontal and vertical calibrations. Thus, rotation, scale, and translations north and east are involved using a process of horizontal calibration with at least two points with know coordinate values in both systems. If more than two points are included, a least-squares best fit can be applied. GPS height differences, however, are based on the ellipsoid, which must be equated to an orthometric height for use in most land surveying elevation applications. Geoid heights are available using a model (e.g., GEOID99); however, most roving units do not have access to this so the differences are referenced to ellipsoid heights. Vertical calibration of the survey must therefore be undertaken to define the relationship of the geoid and the ellipsoid in the area of interest. Since the geoid surface is not planar, at least three points must be occupied so that the slope of the surface can be approximated. Often least-squares techniques are used by the equipment to determine geoid heights from the observation data from which elevations can be obtained. Frequent field checks to known reference points allow the RTK surveyor an opportunity to compare collected data with established control values to ensure precision is being maintained. If problems are evident, then the calibration process must be repeated.

4.10.4 GPS Field Preparation

Planning a GPS survey requires knowledge of where existing survey control points are located relative to within and surrounding the designated survey area. This will ensure that the data will fit not only the local control but also the overall control network. For static differential work, as well as RTK applications, the position of points proposed to be occupied should be where the effects of noise and multipath interference are minimized. Multipath interference can be caused by tree leaves and vegetation where received signals include those reflected from surroundings, not just those directed from the satellite. The proximity of high-tension power poles, power lines, near microwave towers, and radar transmitters can adversely affect data quality and thus should be avoided if possible.

When planning for a typical control survey that will be used for photogrammetric panel ground control, associated with a very large land development, try to set up a 3 to 5-mile baseline using a 10,000 × 10,000-foot grid on the county map. All existing NGS horizontal and vertical control points that exist can be plotted and then reconnoitered. Generally over half of the control points are present, but not all are accessible for GPS observations where a clear line of sight (open canopy) is required. Those points are disregarded, and then new point positions are studied where the monument gaps are filled in on the overall grid, and a good dispersion of existing monuments and new positions are achieved. Use of CORS and HARN stations are recommended to minimize situations where observed GPS field data do not precisely fit published monumentation, as described in Section 4.8.1.

Canopy surveys provide useful information necessary for realistically planning GPS field work sessions because it yields information that can be used to account for blocked-out areas surrounding proposed points of observation. The location and amount of obstructed view is often referred to as curtains created by buildings, trees, and other obstructions. Data obtained in the field during a canopy survey associated with static differential work should be recorded in a field book and include the following:

(a) point name, a description of the mark being occupied;
(b) location of each curtain's beginning and end using true azimuth (magnetic compass plus declination) and altitude from horizon; and
(c) location of intermediate points using true azimuth and altitude from horizon to define its profile.

It must be noted that for realistic session planning the combined canopy effect must be considered where one can determine the combined satellite viewing window that represents the limitations at both the base station and roving station which would control the ability to simultaneously view a particular satellite at a given moment in time.

Computer software available from various equipment vendors (e.g., Trimble) allow the user an opportunity to estimate the best time [lowest position dilution of precision (PDOP)] to occupy various proposed points based on anticipated satellite positions and limitations imposed by receiver sight obstructions. Planning software utilizes projected satellite ephemeris track data downloaded from satellites via GPS receivers, described in Section 4.10.1. These data are used to predict satellite visibility at a particular location and time while considering canopy restrictions. The position of visible satellites minimum of four for three-dimensional positions) to each receiver has an effect on data quality and is measured by the PDOP, which represents a dimensionless relationship between the error associated with the position of the receiver and that of the satellite. The lower the PDOP, the higher the probability of increased positional accuracy. PDOP refers to horizontal and vertical measurement conditions (latitude, longitude, and altitude), where $PDOP^2 = HDOP^2 + VDOP^2$. Dilution of precision can be illustrated by first considering a two-dimensional example where two fictitious satellites simultaneously occupy the same point in space and a receiver determines the range to each satellite to be 10,900 ±0.001 miles. The position of the two satellites creates a situation where a unique receiver location cannot be obtained—from these data the receiver could theoretically be anywhere within the ring formed by range radius of 10,900 +0.001 miles and another of 10,900 − 0.001 miles. Taking this example one step further toward reality, if the two satellites move apart a short distance, the resulting intersection better defines an area (two created, but one is bogus) within which the receiver is probably located, as illustrated by the hatched area in Figure 4.44*a*. This area is relatively large because of the relative satellite positions. If the satellites move further apart as indicated in Figure 4.44*b*, the shaded area is much reduced, thus indicating an improved prediction of the receiver's probable location.

For planning purposes a PDOP of 4 or less yields excellent results (except that the presence of lightning or sunspots during actual observation can offset this), and the time corresponding to these conditions can be obtained from planning software graphs and reports similar to Figure 4.45. For best results, when determining horizontal positions, look for satellites in the north/south/east/west directions at 45 degrees altitudes. For vertical positions, try to have at least one satellite pass the zenith (or as close thereto as possible).

By examining different times of day and periods during the week, the planner can determine the best time for occupying each point. Knowing this and allowing about $1\frac{1}{2}$ hours for each occupation (this will give a least 1 full hour of good simultaneous data collection for static differential work), a coordinated session plan can be developed while taking into account the network's configuration.

To avoid distorting estimates of surveying accuracy when conducting relative positioning using static procedures with carrier-phase GPS units, it is important to consider which lines are included when adjusting the network. Only independent baselines should be included. Referring to Figure 4.43*a*, one can readily see that with two receivers, baseline *A–B* is independent because it is the only line being determined from the observations using relative means. Sometimes additional baselines that are not independent are measured when more than two receivers are simultaneously used, and these are referred to as trivial baselines. Figure 4.46*a* depicts three receivers simultaneously collecting data from four common satellites. Vector *A–B* and vector *A–C* form independent baselines where ΔX, ΔY, and ΔZ between *A–B* and *A–C* are obtained. If the observed data of B and C are processed to determine the vector *B–C*, a trivial baseline results because these data have already been utilized in generating other independent baselines and baseline *B–C* can be readily obtained by inverse calculations.

Figure 4.46*b* shows what happens when four receivers are simultaneously involved. Lines *A–B*, *A–C*, and *A–D* are examples of independent baselines; however, any three lines (one less than the number of receivers) will produce independent baselines because the remaining lines are comprised of points where the data would have been previously used. (*Note:* The total number of baselines per session equals the product of one-half the number of receivers and the number of independent baselines.) If trivial baselines are included when adjusting a network, the resulting accuracy appears to be better than actuality and thus should be avoided. *Geometric Geodetic Accuracy Standards and Specifications for Using GPS Relative Positioning Techniques*, published by the U.S. Federal Geodetic Control Subcommittee, specifically states that only nontrivial lines be included in a geodetic network and conform to the following:

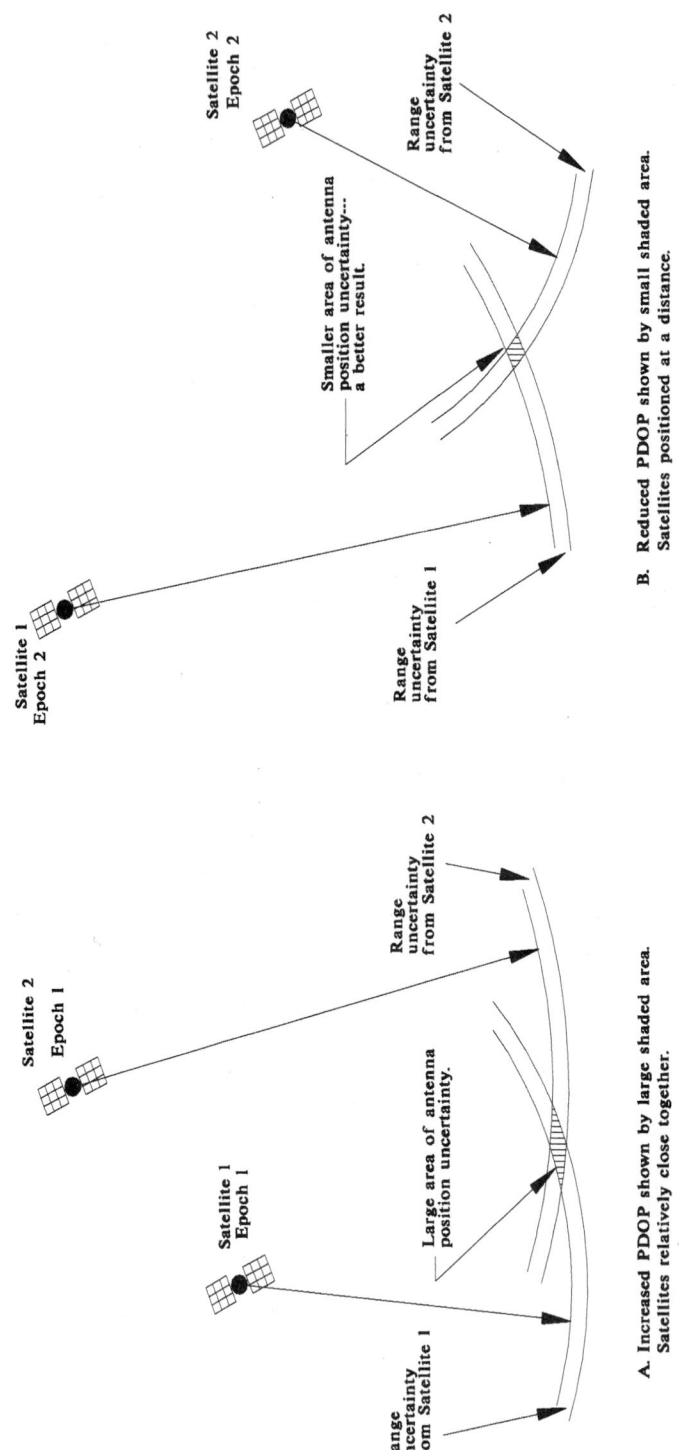

FIG. 4.44 PDOP relationships defined by satellite positions.

A. Increased PDOP shown by large shaded area. Satellites relatively close together.

B. Reduced PDOP shown by small shaded area. Satellites positioned at a distance.

182

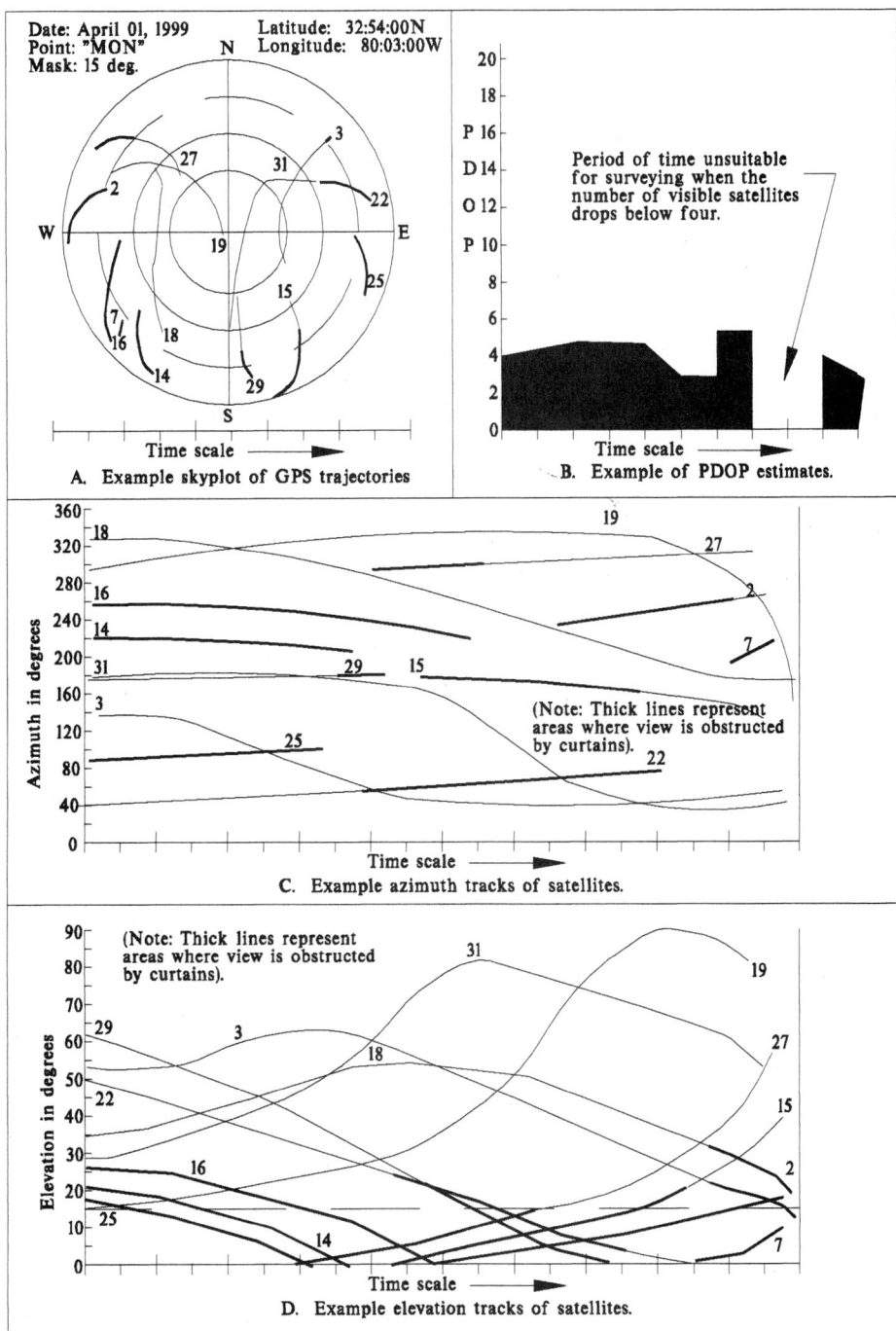

FIG. 4.45   Example GPS planning session results showing expected conditions for a specific time and place at a prescribed location subjected to defined canopy curtain limits. Adapted from *Trimble Geomatics Office™, Trimble Geomatics Office* is a trademark of Trimble Navigation Limited.

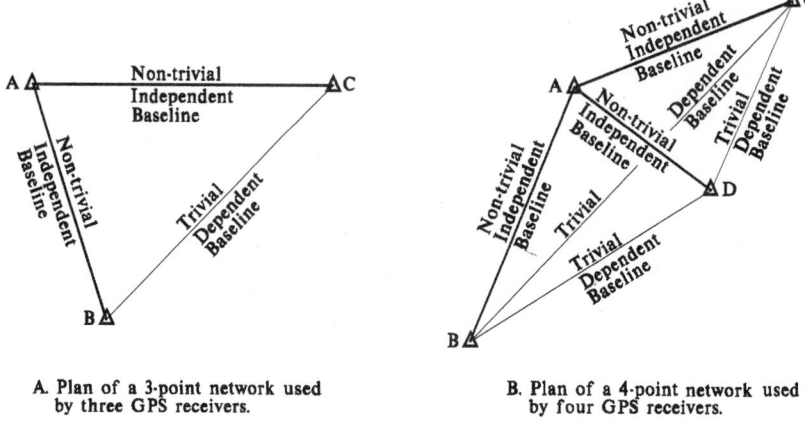

A. Plan of a 3-point network used by three GPS receivers.

B. Plan of a 4-point network used by four GPS receivers.

FIG. 4.46   Trivial and nontrivial baselines.

1. 30 percent of all stations must be occupied twice,
2. 10 percent of all stations must be occupied three times, and
3. there should be no more than 10 lines in a loop.

The network design should reflect a session plan that mobilizes the available GPS equipment in an efficient manner where the least number of sessions are scheduled. This plan should also meet the criteria noted above and at the same time incorporate results of software session planning analysis which considers composite canopy analysis and satellite availability.

4.10.5   GPS Field Procedures

To provide a permanent record of field activities, each static GPS session should be recorded in a survey field book with the following minimum information:

1. project name;
2. point name, a description of the mark being occupied (if possible, a rubbing of the mark's surface is an excellent way to preserve evidence of the actual mark's identification);
3. operator;
4. equipment model/serial number/manufacturer to include GPS receiver, antenna, and tripod;
5. tripod height;
6. antenna type and how measured;
7. Julian date/start time (for each session);
8. Julian date/stop time (for each session), new sessions requiring new start times;
9. weather conditions; and
10. PDOP.

When setting up the instrument, extra care should be taken to ensure the antenna (or instrument if a built-in antenna exists) is correctly plumbed over the point using a calibrated pole leveling device (or tribrach). The antenna height above the occupied point must be measured and recorded accurately for good results. For the most accurate work antennae are mounted on fixed-height range poles (e.g., 2 meters in length) to eliminate any uncertainty in this measurement. Once set up, the GPS receiver operator must be vigilant while the unit is engaged in a recording session to protect the unit from disturbance by outsiders and preserve its security.

Field procedures for conducting RTK GPS-related topographic, construction, and as-built surveys are discussed elsewhere in this text.

4.10.6   GPS Data Analysis and Adjustment

GPS data are obtained in the field and can be downloaded from the receiver or data collector with a cable or infrared interface device to a computer for analysis and adjustment. Once these data are input into processing software, the baselines can be processed and computed. The resulting baseline is said to have a fixed solution if the processor is able to find integer values for the ambiguity terms that are then used to compute final baseline components. If the baseline processor cannot compute an integer solution to resolve the ambiguity (because of atmospheric, orbital, and other errors), then the process must rely on the ambiguities as real values to arrive at a less accurate solution known as a float solution.

Adjustment of each GPS vector included in the surveyed network is usually accomplished using computer software that incorporates least-squares procedures. The initial network adjustment should be performed using only minimal constraints to check the integrity of the field data. To illustrate the process, Figure 4.47 shows a simple six-station network where side A-B is measured twice. In this example there are 18 unknowns (X, Y, Z at each station). The number of lines observed are eight, and the number of observations are 24, based on 3 per line (Δx, Δy, Δz). If point A is initially held as the only station fixed in three dimensions in the network (referred to as a minimally constrained network), there are 5 unknown stations, 15 unknown parameters, and still 24 observations. This configuration having 24 observations and 15 unknowns produces nine degrees of freedom (24 − 15), which is satisfactory to determine all remaining network coordinates. For example, if the reference variance, S_1^2, equals 1.04 for this condition, the data are assumed to be acceptable because the value is close to unity (Wolf and Ghilani, 1997).[*] This indicates that all observed vectors fit well from a geometric point of view. Higher reference variance values indicate the need to make field remeasurements to eliminate poor-quality data, particularly blunders.

Once the minimally constrained data are verified as acceptable where the internal network accuracy looks acceptable, a fully constrained analysis can be undertaken by fixing at least seven of the network's parameters. If an overly determined constrained adjustment (more than seven parameters fixed) indicates a variance of unit weights close to unity, then the network indicates little or no distortion. For this situation, assume that points A, D, and F are fixed using their published values. The resulting adjustment is based now on three unknown stations, where the number of unknown parameters is

---

[*] Taken from Paul R. Wolf and C.D. Ghilani. *Adjustment Computations—Statistics and Least Squares in Surveying and GIS*. © 1997 John Wiley & Sons, Inc. Reprinted by permission of John Wiley & Sons, Inc.

FIG. 4.47 Example six-station GPS network.

nine, the number of observations is still 24, but the reference variance under these constraint conditions, S_2^2, equals 5.7. The ratio of S_1^2/S_2^2 should equal unity if the published data and the observed data are consistent with each other (Wolf and Ghilani, 1997);[*] however, in this case the ratio equals 0.189, indicating that the observed data are better than the published coordinate data for the known points being constrained (see Section 4.8.1 for a related discussion). One way to minimize this situation is to utilize HARN and CORS monuments for network control. Computer software offered by various vendors (e.g., STARPLUS and Trimble Navigation) allows the user the ability to conduct baseline adjustments and statistical analysis. Some offer graphics interfaces where drawings can be generated in formats such as DWG for import into AutoCAD and also allow the user to mix both GPS observations with conventional surveying measurements for simultaneous data adjustment (STAR*NET). The reader is invited to refer to other books that more fully describe adjustment techniques and statistical analysis (Wolf and Ghilani, 1997).

4.11 CADASTRAL SURVEYS

Cadastral surveys are conducted to define a property's legal boundary. This would include identifying corners, bearings, distances, horizontal curve data, if any, and area. In some locales cadastral surveying can be understood to include the elements of topographic surveying contained in Section 4.12.

4.11.1 Property Boundary Determination

A property line survey is required either to establish the true location of the site extremities or to verify current recorded information along with assurance that no encroachments exist. Adjusted results of this survey will yield a mathematically closed figure, called a closed traverse, whereby a person walking the property lines, while

---

[*]Taken from Paul R. Wolf and C.D. Ghilani. *Adjustment Computations—Statistics and Least Squares in Surveying and GIS.* © 1997 John Wiley & Sons, Inc. Reprinted by permission of John Wiley & Sons, Inc.

following the defining boundary data, would travel as far to the north as to the south and as far to the east as to the west such that the ending point is the same as the point of beginning (POB). To obtain a legal boundary survey requires the services of a registered land surveyor who possesses educational and practical experience along with having successfully completed licensing examinations. Arrangements to procure these services would have to be made unless the project's consulting engineer holds dual registration that would allow these activities to be provided as part of the rendered services.

4.11.2   Background Investigation

The land surveyor (or dual registrant) begins a survey by obtaining from public records repositories [possibly the county records office or register mesne conveyances (RMC) office] information pertaining to deeds, mortgages, easements, and rights-of-way of record that are either on or affect the subject property. The tax assessor's office, in conjunction with city or county engineering offices, can provide additional data, while the highway and public transportation departments, along with utility companies, can further augment background information. Once this material has been studied, a base map similar to Figure 4.11 is assembled. This map includes additional details pertaining to property line definitions. These are helpful for measuring property lines on the ground.

If the survey is to be referenced to a state plane system, then the surveyor must locate two separate control marks from which to reference the survey (or a verified mark) where the coordinate values are known. NGS-prepared computer software that can convert NAD 27 to NAD 83 datums, and vice versa, is available under the name CORPSCON.

Analysis of this preliminary information on the base map may indicate potential problems such as discrepancies between measurements recorded for contiguous property lines. This knowledge will alert the field party to anticipate possible problems in questionable areas. The surveyor is cautioned, however, that no judicial authority has been vested through licensing that provides a platform from which to settle boundary disputes. All a surveyor is entitled to do after following such doctrines as senior rights is either to offer suggestions for a solution based on available information or, that failing, to record the conditions that produce the conflict on a plat to show the basis of conflict that can be used for legal proceedings.

Prior to conducting field work, the surveyor must establish the minimum acceptable precision that the work must meet for both angular and linear measurements. Some surveys may be governed by states that have established minimum standards that regulate the conduct of boundary surveys within their jurisdiction. Other surveys may be governed by guidelines created to address particular survey needs, as indicated in Section 4.8.5.

4.11.3   Field Work

Activities undertaken in the field commence by initially walking the site with the current landowner, if possible, to obtain assistance in the initial spotting of property corners. This will reduce the amount of time spent in subsequent recovery efforts required to locate hidden monuments. Additional time spent in interviewing adjacent landowners is helpful, especially in establishing a rapport that could aid in verifying measurements of their intersecting property lines, which, if measured, could help eliminate the possibility of boundary conflicts. Once initial site reconnaissance has been completed, the surveyor should have a fuller understanding of the impending field work, along with the approximate limits of the work area. All information gathered thus far is only preliminary and will be subject to verification with instrument measurements and mathematical calculations. Rules of conduct require the person in charge of the field work to pursue the work in a professional manner. Of primary importance is that a surveyor's services are procured but for one reason: to report through measurements what is seen in the field, be it good or bad. In addition, care must be exerted to avoid damage to landowner's shrubs, fences, and property while at the same time observing applicable laws of trespass or even fence laws that require gates to be secured.

4.11.4   SPC Example Survey

Figure 4.48 illustrates reduced observed field measurements on a potential development's boundary. Local regulatory authorities, for this example, have adopted utilization of the established SPC Lambert projection system as a basis for horizontal control that will provide consistent data necessary for incorporation into the local GIS database. A transverse Mercator projection would follow this same procedure when using published coordinates and convergence angles, except that for transverse Mercator projections, the arc-to-chord relationships would reflect transverse Mercator configurations illustrated in Figure 4.28, and $(t - T)$ values would be determined using Equation 4.3 as applied to transverse Mercator projections. The regulations are also assumed to require that the final plat will show:

(a) a bearing and distance tie between an SPC monument and the property boundary using grid values,

(b) state plane coordinates for the property corner located by the tie obtained above,

(c) grid bearing denoting each property line's direction,

(d) project (or ground) distances for each property line, and

(e) a combined factor for the survey that will allow property line distances (shown as project distances) to be reduced to equivalent grid distances (end user can determine SPC values for each property corner by using recorded grid bearings and reduced grid distances).

It is emphasized that this example is intended only to demonstrate an activity process when conventional surveying techniques are used. GPS is assumed not to be available in this example; however, if it were available, relative positioning procedures described in Section 4.10.2 would be followed after proper analysis of the network configuration (Section 4.10.6) and session planning. Actual procurement of field data would vary to accommodate the presence of several short lines along with the included limitation that the assumed next close SPC monument exists several miles away and the one utilized has been recently verified by another survey. Sequential lettering indicates the order in which points were occupied during traverse operations, beginning with point "A"

FIG. 4.48 Example of field angles and distances for a boundary survey of a land development site.

189

being closest upon site arrival. Precision requirements for conducting this survey are assumed to be those specified for local suburban-type surveys.

Initially, astronomic observations were made by occupying both the existing SPC monument (MONUMENT)—a fictitious monument for illustration only—and point L, which produced correlative data that differed insignificantly after application of La Place corrections. Concurrent with each astronomic observation, a series of horizontal line measurements were made between these points, resulting in the most probable value of the observed horizontal length of line MONUMENT–L, as being 1496.88 feet. Calculations for determining the grid bearing of line MONUMENT–L using "published" data pertaining to MONUMENT are shown in Figure 4.49.

Field angles can now be reduced to grid by first determining the $(t - T)$ corrections. On a traverse the size of this example, all $(t - T)$ corrections are negligible; however, $(t - T)$ corrections should be considered on larger surveys. The dotted lines in Figure 4.48 represent observed geodetic lines, while the straight lines represent equivalent

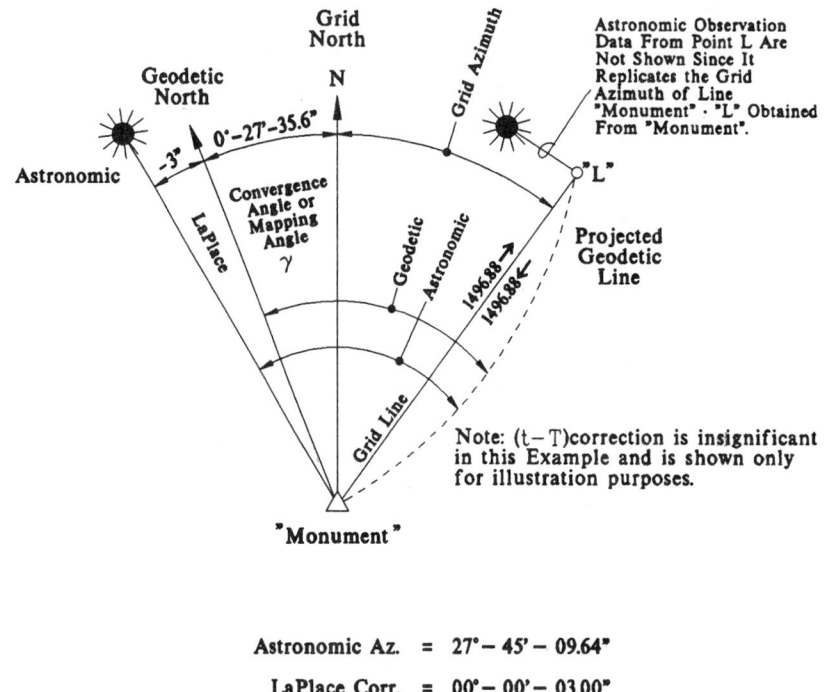

Astronomic Az.	=	27° – 45' – 09.64"
LaPlace Corr.	=	00° – 00' – 03.00"
Geodetic Az.	=	27° – 45' – 06.64"
Convergence Angle	=	00° – 27' – 35.60"
Convergence Adjustment	=	27° – 17' – 31.04"
(t– T) Corr.	=	00° – 00' – 00.04"
Grid Az.	=	27° – 17' – 31.00"

FIG. 4.49 Example of a grid bearing determination for a boundary survey of a land development site.

grid counterparts. In this example the difference between them is less than 0.1 second and therefore is ignored for the rest of this illustration.

The angular error of closure is determined using equivalent grid values and checked against the appropriate summation for either interior angles $(N - 2)180°$, exterior angles $(N + 2)180°$, or deflection angles (summation angles equal $360°$ while accounting for directional signs). In this case, the sum of the exterior angles should be $(12 + 2) \times 180° = 2520°$; therefore, the angular closure is 26", which would be checked with the local survey standards and is assumed to be within the allowable limits. For this example, the angles are adjusted by assuming equal weight and reliability given that field conditions remain fairly constant. Figure 4.50 shows these steps.

The observed horizontal project distances can be reduced to grid by applying both an elevation factor and a grid scale factor. When combined, these form what is known as the combined grid factor. Since the geoid height for the example's location is -90.00 feet, and the average project height is approximately 100 feet (NGVD 29), using information contained in Figure 4.26*d*,

$$\text{Elevation factor} = \frac{20{,}906{,}000}{20{,}906{,}000 + 100 - 90} = 0.9999995$$

The combined factor is then calculated (using an average grid scale factor of 0.9998562 for this example) as

$$(0.9999995)(0.9998562) = 0.9998557$$

This correction will be shown on the final plat and is the amount each measured

Point	Obs. Angle	Angle Corr'n.	Adjusted Grid Angle	Line	Grid Bearing
L	125-21-44	-2"	125-21-42	MON - L	N 27-17-31 E
A	179-58-39	-2"	179-58-37	L - A	N 27-20-47 W
B	180-01-12	-2"	180-01-10	A - B	N 27-22-10 W
C	269-58-22	-2"	269-58-20	B - C	N 27-21-00 W
D	089-15-00	-2"	089-14-58	C - D	N 62-37-20 E
E	271-03-20	-2"	271-03-18	D - E	N 28-07-42 W
F	269-23-10	-2"	269-23-08	E - F	N 62-55-36 E
G	090-02-06	-2"	090-02-04	F - G	S 27-41-16 E
H	270-01-49	-2"	270-01-47	G - H	N 62-20-48 E
I	179-59-33	-2"	179-59-31	H - I	S 27-37-25 E
J	179-58-54	-2"	179-58-52	I - J	S 27-37-54 E
K	270-15-44	-2"	270-15-42	J - K	S 27-39-02 E
L	144-40-53	-2"	144-40-51	K - L	S 62-36-40 W
Sum	2520-00-26	-26"	2520-00-00		

Grid Angle Adjustment Calculations

Note: 1. All angles are shown as degrees, minutes, and seconds unless otherwise noted.

2. The observed angles are taken as grid in this example because (t-T) corrections are less than 0.1 second.

FIG. 4.50 Example of angular adjustment to grid.

horizontal distance must be corrected to obtain equivalent grid distances. Since it is assumed that local regulations require the plat to include the SPC values of only one of the property corners (along with a tie bearing and distance), the equivalent grid distance for line MONUMENT–L must be determined by 1496.88 × 0.9998557, or 1496.66 feet.

Using the verified grid bearing and distance of line MONUMENT–L, the SPCs of point L can be determined by first obtaining the latitude and departure of line MONUMENT–L, as shown in Figure 4.51.

Point L now serves as the beginning for traverse closure calculations. Figure 4.52 contains the closing calculations and demonstrates that the actual precision obtained

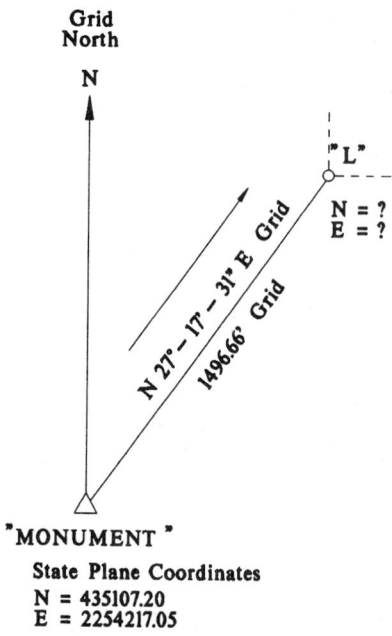

Grid North

N

"L"

N = ?
E = ?

N 27° – 17 – 31" E Grid
1496.66' Grid

"MONUMENT"
State Plane Coordinates
N = 435107.20
E = 2254217.05

The latitude from "MONUMENT" to point "L" calculation :

1496.66 x Cos (27° – 17 – 31") = +1330.05'

The departure from "MONUMENT" to point "L" calculation :

1496.66 x Sin (27° – 17 – 31") = +686.26'

The northing of "L" = 1330.05 + 435107.20 = 436,437.25 feet
The easting of "L" = 686.26 + 2254217.05 = 2,254,903.31 feet

FIG. 4.51 Example of calculations for coordinate determination from latitudes and departures.

meets minimum specifications set forth under local survey standards of 1 : 10,000. If the calculated precision was out of compliance, remedial field work would be required.

Although adjustment of traverse closure errors can be accomplished by several methods, the compass or Bowditch rule is utilized in this example. The compass rule states that the magnitude of correction applied to any traverse leg shall be allocated as part of the total error proportioned at a ratio of each side's length to the sum of all sides; more specifically,

$$\text{Correction in latitude per side} = (\text{sum of all latitudes})\,(L)/(P) \qquad (4.5)$$

$$\text{Correction in departure per side} = (\text{sum of all departures})\,(L)/(P) \qquad (4.6)$$

where L is the length of each traverse side and P is the traverse perimeter.

Figure 4.53 shows the necessary calculations to produce adjusted distances and directions pertaining to each property line that will be included on the final plat.

Final area calculations are made by utilizing either the coordinate or the double-meridian distance (DMD) method, both of which are illustrated in Figure 4.54 using a fictitious 10×10-unit boundary. Figure 4.55 contains a tabulation of the area calculations for this example using the coordinate procedure.

If SPC values are desired for each property corner, then each adjusted boundary distance must first be reduced to grid using the combined factor. The adjusted boundary grid bearing and this adjusted grid distance will provide grid latitudes and grid departures necessary to calculate SPCs, as indicated in Figure 4.56.

This example utilizes a field procedure that consists of occupying each sequential point that in fact is a true property corner; therefore, no ancillary calculations are necessary to solve for unknown sides using a process called inversing. Tie line MONUMENT–L can be the tie reference on the plat; however, to provide a direct

Error of Closure Calculations					
Side	Quadrant	Bearing	Distance	Latitude	Departure
L - A	N W	27.34639	145.04	128.83114183	-66.62686016
A - B	N W	27.36944	50.03	44.42967606	-23.00010403
B - C	N W	27.35000	145.17	128.94241066	-66.69470469
C - D	N E	62.62222	230.26	105.88630753	204.46945365
D - E	N W	28.12833	139.86	123.34167258	-65.93672273
E - F	N E	62.92667	140.19	63.80474884	124.82864305
F - G	S E	27.68778	140.00	-123.96898682	65.05144354
G - H	N E	62.34667	230.16	106.82203064	203.86927029
H - I	S E	27.62361	145.22	-128.66674752	67.33287818
I - J	S E	27.63167	50.16	-44.43912074	23.26349388
J - K	S E	27.65056	144.96	-128.40476150	67.27271976
K - L	S W	62.61111	601.40	-276.66060335	-533.98583366
Summation			2162.45	-0.08223178	-0.15632292
Linear Error of Closure:			0.176632		
Precision:		1 to	12242.67		

Note: 1. Grid bearings are shown in degrees and decimal degrees.
      2. Distances are horizontal project distances in feet.

FIG. 4.52 Example of traverse error of closure calculations.

Traverse Adjustments Using The Compass Rule

Side	Length	Correction for Latitude	Correction for Departure	Adjusted Latitude	Adjusted Departure	Adjusted Distance	Adjusted Direction	Bearing
L - A	145.04	0.00551546	0.01048490	128.83665729	-66.61637525	145.04	-27.341709	N 27-20-30 W
A - B	50.03	0.00190250	0.00361666	44.43157855	-22.99648738	50.03	-27.364764	N 27-21-53 W
B - C	145.17	0.00552040	0.01049430	128.94793106	-66.68421039	145.17	-27.345320	N 27-20-43 W
C - D	230.26	0.00875613	0.01664543	105.89506366	204.48609909	230.28	62.622192	N 62-37-20 E
D - E	139.86	0.00531848	0.01011044	123.34699106	-65.92661229	139.86	-28.123653	N 28-07-25 W
E - F	140.19	0.00533102	0.01013430	63.81007987	124.83877734	140.20	62.926612	N 62-55-36 E
F - G	140.00	0.00532380	0.01012056	-123.96366302	65.06156411	140.00	-27.692458	S 27-41-33 E
G - H	230.16	0.00875233	0.01663820	106.83078297	203.88590850	230.18	62.346659	N 62-20-48 E
H - I	145.22	0.00552230	0.01049791	-128.66122522	67.34337609	145.22	-27.628291	S 27-37-42 E
I - J	50.16	0.00190744	0.00362605	-44.43721330	23.26711994	50.16	-27.636347	S 27-38-11 E
J - K	144.96	0.00551241	0.01047912	-128.39924909	67.28319888	144.96	-27.655236	S 27-39-19 E
K - L	601.40	0.02286952	0.04347504	-276.63773383	-533.94235862	601.35	62.611140	S 62-36-40 W
Sum	2162.45	0.08223178	0.15632292	-0.00000000	0.00000000			

Note: 1. Distances are horizontal project distances in feet.
   2. Adjusted directions are shown in degrees and decimal degrees.
   3. Bearings are shown in degrees, minutes, and seconds.

FIG. 4.53 Example of traverse adjustment using the compass rule.

194

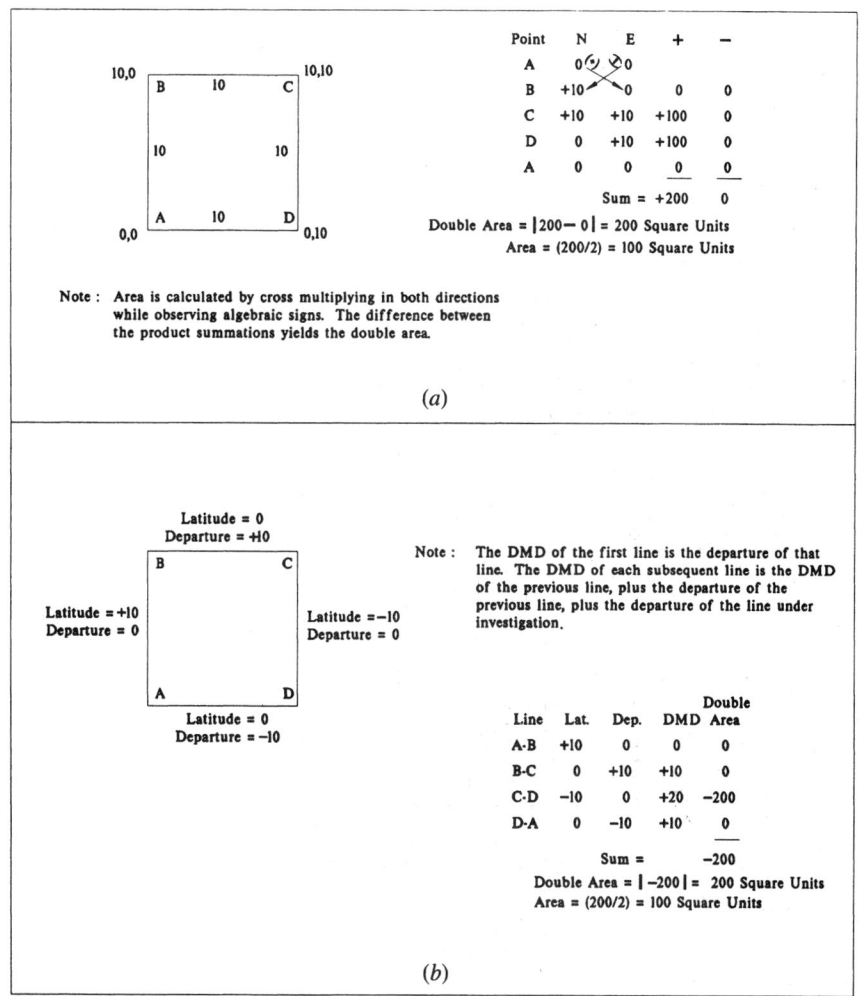

FIG. 4.54 Examples of area calculations: (*a*) by coordinates using a square traverse with sides north, south, east, and west and lengths of ten units; (*b*) by DMDs using a square traverse with sides north, south, east, and west and lengths of ten units.

reference tie to MONUMENT from point A, which is the most obvious property corner spotted upon first arrival, an inverse grid bearing and grid distance must be computed, as shown in Figure 4.57. This is the tie that will be included on the plat.

After the plat has been drafted into its final form, as shown in Figure 4.58, either by manual or CAD methods, and after all remedial field stake-out data have been determined, if resetting obliterated or lost monuments is required, then a return trip to the site is necessary. The plat should be field checked by rewalking the boundaries to ensure all pertinent features, including encroachments, are properly reflected on the plat. In addition, the plat's final boundary information should be verified independently in the office for notation errors. This is done by double checking each line's bearing and distance to ensure that in fact a close traverse is represented by the shown data. This must be done prior to disseminating copies of the plat.

Area Calculations by Double Meridian Distances				
Side	Adjusted Latitude	Adjusted Departure	D M D	Double Area
L - A	128.83665729	-66.61637525	-66.616	-8582.631
A - B	44.43157855	-22.99648738	-156.229	-6941.512
B - C	128.94793106	-66.68421039	-245.910	-31709.577
C - D	105.89506366	204.48609909	-108.108	-11448.109
D - E	123.34699106	-65.92661229	30.451	3756.093
E - F	63.81007987	124.83877734	89.364	5702.299
F - G	-123.96366302	65.06156411	279.264	-34618.582
G - H	106.83078297	203.88590850	548.211	58565.855
H - I	-128.66122522	67.34337609	819.441	-105430.245
I - J	-44.43721330	23.26711994	910.051	-40440.139
J - K	-128.39924909	67.28319888	1000.602	-128476.484
K - L	-276.63773383	-533.94235862	533.942	-147708.604
			Sum	-447331.635
			Area, s.f.:	-223665.817
			Area, Ac.:	-5.135

Note: The absolute value of the area is used.

FIG. 4.55 Example of area determination by DMDs using adjusted latitudes and departures.

State Plane Coordinate Calculations								
Side	Adjusted Bearing	Adjusted Project Distance	Grid Distance	Grid Latitude	Grid Departure	Point	Grid North	Grid East
L - A	-27.34171	145.04008	145.01915	128.81806616	-66.60676251	L	436437.25	2254903.31
A - B	-27.36476	50.03003	50.02281	44.42516708	-22.99316899	A	436566.07	2254836.70
B - C	-27.34532	145.17008	145.14913	128.92932387	-66.67458786	B	436610.49	2254813.71
C - D	62.62219	230.27881	230.24558	105.87978300	204.45659174	C	436739.42	2254747.04
D - E	-28.12365	139.85992	139.83974	123.32919209	-65.91709908	D	436845.30	2254951.49
E - F	62.92661	140.20145	140.18122	63.80087207	124.82076311	E	436968.63	2254885.57
F - G	-27.69246	139.99999	139.97979	-123.94577506	65.05217572	F	437032.43	2255010.40
G - H	62.34666	230.17880	230.14558	106.81536729	203.85648776	G	436908.49	2255075.45
H - I	-27.62829	145.21998	145.19902	-128.64265940	67.33365844	H	437015.30	2255279.30
I - J	-27.63635	50.15999	50.15275	-44.43080101	23.26376249	I	436886.66	2255346.64
J - K	-27.65524	144.95998	144.93906	-128.38072107	67.27348991	J	436842.23	2255369.90
K - L	62.61114	601.35088	601.26410	-276.59781501	-533.86531074	K	436713.85	2255437.18
				-0.00000000	0.00000000	L	436437.25	2254903.31

Note: 1. Adjusted grid bearings are shown in degrees and decimal degrees.
　　　2. Distances are shown in feet.
　　　3. The combined grid factor used is 0.999856 .

FIG. 4.56 Example SPC calculation.

Inverse bearing determination for line "MONUMENT" to "A" = :

$$\text{Tan}^{-1}\left(\frac{E_2-E_1}{N_2-N_1}\right) = \left[\frac{2254836.70 - 2254217.05}{436566.07 - 435107.20}\right]^{\text{Tan}^{-1}} = \left[0.4247466\right]^{\text{Tan}^{-1}}$$

Bearing from "MONUMENT" to "A" = N 23° – 00' – 47" E

Inverse distance determination for line "MONUMENT" to "A" =

$$\sqrt{(E_2-E_1)^2 + (N_2-N_1)^2} = \sqrt{(619.65)^2 + (1458.87)^2} = 1585.01 \text{ feet}$$

FIG. 4.57   Example of inverse calculations or solving for an unknown side.

4.11.5   Property Boundary Description

The surveyor, following requirements set forth in Section 2.7.7, is now able to draft a legal description of the property boundary based on final plat data. Much has been written and said about property descriptions. Nevertheless, the essence of these

FIG. 4.58 Example of a finished land boundary plat.

198

activities centers on one issue: describing property with clarity. Several methods are utilized, including descriptions by lot and block, public land surveys, and metes and bounds. A metes-and-bounds description provides exact boundary information that replicates the plat information using written words. References to terms such as *left* or *right* for direction can be confusing to a nonsurveyor or nonengineer and should be avoided by using substitute compass-related terms like *northeasterly* or *southerly*. The metes-and-bounds method provides a means to describe any shape parcel, including ones bounding on curves, by utilizing phrases such as "Thence, along the arc of a curve, whose segment area, being defined by the arc and chord, is (*part* or *not part*) of the include tract, subtended by a radius of ____ feet, in a southwesterly direction for a distance of ____ feet and having an included chord bearing of ____ and distance of ____, to a ____ monument (*found* or *set*), point ____."

To illustrate the metes-and-bounds process, the 5.135-acre tract depicted in Figure 4.58 will be described. Each description, which includes a formal title for identification, commences by identifying and describing a beginning point, called the point of beginning (POB), from which all references are based, such that upon completion a closed traverse has been described. The POB should be one of the most accessible points that a casual site visitor would most easily locate when orienting the property on the ground. From the purest of interpretations, the POB is exactly that—the point that begins the description of the closed traverse. It is from this point, which is part of the described boundary, that back identification references are made. An easy way to facilitate clarity in writing descriptions includes the use of identification letters being placed at each property corner on the final boundary plat that is cross-referenced in the description, as shown in Figure 4.59.

Copies of the completed plat and description can now be distributed to end users such as the developer, the project attorney, the appraiser, and local jurisdictional authorities that approve plats prior to recording.

4.12 TOPOGRAPHIC SURVEYS

Once the boundary, or cadastral, survey has been completed, a topographic survey may be undertaken. This survey shows the natural features of the property that will influence the project planning activities. Several methods for conducting topographic surveys are used, including the grid, radial, and traverse methods when conventional surveying techniques are employed. RTK GPS and the use of aerial photographs are alternate means to obtain data useful for topographic mapping.

Topographic information provides a viewer in-depth knowledge of the land's surface features. To visualize these features, an understanding of contour lines is required. A contour line is a line that depicts connected points of equal elevation. All contour lines sooner or later close on themselves. Natural contour lines depicting areas of natural curvature are rounded and free flowing. Man-made improvements can result in contour lines that usually appear straight with sharp bends. Contour lines are shown at some predetermined interval and must be positioned in logical sequence. The lines do not touch or cross each other unless contiguous to a wall or part of an overhanging cliff. Figure 4.60 contains various contour line characteristics.

Topographic maps normally show, in addition to contour lines, all site improvements including those above and below ground if definable. Trees and their canopy spreads are usually shown if of a certain type or size.

LEGAL PROPERTY DESCRIPTION OF A 5.135 ACRE TRACT OF LAND
OWNED BY _____, BEING LOCATED AND CONTIGUOUS
TO THE EAST OF SUBDIVISION "A" AND
TO THE WEST OF SUBDIVISION "B'
SITUATED NORTHWEST OF _____TOWN, _____ COUNTY,
STATE OF _____

Beginning at an old iron found in the southern edge of the right of way of Road "A",
Point A, the Point of Beginning (POB), from whence State Plane Monument
"MONUMENT" bears S 23°-00'-47" W, at a distance of 1585.24 feet;

Thence N 27°-21'-53" W crossing the right of way of Road "A" for a distance of 50.03
feet to an old iron found, Point B;

Thence N 27°-20'-43" W along the eastern side lot line of Lot 22, Block C-C of existing
Subdivision "A", for a distance of 145.17 feet, to an old iron found, Point C;

Thence N 62°-37'-20" E along the lands of Smith for a distance of 230.28 feet to an old
iron found, Point D;

Thence N 28°-07'-25" W along the lands of Smith for a distance of 139.86 feet to an old
iron found, Point E;

Thence N 62°-55'-36" E along the lands of Smith for a distance of 140.20 feet to an old
iron found, Point F;

Thence S 27°-41'-33" E along the lands of Smith for a distance of 140.00 feet to an old
iron found, Point G;

Thence N 62°-20'-48" E along the lands of Smith for a distance of 230.18 feet to an old
iron found, Point H;

Thence S 27°-37'-42" E along the western side lot line of Lot 15, Block "B" in Subdivision
"B", for a distance of 145.22 feet to an old iron found in the northern edge of the right of
way of Road "B", Point I;

Thence S 27°-38'-11" E across the right of way of Road "B" for a distance of 50.16 feet,
to the southern edge of the right of way of Road "B" to an iron pipe found, Point J;

FIG. 4.59 Example of a property boundary description of a parcel of land using metes and
bounds.

Thence S 27°-39'-19" E along the western side lot line of Lot 44, Block "A" contained within Subdivision "B", for a distance of 144.96 feet to an old iron found, Point K;

Thence S 62°-36'-40" W along the lands of Jones for a distance of 601.35 feet to an old iron found, Point L;

Thence N 27°-20'-30" W along the eastern side of Lot 17 , Block D-D contained within Subdivision "A" for a distance of 145.04 feet to an iron pipe found, Point A, the Point of Beginning.

Said tract measures and contains 5.135 Acres more or less, and is more clearly shown on a Plat by _____surveyor, dated 6 June 1992, and recorded in Plat Book _____ page _____ , RMC Office, _____ County, in the State of _____-, and is subject to a contiguous 50 foot storm drainage easement whose centerline is located at midpoint on side E-F, which then proceeds from the property along a bearing of N 27°-04'-24" W with 25 feet along each side of the centerline.

Date Of Preparation: Certified to be Correct:

Seal Impression   Surveyor's Name and Signature
                                         Surveyor's Address

FIG. 4.59   (continued)

Prior to starting the topographic survey using conventional techniques, a benchmark, or point of known elevation, must exist on or near the site based on the required vertical datum. Often verified benchmarks exist some distance from the site. In such cases a level loop must be run using differential leveling techniques from the known elevation point to the job site and then, reversing the process, to tie back to the starting point. To help eliminate mistakes and thus reduce field time consumed by remedial action while searching for leveling errors, the field crew should establish along the level run turning points that can easily be identified. These same points are utilized both ways when performing differential leveling, which provides a means to check accumulated field data by verifying both differences in elevation between adjacent turning points. If the difference between the beginning elevation and the computed closing elevation for the starting point (the vertical error of closure) exceeds the allowable tolerance, a check of the turning point differences likely will indicate where the problem exists in the field, allowing efficiency in eliminating unreliable data. It is also helpful to use the three wire leveling process to conduct a control level loop that will help eliminate possible data errors (see Figure 4.33). Although Figures 4.33 and 4.61 illustrate these procedures, the reader is referred to other sources for leveling adjustments using least squares as an alternative method (Davis et al., 1981; Moffitt and Bouchard, 1987).

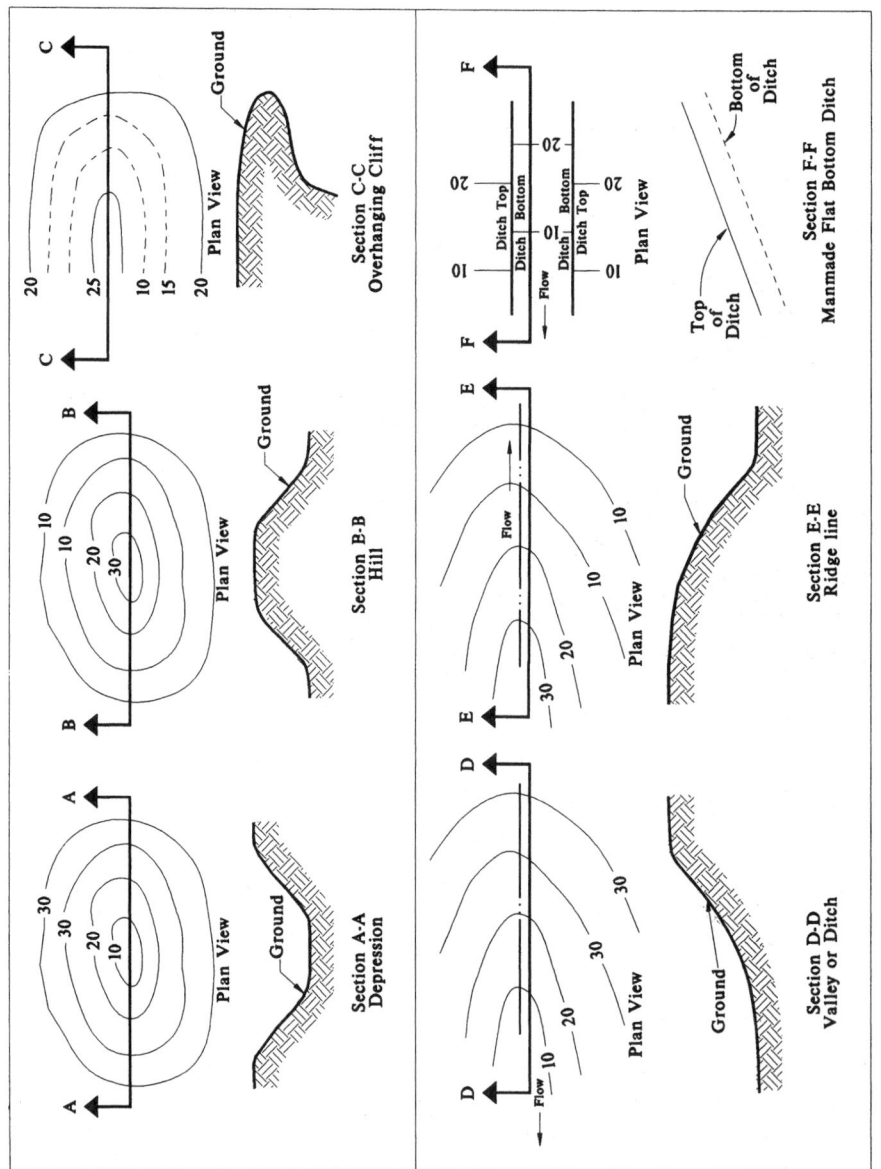

FIG. 4.60 Examples of various contour line configurations.

FIG. 4.61 Examples of vertical control adjustments: (a) adjustment of a level loop; (b) adjustment of multiple level lines; (c) adjustment of a level net. Source: Raymond E. Davis and Francis S. Foote. *Surveying Theory and Practice,* 4th ed. Copyright © 1962 by McGraw-Hill Companies. Reproduced by permission of McGraw-Hill Companies.

Of the various methods available to conduct a conventional topographic survey, the one most utilized is the grid method. This process uses an imaginary grid of predetermined size on the property with spot elevations (which are the ground elevations at a particular spot) being determined at each grid intersection using differential leveling. It is assumed that a uniform slope exists between adjoining spot elevations; therefore, if surface irregularities occur, additional intermediate spot elevations must be obtained.

Prior to laying out a project grid in the field, it is helpful to establish a series of temporary benchmarks (TBMs) along the property boundary that can be used later as references to tie each line of differential levels into when spot elevations are measured. Figure 4.62 shows an example of reduced field data, upon which contour lies must be located. Digital computer software application programs can be utilized to perform this task; however, other methods are often utilized, including mathematical or graphical interpolation. The interpolation process determines intermediate positions of various contour lines that must be shown on the final topographic map based on the interval used.

1. The reduced spot elevation of point "X" is 100.1 NGVD

2. The reduced spot elevation at point "Y" is 99.9 NGVD

3. The horizontal distance from either spot elevation must be determined for all contour lines that will be shown on the topographic map. Using a one foot contour interval, there will be only one contour line having an elevation of 100 feet between these spot elevations.

4. To mathematically proportion the distance use:

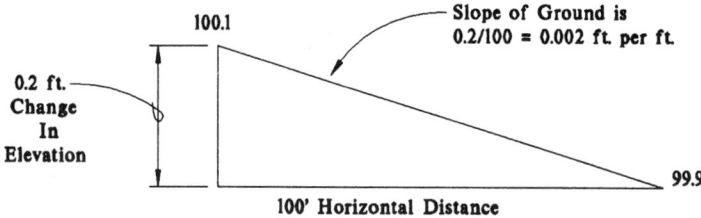

5. From the 99.9 foot spot elevation, the 100 foot contour crosses at (100−99.9)/0.002, or 50 feet.

FIG. 4.62 How to determine contour line placement.

If the graphical method is utilized, a transparent paper sheet with a series of parallel lines must be available. This overlay is positioned over two adjoining spot elevations within the grid. Each parallel line represents an integer value elevation with the distance between lines being proportional subdivisions. The resulting positions of intermediate contour lines are located as illustrated in Figure 4.63.

The process usually begins at the lowest or highest spot elevation in an area and systematically progresses outward in order to maintain contour line integrity. A finished working drawing from which the topographic map will be derived is shown in Figure 4.64. The final drawing should be field verified to ensure accurate information is included and that no details are overlooked prior to design use.

Today conventional topographic surveying often employs the use of a total station and data collector to obtain and store data in the field. These data are usually downloaded in the office using a cable connected to a personal computer or can even be downloaded in the field to a laptop computer where the downloaded data are relayed via wireless modem (cell telephone) to a distant office computer for processing. If a total station is used for this purpose, the surveyor must run what is called a three-dimensional survey. Not only are horizontal angles observed to various points that define the site's surface features, slope distances as well as vertical angles must be included. These data define Δx, Δy, and Δz between the total station and reflecting prism. The height of the instrument must be determined through a backsight (or some type of measurement) and the height of prism must also be known. Using the instrument's height, Δz, and the prism's height, the ground elevation where the prism is located is obtained. Since the observed horizontal angles and slope

1. Contour lines are located graphically between spot elevations 99.9 and 102.5 .
2. This process is repeated between all adjacent pairs of spot elevations.
3. Contour lines are drawn between logical intercept points.

FIG. 4.63 Graphical interpolation for contour line positions.

1. Spot elevation values are not shown for clarity.
2. Grid interval is 100' x 100'.
3. Points "A", "B", "C" and "D" refer to Section 7.5.2

FIG. 4.64 Example of a working drawing for a topographic map.

distances can be used to determine the prism's northing and easting (using latitudes and departures), a three-dimensional coordinate set can be generated for every point needed to define the ground surface being surveyed. Care must be taken to ensure that there are sufficient points gathered in the field to fully define the surface, especially where breaks occur in the topography; otherwise the final map will show erroneous information.

Computer software can be utilized to generate a TIN using the three-dimensional coordinate values. From this contour lines can be automatically generated and even a DTM. An example is shown in Figure 4.65. With a DTM an engineer can select any point on the modeled surface and determine its elevation or generate a cross section using modeling software.

RTK GPS (Section 4.10.3) can be used for topographic surveying if the site is fairly free from overhead obstructions. Prior to commencing GPS observations, various reference control points must be established on the site for calibration and periodic quality control field verification checks.

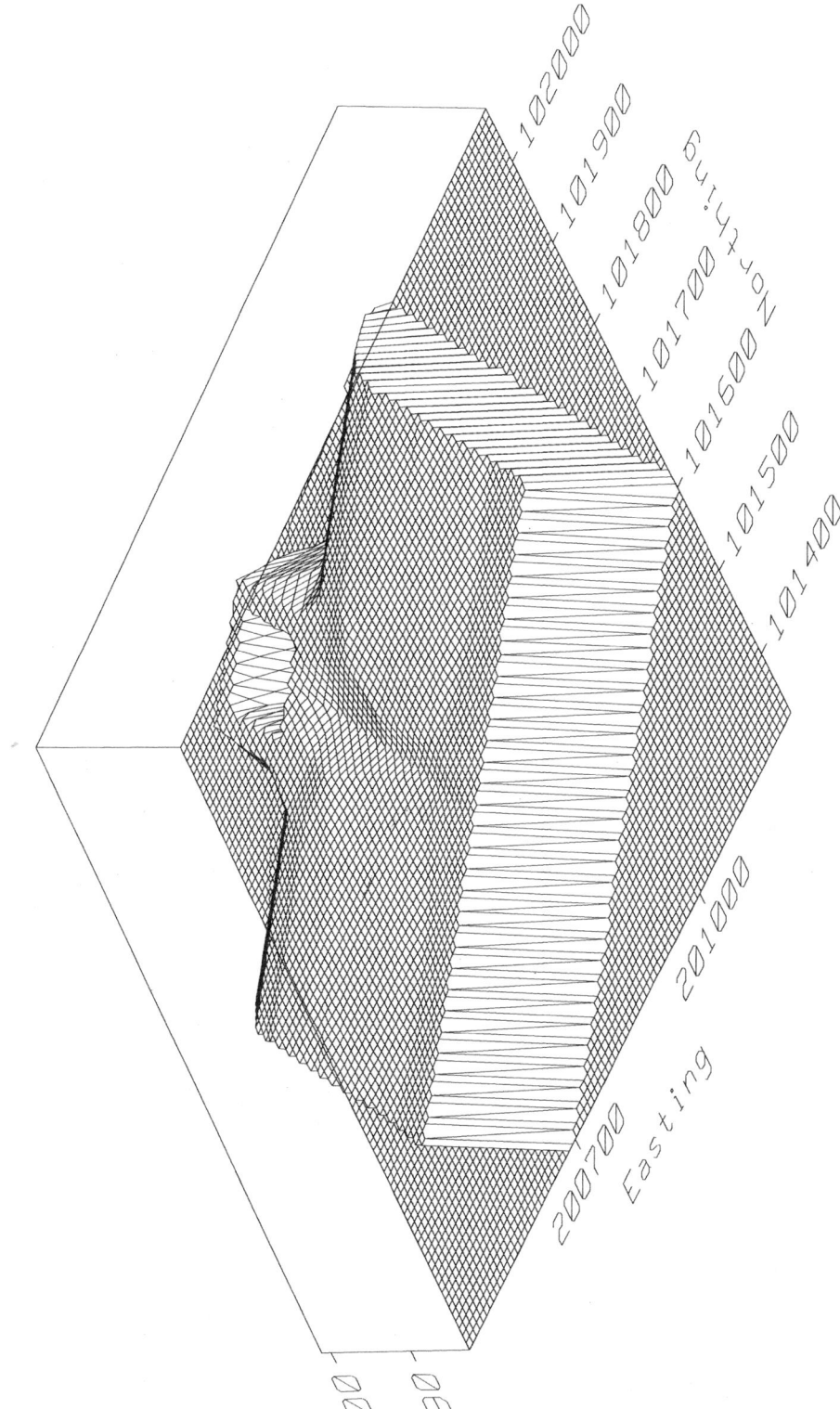

FIG. 4.65 Example DTM view of a small land development site generated by SMI plot. Reprinted by permission of Surveyors Module International, LLC, 2000. (www.smi.com).

4.13   UTILITY EXTENSION SURVEYS

Often it is necessary to include as part of the project some utility line connections that join existing off-site facilities. These improvements could, for example, include installation of gravity or pressure sanitary sewer mains or an extension to the existing portable water system. Field data about the area between the project site and the point of connection are required for both horizontal and vertical control. These data are necessary for the preparation of a plan and profile that show the proposed extension and that are included with the other construction drawings. Best results can be obtained when a foreknowledge of the proposed utility system's general features allows positioning of the field control point during the survey operations on which the design is based. Most utility line extensions occur within public rights-of-way that require a permit to be issued prior to construction. Often other buried facilities exist in the right-of-way where the proposed improvement needs to be located, including possibly storm drains. It is prudent to request underground location services and to excavate test holes so control point can be properly positioned to miss these features prior to making any field measurements. The resulting survey will produce measured design stationing that will result in construction plans that contain a minimum of conflicts. This translates into less contractor change order requests being submitted during construction. Figure 4.66 provides an example of how the process works in the field when the position of a future sanitary sewer manhole is being tentatively located. In this example, the existing storm drainage system is already in place. The new manhole, after being staked and measured, will produce excellent horizontal and vertical control data that will allow the engineer to show completed information on the plans, including the final elevation of the manhole's rim and cover.

4.14   PHOTOGRAMMETRIC SURVEYS

Photogrammetry is the science of making measurements of images on photographs. Although the photographs can be obtained using a fixed-position camera (called terrestrial photogrammetry), the practice of using aircraft-mounted aerial cameras (called aerial photogrammetry, or just photogrammetry in this discussion) is the more likely application associated with the development of land. It is one member of the family called remote sensing where differences in light patterns are sensed by the camera. Another member of the remote sensing family is satellite imaging, which is useful for creating a large-scale base map that could be incorporated into some GIS applications. Two other emerging remote sensing applications that could be used for land development mapping include airplane-mounted light detection and ranging (LIDAR) and interferometric synthetic aperture radar (IFSAR) systems.

4.14.1   Existing Aerial Photography

Available aerial photographs possess a vast amount of information helpful to the design professional; however, the photograph's scale must be known. Because a photograph's scale is determined by the ratio of photo distances to corresponding ground distances, if the photographed surface is fairly level, then a constant scale exists for that ground elevation, as indicated in Figure 4.67a. If, however, the photographed terrain is not

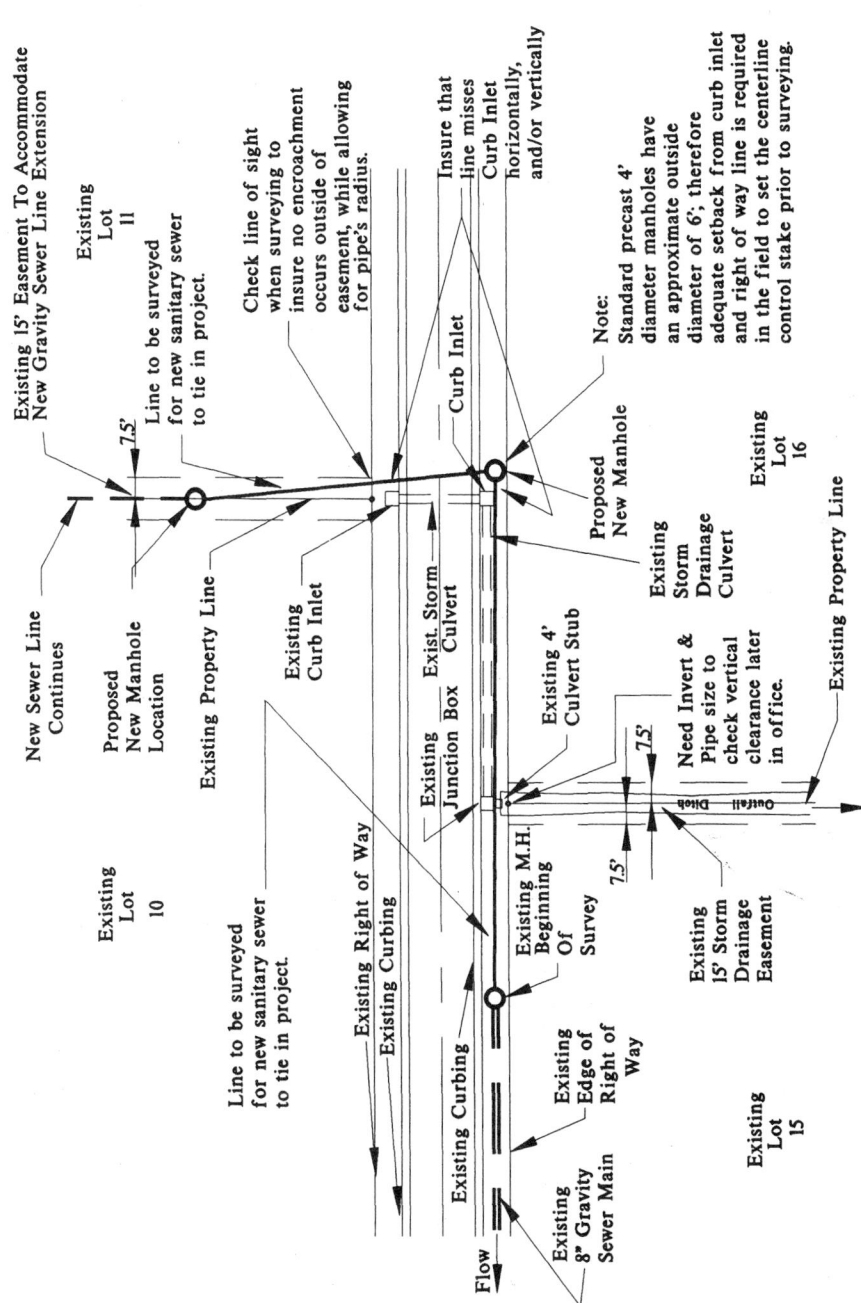

FIG. 4.66   Example of the field survey techniques for an extension of an existing utility system.

209

level, then the average scale can be determined knowing the focal length of the camera, the average elevation of the ground, and the flying height above datum, as illustrated in Figure 4.67*b*. Datum scale, illustrated in Figure 4.67*c* is the scale obtained by projecting all image points vertically to the datum surface.

If the land development professional has access to an existing aerial photograph that pertains to a selected site and the flight height of the photograph is unknown (*f* must be known), the following iterative procedure can be used to determine the flight height *H* above datum if one control line (line *L–M* for this example) is identified and the elevation of point *L* is different from that of point *M*:

Step 1. Obtain the ground distance *LM* of the identified control line through field measurements or using known ground coordinate values.

Step 2. Obtain the elevation of both ends of points h_l and h_m through field measurements or using known ground coordinate values. The average elevation used for the initial approximation of *H* is obtained using

$$h_{\text{ave}} = \tfrac{1}{2}(h_l + h_m) \tag{4.7}$$

Step 3. Measure the photographic coordinates (e.g., x_l, y_l, x_m, y_m) and straight-line distance (*lm*) for a check of the identified control line image and then calculate its photographic length (check against the straight-line measurement) using the inverse procedure described in Section 4.11.4.

Step 4. Solve for a first trial the approximate height *H* using the following relationship:

$$H = \frac{(f)(LM)}{lm} + h_{\text{ave}} \tag{4.8}$$

Step 5. Using this first trial approximation of *H*, compute corresponding ground coordinate values for points *L* and *M* with the following equations:

$$X_L = \left[\frac{H - h_l}{f}\right](x_l) \qquad X_m = \left[\frac{H - h_m}{f}\right](x_m)$$

$$Y_L = \left[\frac{H - h_l}{f}\right](y_l) \qquad Y_m = \left[\frac{H - h_m}{f}\right](y_m) \tag{4.9}$$

Step 6. Calculate, using the ground coordinate values obtained in Step 5, the distance *L–M* by inversing:

$$LM = \sqrt{(X_L - X_M)^2 + (Y_L - Y_M)^2} \tag{4.10}$$

Step 7. Compare the computed distance to the measured ground distance using

$$\frac{LM_{\text{measured ground distance}}}{LM_{\text{computed in Step 6}}} = \frac{H_{\text{improved}}}{H_{\text{trial}}} \tag{4.11}$$

If the first trial *H* is correct, the *LM* ground measurement would be the same as the *LM* value computed in Step 6, and the procedure is completed. However, chances

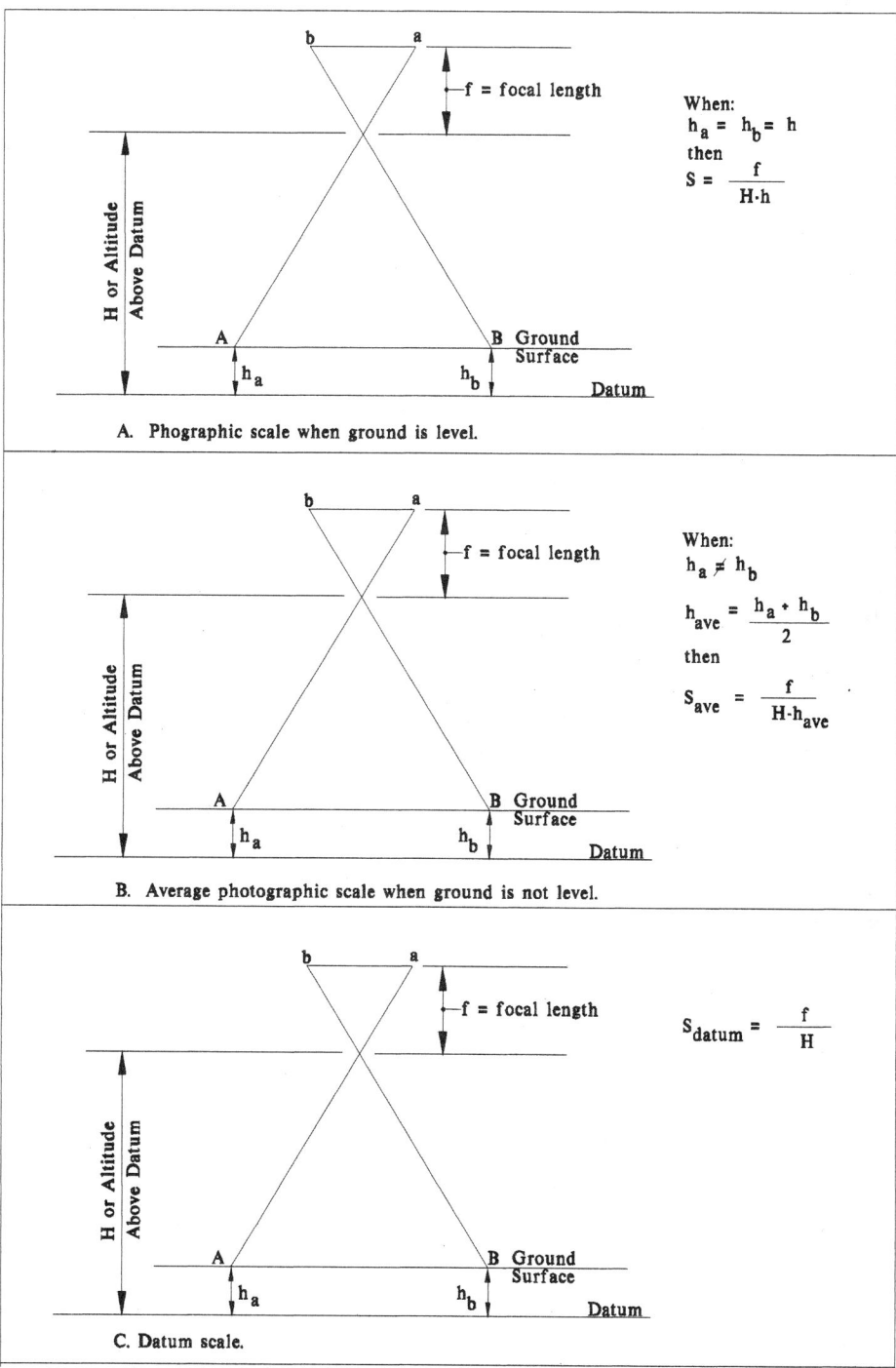

A. Phographic scale when ground is level.

When:
$h_a = h_b = h$
then
$$S = \frac{f}{H \cdot h}$$

B. Average photographic scale when ground is not level.

When:
$h_a \neq h_b$
$$h_{ave} = \frac{h_a + h_b}{2}$$
then
$$S_{ave} = \frac{f}{H \cdot h_{ave}}$$

C. Datum scale.

$$S_{datum} = \frac{f}{H}$$

FIG. 4.67 Scales associated with a vertical photograph.

are that the first trial H is not exactly correct, so the computed H_{improved} value is now used in the equations shown in Step 5 and the process is repeated until the ground and computed LM measurements coincide.

Once the flight height is determined, ground distances and directions can be computed between any points if the elevation of these new points are known. The relationships shown in Step 5 would define the coordinate values for new points being substituted for L, l, M, and m.

Should the mapping professional have access to a pair of stereoscopic vertical photographs, instead of a single photograph, the distance between exposures (called the air base B) and the flight height H above the datum can be determined using parallax formulas when one ground control line (C–D for this example) can be identified and one of the line's terminal elevations (point C) is known. As before, the camera focal length must be known to use the following procedure. Before starting, the x axis (usually parallel to the flight line) and y axis (perpendicular to the flight line and through the photograph's principal point (geometric center) for each photograph must be established by stereoscopically transferring each principal point to the paired photograph. The resulting x and y photographic axes for both the left and right (the stereo pair) photographs can be used to obtain photographic coordinate values for the photographic image of the ground control line being utilized. The reason both photographs are used is to determine the apparent movement (parallax) of the line endpoint that corresponds to the point of known elevation; however, determining the parallax for both endpoints is useful when starting this trial-and-error procedure. Since the aircraft is moving along the flight line (x axis), parallax p is computed by using the photographic x values with regard to algebraic signs from each photograph for the point of known elevation, where

$$p = x_{\text{left}} - x_{\text{right}} \tag{4.12}$$

Once the parallax is determined, one of the two stereo pair photographs is discarded, and the photograph x and y coordinate values off of the remaining photograph for control line are used to compute corresponding ground coordinate distances as follows:

Step 1. Compute an approximate first trial air base value using

$$B_{\text{first approximation}} = \frac{(CD)(p_{\text{ave}})}{cd} \tag{4.13}$$

where cd is the photographic distance of the control line, CD is the ground distance of the control line, and P_{ave} is the average parallax of points C and D.

Step 2. Using this first trial approximation of B, compute corresponding ground coordinate values for points C and D with the following parallax equations:

$$X_C = \left[\frac{B}{p_c}\right](x_c) \qquad X_D = \left[\frac{B}{p_d}\right](x_d)$$

$$Y_C = \left[\frac{B}{p_c}\right](y_c) \qquad Y_D = \left[\frac{B}{p_d}\right](y_d) \tag{4.14}$$

Step 3. Calculate, using the ground coordinate values obtained in Step 2, the distance C–D by inversing:

$$CD = \sqrt{(X_C - X_D)^2 + (Y_C - Y_D)^2} \tag{4.15}$$

Step 4. Compare the computed distance to the measured ground distance using

$$\frac{CD_{\text{measured ground distance}}}{CD_{\text{computed in Step 3}}} = \frac{B_{\text{improved}}}{B_{\text{trial}}} \tag{4.16}$$

If the first trial B is correct, the CD ground measurement would be the same as the CD values computed in Step 3, and the procedure is completed. However, chances are that the first trial B is not exactly correct, so the computed B_{improved} value is now used in the equations shown in Step 2 and the process is repeated until the ground and computed CD measurements coincide.

Once the air base is determined, the flight height can be computed using

$$H = \left[\frac{B}{p_c}\right](f) + h_c \tag{4.17}$$

Now knowing the flight height, the elevation of any random point on the photograph can be determined simply by measuring that point's parallax using the stereoscopic photograph pair and substituting that value for p_c in Equation (4.17) to obtain a new value of h, the point's elevation.

4.14.2 Requirements for New Photography

If aerial photographs are desired for a particular land development site but are not available, a flight plan must be designed for features desired in the finished photographs. The resulting images are often captured on black and white, color, or infrared film. The accuracy of photographic measurements made with the captured images depends on relative errors associated with the camera, film, and measurement equipment. In addition, the scale of the stereo model greatly influences measurement errors. An increase in flight height produces photographs that show an increase in land area at the expense of reducing the corresponding accuracy in any photographic measurements.

The relief displacement depicted on aerial photographs refers to any radial position shift of features shown on the photograph. This is a result of the object having an elevation above or below the set datum. Relief displacement causes the image to radiate outward for points located above the datum and inward for points located below the datum. Relief displacement causes any straight-line features such as a road, utility right-of-way, or fence segment located on rolling terrain to appear bent on a vertical photograph. In addition, relief displacement could affect what flight height to use when designing the flight plan, because an intended feature may not be captured if the relief displacement positions the photographic image off the photograph.

Photographs should be taken on clear days during late fall or early spring (minimum foliage and snow cover) and at low tide for best results. The sun should be at 30 degrees above the horizon to minimize the effects of shadows. When new aerial photography is desired, the basic components that must be considered include the following five components that are examined in the following sections: (a) equipment characteristics, (b) flight planning, (c) ground control, (d) data translations, and (e) final products.

Typically for topographic mapping, the following sequencing of activities occurs:

flight planning, establishment of ground control points, acquisition of aerial photography, aerial triangulation, numerical adjustment, map compilation/editing, revisions, quality control, and completion of the job product. Today the final product can be in hard-copy format or stored in digital files.

4.14.3 Photogrammetric Equipment Characteristics

There are currently many different types of film cameras available for aerial photography; however, the most common type used for land development applications is a large-format negative (9 × 9 inches) with focal lengths of 12, $8\frac{1}{4}$, or 6 inches. The camera consists of a magazine for storing and moving the film, a platen—or focal plane—upon which the film is positioned for exposure, a series of lenses with a shutter and diaphragm, and filters. The camera is designed so that the optical axis (axis through the center of the lens system) is oriented vertically while the platen is positioned horizontally. Pictures taken while the camera is aligned in this intended manner result in what is known as a vertical photograph. Photographs obtained when the optical axis is unintentionally inclined from the vertical are called tilted photographs and must be corrected. Gyro-stabilized camera mounts are often used to compensate for tip or tilt of the aircraft to help promote vertical alignment of the optical axis. In some cases, however, photographs are obtained when the optical axis is intentionally tilted at some established angle and are called oblique photographs (a subject beyond this discussion).

Section 4.14.1 illustrated how the scale of a vertical photograph was affected by the plane's altitude. Because aerial cameras have a fixed focal length, the scale is determined by the height of the plane above the terrain being photographed

State-of-the-art cameras used today have improved optics where their resolving power is determined through an area-weighted average resolution (AWAR), that is, the number of distinguishable lines per linear millimeter of a photograph. A decade ago typical camera AWAR values ranged from 50 to 60 lines per millimeter, but today's improved cameras have AWAR values ranging between 80 and 100. Another camera improvement, called Forward Motion Compensation (FMC) technology, eliminates any image blurring caused by the aircraft's forward movement during the time of exposure. This technology senses the ground speed of the aircraft and then moves the film across the platen at a speed to correspond with that movement while the shutter is open. Both FMC and gyro stabilization are critical for the production of high-quality digital orthophotographs (see Section 4.14.7) because the clarity of the original imagery proportionally affects any derivations.

State-of-the-art aerial cameras are being constructed that operate on digital technology. Remotely sensed satellite imagery has made use of similar sensor devices for many years; however, its use in the photogrammetric field is emerging. The digital camera offers speed in data availability since film processing is eliminated. However, these cameras do not produce the same quality images as conventional film cameras and their format is also much smaller. Digital camera data transmission and storage can also create problems; however, with the rapid progress being made in improved computing power and data-handling systems, a number of improvements are underway that could eventually allow digital technology to replace conventional film utilization. Stereoplotters have traditionally produced maps using adjacent stereoscopic photographs and three-dimensional positional values of known points on each photograph. From these known control points, the stereomodel can be used to determine any number of unknown positions located within the model. Often control points are established using analytical aerotriangulation where a small number of known control points found on some of the photographs are used as a basis to create a dense network of control points. Today stereoplotters include both analog and analytical types. Analog plotters rely on manual (mechanical) methods and regular optics while the analytical plotters utilize computer technology. However, many analog plotters incorporate digital encoders that allow the direct input of measured distances directly into a computer and at the same time allow one to graphically view what is being stored usually with some type of CAD program.

Analytical stereoscopic plotting instruments give a higher quality finished product than the analog types because a computer is used to calculate the stereomodel's geometry. This allows the observer to view stereoimages created by the photographs where the errors associated with the machine's mechanical and optical systems are eliminated. This type of instrument uses scanned photographs and is fully digital where three-dimensional images are processed by computer from which measurements can be recorded. Because of current limitations in computer storage and the often large data files generated by scanning images, the full potential of this technology will be realized in the near future.

Stereoplotters are rated by their ability to enlarge from a given photograph scale to some optimum map scale (enlargement ratio) and also by their ability to discern contour lines as indicated by a C factor. In the case of the enlargement ratio, the product of the desired map scale, the camera focal length, and the enlargement ratio yields the maximum height that the plane can fly to gather data for a particular plotter:

$$\text{Maximum flight height} = \text{map scale} \times \text{focal length} \times \text{enlargement ratio} \quad (4.18)$$

C factors are the ratio of flying height to the contour interval that can be accurately determined for a photograph. Most modern analog plotters have C factors that range from 1400 to 2000 while analytical plotter values fall around 2400. These factors directly affect flight planning because if a particular plotter is to produce the final map using a desired contour interval, then

$$\text{Flight height} = C \text{ factor} \times \text{contour interval} \quad (4.19)$$

4.14.4 Flight Planning

A sample flight plan design follows for a proposed development site where the average elevation is 100 feet NAVD 88. The site consists of approximately 5000 acres in a rectangular boundary of 20,800 × 10,400 feet. A camera with a 12-inch focal length will be used.

(a) Desired photographic scale: 1 inch = 500 feet.
(b) Size of photographs: 9 × 9 inches.
(c) Flight map scale: 1 : 24,000.
(d) Minimum sidelap: 30 percent.
(e) Minimum overlap: 65 percent.
(f) An intervalometer is to be used.
(g) The first and last flight lines will coincide with the larger boundaries.

1. First determine the flight height above the datum the aircraft must fly to produce the desired photographic scale:

$$\frac{1 \text{ inch}}{500 \text{ feet}} = \frac{f}{H - h_{\text{ave}}} = \frac{12}{H - 100}$$

$$H - 100 = 6000 \qquad H = 6100 \text{ feet NAVD 88}$$

2. Next, a determination of the ground distance between exposures must be undertaken. If the minimum overlap is given as 65 percent, then the net gain per picture is 35 percent, resulting in an effective photographic distance of 0.35×9 inches $= 3.15$ inches. This would equate to a corresponding ground distance of 3.15 inches $\times$ 500 feet per inch $= 1575$ feet on the ground.

3. If the anticipated flight speed is to be maintained at 100 miles per hour, the velocity of transit can be computed as

$$\frac{100 \text{ miles per hour} \times 5280 \text{ feet per mile}}{3600 \text{ seconds per hour}} = 146.7 \text{ feet per second}$$

Because an intervalometer will be used to activate the camera, the exposure interval can be computed by 1575 feet/146.76 feet per second $= 10.7$ seconds; however, this must be rounded downward in integer form to 10 seconds. The adjusted ground distance would equal 10 seconds $\times$ 146.7 feet per second $= 1467$ feet of ground coverage between exposures, which equates to an adjusted overlap of 67.4 percent, which is greater than the allowable minimum.

4. The total number of photographs per flight line, with the aircraft flying parallel to the long boundaries, is (20,800 feet/1467 feet) + 4 photographs (2 additional at each end for complete coverage) $= 18.2$, or 19 photographs per flight line.

5. The number of flight lines can be determined by first computing the initial spacing between flight lines using the minimum allowable sidelap of 30 percent. This leaves a net gain of 70 percent between flight lines. The corresponding ground spacing is therefore 0.70×9 inches $\times$ 500 feet per inch $= 3150$ feet. The number of corresponding flight lines is

$$\frac{10{,}400 \text{ feet}}{3150 \text{ feet}} + 1 = 4.3, \qquad \text{but round to 5}$$

The adjusted net gain distance between flight lines accounting for rounding that equalizes the number of trips is

$$\frac{10{,}400}{5 - 1} = 2600 \text{ feet}$$

This distance produces a 42 percent sidelap, which is in excess of the minimum allowed.

6. The total number of pictures required for the mission would be 5×19, or 95 photographs.

7. To properly draw flight lines on the flight map, the spacing must be determined

using the adjusted ground distance of 2600 feet between flight lines. At a scale of 1 : 24,000 the equivalent map spacing for flight lines would be every 1.3 inches.

The completed flight plan would then be:

(a) Flight height = 6100 NAVD 88
(b) Camera focal length = 12 inches.
(c) Speed = 100 miles per hour.
(d) Intervalometer setting = 10 seconds.
(e) Total number of photographs = 95.
(f) Map spacing for drawing flight lines on the flight map is 1.3 inches apart.

Figure 4.68 illustrates some photogrammetric relationships that were alluded to in this example.

The most recent USGS quadrangle or county map should be used for the flight map, especially if airborne GPS equipment is to be utilized. This will allow the flight crew to obtain latitude and longitude coordinates for the end positions of each flight line that can then be used as a basis for navigation during data collection.

4.14.5 Photogrammetric Ground Control

A key element in flight planning is the availability of both horizontal and vertical control points within the flight area usually tied to some adopted reference datum. Photographs that do not have an established relationship between photographic image point coordinate values and their corresponding ground image point coordinate values are of little mapping interest because photographic distances and angles cannot be defined. According to Lillesand and Kiefer (1979, p. 312):

> Accurate ground control is essential to virtually all photogrammetric operations because photogrammetric measurements can only be as reliable as the ground control on which they are based.[*]

If ground control points exist, they must be recovered and, if in open areas where visibility from aircraft is likely, flagged as possible targets. If control monuments do not exist, then new ones must be established using conventional surveying techniques or GPS surveying methods previously described in Section 4.10.3. Because GPS requires little or no canopy obstructions for good satellite signal reception, most control points established using GPS methods are good candidates for use in photogrammetric surveying.

4.14.6 Data Translations

In most cases photogrammetrically compiled maps rarely exist in file formats similar to those commonly used by land surveyors or engineers. Photogrammetrists most often

---

[*] Taken from Lillesand and Kiefer (1979), by permission.

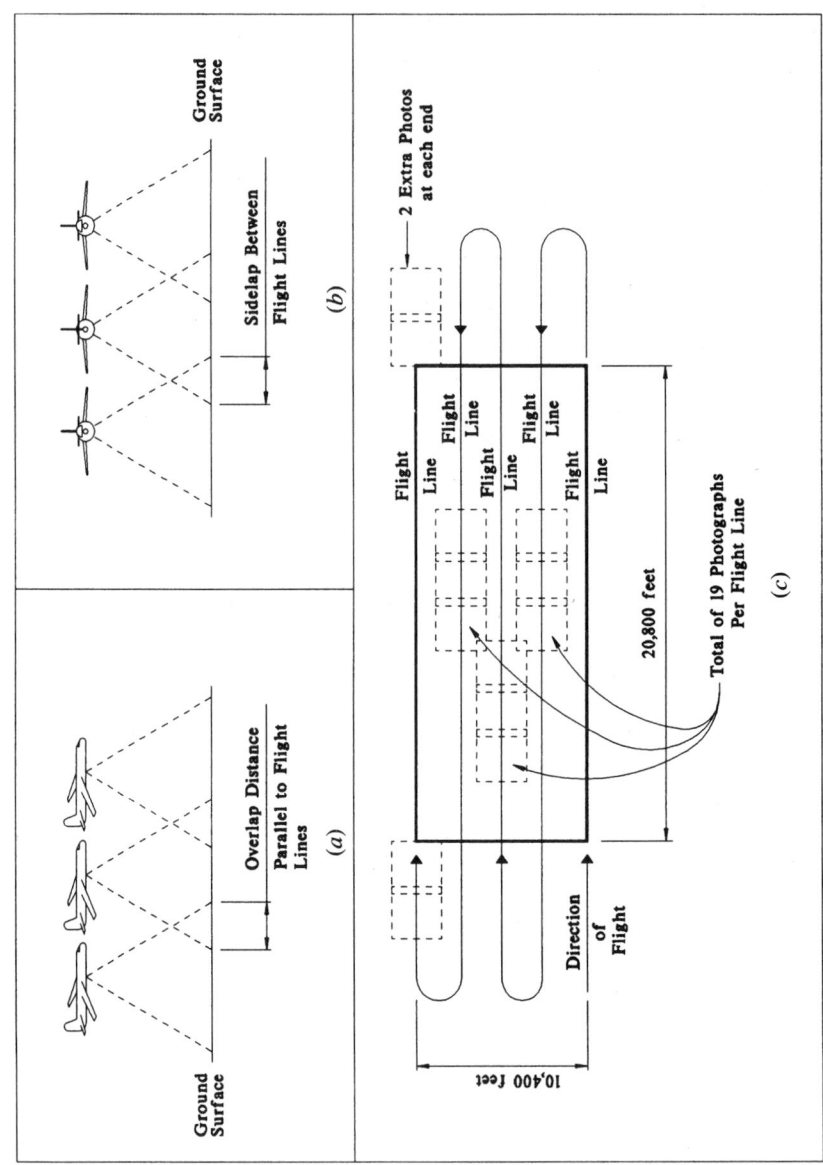

FIG. 4.68 Photogrammetric relationships; (a) photo overlap; (b) photo sidelap; (c) relationships pertaining to the sample flight plan.

218

use a set of programs specifically written for photogrammetric applications and digitizer interface support. Data translation to common land surveying and engineering file formats can be made simple if the photogrammetrist is given layer and other pertinent assignments before beginning the work. Software vendors have excellent software routines to make this translation routine.

4.14.7 Photogrammetric Mapping Products

Typical photogrammetric products produced for end users have evolved over the years from paper products to a combination of paper and digital products that are produced today. Unrectified photographic mosaics (an assemblage of aerial photographs) and planimetric maps have given way to orthophotos and DTM.

An orthophoto is an aerial photograph that depicts the earth's surface as an orthographic plan view where scale and shape distortions have been removed. The process of differential rectification removes the distortions present in "vertical" photographs (including tilt and terrain relief) and then transforms the photographs into a map. In the past, orthophotos were produced with a special photogrammetric stereoplotter. Today digital orthophotos can be produced using a desktop PC with software that is available for this purpose. Digital orthophotography can be used with most CAD systems and combined with traditional CAD vector image properties. To create an orthophoto, the scale and position distortion of each pixel of the aerial photograph must be computed, rescaled, and repositioned for the new computer-generated image. This process is called differential rectification. To accomplish this, a high-quality image scanner is used to create a digital image file from film. This file represents a raster image which is comprised of pixels of information arranged in a grid that represents the original photographed image. Each pixel represents the photographic gray tone of a small section of ground area where its size controls the image resolution. (As an alternative to high-resolution photograph scanning, imagery from digital aerial cameras and satellite sensors will become increasingly available from on-line services over the next few years.)

Next the location of the camera and orientation of each photograph at the time of exposure with respect to some coordinate system must be computed using aerial triangulation based on known ground control point positions. To correct scale distortions attributed to the perspective projection of the image, the elevation of each pixel must be determined. The pixel elevation can be interpolated from a DEM. The most accurate way to obtain elevations is by field surveying; however, the cost associated with developing a DEM (even a small one) using this procedure can be prohibitive. Most data used for DEM are therefore obtained using photogrammetry where the same photographs can be used for both aerial triangulation as well as DEM generation; however, kinematic GPS surveys or digitized contour lines taken from topographic maps can also be used. Once image, orientation, and DEM data are available, each digital photograph can be rectified using software made for this purpose. Since many rectified photographs may be used to provide complete coverage of a development project, individually rectified photographs are pieced together into a seamless and continuous orthophoto through edge matching. Some problems are usually encountered when performing this procedure, such as features that overlap or jump where the rectified photographs join and also inconsistencies in contrast and image brightness due to differences in photographic processing, image scanning, or even atmospheric variations between exposures. Digital orthophoto software is capable of feathering the individual images to ensure consistency in tones and contrasts where any detectable splice lines are eliminated. Once any encountered problems have been corrected, the orthophoto is inspected for incorrect rectification, image-matching difficulties, and missing images attributed to hidden ground problems. Once accepted, the orthophoto is formatted and finalized with appropriate cartographic enhancements. Digital orthophotos are used by design professionals in both hard-copy form and a digital georeferenced image base for GIS systems. Georeferencing means the pixels are assigned real-life coordinate values based on a selected reference datum such as SPCs.

Another mapping product produced (see Section 4.12) through photogrammetric means is the DTM, which has replaced standard contour maps. DTMs contain complete elevation information found on a DEM but also contain break lines and break points which depict ridges, valleys, peaks, and so on. Specialized software uses the DTM to generate topographic maps, ground profiles, or cross sections. Because of this, digital copies of the DTM are often included as part of the final product deliverables for use by other members of the land development planning and design team.

4.14.8   Evolving Photogrammetric Technologies

With the advent of full GPS operation (Section 4.10), the system can be used at any time during the day or night. This availability has opened the door for photogrammetrists to place increased utilization and reliance on airborne GPS navigation using radial survey techniques. This method requires one or more GPS receivers to be positioned over points of known location, while an additional receiver is placed onboard the aircraft. The vector from the receiver located over the known point to the aircraft's receiver is determined using differencing techniques described in Section 4.10.2. This procedure is similar to using a total station for measuring the distance and azimuth to a moving target. Approximate aircraft positional checks can be made by using several ground points to recompute its location; however, independent verification cannot be made unless the aircraft has more than one receiver/antenna onboard.

Aircraft GPS position determination can be achieved using various techniques described in Section 4.10.3. The least accurate method employs the use of differential pseudoranging techniques. A more accurate approach utilizes kinematic techniques where the base carrier-phase measurement of both the L1 and L2 frequencies are utilized for improved accuracy. At the beginning of a kinematic survey, the aircraft's position must be initialized and precisely determined (e.g., maneuvering the aircraft over a presurveyed reference point before takeoff or using the rapid static technique if the GPS base station is close by). The survey can begin when five or more satellites are available for the flight's duration. Both the GPS base station and airborne roving units are set to record the satellite (L1 and L2) data at the 1-second or less interval. The procedure is successful where the collected data can be postprocessed to determine the aircraft's location for every recorded epoch if the aircraft can fly its mission without losing lock on four or more satellites initially used. If satellite lock is lost during flight, the mission must be repeated after the plane's position is reinitialized on the ground.

A variation of airborne kinematic GPS positioning, described in Section 4.10.3 as on-the-fly technology (proposed by Benjamin Remondi of C&GS), makes use of a Kalman filter algorithm to smooth a number of satellite measurements where an improved position can be determined. This improved position in turn is used to resolve integer biases while the roving airborne GPS receiver is moving. The advantage of on-the-fly technology is that the mission can continue even with temporary loss of satellite lock. The only loss of data occurs between the time that loss of lock occurred and the time that reinitialization took place.

When airborne GPS positioning techniques are utilized, the aerial camera must be equipped with a pulsing device that indicates precisely the instant that a photograph is taken. The instant in time for each exposure can be used to determine the photograph's geodetic position by interpolating between subsequent aircraft positions. For more complete information on this subject see the report by Lewis Lapine (1991).

Concomitant with increased utilization of airborne GPS techniques is an increase in utilization of active airborne sensors such as IFSAR, which operates in the microwave band, and LIDAR which operates in the near-infrared and visible spectrum. These technologies rely heavily on the use of internal measuring units (IMUs). An IMU on each airborne sensor provides orientation parameters so that its position and orientation can be determined precisely. Typically these types of data are unknown, and classical procedures using advanced mathematical adjustment techniques (along with known ground control) are used to make these time-consuming determinations. A sensor-mounted IMU that measures the sensor's orientation to 10 seconds accuracy can reduce the need to rely on ground control and mathematical adjustment techniques to rectify the imagery for use in map making.

LIDAR laser systems, which emerged from defense industry efforts, have become affordable for commercial use over the past several years. Its use has also been enhanced through increased computer power, improved GPS techniques, and the emergence of IMU technologies. LIDAR can be used to create DEMs of the earth's surface. The airborne LIDAR system consists of a laser mounted along with positioning and orientation equipment to measure three-dimensional coordinates of passive targets. The laser determines the range to the target (ground surface), and in combination with the position and orientation of the sensor it provides the three-dimensional position of the target which can be ideally used to create DEM rectification models for creation of orthophotos. The accuracy of LIDAR data is relatively insensitive to changes in height, unlike other photogrammetric technologies.

IFSAR, another new tool, utilizes a coherent combination of radar measurements made by two antennae located a short distance apart. The phase difference between two backscatter measurements is used to compute the range difference between the two antennae and then compute the target's location and height. Depending on the relative geometry of the antennae, the combined measurements can be used to derive topographical features.

4.15 GEOTECHNICAL SURVEYS

Four reasons why a subsurface exploration program is necessary are as follows:

(a) The developer must purchase property wisely.

(b) The engineer must locate improvements efficiently on the site.

(c)   The engineer must design site structures properly.

(d)   The engineer must design all site improvements correctly.

Subsurface investigations usually are performed by geotechnical engineering testing firms whose services generally are classified as auxiliary services. A determination of the structural loads, locations of proposed improvements, and proposed sampling sites must be available prior to commencing any subsurface investigation. Investigations can include exploratory borings, extraction of undisturbed soil samples, or employment of sounding rods for crude investigations. Hand augers are excellent for roadbed and floor slab investigations where a visual inspection of the soil can be made; however, these soil samples are totally destroyed and provide no indication of a clay's stiffness or a sand's density. The investigator is limited as to depth, especially in sandy conditions below the groundwater table.

Wash boring samples may be obtained where the sample hole is advanced by the jetting action of water, but again no idea of the soil's stiffness or density is obtained. A better method is to employ the spoon sample boring with a standard penetration test. This process is in part similar to the washed boring method except that at various intervals a standard split casing is driven into the soil using a 140-pound weight that falls 30 inches. The number of blows to drive the standard spoon 1 foot is the standard penetration value from which the apparent soil strength can be estimated. The hole, rather than being advanced by jetting, utilizes a rotary drill in conjunction with a clay slurry to maintain the integrity of the interior wall. If an on-site test pit is available, excellent samples can be obtained from within; nevertheless, many patented samplers including thin-walled tubes are available today to extract subsurface "undisturbed" samples for laboratory testing without requiring a pit facility.

4.15.1   Soils Exploration Program

Because the complexity and proposed uses of sites vary, as does the uniformity of the underlying soil, exploratory activities cannot be based strictly on economics (although economics always play a part). If the soil if fairly uniform and the anticipated structural loads are light, a minimum program that encompasses at least one deep boring every 10 acres will probably provide sufficient preliminary investigative data. Also required are supplemental hand auger borings in locations where improvements such as roads are proposed or where soil problems may exist. Proposed boring locations should be shown with distances on a site plan to allow accurate field positioning. Each boring site will yield a soil log showing the extents of the underlying strata of soil. In addition, soil samples and standard penetration values can be obtained if spoon or thin-walled tube sampling is employed. Finally, the stabilized groundwater table and possibly an indication of the seasonal table fluctuations can be determined. Should results of this exploratory program warrant additional investigative work, arrangements with the developer must be made.

Other procedures used for subsurface exploration may incorporate geophysical methods, which are good for locating bedrock or gravel deposits. One method utilizes seismic shocks that are measured on a calibrated recorder. Another uses electrical resistivity by measuring electrical currents through electrodes driven into the soil. Finally, air photograph interpretation can be utilized to identify soils on the potential site if such photographs are available.

4.15.2   Soil Classification

A soil's composition varies relative to particle size distribution and the amount of plasticity. Two primary systems used for classifying soils for purposes of construction are the American Association of State Highway and Transportation Officials (AASHTO) and the Unified Classification System (UCS). These systems require the soil's grain size distribution and Atterberg limit values, which indicate the amount of cohesive material present.

The AASHTO method M 145–91(1995)[*] classifies soils into seven groups based on grain size distribution and Atterberg limits. Evaluation of soils within each group is made using a "group index" that is obtained with an empirical formula. The group classification and group index can indicate the relative quality of the soil; however, additional soil testing is usually required for design purposes.

Figure 4.69 can be used to classify soils, while Figure 4.70 can be used to further define soils by subgroups. Liquid limit and plasticity index ranges for A-4 through A-7 soils are shown in Figure 4.71.

AASHTO soil classification is accomplished by a process of elimination beginning with entering Figure 4.69 or 4.70 at the left and then proceeding toward the right. The first group that the grain size distribution and Atterberg limits satisfy is the correct classification. Group index values are shown in parentheses after the group symbol. The group index can be obtained by the following equation using the nearest computed integer value:

$$\text{Group index } = (F - 35)\,[0.2 + 0.005(LL - 40)] + 0.01(F - 15)(PI - 10) \quad (4.20)$$

where F is the percentage passing the 0.075-mm (No. 200) sieve, expressed as a whole number [this percentage is based only on the material passing the 75-mm (3-inch) sieve]; LL is the liquid limit; and PI is the plasticity index.

If the calculated group index is negative, then zero is used for the index value.

The UCS uses Figure 4.72 to base the classification of soils into one of 15 major soil groups. Soils are classified using:

(a) percentage of gravel, sand, and fines (fraction passing the No. 200 sieve);

(b) shape of grain size distribution curve using C_u (coefficient of uniformity) and C_g (coefficient of gradation), where D_{10} is the grain size diameter that corresponds to 10 percent of the grain size distribution curve; and

(c) plasticity and compressibility characteristics.

Generally clay (C) type materials are located above the "A" line of the plasticity chart, while silts (M) are below. Silts and clays are further classified based on high (H) or low (L) liquid limit values. The reader is referred elsewhere for a more complete presentation of this soil classification system (Joint Publication of the U.S. Army, Navy, and Air Force, 1971).

---

[*] From *Standard Specifications for Transportation Materials and Methods of Sampling and Testing, Part I Specifications*, Copyright © 1998 by the American Association of State Highway and Transportation Officials, Washington, D.C. Used by permission.

General Classification	Granular Materials (35% or Less Passing 0.075 mm)				Silt-Clay Materials (More than 35% Passing 0.075 mm)			
Group Classification	A-1	A-3[a]	A-2	A-4	A-5	A-6	A-7
Sieve analysis, percent passing:							
2.00 mm (No. 10)	—	—	—	—	—	—	—
0.425 mm (No. 40)	50 max.	51 min.	—	—	—	—	—
0.075 mm (No. 200)	25 max.	10 max.	35 max.	36 min.	36 min.	36 min.	36 min.
Characteristics of fraction passing 0.425 mm (No. 40)							
Liquid limit	—	—		40 max.	41 min.	40 max.	41 min.
Plasticity index	6 max.	N.P.	[b]	10 max.	10 max.	11 min.	11 min.
General rating as subgrade	Excellent to good			Fair to poor			

[a]The placing of A-3 before A-2 is necessary in the "left to right elimination process" and does not indicate superiority of A-3 over A-2.
[b]See Figure 4.70 for values.

FIG. 4.69 Classification of soils and aggregate mixtures. Taken from *Standard Specifications for Transportation Materials and Methods of Sampling and Testing, Part I Specifications*, Copyright © 1998 by the American Association of State Highway and Transportation Officials, Washington, DC. Used by permission.

General Classification

	Granular Materials (35% or Less Passing 0.075 mm)							Silt-Clay Materials (More than 35% Passing 0.075 mm)			
	A-1		A-3	A-2				A-4	A-5	A-6	A-7
Group Classification	A-1-a	A-1-b		A-2-4	A-2-5	A-2-6	A-2-7				A-7-5, A-7-6
Sieve analysis, percent passing:											
2.00 mm (No. 10)	50 max.	—	—	—	—	—	—	—	—	—	—
0.425 mm (No. 40)	30 max.	50 max.	51 min.	—	—	—	—	—	—	—	—
0.075 mm (No. 200)	15 max.	25 max.	10 max.	35 max.	35 max.	35 max.	35 max.	36 min.	36 min.	36 min.	36 min.
Characteristics of fraction passing 0.425 mm (No. 40)											
Liquid limit	—		—	40 max.	41 min.	40 max.	41 min.	40 max.	41 min.	40 max.	41 min.
Plasticity index	6 max.		N.P.	10 max.	10 max.	11 min.	11 min.	10 max.	10 max.	11 min.	11 min.[a]
Usual types of significant constituent materials	Stone fragments, gravel and sand		Fine sand	Silty or clayey gravel and sand				Silty soils		Clayey soils	
General Ratings as Subgrade	Excellent to Good							Fair to poor			

[a]Plasticity index of A-7-5 subgroup is equal to or less than LL minus 30. Plasticity index of A-7-6 subgroup is greater than LL minus 30 (see Figure 4.71).

FIG. 4.70 Classification of soils and aggregate mixtures (continued). Taken from *Standard Specifications for Transportation Materials and Methods of Sampling and Testing, Part I Specifications*, Copyright © 1998 by the American Association of State Highway and Transportation Officials, Washington, DC. Used by permission.

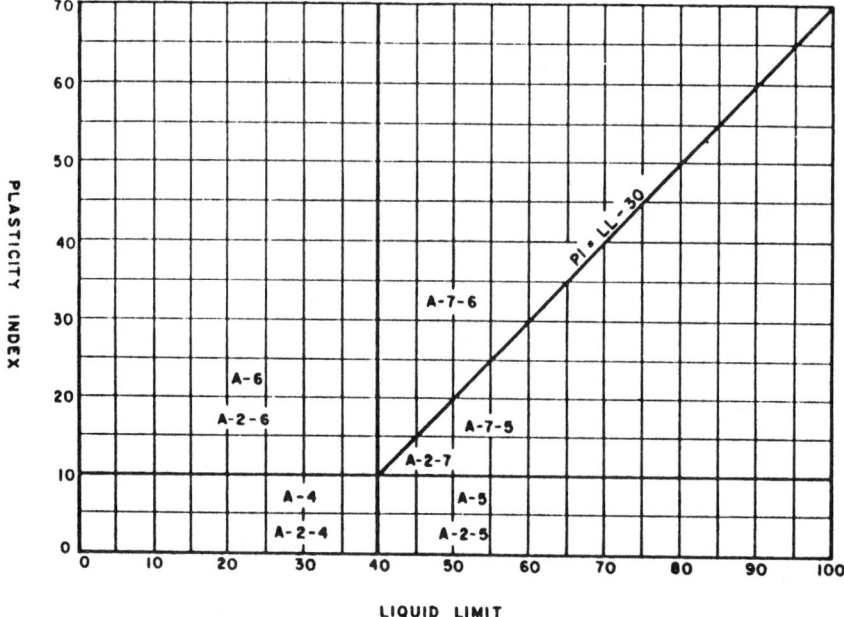

FIG. 4.71 Liquid limit and plasticity index ranges for silt-clay materials. Taken from *Standard Specifications for Transportation Materials and Methods of Sampling and Testing, Part I Specifications,* Copyright © 1998 by the American Association of State Highway and Transportation Officials, Washington, DC. Used by permission.

A third method of soil classification utilized by the U.S. Department of Agriculture is illustrated in Figure 4.73.

4.15.3 Combined Mechanical Analysis of a Soil

When a soil contains particle sizes ranging from clay to gravel, it is necessary to run both the dry (sieve) and wet (hydrometer) analyses to determine the grain size distribution curve. Sieve analysis is suitable for particle sizes down to the No. 200 sieve, or 0.074 mm, whereas hydrometer analysis is suitable for particles ranging from 0.2 mm down to 0.2 μm, which provides some redundant information in the range between 0.2 and 0.07 mm.

A dry representative soil sample is carefully weighed and then washed (without screening) with wash water containing a dispersing agent. The muddy wash water is retained for hydrometer analysis. The washed soil is dried and then used in the sieve analysis. The mechanical analysis is conducted by placing the dry soil sample in the top of a nest of sieves, which range usually from 1.5 inches through the No. 200 sieve, for the purpose of separating the soil particles to determine the percent weight retained on and passing through each sieve. The hydrometer analysis, based on Stokes's law where "particles of equal specific gravity settle in water at a rate which is in proportion to the size of the particle" (Seelye, 1960), is conducted on the wash water using a hydrometer for sequentially timed readings (Lambe, 1951).

Major Divisions			Group Symbols	Typical Name	Field Identification Procedures (Excluding Particles Larger Than 3 Inches and Basing Fractions on Estimated Weights)		
1	2		3	4	5		
Coarse-grained Soils — More than half of material is larger than No. 200 sieve size. The No. 200 sieve size is about the smallest particle visible to the naked eye.	Gravels — More than half of coarse fraction is larger than No. 4 sieve size.	Clean gravels (little or no fines)	GW	Well-graded gravels, gravel-sand mixtures, little or no fines	Wide range in grain sizes and substantial amounts of all intermediate sizes		
			GP	Poorly-graded gravels, gravel-sand mixtures, little or no fines	Predominantly one size or a range of sizes with some intermediate sizes missing		
		Gravels with fines (appreciable amount of fines)	GM	Silty gravels, gravel-sand-silt mixtures	Nonplastic fines or fines with low plasticity (for identification procedures see ML below)		
			GC	Clayey gravels, gravel-sand-clay mixtures	Plastic fines (for identification procedures see CL below)		
	Sands — More than half of coarse fraction is smaller than No. 4 sieve size. (For visual classification, the ¼-inch size may be used as equivalent to the No. 4 sieve size.)	Clean sands (little or no fines)	SW	Well-graded sands, gravelly sands, little or no fines	Wide range in grain size and substantial amounts of all intermediate particle sizes		
			SP	Poorly-graded sands, gravelly sands, little or no fines	Predominantly one size or a range of sizes with some intermediate sizes missing		
		Sands with fines (appreciable amount of fines)	SM	Silty sands, sand-silt mixtures	Nonplastic fines or fines with low plasticity (for identification procedures see ML below)		
			SC	Clayey sands, sand-clay mixtures	Plastic fines (for identification procedures see CL below)		

Major Divisions			Group Symbols	Typical Name	Identification procedures on fraction smaller than No. 40 sieve size		
					Dry strength (crushing characteristics)	Dilatancy (reaction to shaking)	Toughness (consistency near PL)
Fine-grained Soils — More than half of material is smaller than No. 200 sieve size.	Silts and Clays — Liquid limit less than 50		ML	Inorganic silts and very fine sands, rock flour, silty or clayey fine sands with slight plasticity	None to slight	Quick to slow	None
			CL	Inorganic clays of low to medium plasticity, gravelly clays, sandy clays, silty clays, lean clays	Medium to high	None to very slow	Medium
			OL	Organic silts and organic silty clays of low plasticity	Slight to medium	Slow	Slight
	Silts and Clays — Liquid limit greater than 50		MH	Inorganic silts, micaceous or diatomaceous fine sandy or silty soils, elastic silts	Slight to medium	Slow to none	Slight to medium
			CH	Inorganic clays of high plasticity, fat clays	High to very high	None	High
			OH	Organic clays of medium to high plasticity, organic silts	Medium to high	None to very slow	Slight to medium
Highly organic soils			Pt	Peat and other highly organic soils	Readily identified by color, odor, spongy feel and frequently by fibrous texture		

(1) *Boundary classifications* Soils possessing characteristics of two groups are designated by combinations of group symbols. For example. Adopted by Corps of Engineers and Bureau of Reclamation, January, 1952.

FIG. 4.72 Unified Classification System (UCS) for soils. Source: Joint Publication of the U.S. Army, Navy, and Air Force, 1971.

Figure 4.74 indicates a graph of the combined analysis results from which the D_{60} and D_{10} UCS values can be obtained.

4.15.4 Atterberg Limits Testing

The Atterberg limits indicate the amount of cohesive material present in the soil and consist of the liquid limit, plastic limit, and shrinkage limit (Lambe, 1951).

The liquid limit is the minimum moisture content, expressed as a percentage of the oven dry soil weight, at which the soil will begin to flow when subjected to a small shearing force. This test is conducted using a standard liquid limit device, which

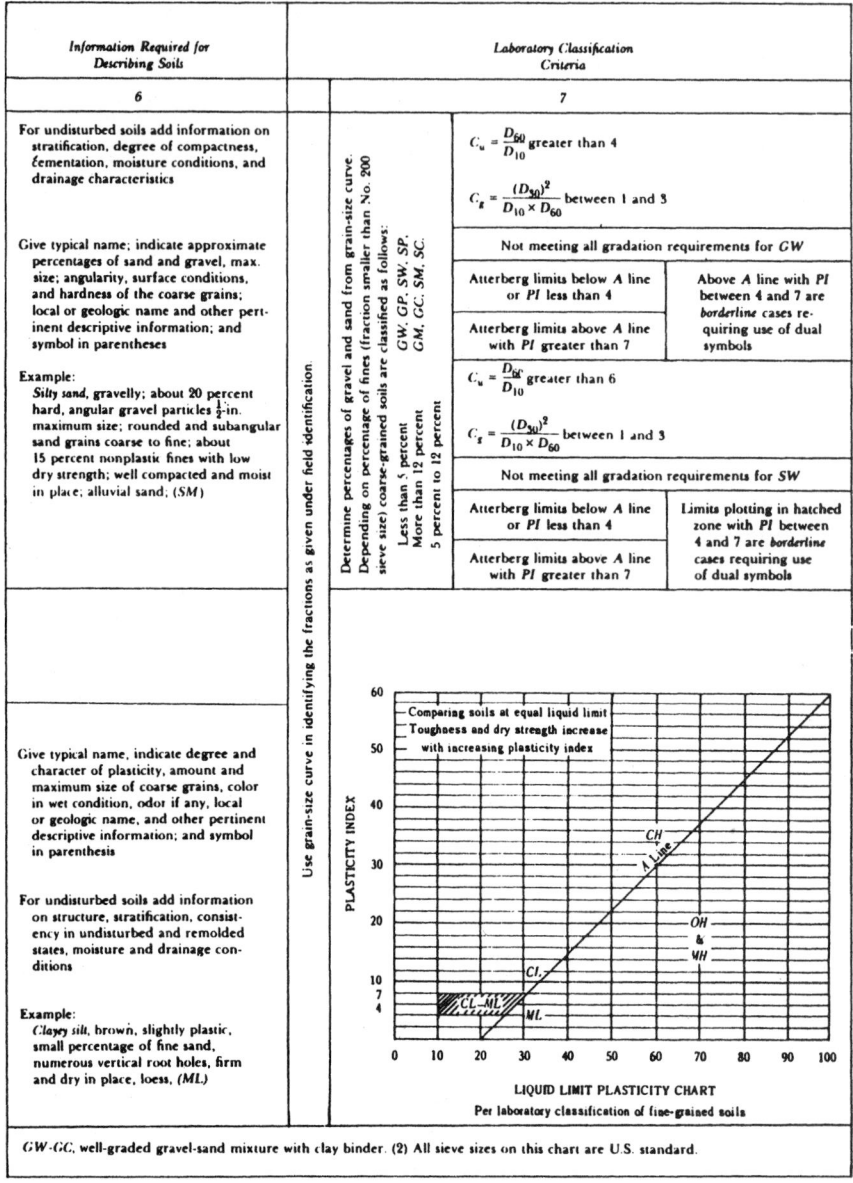

FIG. 4.72 (continued)

consists of a small cup fastened on a pivot that can be lifted and dropped. The soil sample is placed in the shallow cup and grooved with a special tool. The cup and sample are then dropped under the action of a cam that is turned at a rate of two rotations per second until the groove closes for a distance of 0.5 inch. This process is repeated several times while altering the soil's moisture content. A plot of the data is then made to determine the moisture content that would allow groove closure using exactly 25 blows, as illustrated in Figure 4.75.

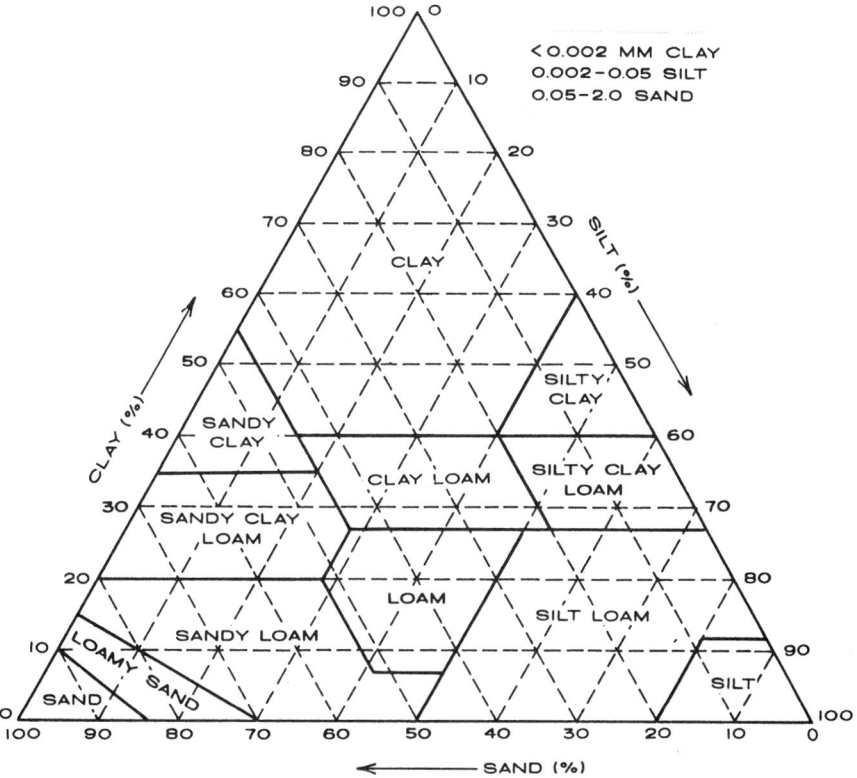

FIG. 4.73 Textural classification chart for soils. Source: Soil Survey Staff, Bureau of Plant Industry, Soils, and Agricultural Engineering. *Soil Survey Manual—USDA Agricultural Handbook No. 18.* Washington: U.S. Govt. Printing Office, 1951.

The plastic limit is the minimum moisture content at which the soil can be rolled into a thread one-eighth of an inch in diameter without crumbling. It is determined by trial and error. Finally, the shrinkage limit is the moisture at which a reduction in moisture will not cause a volume decrease in the soil mass but at which an increase in the moisture will result in a corresponding increase in soil mass. The plasticity index is the difference between the liquid limit and the plastic limit.

4.15.5 Compaction Testing

The requirements of constructing excavations or placing controlled fills within a development to reduce settlement and other considerations necessitate adequate assessment of a soil's compaction capability. Granular soils usually rely on vibration, water, and rolling for best results. Soils containing more cohesive materials rely on pneumatic rubber-tired rollers for silty soils or sheeps foot rollers for plastic soils. Factors affecting a soil's capability of being densified include moisture content, compactive effort exerted, and soil types. Standard and modified Proctor testing will indicate the optimum moisture content (OMC) of a soil passing the No. 10 sieve using a given compactive effort. Using either designated test procedure, samples are compacted at

FIG. 4.74 Example of a grain size distribution curve.

230

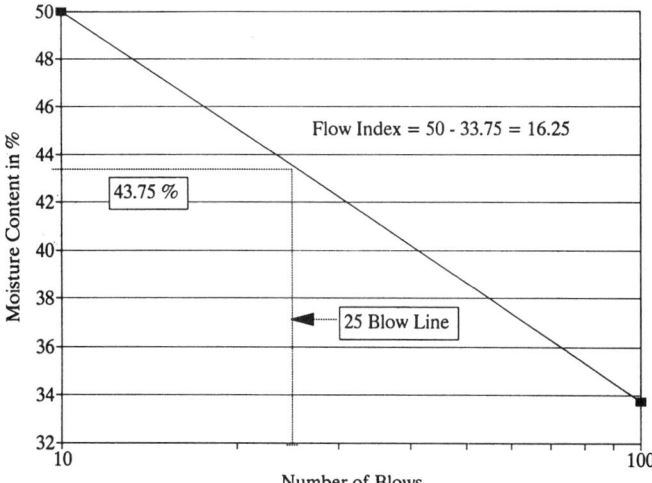

FIG. 4.75 Sample graph for liquid limit determination.

various moisture contents with the resulting equivalent dry unit weights being plotted. The OMC then can be determined at the point of maximum dry unit weight, as shown in Figure 4.76.

Results of the compaction test provide the engineer with information for proper field control during construction.

4.15.6 Other Soil Tests

Although there are many additional tests available to analyze soils, four of them are applicable to land developments that do not include structures requiring deep foundations. These tests are:

(a) The field density test, which is used to determine a soil's density either in its natural state or in a fill section. The test provides an estimate of compaction strength or checks results of field compaction. This test typically is conducted using a nuclear density meter or a sand cone apparatus.

(b) The California bearing ratio (CBR) test can be conducted either in the field or in the laboratory. In laboratory tests the soil is compacted to anticipated field conditions and then soaked to duplicate field moisture. A surcharge weight is placed on top of the sample to simulate overburden pressures; then a 1.95-inch-diameter piston is forced into the prepared soil sample at a rate of 0.05 inch per minute. Corresponding loads for 0.1-, 0.2-, 0.3-, 0.4-, and 0.5-inch penetrations are recorded and plotted. From this information a corrected load value representing the 0.1-inch penetration is obtained. The CBR value is then defined as

$$CBR(\%) = \frac{\text{unit load for 0.1-inch penetration} \times 100}{1000} \qquad (4.21)$$

FIG. 4.76 Example of moisture density relationships for a given compaction effort. Source: Paul H. Wright, with contributions by James S. Lai, Peter S. Parsonson, and Michael Meyer. *Highway Engineering*, 6th ed. Copyright © 1996 by John Wiley & Sons, Inc. Reprinted by permission of John Wiley & Sons, Inc.

Thus the relative shear strength of a soil can be compared to a standard material's strength, which in this case is crushed stone with a corresponding bearing strength of 1000 pounds at 0.1-inch penetration (Wright and Paquette, 1987).

(c)   The resistance R value of compacted soils can be determined in the laboratory using procedures contained in AASHTO T 190-93 (AASHTO, 1997b). Four soil specimens 101.6 mm (4 in.) in diameter by 63.5 mm (2.5 in.) high are formed by compacting them at various moisture contents to anticipated field conditions using a mechanical kneading compactor that simulates the kneading action soil is subjected to during field compaction. Each sample is then subjected to a uniformly increasing pressure using a compression testing machine. The pressure required to cause exudation of water from the sample is the "exudation pressure." The moisture content used to prepare the specimens should yield exudation pressures that bracket a value of 2068 kPa (300 psi). Each sample is allowed to rebound and then is placed in an expansion pressure device for 16–24 hours. This device measures the expansive force exerted by the swelling soil. The resistance R value can then be obtained for each specimen by placing it in an Hveem stabilometer where lateral specimen deformation caused by vertical loadings results in compressive lateral pressures being developed. Materials can be rated from 0 to 100 based on their ability to limit lateral pressure from developing. For example, an R value of 0 indicates a liquid.[*] [The R value used for pavement thickness design can be estimated from either the R value that corresponds to an exudation pressure of 300 psi or the R value based on the thickness of pavement required to maintain the soil's compaction as described in MS-10 (Asphalt Institute, 1991)].

(d) Permeability testing can yield an indication of how quickly water is capable of flowing through the soil structure.

4.15.7 Additional Considerations

The subsurface site investigation should include an examination of soil conditions that will allow the determination of safe slope angles for stability, if applicable. Volume swelling and shrinkage factors also are required for earthwork mass diagram balances. Some sites contain expansive clays that could be detrimental to foundations and road subgrades, while other sites may have sinkholes. Other concerns might focus on the investigation of previously low areas where logs and stumps are covered with soil, which could cause problems. The potential for soil liquefaction resulting in mudslides must also be investigated, along with earthquake fault data.

4.16 TRAFFIC STUDIES

Most transportation systems are studied on a local or regional basis, as indicated in Chapter 1, consistent with federal and state planning. Included is a classification inventory of existing streets and highways into usage categories such as arterial, collector, or local streets. Existing traffic studies include traffic volume counts, travel time studies, street capacities, accident studies, parking studies, and traffic control device studies. Land use changes resulting from development cause a change in traffic activity in terms of trip generating and reduced levels of service on existing facilities. Figure 4.77 illustrates the interrelationship between land use and traffic activity. One of the most important goals of a designer is to minimize the degradation of service levels on transportation networks.

4.16.1 Traffic Associated with Development

Land developments include improvements that act as traffic generators affecting servicing traffic facilities. Multifamily developments containing commercial areas are of concern and may require the anticipated traffic demands to be evaluated to determine how the facilities will perform.

---

[*] From *Standard Specifications for Transportation Materials and Methods of Sampling and Testing, Part II— Tests*, Copyright © 1997, by the American Association of State Highway and Transportation Officials, Washington, D.C. Used by permission.

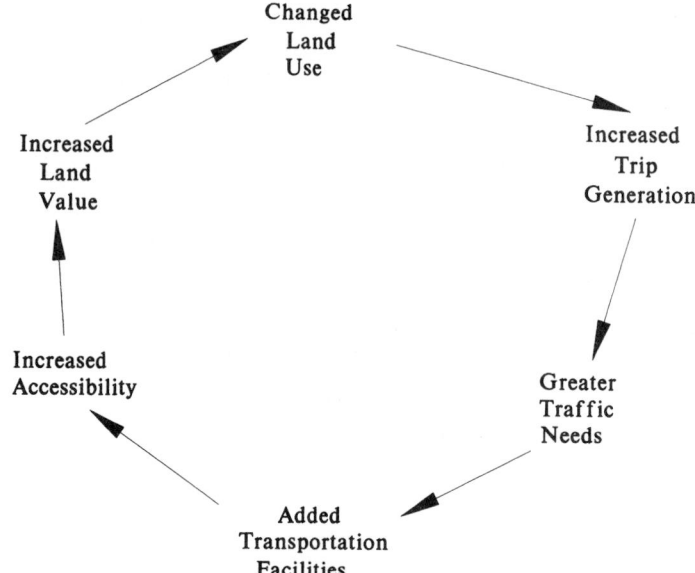

FIG. 4.77 Land use transportation cycle. Source: Paul H. Wright, Norman J. Ashford, and Peter J. Stammer Jr. *Transportation Engineering, Planning, and Design*, 4th ed. Copyright © 1998 by John Wiley & Sons, Inc. Reprinted by permission of John Wiley & Sons, Inc.

Proper evaluation techniques require the determination of existing and future traffic conditions, including existing peak hour levels of service within the affected area. A development's anticipated contribution would be included when analyzing the effect of traffic on the system's carrying capability. Should the existing system indicate inadequate future capacity, various improvement alternatives must be investigated to determine the best course of action.

Some information required to undertake an analysis such as this includes

(a) proposed development site plan showing the proposed system and internal street layouts and the anticipated average daily traffic (ADT) volumes obtained from the size and type of development and trip generation factors;

(b) location and characteristics of all traffic accesses to the proposed site, including intersections used currently and those planned; and

(c) an analysis of service levels using conditions given to various system inputs to accommodate future requirements.

4.16.2   Highway Capacity Studies

Highway capacity is defined as the maximum hourly rate at which vehicles reasonably can be expected to pass through a section of roadway during a given time period under prevailing roadway, traffic, and control conditions (TRB, 1994).* Road-

---

*Taken from Special Report 209: Highway Capacity Manual, Third Edition, Copyright 1994 by the Transportation Research Board, National Research Council, Washington, DC.

way conditions usually refer to street geometry, including alignment, geometric cross section, and geographic location, whereas traffic conditions refer to the mixture and characteristics of the flow stream. Control conditions refer to traffic regulatory devices. These conditions affect the level of service of a given roadway ranging from "A," which consists of free-flow conditions, to "F," which denotes congestion. The service flow rate is the facility's capacity when a designated level of service is maintained in hourly flow measured at 15-minute intervals. Service flow rates that have no traffic signals or intersections operate as uninterrupted flows, whereas the presence of signals and intersections results in interrupted flows. Uninterrupted flows are affected by various roadway factors such as alignment, grades, surface riding quality, shoulder extent, and lane widths. Uninterrupted flows also are affected by various traffic factors including the mixture and direction of trucks, buses, and passenger vehicular volumes, including driver characteristics. Urban streets similar to those associated with land developments would operate under interrupted flow conditions and would be affected by the surrounding land use. The number of available traffic lanes and their widths, street alignment, and existence of turning lanes or parking characteristics affect the flow. Interrupted-flow traffic factors could be dependent on traffic volumes and peaking factors, pedestrian interfaces, public transportation facilities, and parking activities. Control factors usually include signal timing (Transportation Research Board, 1994).

A Policy on Geometric Design of Highways and Streets—1994 (AASHTO, 1994) provides characteristics associated with the six levels of service associated with several highway types, of which the urban and suburban arterial characteristics are shown in Figure 4.78.

Level Service	Characteristics of Traffic
A	Average travel speed of about 90 percent of free flow speed. Stopped delay at signalized intersections is minimal.
B	Average travel speeds drop due to intersection delay and intervehicular conflicts, but remain at 70 percent of free flow speed. Delay is not unreasonable.
C	Stable operations. Longer queues at signals result in average travel speeds of about 50 percent of free flow speeds. Motorists will experience appreciable tension.
D	Approaching unstable flow. Average speeds down to 40 percent of free flow speed. Delays at intersections may become extensive.
E	Average travel speeds 33 percent of free flow speed. Unstable flow. Continuous backup on approaches to intersections.
F	Average travel speed between 25 and 33 percent of free flow speed. Vehicular backups, and high approach delays at signalized intersections.

FIG. 4.78 Level-of-service characteristics for urban and suburban arteries. Taken from *A Policy on Geometric Design of Highways and Streets*. Copyright © 1994 by the American Association of State Highway and Transportation Officials, Washington, DC. Used by permission.

During the preliminary stages of project planning, only an indication of potential traffic problems can be ascertained since the project's scope has not been fully established. Once the preliminary design has been refined, a definitive impact analysis of the traffic can be made as described in Chapter 6.

4.17   MATERIALS EVALUATION

On occasion, a development might include the use of novelty or specialty materials that may require testing and evaluation to ensure adequate performance. Various materials testing firms are available to undertake such evaluations as auxiliary project consultants.

4.18   ANALYSIS AND REPORT ON PROSPECTIVE SITES

Background data pertaining to all possible sites must be gathered and assembled in a manner conducive to thorough analysis. Information from site inspections and other background data that could be in various stages of completion are useful to the design professional when generating a proposed site plan that satisfies the developer's needs and regulatory requirements.

4.18.1   Developing a Schematic Site Plan

The design professional starts by taking a piece of tracing paper and placing it over a scaled display of the site boundary that also contains the topography. Significant areas are delineated to indicate wetlands, drainage facilities, positions for improved views, areas for potential open space, and so on. The site sketch is useful to assist the designer in this endeavor, along with other information.

Once controlling factors have been indicated on this initial overlay, another tracing paper sheet is placed as before and the design professional sketches to approximate scale a layout as specified by the developer. There are probably many things that appear on the sketch that indicate a poor design. If so, another fresh piece of tracing paper must be substituted to allow sketching yet another configuration. In the resketching process insufficiencies may again develop, necessitating even yet another approach. This process requires skill and knowledge, and it may take 10–20 tries before a product of quality is otained. Figure 4.79 demonstrates two plans for a particular site. A precise drawing can be made once a desirable solution is attained. During this planning process colored pencils are used to great advantage to indicate various utilities or areas of different land use. They add perception to the drawing and often, because of the resulting contrasting effect, flaws are noticed.

Utilization of CAD systems to generate schematic plans can parallel the manual approach just described. The major difference is that the tracing paper is replaced by an electronic drawing "layer."

The need for drawing these preliminary plans to scale is apparent when one counts the number of lots or dwelling units contained in a resulting configuration. If no scale were used, a piece of leftover land thought to be a great recreational area may not even exist because the actual lot sizes or road rights-of-way might result in no excess area. At true scale, such fallacies in the plan are revealed. Additional complexities resulting from

FIG. 4.79 Examples of several site layout studies for a single-family development: (*a*) condominium concept plan; (*b*) villa concept plan. Courtesy of Post, Buckley, Schuh, and Jernigan, Inc., Tampa, FL, 1992.

237

FIG. 4.79 (continued)

238

interrelationships of roads to car storage to dwelling unit to recreational areas necessitate the use of specialists such as architects and land planners. In group developments or multifamily developments problems may arise concerning space relationships, solar lighting, and view. These are prime considerations a specialist must address. By doing so, the specialist will upgrade the project from mediocre to excellent.

Each potential site's conceptual plan will be assessed by the engineer regarding how well it meets assigned criteria. All beneficial features as well as potential problems are summarized in a comprehensive report to the developer that includes rough cost estimates. The developer can study the various alternatives to establish a site of choice. Once the project's site is chosen, the design engineer can finalize a preliminary development plan by improving the schematics that pertain to that site. This plan will be submitted to the jurisdictional regulatory review agencies in order to seek permission to proceed with the design of the project. The estimated construction cost is prepared from the preliminary plan and used as a basis for financial feasibility analysis by the developer.

REFERENCES

American Association of State Highway and Transportation Officials (AASHTO). *A Policy on Geometric-Design of Highways and Streets—1994.* Washington: AASHTO, 1994, pp. 90–91.

American Association of State Highway and Transportation Officials (AASHTO). *Recommended Practice for Classification of Soils and Soil-Aggregate Mixtures for Highway Construction Purposes—AASHTO M145-91 (1995). Standard Specifications for Transportation Materials and Methods of Sampling and Testing—Part I, 18th.* Washington: AASHTO, 1997a, pp. 158–162.

American Association of State Highway and Transportation Officials (AASHTO). *Standard Specification for Resistance R-Value and Expansion Pressure of Compacted Soils—AASHTO T190-93). Standard Specifications for Transportation Materials and Methods of Sampling and Testing—Part II, 18th.* Washington: AASHTO, 1997b, pp. 584–590.

American Congress on Surveying and Mapping and American Land Title Association (ACSM/ALTA). *Accuracy Standards for ALTA-ACSM Land Title Surveys.* Bethesda: ACSM, 1999a.

American Congress on Surveying and Mapping and American Land Title Association (ACSM/ALTA). *Minimum Standard Detail Requirements for ALTA-ACSM Land Title Surveys.* Bethesda: ASCM, 1999b.

American Society for Testing and Materials (ASTM). *E1527-97 Standard Practice for Environmental Site Assessments: Phase 1 Environmental Site Assessment Process.* Philadelphia, ASTM, 1997a.

American Society of Civil Engineers (ASCE). *Environmental Site Investigation Guidance Manual—ASCE Manuals and Reports on Engineering Practice No. 83.* Reston: ASCE, 1996.

Asphalt Institute. *Soils Manual* (MS–10). 5th ed. Lexington, KY: Asphalt Institute, 1991.

Davis, R. E., F. S. Foote, J. M. Anderson, and E. M. Mikhail. *Surveying Theory and Practice*, 6th ed. New York: McGraw-Hill, 1981.

Dorchester County Tax Assessor. *Tax Map Showing a Portion of the Town of Summerville, S.C.* St. George, SC: Dorchester County Government, 1980.

Dunkle, F., R. Hanmer, R. W. Page, and W. Scaling. *Federal Manual for Identifying and Delineating Jurisdictional Wetlands.* Washington: U.S. Government Printing Office, 1989, pp. 5, 23–25, B-2.

Federal Emergency Management Agency (FEMA). *Flood Insurance Rate Map: 450073—Town of Summerville, South Carolina.* Washington: Federal Insurance Administration, 1985.

Federal Emergency Management Agency (FEMA). *Guide to Flood Maps—How to Use a Flood Map to Determine Flood Risk for a Property (FEMA 258/May 1995)*. Washington: FEMA, 1995.

Hothem, L. D., D. B. Zilkoski, and R. J. Fury. *Orthometric Height Determination with GPS Satellite Surveys*. Rockville, MD: National Geodetic Survey.

Joint Publication of the U.S. Army, Navy, and Air Force. *Materials Testing—TM 5-530/ NAV-FAC MO-330/AFM 89-3*. Washington: U.S. Government Printing Office, 1971.

Keen, J. E. *ALTA/ACSM Land Title Surveys—The Cadillac of Surveys*. Spruce Pine, NC: Land Surveyor's Workshops, 1999.

Lambe, T. W. *Soil Testing for Engineers*. New York: Wiley, 1951, pp. 30, 43–49, 152.

Lapine, L. A. *Analytical Calibration of the Airborne Photogrammetric System Using A Priori Knowledge of the Exposure Station Obtained from Kinematic Global Positioning System Techniques*. Report No. 411. Columbus: Ohio State University, 1991.

Lillesand, T. M., and R. W. Kiefer. *Remote Sensing and Image Interpretation*. New York: Wiley, 1979, p. 312.

Moffitt, F. H., and H. Bouchard. *Surveying*. 8th ed. New York: Harper and Row, 1987.

Moskowitz, J. S. *Environmental Liability and Real Property Transactions: Law and Practice*. New York: Wiley, 1989, pp. 313–330.

National Oceanic and Atmospheric Administration (NOAA). *State Plane Coordinate Systems of 1983: NOAA Manual NOS NGS 5*. Rockville: National Geodetic Information Center, 1990, pp. 1–7, 18–20, 46–53.

National Oceanic and Atmospheric Administration (NOAA). *Guidelines for Establishing GPS-derived Ellipsoid Heights—NOAA Technical Memorandum NOS NGS-58*. Rockville: National Geodetic Information Center, 1997.

Seelye, E. E. *Data Book for Civil Engineers—Design Volume I*. New York: Wiley, 1960, pp. 9–25 to 9–27. Copyright © 1960 by John Wiley & Sons, Inc. Reprinted by permission of John Wiley & Sons, Inc.

Transportation Research Board (TRB). *Highway Capacity Manual*, 3rd. ed. Special Report 209. Washington: TRB, National Research Council, 1994, pp. 1-1 to 1-14.

U.S. Army Corps of Engineers (COE) and South Carolina Coastal Council. "Developer's Handbook for Freshwater Wetlands." An unpublished handbook. Charleston: South Carolina Coastal Council, 1991.

U.S. Department of Housing and Urban Development (HUD). *The Noise Guidebook, HUD-953-CPD(1)*. Washington: HUD, 1991.

U.S. Department of Interior, Bureau of Land Management. *Manual of Instruction for the Survey of Public Lands of the United States—1973*. Washington: U.S. Government Printing Office, 1988, pp. 1–11, 59–76.

U.S. Department of Transportation. *Highway Traffic Noise Analysis and Abatement Policy and Guidance*. Washington: USDOT, 1995.

U.S. Environmental Protection Agency (EPA). *Protective Noise Levels—Condensed Version of EPA Levels Document*. Washington: EPA, 1978.

Weldon, C., Frank, Kolter, Dannemeyer, Bliley, Chapman, Lagomarsino, and James. *H.R. 2787—A Bill to Amend the Comprehensive Environmental Response, Compensation and Liability Act of 1980*. 101st Congress, 1989.

Wolf, P. R., and C. D. Ghilani. *Adjustments Computations—Statistics and Least Squares in Surveying and GIS*. New York: Wiley, 1997, pp. 156, 157, 299, 407.

Zilkoski, D. B. "North American Vertical Datum and International Great Lakes Datum: They Are Now One and the Same." Publication No. 28, U.S. Hydrographic Conference 1992. Baltimore: Hydrographic Society, 1992, pp. 68–74.

Zilkoski, D. B., J. H. Richards, and G. M. Young. "Results of the General Adjustment of the North American Vertical Datum of 1988." *Journal of the American Congress on Surveying and Mapping—ACSM Surveying and Land Information Systems*, September 1992.

SUGGESTED READINGS

American Association of State Highway and Transportation Officials (AASHTO). Highway Subcommittee on Design—Task Force for Environmental Design. *Guide On Evaluation and Abatement of Traffic Noise*. Washington: AASHTO, 1993.

American Society for Testing and Materials (ASTM). *ASTM Standards on Environmental Site Assessments for Commercial Real Estate, 3rd. E 1527-97 and E 1528-96*. Philadelphia: ASTM, 1997b.

American Society of Civil Engineers (ASCE). *Engineering Surveying Manual—ASCE Manuals and Reports on Engineering Practice No. 64*. Reston: ASCE, 1985.

American Society of Civil Engineers (ASCE). *Right-of-Way Surveying Manual—ASCE Manuals and Reports on Engineering Practice No. 75*. Reston: ASCE, 1991.

Asante-Duah, D. K. *Managing Contaminated Sites: Problems, Diagnosis, and Development of Site Restoration*. New York: Wiley, 1996.

Crawford, W. G. *Construction Surveying and Layout—A Step-By-Step Field Engineering Methods Manual*. 2nd. ed. West Lafayette, IN: Creative Construction Publishing, 1995.

Crazier, L. *Surveys and Surveyors of the Public Domain 1785–1975*. Washington: U.S. Government Printing Office, 1975.

Cuniff, P. F. *Environmental Noise Pollution*. New York: Wiley, 1977.

Davis, Raymond E., Francis S. Foote. *Surveying Theory and Practice, 4th*. New York: McGraw Hill, 1962, pp. 198–203.

DeMers, M. N. *Fundamentals of Geographic Information Systems*. New York: Wiley, 1997.

Dion, T. R., T. W. Bailey, R. D. Branton, J. B. Priestner, J. E. Chellis, S. B. Jones, D. Sherwood Miler, R. Still, S. Seegers, and B. Dukes. *1997 Detmold Baseline Readjustment Map*. Summerville, SC: Town of Summerville, 1997.

Graphic Mapping Group, INC. *Utilimap*. An unpublished flyer. 2001 Asheville Highway, Hendersonville, N.C.: Graphic Mapping Groups.

Leick, A. *GPS Satellite Surveying*, 2nd ed. New York: Wiley, 1995.

Lillesand, T. M., and R. W. Kiefer. *Remote Sensing and Image Interpretation*, 3rd ed. New York: Wiley, 1994.

MapInfo Corporation. One Global View, Troy, NY.

Marriott, B. B. *Practical Guide to Environmental Impact Assessment*. New York: McGraw-Hill, 1997.

McCormac, J. C. *Surveying Fundamentals*. 2nd ed. Englewood Cliffs, NJ: Prentice-Hall, 1991.

Moffitt, F. H., and E. M. Mikhail. *Photogrammetry*. 3rd ed. New York: Harper and Row, 1980.

Muskett, J. *Site Surveying*. 2nd ed. Oxford: Blackwell Science, 1995.

Payne, S. M. *Strategies for Accelerating Cleanup at Toxic Waste Sites—Tracking Environmental Actions and Decision Making*. New York: Lewis, 1998.

Reilly, J. P. " 'The GPS Observer'—A Series of Articles Published in the POB Magazine." In *POB Point of Beginning*. Troy, MI: Business News Publishing Company, 1994–2000.

Sellers, K. *Fundamentals of Hazardous Waste Site Remediation*. New York: Lewis, 1999.

Sowers, G. F. *Building on Sinkholes—Design and Construction of Foundations in Karst Terrain*. Reston: ASCE, 1996.

Spielberg, F. "The Traditional Neighborhood Development: How Will Traffic Engineers Respond?" *ITE Journal*, September 1989, pp. 17–18. Reference material is reprinted by permission of The Institute of Transportation Engineers.

Starplus Software. *STAR*NET™, Version 5.0*. Oakland, CA; Starplus Software, 1997.

Terzaghi, K., and R. B. Peck. *Soil Mechanics in Engineering Practice*, 2nd ed. New York: Wiley, 1967.

U.S. Army Corps of Engineers (COE). *Photogrammetric Mapping—Technical Engineering and Design Guidelines as Adopted from the U.S. Army Corps of Engineers, No. 14*. Reston: ASCE, 1996.

U.S. Department of the Interior. *National Wetlands Inventory Map (NWI)—Bennetts Point, S.C.* Atlanta: U.S. Fish and Wildlife Service, 1989.

U.S. Department of Transportation. *Federal Aid Highway Program Manual*, Vol. 7, Ch. 7 Sect. 3, Transmittal 348, August 9, 1982.

U.S. Environmental Protection Agency (EPA). *Protecting Natural Wetlands—A Guide to Stormwater Best Management Practices*. Washington: EPA, 1996.

U.S. Geological Survey. *7.5 Minute Series Map of Summerville, South Carolina*. Washington: U.S. Geological Survey, 1990.

Van Sickle, J. *GPS for Land Surveyors*. Chelsea: Ann Arbor Press, 1996.

Vernon, R. C. *Professional Surveyor's Manual*. New York: Mc-Graw Hill, 1997.

Ward, C. H., J. A. Cherry, and M. R. Scalf. *Subsurface Restoration*. Chelsea: Ann Arbor Press, 1997.

Wolf, P. R. *Elements of Photogrammetry*, 2nd ed. New York: McGraw-Hill, 1983.

5 Preliminary Design and Costs

Developers often question information in the report that summarizes various alternative conceptual plans and costs. This largely results from a lack of the details and complete information expected in final construction documents. As a result, the engineer can be expected to clarify and amplify information contained in the report, allowing the developer to analyze and consider alternatives. On occasion, preliminary inquiries made by the engineer to jurisdictional authorities while seeking initial conceptual guidance may trigger follow-up agency clarification inquiries requiring the engineer to respond. Resulting information can affect various conceptual assumptions. The developer can choose the most appealing project conceptual plan after reviewing the report. Design professionals can then be directed by the developer to prepare all required preliminary project documents necessary for regulatory and financial institution approval.

5.1 DETERMINATION OF ADDITIONAL AUXILIARY SERVICE

During the report review, the developer and/or engineer may determine a need for additional background studies. The scope of these additional services must be defined in a manner that will result in the studies yielding adequate information. The summary report may need to be modified by the engineer as a result of these activities; nevertheless, the developer normally continues analysis until the best conceptual plan and/or site is determined.

5.2 CONCEPTUAL PLAN APPROVAL

Once the developer decides on the project's conceptual plan, a conference usually follows that culminates with the developer, engineer, land planner, architect, landscape architect, marketing specialists, financial analysts, and lenders reaching agreement on the project's preliminary design features and phasing requirements. This agreement usually is put in writing to eliminate possible misunderstandings, especially if variable conditions prevail that could alter the project's scope. If the project is small, the conceptual agreement can allow the developer to commence activities to procure title to the desired project site. However, if the project is over a certain threshold where a large portion of the locale has the potential of being affected by the proposed development, jurisdictional regulations might require the developer to supply a regional impact report for approval. This report will include information pertaining to the development such as maps, environmental factors, transportation factors, and factors pertaining to housing, education, recreation, health, police, fire, natural hazard mitigation, and

243

preservation. Once this report is approved, the preparation of the preliminary project documents can begin.

5.3 PREPARATION OF PRELIMINARY DOCUMENTS

The engineer can now refine the applicable conceptual scheme and make additional preliminary studies or analyses necessary to formulate a preliminary development plan, or preliminary plat, suitable for regulatory authority reviews. Approval of the preliminary plan allows the developer to pursue preparation of construction plans through the action of the design professional.

5.3.1 Preliminary Development Plans

Depending on the site size, availability of financing, and market absorption potential, the approach used by developers can vary from project to project. For a small project a single preliminary plan may be sufficient for regulatory and financial review; however, on large projects that are phased developments, several plans may be required. These might include a comprehensive master development plan, neighborhood plans, or preliminary development plans for each phased section. Typical plans are illustrated in Figures 5.1–5.3. Supplemental plans might contain master development plans for storm drainage, wastewater, transportation, and potable water systems. These plans usually address major system components that will allow subsequent development phases. Many regulatory agencies want information indicating the interrelationship between the various proposed development improvements that are contained in the preliminary plan. Local planning agencies usually require this plan to show the property's metes and bounds along with all easements and rights-of-way. The property area and adjacent landowners are shown with information indicating how their boundary intersects the development. All proposed land divisions within the development must be drawn to scale to indicate new road rights-of-way, roadway widths, street names, locations and approximate sizes of proposed utilities, lot lines, dimensions, reserved areas, floodways, wetlands, and areas for recreational, religious, educational, or public use. Land use area summaries usually are included for land use density analysis. The plat should include a title, graphic and numeric scale, north arrow with reference meridian, date, and names of the developer, engineer, and land surveyor.

The preliminary plan must delineate all existing physical features, including contour lines referenced to an approved datum, drainage features, and wooded areas. The plan also should contain all proposed storm drainage facilities with the anticipated finished surface flows being delineated with directional arrows, in addition to flows in proposed ditches and retention/detention facilities.

5.3.2 Preliminary Cost Estimate

On completion of the preliminary plans, a more exact, but still preliminary, cost estimate can be prepared by the engineer. This estimate is intended to reasonably ascertain the cost of the development; however, it is subject to several variables that make accuracy difficult. For example, larger projects can offer cost savings attributed to economy of scale resulting in the estimated construction costs being higher than actual costs.

FIG. 5.1   Example of a project's conceptual master plan. Courtesy of Post, Buckley, Schuh, and Jernigan, Inc., Tampa, FL, 1992.

FIG. 5.2 Example of a preliminary site development plan. Courtesy of Post, Buckley, Schuh, and Jernigan, Inc., Tampa, FL, 1992.

FIG. 5.3 Example of a final plat. Courtesy of Post, Buckley, Schuh, and Jernigan, Inc., Tampa, FL, 1992.

247

Also, the site's location and condition can affect actual costs of improvements, as can various seasonal restrictions placed on certain types of construction. The uncertainty of regulatory review comments could result in required plan revisions that might alter the cost of the project. The prevailing economic climate also can affect final construction costs resulting in large pricing fluctuations.

This estimate should specify any anticipated expenses not included as part of the estimate to alert the developer before they occur. Procedures for preparing a preliminary cost estimate parallel those contained in Chapter 14 for the final cost estimate.

5.3.3 Specification Outline

Preliminary activities include preparation of a specification outline for the proposed project. This outline should contain significant specification elements to be included in the final draft. Construction specifications are more fully covered in Chapter 13.

5.3.4 Project Synopsis

Most preliminary documents include a synopsis that describes various aspects of the proposed project. The synopsis is useful to the developer, regulatory review agencies, and financial lending institutions. It can include various components of the previously prepared summary report; however, more explicit detail is presented to focus on only one plan, as indicated by Figure 5.4.

5.4 PRELIMINARY PROJECT SUBMITTALS

Prior to releasing copies of the preliminary plan, construction specification outline, cost estimate, and proposed project synopsis, the engineer should double-check every document to ensure completeness and to verify that no obvious errors exist.

5.4.1 Developer Approval

The first outside review of the preliminary documents is usually performed by the developer. This review is important to ensure that the preliminary project appeals to the developer with regard to theme, estimated cost, and anticipated regulatory compliance. Any concerns expressed by the developer during this review are usually minor because previous communications have allowed prior concerns to be incorporated in these documents.

Once the developer has satisfactorily reviewed the documents, the engineer, acting for the developer, can petition all required regulatory authorities and request preliminary project approval. Application fees for petitions are paid by the developer.

5.4.2 Regulatory Agency Approval

Regulatory agencies that review development plans may include the local planning board, the health department, municipal engineer, building inspector, zoning officer, and jurisdictional utility companies. A typical procedure for local project approval, shown previously in Figure 2.28, varies from locale to locale; however, all regulatory

PROJECT NAME

Preliminary Report
1. Property Identification
    (a) Current owner.
    (b) General tract identification (e.g. tract number, tax map number, etc.).
    (c) General location.
2. Background Information
    (a) Reduced copy of boundary plat.
    (b) Reduced copy of topographic map.
    (c) Description of the property.
    (d) Zoning classification.
    (e) Wetland findings.
    (f) Phase I Environmental Assessment results.
    (g) Historical and Natural Resource findings.
    (h) Potable water and wastewater factors.
    (i) Soil and geologic conditions.
    (j) Transportation factors.
    (k) Other factors:
        (1) Natural disasters (earthquakes, hurricanes, etc.).
        (2) Landslide and mudslide potential.
        (3) Noise.
        (4) Air pollution.
3. Proposed Development Strategy
    (a) Theme.
    (b) Project configuration and phasing.
    (c) Required improvements.
    (d) Estimated construction costs.
4. Project Summary
    (a) Potential problems.
    (b) Presubmission regulatory comments.
    (c) Estimated schedule for construction document preparation.
    (d) Suggestions.

FIG. 5.4 Example of a synopsis of a land development project.

review processes require time. Maintaining lines of communication is helpful to allow timely discussions for satisfactory reconciliation of problems.

5.4.3 Financial Lending Agencies Reviews

Information in the preliminary documents is important to a financial lending institution's review process. The engineer usually provides copies of these documents to the institution while the developer makes all other arrangements. Revisions necessitated by regulatory agency comments must be reflected in an updated set of documents that would be redistributed to the lender.

5.5 PRELIMINARY PROJECT APPROVAL

Approval of the preliminary documents by regulatory agencies establishes a common understanding of the project's scope and direction, allowing the developer to secure financing. Preliminary approval enables jurisdictional regulatory authorities to anticipate improvements that are compatible with established systems. This approval also allows the engineer to move into the final project design phase when construction plans and specifications are prepared.

PART III
Final Design-Related Activities

Transforming information on the preliminary plan into construction drawings is called design. Design is based on theory, technical knowledge, and practical experience. The ability to design is acquired by personal involvement and the willingness of an experienced designer to provide guidance. Many project designs can be formulated for a site; however, there are usually a limited number that best satisfy regulatory, financial, and project goal constraints.

The design process is iterative, with various elements being incorporated into a central project core, which then provides the framework for subsequent attachment of design details. Adjustments may be required to attain the proper relationship between the various designed components. This requires the designer to constantly check and reevaluate the overall project. At any stage during the project's design, other system alternatives that are subsequently designed may cause adverse impacts on previously designed components. This requires the designer either to make adjustments to the previously designed features or to modify the features currently being designed (or both). Refinements are therefore required as the process continues to completion.

Design activities also include preparation of construction specifications. (Project Manual), a final cost estimate, and document review by regulatory agencies for construction approval.

6 Transportation System Design

The limiting feature of a land development is the site's property boundary. It is within this boundary that the designer must work to fabricate individual system components and appurtenances. If the boundary is not properly defined as described in Section 4.11, then the designed systems will not fit when the project is constructed, resulting in problems.

The common thread that weaves through the entire project is the transportation system. According to the Institute of Transportation Engineers (ITE, 1993), transportation considerations in land developments include the layout of streets and pedestrian systems (as related to land use) and also the engineering dimension for vehicular, pedestrian, and bicycle facilities. Both must be addressed during design whether the project is categorized as a suburban or neotraditional type development. [Note: ITE's (1999) *Traditional Neighborhood Development—Street Design Guidelines* should be referred to for neotraditional applications.] The importance of designing the transportation system is that all building sites are contiguous to it, most utilities are located within it, and the storm drainage is usually affected by it. It is therefore necessary, in most cases, to initially establish horizontal and vertical control of the streets prior to undertaking the sequential design of the other site improvements, including the storm drainage system, followed by the sanitary sewer system, and finally the potable water system. An example of an exception to this normal sequencing would be if there is a limiting tie-in elevation critically affecting the design of the storm drainage or sanitary sewer system.

6.1 TRAFFIC-RELATED SITE IMPACT STUDIES AND ACCESS

Traffic-related site impact studies undertaken by qualified persons can be conducted using guidelines similar to those included in ITE's (1991) *Traffic Access and Impact Studies for Site Development*. These recommended practices are presented in this section. [Other guidelines are available, including the U.S. Department of Transportation–Federal Highway Administration's (1985) *Site Impact Traffic Evaluation (S.I.T.E.) Handbook*.] Some reasons for undertaking studies such as this include safety, accessibility, and determination of street improvement needs. Increased trip generation and parking demands are additional reasons to conduct a study. Study results should help maintain satisfactory levels of transportation service; however, site-related studies are not satisfactory as substitutes for areawide or regional transportation studies where many sites must be evaluated over long time periods. The overall purpose of a site study is therefore to determine what effect the proposed development will have on the transportation plans that exist to accommodate future growth. This includes determining the time when peak traffic flow is expected to occur both on and contiguous to

253

the site, because peak site traffic can vary from that of the adjoining road. According to ITE guidelines, one reason a traffic impact study may be required occurs when the site's change in land use results in at least 100 additional peak direction trips to or from the site during the abutting street's peak hour or the proposed development's peak hour (ITE, 1991). Other reasons for conducting the study may be associated with applications for zoning changes, driveway encroachment permits, and building permits. Such accumulated traffic data are necessary not only for geometric design but also for structural design on the pavement. Safety, traffic flow, and structural integrity are basic objectives of a comprehensive traffic analysis program.

Section 4.16 indicates that a definitive traffic impact study can be undertaken once the preliminary plan is formulated. This is because the various on-site land uses are delineated. The traffic engineer should begin the actual study process, however, during the inventory, analysis and reporting phase (Chapter 4) when other factors associated with the proposed development are being investigated. The scope of the traffic impact study should be established and agreed to by the jurisdictional regulatory review agency because not all studies require the same magnitude of detail. Initial study activities nevertheless should include the establishment of study area limits based on abutting streets, major intersections, and street segments up to the first signalized intersection in all directions from the site, if that distance is not unreasonable (possibly up to one-half mile). Other areas can be included for study depending on actual study area conditions. Figure 6.1 illustrates a typical study area.

Evaluation of an appropriate project study horizon (time frame for analyzing traffic

FIG. 6.1 Example of a traffic study area for a typical land development. Courtesy of Post, Buckley, Schuh, and Jernigan, Inc. Tampa, FL, 1992.

conditions) must also be undertaken. This horizon can consist of development phasing periods or it can be synchronized with local planning horizons, areawide transportation planning horizons, or even governmental capital improvements plans. The development shown in Figure 6.1 consists of three developmental phases.

The site impact study should include not only the effects of changing land use attributed to the study site but also any effects attributed to other sites experiencing land use changes within the study area. In addition, the effects of through-traffic growth (both short and long range) and proposed off-site traffic system improvements must be included. This information can be requested from local regulatory authorities having jurisdiction over the system located in the study area. The engineer must conduct a field investigation of the study area and especially the proposed site to inventory traffic control devices, signal phasing and timing, roadway configurations, parking, street lighting, right-of-way widths, speed limits, areas involving accidents at high incident rates, travel lanes, driveway access close to the site, public transportation facilities, adjacent land uses, and turning movements at or near the proposed site. Field studies should be conducted using methods contained in ITE's (Box, 1976) *Manual of Traffic Engineering Studies.*

6.1.1   Nonsite Traffic

Nonsite traffic can include traffic moving through the study area that originates and terminates outside of the study area. This is sometimes called "background" traffic. Nonsite traffic also includes traffic generated within the study area but not as a result of contributions made by the site under study. This can include present and future traffic expected in the study area during the established project study horizon that is independent of traffic generated by the study site. Figure 6.2 illustrates current daily volumes while Figure 6.3 shows initial peak-hour volumes for this example.

According to ITE (1991), nonsite traffic can be estimated using the built-up method, an existing transportation plan, the projection of historical trends, or possibly a combination of these methods.

The build-up method is usually used in areas experiencing moderate growth rates and in conjunction with a study site that has a study horizon of usually no more than 10 years. This method is used when several major projects are being developed over the same time period where traffic information exists on approved (but yet-to-be-constructed) developments and other anticipated developments. The build-up method also includes investigating the impact of proposed traffic system improvements that are scheduled for implementation during the project study horizon. These factors, coupled with estimated new traffic volumes associated with other off-site land use changes, must be included.

The use of an existing transportation plan requires that the plan include the area under study. Most areawide transportation plans reflect average daily traffic (ADT): however, some include peak-hour values. Use of this method often occurs in areas experiencing rapid growth where regional-type project sites having phased construction are being studied. The local transportation plan must be adaptable to the project's study horizon. This method requires refinement and verification of results. Most area forecasts are focused on future traffic volumes with emphasis being placed on rights-of-way and roadway lane requirements. Care must be exercised to obtain the desired results relating to intersection service levels and the impact of site traffic.

FIG. 6.2 Example of daily traffic volumes for the sample traffic study area. Courtesy of Post, Buckley, Schuh, and Jernigan, Inc. Tampa, FL, 1992.

The use of projected historical trends occurs when short study horizons exist. This method is used on smaller projects when locally collected data over at least a five-year period are available. This method may not provide accurate results if the data are of poor quality. Average daily traffic is used to develop growth rates that can project nonsite traffic. This method, however, does not necessarily reflect localized changes or peak-hour traffic trends.

Whichever method of determining nonsite traffic is chosen, it should be acceptable to the reviewing jurisdictional authority and should provide reasonable results. Projected nonsite daily traffic volume for Phase I is shown in Figure 6.4.

6.1.2 Site-Generated Traffic

Another important element of the traffic-related site impact analysis is the estimation of the amount of site-generated traffic (new traffic). ITE defines a trip as a single or one-directional movement that either originates or terminates within a designated area (ITE, 1991). In residential land developments the number of trips can be estimated from the number of included dwelling units or from other parameters. Estimates of trip generation can be obtained by using either trip generation references or results of local study data on comparable sites or by conducting special field studies. Often many state and local transportation planning agencies have available trip generation data collected from within their jurisdictional area that can be used. The engineer must also determine what design level of traffic to use in the analysis, whether it is peak hour, peak season, or something else. Figure 6.5 illustrates typical peak traffic flow times for various land uses.

FIG. 6.3 Existing P.M. peak-hour traffic conditions for the sample traffic study area. Courtesy of Post, Buckley, Schuh, and Jernigan, Inc. Tampa, FL, 1992.

One technical reference (ITE, 1997c), contains information obtained primarily from suburban sites. A reference such as this provides the engineer with a starting point to begin trip generation analysis; good judgment must also be used, since average rates, best-fit equations or a combination of these confronts the user. The best estimate is obtained by comparing average trip rates and the results from the equations, while adjusting the results to represent local conditions. An adjustment to this estimate of generated traffic may be required to account for pass-by trips, transit trips, internal site trips resulting from the presence of mixed land use on-site, and other factors. The need for trip generation adjustment is especially true when considering a neotraditional development because one of the primary goals of this type of development is to reduce reliance on automobiles by promoting pedestrian and bicycle modes of travel to neighborhood centers for on-site shopping, recreation, and employment.

The study can include a summary table where different land uses within the proposed development are classified along with the chosen method of estimating trip generation for each classification and any associated adjustments.

6.1.3 Site-Related Traffic Distribution and Assignment

Estimated new traffic generated by the development must be distributed and assigned along with any pass-by (captured) trips to the various study area roadways for analysis. This establishes the amount of traffic that is expected on each route and at intersections

FIG. 6.4 Projected nondevelopment daily traffic during Phase I for the sample traffic study area. Courtesy of Post, Buckley, Schuh, and Jernigan, Inc., Tampa, FL, 1992.

where system capacities, signalization times and needs, circulation patterns, parking needs, and safety can be analyzed. Traffic distribution is a function of the type of proposed development and its size. Variable land uses within the study area and population distributions are also factors affecting traffic distribution, as well as the condition of the roadway system. Before distributing the traffic, areas contributing to large percentages of the generated trips should be identified and their areas of influence should be delineated within the study boundary.

According to ITE (1991), two of the most common methods to establish influence areas are either to use a reasonable maximum travel time to the site or to delineate the boundary based on the proximity of competing developments. Once the influence area is defined, the distribution of trips can be determined by analogy, modeling, or the use of other reference data.

The analogy method utilizes trip distributions that exist at similar developments near the study site. These data can include turning movements, origin-destination surveys, and driver response surveys.

A trip distribution model, such as the gravity model, can be used assuming that potential trips between two areas are proportional to the magnitude of each area and simultaneously inversely proportional to the separating distance or travel time.

Land Use	Typical Peak Hours[a]	Peak Direction
Residential	7:00–9:00 A.M. weekdays	Outbound
	4:00–6:00 P.M. weekdays	Inbound
Regional shopping	5:00–6:00 P.M. weekdays	Total[b]
	12:30–1:30 P.M. Saturdays	Inbound
	2:30–3:30 P.M. Saturdays	Outbound
Office	7:00–9:00 A.M. weekdays	Inbound
	4:00–6:00 P.M. weekdays	Outbound
Industrial	Varies with employee shift schedule	
Recreational	Varies with type of activity	

Notes:

[a]Hours may vary based on local conditions.

[b]Period of maximum weekday traffic impact.

FIG. 6.5 Typical peak traffic flow hours for selected land uses. Taken from *Traffic Access and Impact Studies for Site Development*. Copyright © 1991 by the Institute of Transportation Engineers. Reproduced by permission.

Should the engineer elect to use other reference data, which can take the form of demographic, socioeconomic, or other related census survey data, trip distributions can also be estimated. Development sites connected with multiple land use areas can require traffic to be distributed and assigned several times.

Trip assignment is focused on the amount of traffic that will travel over specific routes within the study area's roadway network. Assigning trips is dependent upon street capacity, intersection capacity, and travel times. Commonly, traffic is assigned over parallel routes even though some routes require less travel time. Traffic assignment also includes pass-by trips. Pass-by trips result when traffic initially having trip destinations other than to the study site temporarily use the study site as an unplanned intermediate destination. Traffic generated from pass-by trips temporarily exits adjacent street traffic until being returned to continue travel on the adjacent street to the original destination. This results in additional trips being generated by the study site in addition to the new trips that originate with the study site being the intended destination.

Should anticipated pass-by trips be a significant factor, then the percentage of pass-by trips must be estimated as a part of the total trips. This can be determined from driver response surveys. This will separate the traffic into pass-by trips and new trips. Both categories of trips can be distributed and assigned, with caution being applied to account for pass-by departure routes that may not coincide with the route used for entrance.

Traffic assignment reflects peak morning and afternoon volumes where two-way trips are usually used. Travel times are important and are included as a "function factor" in gravity model analysis. If a gravity model is not used, trip length frequency distribution curves obtained from local trends are needed for traffic assignment. Results from analyzing trip generation include the number of generated trips that are expected at each of the site's access locations. Assigned traffic volumes should be checked to ensure that logical assignment has been made. Figure 6.6 illustrates projected P.M. peak-hour traffic for the study in the example.

FIG. 6.6 Projected P.M. peak-hour traffic for Phase I for the sample traffic study area. Courtesy of Post, Buckley, Schuh, and Jernigan, Inc., Tampa, FL, 1992.

6.1.4 Impact Analysis and Alternatives

An analysis of system performance can be made after the traffic has been estimated, distributed, and assigned. This usually includes evaluation of performance levels for each major roadway and intersection. Figure 6.7 illustrates typical intersection turning movements. Other analysis considerations can include parking, pedestrian and bicycle activities, safety, and traffic control considerations. The results of this analysis should provide an indication of potential problem areas and possible ways to eliminate or mitigate them.

The most common method to evaluate transportation system performance uses the level-of-service concept (Section 4.16.2). Levels of service for signalized intersections are shown in Figure 6.8a and for two-way stop-controlled (TWSC) intersections in Figure 6.8b.

The engineer can use the *Highway Capacity Manual* (TRB, 1994), the intersection capacity analysis, the critical movement analysis, or other methods to evaluate the roadway segments. Safety analysis includes evaluating sight distances, areas experiencing high accident rates, and driveway turning speeds.

Providing access to a site requires the traffic engineer to consider providing access availability to adjoining properties, preserving roadway capacities, and maintaining

FIG. 6.7   Typical intersection P.M. peak-hour turning movements for the sample traffic study area, Phase III. Courtesy of Post, Buckley, Schuh, and Jernigan, Inc. Tampa, FL, 1992.

safety and traffic flows. Most residential areas front on local streets (Section 2.5.7), while nonresidential developments usually access major streets that are contiguous to the development site. A limited number of driveway openings are usually provided to allow controlled and efficient ingress and egress.

The location and spacing of ingress-egress access depends on corner clearances, property clearances, median openings, traffic controls, sight distances, adjacent property access, traffic queues, traffic gaps, weaving distances, and internal traffic circulation patterns. Off-site improvements necessary to accommodate anticipated traffic demands include possible improvements to intersections, roadway segments, interchanges, and site access. If off-site improvements prove inadequate, then the scope of the proposed project may have to be altered.

Impact analysis also should include the effects of future traffic passing through roadway areas that have experienced high accident rates. Improved safety conditions should be considered as part of the overall improvement plan.

Level of Service	Stopped Delay Per Vehicle (sec)	Qualitative Description
A	≤ 5.0	Good progression, few stops, and short cycle lengths.
B	>5.0 and ≤ 15.0	Good progression and/or short cycle lengths; more vehicle stops.
C	>15.0 and ≤ 25.0	Fair progression and/or longer cycle lengths, some cycle failures; significant portion of vehicles must stop.
D	>25.0 and ≤ 40.0	Congestion becomes noticeable; high volume-to-capacity ratio, longer delays, noticeable cycle failures
E	>40.0 and ≤ 60.0	At or beyond limit of acceptable delay; poor progression, long cycles, high volumes, long queues
F	> 60.0	Unacceptable to drivers. Arrival volumes greater than discharge capacity; long cycle lengths, unstable and unpredictable flows.

(a)

Level of Service	Average Total Delay (sec/veh)	Qualitative Description
A	≤ 5	Little or no delay
B	>5 and ≤ 10	Short traffic delays
C	>10 and ≤ 20	Average traffic delays
D	>20 and ≤ 30	Long traffic delays
E	>30 and ≤ 45	Very long traffic delays
F	> 45	When demand exceeds the capacity of the lane, extreme delays will be encountered with queuing, which may cause severe congestion affecting other traffic movements in the intersection; this condition usually warrants improvement to the intersection

(b)

FIG. 6.8 Levels of service criteria for intersections: (a) signalized intersections; (b) two-way stop-controlled (TWSC) intersections. Source: Special Report 209: Highway Capacity Manual, 3rd ed., © 1994 by the Transportation Research Board, National Research Council, Washington, DC. Qualitative description adapted from ITE, 1991. Copyright © 1991 by the Institute of Transportation Engineers. Used by permission.

Should the impact analysis indicate that adequate levels of service are attainable, then no additional improvements need to be made. However, should potential traffic problems manifest during the analysis process, an assessment of various on-site and off-site improvements must be undertaken to identify cost-effective means of providing adequate levels of service. This usually requires coordination with the jurisdictional regulatory reviewing authority.

The minimal acceptable level of service could be established at "D" during the design peak-hour traffic. The alternatives can include making improvements adjacent to or close to the study site by improving site access for ingress-egress, parking, and traffic control. The proposed improvements can be scheduled for implementation either over a short or long period of time, depending on traffic growth and system needs.

6.1.5 On-Site Traffic Planning

Interior traffic patterns will affect ingress-egress locations and their performance. According to the ITE, accesses should be treated as intersections and designed accordingly, with the location being determined by

(a) allowing adequate distance between the abutting street and interior driveway intersections to provide space where the intersection or traffic lanes do not become blocked,

(b) providing an adequate number of lanes and associated traffic controls to reduce conflicts in traffic flow patterns, and

(c) providing facilities to accommodate handicapped persons and public transit users.

Interior traffic movements should be accomplished safely and efficiently. This includes the movements of automobiles, service and emergency vehicles, pedestrians, and bicyclists. Parking areas provided should be as convenient as possible to the intended point of destination while including safety lanes for emergency vehicles, adequate space for vehicle maneuvering, and adequate space for passengers to enter and exit parked vehicles without damaging adjacent property.

6.1.6 Summary Traffic Impact Report

Usually the findings of the traffic-related site impact studies and access are placed in a final report. This report can include a description of the study areas, study site, and existing conditions. Future conditions are also described along with the system's analysis. Findings and recommendations for various highway improvements resulting from the study are presented as indicated in Figure 6.9. A favorable review of this report allows the engineer to include applicable improvements in the project's construction drawings. Other improvements can be scheduled for future implementation in accordance with proposed phasing schedules.

6.2 STREET AND ROAD DESIGN CRITERIA

In designing a project's street system, the design professional faces the same constraints as the developer, the constraints described in Section 2.5. In addition, aspects other than just providing site access and conveying traffic must also be included. This is because residential streets provide movement for pedestrians and bicycles and areas for neighborhood children to play and at the same time give the development continuity

FIG. 6.9 Projected highway improvements for the sample study area. Courtesy of Post, Buckley, Schuh, and Jernigan, Inc., Tampa, FL, 1992.

and neighborhood cohesiveness. Streets associated with suburban-type land developments are primarily classified as local and collector streets, while arterial types are not usually included.

According to AASHTO (1994),[*] the urban principal arterial system accommodates most trips entering and leaving the area while serving major centers of activity and accommodates the longest trip desired. A minor arterial system interconnects and augments the principal arterial system by accommodating trips of moderate lengths. The urban collector system can go through residential areas to distribute trips from arterials and collect traffic from urban local streets and channel it to the arterial system. Urban local streets permit access to individual parcels of land (land service) and to higher order street systems.

Traditional neighborhood development (TND) streets can be classified according to the following six ITE (1999)[**] categories:

A. Boulevards—larger multilane and generally urban corridors, with central, planted medians.

B. Avenues—connectors that may be multilaned and run through a TND, terminating at a civic building or monument.

C. Streets—provide for two-way vehicular travel using curb and gutter design, most often with sidewalks on both sides, and have no central median.

[*] From *A Policy on Geometric Design of Highways and Streets*. Copyright © 1994 by the American Association of State Highway and Transportation Officials, Washington, D.C. Used by permission.
[**] From *Traditional Neighborhood Development—Street Design Guidelines*. Copyright © 1999 by the Institute of Transportation Engineers, Washington, D.C. Used by permission.

D. Roads—provide access to neighborhoods of lower densities, may be constructed as a curb and gutter or open cross section, and seldom has striped street-side parking.

E. Drives—constructed between urban and green areas with curbing, gutters, sidewalks, and striped parking on one side while the other may lend itself to the rural green area setting.

F. Alleys—constructed usually with inverted crowns with center drainage, they are found at the rear of lots, without sidewalks or setbacks, and designed as a service route for the lot it abuts.

The design of local urban streets is governed more by practical limitations than that of local rural streets. Dominant design controls include the type and extent of urban development and zoning (or regulatory) restrictions. In residential areas, many of the streets are "land service streets," where the primary consideration is to promote a safe and pleasant environment while convenience of the motorist is secondary. The other streets in residential areas can provide some "land service," while "traffic service" dominates.

Local urban land service streets in suburban residential areas do not usually depend on traffic volumes as a criterion for design. Usually design speeds are low (ranging from 20 to 30 miles per hour) for local urban streets in residential areas, and passing sight distances are not usually a consideration because of the street's character and function. The TND typically has design speeds ranging from 15 to 20 miles per hour where the streetscape is used (e.g., narrow streets, parallel parking, rotaries, landscaped medians, T intersections) to help limit motorists from exceeding these speeds. Because low speeds are promoted in residential areas, superelevation is not usually provided on local urban streets.

A primary concern is to provide the motor vehicle operator with adequate sight distance to allow safe vehicle stops under inclement conditions regardless of the street classification. On a level roadway, the estimated braking distance can be calculated using

$$d = \frac{V^2}{30f} \quad \text{(Imperial units)} \quad \text{or} \quad d = \frac{V^2}{254f} \quad \text{(metric units)} \tag{6.1}$$

where d is the braking distance in feet (Imperial) or meters (metric), V is the initial vehicle speed in miles per hour (Imperial) or kilometers per hour (metric), and f is the coefficient of friction between the tires and the roadway. If the road is not level, the braking distance must account for the road's gradient by

$$d = \frac{V^2}{30(f \pm G)} \quad \text{(Imperial units)} \quad \text{or} \quad d = \frac{V^2}{254(f \pm G)} \text{(metric units)} \tag{6.2}$$

where G is the road's slope in feet per foot (Imperial) or meters per meter (metric).

Braking distances adjusted for uphill slopes are shorter than those required when braking downhill. Stopping distances include braking distance and the distance traveled while the operator reacts to commence application of the brakes. The AASHTO

(1994)[*] proposes that a reasonable reaction time for the driver is 2.5 seconds. The distance traveled during this time to perceive and react is

$$d_r = V \times 2.5 \times 1.47 \quad \text{(Imperial units)} \quad \text{or} \quad (0.278)(2.5)V \quad \text{(metric units)} \quad (6.3)$$

where d_r is the distance traveled while the driver is perceiving and reacting in feet (Imperial) or meters (metric units), V is the vehicle speed during the reaction time in miles per hour (Imperial) or kilometers per hour (metric), and 2.5 is the recommended reaction time in seconds. The sum of the braking distance and d_r yields the required safe stopping distance. Figure 6.10 shows various recommended stopping sight distances for various design speeds on level wet pavements. Sight distances required for safe stopping along horizontal curves can be determined using the geometric relationships contained in Figure 6.11.

Most streets are designed to meet standards that will allow their dedication to the jurisdictional governing body once the streets are installed and approved. Streets that are not dedicated will remain private for the development or homeowner's association to own, operate, maintain, and control. Regardless of the street's final disposition, it must be safe, serviceable (including provisions for the operation of refuse and fire trucks, moving vans, and snow removal vehicles), economical, and attractive. The design professional must consider criteria set forth in this section while formulating the preliminary development plan (Chapter 5) and during the design process discussed in Section 6.4.

6.2.1   Road Cross Sections

Local streets in residential areas can permit parking along one or both sides of the street, resulting in possibly one usable lane for passage. Collector roads, however, usually provide two or more active traffic lanes to function properly. Moving lanes should be designed for a minimum width of 10 feet (12 feet if cars will be allowed to park along both sides), while parking lanes should not be less than 8 feet wide.

Design Speed (km/h)	Assumed Speed for Condition (km/h)	Brake Reaction		Coefficient of Friction f	Braking Distance on Level (meters)	Stopping Sight Distance for Design (meters)
		Time (sec.)	Distance (meters)			
30	30-30	2.5	20.8-20.8	0.40	8.8-8.8	29.6-29.6
40	40-40	2.5	27.8-27.8	0.38	16.6-16.6	44.4-44.4
50	47-50	2.5	32.6-34.7	0.35	24.8-28.1	57.4-62.8
60	55-60	2.5	38.2-41.7	0.33	36.1-42.9	74.3-84.6
70	63-70	2.5	43.7-48.6	0.31	50.4-62.2	94.1-110.8

FIG. 6.10 Stopping sight distances for wet pavements. Taken from *A Policy on Geometric Design of Highways and Streets*. Copyright © 1994 by the American Association of State Highway and Transportation Officials, Washington, DC. Used by permission.

[*] From *A Policy on Geometric Design of Highways and Streets*. Copyright © 1994 by the American Association of State Highway and Transportation Officials, Washington, D.C. Used by permission.

FIG. 6.11 Range of upper values: relation between radius and value of middle ordinate necessary to provide stopping sight distance on horizontal curves. Taken from *A Policy on Geometric Design of Highways and Streets.* Copyright © 1994 by the American Association of State Highway and Transportation Officials, Washington, DC. Used by permission.

Typical TND streets consisting of two lanes of traffic and one lane of parking can range anywhere from 28 to 30 feet in width.

The typical road cross section includes all features contained within the street right-of-way. Within this strip of property are located not only the driving and parking lanes but also utilities, storm drainage facilities, sidewalks, bicycle facilities, and road shoulders. Most modern developments at medium or higher densities include curb and gutter road cross sections in the design of local streets similar to those shown in Figure 6.12c. The minimum curb radius at intersections is usually 20 feet on local streets and 30 feet on collector streets. In the case of curb and gutter road sections, care must be exercised during design to incorporate the section in a "cut" environment, or a place where excavation is required, to allow positive drainage from the contiguous property. If a fill section is encountered, special provisions must be incorporated to remove otherwise trapped runoff that accumulates in abutting areas having elevations lower than the curb top.

Roadside ditch sections (Figure 6.12a) have been used in both cut and fill environments; however, this type of section may not be as attractive or as easy to maintain as

FIG. 6.12 Typical street cross sections: (*a*) roadside ditch section; (*b*) asphalt valley section; (*c*) rolled curb and gutter section; (*d*) typical divided entrance.

a curb and gutter section when used with local streets. In some areas, however, a more "rural" look of no curbs may be preferred or the design engineer may want to provide an "open" drainage system using roadside ditches to promote stormwater runoff infiltration on site.

6.2.2 Local Street Design Criteria

The design of local streets located in urban settings is usually predicated by two factors. One factor is to allow street access while promoting safety and a pleasing effect. The other factor is to promote traffic service. Some local streets primarily serve the first factor while others serve the latter. Typical guidelines for local street design contained in ITE (1993) are shown in Figure 6.13.

The magnitude of traffic volumes on local streets usually does not require a capacity analysis since adequate capacity is provided by using a two-lane street cross section. Other cross sections, such as divided roadway entrances, can be utilized if required by local regulations or desired by the developer for aesthetic reasons. Some types of developments, however, generate enough traffic volume to require a capacity analysis.

For land developments located in rural settings, minimum design speeds for local roads and streets that are not strictly for land service are affected by various traffic volumes, as shown in Figure 6.14a. Regardless of the volume of anticipated traffic, the design engineer must provide adequate sight distance for safety. While Figure 6.10 provides stopping sight distances on level pavements under wet conditions, minimum stopping sight distances for wet-pavement conditions for vertical curves are included in Figure 6.14b.

To determine the minimum length of a vertical curve to allow safe stopping, the algebraic slope difference of the road on either side of the point of vertical intersection is multiplied by the tabulated K value, as shown in Figure 6.14b. Current sight distance criteria used by the AASHTO (1994)[*] for stopping utilize a height of driver's eye of 3.5 feet (1070 mm) above the pavement and a roadway object of 0.5 foot (150 mm) high. The minimum lengths for a crest vertical curve are

$$L_{min} = 2S - \frac{1329}{A} \quad \text{(Imperial)} \quad \text{or} \quad L_{min} = 2S - \frac{404}{A} \quad \text{(metric)} \quad \text{for } S > L \quad (6.4)$$

$$L_{min} = \frac{AS^2}{1329} \quad \text{(Imperial)} \quad \text{or} \quad L_{min} = \frac{AS^2}{404} \quad \text{(metric)} \quad \text{for } S < L \quad (6.5)$$

where L is the length of vertical curve in feet (Imperial) or meters (metric), S is the sight distance in feet (Imperial) or meters (metric), and A is the algebraic difference of tangent grades in percent.

For sag vertical curves, the minimum length can be obtained where the sight distance is governed by night driving conditions. If the headlight height is 2 feet (Imperial) or 600 mm (metric) and a 1-degree upward divergence of the light beam from the vehicle's longitudinal axis is used as criterion, the following equations for sag vertical curves are obtained:

$$L_{min} = 2S - \frac{400 + 3.5S}{A} \quad \text{(Imperial)} \quad \text{or}$$

$$L_{min} = 2S - \frac{120 + 3.5S}{A} \quad \text{(metric)} \quad \text{for } S > L \quad (6.6)$$

$$L_{min} = \frac{AS^2}{400 + 3.5S} \quad \text{(Imperial)} \quad \text{or}$$

$$L_{min} = \frac{AS^2}{120 + 3.5S} \quad \text{(metric)} \quad \text{for } S < L \quad (6.7)$$

[*] From *A Policy on Geometric Design of Highways and Streets.* Copyright © 1994 by the American Association of State Highway and Transportation Officials, Washington, D.C. Used by permission.

Terrain Classification	Level			Rolling			Hilly		
Development Density	Low	Medium	High	Low	Medium	High	Low	Medium	High
Right-of-Way Width (feet)	50	60	60	50	60	60	50	60	60
Pavement Width (feet)	20-28	28-34	36	20-28	28-34	36	28	28-34	36
Type of Curb V = vertical face R = roll-type O = none	O/R	V	V	V	V	V	V	V	V
Sidewalks and Bicycle Paths (feet)	0	4-6	5-6	0	4-6	5-6	0	4-6	5-6
Sidewalk Distance from Curb Face (feet)	—	5-6	5-6	—	5-6	5-6	—	5-6	5-6
Minimum Sight Distance (feet)	200	200	200	150	150	150	125	125	125
Maximum Grade	4%	4%	4%	8%	8%	8%	15%	15%	15%
Maximum Cul-de-Sac Length (feet)	1500	1000	700	1500	1000	700	1500	1000	700
Minimum Cul-de-Sac Radius (right-of-way)(feet)	60	60	60	60	60	60	60	60	60
Design Speed (mph)	30	30	30	25	25	25	20	20	20
Minimum Centerline Radius of Curves (feet)	300	300	300	180	180	180	100	100	100
Minimum Tangent Between Reverse Curves (feet)	50	50	50	50	50	50	50	50	50

FIG. 6.13 Proposed local street design guidelines. Taken from *Guidelines for Residential Subdivision Street Design, A Recommended Practice.* Copyright © 1993 by the Institute of Transportation Engineers. Used by permission.

270

Speeds (Km/h) For Design Volumes Of			
Type of Terrain	ADT Under 50	ADT 50-250	ADT 250-400	ADT 400-1500
Level	50	50	60	80
Rolling	30	50	50	60
Mountainous	30	30	30	50

(a)

Design Speed (km/h)	Assumed Speed for Condition (km/h)	Stopping Sight Distance for Design (meters)	K Value For Crest Vertical Curves (rounded)	K Value For Sag Vertical Curves (rounded)
30	30-30	29.6-29.6	3-3	4-4
40	40-40	44.4-44.4	5-5	8-8
50	47-50	57.4-62.8	9-10	11-12
60	55-60	74.3-84.6	14-18	15-18
70	63-70	94.1-110.8	22-31	20-25

(b)

Note: K value is a coefficient by which the algebraic difference in

FIG. 6.14 Street design standards: (a) minimum design speeds for local rural roads; (b) minimum stopping sight distances for local roads and collectors (wet pavements). Taken from *A Policy on Geometric Design of Highways and Streets.* Copyright © 1994 by The American Association of State Highway and Transportation Officials, Washington, DC. Used by permission.

where L is the length of vertical curve in feet (Imperial) or meters (metric), S is the light beam distance in feet (Imperial) or meters (metric), and A is the algebraic difference of tangent grades in percent.

For example, consider a design speed of 70 km/h (43.5 mph) for a symmetrical vertical curve between a +1.5 percent slope and a −3.5 percent slope. The minimum downhill braking distance using an average running speed of 63 km/h according to Figure 6.10 (approximately 40 mph) and the −3.5 percent slope in conjunction with Equation 6.2 is

$$\text{Braking distance} = \frac{63^2}{254(0.31 - 0.035)} = 56.8 \text{ meters, or } 86 \text{ feet}$$

Although motor vehicle passing conditions do not usually prevail on local residential streets, certain situations may allow safe vehicle passing to occur. Typical minimum passing sight distances for safe maneuvering are shown in Figure 6.15a along with K factors that produce the minimum length of vertical crest curves to allow safe passing.

Design considerations for establishing vertical control of local streets include allowable centerline grades. Local regulations usually prescribe these limiting gradients;

Design Speed (km/h)	Min. Passing Sight Distance For Design (meters)	K Value for Crest Vertical Curves (rounded)
30	217	50
40	285	90
50	345	130
60	407	180
70	482	250

Note: K value is a coefficient by which the algebraic difference in grade may be multiplied to determine the length in meters of the vertical curve which will provide minimum sight distance.

(a)

Type of Terrain	Design Speed (km/h)				
	30	40	50	60	70
Level	8	7	7	7	7
Rolling	11	11	10	10	9
Mountainous	16	15	14	13	12

(b)

Surface Type	Range in Rate of Cross Slope (%)
High	1.5 to 2.0
Intermediate	1.5 to 3.0
Low	2.0 to 6.0

(c)

FIG. 6.15 Street design standards: (a) minimum passing sight distance for local rural roads and collector streets (rural and urban); (b) maximum grades for local rural roads; (c) normal cross slopes for local rural roads and local urban streets. Taken from *A Policy on Geometric Design of Highways and Streets*. Copyright © 1994 by the American Association of State Highway and Transportation Officials, Washington, DC. Used by permission.

however, typical maximum values are shown in Figure 6.15*b* for local rural roads along with typical values for pavement cross slopes in Figure 6.15*c*. The minimum centerline slope is usually not less than 0.3 percent.

Horizontal alignment considerations include establishing a minimum centerline radius value. TND criteria are shown in Figure 6.16. For suburban developments where minimum radii are based on design speeds (and possible superelevation), minimum requirements are shown in Figure 6.17.

Design Speed*	Minimum Centerline Radius (feet)
10 mph	22
15 mph	50
20 mph	89
25 mph	166

* Design speed = posted speed + 5 mph.
Note: No superelevation used on any of these curves.

FIG. 6.16 Minimum centerline radii for TND applications. Taken from ITE, 1999. Copyright © 1999 Institute of Transportation Engineers. Used by permission.

Design Speed (km/h)	Maximum e/100	Maximum f	Total (e/100+f)	Minimum Radius, m
30	0.04	0.312	0.352	20
40	0.04	0.252	0.292	45
50	0.04	0.214	0.254	80
30	0.06	0.312	0.372	20
40	0.06	0.252	0.312	40
50	0.06	0.214	0.274	70

Note: With design speeds of 30 km/h or less, conditions
may warrant elimination of superelevation.

FIG. 6.17 Minimum radii for different values of maximum superelevation on low-speed urban streets. Taken from *A Policy on Geometric Design of Highways and Streets.* Copyright © 1994 by the American Association of State Highway and Transportation Officials, Washington, DC. Used by permission.

Superelevation Rate in %	Length of Runoff, L (m) for Design Speed ((km/h)			
	30	40	50	60
3.6 meter lanes				
2	20	25	30	35
4	20	25	30	35
6	30	35	35	40
8	40	45	45	50
3.0 meter lanes				
2	20	25	30	35
4	20	25	30	35
6	25	30	30	35
8	35	35	40	40

FIG. 6.18 Minimum length for superelevation runoff for two-lane roads. Taken from *A Policy on Geometric Design of Highways and Streets.* Copyright © 1994 by the American Association of State Highway and Transportation Officials, Washington, DC. Used by permission.

Superelevation incorporates the banking of the riding surface to counteract the centrifugal force that otherwise would be counteracted only by side frictional forces produced by tire contact on the pavement. In land developments, most local streets promote low vehicle speeds and therefore are not superelevated. In some cases, however, superelevation is desired. Typical minimum values for superelevation runoff are also shown in Figure 6.18. The limiting rate of 0.08 ft/ft (m/m) of road pavement width is recommended where ice and snow occur. Superelevation runoff is the horizontal distance required to restore the road cross section from its normal crown into a fully superelevated riding surface. Although there is no exact method to place superelevation runoff when transition curves (spiral) are not included, the usual procedure is to place one-third of the runoff in the curve proper and two-thirds outside of the curve along the tangent beginning from the point of curvature (PC) or the point of tangency (PT). A common practice in land developments is to rotate the cross section about the centerline of the street.

6.2.3   Collector Roads and Streets

Collector roads and streets in urban areas serve two functions. They collect traffic between local roads and arterial roads, and they provide access to contiguous property. These streets accommodate higher traffic volumes than local residential streets, and therefore more emphasis is placed on traffic conveyance.

Typical collector street design guidelines taken from ITE (1993) are shown in Figure 6.19.

Some design criteria for collector streets differ from local streets. Figure 6.20*a* shows minimum design speeds for various collector street traffic volumes.

The minimum stopping sight distance for collector roads is shown in Figure 6.14*b* and the minimum passing sight distance is contained in Figure 6.15*a*.

The maximum recommended roadway gradients for collector roads are shown in Figure 6.20*b*.

Required collector street roadway widths vary for various traffic volumes. Recommended widths are shown in Figure 6.21.

6.2.4   Intersection Design

Typical intersection design guidelines for residential areas included in ITE (1993) are shown in Figure 6.22.

Figures 6.23*a* and *b* show examples of sight distances necessary for safety. Figure 6.24 shows typical intersection configurations that are often used in TND designs. Traffic circles, or rotary intersections (roundabouts), allow for continuous traffic movement at relatively low speeds. When a "skewed T" channel is used, consideration should be given to incorporating channels where vehicles headed from the traffic circle approach the intersection at right angles to the skewed road being intersected.

6.2.5   Traffic Calming

The goal of maintaining low vehicle speeds within residential areas is a constant challenge to project engineers. A number of different techniques are used to help achieve this goal. The most common technique utilized is the employment of speed humps to act as an impediment when vehicles are traveling rapidly. These humps, positioned

Terrain Classification	Level			Rolling			Hilly		
Development Density	Low	Medium	High	Low	Medium	High	Low	Medium	High
Right-of-Way Width (feet)	70	70	80	70	70	80	70	70	80
Pavement Width (feet)	24-36	24-36	40	24-36	24-36	40	24-36	24-36	40
Type of Curb, Vertical Face (VF)	VF	VF	VF	VF	VF	VF	VF	VF	VF
Sidewalk Width (feet)	4-6	4-6	4-6	4-6	4-6	4-6	4-6	4-6	4-6
Sidewalk Distance from Curb Face (feet)	10	10	10	10	10	10	10	10	10
Minimum Sight Distance (feet)	250	250	250	200	200	200	150	150	150
Maximum Grade	4%	4%	4%	8%	8%	8%	12%	12%	12%
Minimum Spacing Along Major Traffic Route (feet)	1300	1300	1300	1300	1300	1300	1300	1300	1300
Design Speed (mph)	35	35	35	30	30	30	25	25	25
Minimum Centerline Radius of Curves (feet)	480	480	480	300	300	3090	180	180	180
Minimum Tangent Between Reverse Curves (feet)	100	100	100	100	100	100	100	100	100

FIG. 6.19   Collector street design guidelines. Taken from ITE, 1993. Copyright © 1993 by the Institute of Transportation Engineers. Used by permission.

Type of Terrain	Minimum Design Speeds (km/h) for Design Volumes of:	
	ADT 0-400	ADT 400-2000
Level	60	80
Rolling	50	60
Mountainous	30	50

(a)

Type Terrain	Design Speed (km/h) Rural Collectors			
	30	40	50	60
Level	7	7	7	7
Rolling	10	10	9	8
Mountainous	12	11	10	10

Type Terrain	Design Speed (km/h) Urban Collectors			
	30	40	50	60
Level	9	9	9	9
Rolling	12	12	11	10
Mountainous	14	13	12	12

(b)

Note: Maximum grades shown for rural and urban conditions of short lengths, (less than 150 m), on one-way down grades and on low-volume rural collectors may be 2% steeper.

FIG. 6.20 Collector street design standards: (a) minimum design speeds (km/h) for various traffic volumes; (b) maximum grades. Taken from *A Policy on Geometric Design of Highways and Streets*. Copyright © 1994 by the American Association of State Highway and Transportation Officials, Washington, DC. Used by permission.

perpendicular to the road centerline, give the vehicle a rapid vertical jolt which helps to reduce travel speeds because at low velocities the resulting vertical displacement is more gentle. Sometimes these humps include cutouts where wide-span bus wheel bases can straddle the hump so that they can proceed at appropriate speeds without creating major bus passenger discomfort. These devices can be positioned near pedestrian crosswalks to help improve safety.

Another approach to promote traffic calming is to narrow the pavement on one side of the road to cause traffic to slow. Should both sides be narrowed at the same point in the roadway, the traffic can be throttled. Another approach is to utilize rumble strips and rumble areas where the pavement is raised about $\frac{1}{2}$ inch and is often painted in a contrasting color to capture the driver's attention to reduce speed.

Design Speed (km/h)	Width (m) For Design Traffic Volumes Of:	
	ADT Under 400	ADT 400-1500
30	6*	6
40	6*	6
50	6*	6
60	6*	6.6
70	6	6.6
Width of Graded Shoulder (m)--Each Side of Pavement		
All Speeds	0.6	1.5**

* 5.4 meters minimum for ADT under 250.

**May be adjusted to achieve a minimum roadway width of 9.0 m for design speeds of 50 km/h or less.

FIG. 6.21 Minimum width of traveled way and graded shoulder for rural collector streets. Taken from *A Policy on Geometric Design of Highways and Streets.* Copyright © 1994 by the American Association of State Highway and Transportation Officials, Washington, DC. Used by permission.

Terrain Classification	Level	Rolling	Hilly
Development Density	All	All	All
Approach Speed (MPH)	25	25	20
Clear Sight Distance (length along each approach leg)(feet)*	90	90	70
Vertical Alignment Within Intersection Area	flat	2%	4%
Minimum Angle of Intersection in Deg. (Rt. Angle Preferred)	70	70	70
Minimum Curb Radius (feet):			
a. Local-local	20	20	20
b. Local-collector	25	25	25
c. Collector-major	30	30	30
Minimum Centerline Offset of Adjacent Intersection (feet):			
a. Local-local	125	125	125
b. Local-collector	150	150	150
c. Collector-collector	200	200	200
Minimum Tangent Length Approaching Intersection (each leg)	50	30	20

Note: * At an alley intersection with a street (or with another alley) a 15-foot minimum clear sight distance leg is recommended along each intersecting property line.

FIG. 6.22 Street intersection design guidelines for suburban-type developments. Taken from ITE, 1993. Copyright © 1993 by the Institute of Transportation Engineers. Used by permission.

Another traffic-calming technique is to utilize gate posts placed near the pavement edge to give the driver the appearance that the pavement narrows. Gateways are often used with rumble strips or humps to help keep traffic speeds in check.

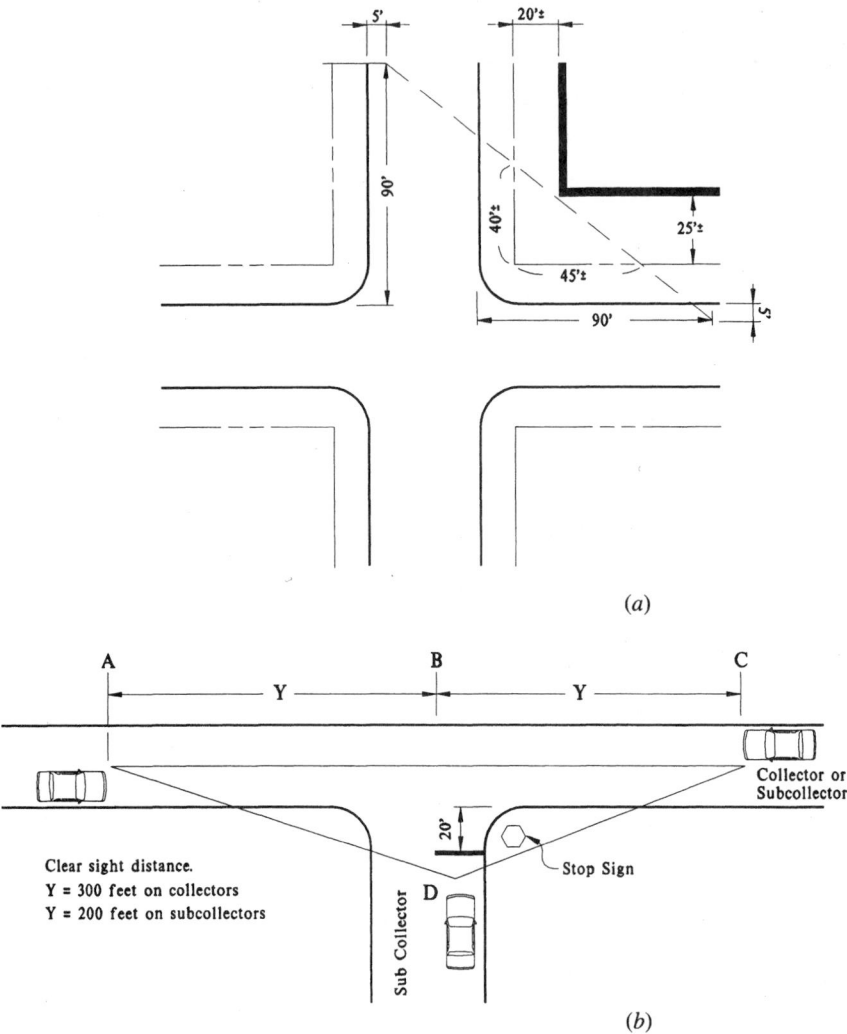

FIG. 6.23 Clear sight distances at typical intersections found in land developments. (*a*) Local-local intersection. Taken from ITE, 1993. Copyright © 1993 by the Institute of Transportation Engineers. Used by permission. (*b*) Higher order street intersection. Source: American Society of Civil Engineers, the National Association of Home Builders, and the Urban Land Institute, *Residential Streets,* 2nd ed. 1990, p. 73. Reproduced by permission of the American Society of Civil Engineers.

6.2.6 Traffic Control Devices: Suburban and Neotraditional

Traffic flow through an intersection is gauged by the prevailing level of service defined in Figure 6.8. In suburban-type neighborhoods, traffic concentrates on collectors and arteries where delays increase and left turns crossing on-coming traffic become increasingly difficult to complete. When a traffic signal is warranted, extra cycle time often must be provided with multiphase signals to allow both vehicles and pedestrians safe transit.

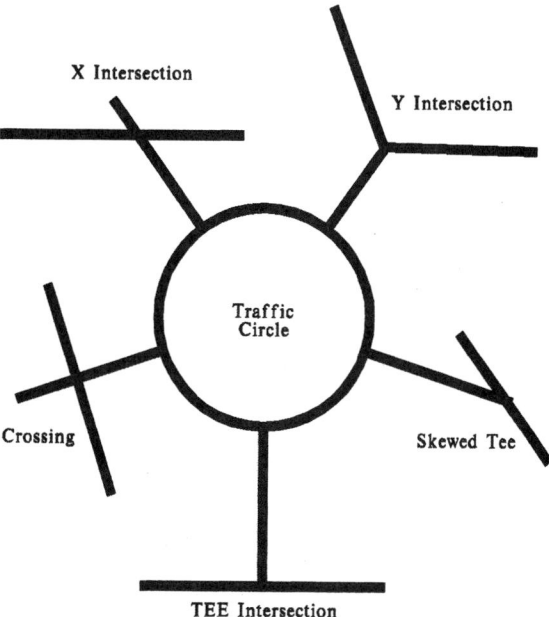

FIG. 6.24   Typical intersection configurations for TND applications.

TNDs provide increased networks where traffic is less concentrated because multiple routes encourage traffic to disperse. In most cases unsignalized intersections or short cycle two-phase signals are used for traffic control.

6.2.7   Alley Design

Alleys are not usually included as part of the design of suburban neighborhood developments; however, neotraditional developments depend on the use of alleys to allow rear-lot access for automobile parking, trash pickup, and places to locate utilities away from the streetscape. In these plans, alleys also provide greater pedestrian connectivity within the neighborhood. According to AASHTO (1994),[*] alleys should be oriented parallel to or concentric with street property lines and should terminate where both ends connect to streets or other alleys. In cases where one end dead-ends, a turning area similar to that shown in Figure 6.25 should be provided.

6.2.8   Driveways

Driveways provide vehicular access to street riding surfaces as well as allow vehicular departure from the riding surface. Driveways must provide for certain rates of travel when exiting the street's traffic pattern and also for smooth ingress-egress. Typical guidelines for driveway design are illustrated in Figure 6.26. Typical residential driveway details are shown in Figure 6.27. Proper clearance must be maintained between the

---

[*] From *A Policy on Geometric Design of Highways and Streets*. Copyright © 1994 by the American Association of State Highway and Transportation Officials, Washington, D.C. Used by permission.

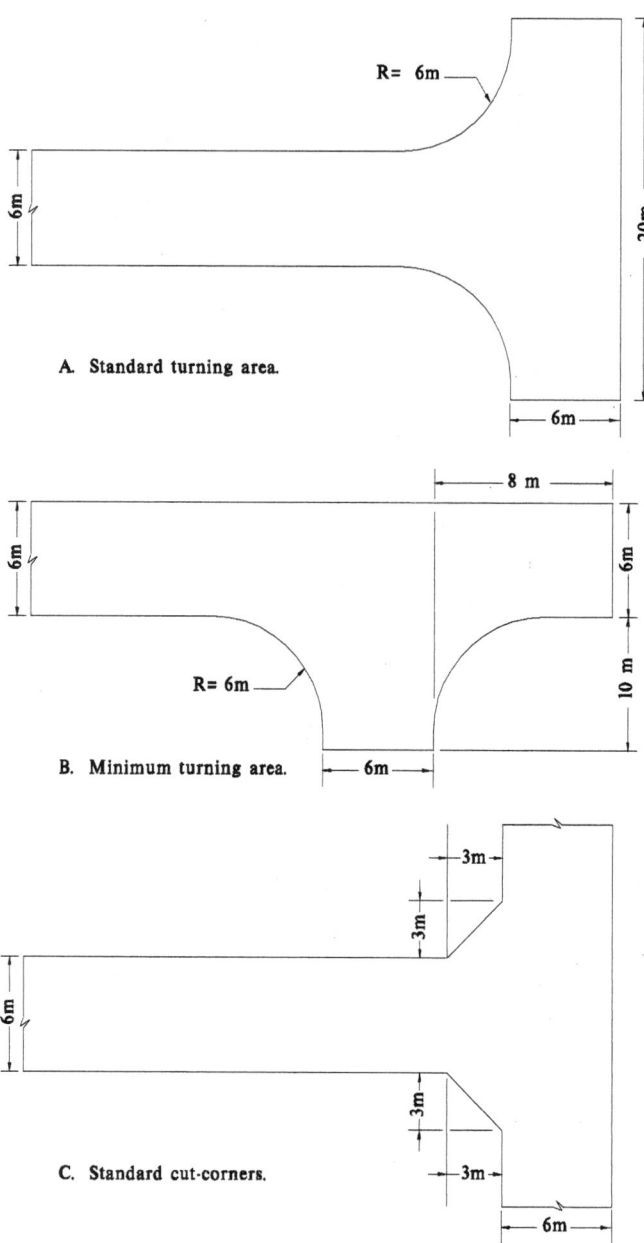

FIG. 6.25   Examples of turning areas for dead-end alleys. Taken from *A Policy on Geometric Design of Highways and Streets.* Copyright © 1994 by the American Association of State Highway and Transportation Officials, Washington, DC. Used by permission.

vehicle's underside and the pavement, especially where driveway slopes are steep. Provisions must be made to keep storm runoff from being encumbered by the driveway. This includes not blocking the throat of the curb, not restraining a roadside ditch's capacity with fill, and not installing any driveway culvert improperly.

6.3. PARKING CONSIDERATIONS

Most residential developments provide for off-street parking where residential owners can place their vehicles out of the street right-of-way. Often in single-family detached suburban-type developments, the parking area is the driveway. In multifamily developments designated parking areas are normally provided. These areas must be convenient to the user, especially when considering the need to manually transport purchased goods from the parked vehicle to the dwelling. TNDs can allow on-street parking while resident's vehicles are often parked behind dwellings using a back-alley access.

When parking is provided, stalls should provide adequate space to allow vehicle

Recommended Basic Driveway Dimension Guidelines				
	Dimension Reference	Residential	Commercial	Industrial
Nominal Width (See Note 1)	W			
One-way		10	15	20
Two-way		10	30	40
Right turn radius or flare	R			
Minimum (See Note 2)		5	15	20
Minimum Spacing (See Note 3)				
From property line	P	0	0	-R
From street corner	C	5	10	10
Between driveways	S	3	3	10
Minimum Angle (See Note 4)	A	45^0	45^0	30^0

Notes:

1. Residential driveway widths typically should not exceed about 24 feet. Commercial driveway widths may vary from about 24 feet for low volume activity (providing that 20 foot radii are used) to a maximum of 36 feet for undivided design, higher volume activity. A 36 foot driveway is usually marked with two exit lanes of 10 to 11 foot width, with the balance used for a single, wide entry lane. Industrial driveway widths should not exceed 50 feet.
2. On the side of a driveway exposed to entry or exit by right turning vehicles. The radii for major generator driveways should be much higher than the values shown.
3. Measured along the curb or edge of pavement from the roadway end of the curb radius or flare. For individual properties, a suggested limitation on the number of driveways is: 1 for 0-50 foot frontage, 2 for 51-150 foot frontage 3 for 151-500 foot frontage, and 4 for over 500 foot frontage.
4. Minimum acute angle measured from edge of pavement, and generally based on one-way operation. For two-way driveways and high pedestrian activity areas, the minimum angle should be 70 degrees.

FIG. 6.26 Driveway layout guidelines. Taken from ITE, 1987b, p. 19. Copyright © 1987 by the Institute of Transportation Engineers. Reproduced with permission.

FIG. 6.27 Residential driveway details. Taken from ITE, 1993. Copyright © 1993 by the Institute of Transportation Engineers. Used by permission.

doors to open when embarking or disembarking a vehicle parked adjacent to other vehicles. The parking facility should provide an adequate number of stalls for occupants, visitors, and possibly space for some recreational vehicles such as boats and their trailers or campers. Parking for handicapped persons requires additional provisions (see Section 2.5.9).

The geometric layout of all parking areas (residential and commercial) should allow the vehicle operator ease in positioning the vehicle in the parking stall by providing adequate radii on curb turnouts and islands. Drainage is also important to prevent standing surface water and degradation of the parking lot pavement and subgrade from subsurface water. Arrangements also must be made for security, night illumination, and landscaping.

6.3.1 Parking Generation

Estimates of the volume of parked vehicles in various settings can be made using references such as ITE (1987a). As in the case of traffic generation, the designer must use this source as a beginning point because local adjustments are required. Should the parking facility accommodate "shared parking" (uses that have different peak parking periods), the total number of needed parking spaces is not necessarily the sum of the individual needs attributed to each use.

6.3.2 Parking Lot Configurations

Numerous configurations for parking lots are available; however, the most common relationships are shown in Figure 6.28. A percentage of the provided parking spaces can be allocated to smaller sized cars in designated areas using relationships shown in Figure 6.29.

Other forms of parking facilities include parking garages and underground parking structures. The reader is referred elsewhere for information pertaining to these alternatives (ULI and NPA, 1990).

6.3.3 Parking Aesthetics

Parking lots can be unattractive if inadequate attention is given to their appearance during design. Without attention they can be extremely hot in the summer as a result of absorbed heat and reflected energy from vehicles. In the winter they can be cold because of cross winds. Repetitive space configurations can provide mundane views for an observer, resulting in visual unattractiveness.

With landscaping, appeal can be obtained. Plants and trees can provide shade, protection from cross winds, and visual delineation to mask parked vehicles. Vegetation also helps by reducing adsorbed heat and glare and filtering noises to help maintain quiet.

Planting screens and buffer zones can be placed around the parking lot's exterior to act as a visual screen, and interior medians can provide space for other plantings.

6.4 STREET ALIGNMENT

In addition to adhering to proper geometric relationships necessary for safe vehicular operation, street and roadway design also includes the establishment of horizontal control using bearings and distances and vertical control with stations, elevations, and grades. The design procedure presented in this section includes the following sequential steps:

(a) determining centerline bearings and distances from point of centerline intersection (PI) to PI for the entire road system;

(b) calculating the horizontal curve nomenclature for each included curve;

(c) calculating the road's stations using the PI-to-PI distances and each horizontal curve's nomenclature;

Θ = Parking Angle
W1= Parking module width (wall to wall),
     single loaded aisle
W2= Parking module width (wall to wall),
     double loaded aisle
W3= Parking module width (wall to inter-
     lock), double loaded
W4= Parking module width (interlock to
     interlock), double loaded aisle
AW= Aisle width
WP= Stall width parallel to aisle
VPl= Projected vehicle length from
     interlock
VPw= Projected vehicle length from
     wall measured perpendicular to
     aisle
Sl= Stall length
Sw= Stall width

Stall Width Classifications

Class	Width	Typical Uses
A	9.00	Retail customers, banks, fast foods, other very high turnover
B	8.75	Retail customers, visitors
C	8.50	Visitors, office employees, residential, airports, hospitals
D	8.25	Industrial, commuter, univesity

Note: Width is measured for large-size vehicle, at right angles to stall

Parking Class	Sw Basic Stall Width,ft.	WP Stall Width Parallel To Aisle, ft.	VPw Stall Depth to wall, ft.	VPl Stall Depth to interlock,ft.	AW Aisle Width,ft.	Modules W2 Wall to Wall,ft.	Modules W4 Interlock to Interlock,ft.
	Two-Way Aisle- 90 Degrees						
A	9.00	9.00					
B	8.75	8.75					
C	8.50	8.50	17.5	17.5	26.0	61.0	61.0
D	8.25	8.25					
	Two-Way Aisle- 60 Degrees						
A	9.00	10.4					
B	8.75	10.1	18.0	16.5	26.0	62.0	59.0
C	8.50	9.8					
D	8.25	9.5					
	One-Way Aisle-75 Degrees						
A	9.00	9.3					
B	8.75	9.0	18.5	17.5	22.0	59.0	57.0
C	8.50	8.8					
D	8.25	8.5					
	One-Way Aisle-60 Degrees						
A	9.00	10.4					
B	8.75	10.1	18.0	16.5	18.0	54.0	51.0
C	8.50	9.8					
D	8.25	9.5					
	One-Way Aisle- 45 Degrees						
A	9.00	12.7					
B	8.75	12.4	16.5	14.5	15.0	48.0	44.0
C	8.50	12.0					
D	8.25	11.7					

FIG. 6.28 Proposed parking layout criteria for large-size vehicles. Taken from *A Proposed Recommended Practice—ITE Guidelines for Parking Facility Location and Design.* Copyright © 1994 by the Institute of Transportation Engineers. Used by permission.

Parking Class	Sw Basic Stall Width,ft.	WP Stall Width Parallel To Aisle, ft.	VPw Stall Depth to wall, ft.	VPl Stall Depth to interlock,ft.	AW Aisle Width,ft.	Modules	
						W2 Wall to Wall,ft.	W4 Interlock to Interlock,ft.
	Two-Way Aisle- 90 Degrees						
A/ B	8.0	8.0					
C/ D	7.5	7.5	15.0	15.0	21.0	51.0	51.0
	Two-Way Aisle- 60 Degrees						
A/ B	8.0	9.3					
C/ D	7.5	8.7	15.4	14.0	21.0	52.0	50.0
	One-Way Aisle-75 Degrees						
A/ B	8.0	8.3					
C/ D	7.5	7.8	16.0	15.1	17.0	49.0	47.0
	One-Way Aisle-60 Degrees						
A/ B	8.0	9.3					
C/ D	7.5	8.7	15.4	14.0	15.0	46.0	43.0
	One-Way Aisle- 45 Degrees						
A/ B	8.0	11.3					
C/ D	7.5	10.6	14.2	12.3	13.0	42.0	38.0

FIG. 6.29 Proposed parking layout dimension guidelines for smaller sized vehicles. Taken from *A Proposed Recommended Practice—ITE Guidelines for Parking Facility Location and Design.* Copyright © 1994 by the Institute of Transportation Engineers. Used by permission.

(d) plotting the plan view of the centerline and features that are related to center-line controls;

(e) laying out the corresponding profile plot area using previously determined sta-tioning along the profile bottom while providing adequate space to accommo-date the maximum and minimum expected elevations along the profile's sides;

(f) plotting the existing ground profile using contour lines contained on the pre-liminary approved plan;

(g) laying out the proposed road profile on a profile drawing; and

(h) calculating grades for each desired station once the storm and sewage systems are evaluated to complete the profile drawing.

6.4.1 Horizontal Control

Horizontal control for streets and roads consists of establishing the horizontal rela-tionship of the road's centerline to the property's boundary. This can be accomplished by scaling graphical relationships from the approved preliminary plan, computing re-lationships using coordinate geometry, or assigning relationships using a desired dis-tance or angle. An example of a scaled relationship is a scaled tangent (radius or chord) distance of a horizontal curve, while a computed relationship can be represented by a calculated radius for some other horizontal curve where the tangent length is scaled and the tangent bearings are assigned.

Once the proposed road's centerline bearings and distances are defined from PI to the next PI, circular horizontal curves having constant radii (and meeting design criteria similar to those contained in Section 6.2.2 or 6.2.3) can be superimposed on the centerline's bearings and distances. Road stationing is then calculated, where a station represents (in hundreds of feet) the distance obtained by continuous measurement along the centerline from the beginning point to the point under investigation.

To further illustrate a procedure to establish alignment for horizontal control, the approved preliminary plan shown in Figure 6.30 is utilized. Following the approach contained in ITE (1997b), where Imperial units are given preference, this illustration follows suite. In this example, the designer decides that the property line between lots 5 and 6 will be 200 feet long and that the back lot line of lot 6 will be 140 feet. These two assignments can result from the designer rounding scaled distances on the approved preliminary plat. Property line C–D has a total distance of 176.27 feet. If 140.00 feet is allocated to lot 6, then the remaining 36.27 feet must be included as part of lot 5's back property line.

The centerline of the proposed new road leading to the cul-de-sac is to be made parallel to the back property line (C–D). The 200-foot lot line between lots 5 and 6 (line X–Z) is to be made perpendicular to the proposed road's alignment (by assignment). The corresponding bearing of line X–Z is obtained by subtracting the assigned bearing from 90° and giving this angular difference an adjoining bearing quadrant (northwest or southeast) by

$$
\begin{array}{lllll}
 & 89° & 59' & 60'' & \\
S & 57° & 40' & 25'' & W \\
\hline
N & 32° & 19' & 35'' & W \\
\end{array}
$$

FIG. 6.30   Example of road layout for centerline calculations.

The distance from point X to the centerline of the proposed road is 200.00 feet plus one-half of the width of the road's total right-of-way of 50 feet, or 225.00 feet to point W. Point W is a control point used to calculate the horizontal alignment back to the existing street right-of-way at point U. Using the preliminary plan (Figure 6.30) for reference, the scaled distance from point H to the western edge of the right-of-way of the proposed road is 262 feet. If 262.00 feet plus 25.00 feet (the proposed road intersects the existing road at 90°) are used to locate point U relative to point H, coordinate geometry calculations can proceed.

The centerline alignment of the proposed road is determined by calculating the inverse bearing and distance between points U and W (with latitudes, departures, and coordinates as described in Chapter 4), then finding the point of intersection of the proposed road's tangents (point Y) based on a bearing of S 57°40'25" W from point W and a perpendicular bearing to line H–U of N 05°3'53" E from point U. Inverse calculations between points U and W are shown in Figure 6.31.

The inverse bearing of line U–W is determined as N 34°20'27" E and the inverse distance as 364.66 feet. The proposed tangents can now be intersected mathematically to determine the distances from the PI (point Y) to both points U and W. Figure 6.32 illustrates this procedure using the law of sines.

The horizontal curve's nomenclature can now be computed based on applicable criteria. Should a particular degree of curve, or "D" (the central curve angle between two radii representing 100 feet of corresponding arc distance for arc definition), be required, the corresponding curve nomenclature can be calculated using the angular relationship between the curve's tangent bearings. In this example, however, the centerline radius (R) is assumed to be scaled from the approved preliminary plan where the approximate value of 110 feet is obtained. Since the centerline bearings have been assigned, the curve's central angle (I) can be calculated as shown in Figure 6.33.

Line	Bearing	Dist,ft.	Pt.	Coordinates North,ft.	East,ft.
W · X	N 32°-19'35" W	225.00	W	0.00	0.00
X · C	S 57°-40'-25" W	36.27	X	190.13	-120.32
C · B	S 57°-42'-39" W	250.60	C	170.73	-150.97
B · A	S 61°-21'-05" W	217.05	B	36.86	-362.81
A · H	S 16°-28'-51" E	217.50	A	-67.20	-553.29
H · U	S 84°-56'-07" E	287.00	H	-275.76	-491.59
			U	-301.10	-205.71

$$\text{Inverse Bearing} \atop (U \cdot W) = \text{Tan}^{-1} \left[\frac{E_2 \cdot E_1}{N_2 \cdot N_1} \right] = \left[\frac{\cdot(205.71) \cdot 0}{\cdot(301.10) \cdot 0} \right]$$

$$\text{Inverse Bearing} \atop (U \cdot W) = \text{N } 34°\text{-}20'\text{-}27" \text{ E}$$

$$\text{Inverse Distance} = \left[\left(E_2 \cdot E_1\right)^2 + \left(N_2 \cdot N_1\right)^2 \right]^{0.5} = 364.66 \text{ ft.}$$

FIG. 6.31 Example of inverse calculations to determine horizontal control for a street.

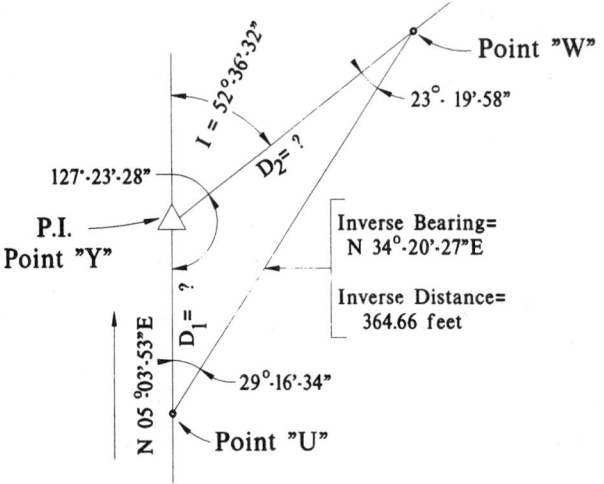

$$\frac{a}{\text{Sin A}} = \frac{b}{\text{Sin B}} \quad ; \quad \frac{364.66'}{\text{Sin } 127°\cdot23'\cdot28''} = \frac{D_1}{\text{Sin } 23°\cdot19'\cdot58''}$$

Distance D_1 = 181.79 feet.

$$\frac{a}{\text{Sin A}} = \frac{c}{\text{Sin C}} \quad ; \quad \frac{364.66'}{\text{Sin } 127°\cdot23'\cdot28''} = \frac{D_2}{\text{Sin } 29°\cdot16'\cdot34''}$$

Distance D_2 = 224.45 feet.

FIG. 6.32 Horizontal control relationships for the sample road centerline.

Other centerline curve nomenclature is calculated using R and I as follows:

(a)   Tangent distance $(T) = R\tan(I/2)$ (6.8)

      or 110 feet × tan 26°18'16". The curve's tangent is therefore 54.38 feet.

(b)   Curve arc length $(L) = 3.14 \times 2 \times R \times I/360$ (6.9)

      or 3.14 × 2 × 110 × 52.508889/360. The curve's arc length is 101.00 feet. (Note: I is in decimal degrees.)

(c)   Chord length $(CH) = 2R\sin(I/2)$ (6.10)

      or 2 × 110 × sin 26°18'16". The curve's chord distance is therefore 97.49 feet.

(d)   Degree of curve $(D) = (100 \times I)/L$ (6.11)

      or 100 × 52.608889/101.00. The corresponding degree of curve is 52.088009°, or 52°05'17".

Station calculations for key control points along the road centerline can now begin starting with station 0 + 00. Usually key control points consist of PIs, PCs, and PTs

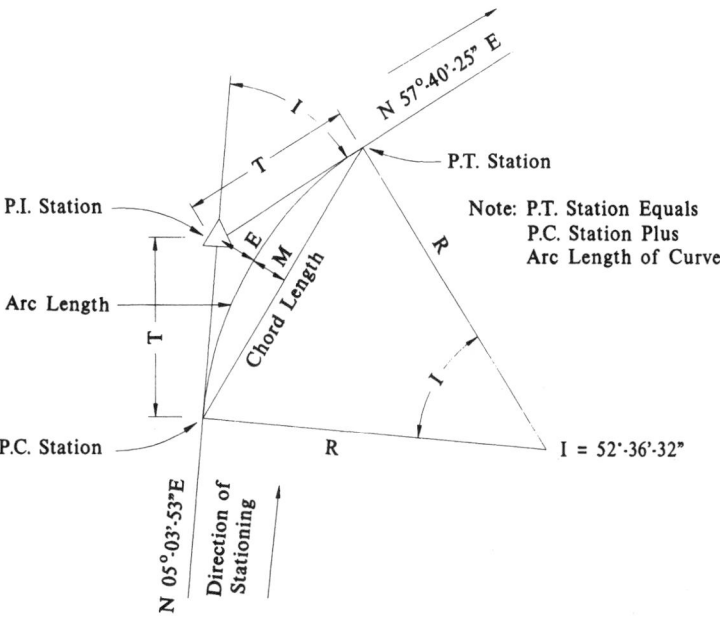

Curve Nomenclature

$L = (100 \times I)/D$, or $(2 \times R \times \pi \times I)/360$

$T = R \ \text{Tan} \ (I/2)$

$E = (T^2 + R^2)^{0.5} \ R$

$M = R \cdot (R^2 - (\text{Chord Dist.}/2)^2)^{0.5}$

Chord Length $= 2 \ R \ \text{Sin} \ I/2$

Where D = Degree of Curve by Arc Definition

FIG. 6.33   Horizontal highway curve nomenclature.

[also points of vertical intersection (PVIs), points of vertical curvature (PVCs), and points of vertical tangency (PVTs) and high/low points for vertical curves described in Section 6.4.2]. The beginning point (station 0 + 00) for this example is defined by the centerline intersection of the existing and proposed roads. To establish this point relative to the proposed road's alignment, one-half of the existing road's total right-of-way (50 feet) must be added to the distance of line U–Y. In the field, it is helpful to install a marker at this beginning point by measuring from the site's existing property lines and monuments. This installed marker locates the proposed road's beginning, which is useful for staking out the road for construction.

Point U would be assigned a station value of station 0 + 25 and point Y would have a PI station value of station 2 + 06.79 (181.79 feet + 25.00 feet). The corresponding PC station for this horizontal curve is station 1 + 52.41 (206.79 − 54.38), and the PT station is 2 + 53.41 (PC station plus the arc length, or 152.41 + 101.00). Point W therefore has a station value of station 4 + 23.48 (PT station + D_2 − T, or 253.41 + 224.45 − 54.38).

In this case the scaled distance on the preliminary plan from point W to the point where the cul-de-sac centerline is offset is approximately 95 feet. Therefore, the PI station of the cul-de-sac is $4 + 23.48 + 95.00$, or station $5 + 18.48$, and the offset station of the cul-de-sac's center is station $5 + 43.48$ ($518.48 + 25.00$). If the proposed road's geometry does not lend itself to scaling the last segment because some geometric relationship is desired, such as having the cul-de-sac tangent to a property line, a procedure to determine this distance is presented in Section 6.4.3.

If the road under design includes several horizontal curves, the PI-to-PI bearings and distances for the entire roadway should first be determined. Then each horizontal curve's nomenclature can be calculated and used to obtain road stations. Figure 6.34 illustrates this situation where the depicted road's centerline consists of four horizontal curves (one at each PI) and associated tangent segments. Should the proposed road not end with a cul-de-sac because it intersects another road (or itself by looping), the resulting centerline horizontal alignment (bearings and distances) must represent a closed traverse. This should be verified by the designer prior to beginning vertical control design. Once this has been verified, a plan view of the road can be plotted by using the completed horizontal alignment and by offsetting right-of-way lines and pavement edges (or curbing) from the centerline geometry since they are normally parallel to the centerline. In some TND designs the outside arc forming the edge of the right-of-way is omitted. The right-of-way is then defined by the tangents intersecting at a point, resulting in a nonsymmetrical right-of-way being created.

6.4.2 Vertical Control

Control for street and road vertical alignment depends on establishing the relationship of the road's grade and elevation to the site's topography. This process can begin once the road's horizontal control is established because it is necessary to horizontally locate existing contour lines where they intersect the proposed road's centerline in order to draw the ground profile.

The ground profile shows in an elevation view the surface characteristics of the existing ground along the proposed road centerline. This profile line indicates whether the ground (before construction) is level and uniformly sloped or consists of irregular features. To construct this ground profile, the existing ground elevations are obtained where contour lines cross the proposed road's centerline using the approved preliminary plan. Intercept stations for each contour line must be defined by scaling the preliminary plan. On the profile drawing, road stationing is plotted along the bottom (abscissa) while elevators are plotted vertically (ordinates). Figure 6.35 illustrates a sample plan view of a portion of a proposed road and the corresponding existing ground profile.

Once the existing ground profile is plotted, a preliminary proposed road grade can be superimposed on the profile beginning at the edge of the existing pavement for uniform transition between the existing road surface and that of the proposed road. This proposed road grade usually represents the finished road surface in land developments, although the road's subgrade can be shown as an alternative. Regardless, the proposed road grade must be clearly labeled to indicate what it represents. This is important during construction because field grade stakes used for construction control must reflect intended design elevations, requiring the earthwork contractor to account for pavement thickness when grading the roadbed.

FIG. 6.34 Example of a layout for horizontal control calculations using a road with four horizontal curves.

FIG. 6.35   Plotting the existing ground profile on a road plan and profile sheet using contour lines.

The direction of proposed roadway slope should be consistent with the approved preliminary development plan. For example, catch basins shown on the approved preliminary plan that are located in low areas must guide the designer to provide low profile elevations (low point on a sag vertical curve) there. This will allow stormwater runoff to concentrate in accordance with the approved preliminary plan. For developments utilizing curb and gutter sections, the roadbed should be placed in a cut environment, if possible, to allow runoff from adjacent property to flow across the top of the curb and into the street gutter for removal. Cut sections are represented on the profile wherever the existing ground is located above the proposed road grade.

Other criteria for laying out the preliminary road profile include minimizing earthwork (Chapter 9), providing necessary sight distances, and providing road grades that are within acceptable regulatory limits. The centerline road grade for vertical control from PVI to PVI is straight; therefore, initially the designer only needs to establish PVI stations and corresponding elevations to verify that the computed tangent gradients (or grades) are acceptable. At this design stage, the corresponding elevations of each station and included elevations within vertical curves are not necessarily needed (see Figure 6.36) because modification to the road's vertical alignment might be required later to accommodate design requirements for storm drains or sanitary sewers.

As in the case of horizontal control, if the road system consists of a closed centerline loop, then the vertical control network must also be a closed traverse relative to stations, elevations, and included grades. The design engineer cannot begin design of the storm drainage system if the proposed road grades are greater than the maximum allowed or less than the minimum allowed. PVI stations or elevations may have to

FIG. 6.36 Preliminary design of a road profile.

be altered to give acceptable results. Completion of the system includes calculating vertical curve elevations and tangent elevations outside of the vertical curves once the storm and sanitary sewer systems have been evaluated.

When completing the design, symmetrical parabolic vertical curves are most commonly used to join vertical tangents in land developments, where $l_1 = l_2$. Vertical curves can either be crest types or sag types.

Figure 6.37 illustrates a typical 400-foot crest vertical curve. Intermediate elevations for stations included within the vertical curve and the low or high point data can be obtained, as shown in Figure 6.37, using the following procedure:

(a) Determine the total change of grade in the vertical curve (A) using $g_2 - g_1$, where a negative A value denotes a crest vertical curve. Data included in Figure 6.37 indicate that $A = -16$ percent, that is -7 percent $- (+9$ percent).

(b) Determine the vertical offset (E) from the PVI to the curve (plus for sag curves, minus for crest curves) by multiplying the total change in grade (in percent) by the curve's total length (in stations) and dividing the product by the constant value 8. In this case $E = (-16 \times 4)/8$, or -8 feet. Vertical offsets for intermediate points between the PVC station and the PVI or the PVI station and the PVT can be determined using

$$y = E \left(\frac{x}{l}\right)^2 \tag{6.12}$$

where x is the horizontal distance from the PVC or PVT to the point under investigation in feet, l is the horizontal distance from the PVC to the PVI or the PVI to the PVT in feet, and E is the vertical offset at the PVI in feet.

FIG. 6.37   Example of a 400-foot symmetrical crest vertical highway curve.

Station	Tan. Elev.	(x/l)	(x/l)(x/l)	Vert. Off.	Curve El.
28+00 PVC	21.12	0	0	0.00	21.12
+50	25.62	1/4	1/16	0.50	25.12
29+00	30.12	1/2	1/4	2.00	28.12
+50	34.62	3/4	9/16	4.50	30.12
30+00 PVI	39.12	1	1	8.00	31.12
+50	35.62	3/4	9/16	4.50	31.12
31+00	32.12	1/2	1/4	2.00	30.12
+50	28.62	1/4	1/16	0.50	28.12
32+00 PVT	25.12	0	0	0.00	25.12

FIG. 6.38   Example of vertical curve calculations.

Figure 6.38 includes the calculations for points between stations $28 + 00$ and $32 + 00$.

To determine the high or low point relative to the PVC or PVT station and elevation, the following can be used:

Distance to high or low point from PVC:

$$\frac{L(g_1)}{g_1 - g_2} \tag{6.13}$$

Elevation change to high or low point from PVC elevation:

$$\frac{L(g_1)^2}{2(g_1 - g_2)} \tag{6.14}$$

Distance to high or low point from PVT:

$$\frac{L(g_2)}{g_1 - g_2} \tag{6.15}$$

Elevation change to high or low point from PVT elevation:

$$\frac{L(g_2)^2}{2(g_1 - g_2)} \tag{6.16}$$

In this example the high point from the PVT is determined to be at station $30 + 25$ using the results of Equation 6.15. The corresponding difference in elevation between the PVT's elevation and that of the high point is found to be 6.13 feet; therefore, the elevation of the high point is 31.25 feet (6.13 feet + 25.12 feet).

Road elevations corresponding to other stations along the vertical tangent can be calculated by determining the horizontal distance between a station of known elevation and the station under investigation and multiplying this distance by the intervening slope. This product gives the difference in elevation between the point of known elevation and the one under investigation. The desired elevation is obtained by adding or subtracting this difference from the referenced elevation. For example, to determine the road elevation at station $27 + 00$ (using the PVI at station $30 + 00$ for reference), the horizontal distance between the two stations is 300 feet. The difference in elevation between the two stations is 27 feet (300 feet $\times$ 0.09). The required elevation at station $27 + 00$ is 12.12 feet (39.12 $-$ 27.00). Elevations at any station can be obtained using this procedure.

A completed plan and profile sheet showing part of a road system are shown in Figure 6.39.

6.4.3 Special Geometric Conditions

Occasionally, the design professional must determine horizontal control relationships for roads, parking areas, and boundaries for parcels of land that require special techniques. This can include applications of a compound or reverse curve tangent to two lines or a single curve tangent to a single line. Figure 6.40 contains illustrations of each of these conditions.

When a road system contains a reverse curve (Figure 6.40b) that is to be made tangent to a property line (where the road's right-of-way edge must not encroach on the adjoining property), a beginning point, pont B (the PC), must be assigned or known, as shown in Figure 6.41, as well as the magnitude of the two radii (R_1 and R_2). Not known are the curves' remaining nomenclature and the distance from reference point D to the point of tangency along the property line (point H). Figure 6.41 shows these conditions where the right-of-way edge is used because it joins the property line at point H (not the road's centerline). Centerline control can be determined once I_1 and I_2 are known since I_1 and I_2 are the same for the edge of right-of-way and the centerline. Centerline curve data for the road can be calculated by adding or subtracting one-half of the total right-of-way distance to R_1 and R_2.

The solution to the reverse curve problem shown in Figure 6.41 is as follows:

(a) Establish the relationship of point A, the first arc's center, to the property line B–C by making the bearing of line A–B perpendicular to the bearing of line B–C at point B (the given PC), and assign the appropriate quadrant to reflect the direction from point A to point B. The distance A–B is assigned as R_1 (given). Half the problem is now solved because the location of one of the arc centers is now established.

FIG. 6.39 Example of a completed plan and profile for a street.

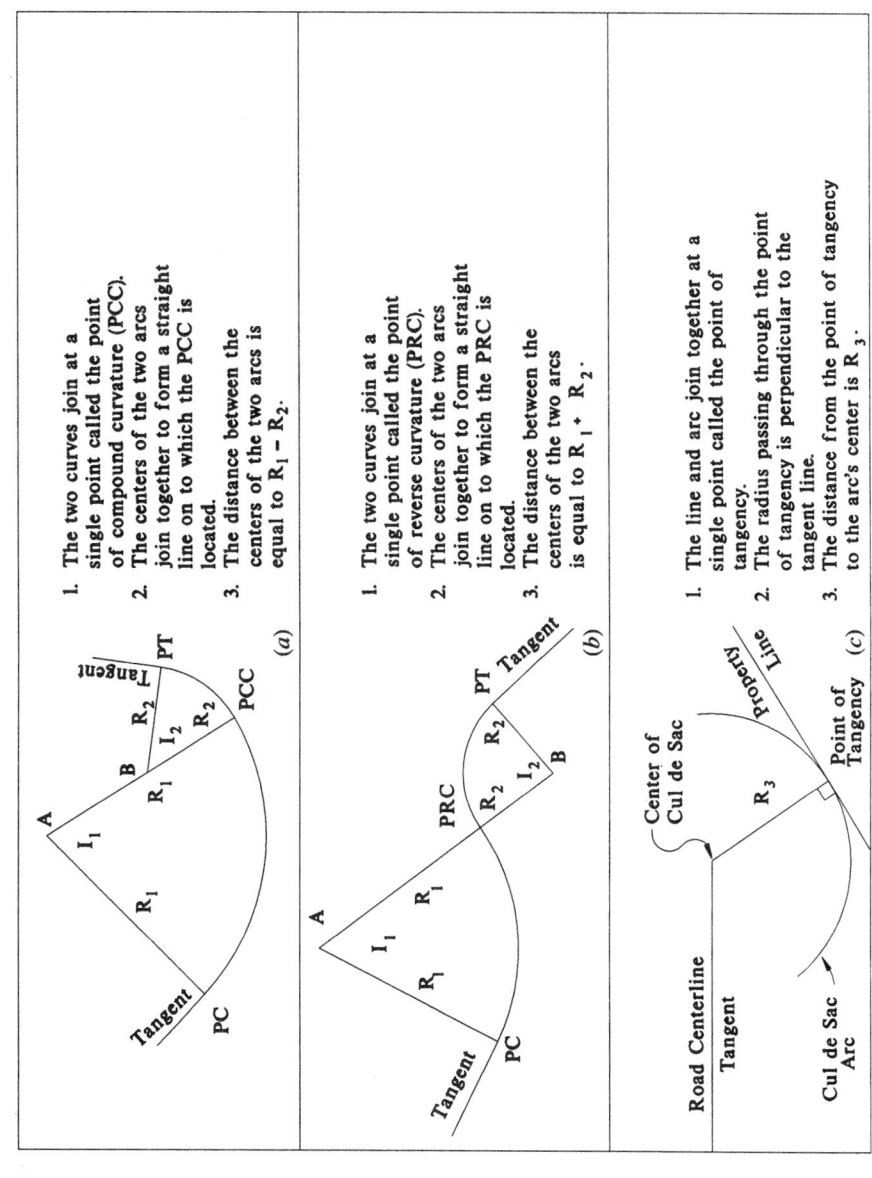

FIG. 6.40 Special geometric conditions for tangent horizontal curves: (a) compound curves; (b) reverse curves; (c) single curve tangent to a straight line.

297

FIG. 6.41   Reverse highway curve tangent to two lines.

To locate the center of the second arc (point G), two geometric conditions must prevail: First, the position of point G must be such that when an arc having a radius of R_2 is constructed, it will be tangent to line $D-H$ (the distance $D-H$ is unknown at this time). Second, the distance between the centers of the two arcs (points A and G) must be $R_1 + R_2$. These two conditions can be used to obtain the location of the second arc's center by continuing with the following solution process:

(b) Establish an imaginary line that represents a locus of points, any of which could represent the center of the second arc that would be tangent to line $D-H$. This is achieved by assigning some value to "X" (possibly 10 feet) to locate an intermediate point E. Through point E a line representing a fictitious radius is positioned, having a bearing perpendicular to side $D-H$ and a magnitude equal to R_2 (line $E-F$). A line $F-G$ passing through F that is parallel to line $D-H$ represents a locus of points, any of which, if used as a center, would produce an arc that is tangent to line $D-H$ if a radius of R_2 is used. The remaining problem is that distance $F-G$ is still unknown.

(c) The only available relationship that can be used to locate the second arc's center is that the distance between the centers of the two arcs is equal to $R_1 + R_2$. Therefore, an additional arc equal to $R_1 + R_2$ is used with point A as its center. This additional arc intersects line $F-G$ at point G, the desired point that locates the second arc's position.

(d) The point of reverse curvature (PRC) is located along the line $A-G$ (R_1 distance from A or R_2 distance from point G. Here, I_1 can be determined using bearings $A-B$ and $A-$PRC. Since R_1 and I_1 are known, the remaining nomenclature for curve 1 can be computed.

(e) Point H (the point of tangency) is located from point G on a line perpendicular to line $D-H$ and at a distance equal to R_2. Now, I_2 can be determined by using bearings $G-$PRC and $G-H$. The rest of the second arc's nomenclature can now be determined.

(f) The distance $D-H$ equals X plus the length of line $F-G$. The road centerline alignment can now be determined from the edge of right-of-way data for this portion of the road.

To solve this problem mathematically using coordinate geometry, coordinate values for points A, B, C, D, E, and F must be obtained using latitudes and departures. Line $A-F$ is inversed to obtain its corresponding bearing and distance. The bearing of line $F-G$ is known (same as the property line) and the distance $A-G$ is known ($R_1 + R_2$). The unknown values are the distance of line $F-G$ and the unknown direction of line $A-G$. These unknowns can be determined once triangle AFG is solved using first the law of cosines and then the law of sines. The resulting interior angles of this triangle can be used to determine the remaining bearing. From this, I_1 and I_2 can be calculated along with both curves' remaining nomenclature. The distance $D-H$ is obtained by adding the value of X to the length of the triangle's side $F-G$.

If a compound curve is being designed instead of a reversed curve, each step just described is followed, except that instead of using an arc equal to $R_1 + R_2$ from point A to help find the location of the second curve as described in step (c), the arc's magnitude must be $R_1 - R_2$. Figure 6.42 illustrates a compound curve tangent to a property line. In this case the outside right-of-way edge is used for calculations because it is this outside edge that must be tangent to the property line.

Another special problem sometimes encountered by the designer is determining the location of a cul-de-sac's center when a cul-de-sac's edge is tangent to a property line.

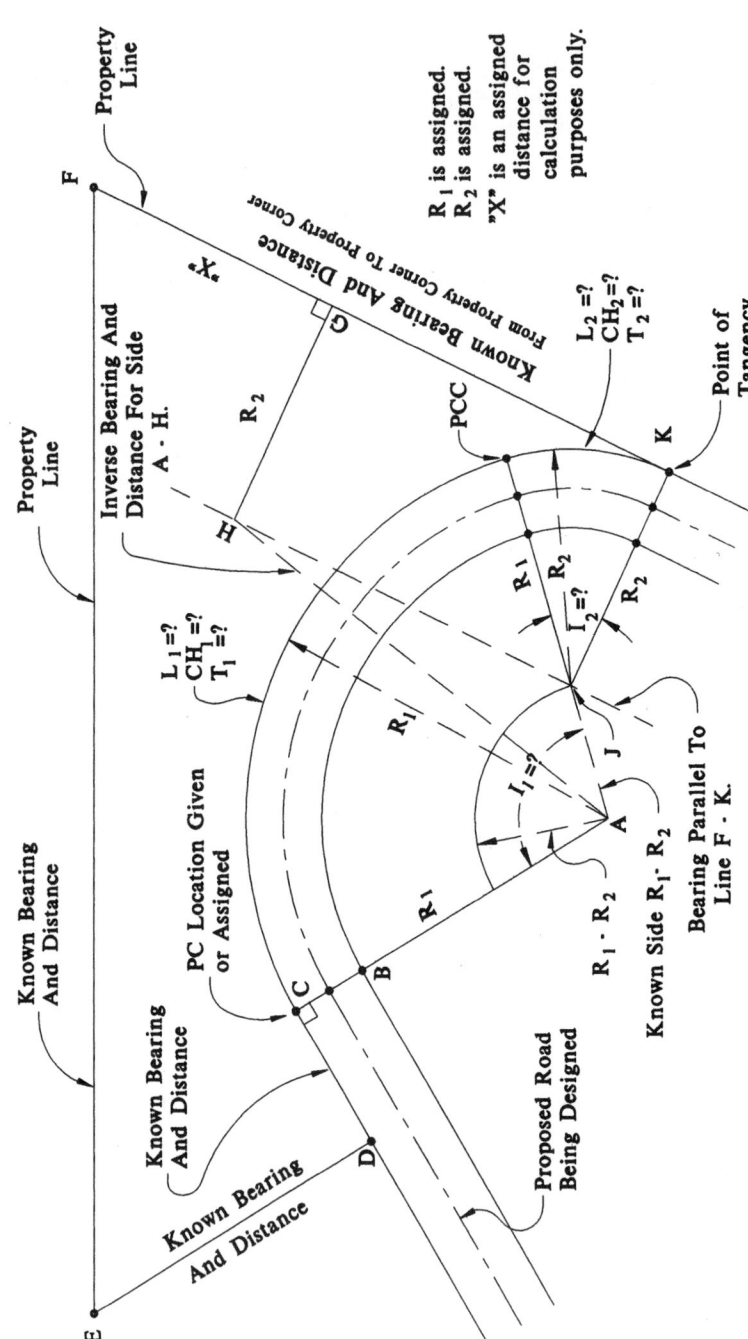

FIG. 6.42 Compound highway curve tangent to two lines.

300

To solve this problem, the bearing of the street's tangent where the cul-de-sac's center is located (line *A–G*) must be assigned or known (the distance is unknown). Also, the cul-de-sac's radius must be known. Figure 6.43 illustrates this situation.

In this example, the cul-de-sac's center is located along bearing *A–G*. The center must also be located where an arc of radius R_3 will be tangent to line *D–H*. The distance of *D–H* is unknown. The solution can be obtained by intersecting two bearings formed by lines *A–G* and *F–G*. As in the case with compound or reverse curves, line *F–G* must be established where every point along that line could represent the cul de sac's center, such that if an arc having a radius equal to R_3 were constructed, it would be tangent to line *D–H*. The arc's center, however, must also be located along line *A–G*. In this case, coordinates for points *A*, *B*, *C*, *D*, *E*, and *F* must be obtained using latitudes and departures. Line *A–F* is inversed to obtain its corresponding bearing and distance. The bearing of line *A–G* is known, as is the bearing of line *F–G* (same as the back property line). The unknown distances of two sides of triangle *AFG* can be solved using the law of sines, as the three interior angles can be obtained from bearings and one distance is known from inversing.

6.4.4 Computer Applications For Civil Design

Computer software available for collaborative use by today's design professionals automate much of the previously described steps that must be taken to produce a completed road plan. AutoCAD® Land Development Desktop (LDDT), for example, allows the user to utilize surveying data to create civil designs involving plan and profile sheets, cross sections, and more. The graphical user interface for civil design is illustrated in Figure 6.44. Creation of detention ponds, hydrology analysis, and culvert design (see

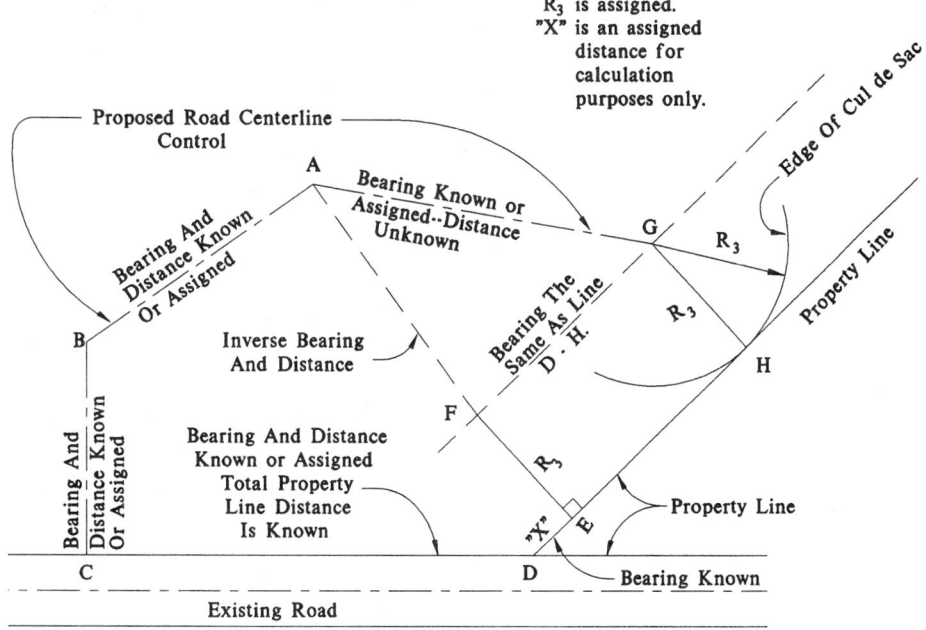

FIG. 6.43 Design of a street having a cul-de-sac tangent to a property line.

Chapter 7) along with site grading and earthwork (see Chapter 9) are also available with this software package.

6.5   PAVEMENT DESIGN

According to the Asphalt Institute's (1991) manual on thickness design, pavement design involves three basic steps: selecting construction materials, determining thickness requirements for each material used, and establishing construction-related requirements (usually covered in the project's construction specifications). Construction-related requirements have undergone major revisions during the past decade resulting from the Strategic Highway Research Program (SHRP) that was established by Congress in 1987. For example, Superpave (Superior Performing Asphalt Pavements), a National Academy of Sciences trademark, incorporates both performance based materials and design environmental conditions to improve control of rutting, fatigue, and low-temperature cracking where increased performance results. Superpave

FIG. 6.44   Example civil design graphical user interface. Source: Autodesk, Inc. *Autodesk® Land Development Desktop—Civil Design*, 1998. Reproduced by permission of Autodesk, Inc., San Rafael, CA.

is comprised of three major components: (1) asphalt binder specifications, (2) mixture design and analysis, and (3) a computer software system. The Superpave mix design and analysis system is being widely accepted and implemented by federal, state, and some local governments throughout the country. For more information on Superpave, the reader is referred to publications of the Asphalt Institute (1996, 1997) and other available sources.

Land developments require pavement as part of the incorporated site improvements to produce a finished product. Pavements can be rigid (portland cement concrete) or flexible (asphaltic concrete). The flexible pavement may be of two types—full-depth asphalt or a composite pavement. A composite pavement consists of a granular base with a hot-mix asphalt surface. Parking lots and residential streets often are constructed of flexible-type pavements, where maintenance crews find it relatively easy to cut pavement when necessary for effecting utility repairs. Also, resurfacing is facilitated by accommodating direct overlays or, if the existing pavement is unsound, by milling and hot-mix recycling. The reader is referred to other sources for rigid pavement design and alternate methods of flexible pavement design (such as the approach used by the AASHTO). The Asphalt Institute's pavement thickness design procedure follows.

6.5.1   The Asphalt Institute Method

The Asphalt Institute's (1991) manual on thickness design contains procedures based on test data, theory, and experience for structural design of various asphalt pavements. Two stress-strain relationships shown in Figure 6.45 are incorporated into this method as follows:

(a)   Wheel loads are transferred to the pavement's structure at the tire/pavement contact interface as a uniform vertical load P_0. Stresses are spread through the pavement where a reduced maximum vertical stress, P_1, results at the subgrade surface, as indicated in Figure 6.45a.

(b)   The wheel load depicted in Figure 6.45b also causes pavement deflection where compressive stress occurs near the pavement's surface and tension near the pavement's bottom. The resulting strains are used to determine required pavement thicknesses for ten different pavement structures as a result of computer analysis using a program called DAMA (available from the Asphalt Institute). The included design nomographs address the following pavement structures:

(1)   full-depth asphaltic concrete;

(2)   emulsified asphalt mix, Type I, comprised of densely graded aggregate and emulsified asphalt;

(3)   emulsified asphalt mix, Type II, comprised of semiprocessed, crusher run, bank run, or pit run aggregates and emulsified asphalt;

(4)   emulsified asphalt mix, Type III, comprised of sands or silty sands and emulsified asphalt; and

(5)   untreated aggregate base thicknesses of 6 and 12 inches.

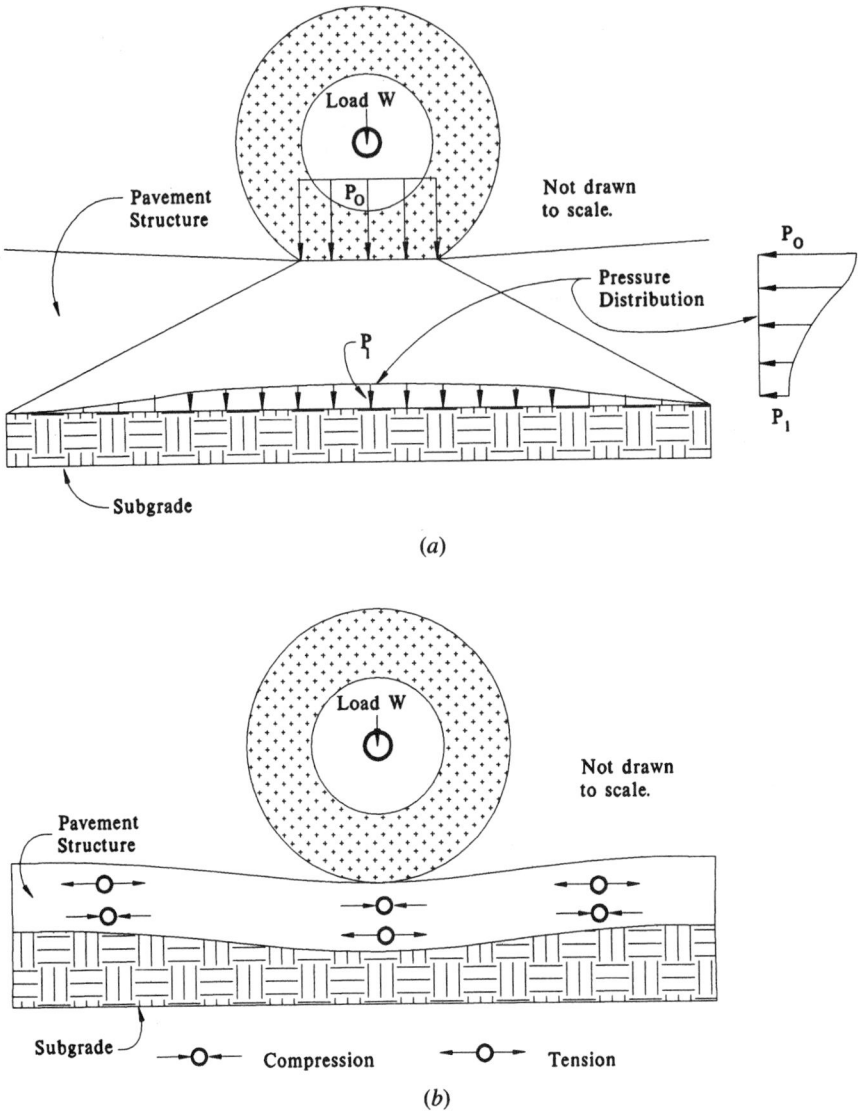

FIG. 6.45 Load distribution and pavement stresses for pavement thickness design: (a) spread of wheel-load pressure through pavement structure; (b) pavement deflection results in tensile and compressive stresses in pavement structure. Source: Asphalt Institute, 1991. Reprinted by permission of the Asphalt Institute, Lexington, KY.

6.5.2 Traffic Analysis

Estimated initial and future traffic mixtures and volumes are required to design a pavement's thickness. Typically a pavement analysis horizon will extend to 20 years for residential areas. The analysis horizon is important for establishing the number of anticipated 18,000-pound equivalent single-axle loads (ESALs) expected to pass over the pavement. Each axle load is presumed to be applied to the pavement with two sets of dual tires that are represented by two 4.52-inch circular plates spaced 13.57 inches apart while providing a surface contact pressure of 70 lb/in.^{2}

Due to the mixture of motor vehicle types, an estimate of the traffic's composition must be made to equate anticipated traffic axle loads to 18,000-pound ESALs. Should vehicular traffic volumes be expected to grow throughout the analysis horizon, appropriate growth factors must be applied. For local streets within residential developments, the estimated phased population or numbers of dwelling units can be used to determine incremental increases in traffic if off-site traffic is not involved with the maximum being generated at project build-out.

According to the Asphalt Institute (1991), national average traffic growth rates range from 3 to 5 percent compounded each year, however, rates ranging from 4 to 9 percent have been suggested for rural roads and from 8 to 10 percent for some interstate highways. Figure 6.46 can be used to obtain appropriate factors that can be used to multiply with the initial ESAL to estimate future traffic with the percentage used to reflect local growth trends. Future traffic that reflects growth should be checked against overloading the facility's capacity, thus promoting reasonable results.

ESAL determination requires knowledge of the number of various vehicle types (passenger, single-unit trucks, and multiple-unit trucks). The distribution of vehicle types can be estimated from state or local data associated with similar facilities, although Figure 6.47 can be used if local information does not exist.

According to the Asphalt Institute (1991), overall, heavy trucks (two-axle six-tire trucks and larger) average 7 percent of the total traffic volume on all highway classes,

Design Period, Years (n)	Annual Growth Rate, Percent							
	No Growth	2	4	5	6	7	8	10
1	1	1.00	1.00	1.00	1.00	1.00	1.00	1.00
2	2	2.02	2.04	2.05	2.06	2.07	2.08	2.10
3	3	3.06	3.12	3.15	3.18	3.21	3.25	3.31
4	4	4.12	4.25	4.31	4.37	4.44	4.51	4.64
5	5	5.20	5.42	5.53	5.64	5.75	5.87	6.11
6	6	6.31	6.63	6.80	6.98	7.15	7.34	7.72
7	7	7.43	7.90	8.14	8.39	8.65	8.92	9.49
8	8	8.58	9.21	9.55	9.90	10.26	10.64	11.44
9	9	9.75	10.58	11.03	11.49	11.98	12.49	13.58
10	10	10.95	12.01	12.58	13.18	13.82	14.49	15.94
11	11	12.17	13.49	14.21	14.97	15.78	16.65	18.53
12	12	13.41	15.03	15.92	16.87	17.89	18.98	21.38
13	13	14.68	16.63	17.71	18.88	20.14	21.50	24.52
14	14	15.97	18.29	19.16	21.01	22.55	24.21	27.97
15	15	17.29	20.02	21.58	23.28	25.13	27.15	31.77
16	16	18.64	21.82	23.66	25.67	27.89	30.32	35.95
17	17	20.01	23.70	25.84	28.21	30.84	33.75	40.55
18	18	21.41	25.65	28.13	30.91	34.00	37.45	45.60
19	19	22.84	27.67	30.54	33.76	37.38	41.45	51.16
20	20	24.30	29.78	33.06	36.79	41.00	45.76	57.28

If Annual Growth is zero, Growth Factor = Design Period.

FIG. 6.46 Traffic growth factors for pavement thickness design. Source: Asphalt Institute, 1991. Reprinted by permission of the Asphalt Institute, Lexington, KY.

whereas on a regional basis heavy trucks can range from 2 to 25 percent or more, and on urban highways they range from 5 to 15 percent or more. Peak-hour truck traffic can be estimated at one-half the daily average percentage of trucks on urban arteries and between one-half and two-thirds on rural roads. Again, local traffic characteristics should be used if available.

Of primary importance is the determination of the design lane. On multilane highways the design lane is usually the outside lane, while on two-lane highways and streets it can be either lane. Local data can reveal the relative proportions of trucks expected to use the design lane; however, Figure 6.48 can be used if local data are not available.

To estimate ESALs from the anticipated traffic, either the number of various axle loads or the number of vehicles in each weight class must be known. Figure 6.49 can be used to convert different axle loads to equivalent 18,000-pound axle loads. When the axle loads of the vehicles are not known, Figure 6.50 can be used to estimate truck factors (TFs). Truck factors, when multiplied by the number of trucks, yield the equivalent number of 18,000-pound axle loads generated by those trucks. Therefore, the TF is the number of 18,000-pound single-load applications caused by a single passage of a vehicle.

Traffic on residential streets and parking lots usually is comprised of automobile

Truck Class	Percent Trucks *						
	Rural Systems				Urban Systems		
	Minor Arterial	Collectors		Range ***	Minor Arterial	Collectors	Range ***
		Major	Minor				
Single-unit trucks							
2-axle, 4-tire	71	73	80	43-80	84	86	52-86
2-axle, 6-tire	11	10	10	8-11	9	11	9-15
3-axle or more	4	4	2	2-4	2	<1	<1-4
All single-units	86	87	92	53-92	95	97	66-97
Multiple Unit Trucks							
4-axle or less	3	2	2	2-5	2	1	1-5
5-axle **	11	10	6	6-41	3	2	2-28
6-axle or more **	<1	1	<1	<1-1	<1	<1	<1-1
All multiple units	14	13	8	8-47	5	3	3-34
All trucks	100	100	100		100	100	

   * Compiled from data supplied by the Highway Statistics Division, U.S. Federal Highway Administration.
  ** Including full-trailer combinations in some states.
*** These values represent truck distributions comprised of the highway classes included in this figure and for Interstate and other principal highway classes not shown.

FIG. 6.47 Distribution of trucks on different classes of highways, United States. Source: Asphalt Institute, 1991. Reprinted by permission of the Asphalt Institute, Lexington, KY.

Number of Traffic Lanes (Two Directions)	Percentage of Trucks in Design Lane
2	50
4	45(35-48)*
6 or more	40(25-48)*

*Probable range.

FIG. 6.48 Percentage of total truck traffic expected in the design lane. Source: Asphalt Institute, 1991. Reprinted by permission of the Asphalt Institute, Lexington, KY.

Gross Axle Load		Load Equivalency Factors *		
kN	lb	Single Axle	Tandem Axles	Tridem Axles
8.90	2000	0.00018		
17.80	4000	0.00209	0.00030	
26.70	6000	0.01043	0.00100	0.0003
35.60	8000	0.03430	0.00300	0.0010
44.50	10000	0.08770	0.00700	0.0020
53.40	12000	0.18900	0.01400	0.0030
62.30	14000	0.36000	0.02700	0.0060
71.20	16000	0.62300	0.04700	0.0110
80.00	18000	1.00000	0.07700	0.0170
89.00	20000	1.51000	0.12100	0.0270
97.90	22000	2.18000	0.18000	0.0400
106.80	24000	3.03000	0.26000	0.0570
115.60	26000	4.09000	0.36400	0.0800
124.50	28000	5.39000	0.49500	0.1090
133.40	30000	6.97000	0.65800	0.1450
142.30	32000	8.88000	0.85700	0.1910
151.20	34000	11.18000	1.09500	0.2460
160.10	36000	13.93000	1.38000	0.3130
169.00	38000	17.20000	1.70000	0.3930
178.00	40000	21.08000	2.08000	0.4870
222.40	50000	52.88000	4.86000	1.2200
267.00	60000		9.59000	2.5100
311.50	70000		17.19000	4.5200
356.00	80000		29.00000	7.4500

* From Appendix D of AASHTO Guide for Design Of Pavement Structures, American Association of State Highway And Transportation Officials, Washington, D.C. 1986.

FIG. 6.49 Axle load equivalency factors for pavement thickness design. Taken from *Guide for Design of Pavement Structures.* Copyright © 1986 by the American Association of State Highway and Transportation Officials, Washington, DC. Used by permission. Source: Asphalt Institute, 1991. Reprinted by permission of the Asphalt Institute, Lexington, KY .

traffic. Pavements designed using ESALs based on automobiles will not provide adequate pavement depths to withstand occasional traffic resulting from trash and refuse trucks, moving vans, and ready-mix concrete trucks. Estimates of use by these occasional vehicles must be included; otherwise minimum recommended pavement thicknesses must be used.

To illustrate estimating ESALs for a 330-unit single-family detached subdivision, consider these characteristics:

(a) Total number of dwelling units $= 330$.

(b) Average trips/day/unit $= 10$.

(c) Garbage pickup—two times a week.

(d) Delivery trucks—three times a week.

(e) Trash pickup—two times a week.

Vehicle Type	Rural Systems *				Urban Systems *		
	Minor Arterial	Collectors		Range ***	Minor Arterial	Collectors	Range ***
		Major	Minor				
Single-unit trucks							
2-axle, 4-tire	0.003	0.017	0.003	0.003-0.017	0.006	---	0.006-0.015
2-axle, 6-tire	0.280	0.410	0.190	0.19-0.41	0.230	0.13	0.13-0.24
3-axle or more	1.060	1.260	0.450	0.45-1.26	0.760	0.72	0.61-1.02
All single-units	0.080	0.120	0.030	0.03-0.12	0.040	0.16	0.04-0.16
Tractor semi-trailers							
4-axle or less	0.620	0.370	0.910	0.37-0.91	0.460	0.40	0.04-0.98
5-axle **	1.050	1.670	1.110	1.05-1.67	0.770	0.63	0.63-1.17
6-axle or more **	1.040	2.210	1.350	1.04-2.21	0.640	---	0.64-1.19
All multiple units	0.970	1.520	1.080	0.97-1.52	0.670	0.53	0.53-1.05
All trucks	0.210	0.300	0.120	0.12-0.52	0.070	0.24	0.07-0.39

* Compiled from data supplied by the Highway Statistics Division, U.S. Federal Highway Administration.

** Including full-trailer combinations in some states.

*** These values represent truck factors comprised of the highway classes included in this figure and for Interstate and other principal highway classes not shown.

FIG. 6.50 Distribution of truck factors (TFs) for different classes of highways and vehicles, United States. Source: Asphalt Institute, 1991. Reprinted by permission of the Asphalt Institute, Lexington, KY.

(f) Moving vans—in a young neighborhood, there is a four-year average turnover rate.

Determination of ESALs on a weekly basis is as follows:

1. For passenger cars, assume two each, 2000-pound axle loads with a load equivalency factor of 0.00018 (see Figure 6.49); this yields

$$330 \text{ units} \times 10 \text{ trips per unit per day} \times 0.00018 \times 2 \text{ axles} \times$$
$$7 \text{ days per week, or } 8.32 \text{ ESALs per week}$$

2. For garbage trucks assume three-axle, 8000-pound axle loads with a load equivalency factor of 0.03430 and allow for four passes per week; this yields

$$3 \text{ axles} \times 0.03430 \times 4 \text{ passes, or } 0.41 \text{ ESALs per week}$$

3. For delivery trucks assume two-axle, 6000-pound axle loads with a load equivalency factor of 0.01043 and allow for six passes per week; this yields

$$2 \text{ axles} \times 0.01043 \times 6 \text{ passes per week, or } 0.13 \text{ ESALs per week}$$

4. For trash and refuse pickup, assume: rear axle, 16,000 pounds with a load equivalency of 0.623; front axle, 6000 pounds with a load equivalency of 0.01043. Allowing four passes per week, this yields

$$(0.623 + 0.01043) \times 4 \text{ passes per week, or } 2.53 \text{ ESALs per week}$$

5. For moving vans assume: rear tandem axle, 34,000 pounds with a load equivalency factor of 1.095; middle tandem axle, 34,000 pounds with a load equivalency factor of 1.095; front axle, 12,000 pounds with a load equivalency factor of 0.189. Allowing for occupants consisting of young adults with an average estimated turnover rate of four years, the number of estimated dwellings being vacated or occupied per week is 330 dwellings/(4 years × 52 weeks per year), or 1.6 dwellings being turned over each week. Assuming two van passes for vacating the premises and two van passes for delivery of the new occupant's goods, the total estimated passes per week are 1.6 × 4, or 6.4 passes per week. The estimated ESALs are 6.4 × (1.095 + 1.095 + 0.189), or 15.23.

Determination of ESALs for the 20-year analysis horizon yields

$$20 \text{ years} \times 52 \text{ weeks per year} \times (8.32 + 0.41 + 0.13 + 2.53 + 15.23),$$

$$\text{or } 27{,}685 \text{ ESALs}$$

The percentage of traffic in the design lane is assumed to be 50 percent (see Figure 6.48); therefore, the design ESAL is 27,685 × 0.5, or 13,843 ESALs.

6.5.3 Materials Evaluation

The pavement thickness required to perform under anticipated traffic conditions is a function of the subgrade soil's strength. The strength of the subgrade is best measured by determining the resistance value of the soil (R) using AASHTO T 190-93 or by determining the resilient modulus (M_r) using the Asphalt Institute's *Soils Manual (MS-10)*. According to the Asphalt Institute (1991), an approximate resilient modulus can be estimated by

$$M_r \text{ in psi} = 1500 \text{ CBR} \tag{6.17}$$

$$M_r \text{ in psi} = 1155 + 555 \, (R \text{ value}) \tag{6.18}$$

Equations 6.17 and 6.18, relating the California bearing ratio (CBR) and R values to the subgrade soil resilient modulus, do not apply to granular untreated aggregate base and subbase materials according to the Asphalt Institute (1991).

To obtain a design value for the subgrade's representative resilient modulus, the following procedure can be used in conjunctions with Figure 6.51.

Traffic Level EAL	Design Subgrade Value, Percent
10,000 or less	60
Between 10,000 and 1,000,000	75
1,000,000 or more	87.5

FIG. 6.51 Subgrade design limits (EAL = equivalent axle load). Source: Asphalt Institute, 1991. Reprinted by permission of the Asphalt Institute, Lexington, KY.

(a) Obtain M_r values from various soil tests conducted to determine the subgrade's resilient modulus. These data can be obtained by converting CBR or R values to M_r values.

(b) List the test results in descending numerical order.

(c) Calculate for each M_r (beginning with the lowest value) the percentage of total values that it is equal to or greater than.

(d) Plot the results and fit a smooth curve through the points.

(e) From the graph prepared in step (d) obtain the design subgrade resilient modulus using Figure 6.51 as a guide.

For example, five random soil tests obtained from the proposed roadbed indicate the following M_r values: 6000, 7500, 8800, 9100, and 14,700 psi. Determination of the design M_r is shown in Figures 6.52 and 6.53, following steps (a)–(e) above. The Asphalt Institute's (1991) pavement design procedures assume the following:

(a) Cohesive subgrades will be compacted to a minimum of 95 percent of maximum density determined by AASHTO T 180-90, Method D, for the top 12 inches of fill and a minimum of 90 percent in fill areas below the top 12 inches.

Test Value, psi.	No Values Equal or Greater	Percent Equal to or Greater
14700	1	1/5 x 100 or 20%
9100	2	2/5 x 100 or 40%
8800	3	3/5 x 100 or 60%
7500	4	4/5 x 100 or 80%
6000	5	5/5 x 100 or 100%

FIG. 6.52   Determining the M_r value for pavement thickness design.

FIG. 6.53   Plot of field M_r values for pavement thickness design.

(b) Cohesionless subgrade soils will be compacted to 100 percent of the maximum density determined by AASHTO T 180-90, Method D, for the top 12 inches of fill and a minimum of 95 percent in fill areas below the top 12 inches.

(c) Asphalt concrete mix designs for both surface and base courses will be for dense-graded aggregates and made in accordance with the Asphalt Institute's (1988) *Mix Design Methods for Asphalt Concrete and Other Hot Mix Types (MS-2)*. Coarse aggregate for surface mixes should be comprised of at least 50 percent crushed material with two or more fractured faces.

(d) The prepared asphalt mix will be properly placed and compacted in the field to at least 96 percent of the average laboratory density (based on Marshall compacted specimens) or will achieve 92 percent of the maximum theoretical specific gravity.

(e) Emulsified asphalt base mixes will meet requirements set forth in Chapter VII of the Asphalt Institute's (1987) *A Basic Asphalt Emulsion Manual (MS-19)*. For dense graded emulsified aggregate base mixes, plant mixing is required to ensure uniform aggregate blending.

(f) Field compaction of emulsified asphalt base mixes shall be accomplished in accordance with procedures set forth by the Asphalt Institute (1987).

(g) Untreated aggregate base and subbase materials will be compacted to 100 percent of their maximum density as determined by AASHTO T 180, Method D, and will conform to the requirements contained in Figure 6.54; otherwise ASTM D 2940 will be used.

(h) Appropriate grades of asphalt cement will be incorporated into the mix design using guidelines included in Figure 6.55. This will help reduce low-temperature cracking and rutting in the installed pavement.

6.5.4   Pavement Structural Design Procedures

Recommended procedures for designing a pavement's thickness include initially determining the design traffic in terms of ESALs, the subgrade's design strength, M_r, and the types of surface and base courses potentially available for use.

The pavement thickness can be determined from these initial data. If incremental project phasing is proposed, stage construction can be analyzed and evaluated for

Test	Test Requirements	
	Subbase	Base
CBR, minimum*	20	80
or		
R-value, minimum*	55	78
Liquid Limit, maximum	25	25
Plasticity Index, maximum, or	6	NP
Sand Equivalent, minimum	25	35
Passing No. 200 sieve, maximum	12	7

Equations 6-17 and 6-18 relating CBR and R-value to subgrade soil resilient modulus do not apply to untreated aggregate base and subbase.

FIG. 6.54   Untreated aggregate base and subbase quality requirements for pavement thickness design. Source: Asphalt Institute, 1991. Reprinted by permission of the Asphalt Institute, Lexington, KY.

Temperature Condition	Asphalt Grades *	
Cold, mean annual air temperature less than and including 45-deg F.	AC-5 AR-2000 120/150 pen.	AC-10 AR-4000 85/100 pen.
Warm, mean annual air temperature between 45-deg F and 75-deg F.	AC-10 AR-4000 85/100 pen.	AC-20 AR-8000 60/70 pen.
Hot, mean annual temperature greater than and including 75-deg F.	AC-20 AR-8000 60/70 pen.	AC-40 AR-16000 40/50 pen.

* Both medium setting (MS) and slow setting (SS) emulsified asphalts are used in emulsified asphalt base mixes. They can be either of two types: cationic (ASTM D 2397 or AASHTO M 208) or anionic (ASTM D 977 or AASHTO M 140).

The grade of emulsified asphalt is selected primarily on the basis of its ability to satisfactorily coat the aggregate. This is determined by coating and stabilty tests (ASTM D 244, AASHTO T Other factors important in the selection are the water availability at the jobsite, anticipated weather at the time of construction, the mixing process to be used, and the curing rate.

FIG. 6.55 Selecting asphalt grade for pavement thickness design. Source: Asphalt Institute, 1991. Reprinted by permission of the Asphalt Institute, Lexington, KY.

economical results. Stage construction in some areas can take the form of initially installing the pavement's base or binder course, then waiting until near the end of the regulatory warranty period to install the remaining surface courses. This allows repairs to be made to distressed areas and at the same time provides a quality riding surface that will not be subjected to heavy construction traffic during the intervening time period.

Using procedures set forth by the Asphalt Institute (1991), the resulting pavement thickness satisfies two strain design criteria (the vertical compressive strain at the subgrade's surface and the horizontal tensile strain on the underside of the lowest bituminous layer). When asphaltic concrete wearing courses are placed over emulsified asphalt base materials classified as Type II or III (see Section 6.5.1), the minimum thickness of asphaltic concrete to be used is shown in Figure 6.56.

Emulsified asphalt mix Type I can be substituted in accordance with Figure 6.56; however, when the Type I base course is used, a surface treatment must be provided.

Nomographs for asphalt pavement design using emulsified asphalt base courses (allowing six-month curing periods) include the following for three sets of environmental conditions where the mean annual air temperature (MAAT) is less than or equal to 45°F, 60°F, or greater than or equal to 75°F:

(a) Type I—emulsified asphalt mixes made with processed, dense graded aggregates.

(b) Type II—emulsified asphalt mixes made with semiprocessed, crusher run, or bank run aggregates.

(c) Type III—emulsified asphalt mixes made with sands or silty sands.

Traffic Level EAL	Type II and Type III * Inches
10,000	2
100,000	2
1,000,000	3
10,000,000	4

* Asphalt concrete, or Type I
Emulsified asphalt mix with
a surface treatment, may be used
over Type II or Type III emulsified
asphalt base courses.

FIG. 6.56 Minimum thickness of asphalt concrete over emulsified bases. Source: Asphalt Institute, 1991. Reprinted by permission of the Asphalt Institute, Lexington, KY.

Figure 6.57 illustrates a typical design chart for emulsified pavements where Type I is shown. This and other included charts provide combined thicknesses of asphaltic concrete surface and base or emulsified asphalt surface with surface treatment and emulsified asphalt base.

For example, if the design ESAL is 10^5 and the design subgrade modulus, M_r is 6000 psi, the required pavement thicknesses (MAAT = 60°F) are shown in Figure 6.58 for the three types of emulsified bases.

For full-depth asphaltic concrete pavement, minimum thickness requirements can be obtained using a similar procedure with other design charts contained in the Asphalt Institute's MS-1 (1991). Figure 6.59 illustrates one of these chart types.

Full-depth pavement thicknesses can be obtained by entering the appropriate nomograph with the design ESAL and M_r values. For example, the minimum full-depth asphaltic concrete pavement for a design ESAL of 10^5 and M_r equal to 6000 psi (MAAT = 60°F) is 6.5 inches of asphaltic concrete surface and base.

Thickness determination for asphaltic concrete surface and base courses placed over untreated aggregate bases can be obtained by using nomographs similar to the one for 6 inches of untreated base shown in Figure 6.60. Minimum allowed thickness of asphaltic concrete over untreated base material is shown in Figure 6.61. Charts for pavement thicknesses using other depths of untreated base are contained in the Asphalt Institute's MS-1 (1991). To illustrate the use of Figure 6.61, a design of ESAL of 10^5 and an M_r value of 6000 psi (MAAT = 60°F) can be used to determine that the minimum thickness of asphaltic concrete pavement is 4.8 inches. The total pavement depth is the 6 inches of untreated aggregate base and the 4.8 inches of asphaltic concrete, or approximately 11 inches total. The engineer should develop several pavement structural designs and make an economic analysis of each. In many cases, alternate bids are invited to determine the least expensive design. Other factors, such as ease of construction and traffic conditions, should be considered in the final determination of pavement courses.

Design conditions not covered by the supplied monographs in the Asphalt Institute's MS-1 (1991) require use of computer analysis to obtain the desired results. The reader is referred to the Asphalt Institute IS-91 and IS-96 (1981a) for additional information pertaining to pavement thickness design.

FIG. 6.57 Emulsified asphalt mix Type I. Source: Asphalt Institute, 1991. Reprinted by permission of the Asphalt Institute, Lexington, KY.

Emulsified Asphalt Pavement Thickness Design				
Type Base	Minimum Thickness, In.	Surface Course	Final Design in Inches	
			Emul. Base	Surface
I	6.5	Surface Treatment	6.5	Surface Treatment
II	7.0	2 Inches of Asphaltic Conc.	5.0	2 Inches A.C.
III	9.5	2 Inches of Asphaltic Conc.	7.5	2 Inches A.C.

FIG. 6.58   Examples of emulsified asphalt pavement thickness design.

6.6   PEDESTRIAN FACILITY DESIGN

The most common pedestrian facilities in land developments are sidewalks. They are usually 4 or 5 feet wide in suburban-type developments unless unusually high usage is anticipated, such as near schools and public transportation stops. TND designs usually include at least a 5-foot-wide sidewalk on both sides of the street. *The Highway Capacity Manual* (TRB, 1994) contains information on determining various levels of service for pedestrian travel.

In suburban areas more often than in urban ones sidewalks are usually offset from the curb several feet to allow a planting area. This offset of several feet also allows protection by placing the pedestrian away from vehicular movements. Considerations for handicap access must be included, as indicated in Section 2.5.9, in addition to other pedestrian considerations contained in Section 2.5.8.

6.7   BICYCLE FACILITY DESIGN

Depending on the project's scope, the design professional can include various site improvements to accommodate bicycling. Section 2.5.10 includes some of these considerations.

Design of bicycle facilities includes provisions for safety and access. Consideration for separation of bicyclists from pedestrian and motor vehicle movements is warranted to help prevent accidents, especially when the travel way is shared. Properly designed bicycle paths are one approach that promotes safe travel on a shared basis.

6.7.1   Shared-Use Bicycle Path Design

Two-way bicycle paths require a 10-foot paved area with 2 feet of graded shoulder along each edge. Pavements should be smooth and free from depressions, bumps, and irregular surface patterns. The minimum pavement radii that can be safely used for horizontal control are shown in Figure 6.62, where a 2 percent superelevation is assumed to exist (AASHTO, 1999).*

Most bicycle paths are kept to less than a 5 percent gradient to minimize downslope speeds and up-hill fatigue. Of concern to the designer is the distance required to allow

---

* From *Guide for the Development of Bicycle Facilities*, Copyright © 1999 by The American Association of State Highway and Transportation Officials, Washington, D.C. Used by permission.

FIG. 6.59 Full-depth asphalt concrete. Source: Asphalt Institute, 1991. Reprinted by permission of the Asphalt Institute, Lexington, KY.

316

MAAT 60°F

FIG. 6.60 Untreated aggregate base, 6 inches thickness. Source: Asphalt Institute, 1991. Reprinted by permission of the Asphalt Institute, Lexington, KY.

Traffic EAL	Traffic Condition	Minimum Thickness of Asphalt Concrete
10,000 or less	Light traffic *	3 inches **
Between 10,000 and 1,000,000	Medium truck traffic	4 inches
1,000,000 or more	Heavy truck traffic	5 inches, or greater

* Includes parking lots, and driveways.

** For Full-Depth asphalt concrete or emulsified asphalt pavements
   a minimum thickness of 4-inches applies in this traffic region, as shown
   in Figures 6.53 and 6.55 .

FIG. 6.61 Minimum thickness of asphalt concrete over untreated aggregate base. Source: Asphalt Institute, 1991. Reprinted by permission of the Asphalt Institute, Lexington, KY.

Design Speed-V		Friction Factor (f)*	Minimum Radius, R	
km/h	(mph)	paved surface	meters	(feet)
20	(12)	0.31	10	(30)
30	(20)	0.28	24	(90)
40	(25)	0.25	47	(155)
50	(30)	0.21	86	(260)

* e = 2% with 20 degree lean angle

FIG. 6.62 Design radii for paved bicycle paths. Taken from *Guide for the Development of Bicycle Facilities.* Copyright © 1999 by The American Association of State Highway and Transportation Officials, Washington, DC. Used by permission.

a cyclist to stop. Figure 6.63 shows various stopping sight distances for various bicycle speeds and pavement gradients.

Stopping sight distances for bicycles can be limited by vertical curves and horizontal curves. Figures 6.64 and 6.65 depict examples of design constraints for each condition.

6.7.2 Miscellaneous Considerations

Bicycle lanes and paths require pavement marking to separate opposing traffic and to provide safety. Proper route markers also are necessary to warn bicyclists of potential hazards and general information. Specific marking applications are contained in the *Manual of Uniform Traffic Control Devices* (Federal Highway Administration, 1988).

Night lighting can reduce chances for accidents and provide user safety. This is essential if high volumes of bicycle traffic are expected during certain nighttime periods. Restraints and barriers are needed to prevent accidents from falls when crossing bridges or traveling close to a steep embankment or near the edge of an area being supported by a retaining wall.

The bicycle facilities should also include drainage improvements to remove potential runoff while avoiding grates and manhole covers that are not compatible with narrow bicycle tires.

$$S = \frac{V^2}{254(f \pm G)} + \frac{V}{1.4}$$

Where: S = stopping sight distance (m)
V = velocity (km/h)
f = coefficient of friction (use 0.25)
G = grade (m/m)(rise/run)

——— Descend
━━━ Ascend

FIG. 6.63 Minimum bicycle stopping sight distances, vs. grade for various design speeds. Taken from *Guide for the Development of Bicycle Facilities.* Copyright © 1999 by The American Association of State Highway and Transportation Officials, Washington, DC. Used by permission.

REFERENCES

American Association of State Highway and Transportation Officials (AASHTO). *A Policy on Geometric Design of Highways and Streets.* Washington: AASHTO, 1994, pp. 8, 13–15, 88–89, 119–124, 179, 193, 220, 274–295, 419–421, 428–442, 461–465.

American Association of State Highway and Transportation Officials (AASHTO). *Guide for the Development of New Bicycle Facilities.* Washington: AASHTO, 1999, pp. 38–39, 41, 43, 45.

Asphalt Institute. *Soils Manual* 5th ed. (MS-10). Lexington, KY: Asphalt Institute, 1991.

Asphalt Institute. *Full-Depth Asphalt Pavements for Parking Lots, Service Stations, and Driveways (IS-91).* College Park, MD: Asphalt Institute, 1981a.

A (%)	S = Stopping Sight Distance (m)																		
	10	15	20	25	30	35	40	45	50	55	60	65	70	75	80	85	90	95	100
2														10	20	30	40	50	60
3									7	17	27	37	47	57	67	77	87	97	107
4							10	20	30	40	50	60	70	80	91	103	116	129	143
5					4	14	24	34	44	54	64	75	88	100	114	129	145	161	179
6				3	13	23	33	43	54	65	77	91	105	121	137	155	174	193	214
7				10	20	30	40	51	63	76	90	106	123	141	160	181	203	226	250
8			5	15	25	35	46	58	71	86	103	121	140	161	183	206	231	258	286
9			9	19	29	39	51	65	80	97	116	136	158	181	206	232	260	290	321
10		2	12	22	32	44	57	72	89	108	129	151	175	201	229	258	289	322	357
11		5	15	25	35	48	63	80	98	119	141	166	193	221	251	284	318	355	393
12		7	17	27	39	53	69	87	107	130	154	181	210	241	274	310	347	387	429
13		8	18	29	42	57	74	94	116	140	167	196	228	261	297	335	376	419	464
14		10	20	31	45	61	80	101	125	151	180	211	245	281	320	361	405	451	500
15	1	11	21	33	48	66	86	108	134	162	193	226	263	301	343	387	434	483	536
16	3	13	23	36	51	70	91	116	143	173	206	241	280	321	366	413	463	516	571
17	4	14	24	38	55	74	97	123	152	184	219	257	298	342	389	439	492	548	607
18	4	14	26	40	58	79	103	130	161	194	231	272	315	362	411	464	521	580	643
19	5	15	27	42	61	83	109	137	170	205	244	287	333	382	434	490	550	612	679
20	6	16	29	45	64	88	114	145	179	216	257	302	350	402	457	516	579	645	714
21	7	17	30	47	68	92	120	152	188	227	270	317	368	422	480	542	608	677	750
22	7	18	31	49	71	96	126	159	196	238	283	281	385	442	503	568	636	709	786
23	8	18	33	51	74	101	131	166	205	248	296	347	403	462	526	593	665	741	821
24	8	19	34	54	77	105	137	174	214	259	309	362	420	482	549	619	694	774	857
25	9	20	36	56	80	109	143	181	223	270	321	377	438	502	571	645	723	806	893

when S > L $L = 2S - \dfrac{280}{A}$

when S < L $L = \dfrac{AS^2}{280}$

Height of cyclist eye - 1400 mm
Height of object - 0 mm

Shaded area represents S = L

L = Minimum Length of Vertical Curve (m)
A = Algebraic Grade Difference (%)
S = Stopping Sight Distance (m)

Minimum Length of Vertical Curve = 1 m

FIG. 6.64 Minimum length of crest vertical curve based on stopping sight distance for bicycles. Taken from *Guide for the Development of Bicycle Facilities.* Copyright © 1999 by The American Association of State Highway and Transportation Officials, Washington, DC. Used by permission.

Asphalt Institute. *How to Design Full-Depth Asphalt Pavements for Streets (IS-96).* College Park, MD: Asphalt Institute, 1981b.

Asphalt Institute. *Basic Asphalt Emulsion Manual,* 2nd ed. (MS-19). College Park, MD: Asphalt Institute, 1987.

Asphalt Institute. *Mix Design Methods for Asphalt Concrete and Other Hot-Mix Types,* 4th ed. (MS-2). College Park, MD: Asphalt Institute, 1988.

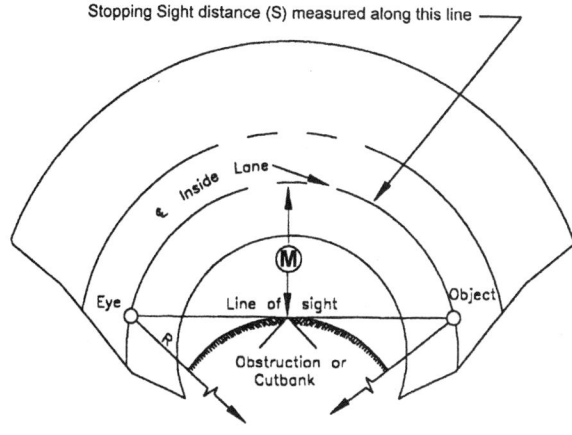

Stopping Sight distance (S) measured along this line ⎯

S = Stopping Sight Distance (m)
R = Radius of centerline of lane (m)
M = Distance from centerline of lane to obstruction (m)

Angle is expressed in degrees

$$M = R\left[1-\cos\left(\frac{28.65S}{R}\right)\right]$$

$$S = \frac{R}{28.65}\left[\cos^{-1}\left(\frac{R-M}{R}\right)\right]$$

Formula applies only when S is equal to or less than length of curve.

Line of sight is 700 mm above centerline of inside lane at point of obstruction.

For Metric Units:

S = Stopping Sight Distance (m)

R = Radius of centerline of lane (m)

M = Distance from centerline of lane to obstruction (m)

For English Units:

S = Stopping Sight Distance (ft)

R = Radius of centerline of lane (ft)

M = Distance from centerline of lane to obstruction (ft)

R (m)	S = Stopping Sight Distance (m)																		
	10	15	20	25	30	35	40	45	50	55	60	65	70	75	80	85	90	95	100
10	1.2	2.7	4.6	6.8	9.3														
15	0.8	1.8	3.2	4.9	6.9	9.1	11	14											
20	0.6	1.4	2.4	3.8	5.4	7.2	9.2	11	14	16	19								
25	0.5	1.1	2	3.1	4.4	5.9	7.6	9.5	11	14	16	18	21	23					
50	0.3	0.6	1	1.6	2.2	3	3.9	5	6.1	7.4	8.7	10	12	13	15	17	19	21	23
75	0.2	0.4	0.7	1	1.5	2	2.7	3.4	4.1	5	5.9	6.9	8	9.2	10	12	13	15	16
100	0.1	0.3	0.5	0.8	1.1	1.5	2	2.5	3.1	3.8	4.5	5.2	6.1	7	7.9	8.9	10	11	12
125	0.1	0.2	0.4	0.6	0.9	1.2	1.6	2	2.5	3	3.6	4.2	4.9	5.6	6.3	7.2	8	8.9	9.9
150		0.2	0.3	0.5	0.7	1	1.3	1.7	2.1	2.5	3	3.5	4.1	4.7	5.3	6	6.7	7.5	8.3
175		0.2	0.3	0.4	0.6	0.9	1.1	1.4	1.8	2.2	2.6	3	3.5	4	4.6	5.1	5.8	6.4	7.1
200		0.1	0.3	0.4	0.6	0.8	1	1.3	1.6	1.9	2.2	2.6	3.1	3.5	4	4.5	5	5.6	6.2
225		0.1	0.2	0.3	0.5	0.7	0.9	1.1	1.4	1.7	2	2.3	2.7	3.1	3.5	4	4.5	5	5.5
250		0.1	0.2	0.3	0.5	0.6	0.8	1	1.2	1.5	1.8	2.1	2.4	2.8	3.2	3.6	4	4.5	5
275		0.1	0.2	0.3	0.4	0.6	0.7	0.9	1.1	1.4	1.6	1.9	2.2	2.6	2.9	3.3	3.7	4.1	4.5
300			0.2	0.3	0.4	0.5	0.7	0.8	1	1.3	1.5	1.8	2	2.3	2.7	3	3.4	3.8	4.2

FIG. 6.65   Minimum lateral clearances for bicycles on horizontal curves. Taken from *Guide for the Development of Bicycle Facilities*. Copyright © 1999 by The American Association of State Highway and Transportation Officials, Washington, DC. Used by permission.

Asphalt Institute. *Thickness Design—Asphalt Pavements for Highways and Streets MS-1*. Lexington, KY: Asphalt Institute, 1991.

Asphalt Institute. *Superpave Mix Design. Superpave Series No. 2 (SP-2)*. Lexington, KY: Asphalt Institute, 1996.

Asphalt Institute. *Performance Graded Asphalt Binder Specification and Testing. Superpave Series No. 1 (SP-1)*. Lexington, KY: Asphalt Institute, 1997.

Federal Highway Administration (FHWA). *Manual of Uniform Traffic Control Devices (MUTCD)*. Washington: FHWA, 1988.

Institute of Transportation Engineers (ITE). *Parking Generation*. Washington: ITE, 1987a.

Institute of Transportation Engineers (ITE). *Guidelines for Driveway Location and Design*. Washington: ITE, 1987b, pp. 18–19.

Institute of Transportation Engineers (ITE). *Traffic Access and Impact Studies for Site Development—A Recommended Practice*. Washington: ITE, 1991, pp. 1–50.

Institute of Transportation Engineers (ITE). *Guidelines for Residential Subdivision Street Design, a Recommended Practice*. Washington: ITE, 1993, pp. 6, 12, 13, 15, 16.

Institute of Transportation Engineers (ITE). *Guidelines for Parking Facility Location and Design—A Recommended Practice*. Washington: ITE, 1994, pp. 6–9.

Institute of Transportation Engineers (ITE). *Traditional Neighborhood Development—Street Design Guidelines*. Washington: ITE, 1997b, pp. iii, 26.

Institute of Transportation Engineers (ITE). *Trip Generation*, 6th ed. Washington: ITE, 1997c.

Institute of Transportation Engineers (ITE). *Traditional Neighborhood Development—Street Design Guidelines*. Washington: ITE, 1999, pp. 8–9, 25.

Transportation Research Board (TRB). *Highway Capacity Manual, Special Report No. 209*, 3rd ed. Washington: National Academy Press, 1994, pp. 9–6, 9–7 and 10–12.

Urban Land Institute (ULA) and National Parking Association (NPA). *The Dimensions of Parking*, 2nd ed. Washington: Urban Land Institute, 1990.

U.S. Department of Transportation–Federal Highway Administration. *Site Impact Traffic Evaluation (S.I.T.E.) Handbook*. Washington: Department of Transportation, 1985.

SUGGESTED READINGS

American Association of State Highway and Transportation Officials (AASHTO). *Guide for Design of Pavement Structures*. Washington: AASHTO, 1993.

American Association of State Highway and Transportation Officials (AASHTO). *Standard Specifications for Transportation Materials and Methods of Sampling and Testing—AASHTO T 180-90 Standard Method of Test for Moisture-Density Relations of Soils Using a 10-Lb. [4.54 kg] Rammer and an 18-inch [457 mm] Drop*. Washington: AASHTO, 1990b, pp. 455–459.

American Association of State Highway and Transportation Officials (AASHTO). *Standard Specifications for Transportation Materials and Methods of Sampling and Testing—AASHTO T 190-90 Standard Method of Test for Resistance R-Value and Pressure Expansion of Compacted Soils*. Washington: AASHTO, 1990c, pp. 472–476.

American Society for Testing and Materials (ASTM). *Standard specification for Graduated Aggregate Material for Bases or Subbases for Highways on Airports (ASTM D2940-74)*. Philadelphia: ASTM, 1974.

American Society of Civil Engineers (ASCE). *Bicycle Transportation—A Civil Engineer's Notebook for Bicycle Facilities*. New York: ASCE, 1980.

Barton-Aschman Associates. *Shared Parking*. Washington: Urban Land Institute, 1983.

Box, P. C., and J. C. Opperlander. *Manual of Traffic Engineering Studies*, 4th ed. Washington: Institute of Transportation Engineers, 1976.

Corwin, M. A. *Parking Lot Landscaping*. Chicago: American Society of Planning Officials, 1978.

Homburger, W. S., E. A. Deakin, P. C. Bosselmann, D. T. Smith, Jr., and B. Beukers. *Residential Street Design and Traffic Control (ITE)*. Englewood Cliffs, NJ: Prentice-Hall, 1989.

Homburger, W. S., ed. L. E. Keefer, and W. R. McGrath, (Eds.). *Transportation and Traffic Engineering Handbook*, 2nd ed. Englewood Cliffs, NJ: Prentice-Hall, 1982.

Institute of Transportation Engineers (ITE). *Proper Location of Bus Stops*. Washington: ITE, 1986.

Institute of Transportation Engineers (ITE). *Guidelines for Planning and Designing Access Systems for Shopping Centers*. Washington: ITE, 1988.

Institute of Transportation Engineers (ITE). *Traffic Engineering for Neo-Traditional Neighborhood Design—An Informational Report*. Washington: ITE, 1994.

Institute of Transportation Engineers (ITE). *Neighborhood Traffic Control*. Washington: ITE, 1996.

Institute of Transportation Engineers (ITE). *Guidelines for the Design and Application of Speed Humps, a Recommended Practice*. Washington: ITE, 1997a.

Johnson, D. E. *Residential Land Development Practices*. Reston: American Society of Civil Engineers, 1997.

Kell, J. H., and I. J. Fullerton. *Manual of Traffic Signal Design*, 4th ed. Englewood Cliffs, NJ: Prentice-Hall, 1991.

Leisch, J. E., and Associates. "Planning and Design Guide At-Grade Intersections." An unpublished report. Evanston, IL: Jack E. Leisch and Associates, 1990.

Mannerling, F. L., and W. P. Kilareski. *Principles of Highway Engineering and Traffic Analysis*, 2nd ed. New York: Wiley, 1997.

McShane, W. R., and R. P. Roess. *Traffic Engineering*, 2nd ed. Englewood Cliffs, NJ: Prentice-Hall, 1998.

National Cooperative Highway Research Program, Transportation Research Board. *Modern Roundabout Practice in the United States—A Synthesis of Highway Practice—NCHRP Synthesis 264*. Washington: National Academy Press, 1998.

Smith, T. P. *The Aesthetics of Parking*. Chicago: American Planning Association, 1988.

Asphalt Institute. *HMA Construction—Construction of Hot Mix Asphalt Pavements Manual Series No. 22*, 2nd ed. Lexington, KY: Asphalt Institute, 1998.

Transportation Research Board (TRB). *Driveway and Street Intersection Spacing*. Washington: National Academy Press, 1996.

Transportation Research Board (TRB). *Bicycle and Pedestrian Research 1998—Transportation Research Record No. 1636*. Washington: National Academy Press, 1998.

Urban Land Institute and International Council of Shopping Centers. *Parking Requirements for Shopping Centers*. Washington: Urban Land Institute, 1982.

Wright, P. H., with contributions by J. S. Lai, P. S. Parsonson, and M. Meyer. *Highway Engineering*, 6th ed. New York: Wiley, 1996.

Wright, P. H., N. J. Ashford, and R. J. Stammer, Jr. *Transportation Engineering*, 4th ed. New York: Wiley, 1997.

7 Drainage System Design

Once the road system is fairly well defined, design of the drainage system can begin. A civil engineering professor once said, "There are three cardinal rules to remember when dealing with a construction project. Rule 1: Get rid of the water. Rules 2 and 3: Get rid of the water."

Reasons for wanting to remove water, in addition to minimizing flooding, are as follows:

(a) Excess moisture can weaken a soil's strength and even make it plastic if it is cohesive material.

(b) Excess moisture can elevate the soil's moisture content to a level that makes it impossible to attain proper field compaction.

(c) Excess moisture can result in seepage problems.

(d) High groundwater can complicate excavation operations by destabilizing the soil.

(e) High groundwater can cause pavement to fail as a result of hydrostatic pressure and loss of subgrade stability.

(f) Excess water resulting from inadequate permanent drainage facilities can result in standing water in yards or water in basements and crawl spaces, providing an environment that invites mildew, rot, and possibly insect damage.

Today, it is not a simple task to control and remove water associated with a construction project.

The water to which these cardinal rules are referring contributes to the hydrologic cycle. This cycle includes water evaporating from oceans and surface waters, along with that which is transpiring from plants to form clouds. Water vapor carried by the air masses can cool into droplets under certain meteorological conditions, resulting in precipitation. Some of the precipitation that falls on the earth can temporarily be stored by surface soil until evaporation or transpiration occurs. Other water can become surface runoff by flowing into streams and lakes, while the remainder can percolate into the soil and become groundwater. The hydrologic cycle plays an important role in determining the water supply demands attributed to lawn sprinkling. (See Chapter 11.)

Drainage design involves surface and subsurface (ground) water. Surface water resulting from stormwater runoff, snow melt, rainfall, hail, or sleet must be controlled both during and after construction. Erosion mitigation and storm drainage collection systems can be effectively used to remove excess water. Since 1992 the U.S. Environmental Protection Agency under the Clean Water Act has regulated stormwater from construction sites having 5 or more acres of disturbed land area using its NPDES General Permit for Stormwater. In 1999 a proposed EPA plan to reduce the regulated area from 5 acres to 1 acre was announced with the first permits possibly being issued in 2002. Biological, chemical, and physical contaminants contained in construction site runoff are targeted through best management practices (BMPs) to curtail sedimentation and erosion (Chapter 8) and minimize the delivery of nutrients, metals, and organic compounds to downstream waterways.

Each drainage system is designed for a particular storm intensity and duration based on a certain return period. During that time period, one storm having a particular magnitude can be expected to occur, on the average. This does not mean that another storm of the same magnitude cannot occur within the same period of time, but chances are small. Typically, land development design is based on a storm of a magnitude associated with a 10-year return period. This gives a 10 percent probability of a storm that size occurring in any given year. Since drainage systems are usually designed to flow at capacity under open-channel conditions for return periods ranging from 5 to 10 years, it is advisable to analyze the designed system's performance using larger return periods to investigate potential flooding. A 25-year return period can be used to investigate the impact of a sizable storm on the development, while using a 50- or 100-year period allows insight into drainage conditions while the system is under duress.

Local regulations usually prescribe which return periods must be utilized for storm system design. The design professional would have to obtain this information while conducting preliminary engineering activities.

7.1 PRECIPITATION AND DURATION

To design a drainage system, the expected amount of rainfall must first be known. Procedures for determining this are now given.

7.1.1 Precipitation Frequency Determination

The U.S. Weather Bureau, a part of the National Oceanic and Atmospheric Administration (NOAA), has gathered rainfall data pertaining to this country for many years. Statistical procedures developed by E. J. Gumbel were used to fit "annual series" values to the extreme value Type I distribution (Frederick et al., 1977). The one largest event occurring during each year of a particular duration was identified, and the group of these events constituted the annual series. The annual series values were converted into equivalent partial duration values in those cases where a single year in a series contained several significant storm events that were greater than the largest event during other years in the series. This information has been recorded in various publications associated with the probability distribution of precipitation in the United States. Information pertaining to geographical areas located generally west of 103° W longitude is included in the NOAA Atlas 2 (Miller et al., 1973). For areas generally east of 103° W longitude, NWS Hydro-35 (Frederick et al., 1977) contains precipitation data for durations of 1 hour or less, and USWB TP-40 (Hershfield, 1961) includes durations of up to 24 hours. The amount of precipitation is depicted using isopluvial lines that show lines of equal precipitation depths for specified return periods and durations. For example, Figure 7.1 depicts precipitation data adjusted to the partial duration series from USWB TP-40.

FIG. 7.1 A 100-year, 24-hour rainfall. Source: Hershfield, 1961.

Space limitations prohibit the inclusion of the other rainfall charts; therefore, the reader is referred to the referenced sources for this information.

Precipitation data in inches for a sample shown in Figure 7.2 are obtained from Figure 7.1 and the other precipitation charts referenced previously. The design professional, however, must use specific data corresponding to the actual geographic location of the development site.

Partial duration values for return periods not included in the published data and for areas generally east of 103° W longitude can be computed using procedures and charts contained in NWS Hydro-35 and USWB TP-40 (Frederick et al., 1977; Hershfield, 1961).

To calculate these values, the following equations, contained in NWS Hydro-35, can be used:

$$\text{5-year precipitation} = 0.278(100 \text{ years}) + 0.674(2 \text{ years}) \qquad (7.1)$$

$$\text{10-year precipitation} = 0.449(100 \text{ years}) + 0.496(2 \text{ years}) \qquad (7.2)$$

$$\text{25-year precipitation} = 0.669(100 \text{ years}) + 0.293(2 \text{ years}) \qquad (7.3)$$

$$\text{50-year precipitation} = 0.835(100 \text{ years}) + 0.146(2 \text{ years}) \qquad (7.4)$$

To empirically compute values for the intervening durations shown in Figure 7.3, the following equations, from NWS Hydro-35, can be used:

$$\text{10-minute value} = 0.59(15 \text{ minute value}) + 0.41(5 \text{ minute value}) \qquad (7.5)$$

$$\text{30-minute value} = 0.49(60 \text{ minute value}) + 0.51(15 \text{ minute value}) \qquad (7.6)$$

Return	Duration	Inches
2 Year	60 Minute Precipitation	2.20
100 Year	60 Minute Precipitation	4.25
2 Year	5 Minute Precipitation	0.51
100 Year	5 Minute Precipitation	0.80
2 Year	15 Minute Precipitation	1.13
100 Year	15 Minute Precipitation	1.85
2 Year	24 Hour Precipitation	4.50
100 Year	24 Hour Precipitation	9.90

FIG. 7.2 Precipitation data for a sample project site.

Period	Duration									
Years	5 min.	10 min.	15 min.	30 min.	1 hr.	2 hr.	3 hr.	6 hr.	12 hr.	24 hr.
2	0.51	0.88	1.13	1.65	2.20	2.45	2.71	3.35	3.93	4.50
5	0.57	0.98	1.28	1.96	2.66	3.00	3.35	4.43	5.01	5.79
10	0.61	1.07	1.39	2.18	3.00	3.40	3.81	4.84	5.76	6.68
25	0.68	1.21	1.57	2.51	3.49	3.98	4.47	5.72	6.83	7.94
50	0.74	1.31	1.71	2.77	3.87	4.43	4.98	6.40	7.66	8.92
100	0.80	1.42	1.85	3.03	4.25	4.87	5.49	7.08	8.49	9.90

FIG. 7.3 Completed depth-duration frequency table in inches.

The remaining durations can be empirically calculated using the following equations adapted from Figure 2 of USWB TP-40 (Bradford et al., 1979):

$$\text{2-hour value} = 0.11(24 \text{ hour value}) + 0.89(1 \text{ hour value}) \qquad (7.7)$$

$$\text{3-hour value} = 0.22(24 \text{ hour value}) + 0.78(1 \text{ hour value}) \qquad (7.8)$$

$$\text{6-hour value} = 0.50(24 \text{ hour value}) + 0.50(1 \text{ hour value}) \qquad (7.9)$$

$$\text{12-hour value} = 0.75(24 \text{ hour value}) + 0.25(1 \text{ hour value}) \qquad (7.10)$$

The completed depth duration values for this example are included in Figure 7.3.

7.1.2 Defining a Storm

If actual data are not available, a design storm can be defined by using either an intensity-duration curve or a depth-duration relationship with synthetic rainfall distribution. The intensity-duration curve does not show rainfall sequentially according to time; it shows average intensities for a particular duration. A rainfall intensity curve can be constructed in the following manner:

(a) Divide each rainfall depth included in Figure 7.3 by its corresponding duration (in terms of hours) to obtain average rainfall intensity values. For example, if a 100-year return period and corresponding 5-minute duration are accompanied by a rainfall depth of 0.80 inch, the average intensity is calculated by (0.8 inch/5 minutes) × 60 minutes per hour, or 9.6 inches per hour. Figure 7.4 contains the calculation results.

(b) Plot each average intensity value based on its duration and return period to produce a series of curves similar to Figure 7.5. For clarity only the 10- and 100-year return periods are included.

Although a number of ways to distribute rainfall are available, the Natural Resources Conservation Service (NRCS) has developed four synthetic rainfall distribution curves that can be used to distribute storm precipitation having periods of 24 hours. These curves have been constructed in a manner that allows use for determining not only peak runoff discharges but also runoff volumes for a variety of drainage area sizes (NRCS TR-55, 1986a). Figure 7.6 shows the four NRCS rainfall distribution curves and their geographic application.

Period	Duration									
Years	5 min.	10 min.	15 min.	30 min.	1 hr.	2 hr.	3 hr.	6 hr.	12 hr.	24 hr.
2	6.12	5.28	4.52	3.30	2.20	1.23	0.90	0.56	0.33	0.19
5	6.84	5.88	5.12	3.92	2.66	1.51	1.12	0.70	0.42	0.24
10	7.32	6.42	5.56	4.36	3.00	1.70	1.27	0.81	0.48	0.28
25	8.16	7.26	6.28	5.02	3.49	1.99	1.49	0.95	0.57	0.33
50	8.88	7.86	6.84	5.54	3.87	2.22	1.66	1.07	0.64	0.37
100	9.60	8.52	7.40	6.06	4.25	2.44	1.83	1.18	0.71	0.41

FIG. 7.4 Example of rainfall intensity values.

FIG. 7.5 Example of rainfall intensity curves for 10- and 100-year return periods.

A 10-year design storm using the NRCS Type II curve (when the 24-hour rainfall is 6.68 inches) yields the rainfall distribution contained in Figure 7.7 for the 24-hour period.

7.2 RAINFALL ABSTRACTIONS

Once rainfall is defined, an estimate must be made as to what portion will be retained and what will become excess precipitation, or runoff. Some of the initial rainfall may land in channelized areas, contributing to the initial runoff flows, while the majority is likely to fall on soil, vegetation, and man-made surfaces. Rainfall occurring on dry porous soil will initially be absorbed to a point where soil voids near the surface become saturated, causing depressed areas to fill. It is not until the surface depressions reach capacity that runoff begins to occur. Rain falling on vegetation can be held by leaves and returned to the atmosphere by evaporation and transpiration rather than becoming runoff.

When rainfall is no longer capable of being abstracted, it becomes runoff. Of particular concern is the amount of moisture present in the soil at the storm's beginning. A storm having a particular return period can provide varying amounts of excess rainfall, depending on whether the exposed soil was initially dry, moist, or saturated. The saturated condition produces the most excess precipitation.

7.2.1 Coefficient of Runoff

One method of defining the estimated amount of runoff is by using a coefficient of runoff, which is utilized in the rational method. The coefficient of runoff is the ratio of the amount of excess precipitation expected to total amount of expected rainfall. Figure 7.8 shows typical coefficient values.

Should the contributory area be comprised of various land uses, a composite coefficient of runoff must be obtained based on the ratio of individual contributory subbasin

(a)

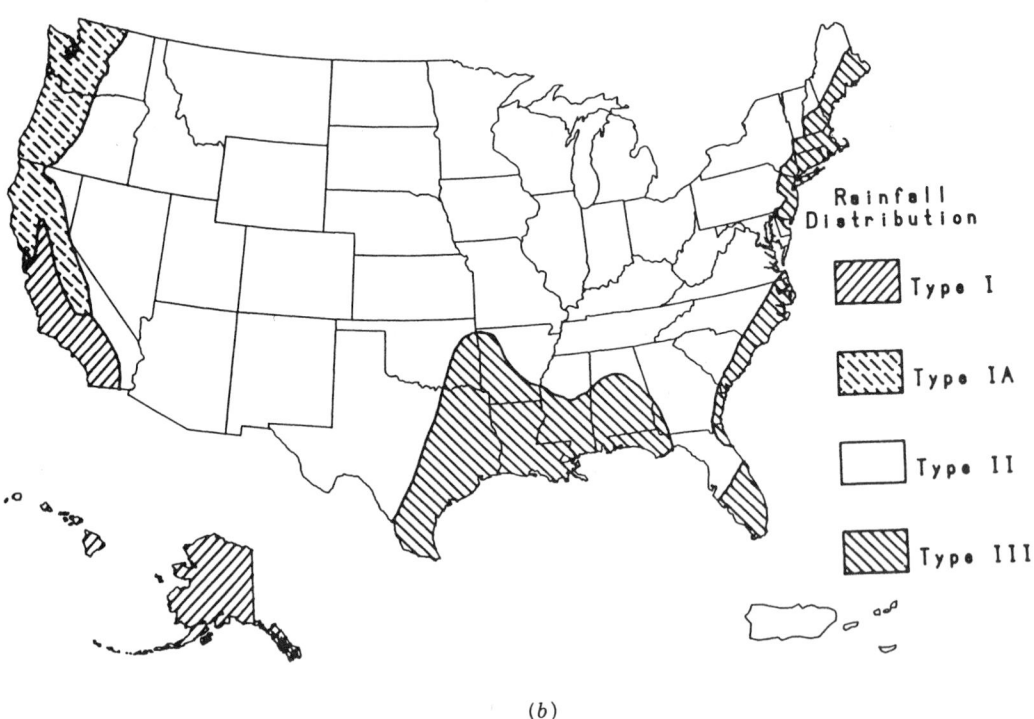

(b)

FIG. 7.6   (a) NRCS 24-hour rainfall distributions. (b) Approximate geographic boundaries for NRCS rainfall distribution. Source: Natural Resources Conservation Service, 1986a.

Time,hrs.	Px/P24	Depth, in.
0.00	0.0000	0.0000
1.00	0.0106	0.0708
2.00	0.0220	0.1470
3.00	0.0344	0.2298
4.00	0.0480	0.3206
5.00	0.0631	0.4215
6.00	0.0800	0.5344
7.00	0.0982	0.6560
8.00	0.1200	0.8016
9.00	0.1470	0.9820
10.00	0.1810	1.2091
10.50	0.2040	1.3627
11.00	0.2350	1.5698
11.50	0.2830	1.8904
12.00	0.6630	4.4288
12.50	0.7350	4.9098
13.00	0.7720	5.1570
13.50	0.7990	5.3373
14.00	0.8200	5.4776
15.00	0.8542	5.7061
16.00	0.8800	5.8784
17.00	0.9029	6.0314
18.00	0.9220	6.1590
19.00	0.9381	6.2665
20.00	0.9520	6.3594
21.00	0.9644	6.4422
22.00	0.9760	6.5197
23.00	0.9876	6.5972
24.00	1.0000	6.6800

FIG. 7.7 Example of rainfall distribution for a 10-year Type II storm where P_x/P_{24} is the ratio of accumulated rainfall to the 24-hour total.

characteristics to that of the total area. The composite coefficient of runoff can be calculated by

$$C_{comp} = \frac{C_1 A_1 + C_2 A_2 + C_3 A_3 + \cdots + C_n A_n}{A_1 + A_2 + A_3 + \cdots + A_n} \qquad (7.11)$$

where C_n is the coefficient of runoff associated with subbasin area A_n and A_n is the subbasin drainage area.

7.2.2 NRCS Runoff Curve Number

The NRCS estimates the amount of runoff by a runoff curve number (CN). The CN is a function of land use and prevailing soil conditions; the higher the CN, the higher the runoff potential if the soil is not frozen.

Soil properties affect the amount of runoff as a result of varying infiltration rates. The NRCS has defined four hydrologic soil groups:

Group A Soils These have low runoff potential and high infiltration rates even when thoroughly wetted. They consist chiefly of deep, well to excessively drained sands

Slope	Land-Use	Sandy Soils **			Clay Soils **		
		Min.		Max.	Min.		Max.
Flat (0-2%)	Woodlands	0.10		0.15	0.15		0.20
	*Pasture, grass & farmland	0.15		0.20	0.20		0.25
	Rooftops and pavement		0.95			0.95	
	Single family residential:						
	1/2 acre lots & larger	0.30		0.35	0.35		0.45
	Smaller lots	0.35		0.45	0.40		0.50
	Multi-family residential:						
	Duplexes	0.35		0.45	0.40		0.50
	Apartments, townhouses, and condominiums	0.45		0.60	0.50		0.70
	Commercial and Industrial	0.50		0.95	0.50		0.95
Rolling (2%-7%)	Woodlands	0.15		0.20	0.20		0.25
	*Pasture, grass & farmland	0.20		0.25	0.25		0.30
	Rooftops and pavement		0.95			0.95	
	Single family residential:						
	1/2 acre lots & larger	0.35		0.50	0.40		0.55
	Smaller lots	0.40		0.55	0.45		0.60
	Multi-family residential:						
	Duplexes	0.40		0.55	0.45		0.60
	Apartments, townhouses, and condominiums	0.50		0.70	0.60		0.80
	Commercial and Industrial	0.50		0.95	0.60		0.95
Steep (7%+)	Woodlands	0.20		0.25	0.25		0.30
	*Pasture, grass & farmland	0.25		0.35	0.30		0.40
	Rooftops and pavement		0.95			0.95	
	Single family residential:						
	1/2 acre lots & larger	0.40		0.55	0.50		0.65
	Smaller lots	0.45		0.60	0.55		0.70
	Multi-family residential:						
	Duplexes	0.45		0.60	0.55		0.70
	Apartments, townhouses, and condominiums	0.60		0.75	0.65		0.85
	Commercial and Industrial	0.60		0.95	0.65		0.95

* Coefficients assume good ground cover and conservation treatment.
** Weighted coefficient based on percentage of impervious surfaces and green areas must be selected for each site.

FIG. 7.8 Runoff coefficients for various land uses. Source: De Kalb County Drainage Procedures Manual, De Kalb County, GA. Copyright © 1976 by W.L. Jorden & Company, Inc. Reproduced with permission.

or gravels and have a high rate of water transmission (greater than 0.30 inch per hour).

Group B Soils   These have moderate infiltration rates when thoroughly wetted and consist chiefly of moderately deep to deep, moderately well to well-drained soils with moderately fine to moderately coarse textures. These soils have a moderate rate of water transmission (0.15–0.30 inch per hour).

Group C Soils   These have low infiltration rates when thoroughly wetted and consist chiefly of soils with a layer that impedes downward movement of water and soils with moderately fine to fine texture. These soils have a low rate of water transmission (0.05–0.15 inch per hour).

Group D Soils   These have high runoff potential. They have very low infiltration rates when thoroughly wetted and consist chiefly of clay soils with a high swelling potential, soils with a permanently high water table, soils with a claypan or clay layer at or near the surface, and shallow soils over nearly impervious material. These soils have a very low rate of water transmission (0.00–0.05 inch per hour).

Compaction and soil mixing attributed to construction can affect soil conditions in a land development and can alter the runoff curve number. Figure 7.9 contains various runoff curves for various land uses.

Soil types can be identified from NRCS soils maps (referred to in Section 4.2.1) containing the project area. Figure 7.10*a* shows a typical soils map used to identify the predominant soil type, while Figure 7.10*b* provides the hydrologic grouping and other characteristics.

The factors affecting CN are the hydrologic soil group, cover type, treatment, hydrologic condition, antecedent runoff condition (ARC), and land use. Should the

Cover description		Curve numbers for hydrologic soil group[1]–			
Cover type and hydrologic condition	Average percent impervious area	A	B	C	D
Fully developed urban areas (vegetation established)					
Open space (lawns, parks, golf courses, cemeteries, etc.)[b]:					
Poor condition (grass cover < 50%)		68	79	86	89
Fair condition (grass cover 50% to 75%)..........		49	69	79	84
Good condition (grass cover > 75%)		39	61	74	80
Impervious areas:					
Paved parking lots, roofs, driveways, etc.					
(excluding right-of-way).		98	98	98	98
Streets and roads:					
Paved; curbs and storm sewers (excluding					
right-of-way)..................................		98	98	98	98
Paved; open ditches (including right-of-way)		83	89	92	93
Gravel (including right-of-way)		76	85	89	91
Dirt (including right-of-way)		72	82	87	89
Western desert urban areas:					
Natural desert landscaping (pervious areas only) ...		63	77	85	88
Artificial desert landscaping (impervious weed					
barrier, desert shrub with 1- to 2-inch sand					
or gravel mulch and basin borders).		96	96	96	96
Urban districts:					
Commercial and business..........................	85	89	92	94	95
Industrial......................................	72	81	88	91	93
Residential districts by average lot size:					
1/8 acre or less (town houses).....................	65	77	85	90	92
1/4 acre	38	61	75	83	87
1/3 acre	30	57	72	81	86
1/2 acre	25	54	70	80	85
1 acre	20	51	68	79	84
2 acres	12	46	65	77	82
Developing urban areas					
Newly graded areas (pervious areas only,					
no vegetation)		77	86	91	94

[1]Average runoff condition, and $I_a = 0.2S$.

FIG. 7.9 Runoff curve numbers for urban areas. Source: Natural Resources Conservation Service, 1986a.

(a)

FIG. 7.10 (a) NRCS soils map, Orangeburg County, SC. Source: Natural Resources Conservation Service. Soil Survey of Orangeburg County, South Carolina (A National Cooperative Survey) Washington: USDA, 1988. (b) Example of soil groupings found in typical NRCS soil surveys. Source: Natural Resources Conservation Service and Forest Service, 1980. Soil Survey of Berkeley County, South Carolina (A National Cooperative Soil Survey) Washington: USDA, 1980.

Soil name and map symbol	Hydro-logic group	Flooding			High water table		
		Frequency	Duration	Months	Depth	Kind	Months
					Ft		
Aquic Udifluvents: AU	D	Common	Long	Dec-Mar	1.0-3.0	Apparent	Dec-Mar
Bayboro: Ba	D	Common	Brief	Dec-Mar	0-1.0	Apparent	Nov-Apr
Bethera: Be	D	Common	Brief	Dec-Apr	0-1.5	Apparent	Dec-Apr
Bohicket: [1]BH	D	Frequent	Very brief	Jan-Dec	(3)-0	Apparent	Jan-Dec
Bonneau: BoA, BoB	A	None	---	---	3.5-5.0	Apparent	Dec-Apr
Borrow pits: Bp.							
Byars: By	D	Common	Brief	Dec-Mar	0-1.0	Apparent	Nov-Apr
Cainhoy: CaB	A	None	---	---	>6.0	---	---
Capers: [2]CP	D	Frequent	Very long	Jan-Dec	(1)-1.0	Apparent	Jan-Dec
Caroline: CoA, CoB	C	None	---	---	>6.0	---	---
Chastain: [2]CS	D	Common	Very long	Dec-Apr	0-1.0	Apparent	Nov-May
Chipley: [2]Ct: Chipley part	C	None	---	---	2.5-5.0	Apparent	Nov-Apr
Echaw part	B	None	---	---	2.5-5.0	Apparent	Dec-Apr
Coxville: Cu	D	Rare	---	---	0-1.5	Apparent	Nov-Apr
Craven: CvA, CvB	C	None	---	---	2.0-3.5	Apparent	Dec-Mar
Duplin: DuA, DuB	C	None	---	---	2.0-3.5	Apparent	Dec-Mar
Goldsboro: GoA	B	None	---	---	2.5-3.5	Apparent	Dec-Mar
Lenoir: Le	D	None	---	---	1.0-2.5	Apparent	Dec-Mar
Leon: Lo	A/D	None	---	---	0-1.0	Apparent	Nov-Mar
Lucy: LuB	A	None	---	---	>6.0	---	---
Lynchburg: Ly	B/D	None	---	---	0.5-1.5	Apparent	Nov-Apr
Meggett: Mg, Mp	D	Common	Brief	Dec-Mar	0-1.0	Apparent	Jun-Apr
Norfolk: NoA, NoB	B	None	---	---	>6.0	---	---

(b)

FIG. 7.10   (continued)

contributory area contain different land uses, a weighted CN value must be obtained. For example, a drainage basin comprised of woods, residential area, and open space has a weighted CN value of:

Land Use	Percent	Soil Type	Curve	Product
Woods (good)	25	C	70	$0.25 \times 70 = 17.5$
Residential ($\frac{1}{3}$ acre)	60	A	57	$0.60 \times 57 = 34.2$
Open space (good condition)	15	D	80	$0.15 \times 80 = 12.0$
			Sum	63.7

The weighted CN is taken as 64.

The CN value is altered by the amount of moisture present in the soil. The ARC reflects estimated variations in CN from storm to storm; however, for design purposes in urban areas, NRCS recommends using the average CN value.

7.3   TIME OF CONCENTRATION

The time of concentration (T_c) is the time required for excess precipitation to travel hydraulically from the remotest point within the drainage basin to the point under investigation. This time is influenced by land use, slope, channel conditions, and ponding. In hydrograph analysis, however, T_c is defined as "the time from the end of excessive rainfall to the point on the falling limb of the hydrograph (point of inflection) where the recession curve begins" (NRCS, 1972, Chapter 15, p. 15-4).

7.3.1   Travel Time

Most drainage basins are comprised of various elements that affect the time required for runoff to travel to a specified downstream point. This requires the contributory area to be analyzed by subbasins where elements affecting the time of concentration within each subbasin are fairly uniform. The individual subbasin travel times (T_n) are then summed for the contributory area's time of concentration. More specifically,

$$T_c = T_1 + T_2 + T_3 + \cdots + T_n \qquad (7.12)$$

Within each subbasin that is analyzed the surface roughness greatly affects the velocity of runoff. For example, areas comprised of grass retard runoff movement, while paved areas promote more rapid movement. The type of flow also affects travel times. Velocities associated with sheet flow are usually less than those associated with channelized flow, because hydraulic features associated with channels promote the higher velocities. The other factor that affects travel time is the slope of the surface carrying the runoff. The steeper the slope, the shorter the time of travel.

7.3.2   Overland Flow

Water flowing freely across plane surfaces is usually referred to as sheet flow. Depths may approach 1 inch, and the time of travel associated with sheet flow is dependent upon a roughness coefficient that includes the effect of erosion, sedimentation transport, drag, and rainfall impact. (See Chapter 8.) According to NRCS (1986b), the time of travel for overland flow can be approximated using the Manning kinematic solution, where the maximum flow level can be up to 300 feet, but is more likely to be 100 feet, using the following equation:

$$T_t = \frac{0.93 \, (nL)^{0.6}}{i^{0.4} \, S^{0.3}} \qquad (7.13)$$

where T_t is travel time in minutes; n a friction factor, Manning's n for sheet flow; L the slope length in feet; i the average rainfall excess intensity for a storm duration equal to T_c (inches per hour); and S the slope in feet per foot. According to the NRCS (1986b), the assumptions for using Equation 7.13 include:

Those for Manning's equation (steady uniform turbulent flow), a flat, wide plane surface, constant rainfall excess intensity, and duration of rainfall equaling T_t. Overton (1972) found that the Manning-kinematic solution for the rising overland flow hydrograph produced a 15 percent standard of error in fitting the observed data. Use of this equation requires an estimate of the rainfall excess intensity and trial and error calculations until T_t equals rainfall duration.[*]

According to the NRCS (1986a) the overland time of travel can be estimated using Equation 7.14, where the rainfall excess intensity term in Equation 7.13 has been replaced by a 24-hour precipitation term:

$$T_t = \frac{0.007 \, (nL)^{0.8}}{(P_2)^{0.5} \, (S)^{0.4}} \tag{7.14}$$

where T_t is the travel time in hours for sheet flow for distances up to 300 feet (although 50 feet is probably more representative); n is Manning's coefficient of roughness as shown in Figure 7.11 for sheet flow conditions; L is the flow length in feet; P_2 is the 2-year, 24-hour rainfall in inches; and S is the slope of the drainage surface in feet per foot.

According to the NRCS (1986a), once sheet flows become shallow concentrated flows, the following equation, based on Manning's formula, can be used to estimate flow velocities:

$$\text{Unpaved} \quad V = 16.1345(S)^{0.5} \tag{7.15}$$

$$\text{Paved} \quad V = 20.3282(S)^{0.5} \tag{7.16}$$

where V is the average velocity in feet per second and S the surface slope in feet per foot.

Surface Description	Value of n
Smooth surfaces (concrete, asphalt, or bare soil)	0.011
Fallow (no residue)	0.050
Cultivated soils:	
Residue cover $\leq$ 20%	0.060
Residue cover $\geq$ 20%	0.170
Grass:	
Short grass prairie	0.150
Dense grasses	0.240
Bermuda grass	0.410
Range (natural)	0.130
Woods:	
Light underbrush	0.400
Dense underbrush	0.800[a]

[a]May be revised to 0.600.

FIG. 7.11 Manning's coefficient of roughness, n, for sheet flow. Source: Natural Resources Conservation Service, 1986a.

[*]Taken from Natural Resources Conservation Service, Technical Note: *Hydrology No. 4* (1986).

The assumptions used for these two equations include Manning's roughness coefficient values of 0.05 for unpaved areas and 0.025 for paved areas. In addition, the hydraulic radius (included in the numerical coefficients in Equations 7.15 and 7.16) is assumed to be 0.4 for unpaved areas and 0.2 for paved areas.

Once concentrated shallow flows enter channelized areas, velocities of travel can be determined utilizing Manning's equation for open-channel flow, which is

$$V = \frac{1.49 r^{0.67} S^{0.5}}{n} \tag{7.17}$$

where V is the average flow velocity in feet per second; r is the hydraulic radius in feet defined by the ratio of the cross-sectional flow area to the corresponding wetted perimeter, the wetted perimeter being the exposed length of a channel cross section where contact is made with the flow; S is the slope of the channel in feet per foot; and n is Manning's roughness coefficient for open channel flow. Typical values are shown in Figure 7.12.

Often Manning's equation is rearranged for use in design by incorporating the continuity equation relationship with

$$Q = AV \tag{7.18}$$

where Q is the rate of flow in cubic feet per second, A is the cross-sectional flow area in square feet, and V is the velocity of flow in feet per second.

The rearranged equation is

$$Q = \frac{1.49 A r^{0.67} S^{0.5}}{n} \tag{7.19}$$

When travel velocities are obtained for shallow concentrated flows or open-channel flows using Equations 7.15–7.17, the corresponding time of travel can be obtained using

$$T_n = \frac{L}{3600 \times V} \tag{7.20}$$

where T_n is the travel time in hours, L is the length of flow in feet, V is the average velocity in feet per second, and 3600 is the factor to convert seconds into hours.

Another method to determine the time of concentration is the lag method. The lag equation was developed from limited agricultural watershed data. It was later adjusted and now has an upper drainage area limit of 2000 acres. Use of the lag method should be confined to agricultural watersheds, while the velocity method included here is better suited in urban areas.

7.4 RUNOFF

After the design storm is established, rainfall abstractions are addressed, and the time of concentration is determined, the expected runoff can be calculated. If the drainage area is 100 acres or less, the rational method is satisfactory for computing peak watershed

Type of Channel or Structure	Value of n
Open Channels for Type of Lining Shown	
Smooth concrete	0.013
Rough concrete	0.022
Riprap	0.03–0.04
Asphalt, smooth texture	0.013
Good stand, any grass—depth of flow	
More than 6 inches	0.09–0.30
Less than 6 inches	0.07–0.20
Earth, uniform section, clean	0.016
Earth, fairly uniform section, no vegetation	0.022
Channels not maintained, dense weeds	0.08
Natural Stream Channels	
(Surface Width at Flood Stage is 100 feet)	
Fairly regular section	
Some grass and weeds, little or no brush	0.030–0.035
Dense growth of weeds, depth of flow materially greater than weed height	0.035–0.05
Some weeds, light brush on banks	0.035–0.05
Some weeds, heavy brush on banks	0.05–0.07
Some weeds, dense willows on banks	0.06–0.08
For trees within channel, with branches submerged at high stage, increase all values above by	0.01–0.02
Irregular sections with pools, slight channel meander: increase values given above by approximately	0.01–0.02
Culverts	
Concrete pipe and boxes	0.012
Corrugated metal	
Unpaved	0.024–0.027
25% paved	0.021–0.026
Fully paved	0.012

FIG. 7.12 Typical values for Manning's roughness coefficient. Source: Paul H. Wright with contributions by James S. Lai, Peter S. Parsonson, and Michael Meyer. *Highway Engineering*, 6th. Copyright © 1996 by John Wiley & Sons, Inc. Reprinted by permission of John Wiley & Sons, Inc.

runoff if care is exercised. The NRCS method may be used for small and medium watersheds up to 25,000 acres. However, NRCS TR-55 (1986a) is limited to a T_c that does not exceed a maximum value of 10 hours, rather than being limited by a maximum drainage area.

7.4.1   Rational Method

According to the American Society of Civil Engineers (ASCE) and the Water Pollution Control Federation, the rational method is based on the following three assumptions (ASCE, 1969):

(a)   The peak rate of runoff at any point is a direct function of the average rainfall intensity during the time of concentration to that point.

(b)   The frequency of the peak discharge is the same as the frequency of the average rainfall intensity.

(c)   The time of concentration is the time required for the runoff to become established and flow from the most remote part of the drainage area to the point under design.

The rational formula takes the form of

$$Q = CIA \qquad (7.21)$$

where Q is the maximum rate of runoff, in cubic feet per second, from a given drainage basin; C the coefficient of runoff, which relates runoff to rainfall; I the intensity in inches per hour, for a duration equal to the time of concentration; and A the contributory drainage basin in acres.

This formula is not dimensionally correct; however, a 1-inch depth of rainfall applied at a uniform rate for 1 hour on an area of 1 acre yields 1.008 ft^3/s of runoff if no loss occurs.

To illustrate the use of the rational method, the peak runoff from a 10-acre drainage basin will be determined. For this example, a coefficient of runoff of 0.45 is used, along with a time of concentration of 30 minutes. If a 10-year return period is specified and Figure 7.5 is assumed to contain rainfall data applicable to this site's geographical area, the rainfall intensity of 4.36 in./h can be obtained. The peak discharge using Equation 7.21 is $Q = CIA = 0.45 \times 4.36 \times 10 = 19.62$, or 20 ft^3/s.

7.4.2   Peak Discharge Using the NRCS Graphical Method

The NRCS has developed a method, contained in TR-55 (NRCS, 1986a), to estimate peak runoff using graphical means; this method takes into account precipitation and initial abstractions. The generalized NRCS runoff equation is

$$Q = \frac{(P - I_a)^2}{(P - I_a) + S} \qquad (7.22)$$

where Q is runoff in inches; P is rainfall in inches, S is potential maximum retention after runoff begins in inches, and I_a is initial abstractions in inches. This term has been further defined empirically from field observations by

$$I_a = 0.2S \qquad (7.23)$$

Therefore the combined equation is

$$Q = \frac{(P - 0.2S)^2}{(P + 0.8S)} \tag{7.24}$$

where P is total precipitation in inches, S is watershed retention factor, and Q is excess rainfall in inches. Also,

$$S = \frac{1000}{\text{CN}} - 10 \tag{7.25}$$

where CN is the NRCS curve number defined in Section 7.2.2.

Figure 7.13 or Figure 7.14 can be used, depending on the type of rainfall distribution, to graphically obtain the unit peak discharge, q_u, knowing the time of concentration and the I_a/P ratio. Once the unit peak discharge is obtained, the peak discharge of the drainage basin can be calculated using

$$q_p = q_u A_m Q F_p \tag{7.26}$$

where q_p is the peak discharge in cubic feet per second; q_u is the unit peak discharge (CSM per inch), CSM = cubic feet per second per square mile; A_m is the drainage area in square miles; Q is runoff in inches; and F_p is pond or swamp adjustment factor.

In summary, peak discharges using the NRCS graphical method are obtained by

(a) establishing a storm return period and obtaining the corresponding 24-hour precipitation (see Figure 7.1);
(b) calculating the initial abstractions, I_a, using Equations 7.23 and 7.25;
(c) calculating I_a/P to obtain the unit peak discharge from either Figure 7.13 or 7.14;
(d) calculating the runoff, Q, using Equation 7.24;
(e) obtaining the pond or swamp factor, if applicable, using Figure 7.15; and
(f) calculating the peak discharge, q_p, using Equation 7.26.

According to TR-55 (NRCS, 1986a), the graphical method has the following limitations:

(a) Watershed areas must be hydrologically homogeneous with uniform soils, cover, and land use.
(b) Although the watershed can have more than one stream or branch, all must have approximately the same T_c.
(c) F_p applies only to ponds or swamps not in the T_c flow path.

7.4.3 The NRCS Unit Hydrograph

A unit hydrograph is a hydrograph that represents 1 inch of runoff produced over a specified duration. The assumption is made that the rate of excess precipitation is constant. The duration allowing the 1 inch of excess precipitation is important because a different duration for the same watershed will produce a different unit hydrograph. Figure 7.16 illustrates the NRCS dimensionless unit hydrograph.

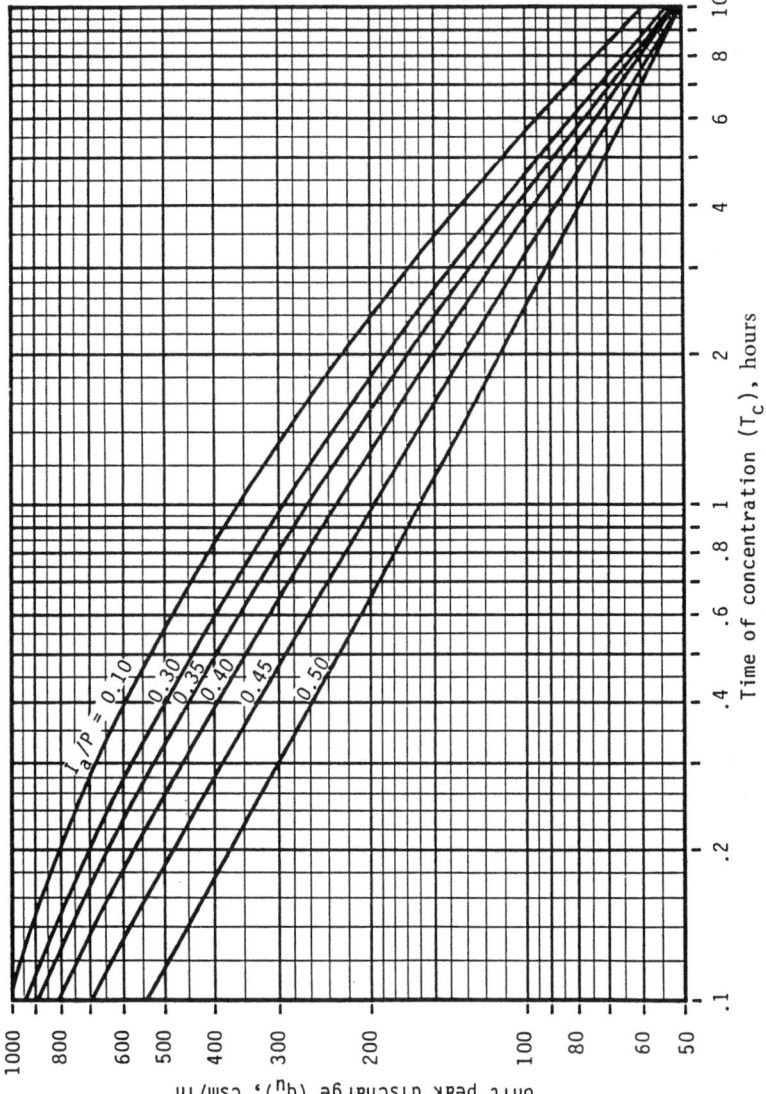

FIG. 7.13 Unit peak discharges: (*a*) for NRCS Type II rainfall distribution; (*b*) for NRCS Type III rainfall distribution. Source: Natural Resources Conservation Service, 1986a.

FIG. 7.13 (continued)

(b)

Time of concentration (T_c), hours

Unit peak discharge (q_u), csm/in

$I_a/P = 0.10$

0.30

0.35

0.40

0.45

0.50

FIG. 7.14 Unit peak discharges: (a) for NRCS Type I rainfall distribution; (b) for NRCS Type IA rainfall distribution. Source: Natural Resources Conservation Service, 1986a.

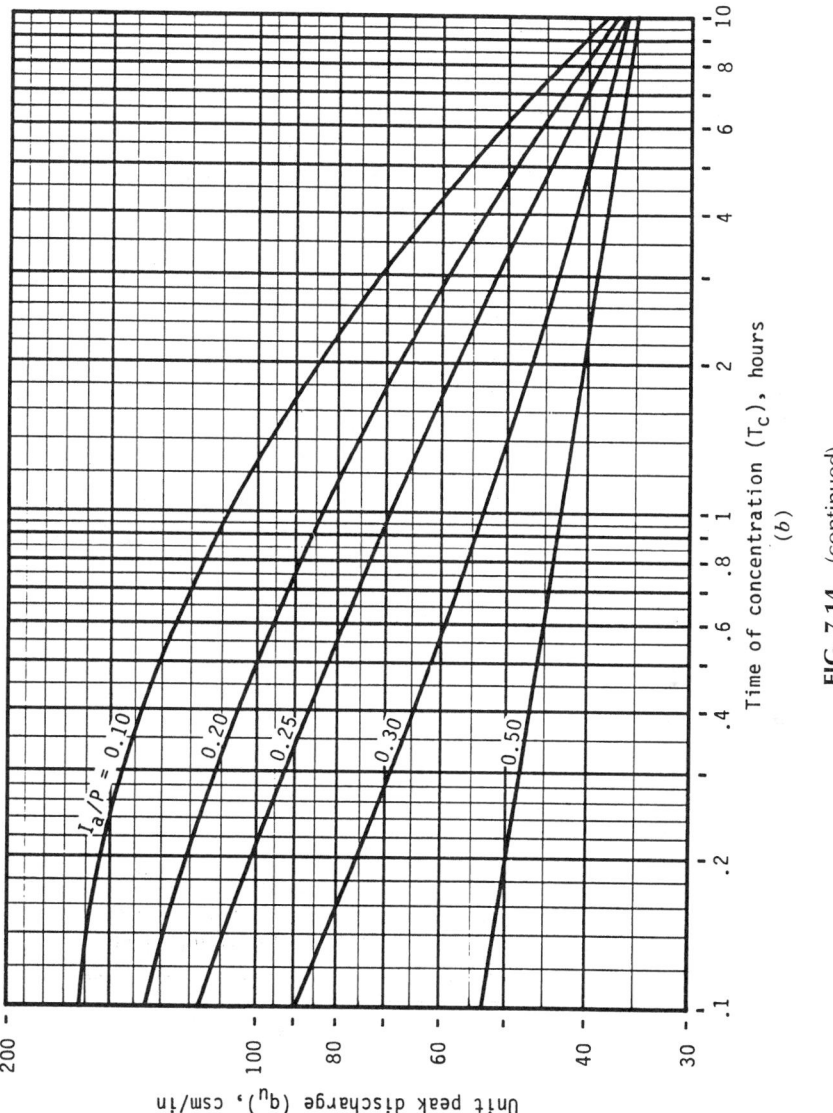

FIG. 7.14 (continued)

Percentage of Pond and Swamp Areas	F_p
0	1.00
0.2	0.97
1.0	0.87
3.0	0.75
5.0	0.72

FIG. 7.15   Adjustment factor, F_p, for ponds and swamp area that are spread throughout the watershed. Source: Natural Resources Conservation Service, 1986a.

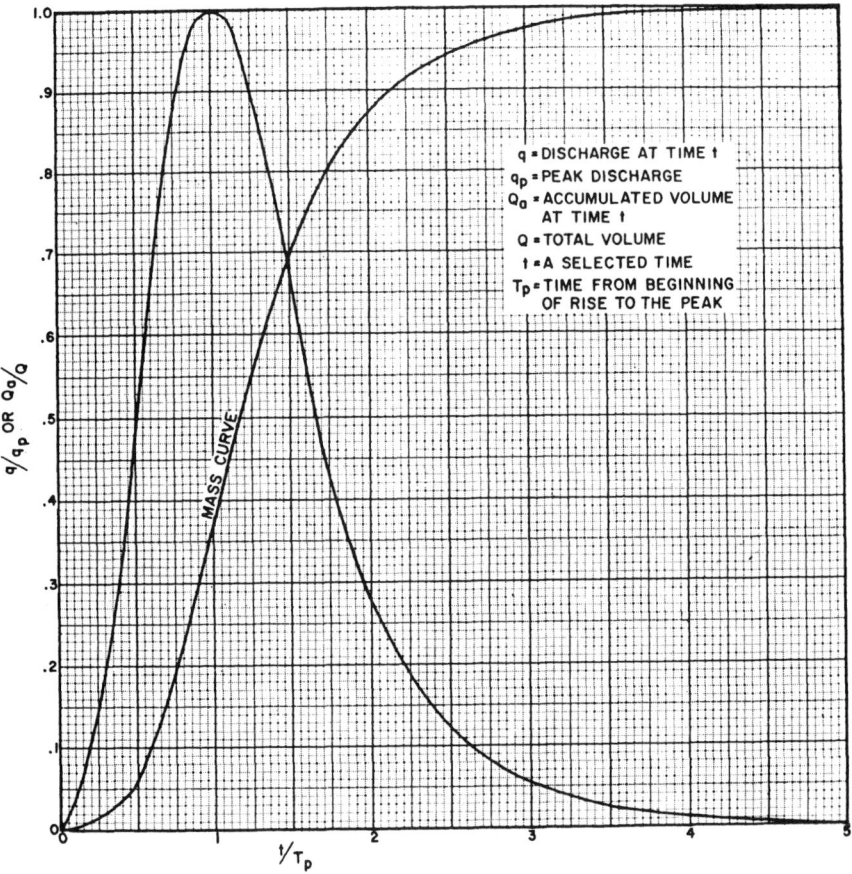

FIG. 7.16   NRCS dimensionless unit hydrograph and mass curve. Source: Natural Resources Conservation Service, 1972, Chapter 16.

This dimensionless hydrograph has the following characteristics:[*]

$$L = 0.6T_c \tag{7.27}$$
$$\tfrac{1}{2}D + L = T_p \tag{7.28}$$
$$D + T_c = 1.7T_p \tag{7.29}$$
$$T = 5T_p \tag{7.30}$$

where L is the time of lag from the centroid of the inch of excess precipitation to the peak of the unit hydrograph; D is the duration of excess precipitation; T_p is the time from the beginning of excess to the peak of the unit hydrograph in hours; T is the total time of direct runoff in hours; and T_c is the watershed time of concentration in hours. The peak flow q_{up} of the unit hydrograph can be determined by

$$q_{up} = \frac{484AQ}{T_p} \tag{7.31}$$

where 484 is a constant for a hydrograph with three-eighths of its included area under its leading edge, called the PFF (if the PFF is modified, then the unit hydrograph shape will change); A is the drainage basin area in square miles; Q is the excess precipitation in inches; and T_p is the time for the hydrograph to peak in hours, $\tfrac{1}{2}D + L$ or $\tfrac{1}{2}D + 0.6T_c$.

If the duration, D, is $0.133 \times T_c$, the unit hydrograph can be constructed. The following example illustrates the process using an assumed $T_c = 9$ minutes and a drainage basin area of 25 acres:

(a) $L = 0.6T_c$,   or $0.6 \times (9/60) = 0.09$ hour.

(b) $D = 0.133T_c$,   or $0.133 \times (9/60) = 0.02$ hour.

(c) $T_p = (0.02/2 + 0.09) = 0.10$ hour.

(d) $q_{up} = \dfrac{(484)(25 \text{ acres}/640 \text{ acres/square mile})(1 \text{ inch})}{0.10}$
   or 189.1 ft^3/s.

(e) The completed unit hydrograph is calculated by multiplying the q/q_p factors obtained from Figure 7.16 with the peak flow of 189.1 ft^3/s. The results are included in Figure 7.17.

Figure 7.18*a* illustrates a unit hydrograph with an assumed unit duration of 1 hour. If 1 inch of excess precipitation falls on the watershed during the period of 1 hour, the unit hydrograph also represents the direct runoff hydrograph for that duration. If, however, 2 inches of excess precipitation fall within the hour, Figure 7.18*b* would represent the direct runoff hydrograph where each unit hydrograph is added lienarly. Figure 7.18*c* shows a different situation, where 1 inch of direct runoff during the period of 1 hour is followed by another 1 inch of runoff during the subsequent hour. In this case the direct runoff hydrograph for the two durations is the sum of the two unit hydrographs with the second being lagged by 1 hour. Each direct runoff hydrograph for each time increment is therefore a ratio of the excess precipitation for a particular duration to that of 1 inch.

---

[*]Only if the peak flow factor (PFF) is 484.

t (hours)	t/t_p	q/q_p	q (cfs)
0.00	0.0	0.000	0.00
0.02	0.2	0.100	18.91
0.04	0.4	0.310	58.62
0.06	0.6	0.660	124.81
0.08	0.8	0.930	175.86
0.10	1.0	1.000	189.10
0.12	1.2	0.930	175.86
0.14	1.4	0.780	147.50
0.16	1.6	0.560	105.90
0.18	1.8	0.390	73.75
0.20	2.0	0.280	52.95
0.22	2.2	0.207	39.14
0.24	2.4	0.147	27.80
0.26	2.6	0.107	22.23
0.28	2.8	0.077	14.56
0.30	3.0	0.055	10.40
0.32	3.2	0.040	7.56
0.34	3.4	0.029	5.48
0.36	3.6	0.021	3.97
0.38	3.8	0.015	2.84
0.40	4.0	0.011	2.08
0.42	4.2	0.009	1.70
0.44	4.4	0.006	1.13
0.46	4.6	0.004	0.76
0.48	4.8	0.002	0.38
0.50	5.0	0.000	0.00

FIG. 7.17   Unit hydrograph ordinate calculations.

7.4.4   Direct Runoff Calculations

Before the direct runoff hydrograph can be calculated, the duration of the unit hydrograph and the time step used to determine excess precipitation must be the same. This would require the data contained in Figure 7.7 to be recomputed using a suitable time increment. A recalculated portion is shown in Figure 7.19 using a 0.02-hour increment. For this example CN is assumed to be 72; therefore, $S = 3.888$ inches when Equation 7.25 is applied. Runoff, Q, is obtained with Equation 7.24 for each time period and corresponding rainfall depth (in inches). Incremental excess precipitation is shown between each time step as ΔQ.

A direct runoff hydrograph for each time step of the storm can then be calculated by multiplying each incremental excess precipitation (ΔQ) by every ordinate of the unit hydrograph, as shown in Figure 7.20. The 11.98- to 12-hour direct runoff hydrograph where $\Delta Q = 0.138437$ in Figure 7.19 is used in this example.

7.4.5   Composite Direct Runoff Hydrographs

Once each direct runoff hydrograph for each time step has been calculated, all of them can be added and lagged by time increment D, and any base flows can be included, to

FIG. 7.18   (a) Unit hydrograph, 1-hour duration. (b) Direct runoff hydrograph for 2 inches of excess precipitation. (c) Direct runoff hydrograph for subsequent 1 inch of excess precipitation.

Time (hours)	P_x/P_{24}	Depth (inches)	Q (inches)	Delta Q (inches)
11.96	0.6060	4.0481	1.494177	——
				> 0.135675
11.98	0.6345	4.2385	1.629852	
				> 0.138437
12.00	0.6630	4.4288	1.768289	
				> 0.023327
12.02	0.6678	4.4606	1.791616	
				> 0.023397
12.04	0.6725	4.4923	1.815013	

FIG. 7.19 Calculations showing a partial development of the excess precipitation.

t (hours)	q-unit (cfs)	Direct Runoff (cfs)
0.00	0.00	0.0
0.02	18.91	2.6
0.04	58.62	8.1
0.06	124.81	17.3
0.08	175.86	24.3
0.10	189.10	26.2
0.12	175.86	24.3
0.14	147.50	20.4
0.16	105.90	14.7
0.18	73.75	10.2
0.20	52.95	7.3
0.22	39.14	5.4
0.24	27.80	3.8
0.26	22.23	3.1
0.28	14.56	2.0
0.30	10.40	1.4
0.32	7.56	1.0
0.34	5.48	0.8
0.36	3.97	0.5
0.38	2.84	0.4
0.40	2.08	0.3
0.42	1.70	0.2
0.44	1.13	0.2
0.46	0.76	0.1
0.48	0.38	0.1
0.50	0.00	0.0

FIG. 7.20 Direct runoff hydrograph calculations for time increment 11.98–12.00 and $\Delta Q = 0.138437$ inch.

produce the total runoff hydrograph for the storm. Figure 7.18c illustrates the process graphically. This total runoff hydrograph can now be routed through downstream channels or reservoirs.

7.4.6 Hydrograph Routing

Hydrograph routing is associated with either stream or reservoir routing. Streamflow routing involves the process of moving a hydrograph from one upstream point to a downstream location. The stream's channel storage affects the amount of hydrograph attenuation.

Although there are a number of ways to route hydrographs through channels, the convex routing method developed by the NRCS is presented here. Figure 7.21 illustrates the fundamental relationship between the inflow and outflow hydrographs using this routing method. The basic equation for this relationship is

$$Q_{t+dt} = (1 - C)O_t + CI_t \tag{7.32}$$

where O_{t+dt} is the outflow at time $t + dt$, O_t is the outflow at time t, I_t is the inflow at time t, and C is the routing coefficient where $0 \leq C \leq 1$.

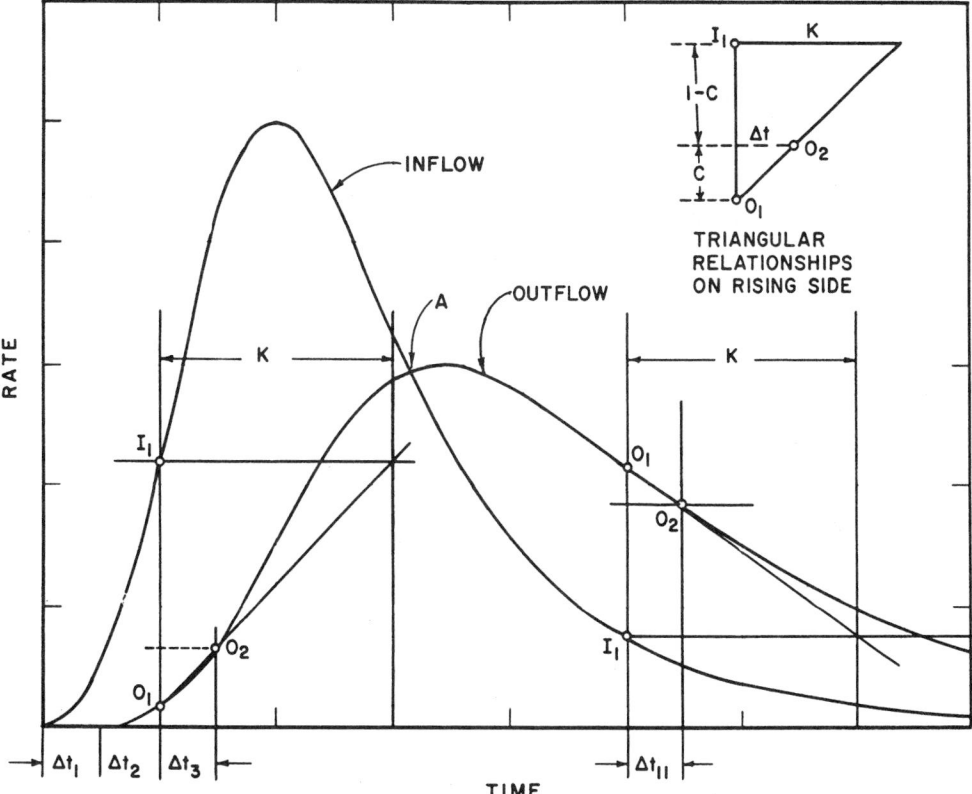

FIG. 7.21 Channel routing using the convex method. Source: Natural Resources Conservation Service, 1972, Chapter 17.

The term dt should be no more than about 20 percent of the time required for the inflow hydrograph to peak. The routing coefficient, C, can be estimated if the steady-state velocity of drainage is known using

$$C = \frac{V}{V + 1.7} \tag{7.33}$$

where V is the velocity of travel in feet per second and $V + 1.7$ is the approximate speed of a kinematic wave traveling through the reach.

We now illustrate the routing of a hydrograph where the inflow hydrograph is assumed to be a direct runoff hydrograph with ordinates the same as the unit hydrograph shown in Figure 7.17. The steady-state channel flow velocity is estimated at 1.5 ft/s. Using Equation 7.33, the routing coefficient $C = 0.46875$. The routing calculations are shown in Figure 7.22.

Reservoir routing, however, considers the following general relationship:

$$\frac{\text{Storage volume change}}{dt} = \text{inflow} - \text{outflow} \tag{7.34}$$

t (hour)	Inflow (cfs)	Routing $(1 - C)Ot + C\,It$				Outflow (cfs)
0.00	0.0	$(1 - 0.46875)$	0.0	$+ (0.46875)$	0.0	0.0
0.02	18.9	$(1 - 0.46875)$	0.0	$+ (0.46875)$	0.0	0.0
0.04	58.6	$(1 - 0.46875)$	0.0	$+ (0.46875)$	18.9	8.9
0.06	124.8	$(1 - 0.46875)$	8.9	$+ (0.46875)$	58.6	32.2
0.08	175.9	$(1 - 0.46875)$	32.2	$+ (0.46875)$	124.8	75.6
0.10	189.1	$(1 - 0.46875)$	75.6	$+ (0.46875)$	175.9	122.6
0.12	175.9	$(1 - 0.46875)$	122.6	$+ (0.46875)$	189.1	153.8
0.14	147.5	$(1 - 0.46875)$	153.8	$+ (0.46875)$	175.9	164.2
0.16	105.9	$(1 - 0.46875)$	164.2	$+ (0.46875)$	147.5	156.4
0.18	73.8	$(1 - 0.46875)$	156.4	$+ (0.46875)$	105.9	132.7
0.20	53.0	$(1 - 0.46875)$	132.7	$+ (0.46875)$	73.8	105.1
0.22	39.1	$(1 - 0.46875)$	105.1	$+ (0.46875)$	53.0	80.7
0.24	27.8	$(1 - 0.46875)$	80.7	$+ (0.46875)$	39.1	61.2
0.26	22.2	$(1 - 0.46875)$	61.2	$+ (0.46875)$	27.8	45.5
0.28	14.6	$(1 - 0.46875)$	45.5	$+ (0.46875)$	22.2	34.6
0.30	10.4	$(1 - 0.46875)$	34.6	$+ (0.46875)$	14.6	25.2
0.32	7.6	$(1 - 0.46875)$	25.2	$+ (0.46875)$	10.4	18.3
0.34	5.5	$(1 - 0.46875)$	18.3	$+ (0.46875)$	7.6	13.3
0.36	4.0	$(1 - 0.46875)$	13.3	$+ (0.46875)$	5.5	9.6
0.38	2.8	$(1 - 0.46875)$	9.6	$+ (0.46875)$	4.0	7.0
0.40	2.1	$(1 - 0.46875)$	7.0	$+ (0.46875)$	2.8	5.0
0.42	1.7	$(1 - 0.46875)$	5.0	$+ (0.46875)$	2.1	3.6
0.44	1.1	$(1 - 0.46875)$	3.6	$+ (0.46875)$	1.7	2.7
0.46	0.8	$(1 - 0.46875)$	2.7	$+ (0.46875)$	1.1	2.0
0.48	0.4	$(1 - 0.46875)$	2.0	$+ (0.46875)$	0.8	1.4
0.50	0.0	$(1 - 0.46875)$	1.4	$+ (0.46875)$	0.4	0.9

FIG. 7.22 Example of stream flow routing using the convex method with $C = 0.46875$.

In an expanded form this equation appears as

$$\frac{I_1 + I_2}{2} - \frac{O_1 + O_2}{2} = \frac{S_2 - S_1}{dt} \tag{7.35}$$

where I_1 is the inflow in cubic feet per second at time t, I_2 is the inflow in cubic feet per second at time $t + dt$, O_1 is the outflow in cubic feet per second at time t, O_2 is the outflow in cubic feet per second at time $t + dt$, S_1 is the storage volume in cubic feet at time t, S_2 is the storage volume in cubic feet at time $t + dt$, and dt is the time interval.

The outflow and storage volumes are dependent upon the surface level of the water in the reservoir. Two curves shown in Figure 7.23a and b represent these relationships based on the reservoir's discharge structure and geometry. These two curves can be combined into a third curve, Figure 7.23c, which represents the relationship between storage and discharge.

One method that can be used for reservoir routing is called the "storage indication" method. This method is often used in land development projects because discharge rates are used for input and output instead of using mass curves.

According to the NRCS (1972), the storage indication method utilizes the following equation:

$$I_{\text{ave}} + \frac{S_1}{dt} - \frac{O_1}{2} = \frac{S_2}{dt} + \frac{O_2}{2} \tag{7.36}$$

where $I_{\text{ave}} = \frac{1}{2}(I_1 + I_2)/2$ represents the values for I that can be taken from regular intervals or at the midpoints of routing intervals.

The working relationship between O_2 and $[(S_2/dt) + \frac{1}{2}O_2]$ must be available in either numerical or graphical form to solve Equation 7.36. Care must be exercised to ensure that an acceptable routing interval is used. According to the NRCS (1972, p. 17–22):

> Trial routings show that negative outflows will occur during recession periods of the outflow whenever dt is greater than $2S_2/O_2$ (or whenever $O_2/2$ is greater than S_2/dt). This also means that rising portions of hydrographs are being distorted.

These problems can be avoided by plotting the working O_2 versus $[(S_2/dt) + \frac{1}{2}O_2]$ values on a log-log graph, along with a second curve where each O_2 value equals each $[(S_2/dt) + \frac{1}{2}O_2]$ value. If any portion of the working O_2 versus $[(S_2/dt) + \frac{1}{2}O_2]$ curve plots above the second curve (representing a plot of equal values), then a sufficiently small dt increment must be utilized to produce a plot that is completely below the curve of equal values.

To illustrate the storage indication reservoir routing procedure, a reservoir that has a square surface will be used. Each side is assumed to be 100 feet long and the reservoir sides are assumed to be vertical (only for simplicity in this example). The outlet structure provides discharge rates that are shown in Figure 7.24. Once the inflow hydrograph and initial reservoir discharges are known, hydrograph routing can be performed.

The inflow hydrograph for this example is the same as the outflow hydrograph contained in Figure 7.22, thus requiring a time step, dt, of 0.02 hour. The working relationship between O_2 and $[(S_2/dt) + \frac{1}{2}O_2]$ is determined using data contained in Figure 7.24. The discharge values in cubic feet per second are taken directly from

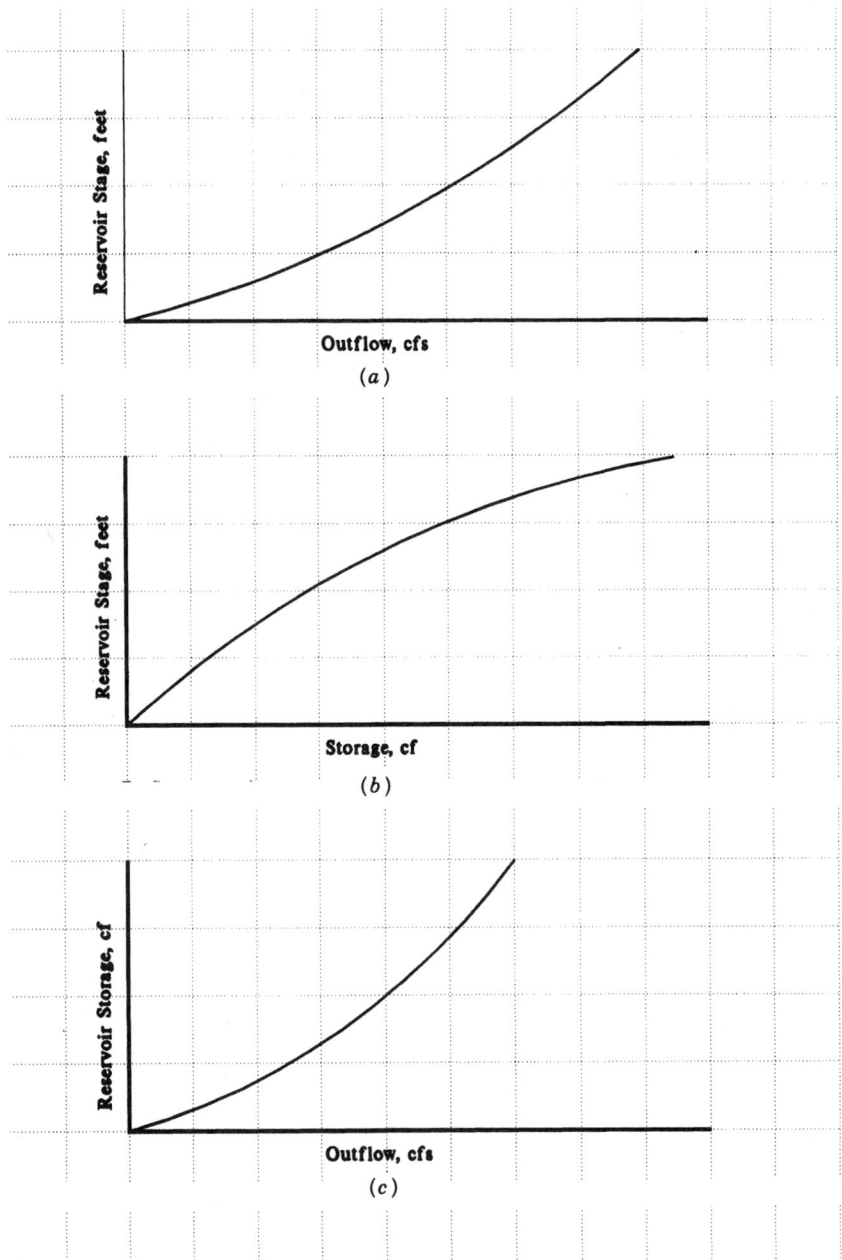

FIG. 7.23 Reservoir outflow and storage relationships: (a) stage versus discharge; (b) stage versus storage; (c) storage versus outflow.

the figure, while the values for S_2 must be in terms of "cfs-hours," thus requiring the storage values shown in cubic feet to first be converted into acre-feet and then into cfs-hours by multiplying each storage value in equivalent acre-feet by a factor of 12.10. Figure 7.25 shows the completed working relationships.

Stage (feet)	Discharge (cfs)	Storage (cft)
0	0	0
2	12	20,000
4	26	40,000
6	34	60,000
8	42	80,000

FIG. 7.24   Example of reservoir stage, discharge, and storage relationships.

The O_2 and $[(S_2/dt) + \frac{1}{2}O_2]$ values shown in Figure 7.25 can be used to numerically solve the storage indication equation or they can be plotted as a curve for graphical use. Equation 7.36 is solved numerically in Figure 7.26. Initially, when solving the routing equation for this example, the beginning time, inflow, average inflow, $[(S_2/dt) + \frac{1}{2}O_2]$, and outflow are known. Also known are other times, inflows, and average inflow values.

To solve for the $[(S_2/dt) + \frac{1}{2}O_2]$ value at time equal to 0.02 hour, subtract the value of column 5 from the value of column 4 for time 0.00 and add to this difference the value of column 3 when the time is 0.02 hour. This yields $0 - 0 + 0 = 0$ for $[(S_2/dt) + \frac{1}{2}O_2]$ when time equals 0.02 hour. According to Figure 7.25, the corresponding value of O_2 is 0.0 ft^3/s.

To solve for the $[(S_2/dt) + \frac{1}{2}O_2]$ value when time equals 0.04 hour, subtract the value of column 5 from the value of column 4 in Figure 7.26 for time 0.02 and add to this difference the value of column 3 when the time is 0.04 hour. This yields $(0 - 0 + 4.45 = 4.45$ ft^3/s for $[(S_2/dt) + \frac{1}{2}O_2]$ when time equals 0.04 hour. The corresponding value of O_2 can be obtained by interpolating (or graphically) using data found in Figure 7.25. In this case $O_2 = (4.45/283.75)(12) = 0.188$ ft^3/s when time equals 0.04 hour.

By repeating these steps, $[(S_2/dt) + \frac{1}{2}O_2]$ for time 0.06 hour is found to be $4.45 - 0.188 + 20.55$, or 24.812 ft^3/s. The corresponding outflow can be found using Figure 7.25 as $(24.812/283.75)(12)$, or 1.049 ft^3/s. The remainder of the outflow hydrograph is obtained using these same procedures.

The reservoir stage and storage values can be obtained for each time step using Equation 7.34 along with the routed hydrograph from the reservoir and the reservoir's geometry. An example is given later in this chapter where computer analysis provides hydrograph routing through a reservoir.

7.5   FACTORS AFFECTING SITE DRAINAGE

The engineer must establish various criteria to follow during the drainage design process including the following:

(a) Projects located on ridge lines that receive no off-site stormwater allow the design engineer to concentrate strictly on drainage design within the site.

(b) Sites located away from ridge lines require the engineer to determine whether off-site storm runoff will contribute to on-site design flows. If so, provisions must be

O_2 (cfs)	$O_2/2$ (cfs)	S_2 (cfs-hours)	S_2/dt (cfs)	$[(S_2/dt) + (O_2/2)]$ (cfs)
0	0	0	0	0
12	6	12.1(20,000/43560) = 5.555	277.75	283.75
26	19	12.1(40,000/43560) = 11.111	555.55	574.55
34	30	12.1(60,000/43560) = 16.667	833.35	863.35
42	38	12.1(80,000/43560) = 22.222	1111.10	1149.10

FIG. 7.25   Example of working relationships between O_2 and $S_2/dt + \frac{1}{2}O_2$.

made to allow the off-site runoff to pass through the development and, if necessary, to intercept these flows for flood prevention.

(c) Generalized lot grading plans (see Chapter 9) must be established to provide positive site drainage.

(d) Regulatory discharge limits for stormwater must be obtained to establish allowable postconstruction site discharge rates. Requirements for stormwater retention or detention facilities are also necessary.

(e) The maximum distance that runoff can be carried in a roadside ditch or curb and gutter before it must be removed must also be determined. Some development regulations specify maximum placement distances for standard inlet devices. These distances can be used when making the preliminary development plan; however, inlet performance should be verified during the final design. If local regulations do not specify a maximum interval, an analysis must be made to determine maximum inlet spacing based on performance.

7.5.1 Inlet Spacing Interval

A maximum spacing interval for curb inlets on a project can be estimated by analyzing a typical road cross section (Figure 7.27a) on minimum grade using a typical inlet device. The simplified road section shown in Figure 7.27b is used in this example for inlet spacing analysis. Manning's equation (Equation 7.19) is used to determine the maximum flow capacity when a predetermined depth of flow is established in the gutter.

If the maximum flow depth in the curb and gutter half section is arbitrarily set for this example at approximately 3.5 inches for a 10-year return period, the corresponding flow cross section would appear as shown in Figure 7.27c for half of the street. If the minimum centerline slope of the road is 0.5 percent, the amount of runoff that can be carried by half of the street section can be found where:

(a) The area of the flowing triangle equals $\frac{1}{2}$ (3.5/12 feet depth of flow)(10.5 feet wide), or 1.53 ft².

(b) The hydraulic radius, A/P, equals $(1.53 \text{ ft}^2)/\{(3.5/12)+[10.5^2+(3.5/12)^2]^{0.5}\}$, or 0.142 foot.

(c) The minimum road slope is given as 0.5 percent, or 0.005 feet per foot of roadway.

(d) Manning's roughness coefficient, n, is taken as 0.013 from Figure 7.12.

T (hours) (1)	I (cfs) (2)	I_{ave} (cfs) (3)	$[(S_2/dt) + (O_2/2)]$ (cfs) (4)	O (cfs) (5)
0.00	0.0	0.00	0.000	0.000
0.02	0.0	0.00	0.000	0.000
0.04	8.9	4.45	4.450	0.188
0.06	32.2	20.55	24.812	1.049
0.08	75.6	53.90	77.663	3.284
0.10	122.6	99.10	173.479	7.337

FIG. 7.26 Storage indication method of hydrograph routing through a reservoir.

Using Equation 7.19, the flow is determined as

$$Q = \frac{1.49(1.53)(0.142)^{0.67}(0.005)^{0.5}}{0.013} = 3.35 \text{ ft}^3/\text{s}$$

This is the capacity of curb flow for the 10-year return period that will be used to estimate inlet spacing.

If a weighted runoff coefficient of 0.54 is used with the rational formula (Equation 7.21), the relationship of rainfall intensity to drainage area is established by

$$Q = CIA$$
$$3.35 = (0.54)(I)(A)$$
$$(I)(A) = 6.2 \text{ acre-inches per hour}$$

The corresponding velocity of flow for a 10.5-foot wide by 3.5-inch deep triangular shaped cross section is computed using Equation 7.18 by

$$Q = AV.$$
$$V = 3.35 \text{ ft}^3/\text{s}/(1.53 \text{ ft}^2)$$
$$V = 2.19 \text{ ft/s}$$

If a typical lot is to be 100 feet wide and the typical lot area is $\frac{1}{3}$ acre, the corresponding lot depth is approximately 145 feet. The contributory drainage area encompassed by one lot and half the fronting road (assuming a 50-foot right-of-way) is $[100 \times (145 + 25)]/43,560$, or 0.39 acre. This assumes that the entire lot drains toward the street. The time of travel from the remotest point in the lot (at the rear lot line) to the street can be estimated using the sheet flow equation 7.13 and Figures 7.5 (for the 10-year return period curve) and 7.11, where

$$T_t = \frac{(0.93)(n \times L)^{0.6}}{(i)^{0.4}(S)^{0.3}} = \frac{(0.93)(0.15 \times 145)^{0.6}}{(i)^{0.4}(0.005)^{0.3}}$$

By trial and error, it is determined that the above equation is satisfied when $i_{10} = 5.73$ inches per hour, and the duration is 14.38 minutes. The amount of time required

FIG. 7.27   Roadway drainage analysis using a rolled curb section. (a) Road cross section for storm drainage analysis. (b) Simplified road cross section for storm drainage analysis. (c) Flow area when 3.5 inches depth of flow occurs in the gutter. This will allow a single 9-foot-wide traffic lane for use during a 10-year storm.

for the runoff to travel 100 feet in the street gutter corresponding to one lot width is (100 feet per lot width)/(2.19 feet per second), or 45.66 seconds. This value is rounded to 0.75 minute for these preliminary calculations. Figure 7.28 is constructed to aid in determining the maximum number of contributory lots. This maximum number provides the largest drainage area that can be handled by this road section using a 10-year return period design.

The $(I)(A)$ based on minimum slopes for this roadway is 6.2 acre-inches per hour; therefore, three lots approximate the maximum contributory drainage area to produce an approximate flood depth of 3.5 inches. This would indicate a maximum spacing between inlet devices of 3×100 feet (typical lot width for this development) or 300 feet, unless a sag vertical curve low point is present, as illustrated in Figure 7.29.

The performance of all inlet structures should be verified to ensure their adequacy. To determine the required length of throat of a curb inlet that depends on weir flow, the generalized weir formula can be utilized (Walesh, 1989):

Number of Lots	T_c (minutes)	I (inches per hour)	A (acres)	$(I)(A)$ (acre-inches per hour)
1	(14.4 + 0.75 = 15.15); use 15.2	5.54	0.39	2.16
2	15.15 + 0.75 = 15.9	5.48	0.78	4.27
3	(15.9 + 0.75 = 16.65); use 16.7	5.42	1.17	6.34
4	16.65 + 0.75 = 17.4	5.37	1.56	8.38
5	(17.4 + 0.75 = 18.15); use 18.2	5.32	1.95	10.37
6	18.15 + 0.75 = 18.9	5.25	2.34	12.29

FIG. 7.28 Inlet spacing capacities for various contributory areas.

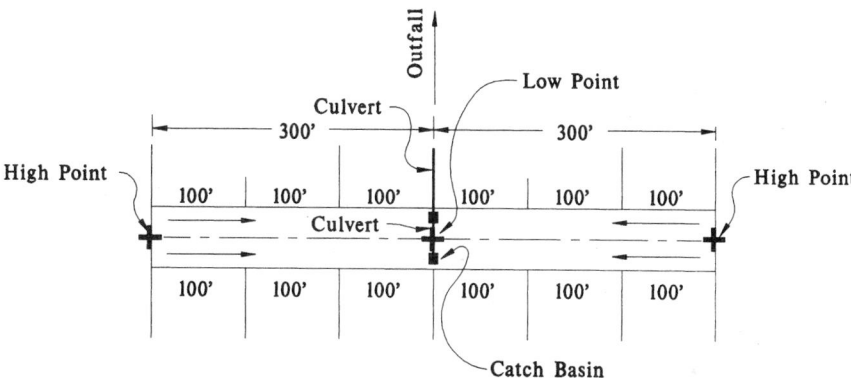

FIG. 7.29 Example of positioning of curb inlets at a vertical sag curve low point.

$$Q = C_w L H^{1.5} \qquad (7.37)$$

where Q is the flow rate in cubic feet per second, C_w is the weir coefficient, L is the throat length in feet, and h is the depth of flow in the gutter in feet.

If an estimated weir coefficient of 3 is assumed, based on free discharge conditions, the corresponding length of inlet throat to accommodate a 10-year return period would be

$$3.35 \ \text{ft}^3/\text{s} = 3 \times L \times (3.5/12)^{1.5}$$

or

$$L = 7.09 \ \text{feet}$$

A curb inlet similar to Figure 7.30a would provide adequate inlet capacity with a 5-foot inlet parallel to the street and another 4-foot inlet perpendicular to the street positioned on the uphill side. If the inlet, however, is located at the low point of a sag curve where the contributory flow totals 6.70 ft³/s (from both directions), a different inlet configuration must be utilized. By including an additional 4-foot inlet perpendicular to the street, a total throat length of 13 feet can be obtained. The required throat length for the 6.70-ft³/s inflow, however, is 14.18 feet. A grate casting that is 2 × 5 feet can be installed as shown in Figure 7.30b to increase the inlet's capacity using orifice flow through the grate. The generalized orifice equation is (Walesh, 1989)

$$Q = C_d A (2gh)^{0.5} \tag{7.38}$$

where Q is the orifice flow in cubic feet per second; C_d is the coefficient of discharge, taken as 0.6 for the grate entrance; A is the cross-sectional area of the opening in the case of a grate the sum of all openings is used; $g = 32.2$ feet per second per second; and h is the depth of flow in feet.

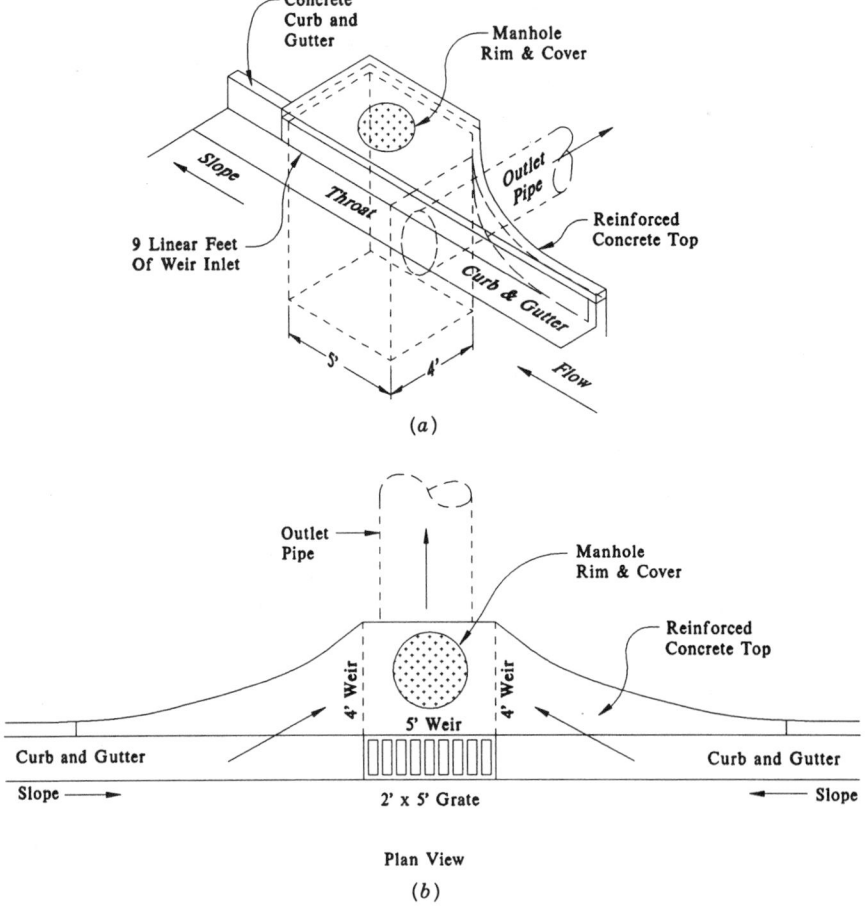

(a)

Plan View

(b)

FIG. 7.30 Examples of curb inlet details: (a) on an incline; (b) at a sag curve low point.

If the proposed grate contains 16 slots, each 1.5 × 16 inches, the total grate opening area is 16 slots × (1.5 × 16 inches)/144 in.^{2}/ft^2, or 2.66 ft^2. Using the approximate flood depth of 3.5 inches for the 10-year return period, the grate inflow may be computed through application of Equation 7.38 as

$$Q = 0.6 \times 2.66 \times [2 \times 32.2 \times (3.5/12)]^{0.5} = 6.92 \text{ ft}^3/\text{s}$$

The curb inlet and grate should be adequate for sag locations. As an alternative to using the weir and orifice equations, the design professional can use inlet capacity charts or nomographs that are available to determine inflow rates. If charts or nomographs are used, the configuration of the inlet used in the design aid must be the same as that proposed for installation in the field.

7.5.2 Retention/Detention Area Requirements

The engineer should determine the approximate size of any required retention/detention facilities based on local ordinances and BMP requirements (Section 7.6) since adequate land area must be set aside for that use. Detention ponds are basins where stormwater runoff is temporarily stored after a rainfall event. Ponds can be designed as "dry" or "wet" facilities. Dry ponds normally have no water present during dry season because they are designed to drain completely. Some dry ponds are designed for multiple use. Neighborhood parks and open-space recreational areas (e.g., ball fields) can sometimes be designed for the pond area where multiple benefits are gained. Wet ponds (see Figure 8.2) normally have a standing pool of water (lake) where the stormwater runoff temporarily increases the water but is lowered over time as the collected runoff discharges until the normal pool level is again achieved. The difference between a retention and detention facility is that a retention facility has zero discharge while a detention facility has a limited discharge.

Some regulatory agencies may require that:

(a) Maximum discharges from storage facilities are to be limited to the peak stormwater runoff rates that prevailed prior to site development.

(b) Stormwater runoff is to be stored and its release controlled when a site's modification results in the peak runoff rate being increased by more than 1 ft^3/s for a 5-year return period.

(c) Some regulatory agencies may require a wet pond facility to store and release the first ½ inch of runoff from the entire site over a 24-hour period [South Carolina Department of Health and Environmental Control (SCDHEC), 1998].

(d) Areas having runoff that is directly channeled into receiving streams are to be required to provide on-site retention volumes equivalent to at least 1½ inches of runoff depth from the built-up portion of the property (SCDHEC, 1998). This volume can include a portion of a detention basin where the basin's outlet structure is elevated to provide the necessary storage. To eliminate standing water in a "dry" basin, a subsurface drain can be installed to allow the water to slowly drain and at the same time be filtered. Natural wetland areas also can be used to provide the required on-site retention volume as discussed in Section 7.6.5 as well as other BMP options.

An approximate detention facility size can be determined once the maximum allowable discharge from the developed site is known. Assuming that the allowable rate will be the predevelopment rate, the allowable discharge is computed using the rational formula equation 7.21. The watershed area, the coefficient of runoff, and the time of concentration must be determined. The rainfall intensity (I) can be obtained from the rainfall intensity curve for the area. Figure 7.5 is assumed to be applicable for the following example.

To illustrate the process, if the contributory area of a development consists of 5.13 acres (Figure 4.31), the predevelopment rate of runoff can be calculated. The site is moderately flat and has an estimated coefficient of runoff of 0.20.

The time of concentration is the maximum time required for a drop of water to travel from the remotest point within the drainage basin to the point under investigation, which in this case is the discharge point for the site. Figure 4.64 shows the existing topography, which can be used as a basis for time of concentration determination.

For this example, the time of concentration is divided into three separate travel times to include sheet flow, shallow concentrated flow, and channel flow. Figure 4.64 indicates where estimated sheet flow will occur between points A and B. Shallow concentrated flows are expected between points B and C, while channelized flows will occur between points C and D.

The time of travel T_1, for sheet flow is calculated using Equation 7.13. Manning's roughness coefficient, n, is estimated from Figure 7.11 as 0.15. The scaled length of travel is approximately 300 feet, and the ground slope is estimated as 0.0033 ft/ft. Time T_1 must be calculated for each return period using the rainfall intensity curve for that particular return period. For the 2-year return period T_1 is estimated, with Equation 7.13 being solved by trial and error, as

$$T_1 = \frac{(0.93)(n \times L)^{0.6}}{(i_2)^{0.4}(S)^{0.3}} = \frac{(0.93)(0.15 \times 300)^{0.6}}{(i_2)^{0.4}(0.0033)^{0.3}}$$

This equation is satisfied when the intensity is 3.227 inches per hour for the 2-year storm, and the corresponding duration is 31.72 minutes. Other T_1 values are shown in Figure 7.31.

The velocity of travel, V_2, for shallow concentrated flow is calculated using Equation 7.15. The travel distance is estimated as 120 feet, and the corresponding ground slope at 0.0125 ft/ft:

$$V_2 = 16.1345(0.0125)^{0.5} = 1.8 \text{ ft/s}$$

The time of travel, T_2, can be obtained by utilizing Equation 7.20:

$$T_2 = \frac{120}{3600 \times 1.8} = 0.02 \text{ hour}$$

Return Period (Years)	2	5	10	25	50	100
T_1 (minutes)	31.7	29.1	27.7	25.9	24.8	23.8

FIG. 7.31   Travel time, T_1, for six return periods.

The velocity of travel, V_3, for channelized flow for line $C\text{–}D$ is calculated using Equation 7.17. The cross-sectional area is estimated at 5 ft^2 and the wetted perimeter is estimated at 10 feet. Manning's roughness coefficient, n, is taken as 0.05 from Figure 7.12 for a ditch bank with weeds and brush, while the slope is estimated as 0.0125 ft/ft:

$$V_3 = \frac{1.49(0.5)^{0.67}(0.0125)^{0.5}}{0.05} = 2.1 \text{ ft/s}$$

The time of travel, T_3, can again be obtained by utilizing Equation 7.20 and the scaled distance between points C and D as 280 feet:

$$T_3 = \frac{280}{3600 \times 2.1} = 0.04 \text{ hour}$$

The time of concentration for the 2-year return period, T_c, is equal to $T_1 + T_2 + T_3$, or $0.53 + 0.02 + 0.04 = 0.59$ hour (or approximately 35 minutes). Other times of concentration are shown in Figure 7.32.

The site's predevelopment discharge rates, shown in Figure 7.33 for various return periods, are found using Equation 7.21, where the coefficient of runoff is estimated to be 0.2, a contributory area to be 5.13 acres, and an intensity for each return period to correspond to the time of concentration shown in Figure 7.32.

The Federal Aviation Administration (FAA) method of ponding can be used to approximate the volume of required detention storage (U.S. Department of Transportation—FAA, 1970). Should the discharge be regulated at zero (for a retention pond), the procedure is identical with zero discharge. The volume of inflow is computed using

$$V_{\text{storage}} = CIAT \tag{7.39}$$

where V_{storage} is the required storage volume in cubic feet, C is the coefficient of runoff,

Return Period (years)	2	5	10	25	50	100
T_c (minutes)	35	33	31	30	28	27

FIG. 7.32   Times of concentration of a sample undeveloped site for six return periods.

Return Period (years)	Predevelopment Q (cfs)
2	3.20
5	3.89
10	4.43
25	5.15
50	5.86
100	6.49

FIG. 7.33   Examples of predevelopment discharge rates for six return periods.

I is the rainfall intensity in inches per hour for time *T*, *T* is the time of runoff volume accumulation in seconds, and *A* is the drainage area in acres.

A coefficient of runoff is assumed for this example to be 0.54 for postconstruction conditions. Using Figure 7.4, the rainfall intensities for durations of 5, 10, 15, 30, 60, 120, 180, and 360 minutes are obtained for the 100-year return period. The corresponding volumes are calculated using Equation 7.39. Volumes of runoff are plotted as ordinates and corresponding times as abscissas. The allowable discharge for the 100-year return period is then plotted as a straight line (starting at time zero with zero discharge). The accumulated discharge volume when time is equal to 360 minutes (360 minutes × 60 s/min × 6.49 ft³/s, or 140,184 ft³) is also plotted, where this point and the initial point are connected. When the curves are completed, the time that provides the maximum vertical separation between the two curves yields the required storage volume. Examples of calculations and curves are shown in Figure 7.34.

An approximate basin surface area can be obtained by scaling the proposed detention area on the preliminary plan (or using a planimeter), as shown in Figure 7.35. This surface area is necessary to compute an approximate basin depth to check outfall elevations.

100 Year Storm- Allowable Discharge: 6.49 cfs.			
Time Min.	Intensity In/hr.	Inflow Volume CFT	Discharge CFT
0	0.00	0.00	0.00
5	9.60	7978.18	1947.00
10	8.52	14161.26	3894.00
15	7.40	18449.53	5841.00
30	6.06	30217.34	11682.00
60	4.25	42384.06	23364.00
120	2.44	48666.87	46728.00
180	1.83	54750.23	70092.00
360	1.18	70606.86	140184.00

FIG. 7.34 FAA method of ponding: preliminary detention pond volume determination for a sample 100-year return period.

FIG. 7.35   Initial detention basin plan for the sample drainage system.

365

The maximum required storage for the 100-year storm is 19,020 ft^3; however, a value of twice this should initially be used to compensate for approximations and to allow flexibility while designing. The estimated basin size (assuming a 90 × 90-foot surface area) is 10 feet deep with 2 : 1 side slopes. The resulting bottom would be 50 × 50 feet. The approximate water depth to provide 38,040 ft^3 of storage is 8.15 feet. Based on the topography, this preliminary depth appears to be satisfactory, based on contour lines shown in Figure 7.35, which indicate an elevation difference of aproximately 13 feet between the pond site and the downstream invert of the outfall ditch. This elevation difference allows proposed culverts to be placed with sufficient cover and slope. If the estimated storage depth is large where gravity flow into the site's stormwater outfall cannot be utilized, then the basin's surface area could be increased to allow a corresponding decrease in detention pond depth. This may accommodate the terrain requirements; otherwise additional basins may be required.

Care must be taken when designing a detention pond to ensure that the inflow velocities are low enough to prevent pond erosion. Should the potential for scour occur, appropriate energy dissipation devices must be installed. The designer should also check the discharge velocities of the pond's outflow for potential erosion problems, as discussed in Section 7.8.5.

7.6 BEST MANAGEMENT PRACTICES

The design of stormwater systems can include provisions for promoting improved water quality runoff from the site. Figure 7.36 illustrates many nonpoint sources of pollution and how their impacts can be controlled through stormwater management facilities. One of these provisions involves sedimentation and erosion control practices featured in Chapter 8. Other provisions can include ways to reduce oxygen demands, bacteria, nutrients, metals, and other toxic chemicals.

7.6.1 Pollutant Concentrations

Pollutant concentrations can be estimated by using average concentrations of pollutants per storm event called event mean concentrations (EMCs). Figure 7.37 shows median EMCs based on various land use categories. The runoff pollutant load for each storm event is the product of the appropriate EMC and the excess rainfall (runoff per average storm event).

Environmental Protection Agency (EPA) studies have determined that the storm duration, runoff rate, and rainfall intensity can affect the types of contaminates removed and the amount of removal. Figure 7.38 shows different rainfall intensities and durations where 90 percent of the particles are removed from impervious surfaces.

According to Wanielista and Yousef (1993, p. 128), average pollutant loadings for various streets and roads can be estimated using Figure 7.39 and other figures; however, one is cautioned that data shown in Figures 7.37 and 7.39 are valuable

> for understanding the magnitude of the problems related to runoff water quality; however, the quantity of pollutants washed from the impervious surfaces have to be determined. Thus mathematical models of both quantity and quality using loading rates, transport functions, degradation effects, and deposition rates are, to a degree, useful in reproducing

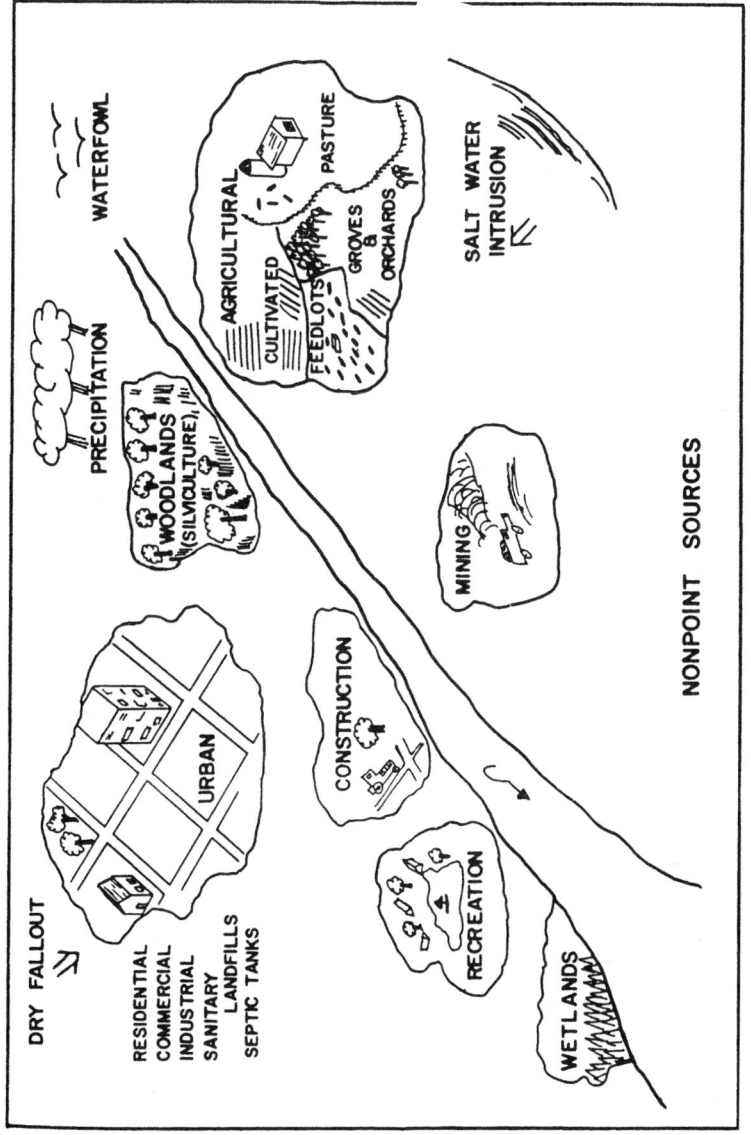

FIG. 7.36 Nonpoint sources with three possible stormwater management location sites. Source: Wanielista and Yousef, 1993. Copyright © 1993 by John Wiley & Sons, Inc. Reprinted by permission of John Wiley & Sons, Inc.

Median EMCs for All Sites by Land-Use Category

Pollutant	Residential Median	CV	Mixed Median	CV	Commercial Median	CV	Open/Nonurban Median	CV
BOD (mg/L)	10.0	0.41	7.8	0.52	9.3	0.31	—	—
COD (mg/L)	73	0.55	65	0.58	57	0.39	40	0.78
TSS (mg/L)	101	0.96	67	1.14	69	0.35	70	2.92
Total lead (μg/L)	144	0.75	114	1.35	104	0.68	30	1.52
Total copper (μg/L)	33	0.99	27	1.32	29	0.81	—	—
Total zinc (μg/L)	135	0.84	154	0.78	226	1.07	195	0.66
Total Kheldahl nitrogen (μg/L)	1900	0.73	1288	0.50	1179	0.43	965	1.00
$NO_2 - N + NO_3 - N$ (μg/L)	736	0.83	558	0.67	572	0.48	543	0.91
Total P (μg/L)	383	0.69	263	0.75	201	0.67	121	1.66
Soluble P (μg/L)	143	0.46	56	0.75	80	0.71	26	2.11

FIG. 7.37 Median EMCs for all sites by land use category. Source: Wanielista and Yousef, 1993. Copyright © 1993 by John Wiley & Sons, Inc. Reprinted by permission of John Wiley & Sons, Inc.

0.01 in./h for 300 min
0.33 in./h for 90 min
0.50 in./h for 60 min
1.00 in./h for 30 min

Source: U.S. Environmental Protection Agency (1974).
[a]in./h × 25.4 = mm/h.

FIG. 7.38 Rainfall intensities and durations for 90 percent particle removal from impervious surfaces. Source: U.S. Environmental Protection Agency, 1974.

existing conditions and in predicting future conditions relative to planning, design, or management of surface water quality and quantity.[*]

To illustrate how one might estimate the concentration of lead in runoff from a curb and gutter lined city street 1500 feet long and 30 feet wide after a 60-minute rainfall event of 0.5 in./h, use the following procedure. For this example it is assumed that it has been 7 days since the last storm completely cleaned the road and there is no evaporation, infiltration, or initial abstractions. Additional lead-free runoff is assumed to be contributed from property adjoining the street as well as the precipitation itself being free from lead. The amount of runoff from the property adjoining the street is assumed to be twice the street's volume.

---

[*] Source: Martin P. Wanielista and Yousef A. Yousef. *Stormwater Management*, 1993, p.128. Copyright © 1993 by John Wiley & Sons, Inc. Reprinted by permission of John Wiley & Sons, Inc.

Parameter	City Street	Rural Road	Highway
BOD_5	0.850	0.140	0.900
COD	5.000	4.300	10.000
PO_4	0.060	0.150	0.080
NO_3	0.015	0.025	0.015
N (total)	0.150	0.055	0.200
Cr	0.015	0.019	0.067
Cu	0.007	0.003	0.015
Fe	1.360	2.020	7.620
Mn	0.026	0.076	0.134
Ni	0.002	0.009	0.038
Pb	0.093	0.006	0.178
Sr	0.012	0.004	0.018
Zn	0.023	0.006	0.070

Source: Wanielista (1976) and Pitt and Amy (1973).

FIG. 7.39 Average loadings for common pollution parameters and certain heavy metals (kg/curb km-day). Sources: Pitt and Amy, 1973, and Wanielista and Yousef, 1993. Copyright © 1993 by John Wiley & Sons, Inc. Reprinted by permission of John Wiley & Sons, Inc.

A. Using Figure 7.39, one can obtain the typical lead loading on a city street as 0.093 kg/km-day. For this drainage basin analysis there are 1500 feet × 2 curbs/3280.83 ft/km, or 0.914 km. The estimated lead concentration is

$$P_b^n = 0.093 \text{ kg/km-day}) \times 0.914 \text{ km} \times 7 \text{ days} = 0.595 \text{ kg}$$

B. A 60-minute rainfall event having an intensity of 0.5 in./h removes 90 percent of the particles on the roadbed (Figure 7.38). The estimated amount of lead in the runoff is 0.9×0.595 kg, or 0.536 kg.

C. The total runoff volume is the sum of the street volume plus that of the surrounding areas:

$$\text{Volume} = \frac{0.5 \text{ in./h}}{12 \text{ in./ft}} \times 1 \text{ hour} \times 30 \text{ feet} \times 1500 \text{ feet} \times \frac{1 \text{ acre}}{43560 \text{ ft}^2}$$

$$\times (1.235 \times 10^6 \text{ L/acre-ft})$$

$$= 0.0532 \times 10^6 \times 3 = 1.596 \times 10^5 \text{ liters}$$

D. The lead concentration is $0.536 \text{ kg}/1.596 \times 10^5$ liters $= 3.36$ mg/L, which is the event mean concentration.

7.6.2   Water Quality Practice Options

Best management practice options (other than for sedimentation and erosion control measures presented in Chapter 8) fall under two recognized categories: (a) nonstructural BMPs and (b) structural BMPs. Some examples are shown in Figure 7.40. Nonstructural BMPs can include improved "housekeeping" procedures relative to street sweeping, control of pesticides/herbicides/fertilizer, and the use of vegetative practices such as wetland applications, grass swale, and the like. Structural BMPs included detention ponds, infiltration trenches, porous pavements, and trap inlets. In the design of a land development typically the nonstructural BMPs utilized as part of the design process are those pertaining to wetland applications, grass swales, and filter strips. Construction specifications (Chapter 13) can be used to address control of herbicides, pesticides, and fertilizers during the construction phase. Nonstructural practices combined with structural practices offer a multitreatment approach to improving the quality of stormwater runoff. On-site structural and nonstructural BMP facilities offer a number of additional benefits. First, they help to recharge groundwater and reduce settlement resulting from groundwater extraction. Next they promote vegetation growth and help to reduce downstream peak flows. Finally, they help reduce the size of storm sewers because runoff volumes are decreased. Typically on-site disposal facilities fall into one of two general categories classified by their function: (a) infiltration devices and (b) impoundments.

7.6.3   Infiltration Devices

Infiltration devices are used to allow polluted stormwater to infiltrate into the soil structure on-site. This can be achieved in a number of ways, including the following options:

A. *Infiltration Basins*   Infiltration basins (beds) are areas where stormwater runoff is diverted and held while it is allowed to infiltrate into the surrounding soil, as illustrated in Figure 7.40*a*. Typically, the infiltration basin is constructed as a depressed area in permeable soils. In areas where the soil has low permeability, a French drain (Section 7.10) can be installed to help increase the rate at which the water subsides in the basin after a rainfall event. The surfaces of these basins are usually protected with a cover growth such as grass.

B. *Infiltration Trenches*   Infiltration trenches are underground reservoirs made from shallow excavations that are backfilled with coarse stone. Examples are shown in Figure 7.40*c*. Collected runoff that is diverted to these trenches is allowed to infiltrate the surrounding soil while removing fine particles and soluble pollutants. Infiltration trenches can be modified where the runoff is piped to and distributed within the stone layer. These modified versions are sometimes called percolation basins or pipe trenches which operate exactly like septic tank absorption fields described in Section 10.3.2 except that stormwater runoff is being conveyed through the perforated pipe instead of septic tank effluent.

C. *Infiltration Ditches*   Roadside ditches (or swales) and ditches adjacent to parking areas can be used to promote infiltration of runoff into on-site soil, as shown in Figures 7.40*c* and 7.40*d*. The stabilized groundwater table must be below the invert of the ditch or swale to work properly. As in the case with all infiltration devices, the rate of infiltration is a function of the surrounding soil porosity, the amount of solids present in the runoff, and the type and amount of vegetation lining the ditch.

D. *Infiltration Ponds*   This option utilizes a depressed area for impounding runoff at greater depths than those involved with infiltration basins. The impoundment can act as a retention or detention pond, further described in the next section. For the pond

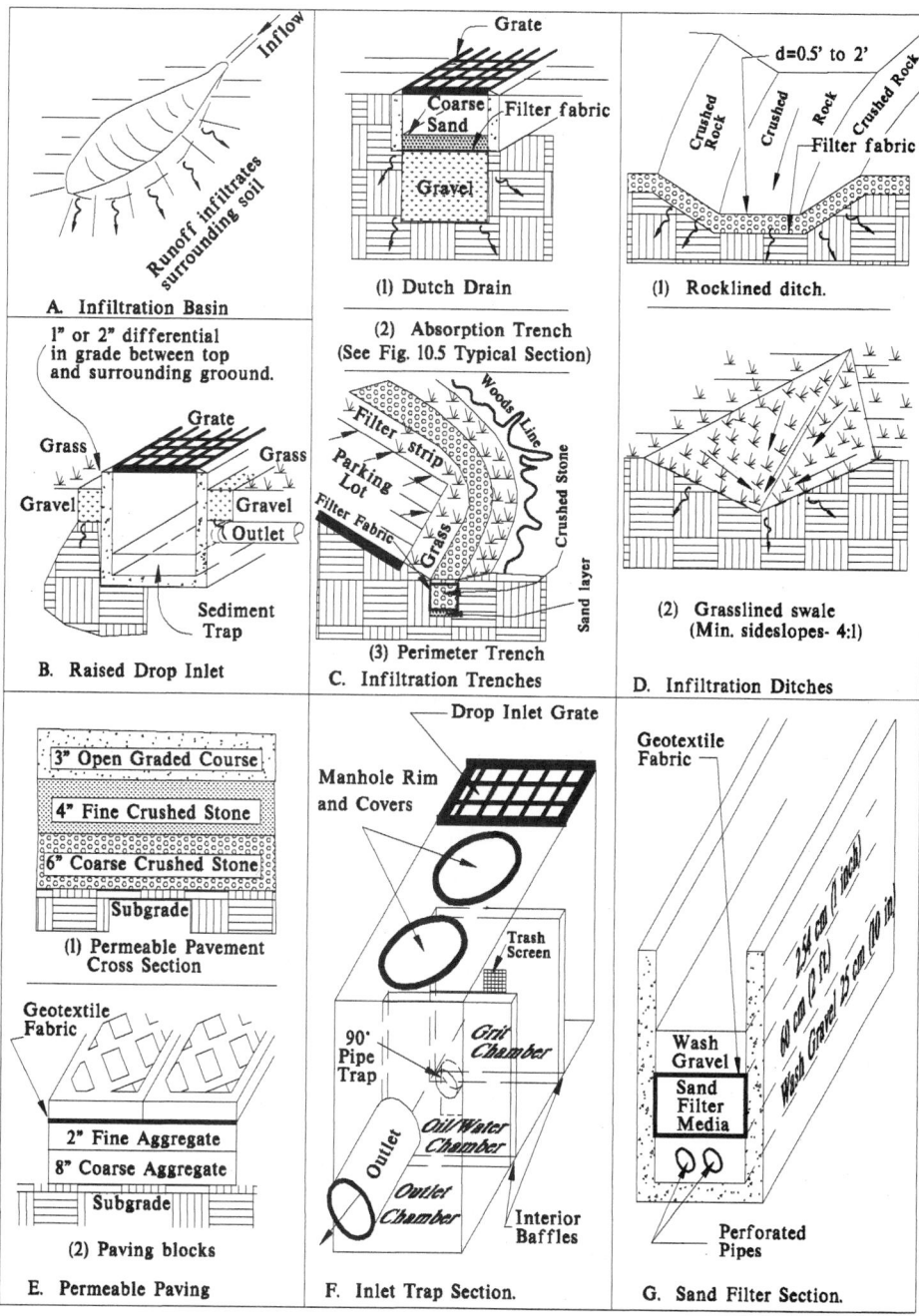

FIG. 7.40 Examples of best management practices (BMPs) that promote improved water quality: (*a*) infiltration basin; (*b*) raised drop inlet; (*c*) examples of infiltration trenches—(1) Dutch drain, (2) reference to an absorption field shown as a detail in Figure 10.5, and (3) perimeter trench; (*d*) infiltration ditches—(1) rocklined and (2) grasslined swale; (*e*) permeable paving—(1) permeable pavement structural cross section and (2) paving blocks; (*f*) inlet trap; (*g*) sand filtration device. Adapted in part from South Carolina Coastal Council, 1988.

to have the infiltration classification, it must be built out of a permeable soil where the runoff can infiltrate into the soil structure. The seasonal groundwater table can affect how infiltration ponds perform, and because larger quantities of runoff are impounded when using this option, the pore structure of the soil tends to clog more quickly than when using the other alternatives.

E. *Permeable Pavement*   Porous pavement made out of asphalt or concrete can be used over a compacted granular base. Typically runoff permeates the pavement structure to an underlying stone reservoir or infiltration trench. Freeze-thaw action can cause porous pavements to deteriorate prematurely in some cold-weather climates, and there are as well cases where these types of pavements have clogged after a short period of service time because the surfaces were not periodically swept. An alternative would be to use interlocking concrete paving blocks which give good service in most geographical areas; however, they can misalign under the action of traffic. Use of paving blocks in areas that accommodate overflow parking make them ideal as well as in areas where they can be confined to resist movement. Figure 7.40e shows two examples of permeable pavement options.

7.6.4   Impoundments

Impoundment areas are used to store runoff for either or both of the following two reasons: (1) to allow pollutants to settle out and/or (2) to allow contaminated water to percolate into the surrounding soil structure so that it does not leave the site. There are a variety of impoundment facilities that can be utilized, including the following:

A. *Detention Ponds*   When considering using a detention pond, discussed in Section 7.5.2, as a BMP alternative, extending the time the stormwater is stored (detained) can allow an opportunity for pollutants to settle (and possibly infiltrate into the surrounding soil) before the stored water leaves the pond. In the case of wet ponds, soluble pollutants can be stabilized biologically.

B. *Underground Caverns*   Underground caverns are stormwater runoff storage areas located underground for the purposes of detention by controlling the rate of release. If the seasonal groundwater table has an elevation low enough below the cavern bottom, these caverns can be built with bottom and side openings where the stored stormwater is allowed to infiltrate the surrounding soil. These caverns can be created using concrete or steel-tank-type structures or a series of interconnected pipes to provide the storage volume required. If sedimentation is to occur in the structure, the inflow and outflow velocities must be kept to a minimum; otherwise previously deposited sediment will become resuspended. Underground caverns are often used on sites where land is at a premium and little space if available for surface storage facilities to be constructed.

C. *Paved Surface Detention*   This approach utilizes areas within parking lots that are purposefully flooded after a rainfall event to take advantage of the existing capacity not normally used for detention purposes. This approach requires the designer to establish a maximum allowable pavement flood depth for a specified design storm. The design would accommodate any storm having a return period in excess of that specified where the additional runoff is removed so as to not increase the flood depth beyond the maximum allowed.

7.6.5 Other BMP Devices

There are many additional BMP devices available for design engineers to consider using (see Chapter 8 for sedimentation and erosion control BMP options); however, three of the most common remaining choices to promote improving water quality are given below.

A. *Inlet Traps*   Inlet traps are inlets that can separate oil and grit from stormwater runoff, as shown in Figure 7.40*f*. These devices, sometimes called "water quality inlets," can be used to pretreat parking lot runoff before routing it to the storm sewer or an infiltration device. Because these devices are usually first in line, they require constant maintenance to operate properly.

B. *Sand Filters*   A sand filter (see Figure 7.40*g*) is a device usually placed underground where the first flush (1–$1\frac{1}{2}$ inches) of runoff is allowed to flow under the force of gravity through a layer of gravel and then through several layers of graded granular filter media (sand and gravel) where heavy metals and suspended solids are trapped. Sand filters are effective; however, they require frequent maintenance to ensure proper operation and should be monitored for offensive odor generation and controlled if necessary. Sand filters can be used in areas where site conditions are not conducive to supporting infiltration-type devices.

C. *Vegetation*   There are different types of vegetation surfaces that can be used as pretreatment devices to help improve the quality of stormwater runoff. Infiltration basins and ditches often utilize vegetative coverings (e.g., grass) as integral component parts. In addition to these applications, vegetation is also used to create filter strips. Filter strips are vegetated areas (can either be grassed or forested areas) constructed where runoff from other surfaces must cross under sheet flow conditions. This requires the strips to have a minimum cross slope with a uniform surface. Sometimes level spreaders (see Section 8.1.5) must be used to promote sheet flow across the strip's surface.

The use of vegetation offers several management benefits. First, runoff is filtered where some of the pollutants and nutrients are removed prior to the runoff typically being routed to another BPM device such as an infiltration trench or retention/detention facility. Second, some of the runoff is trapped by the vegetation allowing biological plant uptake of some of the nutrients contained in the runoff. Third, some of the runoff is allowed to infiltrate into the surrounding soil where pollutants can be adsorbed. And, finally, vegetation practices can be integrated into the overall site's green-belt and natural conservation areas to improve aesthetics and natural habitat.

D. *Stormwater Wetlands*   Freshwater wetlands are natural filters that can often be used as part of the overall stormwater management plan, as indicated in Figure 7.41. When other BMP devices are properly used prior to a freshwater wetland, runoff will not generally damage the wetland. For example, if runoff is pretreated upstream by oil and grease removal, sedimentation, and flow attenuation and then conveyed by an infiltration ditch to the wetland, the flow can then be released under sheet flow conditions. This will help distribute the entering water throughout the wetland and increase residence time where the wetland is able to assimilate the inflow.

FIG. 7.41   Wetlands detention system. Source: South Carolina Coastal Council, 1988.

7.6.6 Ways to Estimate Runoff Water Quality

Each BMP has its own pollutant removal capability. Depending on regulatory requirements, the design engineer may choose one or several BMPs based on individual performance efficiencies indicated in Figure 7.42.

Because of space limitations here, the reader is referred elsewhere to more complete and definitive sources on the subject matter of BMPs (Wanielista et al, 1993; FHWA, 1996; AASHTO, 1991; Urbonas and Stahre, 1993; Schueler, 1987).

7.7 MITIGATING FLOODPLAIN HAZARDS

As mentioned in Section 4.3, any new residential structures built in the floodplain must be positioned above the base flood elevation (BFE). Nonresidential structures must also be elevated to the BFE, or flood proofed to that level, "provided that the structure is with walls that are impermeable to floodwaters," according to the U.S. Army Corps of Engineers (1992).

A structure surrounded by floodwaters is subjected to hydrostatic loads, hydrodynamic loads, and impact loads. A structural engineer is required to ensure that the parts of the structure subjected to floodwater loads are designed for safety. Another approach to ensure that a structure is protected from flood hazards is to elevate the building on fill, where the fill does not constrict floodwater flow where the water surface elevation rises above the allowable surcharge limit (Section 4.3). An alternative to using fill is the use of piles or columns to elevate the structure to the desired elevation. Any outdoor space below the functional floors of the structure could be used for parking, recreation, or open space.

If the developer does not want to flood proof the structure using methods previously cited, consideration can be given to using a dike, levee, or floodwall. This approach attempts to keep the floodwater away from the structure and surrounding property. A geotechnical study would have to be undertaken to ensure that any dike or levee

BMP/design		Pollutant removal efficiency (%)						Overall Removal Capability
		Suspended Sediment	Total Phosphorus	Total Nitrogen	Oxygen Demand	Trace Metals	Bacteria	
Extended detention pond	Design 1	60 - 80	20 - 40	20 - 40	20 - 40	40 - 60	Unknown	Moderate
	Design 2	80 - 100	40 - 60	20 - 40	40 - 60	60 - 80	Unknown	Moderate
	Design 3	80 - 100	60 - 80	40 - 60	40 - 60	60 - 80	Unknown	High
Wet pond	Design 4	60 - 80	40 - 60	20 - 40	20 - 40	20 - 40	Unknown	Moderate
	Design 5	60 - 80	40 - 60	20 - 40	20 - 40	60 - 80	Unknown	Moderate
	Design 6	80 - 100	60 - 80	40 - 60	40 - 60	60 - 80	Unknown	High
Infiltration trench	Design 7	60 - 80	40 - 60	40 - 60	60 - 80	60 - 80	60 - 80	Moderate
	Design 8	80 - 100	40 - 60	40 - 60	60 - 80	80 - 100	60 - 80	High
	Design 9	80 - 100	60 - 80	60 - 80	80 - 100	80 - 100	80 - 100	High
Infiltration basin	Design 7	60 - 80	40 - 60	40 - 60	60 - 80	40 - 60	60 - 80	Moderate
	Design 8	80 - 100	40 - 60	40 - 60	60 - 80	80 - 100	60 - 80	High
	Design 9	80 - 100	60 - 80	60 - 80	80 - 100	80 - 100	80 - 100	High
Porous pavement	Design 7	40 - 60	60 - 80	40 - 60	60 - 80	40 - 60	60 - 80	Moderate
	Design 8	80 - 100	60 - 80	60 - 80	60 - 80	80 - 100	80 - 100	High
	Design 9	80 - 100	60 - 80	60 - 80	80 - 100	80 - 100	80 - 100	High
Water quality inlet	Design 10	0 - 20	Unknown	Unknown	Unknown	Unknown	Unknown	Low
Filter strip	Design 11	20 - 40	0 - 20	0 - 20	0 - 20	20 - 40	Unknown	Low
	Design 12	80 - 100	40 - 60	40 - 60	40 - 60	80 - 100	Unknown	Moderate
Grassed swale	Design 13	0 - 20	0 - 20	0 - 20	0 - 20	0 - 20	Unknown	Low
	Design 14	20 - 40	20 - 40	20 - 40	20 - 40	0 - 20	Unknown	Low

Design 1: First-flush runoff volume detained for 6-12 h. **Design 2**: Runoff volume produced by 25 mm (1.0 in), detained 24 h. **Design 3**: As in Design 2, but with shallow marsh in bottom stage. **Design 4**: Permanent pool equal to 13 mm (0.5 in) storage per impervious acre. **Design 5**: Permanent pool equal to 2.5 (Vr); where Vr = mean storm runoff. **Design 6**: Permanent pool equal to 4.0 (Vr); approx. 2 weeks retention. **Design 7**: Facility exfiltrates first-flush; 13 mm (0.5 in) runoff/imper. acre. **Design 8**: Facility exfiltrates 25-mm (1-in) runoff volume per imper. acre. **Design 9**: Facility exfiltrates all runoff, up to the 2-yr design storm. **Design 10**: 11 m³ (400 ft³) wet storage per impervious acre. **Design 11**: 6-m (20-ft) wide turf strip. **Design 12**: 30-m (100-ft) wide forested strip, with level spreader. **Design 13**: High-slope swales with no check dams. **Design 14**: Low-gradient swales with check dams.

FIG. 7.42 Pollutant removal comparison for various urban BMP designs. Source: Schueler, 1987. Printed with permission.

375

is safely designed and built. The design of a floodwall, similar in some respects to a retaining wall (Section 9.7) except water instead of soil is retained and no weep holes are provided, requires additional professional design services from a structural engineer.

7.8   DRAINAGE SYSTEM DESIGN

After adequate surface area has been reserved for basin sites, design of the storm drainage collection system can begin. The example collection system design included in this section illustrates the customary hydraulic design approach. Any regulatory BMP devices required for a particular site would have to be included, and this design would have to be adjusted to reflect the impact of the hydraulic characteristics of each included device. If the site is subject to stormwater runoff from adjacent property because of the surrounding terrain, then the amount of runoff expected from off-site must be determined. Two situations are possible alone or in combination:

(a)   The site is subject to sheet or overland flow from the adjoining property. If this occurs, a swale (shallow vee bottom ditch) or trapezoidal ditch can be provided parallel and contiguous to the property line where the sheet flows cross to the site for interception and diversion. This ditch system can provide additional benefits by lowering the groundwater table if high groundwater is present.

(b)   The site is subject to channelized flow from the adjoining propety. In this case adequate provisions must be made to allow this off-site water to flow downstream without creating on-site problems.

7.8.1   Contributory Drainage Areas

Interior drainage areas can be identified by using different colored pencils on a transparent paper overlay placed on top of the preliminary plan or by using computer-aided drafting (CAD) techniques similar to this. Either system provides an excellent method to delineate the different contributory areas while determining how each lot will be individually graded to provide positive drainage. The centerline of the road (pavement crown) is usually a limiting boundary for drainage areas where curb and gutter road sections are used. Figure 7.43 shows three contributory areas that generate runoff on the 5.13-acre site. Areas A_1 and A_2 drain into the frontage road, while area A_3 drains into the swale ditch along the rear property line. Pipe 1's inflow comes from contributory area A_1. Pipe 2 has a contributory flow from areas A_1 and A_2, while pipe 3 has a contributory flow from A_1, A_2, and A_3.

7.8.2   The Design Process

The availability of computer software such as HYDROS (Debo, 1987), which utilizes hydrograph generation and routing, allows the design professional to first design the storm drainage system using the rational method and then analyze the designed system's performance using computer-generated hydrographs. Should imperfections surface while routing various hydrographs through the various components with the computer software, adjustments can be made and the hydrographs rerun to refine the design.

FIG. 7.43   Example of a site plan for the drainage system design and analysis.

Initially several rules should be followed that, with experience, can be modified by the designer:

(a) Determine the minimum allowable slopes that produce a self-scouring velocity for all culverts. Some regulations stipulate minimum slopes that should be used to initially establish the culvert alignment throughout the system.

(b) Use a uniform material of construction for all culverts to ease calculations where Manning's roughness coefficient is constant. Alterations can be made later, if required.

(c) Size culverts beginning at the furthest point upstream and progress downstream in logical sequence. The time of concentration must be recalculated for each culvert using the longest time of travel within the basin to the point under consideration. In addition, the composite coefficient of runoff must be determined. If a junction is encountered where other tributary flows enter the system, do not proceed downstream until all other upstream channels have been sized. The process continues until the outfall structure is reached. Pipes are usually sized to flow full while carrying the peak design runoff under gravity flow conditions. If ponding is allowed at culvert entrances, then a capacity chart similar to Figure 7.44 can be utilized to determine flows.

(d) Establish all invert elevations (the elevation of the bottom inside of the channel or pipe where flow first begins) commencing with the outfall, or discharge, and proceed upstream in logical sequence to the most elevated contributory area using the same pipe or channel slopes that were included for capacity determination.

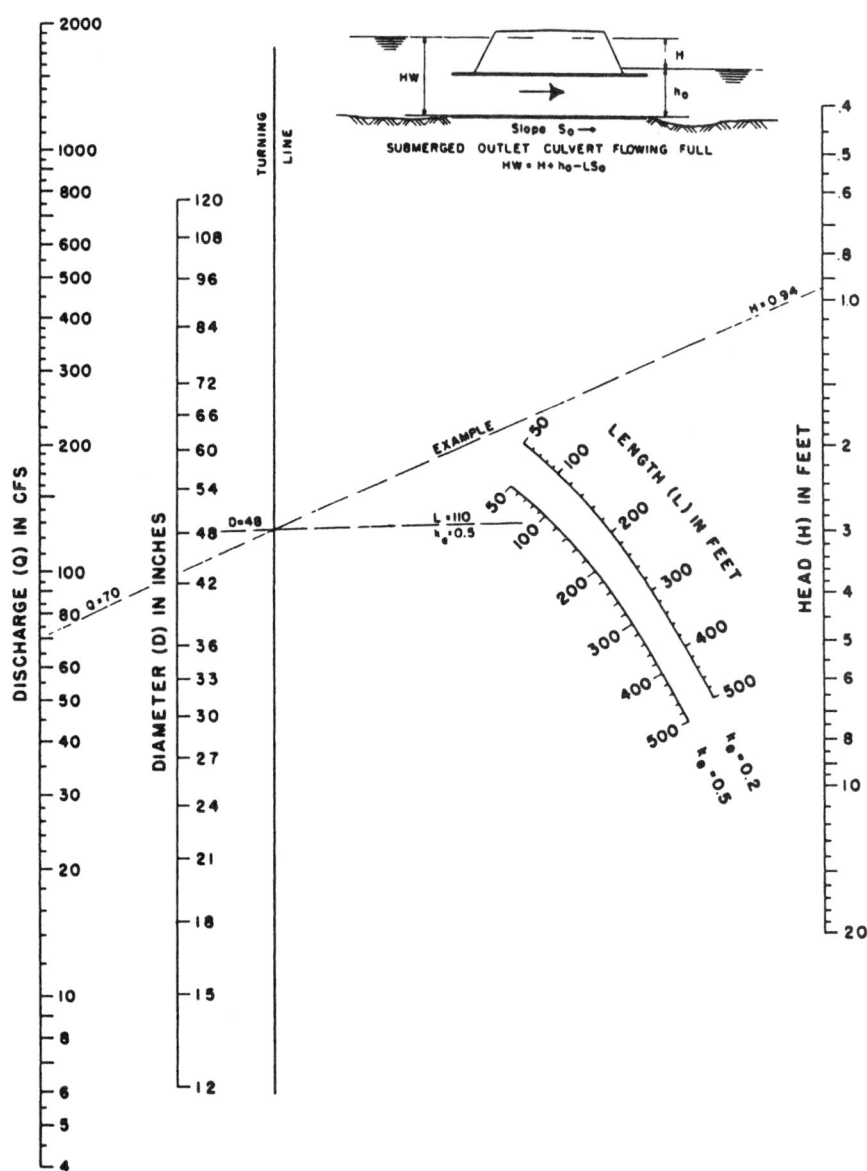

**HEAD FOR
CONCRETE PIPE CULVERTS
FLOWING FULL
n = 0.012**

FIG. 7.44 Culvert capacity chart. Source: Bureau of Public Roads, "Hydraulic Charts for the Selection of Highway Culverts," *Hydraulic Engineering Circular No. 5,* U.S. Department of Commerce, Washington, DC, 1965.

7.8.3   Example of Drainage System Design

To illustrate the design process, an example using the rational method follows using a 10-year return period for design. Figure 7.5 is assumed to represent local rainfall conditions, while Figure 7.43 represents the site conditions.

(a) Compute the composite coefficient of runoff using Equation 7.11 for A_1 and using a typical lot for calculations:

Typical lot size	$= 0.33$ acre
1/2 road R/W	$= 0.05$ acre
Total area	$= 0.388$ acre

APPROXIMATE IMPERVIOUS AREAS

House roof	$= 1600 \text{ ft}^2$
Paved driveway	$= 600 \text{ ft}^2$
Paved patio	$= 400 \text{ ft}^2$
Roadway pavement	$= 1500 \text{ ft}^2$
Total impervious	$= 4100 \text{ ft}^2 = 0.094$ acre

A coefficient of runoff of 0.4 is used for nonimpervious areas and a value of 0.95 is used for impervious areas:

$$C_{w1} = \frac{C_1 A_1 + C_2 A_2}{A_1 + A^2} = \frac{0.094 \times 0.95 + 0.4 \times 0.294}{0.388} = 0.54$$

(b) Compute the composite coefficient of runoff using Equation 7.11 for A_1 and A_2 using a typical lot. Since A_2 has less contributory area, a typical lot in A_2 needs to be evaluated:

Typical lot portion contributing to $A_2 = 90 \times 100$	$= 9000 \text{ ft}^2$
1/2 road R/W (25 feet $\times$ 100)	$= 2500 \text{ ft}^2$
Total	$= 11{,}500 = 0.264$ acre

$$C_{w12} = \frac{0.094 \times 0.95 + 0.094 \times 0.95 + 0.4 \times 0.294 + 0.4 \times 0.17}{0.094 + 0.094 + 0.294 + 0.17} = 0.56$$

(c) Calculate the composite coefficient of runoff for A_1, A_2, and A_3 using Equation 7.11:

$$C_{w123} = \frac{0.094 \times 0.95 \times 2 + 0.4 \times 0.294 + 0.4 \times 0.17 + 0.4 \times 0.126}{0.778} = 0.54$$

Figure 7.45 depicts the contributory area features required for design calculations.

(d) Compute T_c for A_1, where T_c is the time for a drop of water to travel 145 linear feet perpendicular to the road and then 300 linear feet in the road gutter to the curb inlet. With a ground and road slope of 0.5 percent, T_c is obtained from previous calculations included in Section 7.5.1, where the lot flow time for sheet flow is 14.4 minutes and

FIG. 7.45   Delineation of various contributory drainage areas in a sample land development.

the street travel time is 300 feet/2.19 ft/s, or 2.3 minutes. The time of concentration is estimated as 16.7 minutes.

(e) Use Figure 7.5 to determine the intensity corresponding to a time of concentration of 16.7 minutes for a 10-year return period; this intensity is 5.42 in./h.

(f) Compute the peak discharge for A_1 using Equation 7.21:

$$Q_1 = 0.54 \times 5.42 \times 2.34, \text{ or } 6.85 \text{ ft}^3/\text{s}$$

(g) Determine the size of pipe 1 using $S_{min} = 0.3$ percent and Manning's roughness coefficient equal to 0.012 from Figure 7.12. Manning's equation 7.17 can be rearranged to allow a direct solution for the minimum required pipe diameter, in inches, assuming full flow by using

$$d_{inches} = 2 \left[\frac{258.68 \times Q \times n}{S^{0.5}} \right]^{0.375} \tag{7.40}$$

where d_{inches} is the required pipe diameter in inches to carry the required flow for full flow under open-channel conditions, Q is the required flow rate in cubic feet per second, n is Manning's roughness coefficient, and S is the pipe slope in feet per foot. Using Equation 7.40, the required minimum diameter is

$$d_{inches} = 2 \left[\frac{258.68 \times 6.85 \times 0.012}{(0.003)^{0.5}} \right]^{0.375}$$

$$= 18.70 \text{ inches}$$

The closest pipe size locally available is assumed to have a 24-inch diameter. This size will initially be used for pipe 1. The maximum capacity for a full 24-inch pipe using Manning's equation 7.19 is

$$Q = \frac{1.49 A r^{0.67} S^{0.5}}{n}$$

where

$$A = (3.14)(2/2)^2 = 3.14 \text{ ft}^2$$

$$P = (3.14)(2) = 6.28 \text{ feet}$$

$$r = A/P = 0.5 \text{ foot}$$

$$Q = \frac{1.49(3.14)(0.5)^{0.67}(0.003)^{0.5}}{0.012}$$

$$= 13.42 \text{ ft}^3/\text{s}$$

The corresponding velocity at full flow, using Equation 7.18 is

$$V_{\text{full}} = \frac{13.42 \text{ ft}^3/\text{s}}{3.14 \text{ ft}^2}$$

$$= 4.27 \text{ ft/s}$$

Figure 7.46 can be used to graphically determine partial flow relationships when pipe diameters are used that are larger than those that allow full flow to occur. Since a 24-inch pipe is being used, the corresponding velocity for 6.85 ft^3/s is needed; however, Q/Q_{full} must first be computed assuming that Manning's n varies with depth. Here, $Q/Q_{\text{full}} = 6.85$ ft^3/s/13.42 ft^3/s, or 51 percent. Figure 7.46 indicates a percent depth of flow of approximately 56 percent (using the variable n curve for discharge). The corresponding velocity ratio using the same 56 percent depth of flow is approximately 86 percent of full flow velocity. Velocity for the 6.85-ft^3/s flow is therefore 0.86 × 4.27, or 3.67 ft/s.

(h) Compute the contributory area for pipe 2:

$$A_1 + A_2 = 2.34 + 1.58 = 3.92 \text{ acres}$$

(i) Determine the longest time of travel considering the combined areas for T_c. Two possibilities occur. First, the T_c may equal the previous T_c of 16.7 minutes plus the time of travel through pipe 1. The approximate pipe 1 time of travel is 30 feet/3.67 ft/s = 8.2 seconds. Therefore, T_c is approximately 16.8 minutes. The second possibility is the time for a drop of water to travel perpendicular to the road for 90 feet in A_2, and then for 300 feet within the road gutter to the inlet. Based on these conditions, this time would be less than the previously calculated time of 16.8 minutes; therefore, use 16.8 minutes for T_c. The rainfall intensity is estimated as 5.42 in./h.

(j) Compute the peak discharge from A_1 and A_2 using Equation 7.21 and a weighted coefficient of 0.56:

FIG. 7.46 Hydraulic elements for a circular section. Source: American Society of Civil Engineers (ASCE) and the Water Pollution Control Federation (WPCF). *Gravity Sanitary Sewer Design and Construction.* ASCE Manual No. 60; WPCF No. FD-5, 1982. Reproduced by permission of the American Society of Civil Engineers.

$$Q_2 = 0.56 \times 5.42 \times 3.92, \text{ or } 11.9 \text{ ft}^3/\text{s}$$

(k) Size pipe 2 using $S_{\min} = 0.3$ percent and Manning's coefficient of roughness of 0.012 in conjunction with Equation 7.40:

$$d_{\text{inches}} = 2 \left[\frac{258.68 \times 11.9 \times 0.012}{(0.003)^{0.5}} \right]^{0.375}$$

$$= 23.00 \text{ inches.}$$

Due to local supply availability, the closest diameter that can be used for pipe 2 is assumed to be a 24-inch culvert. The corresponding full flow of a 24-inch culvert is the previously calculated 13.42 ft^3/s and the velocity is 4.27 ft/s. Figure 7.46 can again be used to determine the partial flow relationships; however, Q/Q_{full} must first be computed. Here, $Q/Q_{\text{full}} = 11.9$ ft^3/s/13.42 ft^3/s, or 89 percent. Figure 7.46 indicates a percent depth of flow of approximately 82 percent. The corresponding velocity ratio using the same 82 percent depth of flow is approximately 102 percent of full flow velocity. Velocity for the 11.9-ft^3/s flow is therefore 1.02×4.27 or 4.35 ft/s.

(l) Compute the contributory area for pipe 3:

$$A_1 + A_2 + A_3 = 4.68 \text{ acres}$$

(m) Determine the longest time of travel for T_c for pipe 3. First, T_c may be equal to the previous 16.8 minutes plus the time of travel through pipe 2. This time of travel is approximately 145 feet/4.35 ft/s = 0.56 minute:

$$T_c = 16.8 + 0.56 = 17.36, \text{ or } 17.4 \text{ minutes}$$

The second possibility is the time for a drop of water to travel perpendicular to the road for 90 feet in A_2, then in the road gutter for 300 feet to the curb inlet, and finally the time required to travel through pipe 2. This time can be seen as less than the previously computed 17.4 minutes.

The third possibility would be for the time of travel required for a drop of water to go 300 feet in A_3 to the inlet at pipe 3. This is also less than the 17.4 minutes. The time of concentration is therefore 17.4 minutes, and the corresponding rainfall intensity is 5.37 in./h.

(n) Compute the peak discharge for A_1, A_2, and A_3 using Equation 7.21 and a weighted coefficient of runoff of 0.54:

$$Q_{123} = 0.54 \times 5.37 \times 4.68 = 13.57 \text{ ft}^3/\text{s}$$

(o) Size pipe 3 using $S_{min} = 0.3$ percent and Manning's coefficient of roughness of 0.012 using Equation 7.40 by

$$d_{inches} = 2 \left[\frac{258.68 \times 13.57 \times 0.012}{(0.003)^{0.5}} \right]^{0.375}$$
$$= 24.17 \text{ inches}$$

Therefore, it is assumed that the next largest size culvert locally available is 30 inches in diameter.

(p) Compute the preliminary invert elevations of each culvert beginning with the discharge of the detention basin. From the topographic map (Figure 7.35), the outlet invert elevation could be preliminarily set at 83.00. If an elevation of 83.5 is selected for the pond bottom, the estimated maximum height of water based on twice the FAA method of ponding results (8.15 feet) would be 91.65 feet, which can be used for the invert of pipe 3's discharge elevation.

Since pipe 3 is 30 feet long and on a slope of 0.3 percent, the inlet elevation of pipe 3 would be 91.74 feet. To account for the diameter differences in pipes 2 and 3, the outlet of pipe 2 can be set at elevation 91.74 + 0.5, or 92.24 feet.

Pipe 2's inlet can now be calculated as 0.003 × 145 + 92.24 = 92.68. Since pipe 2 is proposed to be 24 inches in diameter, the topography indicates that approximately 5.3 feet of cover over the top of the pipe will be available, which appears acceptable.

Pipe 1's invert elevation at the pipe inlet is computed by 92.68 + 0.003 × 30 = 92.77. The elevation at the centerline of the road where the pipe crosses needs to be checked to ensure that proper cover is available over the crown of the pipe.

To aid the designer in determining pipe cover, drainage information must be plotted on the road plan and profile where the road centerline has been drawn from PVI to PVI, as described in Section 6.4.2. By plotting the elevation view of any culvert that crosses a road (by plotting the station, invert elevation, and pipe diameter), one can view the relationship of the top of the culvert to the road grade. Culverts need a minimum amount of cover for protection from traffic loads (usually a minimum of 12 inches), as indicated in Section 7.9.

If sufficient cover is not available, then either the road grade must be adjusted upward, or the drainage system must be lowered, which means the outfall ditch may need to be lowered and resloped for some distance downstream. Sometimes neither of these options is available, and arched pipe must be utilized in lieu of the circular culvert to obtain the proper clearance. Corrugated metal pipes have a different Manning's coefficient of roughness than reinforced concrete pipe. This must be considered when sizing for a substitute pipe. One other option would be to increase the surface area of the detention pond, thus reducing the amount of working depth required in the pond so that the pond's inlet pipe elevation can be lowered along with upstream contributory pipes.

Should the storm drainage system indicate excessive installment depths, the invert elevations can be increased appropriately and the slopes adjusted accordingly. Sometimes drop structures, similar to the one shown in Figure 7.47, can be utilized to minimize construction costs.

Once the initial refinements have been made and the storm drainage system appears to fit adequately within the road system and terrain, the designer can undertake preliminary design of the sanitary sewer collection system as delineated in Chapter

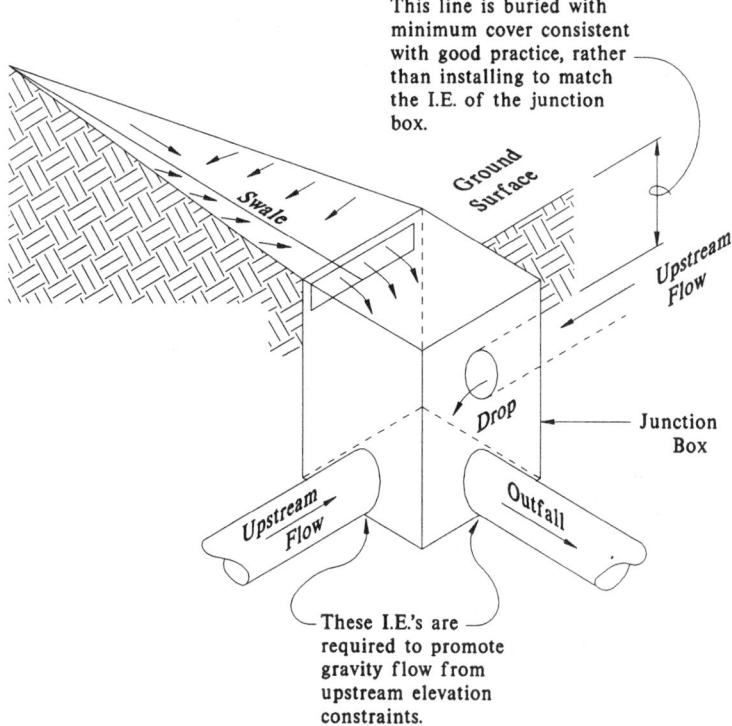

FIG. 7.47   Typical stormwater junction box details with an inlet drop (I.E. = invert elevation).

10. Should future design problems arise from elevation conflicts between the sanitary sewer and storm drainage systems, revisions may need to be made to both systems. This, in turn, will require the engineer to recheck the road alignment, which may also require further adjustment.

The storm drainage hydraulic grade line should be plotted to determine whether local flooding might occur and whether the system will perform as expected. This is presented in the next section.

It is recommended that once the road, storm drainage, and sanitary sewer systems are fairly well established where the most economical costs of construction are obtained, the storm drainage system should then be analyzed using hydrograph generation and routing.

7.8.4   Example of Drainage System Analysis

There are many different computer software packages available (e.g., Haestad Methods StormCAD®, PondPack®, CulvertMaster®, and FlowMaster®; see www.haestad.com) for the practitioner to use when evaluating a storm drainage system. One recently developed application (KY Pipe/FlowNet Systems, Lexington, KY, www.kypipe.com), for example, provides the user a graphical interface where the drainage system can be visually created over an existing map. Pipe sizes, slopes, and materials can be changed where the results can be stored in a database.

To illustrate the analysis of a designed system, HYDROS (Debo, 1987) computer software, which first generates hydrographs and then routes them, will be utilized. In this program NRCS methods are used for hydrograph generation while Muskingum procedures (alternative to the convex method) are used for routing.

The system being analyzed is the same 5.13-acre tract's previously designed storm drainage system; however, A_1, A_2, and A_3 must be further subdivided into A_{1a} and A_{1b}, A_{2a} and A_{2b}, and A_{3a} and A_{3b} because a runoff hydrograph is generated from each of these subareas.

To fully investigate the performance of the storm drainage system under several storm return periods, areas contiguous to inlet structures can be treated as small reservoirs. This gives the designer a chance to investigate potential localized flooding under different storm conditions. The drainage subbasins and flow network are shown in Figure 7.48.

Channel velocities for hydrographs are estimated as those occurring during the design storm. Stage, storage, and discharge relationships for the various reservoirs are used as input data. The program interpolates between data points during calculations to obtain approximate intervening values. The input data are determined as follows:

A. *Calculation of Stage/Storage for Reservoirs 1 and 6*   These reservoirs consist of the swale sections in drainage areas A_{3a} and A_{3b}. Temporary storage is provided there while runoff enters the inlet, as illustrated in Figure 7.49.

Using the average end area method (Equation 9.1) to estimate the storage volume for the tapered sections (see Section 9.3), the volume is calculated as

$$\text{Vol} = [(a_1 + a_2)/2]\,(L)$$
$$= [(0 + 0.5 \times 2 \times 5 \times 2)/2]\,(300) = 1500 \text{ ft}^3$$

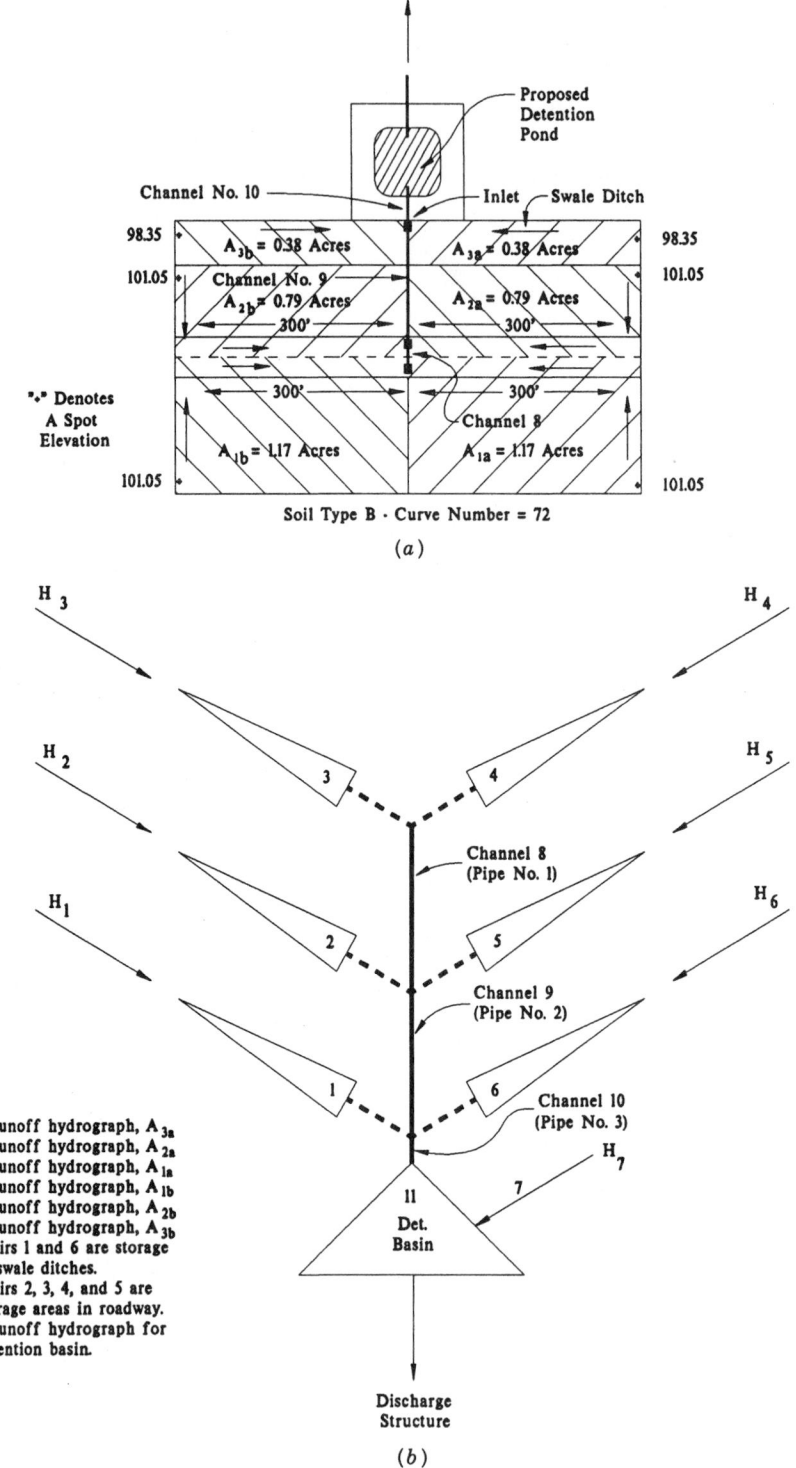

FIG. 7.48 Example of drainage subbasin flow network: (*a*) drainage basins for hydrograph generation; (*b*) network for hydrograph analysis.

FIG. 7.49 Geometry for sample reservoirs 1 and 6: swale ditch sections.

The weir inflow of the inlet is determined using Equation 7.37 with an assumed broad-crested weir coefficient of 3, a length of 3.54 feet, and a maximum flow depth of 2 feet at the junction box entrance because the swale's maximum depth is expected not to be exceeded:

$$Q = 3.0 \times 3.54 \times 2^{1.5} = 30 \text{ ft}^3/\text{s}$$

Therefore two stage, discharge, and storage relationships for reservoirs 1 and 6 are:

Stage (feet)	Discharge ft³/s	Storage ft³
0	0	0
2	30	1500

B. *Calculation of Stage/Storage Relationships for Reservoirs 2–5* Figure 7.50 illustrates the stage and storage relationship in the roadway where maximum flooding 0.41 foot deep is estimated based on the road's cross section and anticipated topography. These reservoirs will drain into curb inlets, then through storm drainage culverts to the detention basis. Additional storage volumes attributed to the space behind the curb where the soil slopes to the road are neglected for this example.

Using a sag-type curb inlet similar to Figure 7.30*b* with an estimated maximum flood elevation of 0.41 feet, the total inflow capacity of the inlet is calculated using the weir and orifice equations by:

Approximate Storage Volume With
Flood Depth At 0.41 Feet is:

$$\left[\frac{0 + 3.1}{2}\right] 82 = 127.1 \text{ Cubic Feet}$$

FIG. 7.50   Street detention volumes for a 5-inch gutter flood depth.

(a)   Weir flow using Equation 7.37 and a constant of 3 for throat flow:

$$Q = 3 \times 13 \text{ feet of weir length } \times (0.41)^{1.5}$$
$$= 10.24 \text{ ft}^3/\text{s}$$

(b)   Orifice flow using Equation 7.38 with a constant equal to 0.6 for grate flow and a cross-sectional area of 2.66 ft²:

$$Q = 0.6 \times 2.66 \times (2 \times 32.2 \times 0.41 \text{ feet})^{0.5}$$
$$= 8.2 \text{ ft}^3/\text{s}$$

The total inlet flow is therefore 18.44 ft³/s. One-half of this inflow is allocated to each contributory area with 9.22 ft³/s being allocated to reservoir 2 and 9.22 ft³/s to reservoir 5. (The same is true for reservoirs 3 and 4.) Two stage, storage, and discharge relationships are shown below that are applicable to each of the four reservoirs:

Stage (feet)	Discharge ft³/s	Storage ft³
0	0	0
0.41	9.22	127.1

C. *Calculation of Stage, Discharge, and Storage Relationships for Detention Pond* This calculation uses procedures contained in Section 7.5.2. For this example, the design objective is to size the detention pond to accommodate the 100-year flood. This may be achieved by placing the pond in the location shown in Figure 7.35, selecting side slopes of 2 : 1, and using 10 feet as the maximum depth. Using bottom dimensions of 50 × 50 feet allows the top to extend to 90 × 90 feet. The limiting discharge out of the pond for the 100-year storm is 6.49 ft³/s; therefore, using 10 feet as the conservative maximum stage value (an 8.15-foot depth is the preliminary value obtained in Section 7.5.2 with the corresponding water surface area being 8.26 × 8.26 feet), an orifice is sized for use as the discharge structure with Equation 7.38. (*Note*: A more representative value of the reservoir's maximum storage height can be obtained after hydrograph generation and routing is accomplished, as indicated in Figure 7.53). The orifice coefficient is taken as 0.58 for a sharp-edged entrance:

$$6.49 = 0.58 \times A \times (2 \times 32.2 \times 10 \text{ feet headwater})^{0.5}$$

$$A = 0.44 \text{ ft}^2 \text{ of opening}$$

The corresponding diameter of the orifice is approximately 9 inches. This can be manufactured from a plate and installed at the discharge entrance. Eleven detention basin stage, discharge, and storage values are determined beginning with 0, 0, 0 at the basin bottom with discharge values corresponding to a 9-inch orifice opening:

Stage (feet)	Discharge (ft³/s)	Storage (ft³)
0	0.00	0
1	2.06	2,708
2	2.91	5,864
3	3.56	9,516
4	4.11	13,712
5	4.60	18,500
6	5.04	23,928
7	5.44	30,044
8	5.82	36,896
9	6.17	44,532
10	6.50	53,000

Computer input data required for each drainage subbasin to perform and be analyzed include the drainage area, high and low elevations, distance of runoff travel including the amount channelized, the amount of impervious area, and the NRCS runoff curve number. Data required for each reservoir include stage, discharge, and storage nomenclature. Channels are identified by contributory areas and are defined by invert slopes, length, and reach (channel) velocities. Lack of space prohibits the inclusion of this input data and intermediate calculations; however, Figure 7.51 shows hydrograph data for the six reservoirs (site symmetry allows dual use of the data) using a 10-year design storm.

These reservoir routed hydrographs are then routed through the system's channels as indicated in Figure 7.52, where detention basin (reservoir 11) routing is also shown.

Using Figure 7.48 as a guide, channel 8 (pipe 1) includes inflow hydrographs from both reservoirs 3 and 4, resulting in a combined inflow hydrograph twice that shown in Figure 7.51. Due to number rounding, however, the inflow into channel 8 varies slightly. Channel 9 (pipe 2) includes inflow hydrographs from channel 8 and reservoirs 2 and 5. Channel 10 (pipe 3) includes inflow from channel 9 and reservoirs 1 and 6.

Time	Reservoir 1 and Reservoir 6				Reservoir 2 and Reservoir 5				Reservoir 3 and Reservoir 4			
	I,cfs.	O,cfs.	V,cft.	H,ft.	I,cfs.	O,cfs.	V,ctf.	H,ft.	I,cfs.	O,cfs.	V,ctf.	H,ft.
0.00	0	0	0.00	0.00	0	0	0.00	0.00	0	0	0.00	0.00
0.04	0	0	13.40	0.02	0	0	4.11	0.01	0	0	1.23	0.00
0.08	1	1	42.18	0.06	1	1	10.51	0.03	0	0	3.46	0.01
0.12	2	1	68.82	0.09	1	1	17.73	0.06	0	0	6.28	0.02
0.16	2	2	87.45	0.12	2	2	26.44	0.09	1	1	10.67	0.03
0.21	2	2	85.84	0.11	3	3	34.81	0.11	1	1	16.43	0.05
0.25	1	1	58.56	0.08	3	3	42.78	0.14	2	2	22.96	0.07
0.29	0	1	27.22	0.04	4	4	48.77	0.16	2	2	31.20	0.10
0.33	0	0	7.56	0.01	4	4	49.62	0.16	3	3	40.13	0.13
0.37	0	0	0.39	0.00	3	3	43.92	0.14	4	4	49.24	0.16
0.41	0	0	0.07	0.00	2	2	33.19	0.11	4	4	56.54	0.18
0.45	0	0	0.01	0.00	2	2	22.18	0.07	5	4	61.91	0.20
0.49	0	0	0.00	0.00	1	1	13.59	0.04	4	4	61.98	0.20
0.53	0	0	0.00	0.00	0	1	6.70	0.02	4	4	58.27	0.19
0.57	0	0	0.00	0.00	0	0	2.30	0.01	4	4	51.61	0.17
0.62	0	0	0.00	0.00	0	0	0.06	0.00	3	3	42.15	0.14
0.66	0	0	0.00	0.00	0	0	0.04	0.00	2	2	32.94	0.11
0.70	0	0	0.00	0.00	0	0	0.03	0.00	2	2	24.18	0.08
0.74	0	0	0.00	0.00	0	0	0.02	0.00	1	1	16.87	0.05
0.78	0	0	0.00	0.00	0	0	0.01	0.00	1	1	10.32	0.03
0.82	0	0	0.00	0.00	0	0	0.01	0.00	0	0	5.81	0.02
0.86	0	0	0.00	0.00	0	0	0.01	0.00	0	0	2.02	0.01
0.90	0	0	0.00	0.00	0	0	0.00	0.00	0	0	0.07	0.00
0.94	0	0	0.00	0.00	0	0	0.00	0.00	0	0	0.05	0.00
0.98	0	0	0.00	0.00	0	0	0.00	0.00	0	0	0.03	0.00
1.03	0	0	0.00	0.00	0	0	0.00	0.00	0	0	0.02	0.00
1.07	0	0	0.00	0.00	0	0	0.00	0.00	0	0	0.02	0.00

FIG. 7.51 Example of reservoir routing: 10-year storm. Source: Debo, 1987.

Reservoir 11, the detention basin, includes inflow from channel 10 and additional inflow resulting from precipitation falling on the basin's surface designated as hydrograph 7 (feeding into a fictitious channel 7 required by the software), as indicated in Figure 7.48. The hydrograph values for fictitious channel 7 are not included here.

The maximum detention basin depth necessary to accommodate a 10-year return period storage requirement is estimated at 4.30 feet.

Figure 7.53 is a summary sheet depicting peak hydrograph flows, maximum storage volumes, and maximum storage depths. For example, the 100-year return period predicts an estimated flood depth in the street at the inlet of pipe 1 to be approximately 0.36 foot, while a flood depth of approximately 0.28 foot is expected in the street at the inlet of pipe 2. Since hydrograph routing indicates that the approximate working depth in member 11 (the main detention pond) is 7.12 feet (rather than the initial estimate of 8.15 feet for the 100-year storm) and pipe 2 has more than the minimum cover, the detention pond's outlet pipe can be raised from its preliminary elevation, which was estimated at 83.00 feet, to an invert elevation of 85.00 feet for the discharge side and 85.60 feet for the inlet where the 9-inch orifice is located. This will allow the discharge invert of pipe 3 to be increased from its preliminary elevation, which was estimated at 91.65 feet, to an elevation of 92.72 feet (85.60 + 7.12).

From Figure 7.53, the computer-generated detention basin discharges are found to be less than the predevelopment release rates for all return periods previously

Time	Channel 8 Routing I,cfs.	Channel 8 Routing O,cfs.	Channel 9 Routing I,cfs.	Channel 9 Routing O,cfs.	Channel 10 Routing I,cfs.	Channel 10 Routing O,cfs.	Reservoir 11 Routing I,cfs.	Reservoir 11 Routing O,cfs	Reservoir 11 Routing V,cft.	Reservoir 11 Routing H,ft.
0.00	0	0	0	0	0	0	0	0	0.00	0.00
0.04	0	0	1	1	1	1	4	0	283.30	0.10
0.08	1	0	2	2	3	3	4	1	788.82	0.29
0.12	1	1	3	3	6	6	6	1	1360.24	0.50
0.16	2	2	5	5	8	8	8	2	2195.56	0.81
0.21	2	2	7	7	10	10	10	2	3277.94	1.18
0.25	3	3	9	9	11	11	11	3	4521.45	1.57
0.29	5	4	12	11	12	12	12	3	5848.55	2.00
0.33	6	6	13	13	13	13	13	3	7251.28	2.38
0.37	7	7	13	13	13	13	13	3	8710.96	2.78
0.41	8	8	13	13	13	13	13	4	10150.60	3.15
0.45	9	9	12	12	12	12	12	4	11486.42	3.47
0.49	9	9	11	11	11	11	11	4	12663.38	3.75
0.53	8	8	9	10	10	10	10	4	13635.78	3.98
0.57	7	8	8	8	8	8	8	4	14369.29	4.14
0.62	6	6	6	7	7	7	7	4	14855.53	4.24
0.66	5	5	5	5	5	5	5	4	15104.82	4.29
0.70	4	4	4	4	4	4	4	4	15148.81	4.30
0.74	2	2	2	3	3	3	3	4	15016.19	4.27
0.78	1	2	2	2	2	2	2	4	14729.59	4.21
0.82	1	1	1	1	1	1	1	4	14319.54	4.13
0.86	0	0	0	0	0	0	0	4	13817.85	4.02
0.90	0	0	0	0	0	0	0	4	13255.01	3.89
0.94	0	0	0	0	0	0	0	4	12668.65	3.75
0.98	0	0	0	0	0	0	0	4	12087.66	3.61
1.03	0	0	0	0	0	0	0	4	11517.87	3.48
1.07	0	0	0	0	0	0	0	4	10959.05	3.34

FIG. 7.52 Example of channel routing: 10-year storm. Source: Debo, 1987.

computed in Figure 7.33. Figure 7.54 illustrates a plot of the 100-year return period inflow and outflow hydrographs in the detention basin. This graph shows the value of detaining runoff to reduce peak discharges.

To evaluate the performance of pipes 1, 2, and 3 over a range of storm return periods, a culvert analysis sheet (Figure 7.55) is constructed. Maximum open-channel flows available for each culvert are recorded. Peak runoff flows generated by various return periods (Figure 7.53) that are less than these maximum available open-channel flows allow open-channel flows to prevail. Analysis of the expected performance of the 24-inch pipe proposed for pipes 1 and 2 indicates that open-channel conditions will prevail for the chosen 10-year design storm. Pipe 3, the 30-inch pipe, will also perform under open-channel flow conditions for the same return period (including the 100-year storm). When return periods greater than 10 years are expected, the analysis indicates that pipe 1 will flow under pressurized conditions during the 50- and 100-year storms. Pipe 2 is also expected to flow under pressurized conditions for storms beginning with the 25-year return period and larger. System adequacy for return periods greater than the 10-year design storm is demonstrated in Figure 7.56, where the hydraulic grade line (HGL) resulting from a 100-year storm is plotted.

In the final analysis, the HGL indicates that all three culverts can be raised approximately 1 foot without adversely affecting the drainage system's performance or

Peak Discharges (CFS)	2-Yr	5-Yr	10-Yr	25-Yr	50-Yr	100-Yr
Sub-Watershed No. 1	1	1	2	2	3	3
Sub-Watershed No.2	2	3	4	5	6	6
Sub-Watershed No.3	2	4	5	6	7	8
Sub-Watershed No. 4	2	4	5	6	7	8
Sub-Watershed No.5	2	3	4	5	6	6
Sub-Watershed No.6	1	1	2	2	3	3
Sub-Watershed No.7	2	3	3	4	4	5
Detention Basin No.1						
Inflow Peak, cfs.	1	1	2	2	3	3
Outflow Peak, cfs.	1	1	2	2	3	3
Maximum Stage,ft.	0.06	0.09	0.12	0.15	0.18	0.21
Maximum Storage, cft.	43	69	87	115	136	157
Detention Basin No.2						
Inflow Peak, cfs.	2	3	4	5	6	6
Outflow Peak, cfs.	2	3	4	5	6	6
Maximum Stage,ft.	0.08	0.13	0.16	0.21	0.25	0.28
Maximum Storage, cft.	25	39	50	65	76	88
Detention Basin No.3						
Inflow Peak, cfs.	2	4	5	6	7	8
Outflow Peak, cfs.	2	4	4	6	7	8
Maximum Stage,ft.	0.1	0.16	0.2	0.26	0.31	0.36
Maximum Storage, cft.	31	49	62	81	96	111
Detention Basin No.4						
Inflow Peak, cfs.	2	4	5	6	7	8
Outflow Peak, cfs.	2	4	4	6	7	8
Maximum Stage,ft.	0.1	0.16	0.2	0.26	0.31	0.36
Maximum Storage, cft.	31	49	62	81	96	111
Detention Basin No.5						
Inflow Peak, cfs.	2	3	4	5	6	6
Outflow Peak, cfs.	2	3	4	5	6	6
Maximum Stage,ft.	0.08	0.13	0.16	0.21	0.25	0.28
Maximum Storage, cft.	25	39	50	65	76	88
Detention Basin No.6						
Inflow Peak, cfs.	1	1	2	2	3	3
Outflow Peak, cfs.	1	1	2	2	3	3
Maximum Stage,ft.	0.06	0.09	0.12	0.15	0.18	0.21
Maximum Storage, cft.	43	69	87	115	136	157
Channel No.7						
Inflow Peak, cfs.	2	3	3	4	4	5
Outflow Peak, cfs.	2	3	3	4	4	5
Channel No.8						
Inflow Peak, cfs.	4	7	9	12	14	16
Outflow Peak, cfs.	4	7	9	12	14	16
Channel No.9						
Inflow Peak, cfs.	7	11	13	18	21	24
Outflow Peak, cfs.	7	11	13	18	21	24
Channel No.10						
Inflow Peak, cfs.	7	11	13	18	21	24
Outflow Peak, cfs.	7	11	13	18	21	24
Detention Basin No. 11						
Inflow Peak, cfs.	7	11	13	18	21	24
Outflow Peak, cfs.	3	3	4	5	5	5
Maximum Stage,ft.	2.11	3.41	4.3	5.48	6.33	7.12
Maximum Storage, cft.	6261	11246	15149	21118	25948	30898

FIG. 7.53 Summary of hydrologic data for the sample 5.13-acre site. Source: Debo, 1987.

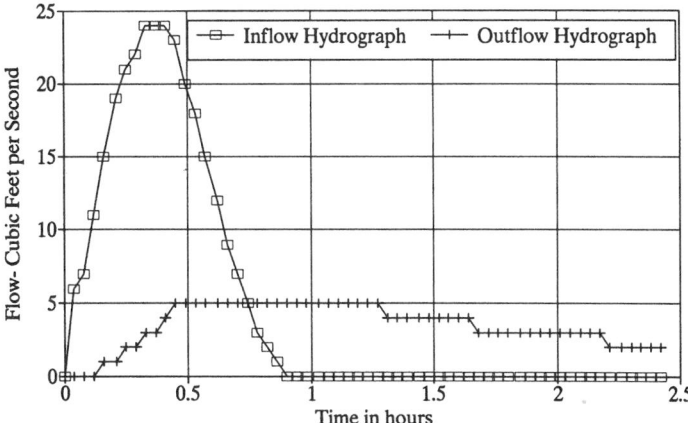

FIG. 7.54 Examples of detention basin hydrographs for 100-year return period. Source: Debo, 1987.

******Pipe Design*******Peak Flow In CFS For Various Storms***										
Pipe	Dia.	n	Slope	Qfull o.c.	2	5	10	25	50	100
1	24"	0.012	0.3%	13.4 cfs.	4	7	9	12	14	16
2	24"	0.012	0.3%	13.4 cfs.	7	11	13	18	21	24
3	30"	0.012	0.3%	24.4 cfs.	7	11	13	18	21	24
[Note: Shaded area signifies pressure flow conditions exist.]										

FIG. 7.55 Example of culvert performance analysis.

providing inadequate cover over the pipes. This assumes that the designer wishes the HGL to be approximately 6 inches below the ground surface when a 100-year storm occurs. Raising the culverts means less excavation will be required during construction. This can result in reduced installation costs. If the invert of pipe 3's outlet is raised (along with that of the emergency spillway shown in Figure 7.57) 1 foot from the shown 92.72-foot elevation to 93.72, and if the maximum headwater for the 100-year storm is estimated at 7.12 feet in the detention basin, the elevation of the basin's bottom can be raised from the shown 85.60 feet to 93.72 − 7.12, or 86.60, feet. This will also provide additional economy by reducing the amount of earthwork associated with building the detention basin. The basin's discharge pipe outlet could then be kept at 85.0 feet to provide adequate outfall, as shown in Figure 7.57.

The stability of the detention basin's slopes should be checked (Section 9.6) to ensure that they will remain safe when subjected to inflows expected for 100-year return periods or greater.

If local regulatory agencies require that a retention volume (the volume of water not allowed to leave the site) equal to 1.5 inches of runoff be provided for this development, a retained volume of 27,932 ft^3 must be provided. This would require

(a) increasing the elevation of the discharge structure to allow 27,932 ft^3 of retained storage below the outlet,

Note:

A) Minor headlosses associated with discharges are assumed to be $V^2/2G$.

B) Minimum headlosses associated with entrance losses are assumed to be $0.2(V^2/2G)$.

C) Friction losses are based on the Hazen-Williams formula (Equation 10-10).

Pipe No.	1	2	3
Flow, CFS	16	24	24
Length, ft	30	145	30
Hazen-Williams Roughness- C	120	120	Open Channel
Diameter, in.	24	24	30
Friction Loss, ft	0.12	1.19	---
Velocity, ft/sec	5.09	7.64	4.88
Velocity Head, $(V^2/2g)$	0.40	0.91	0.37

FIG. 7.56 Example of a hydraulic grade line for a 100-year storm.

FIG. 7.57 Tentative detention basin configuration before final bottom elevation of basin is established.

(b) redesigning the outlet structure to provide the desired discharge flows, and

(c) installing a granular bottom to cover an included subsurface drainage system that will allow the detained water to be filtered and seep into the outfall ditch over a period of as much as 72 hours.

7.8.5 Control Discharge Structures

The types of discharge structures that can be utilized on detention basins vary based on need and site conditions. Some of the most common discharge structures in addition to rectangular weirs and orifices are trapezoidal weirs, V notch weirs, and culverts.

Normally an auxiliary emergency discharge structure is set approximately at the 100-year maximum pond storage elevation to allow free discharge of runoff accumulations in excess of the 100-year storm. Generally, this type of structure is in the form of a broad-crested weir.

The exit velocity of the outfall discharge must be examined to ensure that downstream erosion will not occur. Recommended maximum velocities are shown in Figure 7.58. Normally culvert outfalls have riprap (stones) placed as a blanket between the erodible soil and the exiting water. The riprap also acts as an energy dissipator to slow down the discharge. Baffle devices can also be utilized.

7.8.6 Trapezoidal Ditches for Discharge Outfalls

Often outfall ditches have trapezoidal shapes to minimize bank erosion. Manning's equation can be used to determine the flow properties of this type of ditch; however,

Material	Mean Velocity, Clear Water (feet per second)	Mean Velocity, Silty Water (feet per second)
Fine sand, colloidal	1.50	2.50
Sandy loam, noncolloidal	1.75	2.50
Silt loam, noncolloidal	2.00	3.00
Alluvial silts, noncolloidal	2.00	3.50
Ordinary firm loam	2.50	3.50
Volcanic ash	2.50	3.50
Stiff clay, very colloidal	3.75	5.00
Alluvial silts, colloidal	3.75	5.00
Shales and hardpans	6.00	6.00
Fine gravel	2.50	5.00
Graded loam to cobbles, noncolloidal	3.75	5.00
Graded silts to cobbles, colloidal	4.00	5.50
Coarse gravel, noncolloidal	4.00	6.00
Cobbles and shingles	5.00	5.50

(a)

Material	Loose (feet per second)	Fairly Compact (feet per second)	Compact (feet per second)	Very Compact (feet per second)
Lean clayey soils	1.2	2.5	3.4	4.4
Clays	1.3	2.8	3.9	5.4
Heavy clayey soils	1.5	3.0	4.1	5.8
Sandy clays (sand < 50%)	1.7	3.2	4.3	6.0

(b)

Average Depth (feet)	Correction Factor
1	0.81
2	0.91
3	0.99
4	1.04
5	1.11
6	1.14
7	1.18
8	1.21
9	1.24
10	1.26

(c)

FIG. 7.58 Maximum permissible velocities for various channels. (a) Maximum permissible mean velocities recommended by Fortier and Scobey for straight, aged channels on small slope. Taken from Fortier and Scobey, American Society of Civil Engineers, 1926 *Special Committee on Irrigation*. Reproduced by permission of the American Society of Civil Engineers. (b) Permissible velocities in cohesive soils. (c) Correction factors to be applied to Figure 7.58(b) for cohesive and noncohesive soils. Source: Portland Cement Association, *Handbook of Concrete Pipe Hydraulics*, 1964. Reproduced with permission; however the Portland Cement Association does not verify the accuracy.

uses of nomographs such as the one shown in Figure 7.59 are common. This particular chart is suitable if the trapezoidal ditch has a 3-foot bottom and 2 : 1 side slopes. Direct chart use is available if the channel's coefficient of roughness is 0.03. Other coefficients of roughness require the calculation of Q_n or V_n values prior to entering the chart.

7.9   PIPE BEDDING AND PROTECTION

When considering pipe installations, the type of pipe to be installed and the type of pipe bedding specified or dictated by field conditions are two of the most important factors affecting the overall structural integrity of a buried pipe system. Pipe type may be subdivided into reinforced concrete pipe (RCP), high-density polyethylene pipe (HDPE), polyvinyl chloride pipe (PVCP), and corrugated steel pipe (CSP). To limit the scope of this section, RCP was chosen as the focus of this discussion. Further, pipe bedding may be subdivided into two broad categories: pipes installed in trenches and pipes installed in embankments. This section examines pipes installed in trenches. Pipes installed in embankments have different bedding relationships, and the reader is referred to the American Concrete Pipe Association's (ACPA's) *Concrete Pipe Handbook* (1998) for embankment applications.

During the 1970s the ACPA began a research program to investigate the interaction between buried concrete pipes and soil. As a result of this effort, a finite-element computer program entitled *Soil-Pipe Interaction Design and Analysis* (SPIDA) evolved for use in designing buried concrete pipe installations. Additional research has resulted in four replacement pipe installation classifications and a simplified computer design program called *Standard Installations Direct Design* (SIDD). These four replacement classifications (called standard installations) replace previously used classes (B, C, and D) of soil type bedding developed by M. G. Spangler et al. Class A bedding, consisting of a pipe with a concrete arch or cradle, has not been replaced. Figure 7.60 illustrates standard trench installation types 1–4 along with bedding, haunch, outer bedding, and lower side requirements.

To illustrate the analysis of external loads on a buried culvert, a 24-inch, class II (A) wall-reinforced concrete pipe having the characteristics shown in Figure 7.61 will be used along with standard trench installation type 2 bedding. If the design professional wishes to establish as a design standard the minimum cover over the top of pipes crossing roads as 1 foot ($H = 1$ foot), the pipe should be checked to ensure it will maintain its structural integrity when subjected to wheel loads. An H-20 wheel load, similar to that shown in Figure 7.62, is utilized in this analysis.

First, the earth load on the pipe, W_E, must be determined. In trenches of moderate widths, the resulting earth load equals the weight of material above the culvert minus any forces due to shear along the sides of the trench. Because material used for bedding the pipe generally will settle more than the backfill soil at the interface with the trench sides, the friction along the walls will reduce loads being placed on the pipe. According to ACPA (1998), arching factors for standard installations in trenches have not been developed, and the earth load must be estimated using Marston's formula:

$$W_E = C_d w B_d^2 \tag{7.41}$$

FIG. 7.59 Graphical solution of Manning's equation for a trapezoidal channel with 2 : 1 side slopes and a 3-foot-wide bottom. Source: U.S. Department of Transportation, Federal Highway Administration. *Design Charts for Open Channel Flow. Hydraulic Design Series*, No. 3, 1973.

Installation Type	Bedding Thickness	Haunch and Outer Bedding	Lower Side
Type 1	Do/24 min, not less than 75 mm (3"). If rock foundation, use Do/12 minimum, not less than 150 mm (6").	95% Standard Proctor for SIDD Class Gravelly Sand or AASHTO Classes A1, A3.	90% Standard Proctor for SIDD Class Gravelly Sand or AASHTO Classes A1, A3 or 95% Standard Proctor for SIDD Class Sandy Silt or AASHTO Classes A2, A4 or 100% Standard Proctor for SIDD Class Silty Clay or AASHTO Classes A5, A6
Type 2	Do/24 minimum, not less than 75 mm (3"). If rock foundation, use Do/12 minimum, not less than 150 mm (6").	90% Standard Proctor for SIDD Class Gravelly Sand or AASHTO Classes A1, A3 or 95% Standard Proctor for SIDD Class Sandy Silt or AASHTO Classes A2, A4	85% Standard Proctor for SIDD Class Gravelly Sand or AASHTO Classes A1, A3 or 90% Standard Proctor for SIDD Class Sandy Silt or AASHTO Classes A2, A4 or 95% Standard Proctor for SIDD Class Silty Clay or AASHTO Classes A5, A6
Type 3	Do/24 minimum, not less than 75 mm (3"). If rock foundation, use Do/12 minimum, not less than 150 mm (6").	85% Standard Proctor for SIDD Class Gravelly Sand or AASHTO Classes A1, A3 or 90% Standard Proctor for SIDD Class Sandy Silt or AASHTO Classes A2, A4 or 95% Standard Proctor for SIDD Class Silty Clay or AASHTO Classes A5, A6	85% Standard Proctor for SIDD Class Gravelly Sand or AASHTO Classes A1, A3 or 90% Standard Proctor for SIDD Class Sandy Silt or AASHTO Classes A2, A4 or 95% Standard Proctor for SIDD Class Silty Clay or AASHTO Classes A5, A6
Type 4	Do/24 minimum, not less than 75 mm (3"). If rock foundation, use Do/12 minimum, not less than 150 mm (6").	No compaction required, except if SIDD Silty Clay is used, then 85% Standard Proctor minimum.	No compaction required, except if SIDD Silty Clay is used, then 85% Standard Proctor minimum.

B. Standard trench installations soil and compaction requirements.

Note: (1) Trench top elevation shall be no lower than 0.1 H below finished grade, or for streets, no lower than an elevation of 0.3 m (1') below the bottom of the pavement base.
(2) Soil in bedding and haunch zones shall be compacted to at least same compaction specifed for the majority of soil in the backfill zone.
(3) Trench walls consisting of embankment > 10° of verticle, compact lower side ≥ backfill zone.

Figure 7.60 Standard trench installation and nomenclature. (a) Installation schematic, and (b) Installation requirement showing type of installation, bedding thickness, haunch, outer bedding, and lower side minimum Standard Proctor compaction requirements for different soil types. Source: American Concrete Pipe Association Concrete Pipe Handbook. Irving, Texas, 1998. Reprinted by permission of the American Concrete Pipe Association.

FIG. 7.60 Standard trench installation and nomenclature: (*a*) installation schematic; (*b*) installation requirements showing type of installation, bedding thickness, haunch, outer bedding, and lower side minimum standard Proctor compaction requirements for different soil types. Source: American Concrete Pipe Association, 1998. Reprinted by permission of the American Concrete Pipe Association.

The strength test requirements in pounds-force per linear foot of pipe under the three-edge-bearing method shall be either the D-load (test load expressed in pounds-force per linear foot per foot of diameter) to produce a 0.01-in. crack, or the D-loads to produce the 0.01-in. crack and the ultimate load as specified below, multiplied by the internal diameter of the pipe in feet.

D-load to produce a 0.01-in. crack 1000
D-load to produce the ultimate load[B] 1500

Reinforcement, in.²/linear ft of Pipe Wall

Internal Designated Diameter, in.	Wall A Concrete Strength, 4000 psi				Wall B Concrete Strength, 4000 psi				Wall C Concrete Strength, 4000 psi			
	Wall Thickness, in.	Circular Reinforcement[C] Inner Cage	Circular Reinforcement[C] Outer Cage	Elliptical Reinforcement[D]	Wall Thickness, in.	Circular Reinforcement[C] Inner Cage	Circular Reinforcement[C] Outer Cage	Elliptical Reinforcement[D]	Wall Thickness, in.	Circular Reinforcement[C] Inner Cage	Circular Reinforcement[C] Outer Cage	Elliptical Reinforcement[D]
12	1¾	0.07[E]	—	—	2	0.07[E]	—	—	2¾	0.07[E]	—	—
15	1⅞	0.07[E]	—	—	2¼	0.07[E]	—	—	3	0.07[E]	—	—
18	2	0.07[E]	—	0.07[E]	2½	0.07[E]	—	0.07[E]	3¼	0.07[E]	—	0.07[E]
21	2¼	0.12	—	0.10	2¾	0.07[E]	—	0.07[E]	3½	0.07[E]	—	0.07[E]
24	2½	0.13	—	0.11	3	0.07[E]	—	0.07[E]	3¾	0.07[E]	—	0.07[E]
27	2⅝	0.15	—	0.13	3¼	0.13	—	0.11	4	0.07[E]	—	0.07[E]
30	2¾	0.15	—	0.14	3½	0.14	—	0.12	4¼	0.07[E]	—	0.07[E]
33	2⅞	0.16	—	0.15	3¾	0.15	—	0.13	4½	0.07[E]	—	0.07[E]
36	3	0.14	0.08	0.15	4[F]	0.12	0.07	0.13	4¾	0.07	0.07	0.08
42	3½	0.16	0.10	0.18	4½	0.15	0.09	0.17	5¼	0.10	0.07	0.11
48	4	0.21	0.13	0.23	5	0.18	0.11	0.20	5¾	0.14	0.08	0.15
54	4½	0.25	0.15	0.28	5½	0.22	0.13	0.24	6¼	0.17	0.10	0.19
60	5	0.30	0.18	0.33	6	0.25	0.15	0.28	6¾	0.22	0.13	0.24
66	5½	0.35	0.21	0.39	6½	0.31	0.19	0.34	7¼	0.25	0.15	0.28
72	6	0.41	0.25	0.45	7	0.35	0.21	0.39	7¾	0.30	0.18	0.33
78	6½	0.46	0.28	0.51	7½	0.40	0.24	0.44	8¼	0.35	0.21	0.39
84	7	0.51	0.31	0.57	8	0.46	0.28	0.51	8¾	0.41	0.25	0.46

	9	0.85	0.51	0.46 Plus Elliptical	Inner Circular Plus Elliptical	10	0.76	0.46	0.41 Plus Elliptical	Inner Circular Plus Elliptical	$10^{3}/_{4}$	0.70	0.42	0.37 Plus Elliptical	Inner Circular Plus Elliptical
108					0.34 / 0.51				0.30 / 0.46					0.28 / 0.42	
114	✓	—	—	—	—	✓	—	—	—	—	✓	—	—	—	—
120	✓	—	—	—	—	✓	—	—	—	—	✓	—	—	—	—
126	✓	—	—	—	—	✓	—	—	—	—	✓	—	—	—	—
132	✓	—	—	—	—	✓	—	—	—	—	✓	—	—	—	—
138	✓	—	—	—	—	✓	—	—	—	—	✓	—	—	—	—
144	✓	—	—	—	—	✓	—	—	—	—	✓	—	—	—	—

A For modified or special designs, see Section 7.2 or with the permission of the owner utilize the provisions of M 242. Steel areas may be interpolated between those shown for variations in diameter, loading, or wall thickness. Pipe over 96 in. in diameter shall have two circular cages or an inner circular cage plus one elliptical cage.

B Three-edge-bearing test to ultimate load is not required for any class of pipe 60 in. or less in diameter provided all other requirements of this specification are met.

C As an alternative to designs requiring both inner and outer circular cages the reinforcement may be positioned and proportioned in either of the following manners:

An inner circular cage plus an elliptical cage such that the area of the elliptical cage shall not be less than that specified for the inner circular cage plus the elliptical cage shall not be less than that specified for the outer cage in the table and the total area of the inner circular cage plus the elliptical cage shall not be less than that specified for the inner cage in the table.

An inner and outer cage plus quadrant mats in accordance with Figure 1, or

An inner and outer cage plus an elliptical cage in accordance with Figure 2.

D Elliptical and quadrant steel must be held in place by means of holding rods, chairs, or other positive means throughout the entire casting operation.

E For these classes and sizes, the minimum practical steel reinforcement is specified. The actual ultimate strength is greater than the minimum strength specified for nonreinforced pipe of equivalent diameters in AASHTO M 86.

F As an alternative, single cage reinforcement may be used. The reinforcement area in square in. per linear foot shall be 0.20 for wall B and 0.16 for wall C.

FIG. 7.61 Design requirements for class II reinforced concrete pipe. Taken from *Standard Specification for Transportation Materials and Methods of Sampling and Testing, Part I—Specifications*. Copyright © 1998 by the American Association of State Highway and Transportation Officials, Washington, DC. Used by permission.

FIG. 7.62 AASHTO H-20 wheel loads. Source: American Concrete Pipe Association, 1998. Reprinted by permission of the American Concrete Pipe Association.

where W_E is the load that the culvert is subjected to in pounds per linear foot; C_d is the soil structure factor, which may be obtained from Figure 7.63 using H as the depth of cover over the top of the pipe in feet; w is the unit weight of the backfill soil in pounds per cubic foot; and B_d is the width of the trench at the top of the culvert in feet. Assuming that a typical 3-foot wide excavation bucket (same as trench width, B_d) is used when installing the culvert and that the backfill soil is found to have a unit weight of 120 lb/ft³, C_d can be estimated using Figure 7.63 (where $H/B_d = \frac{1}{3}$ and the soil is found to be sand and gravel) as 0.30. The load on the culvert is then estimated, using Equation 7.41, as

$$W_E = 0.30 \times 120 \times 3^2 = 324 \text{ pounds per linear foot}$$

Next, the live load due to traffic, W_L, must be estimated. According to ACPA (1998), the live load pressure on a buried concrete pipe can be determined by using the average pressure intensity at the top of the pipe. The total live load acting on the pipe can be calculated using the average pressure intensity. The live load pressure per linear foot of pipe can then be obtained by dividing the total live load on the pipe by its effective supporting length. Figure 7.64 shows the approximate live load distribution for a single dual-wheel application.

The average pressure intensity at the top of the pipe is calculated using

$$w_L = \frac{P(1 + I_f)}{A_{\text{LL}}} \tag{7.42}$$

where w_L is the average pressure intensity in pounds per foot², P is the total applied surface wheel load in pounds, A_{LL} is the distributed live load area on the subsoil plane at the outside top of the pipe in square feet, and I_f is the impact factor (use 30 percent in this case). For this analysis, using Equation 7.42, the average pressure is

VALUES OF COEFFICIENT—C$_d$

A = C$_d$ for Kμ and Kμ' = .1924 for Granular Materials Without Cohesion
B = C$_d$ for Kμ and Kμ' = .165 Max. for Sand and Gravel
C = C$_d$ for Kμ and Kμ' = .150 Max. for Saturated Top Soil
D = C$_d$ for Kμ and Kμ' = .130 Ordinary Max. for Clay
E = C$_d$ for Kμ and Kμ' = .110 Max. for Saturated Clay

LOAD PER UNIT OF LENGTH, W$_c$ = C$_d$ w B$_d$^{2}
w = Unit Weight of Fill Materials
B$_d$ = Breadth of Trench at the Top of Conduit
H = Height of Fill Over Top of Conduit

FIG. 7.63 Computation diagram for earth loads on trench conduits (completely buried trenches). Source: American Concrete Pipe Association. Used by permission.

$$w_L = \frac{(16{,}000)(1 + 0.30)}{(0.83 + 1.75)(1.67 + 1.75)} = 2357 \text{ lb/ft}^2 \text{ at 1-foot depth}$$

The total load transferred to the pipe can now be computed based on the affected pipe area. For this example the length, L, parallel to the longitudinal axis of the pipe is $1.67 + 1.75$, or 3.42 feet. The effective width is the projected outside diameter of the pipe or the base area of the frustum pyramid at the 1-foot depth, whichever is smaller. In this case, the 24-inch-diameter pipe has a total wall thickness of 5 inches; therefore, the outside projected distance, as shown in Figure 7.65, would be 29 inches, or 2.42 feet, which will be used in later calculations as variable B_c.

The total live load, W_T, applied to the pipe is equal to

$$W_T = W_L L S_L \tag{7.43}$$

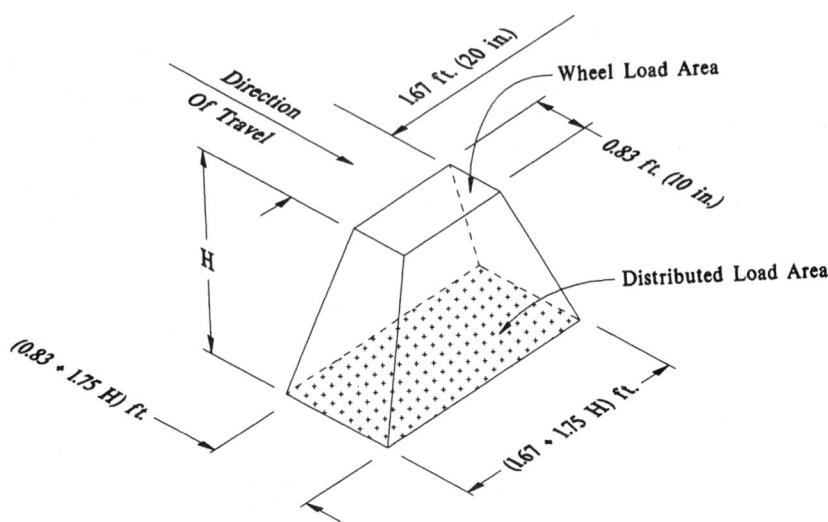

FIG. 7.64 Distribution load area: single dual wheel. Source: American Concrete Pipe Association, 1998. Reprinted by permission of the American Concrete Pipe Association.

where W_T is the total live load in pounds, L is the length of A_{LL} parallel to the longitudinal axis of the pipe in feet, and S_L is the outside horizontal span of pipe or width of A_{LL} transverse to the longitudinal axis of the pipe, whichever is less, in feet. In this example W_T would have a value of

$$W_T = 2357 \times 3.42 \times 2.42, \text{ or } 19,507 \text{ pounds}$$

The live load per foot of pipe, W_L, is equal to the total live load in pounds, W_T, divided by the effective supporting length of pipe in feet shown in Figure 7.65, L_e, where

$$W_L = \frac{19,507}{3.42 + (1.75 \times 0.75 \times 2.42)} = 2957 \text{ pounds per linear foot}$$

The bedding factor must now be determined. Minimum trench bedding factors, B_{f0}, are shown in Figure 7.66. Because some trenches are wide enough to create loading conditions on the buried culverts that are similar to embankment conditions, a variable trench bedding factor, B_{fv}, must be determined using the following equation that has been modified for use with standard installations:

$$B_{fv} = \frac{(B_{fe} - B_{f0})(B_d - B_c)}{B_{dt} - B_c} + B_{f0} \qquad (7.44)$$

where B_c is the outside horizontal span of pipe in feet; B_d is the trench width at top of pipe in feet; B_{dt} is the transition width at top of pipe in feet; B_{fe} is the bedding factor, embankment; B_{f0} is the minimum bedding factor, trench; and B_{fv} is the variable bedding factor, trench.

Variables B_c and B_d have previously been defined in this example. The transition

FIG. 7.65   (*a*) Effective pipe pressure. (*b*) Effective support length of pipe. Source: American Concrete Pipe Association, 1998. Reprinted by permission of the American Concrete Pipe Association.

Standard Installation	Minimum Bedding Factor, B_{fo}
Type 1	2.3
Type 2	1.9
Type 3	1.7
Type 4	1.5

FIG. 7.66   Trench minimum bedding factors, B_{f0}, for pipes installed in trenches. Source: American Concrete Pipe Association, 1998. Reprinted by permission of the American Concrete Pipe Association.

width must now be determined. Transition width is defined as the width at which the trench can be widened where no additional load is transmitted to the pipe (under established height of cover and pipe diameter) because an embankment condition exists. Using a settlement ratio, $r_{sd}p$, equal to 0.5, as recommended by ACPA (1998) for most construction methods, Figure 7.67 can be used to obtain an estimated trench transition width. In this case the factor H/B_c equals 1/2.5, or 0.4. Entering the figure with an ordinate value of 0.4 and then reading horizontally to the right where the 0.5 settlement ratio line ($r_{sd}p$) intersects, the corresponding B_d/B_c value is determined to be 1.18. Since $B_c = 2.5$ feet, the trench transition width value, B_{dt}, equals 2.95 feet (1.18 × 2.5).

Next the bedding factor for embankment conditions, B_{fe} (required to find the trench transition width), must be determined. Using Figure 7.68, a value of 3.0 is obtained.

Using equation 7.44, the variable trench bedding factor can be estimated as

$$B_{fv} = \frac{(4.2 - 2.3)(3 - 2.5)}{2.95 - 2.5} + 1.9 = 4.0$$

The factor of safety for the installed 24-inch culvert being subjected to a single H-20 load can be determined using the following equation:

FIG. 7.67 Trench transition widths. Source: American Concrete Pipe Association, 1998. Reprinted by permission of the American Concrete Pipe Association.

Pipe Diameter	Standard Installation			
	Type 1	Type 2	Type 3	Type 4
300 mm (12 in.)	4.4	3.2	2.5	1.7
600 mm (24 in.)	4.2	3.0	2.4	1.7
900 mm (36 in.)	4.0	2.9	2.3	1.7
1800 mm (72 in.)	3.8	2.8	2.2	1.7
3600 mm (144 in.)	3.6	2.8	2.2	1.7

Notes:

1. *For pipe diameters other than listed above, embankment condition factors, B_{fe} can be obtained by interpolation.*

2. *Bedding factors are based on the soils being placed with the minimum compaction specified in Fig. 7.60 for each standard installation.*

FIG. 7.68 Bedding factors for pipes installed under embankment conditions. Source: American Concrete Pipe Association, 1998. Reprinted by permission of the American Concrete Pipe Association.

$$D - \text{load} = \left(\frac{W_E}{B_{fe}} + \frac{W_L}{B_{f,\text{LL}}} \right) \left[\frac{\text{FS}}{D} \right] \tag{7.45}$$

where

D-load = D load of pipe [three edge bearing test load expressed in pounds per linear foot per foot of diameter to produce a 0.01-inch crack according to AASHTO M 170–95 (1997); for class II pipe, D-load = 1000 pounds].

W_E = earth load on pipe in pounds per foot

W_L = live load on pipe in pounds per foot

B_{fe} = earth load bedding factor; use B_{fe} factor from Figure 7.68 unless B_{fv} value is smaller, in which case substitute with the smaller B_{fv} value

$B_{f,\text{LL}}$ = live load bedding factor; use $B_{f,\text{LL}}$ factor from Figure 7.69 unless B_{fe} value is smaller, in which case substitute with the smaller B_{fe} value; in trench installation where trench width is less than transition width, $B_{f,\text{LL}}$ is compared to variable trench bedding factor, B_{fv}

FS = factor of safety

D = pipe diameter in feet

Solving for the factor of safety, FS, using Equation 7.45,

$$1000 = \left(\frac{324}{3.0} + \frac{2957}{2.2} \right) \frac{\text{FS}}{3}$$

$$\text{FS} \approx 2.1$$

A minimum factor of safety is usually taken as 1.0 when considering the D load that produces a 0.01-inch crack. Based on this analysis, the 24-inch RCP culvert should perform satisfactorily.

Fill Height, Ft.	Pipe Diameter, Inches										
	12	24	36	48	60	72	84	96	108	120	144
0.5	2.2	1.7	1.4	1.3	1.3	1.1	1.1	1.1	1.1	1.1	1.1
1.0	2.2	2.2	1.7	1.5	1.4	1.3	1.3	1.3	1.1	1.1	1.1
1.5	2.2	2.2	2.1	1.8	1.5	1.4	1 4	1 3	1.3	1.3	1.1
2.0	2.2	2.2	2.2	2.0	1.8	1.5	1.5	1.4	1 4	1.3	1.3
2.5	2.2	2.2	2.2	2.2	2.0	1.8	1.7	1.5	1.4	1.4	1 3
3.0	2.2	2.2	2.2	2.2	2.2	2.2	1 8	1 7	1 5	1 5	1 4
3.5	2.2	2 2	2.2	2.2	2 2	2.2	1 9	1.8	1.7	1.5	1.4
4.0	2.2	2.2	2.2	2 2	2.2	2.2	2.1	1.9	1.8	1.7	1.5
4.5	2.2	2.2	2.2	2.2	2.2	2.2	2.2	2.0	1 9	1.8	1.7
5.0	2 2	2.2	2 2	2.2	2 2	2 2	2 2	2.2	2 0	1.9	1.8
5.5	2.2	2.2	2.2	2.2	2.2	2 2	2.2	2.2	2.2	2.0	1.9
6.0	2 2	2.2	2 2	2 2	2.2	2 2	2.2	2.2	2 2	2.1	2.0
6.5	2.2	2.2	2.2	2.2	2.2	2.2	2.2	2.2	2.2	2.2	2.2

Note:

1.   *For pipe diameters other than listed above, B_{fLL} values can be obtained by interpolation.*

FIG. 7.69   HS20 live load bedding factors. Source: American Concrete Pipe Association, 1998. Reprinted by permission of the American Concrete Pipe Association.

7.10   SUBSURFACE WATER

In addition to surface runoff, drainage systems can handle subsurface water removal. Some precipitation becomes subsurface water by infiltration. The amount of infiltration is a function of the permeability of the soil, the slope, and the surface cover characteristics. Subsurface water flows in much of the same manner as surface water does—toward lower elevations. The most common method of determining laminar flow through porous media is by using Darcy's equation (Equation 11.6), contained in Section 11.3.2.

Accurate calculations are difficult because of the nonhomogeneity of soils. Permeability coefficients can be estimated by field pumping, laboratory permeability tests, or gradation analysis as indicated in Chapter 11. Typical permeability values for various soils are included in Figure 7.70.

Normally high groundwater tables in a development can be lowered by open ditching or by installing subsurface drains, commonly called French drains. For example, if the proposed roadbed shown in Figure 7.71 is to be located in an area using a curb and gutter section, and if it is desired to keep the stabilized groundwater table 2 feet below the road surface, a French drain must be properly designed.

AASHO Class.	Liquid Limit Percent	Plasticity Index Percent	Amount Passing # 200 Sieve Percent	Consolidating Load Kips/sq.ft.	Compacted Wet Permeability Coefficient Ft/day	Compacted Dry Permeability Coefficient Ft/day
A - 1	23	8	26	1	0.00045	0.00480
				2	0.00036	0.00200
				4	0.00024	0.00170
A - 2	28	11	40	1	0.00048	0.00470
				2	0.00028	0.00160
				4	0.00018	0.00090
A - 3	NP	NP	Patted with spatula		200.00000	226.00000
A - 4	33	12	99	1	0.00030	0.00075
				2		0.00060
				4	0.00020	0.00024
A - 5	35	6	35	1	0.02100	0.37000
				2	0.01500	0.25000
				4	0.01200	0.17000
A - 6	72	45	86	1	0.00004	0.00011
				2	0.00002	0.00011
				4	0.00001	0.00010
A - 7	67	34	71	1	0.00051	0.26000
				2	0.00028	0.00840
				4	0.00020	0.00120
A - 8	78	8	38	1	0.00390	0.04100
				2	0.00100	0.01800
				4	0.00046	0.01000

FIG. 7.70 Permeability of soils. Source: E. S. Barber and C. L. Sawyer. *Highway Subdrainage Proceeedings—Thirty-First Annual Meeting.* © 1952 by the Highway Research Board, National Research Council, Washington, DC. Used with permission.

To compute the flow per foot of roadbed:

$$Q = \text{coefficient of permeability} \times \text{head}$$
$$\times \left[\frac{\text{number of flow lines}}{\text{number of equipotential lines}} \right] \tag{7.46}$$
$$Q = K_s h(N_f/N_d) = K_s h\left(\tfrac{6}{10}\right)$$

Assuming an A-2 type soil (AASHTO classification), with a permeability of 0.0047 foot per day, the estimated flow for 100 linear feet of perforated pipe is

$$Q = \frac{0.0047 \text{ foot per day} \times 4.3 \text{ feet of head} \times 0.6 \times 100 \text{ feet of pipe}}{24 \times 60 \times 60 \text{ seconds per day}}$$

$$= 0.000014 \text{ ft}^3/\text{s}$$

A 4-inch-diameter perforated pipe on a slope of 0.5 percent can carry 0.12 ft³/s, which would be more than adequate to provide the necessary drainage. Normal installation of the perforated pipe used as a French drain is shown in Figure 7.72.

FIG. 7.71   Example of subsurface flow net development.

FIG. 7.72   French drain installation.

Other subsurface flow conditions may require installation of a layer of graded granular material as part of the road subgrade. It is covered with filter fabric to provide a porous blanket for larger subsurface water removal requirements. For more in-depth coverage on subsurface water and seepage, the reader is referred to Cedergren (1989).

REFERENCES

American Association of State Highway and Transportation Officials (AASHTO). *Model Drainage Manual—1991*. Washington: AASHTO, 1991.

American Association of State Highway and Transportation Officials (AASHTO). *AASHTO M 170–95, Standard Specifications for Reinforced Concrete Culvert, Storm Drain, and Sewer Pipe*. Washington: AASHTO, 1997, p. 284.

American Concrete Pipe Association (ACPA). *Concrete Pipe Handbook*. Irving, TX: ACPA, 1998, pp. 4–7, 4–15, 4–39 to 4–42, 4–47, 4–71 to 4–85. Referenced material reproduced by permission of the American Concrete Pipe Association.

American Society of Civil Engineers (ASCE) and Water Pollution Control Federation. *Design and Construction of Sanitary and Storm Sewers* (ASCE Manual No. 37/MOP-9). New York: ASCE, 1969, pp. 41–62. Referenced material is reproduced by permission of the American Society of Civil Engineers.

Bradford, B. H., N. S. Grigg, and L. Scott Tucker. *Storm Water Management*, Vol. I: *Urban Hydrology*. Atlanta: Water Management Science, 1979, pp. III–5.

Cedergren, H. R. *Seepage, Drainage, and Flow Nets*, 3rd ed. New York: Wiley, 1989.

Debo, T. N. *Hydros*. Computer software. Atlanta: Georgia Institute of Technology, 1987.

Frederick, R. H., V. A. Myers, and E. P. Anciello. *Five to 60 Minute Precipitation Frequency for the Eastern and Central United States, NWS Hydro-35*. Silver Springs, MD: National Weather Service, NOAA, U.S. Department of Commerce, 1977.

Hershfield, P. M. *Rainfall Frequency Atlas of the United States for Durations from 30 Minutes to 24 Hours and Return Periods from 1 to 100 Years*. Technical Paper No 40. Washington: U.S. Weather Bureau, U.S. Department of Commerce, 1961.

Miller, J. F., R. H. Frederick, and R. J. Traley. *Precipitation-Frequency Atlas of the Western United States*, Vol. I–XI: *NOAA Atlas 2*. Silver Springs, MD: National Weather Service, NOAA, U.S. Department of Commerce, 1973.

Natural Resources Conservation Service (NRCS*). National Engineering Handbook, Notice NEH 4-102*. Washington: NRCS, 1972, Chapters 15–17.

Natural Resources Conservation Service (NRCS*). Urban Hydrology for Small Watersheds— (NRCS TR-55)*. Washington: NRCS, 1986a.

Natural Resources Conservation Service (NRCS). "Technical Note: Hydrology No. 4—Time of Concentration." Chester, PA: New England Naturalist Training Center (NENTC), 1986b.

Pitt, R. E., and G. Amy. *Toxic Materials Analysis of Street Surface Contaminants*, USEPA, Washington, DC: U.S. Government Printing Office, 1973.

Schueler, T. R. *Controlling Urban Runoff: A Practical Manual for Planning and Designing Urban BMPs*. Ellicott City, MD: Center for Watershed Protection, 1987.

South Carolina Coastal Council. *Stormwater Management Guidelines*. Columbia: South Carolina Coastal Council, 1988, pp. 25, 28, 31.

South Carolina Department of Health and Environmental Control (SCDHEC). *South Carolina Stormwater Management and Sedimentation control—Handbook for Land Disturbance Activities*. Columbia: SCDHEC, 1998.

U.S. Army Corps of Engineers (COE). *Flood Proofing Regulations*. Washington: COE, 1992, p. 1-1.

U.S. Department of Transportation–Federal Aviation Administration (FAA). *Airport Drainage*. FAA Advisory Circular 150/5320-5B. Washington: FAA, 1970.

Urbonas, B., and P. Stahre. *Stormwater—Best Management Practices and Detention for Water Quality, Drainage, and CSO Management*. Englewood Cliffs, NJ: Prentice-Hall, 1993.

Walesh, S. G. *Urban Surface Water Management*. New York: Wiley, 1989, pp. 195–199. Copyright © 1989 by John Wiley & Sons, Inc. Reprinted by permission of John Wiley & Sons, Inc.

Wanielista, M. P., and Y. A. Yousef. *Stormwater Management*. New York: Wiley, 1993, pp. 109, 126–128, 133–134. Copyright © 1993 by John Wiley & Sons, Inc. Reprinted by permission of John Wiley & Sons, Inc.

SUGGESTED READINGS

American Society of Civil Engineers (ASCE). *Urban Subsurface Drainage—ASCE Manuals and Reports on Engineering Practice No. 95*. Reston: ASCE, 1998.

American Society of Civil Engineers (ASCE) and the Water Pollution Control Federation. *Gravity Sanitary Sewer Design and Construction, ASCE No. 60/MOP FD-5*. New York: ASCE, 1982, pp. 109, 201–204.

Asphalt Institute. *Drainage of Asphalt Pavement Structures—MS-15*. College Park: Asphalt Institute, 1966, pp. 68, 70.

Barber, E. S. and C. L. Sawyer. "Highway Subdrainage" in *Proceedings, Thirty-First Annual Meeting U.S. Highway Research Board*, Vol 31, 1952.

Brown, L.A., Director. "Dekalb County—Drainage Procedures Manual." An unpublished manual. Atlanta: W. L. Jorden & Co., 1976, p. 4.

Campbell, C. S., and M. H. Ogden. *Constructed Wetlands in the Sustainable Landscape*. New York: Wiley, 1999.

Corbet, S., and J. King (eds.), *Geotextiles in Infiltration and Drainage*. London: Telford, 1993.

Federal Highway Administration (FHWA). *Urban Drainage Design Manual—Hydraulic Engineering Circular No. 22—FHWA-SA-96-078*. Washington: U.S. Department of Transportation, 1996.

Fortier and Scobey. *1926 Special Committee on Irrigation*. New York: American Society of Civil Engineers, 1926. Referenced material is reproduced by permission of the American Society of Civil Engineers.

Natural Resources Conservation Service (NRCS). *A Method for Estimating Volume and Rate of Runoff in Small Watersheds—SCS TP 149*. Washington: NRCS, 1973.

Overton, D. E., and M. E. Meadows. *Stormwater Modeling*. New York: Academic, 1976, p. 89.

Peckworth, H. F. *Concrete Pipe Field Manual*. Arlington: American Concrete Pipe Association, 1962, p. 189.

Portland Cement Association (PCA). *Handbook of Concrete Pipe Hydraulics*. Skokie: PCA, 1964, p. 142.

U.S. Department of Agriculture (USDA). *Soil Survey—Orangeburg County South Carolina*. Washington: USDA, 1988, p. 30.

U.S. Department of Transportation–Federal Highway Administration. Design charts for open channel flow. *Hydraulic Design Series No. 3*. Washington: Federal Highway Administration, 1973.

U.S. Department of Transportation–Federal Highway Administration. "Hydraulic Charts for Selection of Highway Culverts." *Hydraulic Eng. Circular. No. 5*, December 1965.

U.S. Environmental Protection Agency (EPA). *Protecting Natural Wetlands—A Guide to Stormwater Best Management Practices*. Washington: EPA, 1996.

Water Environment Federation and American Society of Civil Engineers (ASCE). *Urban Runoff Quality Management. WEF Manual of Practice No. 23 and ASCE Manuals and Reports on Engineering Practice No. 87*. Reston: ASCE, 1998.

8 Sedimentation and Erosion Control

As a corollary to the storm drainage system design, considerations for the control of sedimentation and erosion must be addressed. During construction natural cover is removed, leaving underlying soils exposed to erosive elements. Rain and wind can mobilize this material and transport it great distances. Control of this unwanted activity is necessary to maintain improvements as they are constructed and to eliminate adverse environmental impacts on surrounding areas. The Environmental Protection Agency (EPA), through its National Pollution Discharge Elimination System (NPDES) general permit for stormwater discharges associated with construction activities (Chapter 7), requires that best management practice (BMP) sedimentation and erosion control provisions be utilized on-site. Determining the rate of sediment accumulation in detention or retention facilities is also important for scheduling postconstruction maintenance.

Sedimentation involves three basic processes: erosion, transportation, and deposition. Erosion increases when the protecting vegetation is disturbed, the mantle of the surface soil is destroyed, or the stability of the surface water runoff system is altered.

8.1 GENERAL CONSIDERATIONS

Several topics that should be considered for minimizing or eliminating sedimentation and erosion on construction sites include

- (a) initially selecting the development site where drainage patterns, topography, and soils are suitable for development activities and
- (b) incorporating into the construction specifications (Chapter 13) provisions for
    - (1) exposing the smallest possible area for the least amount of time during construction;
    - (2) retaining topsoil for covering graded areas and protecting natural vegetation;
    - (3) including provisions for temporary planting, mulching, chemical bonding, stone surfacing, and/or structures to control runoff and protect areas subject to erosion during construction;
    - (4) accommodating increased runoff resulting from changed soil and surface conditions during construction;
    - (5) using sediment basins and other forms of silt traps to remove sediment from runoff; and
    - (6) installing permanent vegetative covering and long-term erosion protection structures as soon as practical.

Before commencing work, the contractor, in conjunction with the engineer, should locate storage areas and shops where the potential for erosion is slight. If satisfactory areas are not available, the plans and specifications should include provisions for paving and the use of erosion control practices to minimize potential erosion. In other areas the contractor may be required to saturate the ground or apply suppressors to minimize dust. In addition, if fording of streams is objectionable, the contractor may be required to install and use temporary bridges or culverts.

The contractor and project engineer should give attention to protecting potential borrow areas from erosion after the borrow has been removed. The contractor must also protect streams and the ground from chemicals, fuel, lubricants, sewage, or other pollutants and avoid disposing waste materials in drainage ways or floodplains. Other BMP devices can be utilized during construction to minimize erosion. Several alternatives are presented in the following sections (USDA-NRCS, 1997).

8.1.1   Temporary Construction Entrance

A temporary construction entrance is a stone-filled area located where construction vehicles enter public rights-of-way, streets, or parking lots. This entrance is used to reduce and eliminate mud migration from the construction site. Aggregates up to 3 inches in size can be used to construct the stone-filled mat. Entrances can exceed 50 feet in length to provide adequate space for washing the underside of vehicles prior to their departure. It may be necessary to add new stone periodically to maintain a relatively porous surface. Provisions to divert silt-laden rinse water to an appropriate trap or basin are necessary to help the mat drain properly and also prevent sediment migration.

8.1.2   Temporary Perimeter Dike

Temporary perimeter dikes are usually installed for one year or less to divert sediment laden runoff to trapping facilities. Normally dikes such as these are placed around the perimeter of an area to control runoff from small watersheds of 5 acres or less. The dike can have a 2-foot-wide top with 2 : 1 side slopes and be 18 inches tall, as shown in Figure 8.1a.

8.1.3   Temporary Diversionary Dike

Temporary diversionary dikes, illustrated in Figure 8.1b, are usually in place for one year or less to divert sediment laden runoff from running over unstabilized slopes. Typically these dikes are placed parallel to the top of a cut-or-fill slope and are constructed in a manner similar to perimeter dikes.

8.1.4   Temporary Interceptor Dike

Temporary interceptor dikes are similar to perimeter and temporary diversionary dikes except that they are placed at intervals across disturbed areas (such as cleared rights-of-way) to shorten the distance sediment laden runoff must travel before reaching a stabilized area. Figure 8.1c shows how the interceptor is constructed and placed.

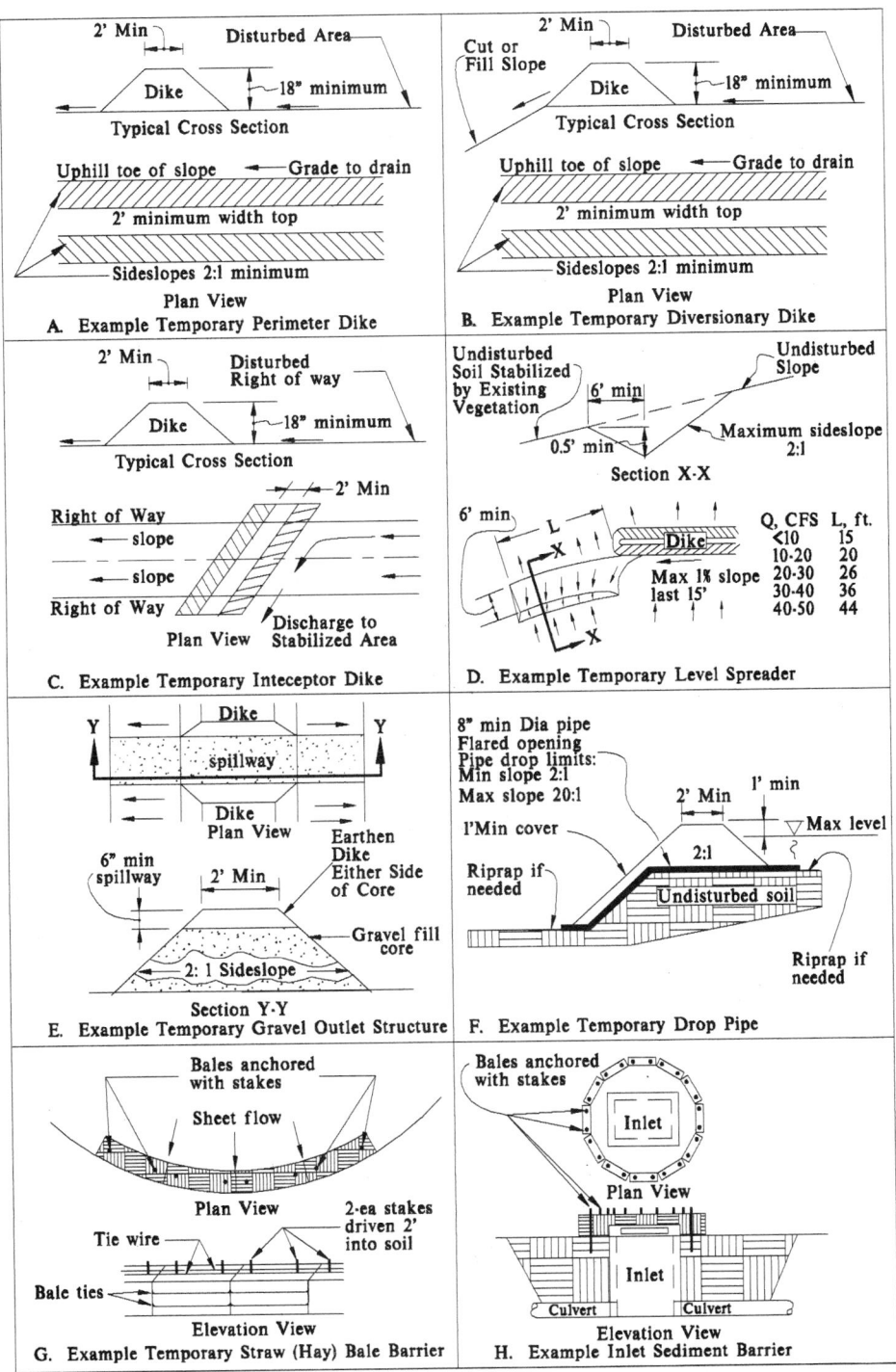

FIG. 8.1 Sedimentation and erosion BMP control devices. Adapted from USDA–NRCS, 1977.

8.1.5   Temporary Level Spreaders

Temporary level spreaders (see Figure 8.1*d*) are stormwater outlets constructed with no slope to spread concentrated discharge runoff where a nonerosive velocity can be obtained. This dispersed water should be prevented from reconcentrating until it reaches a suitable outfall. Level spreader lengths should be based on a ten-year return period for determining the peak runoff generated by the contributory area.

8.1.6   Temporary Gravel Outlet Structure

A temporary gravel outlet structure is an auxiliary facility installed as part of a diversionary, interceptor, or perimeter dike to provide for the removal of storm runoff that collects upstream while retaining the sediment. This outlet structure is usually constructed of gravel that has a maximum size of 3 inches and is well graded. Precautions must be taken to ensure that the gravel does not become clogged with trapped silt, in which case the gravel would have to be replaced. An emergency spillway should be provided to accommodate the runoff peak discharge from at least a ten-year frequency storm. Figure 8.1*e* illustrates these features.

8.1.7   Temporary Pipe Drop

Temporary pipe drops, depicted in Figure 8.1*f*, are used to convey stormwater from the top of a bank to the bottom without causing erosion. The pipe should be at least 8 inches in diameter and have the capability of accommodating the runoff from at least a five-year 24-hour duration storm. Standard flared pipe sections can be installed at each end to decrease entrance and exit velocities.

8.1.8   Straw (or Hay) Bale Barriers

Bale barriers are used to intercept and retain sediment in silt-laden runoff from small drainage areas that are usually less than half an acre in size. These barriers produce excellent results in reducing sheet erosion where large quantities of sediment are not produced. The bales are usually placed in single rows similar to that shown in Figures 8.1*g* and *h*, and anchored by wooden stakes or reinforcing steel rods that are driven into the soil. The bales must be replaced if they become clogged with silt.

8.1.9   Silt Fencing

Silt fencing provides a system of sediment control similar to bale barriers. The fencing material can be made of burlap or geotechnical fabric that is supported by wire stays fastened to posts. Care must be exercised to protect the fence from tearing and ripping during construction.

8.1.10   Temporary Sediment Trap

Temporary sediment traps are small impoundments for trapping sediment being transported by storm runoff. These traps prevent sediment from entering natural drainage systems while the construction area is unstabilized. Usually these temporary traps are constructed at the inlet of catch basins and drop inlets and require periodic cleaning during construction.

8.1.11 Sediment Basins

Sediment basins are used to trap suspended solids in storm runoff to prevent siltation in channels and reservoirs downstream. This improves the quality of stormwater runoff by removing possible pollutants contained in the sediment. The basin can be either temporary or permanent, depending on local regulations and site conditions, as well as for single or multiple purposes. A sediment basin can be designed as an integral part of a detention facility so that runoff control is simultaneous with sediment reduction. The basins can even be placed ahead of retention facilities to reduce sediment pool loadings or after a detention facility to take advantage of fairly controlled flows that will help promote quiescence in the basin. One drawback (or benefit) to placing the basin after the detention facility is that some settling may occur in the detention facility, resulting in a possible reduction in working detention volume. If this is of concern, the detention facility may be overexcavated to provide adequate storage for any sediment and cleaned along with the sediment basin.

The design of sedimentation basins can be based on empirical relationships such as the one set forth in the EPA's NPDES general permit for stormwater discharges. This permit, for example, requires a minimum of 3600 ft^3 of sediment storage be provided per acre drained on construction sites having 10 or more disturbed acres. Sediment basin design, however, usually addresses the trap efficiency of the pool, the volume/weight relationship of the sediment, and the sediment grain size distribution and storage allocation. The trap efficiency of a basin is its ability to allow sediment to settle and remain within the basin, expressed as a percentage of the delivered sediment. Most floodwater-retarding structures with permanent retarding pools have an efficiency of about 85–95 percent. For smaller basins or basins where water is not permanently impounded, a lower efficiency may exist.

When the efficiency of the basin is impaired, trapped sediment must be removed and disposed of properly, as it possibly could contain organic materials high in biochemical oxygen demands (BOD_5) or other pollutants. If the sediment basin is to be incorporated as part of a permanent lake after construction, then a minimum of 6 feet of headwater should be provided at the deep end in addition to any required volume for storage of accumulated sediment. This combined volume can be considered dead storage if the lake is to be used as a detention pond. The working volume for detention plus any flood volume requirement would be added to this dead-storage allocation to determine the total lake volume. Figure 8.2 illustrates these relationships.

Most temporary sediment basins used on construction sites have a perforated vertical pipe spillway riser for conveying normal stormwater runoff discharges and a separate emergency discharge spillway similar to that used in detention basins.

8.2 PREDICTING RAINFALL EROSION LOSS

The erosional relationship between soils and water has been studied since about 1940 and has yielded one generally acceptable method for predicting soil losses with the universal soil loss equation (USLE). According to Wischmeier and Smith (1978), the

FIG. 8.2   Volume relationships for combined flood, detention, permanent pool, and sediment storage.

USLE is an erosion model designed to predict long-term average soil losses in runoff from specific field areas using specific crop management for sheet and rill erosion under certain conditions. (It does not include the effects of concentrated runoff.) It can be applied with due caution to construction sites and other nonagricultural situations, but it does not predict deposition or sediment yields from gully, streambank, and streambed erosion. In 1990s Renard et al. (1997) issued a guide to conservation planning to support use of the revised universal soil loss equation (RUSLE) where the technology for evaluating equation factors along with new data are reflected. According to the *Agricultural Handbook No. 703* (Renard et al., 1997) the improvements include

(a)   new and (in some cases) revised isoerodent maps;

(b)   a time-varying approach to reflect freeze-thaw conditions and consolidation caused by extraction of moisture by a growing crop for the soil erodibility factor;

(c)   use of a subfactor approach for evaluating the crop management factor for cropland, rangeland, and disturbed areas;

(d)   use of a new equation to reflect slope length and steepness where the ratio of rill-to-interrill erosion is considered; and

(e)   new conservation practice values for both cropland and rangeland practices.

RUSLE Version 1.06 (Galetovic et al., 1998), however, specifically addresses soil loss relating to mined lands, construction sites, and reclaimed lands. Calculations using RUSLE are more involved than those connected with USLE; therefore, use of the developed computer program is encouraged. The RUSLE is given as

$$A = RKLSCP \tag{8.1}$$

where

$A = $ The computed spatial average soil loss and temporal average soil loss per unit area, expressed in units selected for K and for period selected for R. In practice, these are usually selected to compute A in tons per acre per year, but other units can be selected. The term A therefore represents the average loss, or detachment of soil particles, but does not determine where the particles are physically located after mobilization. Sediment yield is the amount of eroded soil that is delivered to a point in the watershed that is away from the location where the soil became detached and cannot be estimated directly by RUSLE, although Version 1.06 includes practice factor subroutines which address sediment yield.

$R =$ The rainfall and runoff erosivity factor and is the number of rainfall erosion index units, plus a factor for runoff from snowmelt or applied water where such runoff is significant. Figure 8.3 shows these values throughout the United States.

$K =$ The soil erodibility factor, the soil loss rate per erosion index unit for a specified soil as measured on a unit plot, which is defined as a 72.6-foot length of uniform 9 percent slope continuously in clean-tilled fallow. Figure 8.4 contains typical values available in National Resources Conservation Service (NRCS) soil surveys for undisturbed soils. (Note: T is the soil loss tolerance value, which indicates the rate of soil loss in tons per acre per year.)

$L =$ The slope length factor, the ratio of soil loss from the field slope length to that from a 72.6-foot length under identical conditions.

$S =$ The slope steepness factor, the ratio of soil loss from the field slope gradient to that from a 9 percent slope under otherwise identical conditions.

$C =$ The cover and management factor, the ratio of soil loss from an area with specified cover and management to that from an identical area in tilled continuous fallow.

$P =$ The support practice factor, the ratio of soil loss with a support practice such as contouring, strip cropping, or terracing to that with straight-row farming up and down slope. This factor can be used to take into account soil being deposited when a support device is encountered along the slope (e.g., terrace, sediment barrier, sediment basin) where credit is given for excessive erosion protection.

Each factor will now be examined in detail.

8.2.1 Rainfall and Runoff Factor

In the soil loss equation R represents not only the impact effect of the raindrops but also amounts and rate of runoff. This term is defined as

$$R = \frac{1}{n} \sum_{j=1}^{n} \left[\sum_{k=1}^{m} (E)(I_{30})_k \right] \tag{8.2}$$

where E is the total storm kinetic energy, I_{30} is the maximum 30-minute rainfall intensity, j is the index of number of years used to produce the average, k is the index of number of storms in a year, n is the number of years used to obtain average R (22 years preferred minimum), m is the number of storms in each year, and R is the average annual rainfall erosivity.

The storm erosivity (EI) parameter for any rainstorm is the total energy in the storm (E) times the maximum 30 minute intensity (I_{30}), where E is in hundreds of foot tons

FIG. 8.3 Average annual values of the rainfall erosion index. Source: Renard et al., 1997.

Series	Texture	K	T
Blanton	Sandy, Fine Sand	0.10	5
	Loamy Sand, Loamy Fine Sand	0.15	5
Bohicket	Silty Clay Loam, Silty Clay, Clay	0.28	5
Bonneau	Loamy Sand, Sandy, Fine Sand	0.15	5
Brevard	Loamy, Silty Loam	0.24	5
	Sandy Loam	0.15	4
	Sandy Clay Loam	0.24	4
Brogdon	Loamy Sand, Loamy Fine Sand	0.15	5
	Sandy, Fine Sand	0.10	5
Brookman	Clay Loam, Loam	0.24	4
Buncombe	Loamy Sand, Sandy	0.10	5
Byars	Silty Clay Loam, Clay Loam, Loamy	0.28	5
Cahaba	Silty Loam, Fine Sandy Loam	0.24	5
	Loamy Sand, Loamy Fine Sand	0.15	5
	Loamy	0.28	5
Cainhoy	Sandy Loam, Fine Sandy Loam	0.24	5
Candor	Sandy, Loamy Sand	0.10	5
Cantey	Sandy Loam, Fine Sandy Loam	0.24	5
	Loamy	0.28	5
Cape Fear	Loamy, Silty Loam	0.15	5
Capers	Silty Clay Loam, Clay, Silty Clay	0.28	5

Note: The soil erodibility factor K is the rate of soil loss per erosion unit index. "T" is the soil loss tolerance value which indicates the rate of soil loss in tons per acre per year.

FIG. 8.4 Typical erosion K and T values for various soil types. Source: USDA–NRCS, 1974.

per acre and I_{30} is in inches per hour. The EI is a function of the amount of rain and of all the storm's component intensities express by Laws and Parsons (1943) as

$$e = 916 + 331 \log_{10} I_{30} \qquad I < 3 \text{ in./h} \tag{8.3}$$

or

$$e = 1074 \qquad I > 3 \text{ in./h} \tag{8.4}$$

Later work by Brown and Foster (1987) yielded the following two equations that have been utilized since the publication of *Agricultural Handbook No 703* (Renard et al., 1997):

$$e = 1099 \left[1 - 0.72 \exp(-1.27i)\right] \tag{8.5}$$

where i has units of inches per hour and e has units of foot-tons per acre-inch, and

$$R = \frac{\sum_{i=1}^{j} (E I_{30})_i}{N} \tag{8.6}$$

where $(E I_{30})_i = E I_{30}$ for storm i and j = number of storms in an N-year period.

Local values can be obtained from isoerodent maps produced using Equations 8.3 and 8.4 which show points of equal rainfall erosivity, as illustrated by Figure 8.3;

however, Equations 8.5 and 8.6 (or the city database supplied with the RUSLE computer program) are more frequently used today.

Should one need to determine EI values from recording raingage records, a procedure contained in Appendix B of *Agricultural Handbook No. 703* (Renard et al., 1997) can be used along with new equations being substituted for Equations 8.3 and 8.4.

These parameters account for erosive forces associated with rainfall and runoff but not surface thaws, rain on snow, or snowmelt. In areas where these factors are of consequence (designated geographic areas include eastern Washington, north-central Oregon, northern Idaho, southeastern Idaho, southwestern Montana, western Wyoming, northwestern Utah, and northwestern Colorado), the R_{eq} must be used if the land has been disturbed within approximately the last seven-year period. In areas where the annual precipitation, P, is greater than 7.5 inches, R_{eq} can be estimated using the following relationship; otherwise for annual precipitation values of less than 7.5 inches the usual R value obtained from Equation 8.2 can be used:

$$R_{eq} = -48.4 + 7.78P^* \qquad (8.7)$$

When evaluating the R factor, one must take into account the effects of "ponded" water when dealing with flat slopes where the soil surface tends to be protected. The rainfall erosivity adjustment factor, R_c, can be estimated using Figure 8.5 where the 10-year EI storm value (taken from Figure 8.6) and erosion surface slope in percent are known.

The R factor does not indicate seasonal rainfall patterns affecting soil erosion; therefore the distribution of rain throughout the year is required. This is referred to as the erosion index distribution (EI). Records for each rainfall on specific areas have been recorded over various observation periods from which accumulated percentages of rainfall are determined. An EI distribution curve is a plot of the twice-monthly accumulated rainfall percentages based on an average value. EI distributions can vary based on geographic information contained in Figure 8.7. Twenty-four 15-day periods, corresponding to days 1 and 16 of each month, and the cumulative rainfall percent for the year are shown in Figure 8.8. To determine the amount of erosion that is expected to occur in any 15-day period, calculate the difference between successive entries in Figure 8.8.

RUSLE Version 1.06 (Galetovic et al., 1998) incorporates a city database file where EI distributions and the applicable R factor for many locations throughout the United States are stored for computational utilization. This file also includes data pertaining to average monthly precipitation, temperature, site elevation, and the number of freeze-free days per year. Thus, the city database is used by Version 1.06 to evaluate the C (Section 8.2.4) and P (Section 8.2.5) factors in addition to the R factor.

8.2.2   Soil Erodibility Factor (K)

The K factor is experimentally obtained and represents the rate of soil loss per erosion index unit as measured on a "unit plot" described in Section 8.2. This differs from the erosion rate factor A, because the A variable is a function of rainfall, slope, cover, and

---

*Taken from *Guidelines for the Use of the Revised Universal Soil Loss Equation (RUSLE) Version 1.06 on Mined Lands, Construction Sites, and Reclaimed Lands.* Office of Surface Mining and Reclamation, Denver, CO (Galetovic et al, 1998).

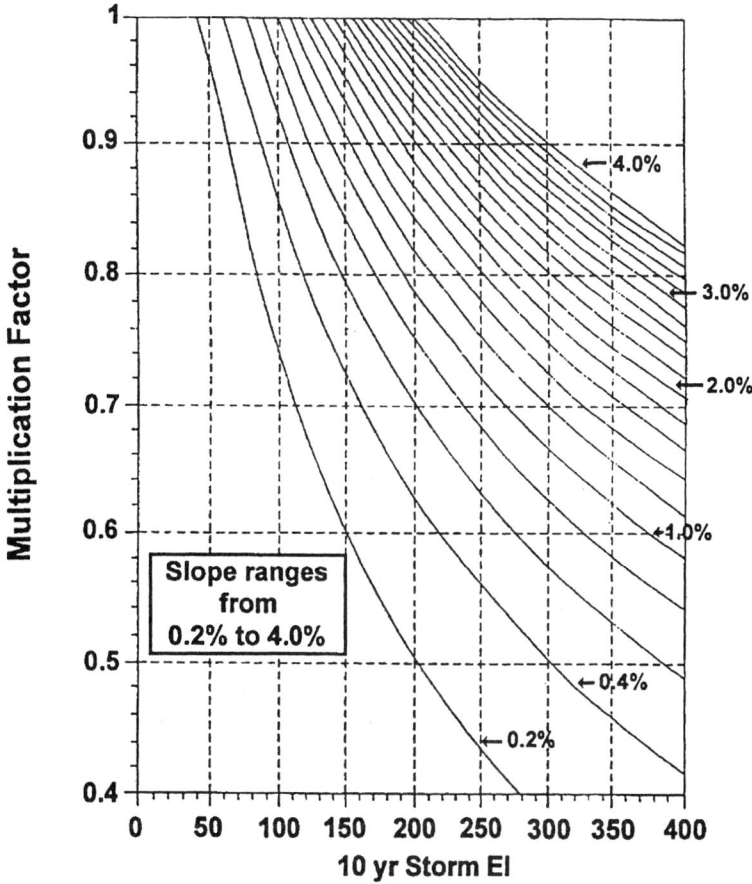

FIG. 8.5 Correction for R factor for flat slopes and large R values to reflect the amount of rainfall on ponded water. Source: Renard et al., 1997, p. 52.

management. The soil erodibility factor denotes the average long-term soil and soil profile response to the erosive effects of rainstorms resulting from soil detachment and transport under the action of rainfall droplet impact and shear created by surface flow. Localized soil deposition, surface roughness, and infiltration can affect soil erodibility.

The K values for most soil types can be obtained from tables or by the use of various computational aids similar to the chart shown in Figure 8.9. The K values available are best for medium-textured soils of the Midwest. Typical K classifications published by the NRCS include the following values: 0.02, 0.05, 0.10, 0.15, 0.17, 0.20, 0.24, 0.28, 0.32, 0.37, 0.43, 0.49, 0.55, and 0.64. For soils having the same properties at two different locations, their initial K values are the same, but the value used in RUSLE could be different due to climate and site differences. For example, soils that exist on a construction site have K values that are altered from the NRCS undisturbed value. *Agricultural Handbook 703* (Renard et al., 1997) recommends that for sodic soils, which have the ability to seal quickly resulting in a decrease in permeability, nomograph values obtained from Figure 8.9 be lowered by one or two classification values previously cited.

FIG. 8.6 Ten year frequency single-storm erosion index for eastern United States. Source: Renard et al., 1997.

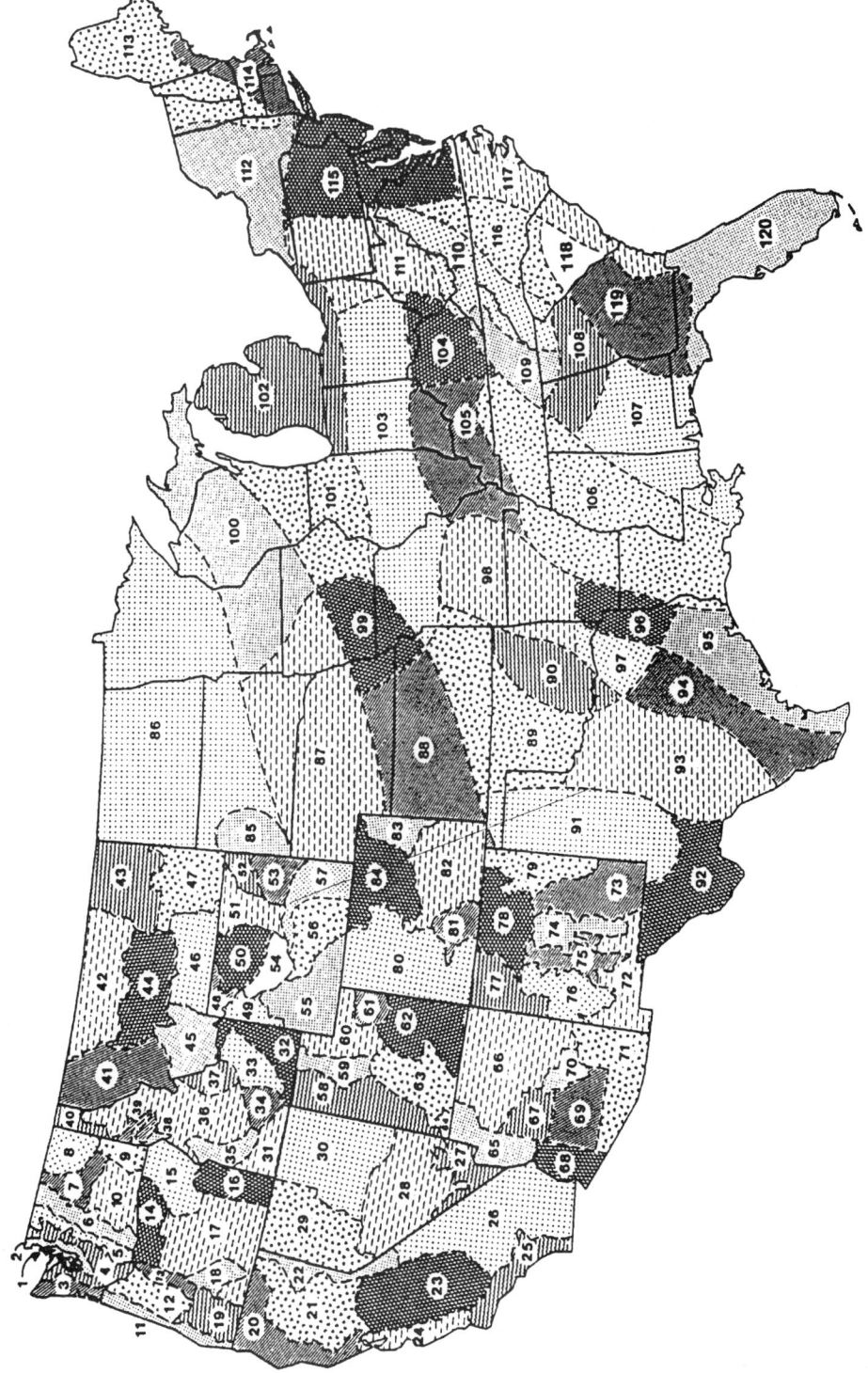

FIG. 8.7 EI distribution zones for contiguous United States. Source: Renard et al., 1997.

Periods

EI number	1	2	3	4	5	6	7	8	9	10	11	12	13	14	15	16	17	18	19	20	21	22	23	24
1	0.0	4.3	8.3	12.8	17.3	21.6	25.1	28.0	30.9	34.9	39.1	42.6	45.4	48.2	50.8	53.0	56.0	60.8	66.8	71.0	75.7	82.0	89.1	95.2
2	0.0	4.3	8.3	12.8	17.3	21.6	25.1	28.0	30.9	34.9	39.1	42.6	45.4	48.2	50.8	53.0	56.0	60.8	66.8	71.0	75.7	82.0	89.1	95.2
3	0.0	7.4	13.8	20.9	26.5	31.8	35.3	38.5	40.2	41.6	42.5	43.6	44.5	45.1	45.7	46.4	47.7	49.4	52.8	57.0	64.5	73.1	83.3	92.3
4	0.0	3.9	7.9	12.6	17.4	21.6	25.2	28.7	31.9	35.1	38.2	42.0	44.9	46.7	48.2	50.1	53.1	56.6	62.2	67.9	75.2	83.5	90.5	96.0
5	0.0	2.3	3.6	4.7	6.0	7.7	10.7	13.9	17.8	21.2	24.5	28.1	31.1	33.1	35.3	38.2	43.2	48.7	57.3	67.8	77.9	86.0	91.3	96.9
6	0.0	0.0	0.0	0.5	2.0	4.1	8.1	12.6	17.6	21.6	25.5	29.6	34.5	40.0	45.7	50.7	55.6	60.2	66.5	75.5	85.6	95.9	99.5	99.9
7	0.0	0.0	0.0	0.0	0.0	1.2	4.9	8.5	13.9	19.0	26.1	35.4	43.9	48.8	53.9	64.5	73.4	77.5	80.4	84.8	89.9	96.6	99.2	99.7
8	0.0	0.0	0.0	0.0	0.0	0.9	3.6	7.8	15.0	20.2	27.4	38.1	49.8	57.9	65.0	75.6	82.7	86.8	89.4	93.4	96.3	99.1	100.0	100.0
9	0.0	0.8	3.1	4.7	7.4	11.7	17.8	22.5	27.0	31.4	36.0	41.6	46.4	50.1	53.4	57.4	61.7	64.9	69.7	79.0	89.6	97.4	100.0	100.0
10	0.0	0.3	0.5	0.9	2.0	4.3	9.2	13.1	18.0	22.7	29.2	39.5	46.3	48.8	51.1	57.2	64.4	67.7	71.1	77.2	85.1	92.5	96.5	99.0
11	0.0	5.4	11.3	18.8	26.3	33.2	37.4	40.7	42.5	44.3	45.4	46.5	47.1	47.4	47.8	48.3	49.4	50.7	53.6	57.5	65.5	76.2	87.4	94.8
12	0.0	3.5	7.8	14.0	21.1	27.4	31.5	35.0	37.3	39.8	41.9	44.3	45.6	46.3	46.8	47.9	50.0	52.9	57.9	62.3	69.3	81.3	91.5	96.7
13	0.0	0.0	0.0	1.8	7.2	11.9	16.7	19.7	24.0	31.2	42.4	55.0	60.0	60.8	61.2	62.6	65.3	67.6	71.6	76.1	83.1	93.3	98.2	99.6
14	0.0	0.7	1.8	3.3	6.9	16.5	26.6	29.9	32.0	35.4	40.2	45.1	51.9	61.1	67.5	70.7	72.8	75.4	78.6	81.9	86.4	93.6	97.7	99.3
15	0.0	0.0	0.0	0.5	2.0	4.4	8.7	12.0	16.6	21.4	29.7	44.5	56.0	60.8	63.9	69.1	74.5	79.1	83.1	87.0	90.9	96.6	99.1	99.8
16	0.0	0.0	0.0	0.5	2.0	5.5	12.3	16.2	20.9	26.4	35.2	48.1	58.1	63.1	66.5	71.9	77.0	81.6	85.1	88.4	91.5	96.3	98.7	99.6
17	0.0	0.0	0.0	0.7	2.8	6.1	10.7	12.9	16.1	21.9	32.8	45.9	55.5	60.3	64.0	71.2	77.2	80.3	83.1	87.7	92.6	97.2	99.1	99.8
18	0.0	0.0	0.0	0.6	2.5	6.2	12.4	16.4	20.2	23.9	29.3	37.7	45.6	49.8	53.3	58.4	64.3	69.0	75.0	86.6	93.9	96.6	98.0	100.0
19	0.0	1.0	2.6	7.4	16.4	23.5	28.0	31.0	33.5	37.0	41.7	48.1	51.1	52.0	52.5	53.6	55.7	57.6	61.1	65.8	74.7	88.0	95.8	98.7
20	0.0	9.8	18.5	25.4	30.2	35.6	38.9	41.5	42.9	44.0	45.2	48.2	50.8	51.7	52.5	54.6	57.4	58.5	60.1	63.2	69.6	76.7	85.4	92.4
21	0.0	7.5	13.6	18.1	21.1	24.4	27.0	29.4	31.7	34.6	37.3	39.6	41.6	43.4	45.4	48.1	51.3	53.3	56.6	62.4	72.4	81.3	88.9	94.7
22	0.0	1.2	1.6	1.6	1.6	1.6	2.2	3.9	4.6	4.6	6.4	14.2	32.8	47.2	58.8	69.1	76.0	82.0	87.1	96.7	99.9	99.9	99.9	99.9
23	0.0	7.9	15.0	20.9	25.7	31.1	35.7	40.2	43.2	46.2	47.7	48.8	49.4	49.9	50.7	51.8	54.1	57.7	62.8	65.9	70.1	77.3	86.8	93.5
24	0.0	12.2	23.6	33.0	39.7	47.1	51.7	55.9	57.7	58.6	58.9	59.1	59.2	59.2	59.2	59.3	59.5	60.0	61.4	63.0	66.5	71.8	81.3	89.6
25	0.0	9.8	20.8	30.2	37.6	45.8	50.6	54.4	56.0	56.8	57.1	57.1	57.2	57.6	58.5	59.8	62.2	65.3	67.5	68.2	69.4	74.8	86.6	93.0

Periods

EI number	1	2	3	4	5	6	7	8	9	10	11	12	13	14	15	16	17	18	19	20	21	22	23	24
26	0.0	2.0	5.4	9.8	15.6	21.5	24.7	26.6	27.4	28.0	28.7	29.8	32.5	36.6	44.9	55.4	65.7	72.6	77.8	84.4	89.5	93.9	96.5	98.4
27	0.0	0.0	0.0	1.0	4.0	5.9	8.0	11.1	13.0	14.0	14.6	15.3	17.0	23.2	39.1	60.0	76.3	86.1	89.7	90.4	90.9	93.1	96.6	99.1
28	0.0	0.0	0.0	0.0	0.2	0.5	1.5	3.3	7.2	11.9	17.7	21.4	27.0	37.1	51.4	62.3	70.6	78.8	84.6	90.6	94.4	97.9	99.3	100.0
29	0.0	0.6	0.7	0.7	0.7	1.5	3.9	6.0	10.5	17.9	28.8	36.6	43.8	51.5	59.3	68.0	74.8	80.3	84.3	88.8	92.7	98.0	99.8	99.9
30	0.0	0.0	0.0	0.0	0.0	0.2	0.8	2.8	7.9	14.2	24.7	35.6	45.4	52.2	58.7	68.5	77.6	84.5	88.9	93.7	96.2	97.6	98.3	99.6
31	0.0	0.0	0.0	0.0	0.0	0.2	1.0	3.5	9.9	15.7	26.4	47.2	61.4	65.9	69.0	77.2	86.0	91.6	94.8	98.7	100.0	100.0	100.0	100.0
32	0.0	0.1	0.1	0.1	0.1	0.6	2.2	4.3	9.0	14.2	23.3	34.6	46.3	54.2	61.7	72.9	82.5	89.6	93.7	98.2	99.7	99.9	99.9	99.9
33	0.0	0.0	0.0	0.0	0.0	0.6	2.3	4.2	8.8	16.1	30.0	46.9	57.9	62.8	66.2	72.1	79.1	85.9	91.1	97.0	98.9	98.9	98.9	98.9
34	0.0	0.0	0.0	0.0	0.0	1.8	7.3	10.7	15.5	22.0	29.9	35.9	42.0	48.5	56.9	67.0	76.9	85.8	91.2	95.7	97.8	99.6	100.0	100.0
35	0.0	0.0	0.0	0.0	0.0	2.5	10.2	15.9	22.2	27.9	34.7	43.9	51.9	56.9	61.3	67.3	73.9	80.1	85.1	89.6	93.2	98.2	99.8	99.8
36	0.0	0.0	0.0	0.0	0.0	0.9	3.4	6.7	12.7	18.5	26.6	36.3	46.0	53.5	60.2	68.3	75.8	82.6	88.3	96.3	99.3	99.9	100.0	100.0
37	0.0	0.0	0.0	0.0	0.0	0.0	0.0	1.0	3.9	9.1	19.1	26.7	36.3	47.9	61.4	75.1	84.5	92.3	96.0	99.1	100.0	100.0	100.0	100.0
38	0.0	0.0	0.0	1.1	4.3	7.2	11.0	13.9	17.9	22.3	30.3	43.1	55.1	65.7	72.1	77.9	82.6	86.3	90.3	93.8	98.4	100.0	100.0	100.0
39	0.0	0.0	0.0	0.0	0.0	1.6	6.5	11.0	17.8	24.7	33.1	42.8	50.3	54.9	59.7	68.9	78.1	83.6	87.5	93.0	96.5	99.2	100.0	100.0
40	0.0	0.0	0.0	0.0	0.0	1.5	6.2	10.1	16.3	23.3	32.5	42.2	50.1	55.6	60.5	67.5	74.3	79.4	84.1	91.1	95.8	99.1	100.0	100.0
41	0.0	0.1	0.2	0.2	0.2	0.2	0.2	0.4	1.1	6.8	22.9	40.1	54.9	63.8	70.7	81.5	89.8	96.3	98.7	99.2	99.3	99.4	99.4	99.7
42	0.0	0.0	0.0	0.0	0.0	0.0	0.0	0.2	0.9	5.2	17.3	33.8	53.2	66.5	75.9	87.6	93.7	97.5	99.0	99.7	100.0	100.0	100.0	100.0
43	0.0	0.0	0.0	0.0	0.0	0.0	0.0	0.1	0.4	2.7	9.5	21.9	42.7	58.6	71.1	84.6	91.9	97.1	99.0	99.8	100.0	100.0	100.0	100.0
44	0.0	1.7	2.3	2.4	2.4	2.4	2.4	2.7	3.5	7.6	18.5	34.3	52.5	64.0	72.3	83.3	90.0	95.1	97.3	98.5	98.9	98.9	98.9	99.2
45	0.0	0.2	0.2	0.3	0.3	0.4	0.6	0.8	1.4	3.7	10.2	22.6	41.8	54.0	64.5	78.7	88.4	96.0	98.7	99.4	99.7	99.7	99.8	99.9
46	0.0	0.0	0.0	0.0	0.0	0.0	0.0	0.6	2.6	7.5	19.6	32.9	48.9	63.0	73.5	83.3	89.5	95.6	98.3	99.6	100.0	100.0	100.0	100.0
47	0.0	0.0	0.0	0.0	0.0	0.0	0.0	0.4	1.6	5.8	17.0	33.0	52.5	66.4	75.7	85.5	91.3	96.5	98.8	100.0	100.0	100.0	100.0	100.0
48	0.0	0.0	0.0	0.0	0.0	0.0	0.0	0.0	0.0	2.0	8.1	15.4	27.8	40.7	52.6	61.1	69.3	82.6	92.0	98.0	100.0	100.0	100.0	100.0
49	0.0	0.0	0.0	0.0	0.0	0.1	0.4	0.7	2.7	8.3	20.0	27.5	35.6	44.6	46.0	70.2	81.3	89.2	93.6	98.5	100.0	100.0	100.0	100.0
50	0.0	0.0	0.0	0.0	0.1	0.1	0.4	2.4	8.2	13.7	23.8	38.8	55.1	66.1	73.6	81.8	87.7	93.8	97.0	99.4	100.0	100.0	100.0	100.0

FIG. 8.8 EI as a percentage of average annual value for selected geographic areas shown in Figure 8.7. Source: Renard et al., 1997.

Periods

EI number	1	2	3	4	5	6	7	8	9	10	11	12	13	14	15	16	17	18	19	20	21	22	23	24
51	0.0	0.0	0.0	0.0	0.0	0.3	1.0	3.1	8.7	18.8	35.8	49.6	60.4	70.2	77.0	84.0	88.8	93.8	96.6	99.1	100.0	100.0	100.0	100.0
52	0.0	0.0	0.0	0.0	0.0	0.0	0.0	0.6	2.5	6.8	17.5	29.8	46.1	60.5	72.7	86.0	92.8	96.8	98.4	99.7	100.0	100.0	100.0	100.0
53	0.0	0.0	0.0	0.0	0.0	0.0	0.0	0.8	3.0	9.5	24.2	35.3	48.0	63.1	76.1	87.7	93.5	97.2	98.6	99.5	99.8	99.9	100.0	100.0
54	0.0	0.0	0.0	0.0	0.0	0.2	0.7	2.4	7.2	14.7	27.2	37.2	47.3	58.8	67.6	74.0	79.2	86.7	92.6	97.9	99.8	99.9	100.0	100.0
55	0.0	0.0	0.0	0.0	0.0	0.0	0.0	1.3	5.4	13.3	25.5	31.6	38.8	52.5	66.8	75.5	81.2	87.9	92.8	98.3	100.0	100.0	100.0	100.0
56	0.0	0.0	0.0	0.0	0.0	0.0	0.0	1.3	5.1	11.4	22.3	29.5	38.5	51.1	65.2	77.8	85.6	91.7	95.0	98.7	100.0	100.0	100.0	100.0
57	0.0	0.0	0.0	0.0	0.0	0.0	0.1	1.0	3.5	9.2	21.5	31.0	43.5	60.4	75.1	86.1	91.6	96.2	98.1	99.4	99.9	100.0	100.0	100.0
58	0.0	0.0	0.0	0.0	0.0	0.2	0.9	2.9	8.0	13.2	21.0	29.1	38.0	45.9	54.5	65.4	74.8	82.1	87.5	95.4	98.8	99.7	100.0	100.0
59	0.0	0.0	0.0	0.0	0.0	0.0	0.0	2.2	8.9	15.6	24.2	31.1	46.0	54.9	64.2	73.2	81.9	88.5	95.7	98.6	99.4	99.7	99.7	100.0
60	0.0	0.0	0.0	0.0	0.0	0.0	0.0	0.4	1.5	4.0	9.5	13.3	20.5	33.6	52.8	66.5	76.7	88.1	94.2	98.6	100.0	100.0	100.0	100.0
61	0.0	0.0	0.0	0.0	0.0	0.0	0.0	1.3	5.0	8.5	15.5	29.8	41.8	46.0	49.2	56.0	65.1	71.6	78.6	91.1	97.3	99.3	100.0	100.0
62	0.0	0.0	0.0	0.1	0.3	0.8	2.1	3.6	6.5	9.7	13.7	16.5	20.8	27.3	40.1	56.9	72.6	83.4	89.4	95.5	98.1	99.6	100.0	100.0
63	0.0	0.0	0.0	0.0	0.0	0.0	0.0	0.9	3.7	7.8	13.3	15.8	19.9	29.0	46.8	64.7	78.3	88.8	93.9	98.5	100.0	100.0	100.0	100.0
64	0.0	0.0	0.0	0.7	2.8	7.4	12.4	14.4	15.6	17.3	19.4	21.0	24.4	32.3	48.0	61.4	72.1	81.9	87.0	90.1	92.4	98.1	100.0	100.0
65	0.0	3.6	7.0	9.6	11.4	13.0	14.4	16.3	17.7	18.4	20.5	23.6	32.0	50.0	66.2	77.2	85.4	88.8	90.4	91.3	92.7	94.8	97.0	97.0
66	0.0	0.0	0.0	0.0	0.1	0.1	0.5	1.1	2.2	3.6	6.0	7.6	11.1	19.8	38.9	59.7	74.4	83.2	88.1	94.6	97.7	99.4	100.0	100.0
67	0.0	0.0	0.0	0.0	0.0	0.1	0.4	0.9	1.6	1.9	2.4	5.0	12.1	24.8	48.3	73.6	86.5	92.0	94.3	96.6	97.9	99.5	100.0	100.0
68	0.0	2.3	4.5	7.8	10.4	12.0	13.3	16.3	17.7	18.1	18.2	18.3	18.4	19.9	24.5	35.0	54.4	69.4	78.6	85.7	89.2	91.9	93.9	97.0
69	0.0	2.0	3.7	5.7	7.8	10.5	12.4	13.7	14.3	14.7	15.1	15.7	17.1	22.7	36.7	50.4	63.6	75.0	81.8	87.8	90.8	93.2	94.9	97.5
70	0.0	0.5	0.7	1.0	1.3	1.7	2.2	2.8	3.4	3.9	4.7	5.4	7.4	15.7	36.5	55.8	70.3	80.9	86.4	90.9	93.4	96.4	98.1	99.4
71	0.0	0.7	1.2	1.6	2.1	2.8	3.3	3.6	4.0	4.5	5.6	6.5	9.1	18.5	40.6	59.7	74.0	86.3	91.7	94.7	96.0	96.7	97.3	98.8
72	0.0	0.0	0.0	0.0	0.0	0.0	0.1	0.2	0.7	0.8	1.3	3.5	9.9	24.7	51.4	71.5	83.6	93.8	97.7	99.2	99.8	99.9	99.9	100.0
73	0.0	0.0	0.1	0.1	0.2	0.2	0.3	0.6	1.3	4.1	11.5	18.1	28.3	40.2	54.1	67.0	77.2	87.7	93.3	97.5	99.1	99.6	99.8	100.0
74	0.0	0.0	0.0	0.0	0.0	0.1	0.2	0.5	1.2	2.7	6.4	10.2	18.4	31.0	50.7	68.7	81.2	91.6	96.1	98.4	99.2	99.8	100.0	100.0
75	0.0	0.1	0.1	0.1	0.2	0.5	1.3	1.9	3.0	4.1	6.6	10.0	17.6	28.3	44.7	59.4	71.6	83.9	90.3	94.7	96.7	98.8	99.6	99.9

Periods

EI number	1	2	3	4	5	6	7	8	9	10	11	12	13	14	15	16	17	18	19	20	21	22	23	24
76	0.0	0.0	0.0	0.0	0.0	0.1	0.2	0.6	1.3	2.0	3.5	4.9	8.4	17.4	37.3	57.5	72.9	83.7	89.5	95.8	98.4	99.6	100.0	100.0
77	0.0	0.2	0.3	0.3	0.4	0.8	1.5	2.0	2.8	3.9	5.9	7.2	10.3	21.5	46.5	66.3	78.3	86.5	90.8	96.0	98.2	99.1	99.5	99.8
78	0.0	0.0	0.0	0.0	0.0	0.0	0.2	0.5	1.6	3.8	8.9	13.2	21.8	35.8	56.6	75.4	86.0	92.9	95.9	98.2	99.2	99.8	100.0	100.0
79	0.0	0.0	0.0	0.0	0.0	0.2	0.7	1.3	2.7	5.8	12.7	18.8	28.8	41.6	58.4	75.7	86.5	94.2	97.3	98.9	99.5	99.9	100.0	100.0
80	0.0	0.6	1.2	1.6	2.1	2.5	3.3	4.5	6.9	10.1	15.5	19.7	26.6	36.4	51.7	67.5	79.4	88.8	93.2	96.1	97.3	98.2	98.7	99.3
81	0.0	0.1	0.2	0.2	0.4	0.5	0.8	0.9	1.5	3.9	9.9	12.8	18.2	30.7	54.1	77.1	89.0	94.9	97.2	98.7	99.3	99.6	99.7	99.9
82	0.0	0.0	0.1	0.1	0.2	0.2	0.5	1.2	3.1	6.7	14.4	20.1	29.8	44.5	64.2	83.1	92.2	96.4	98.1	99.3	99.7	99.8	99.8	99.9
83	0.0	0.0	0.0	0.1	0.1	0.3	0.9	1.6	3.5	8.3	19.4	30.0	44.0	59.2	72.4	84.6	91.2	96.5	98.6	99.5	99.8	99.9	100.0	100.0
84	0.0	0.0	0.0	0.1	0.2	0.3	0.6	1.7	4.9	9.9	19.5	27.2	38.3	52.8	68.8	83.9	91.6	96.4	98.2	99.2	99.6	99.8	99.8	99.9
85	0.0	0.0	0.0	0.0	0.0	0.0	1.0	2.0	3.0	6.0	11.0	23.0	36.0	49.0	63.0	77.0	90.0	95.0	98.0	99.0	100.0	100.0	100.0	100.0
86	0.0	0.0	0.0	0.0	0.0	0.0	1.0	2.0	3.0	6.0	11.0	23.0	36.0	49.0	63.0	77.0	90.0	95.0	98.0	99.0	99.0	100.0	100.0	100.0
87	0.0	0.0	0.0	0.0	1.0	1.0	2.0	3.0	6.0	10.0	17.0	29.0	43.0	55.0	67.0	77.0	85.0	91.0	96.0	98.0	99.0	100.0	100.0	100.0
88	0.0	0.0	0.0	0.0	1.0	1.0	2.0	3.0	6.0	13.0	23.0	37.0	51.0	61.0	69.0	78.0	85.0	91.0	94.0	96.0	98.0	99.0	99.0	100.0
89	0.0	0.0	1.0	1.0	2.0	3.0	4.0	7.0	12.0	18.0	27.0	38.0	48.0	55.0	62.0	69.0	76.0	83.0	90.0	94.0	97.0	98.0	99.0	100.0
90	0.0	1.0	2.0	3.0	4.0	6.0	8.0	13.0	21.0	29.0	37.0	46.0	54.0	60.0	65.0	69.0	74.0	81.0	87.0	92.0	95.0	97.0	98.0	99.0
91	0.0	0.0	0.0	0.0	1.0	1.0	1.0	2.0	6.0	16.0	29.0	39.0	46.0	53.0	60.0	67.0	74.0	81.0	88.0	95.0	99.0	99.0	100.0	100.0
92	0.0	0.0	0.0	0.0	1.0	1.0	1.0	2.0	6.0	16.0	29.0	39.0	46.0	53.0	60.0	67.0	74.0	81.0	88.0	95.0	99.0	99.0	100.0	100.0
93	0.0	1.0	1.0	2.0	3.0	4.0	6.0	8.0	13.0	25.0	40.0	49.0	56.0	62.0	67.0	72.0	76.0	80.0	85.0	91.0	97.0	98.0	99.0	99.0
94	0.0	1.0	2.0	4.0	6.0	8.0	10.0	15.0	21.0	29.0	38.0	47.0	53.0	57.0	61.0	65.0	70.0	76.0	83.0	88.0	91.0	94.0	96.0	98.0
95	0.0	1.0	3.0	5.0	7.0	9.0	11.0	14.0	18.0	27.0	35.0	41.0	46.0	51.0	57.0	62.0	68.0	73.0	79.0	84.0	89.0	93.0	96.0	98.0
96	0.0	2.0	4.0	6.0	9.0	12.0	17.0	23.0	30.0	37.0	43.0	49.0	54.0	58.0	62.0	66.0	70.0	74.0	78.0	82.0	86.0	90.0	94.0	97.0
97	0.0	1.0	3.0	5.0	7.0	10.0	14.0	20.0	28.0	37.0	48.0	56.0	61.0	64.0	68.0	72.0	77.0	81.0	86.0	89.0	92.0	95.0	98.0	99.0
98	0.0	1.0	2.0	4.0	6.0	8.0	10.0	13.0	19.0	26.0	34.0	42.0	50.0	58.0	63.0	68.0	74.0	79.0	84.0	89.0	93.0	95.0	97.0	99.0
99	0.0	0.0	0.0	1.0	1.0	2.0	3.0	5.0	7.0	12.0	19.0	33.0	48.0	57.0	65.0	72.0	82.0	88.0	93.0	96.0	98.0	99.0	100.0	100.0
100	0.0	0.0	0.0	0.0	1.0	1.0	2.0	3.0	5.0	9.0	15.0	27.0	38.0	50.0	62.0	74.0	84.0	91.0	95.0	97.0	98.0	99.0	100.0	100.0

FIG. 8.8 (continued)

Periods

EI number	1	2	3	4	5	6	7	8	9	10	11	12	13	14	15	16	17	18	19	20	21	22	23	24
101	0.0	0.0	0.0	1.0	2.0	3.0	4.0	6.0	9.0	14.0	20.0	28.0	39.0	52.0	63.0	72.0	80.0	87.0	91.0	94.0	97.0	98.0	99.0	100.0
102	0.0	0.0	1.0	2.0	3.0	4.0	6.0	8.0	11.0	15.0	22.0	31.0	40.0	49.0	59.0	69.0	78.0	85.0	91.0	94.0	96.0	98.0	99.0	100.0
103	0.0	1.0	2.0	3.0	4.0	6.0	8.0	10.0	14.0	18.0	25.0	34.0	45.0	56.0	64.0	72.0	79.0	84.0	89.0	92.0	95.0	97.0	98.0	99.0
104	0.0	2.0	3.0	5.0	7.0	10.0	13.0	16.0	19.0	23.0	27.0	34.0	44.0	54.0	63.0	72.0	80.0	85.0	89.0	91.0	93.0	95.0	96.0	98.0
105	0.0	1.0	3.0	6.0	9.0	12.0	16.0	21.0	26.0	31.0	37.0	43.0	50.0	57.0	64.0	71.0	77.0	81.0	85.0	88.0	91.0	93.0	95.0	97.0
106	0.0	3.0	6.0	9.0	13.0	17.0	21.0	27.0	33.0	38.0	44.0	49.0	55.0	61.0	67.0	71.0	75.0	78.0	81.0	84.0	86.0	90.0	94.0	97.0
107	0.0	3.0	5.0	7.0	10.0	14.0	18.0	23.0	27.0	31.0	35.0	39.0	45.0	53.0	60.0	67.0	74.0	80.0	84.0	86.0	88.0	90.0	93.0	95.0
108	0.0	3.0	6.0	9.0	12.0	16.0	20.0	24.0	28.0	33.0	38.0	43.0	50.0	59.0	69.0	75.0	80.0	84.0	87.0	90.0	92.0	94.0	96.0	98.0
109	0.0	3.0	6.0	10.0	13.0	16.0	19.0	23.0	26.0	29.0	33.0	39.0	47.0	58.0	68.0	75.0	80.0	83.0	86.0	88.0	90.0	92.0	95.0	97.0
110	0.0	1.0	3.0	5.0	7.0	9.0	12.0	15.0	18.0	21.0	25.0	29.0	36.0	45.0	56.0	68.0	77.0	83.0	88.0	91.0	93.0	95.0	97.0	99.0
111	0.0	1.0	2.0	3.0	4.0	5.0	6.0	8.0	11.0	15.0	20.0	28.0	41.0	54.0	65.0	74.0	82.0	87.0	92.0	94.0	96.0	97.0	98.0	99.0
112	0.0	0.0	0.0	1.0	2.0	3.0	4.0	5.0	7.0	12.0	17.0	24.0	33.0	42.0	55.0	67.0	76.0	83.0	89.0	92.0	94.0	96.0	98.0	99.0
113	0.0	1.0	2.0	3.0	4.0	5.0	6.0	8.0	10.0	13.0	17.0	22.0	31.0	42.0	52.0	60.0	68.0	75.0	80.0	85.0	89.0	92.0	96.0	98.0
114	0.0	1.0	2.0	4.0	6.0	8.0	11.0	13.0	15.0	18.0	21.0	26.0	32.0	38.0	46.0	55.0	64.0	71.0	77.0	81.0	85.0	89.0	93.0	97.0
115	0.0	1.0	2.0	3.0	4.0	5.0	6.0	8.0	10.0	14.0	19.0	26.0	34.0	45.0	56.0	66.0	76.0	82.0	86.0	90.0	93.0	95.0	97.0	99.0
116	0.0	3.0	5.0	7.0	9.0	12.0	15.0	18.0	21.0	25.0	29.0	36.0	45.0	56.0	68.0	77.0	83.0	88.0	91.0	93.0	95.0	97.0	98.0	99.0
117	0.0	2.0	3.0	3.0	4.0	5.0	7.0	9.0	11.0	14.0	17.0	22.0	31.0	42.0	54.0	65.0	74.0	83.0	89.0	92.0	95.0	97.0	98.0	99.0
118	0.0	2.0	3.0	3.0	5.0	7.0	10.0	14.0	18.0	27.0	32.0	37.0	46.0	58.0	69.0	80.0	89.0	93.0	94.0	95.0	96.0	96.0	97.0	97.0
119	0.0	4.0	6.0	6.0	8.0	12.0	16.0	20.0	25.0	35.0	41.0	47.0	56.0	67.0	75.0	81.0	85.0	87.0	89.0	91.0	93.0	95.0	97.0	97.0
120	0.0	2.0	4.0	4.0	6.0	7.0	9.0	12.0	15.0	23.0	31.0	40.0	48.0	57.0	63.0	72.0	78.0	88.0	90.0	92.0	96.0	97.0	98.0	99.0
[2]121	0.0	8.0	16.0	25.0	33.0	41.0	46.0	50.0	53.0	54.0	55.0	56.0	56.5	57.0	57.8	58.0	58.8	60.0	61.0	63.0	66.5	72.0	80.0	90.0
122	0.0	7.0	14.0	20.0	25.5	33.5	38.0	43.0	46.0	50.0	52.5	54.5	56.0	58.0	59.0	60.0	61.5	63.0	65.0	68.0	72.0	79.0	86.0	93.0
123	0.0	4.0	8.0	12.0	17.0	23.0	29.0	34.0	38.0	44.0	49.0	53.0	56.0	59.0	62.0	65.0	69.0	72.0	75.0	79.0	83.0	88.0	93.0	96.0
124	0.0	4.0	9.0	15.0	23.0	29.0	34.0	40.0	44.0	48.0	50.0	51.0	52.0	53.0	55.0	57.0	60.0	62.0	64.0	67.0	72.0	80.0	88.0	95.0
125	0.0	7.0	12.0	17.0	24.0	30.0	39.0	45.0	50.0	53.0	55.0	56.0	57.0	58.0	59.0	61.0	62.0	63.0	64.0	66.0	70.0	77.0	84.0	92.0

Periods

EI number	1	2	3	4	5	6	7	8	9	10	11	12	13	14	15	16	17	18	19	20	21	22	23	24
126	0.0	9.0	16.0	23.0	30.0	37.0	43.0	47.0	50.0	52.0	54.0	55.0	56.0	57.0	58.0	59.0	60.0	62.0	64.0	67.0	71.0	77.0	86.0	93.0
127	0.0	8.0	15.0	22.0	28.0	33.0	38.0	42.0	46.0	50.0	52.0	53.0	53.0	53.0	53.0	54.0	55.0	57.0	59.0	63.0	68.0	75.0	83.0	92.0
128	0.0	8.0	15.0	22.0	29.0	34.0	40.0	45.0	48.0	51.0	54.0	57.0	59.0	62.0	63.0	64.0	65.0	66.0	67.0	69.0	72.0	76.0	83.0	91.0
129	0.0	9.0	16.0	22.0	27.0	32.0	37.0	41.0	45.0	48.0	51.0	53.0	55.0	56.0	57.0	57.0	58.0	59.0	61.0	64.0	68.0	73.0	79.0	89.0
130	0.0	10.0	20.0	28.0	35.0	41.0	46.0	49.0	51.0	53.0	55.0	56.0	56.0	57.0	58.0	59.0	60.0	61.0	62.0	65.0	69.0	74.0	81.0	90.0
131	0.0	8.0	15.0	22.0	28.0	33.0	38.0	41.0	44.0	47.0	49.0	51.0	53.0	55.0	56.0	58.0	59.0	60.0	63.0	65.0	69.0	75.0	84.0	92.0
132	0.0	10.0	18.0	25.0	29.0	33.0	36.0	39.0	41.0	42.0	44.0	45.0	46.0	47.0	48.0	49.0	51.0	53.0	56.0	59.0	64.0	70.0	80.0	90.0
133	0.0	8.0	16.0	24.0	32.0	40.0	46.0	51.0	54.0	56.0	57.0	58.0	58.0	59.0	59.0	60.0	60.0	61.0	62.0	64.0	68.0	74.0	83.0	91.0
134	0.0	12.0	22.0	31.0	39.0	45.0	49.0	52.0	54.0	55.0	56.0	56.0	56.0	56.0	57.0	57.0	57.0	57.0	58.0	59.0	62.0	68.0	77.0	88.0
135	0.0	7.0	15.0	22.0	30.0	37.0	43.0	49.0	53.0	55.0	57.0	58.0	59.0	60.0	61.0	62.0	63.0	65.0	67.0	70.0	74.0	79.0	85.0	92.0
136	0.0	11.0	21.0	29.0	37.0	44.0	50.0	55.0	57.0	59.0	60.0	60.0	60.0	60.0	61.0	61.0	61.0	62.0	63.0	64.0	67.0	71.0	78.0	89.0
137	0.0	10.0	18.0	25.0	30.0	39.0	46.0	51.0	54.0	57.0	58.0	59.0	59.0	60.0	60.0	60.0	61.0	62.0	63.0	64.0	67.0	72.0	80.0	90.0
138	0.0	11.0	22.0	31.0	39.0	46.0	52.0	56.0	58.0	60.0	61.0	61.0	61.0	61.0	61.0	62.0	62.0	62.0	63.0	64.0	66.0	71.0	78.0	89.0
139	0.0	8.0	14.0	20.0	25.0	32.0	37.0	42.0	47.0	50.0	53.0	55.0	56.0	58.0	59.0	61.0	63.0	64.0	66.0	68.0	71.0	76.0	85.0	93.0
[3]140	0.0	13.0	28.0	43.0	56.0	65.0	69.0	69.4	69.7	70.1	70.4	70.8	71.1	71.5	71.9	72.2	72.6	73.0	73.3	73.6	74.0	76.0	81.0	89.0

[1] Periods are 15-d beginning January 1.
[2] Zones 121-139 are for stations in Hawaii.
[3] Zone 140 is the R_{eq} distribution for Pullman, WA.

FIG. 8.8 (continued)

431

FIG. 8.9 Soil erodibility nomograph. Where the silt fraction does not exceed 70 percent, the equation is $100K = 2.1M^{1.14}(10^{-4})(12 - a) + 3.25(b - 2) + 2.5(c - 3)$, where $M =$ (percent silt + very fine sand) × (100 − percent c), $a =$ percent organic matter, $b =$ soil structure code used in soil classification, and $c =$ profile permeability class. Source: Wischmeier and Smith, 1978.

432

On construction (or reclamation) sites, guidelines for using RUSLE Version 1.06 (Galetovic et al., 1998) indicate that the K value should represent the top 6 inches (last lift) of the final fill material. For K values < 0.1, it is recommended to use only the permeability portion of Figure 8.9, and if K happens to have a value of less than 0.08, to use as a minimum, $K = 0.08$.

Soil containing subsurface rock fragments can affect the amount of infiltration that takes place and can be accounted for through the K factor. In undisturbed soil, rock fragments are usually small and thought to reduce infiltration rates, thereby increasing runoff and erosion. On construction sites, however, if topsoil controls the amount of infiltration, the influence of rock fragments positioned under the topsoil will have little effect on infiltration and the corresponding K value. Guidelines for using RUSLE Version 1.06 (Galetovic et al., 1998) recommend that, for sandy soils having K values (before adjusting for the presence of rock fragments) less than 0.15, K should be increased according to the K factor routine contained in RUSLE Version 1.06. The guidelines also recommend that, for soils having K values equal to or greater than 0.15, the effects of rock fragments in the soil profile should be ignored.

The soil erodibility factor can change throughout the year. Soils fluffed by freeze-thaw conditions and tillage often have high potential for erodibility in early spring and during periods when soil moisture is above average. Generally, K values are lower during the fall after being subjected to a summer of raindrop impact. Temporal variability adjustments to the K value are recommended by the guidelines for only areas generally located east of the 105° W meridian of longitude. For areas west of this, temporal variability relationships that are acceptable have yet to be developed. Knowing when various construction activities will take place on a job site (especially when soil is stripped and exposed during earthwork operations) is important because temporal variability in erodibility factors can alter the amount of soil loss depending on whether K is near its maximum or minimum value.

8.2.3 Topographic Factor (*LS*)

The rate of soil erosion is greatly affected by the length and steepness of the runoff slope. Both factors usually are considered together, and as a result the LS value is the ratio of soil loss from a given field slope to that of a unit plot. The slope length is the distance from the point where overland flows originate to the point where the slope gradient decreases, resulting in the deposition of materials or where the runoff water enters a well-defined channel. In the case where a concave hill slope shows no evidence of deposition along the lower slope, guidelines for the use of RUSLE Version 1.06 offers the following[*]:

> For example, assume a concave hillslope decreases from 18 percent gradient at the upper end to a 2 percent gradient at the lower end. The average gradient is 10 percent, and one-half of the average gradient is 5 percent. Thus, deposition is assumed to begin at the location where the hillslope has flatted to a gradient of 5 percent, and this would mark the endpoint for this particular length segment. On the other hand, consider a concave hillslope that decreases from a gradient of 4 percent on the upper hillslope to a gradient of

[*] Taken from *Guidelines for the Use of the Revised Universal Soil Loss Equation (RUSLE) Version 1.06 on Mined Lands, Construction Sites, and Reclaimed Lands* (Galetovic et al, 1998, pp. 4-3, 4-4).

2 percent on the lower end. In this case the average gradient is 3 percent and one-half of that is 1.5 percent. Because the gradient at the lower end of the hillslope profile (2 percent) is greater than the gradient where deposition would occur (1.5 percent), no deposition is assumed to take place on this hillslope.

The *LS* values predict average soil loss over the entire slope; however, this loss may not be evenly distributed over its total length. Figure 8.10*a* shows *LS* factors for low ratios of rill-to-interill erosion associated with rangeland and other consolidated soil conditions with cover. Figure 8.10*b* can be used to obtain *LS* factors for segments having uniform slopes at construction sites and other areas where soil conditions are greatly disturbed and there is little or no cover but not in a thawing condition and is representative of a high rill-to-interill ratio.[*] Other *LS* factor tables are available in *Agricultural Handbook 703* (Renard et al., 1997) based on

(a) moderate ratios of rill-to-interill erosion associated with row-cropped agricultural and other moderately consolidated soil conditions with little to moderate cover and

(b) thawing soils where most of the erosion is caused by surface (overland) flow.

Soils with a high percentage of silt (>85 percent) often have high rill-to-interill erosion ratios, whereas soils with silt loam can have from high to moderate rill-to-interill erosion ratios. Soils with a high percentage of sand often have a moderate to low rill-to-interill erosion ratio, while soils with a high percentage of clay (>35 percent) exhibit a low rill-to-interill erosion ratio.

The slope length factor, *L*, is defined as

$$L = (\lambda/72.6)^{m} \tag{8.8}$$

where λ is the slope length measured horizontally, not along the ground surface; 72.6 is the RUSLE unit plot length in feet; and *m* is a variable slope length exponent related to the ratio of rill-to-interill erosion by m $= \beta/(1 + \beta)$, where β represents the ratio. Figure 8.11 shows various values for *m* when considering different slope gradients and values of β.

The second part of the *LS* term affects the amount of runoff from a given runoff surface. Soil losses increase much more rapidly than runoff as the slope increases. The slope steepness factor (*S*) can be computed by

$$S = \begin{cases} 10.8 \sin \theta + 0.03 & S < 9\% \tag{8.9} \\ 16.8 \sin \theta - 0.50 & S > 9\% \tag{8.10} \end{cases}$$

where θ is the angle of slope.

For slopes less than 15 feet in length where water drains freely, the following can be used to determine *S*:

$$S = 3.0(\sin \theta)^{0.8} + 0.56 \tag{8.11}$$

---

[*] Rill erosion is caused by overland flow while interill erosion is generally caused by the impact of raindrops.

Horizontal slope length (ft)

Slope (%)	<3	6	9	12	15	25	50	75	100	150	200	250	300	400	600	800	1000
0.2	0.05	0.05	0.05	0.05	0.05	0.05	0.05	0.05	0.05	0.05	0.05	0.05	0.05	0.05	0.05	0.05	0.05
0.5	0.08	0.08	0.08	0.08	0.08	0.08	0.08	0.08	0.09	0.09	0.09	0.09	0.09	0.09	0.09	0.09	0.09
1.0	0.12	0.12	0.12	0.12	0.12	0.13	0.13	0.14	0.14	0.15	0.15	0.15	0.15	0.16	0.16	0.17	0.17
2.0	0.20	0.20	0.20	0.20	0.20	0.21	0.23	0.25	0.26	0.27	0.28	0.29	0.30	0.31	0.33	0.34	0.35
3.0	0.26	0.26	0.26	0.26	0.26	0.29	0.33	0.36	0.38	0.40	0.43	0.44	0.46	0.48	0.52	0.55	0.57
4.0	0.33	0.33	0.33	0.33	0.33	0.36	0.43	0.46	0.50	0.54	0.58	0.61	0.63	0.67	0.74	0.78	0.82
5.0	0.38	0.38	0.38	0.38	0.38	0.44	0.52	0.57	0.62	0.68	0.73	0.78	0.81	0.87	0.97	1.04	1.10
6.0	0.44	0.44	0.44	0.44	0.44	0.50	0.61	0.68	0.74	0.83	0.90	0.95	1.00	1.08	1.21	1.31	1.40
8.0	0.54	0.54	0.54	0.54	0.54	0.64	0.79	0.90	0.99	1.12	1.23	1.32	1.40	1.53	1.74	1.91	2.05
10.0	0.60	0.63	0.65	0.66	0.68	0.81	1.03	1.19	1.31	1.51	1.67	1.80	1.92	2.13	2.45	2.71	2.93
12.0	0.61	0.70	0.75	0.80	0.83	1.01	1.31	1.52	1.69	1.97	2.20	2.39	2.56	2.85	3.32	3.70	4.02
14.0	0.63	0.76	0.85	0.92	0.98	1.20	1.58	1.85	2.08	2.44	2.73	2.99	3.21	3.60	4.23	4.74	5.18
16.0	0.65	0.82	0.94	1.04	1.12	1.38	1.85	2.18	2.46	2.91	3.28	3.60	3.88	4.37	5.17	5.82	6.39
20.0	0.68	0.93	1.11	1.26	1.39	1.74	2.37	2.84	3.22	3.85	4.38	4.83	5.24	5.95	7.13	8.10	8.94
25.0	0.73	1.05	1.30	1.51	1.70	2.17	3.00	3.63	4.16	5.03	5.76	6.39	6.96	7.97	9.65	11.04	12.26
30.0	0.77	1.16	1.48	1.75	2.00	2.57	3.60	4.40	5.06	6.18	7.11	7.94	8.68	9.99	12.19	14.04	15.66
40.0	0.85	1.36	1.79	2.17	2.53	3.30	4.73	5.84	6.78	8.37	9.71	10.91	11.99	13.92	17.19	19.96	22.41
50.0	0.91	1.52	2.06	2.54	3.00	3.95	5.74	7.14	8.33	10.37	12.11	13.65	15.06	17.59	21.88	25.55	28.82
60.0	0.97	1.67	2.29	2.86	3.41	4.52	6.63	8.29	9.72	12.16	14.26	16.13	17.84	20.92	26.17	30.68	34.71

(a)

[1]Such as for rangeland and other consolidated soil conditions with cover (applicable to thawing soil where both interill and rill erosion are significant).

FIG. 8.10 Values for topographic factor, *LS*, for (*a*) low ratio of rill-to-interill erosion and (*b*) high ratio of rill-to-interill erosion. Source: Renard et al., 1997.

435

Horizontal slope length (ft)

Slope (%)	<3	6	9	12	15	25	50	75	100	150	200	250	300	400	600	800	1000
0.2	0.05	0.05	0.05	0.05	0.05	0.05	0.05	0.05	0.05	0.05	0.06	0.06	0.06	0.06	0.06	0.06	0.06
0.5	0.07	0.07	0.07	0.07	0.07	0.07	0.08	0.08	0.09	0.09	0.10	0.10	0.10	0.11	0.12	0.12	0.13
1.0	0.09	0.09	0.09	0.09	0.09	0.10	0.13	0.14	0.15	0.17	0.18	0.19	0.20	0.22	0.24	0.26	0.27
2.0	0.13	0.13	0.13	0.13	0.13	0.16	0.21	0.25	0.28	0.33	0.37	0.40	0.43	0.48	0.56	0.63	0.69
3.0	0.17	0.17	0.17	0.17	0.17	0.21	0.30	0.36	0.41	0.50	0.57	0.64	0.69	0.80	0.96	1.10	1.23
4.0	0.20	0.20	0.20	0.20	0.20	0.26	0.38	0.47	0.55	0.68	0.79	0.89	0.98	1.14	1.42	1.65	1.86
5.0	0.23	0.23	0.23	0.23	0.23	0.31	0.46	0.58	0.68	0.86	1.02	1.16	1.28	1.51	1.91	2.25	2.55
6.0	0.26	0.26	0.26	0.26	0.26	0.36	0.54	0.69	0.82	1.05	1.25	1.43	1.60	1.90	2.43	2.89	3.30
8.0	0.32	0.32	0.32	0.32	0.32	0.45	0.70	0.91	1.10	1.43	1.72	1.99	2.24	2.70	3.52	4.24	4.91
10.0	0.35	0.37	0.38	0.39	0.40	0.57	0.91	1.20	1.46	1.92	2.34	2.72	3.09	3.75	4.95	6.03	7.02
12.0	0.36	0.41	0.45	0.47	0.49	0.71	1.15	1.54	1.88	2.51	3.07	3.60	4.09	5.01	6.67	8.17	9.57
14.0	0.38	0.45	0.51	0.55	0.58	0.85	1.40	1.87	2.31	3.09	3.81	4.48	5.11	6.30	8.45	10.40	12.23
16.0	0.39	0.49	0.56	0.62	0.67	0.98	1.64	2.21	2.73	3.68	4.56	5.37	6.15	7.60	10.26	12.69	14.96
20.0	0.41	0.56	0.67	0.76	0.84	1.24	2.10	2.86	3.57	4.85	6.04	7.16	8.23	10.24	13.94	17.35	20.57
25.0	0.45	0.64	0.80	0.93	1.04	1.56	2.67	3.67	4.59	6.30	7.88	9.38	10.81	13.53	18.57	23.24	27.66
30.0	0.48	0.72	0.91	1.08	1.24	1.86	3.22	4.44	5.58	7.70	9.67	11.55	13.35	16.77	23.14	29.07	34.71
40.0	0.53	0.85	1.13	1.37	1.59	2.41	4.24	5.89	7.44	10.35	13.07	15.67	18.17	22.95	31.89	40.29	48.29
50.0	0.58	0.97	1.31	1.62	1.91	2.91	5.16	7.20	9.13	12.75	16.16	19.42	22.57	28.60	39.95	50.63	60.84
60.0	0.63	1.07	1.47	1.84	2.19	3.36	5.97	8.37	10.63	14.89	18.92	22.78	26.51	33.67	47.18	59.93	72.15

[1]Such as for freshly prepared construction and other highly disturbed soil conditions with little or no cover (not applicable to thawing soil)

(b)

FIG. 8.10 (continued)

Slope (%)	Rill/interrill ratio		
	Low	Moderate	High
0.2	0.02	0.04	0.07
0.5	0.04	0.08	0.16
1.0	0.08	0.15	0.26
2.0	0.14	0.24	0.39
3.0	0.18	0.31	0.47
4.0	0.22	0.36	0.53
5.0	0.25	0.40	0.57
6.0	0.28	0.43	0.60
8.0	0.32	0.48	0.65
10.0	0.35	0.52	0.68
12.0	0.37	0.55	0.71
14.0	0.40	0.57	0.72
16.0	0.41	0.59	0.74
20.0	0.44	0.61	0.76
25.0	0.47	0.64	0.78
30.0	0.49	0.66	0.79
40.0	0.52	0.68	0.81
50.0	0.54	0.70	0.82
60.0	0.55	0.71	0.83

[1]Not applicable to thawing soils

FIG. 8.11 Slope length exponents (*m*) for a range of slopes and rill/interill erosion classes.[1] Source: Renard et al., 1997.

Many construction sites exhibit nonuniform gradients. Some gradients steepen toward the lower end (convex slope) while others flatten toward the lower end (concave slope). An underestimate of soil movement would be attained for the convex slope, and an overestimate of movement would be attained for the concave slope by using the average slope. These slopes can be divided into uniform segments that contain uniform gradients; however, when the runoff flows from one segment to the next, the slopes must be evaluated together.

A complex slope can be divided into equal-length segments to determine a composite *LS* value. First the segments must be listed in their order of progression downstream beginning with the most elevated. Next the *LS* factor must be obtained from Figure

8.10 using the total slope length and then adjusted by multiplying it by factors from Figure 8.12. The sum gives a composite *LS* value; however, before Figure 8.12 can be utilized, the slope length exponent *m* must first be obtained using Figure 8.11.

For example, if a 600-foot convex slope has a gradient of 4 percent for the upper one-third of the slope, an 8 percent gradient for the middle one-third, and a 12 percent gradient for the lower one-third, the computed composite *LS* value would be as shown in Figure 8.13.

If various soil types are encountered, they too can be included in the calculations. If segment 1 has a *K* value of 0.25, segment 2 a *K* of 0.34, and segment 3 a *K* of 0.39, the composite *KLS* value would be as indicated in Figure 8.14.

Number of segments	Sequential number of segments	Slope-length exponent (m)								
		.05	.1	.2	.3	.4	.5	.6	.7	.8
2	1	0.97	0.93	0.87	0.81	0.76	0.71	0.66	0.62	0.57
	2	1.03	1.07	1.13	1.19	1.24	1.29	1.34	1.38	1.43
3	1	0.95	0.90	0.80	0.72	0.64	0.58	0.52	0.46	0.42
	2	1.01	1.02	1.04	1.05	1.06	1.05	1.05	1.04	1.03
	3	1.04	1.08	1.16	1.23	1.30	1.37	1.43	1.50	1.55
4	1	0.93	0.87	0.76	0.66	0.57	0.50	0.44	0.38	0.33
	2	1.00	1.00	0.98	0.96	0.94	0.92	0.88	0.85	0.82
	3	1.03	1.05	1.09	1.13	1.16	1.18	1.2	1.22	1.23
	4	1.04	1.08	1.17	1.25	1.33	1.40	1.48	1.55	1.62
5	1	0.92	0.85	0.73	0.62	0.53	0.45	0.38	0.32	0.28
	2	0.99	0.97	0.94	0.90	0.86	0.82	0.77	0.73	0.69
	3	1.01	1.03	1.04	1.05	1.06	1.06	1.06	1.05	1.03
	4	1.03	1.06	1.12	1.17	1.21	1.25	1.29	1.32	1.35
	5	1.05	1.09	1.17	1.26	1.34	1.42	1.50	1.58	1.65

Soil-loss factors = $[i^{m+1} - (i - 1)^{m+1}]/n^m$ where i = sequential number of segment, m = slope length exponent, and n = number of segments. Values are forced to give a factor total equal to number of segments. Values from RUSLE computer program may differ slightly due to round-off.

FIG. 8.12 Soil loss factor to estimate soil loss on a segment of a uniform slope. Source: Renard et al., 1997.

(1) Segment Number	(2) % Slope for Segment	(3) (Fig. 8.10(b))	(4) (Fig 8.12)	(5) [(3) x (4)]/n	(6) LS for Segment
1	4	1.42	0.56	0.27	0.80
2	8	3.52	1.05	1.23	3.70
3	12	6.67	1.51	3.36	10.07
			Total LS for Slope = 4.86		

FIG. 8.13 Example of *LS* factor determination for a nonuniform slope with a high rill-to-interill ratio.

(1) Segment Number	(2) Soil K Value	(3) (See Fig 8.13) (5) x K	(4) KLS for Segment
1	0.25	0.068	0.204
2	0.34	0.418	1.254
3	0.39	1.310	3.930
	Total KLS for Slope = 1.796		

FIG. 8.14 Example of *KLS* factor determination of a nonuniform slope with varying soil types based, in part, on values shown in Figure 8.13, column 5.

8.2.4   Cover and Management Factor (*C*)

The cover and management factor (*C*) is the ratio of soil loss from land cropped under specified conditions to a corresponding loss from clean-tilled, continuous fallow. The cover and management options instituted become part of the site's conservation plan and reflect the effects of plants, soil covers, roots, residue, and soil-disturbing activities. Because growing seasons change and ground cover varies based on management practices, the *C* factor is usually evaluated along with the seasonal rainfall patterns (EI distribution data).

The soil loss ratio (SLR) refers to the estimated loss of soil under actual conditions as compared to losses experienced under reference conditions. Cover and management practices have been categorized using various subfactors which account for the following:

(a) the impacts of previous cover and management activities: *prior land use* (PLU).

(b) the protection provided by any available vegetative canopy to the soil: *canopy cover* (CC).

(c) the effects of surface cover and roughness on erosion: *surface cover* (SC) and *surface roughness* (SR).

(d) the potential erosion resulting from varying soil moisture: *soil moisture* (SM).

RUSLE assigns a subfactor value to each of these conditions using information obtained from three included databases[*] (vegetation, operations, and city); and when multiplied together, their product gives the SLR as indicated below:

[*] Users might want to contact their respective state NRCS agronomist and climatologist for additional information to expand their databases.

$$SLR = PLU \times CC \times SC \times SR \times SM \qquad (8.12)$$

SLR values are computed over time periods when conditions are assumed to remain constant and are thus weighted by the fractional EI associated with that time period. The sum of all time periods then provide the annual or seasonal value of C.

In land developments where a permanent land cover has been established and stabilized, such as on a golf course, SLR parameters may remain fairly constant over time, where the C factor used to evaluate long-term soil loss can be based on a single annual SLR average. However, this is not the case when the development is being constructed and land covers are being established. During construction, there are periods of time when soil exposure conditions vary a great deal. Clearing and grubbing operations generally remove any canopy, surface vegetation, root systems, and soil consolidation properties. When topsoil is stripped after clearing and grubbing have been accomplished, the exposed soil is unprotected from the elements. During grading operations, the soil remains unprotected. It is not until fine grading is completed that permanent land cover can begin to be established. According to *Agricultural Handbook No. 703* (Renard et al., 1997), if no cover or management practices are employed during the time between clearing and grubbing operations and the completion of fine grading, the C factor would vary according to the following:

(a) Fill:
    (1) Packed smooth: $C = 1$.
    (2) Fresh disking: $C = 0.95$.
    (3) Rough: $C = 0.85$.
(b) Excavation:
    (1) Below the root zone: $C = 0.45$.
    (2) Grubbed with some roots: $C = 0.15$.
    (3) Grubbed with some roots and weeds: $C = 0.42$.

However, BMP activities are required as indicated in Section 8.1, and these practices (along with temporary plantings, mulching, and hydroseeding with emulsified asphalts) can reduce the C factor where soil loss during earthwork operations is much reduced. Corresponding C factors can be obtained utilizing Figures 8.15 and 8.16.

Because RUSLE bases soil loss on rainfall using percent EI distributions based on intended 15-day increments (Figures 8.7 and 8.8), the C factor during earthwork operations can be estimated using the same approach; however, the C factor must remain fairly constant during that 15-day time period. (*Note:* The second time period for February extends from day 16 through the end of the month, and the second period for any month having 31 days extends from day 16 until the end of the month.) If management operations change within the 15-day period, then that period of time must be segmented and the segments analyzed. The C factor can therefore be obtained using the following equation:

$$C = (SLR_1 \, EI_1 + SLR_2 \, EI_2 + \cdots + SLR_n \, EI_n)/EI_t \qquad (8.13)$$

Each SLR subfactor is now examined:

Mulching Factors for Construction Slopes

Type of Mulch	Mulch Rate Tons/Acre	Land Slope Percent	Factor C	Length Feet
None	0	all	1.00	----
Straw or Hay, Anchored	1	1-5	0.20	200
	1	6-10	0.20	100
	1.5	1-5	0.12	300
	1.5	6-10	0.12	150
	2	1-5	0.06	400
	2	6-10	0.06	200
	2	11-15	0.07	150
	2	16-20	0.11	100
Crushed Stone 1/4 to 1.5 in.	135	<16	0.05	200
	135	16-20	0.05	150
	135	21-33	0.05	100
	240	<21	0.02	300
	240	21-33	0.02	200
Wood Chips	7	<16	0.08	75
	7	16-20	0.08	50
	12	<16	0.05	150
	12	16-20	0.05	100
	12	21-33	0.05	75
	25	<16	0.02	200
	25	16-20	0.02	150
	25	21-33	0.02	100

FIG. 8.15 Mulch factors and length limits for construction slopes. Source: Smith, 1978.

A. *Prior Land Use* Prior land use reflects tillage practices on soil consolidation as well as the influence the residual effects of previous plantings have on soil erosion. It is expressed by the relationship

$$\text{PLU} = C_f C_b \exp\left[(-c_{ur} B_{ur}) + (c_{us} B_{us}/C_f^{C_{uf}})\right] \tag{8.14}$$

where
PLU = Prior-land-use subfactor which can range from 0 to 1. On a construction site where topsoil is stripped and stockpiled, this factor would approach unity.

C_f = Surface soil consolidation factor. C_f values vary exponentially ($C_f \approx 0.892^y$) from 1.0 for freshly tilled soil to 0.45 for soils undisturbed for seven years, although some research indicates this duration can vary. (*Note:* If part of the area being evaluated is disturbed due to tillage, then a weighted C_f must be computed by taking into account both the area disturbed, a_1, and that undisturbed, a_2, and their corresponding C_f coefficients: $C_{f\text{-weighted}} = (C_{f1}a_1 + C_{f2}a_2)/(a_1 + a_2)$.

C_b = Consolidation coefficient representing the relative effectiveness of subsurface residue where the recommended value is 0.951.

c_{ur}, c_{us} = Calibration coefficients reflecting the influence of subsurface residues where the recommended values of 0.00199 acre-in./lb for c_{ur} and 0.000416 acre-in./lb for c_{us} should be used. In areas where rainfall

C Values For Various Cover Treatments

Type	Annual Yield	C-Value
Grasses	3-Tons/Ac	0.004
Clover	2-Tons/Ac	0.006
Clover	1-Ton/Ac	0.010
Lawns-Good Condition		0.008
Lawns-Fair Condition		0.010
Lawns-Poor Condition		0.020

Type and Height of Raised Canopy	% Canopy Cover	Type*	Percent of Perennial Ground Cover					
			0	20	40	60	80	95-100
No appreciable canopy of weeds		G	0.450	0.200	0.100	0.042	0.013	0.003
		W	0.450	0.240	0.150	0.090	0.043	0.011
Canopy of tall weeds or short brush- 20 inches fall height	25	G	0.360	0.170	0.090	0.038	0.012	0.003
	25	W	0.360	0.200	0.130	0.082	0.041	0.011
	50	G	0.260	0.130	0.070	0.035	0.012	0.003
	50	W	0.260	0.160	0.110	0.075	0.039	0.011
	75	G	0.170	0.100	0.060	0.031	0.011	0.003
	75	W	0.170	0.120	0.090	0.067	0.038	0.011
Appreciable brush or bushes 6.5 feet fall height	25	G	0.400	0.180	0.090	0.040	0.013	0.003
	25	W	0.400	0.220	0.140	0.085	0.042	0.011
	50	G	0.340	0.160	0.085	0.038	0.012	0.003
	50	W	0.340	0.190	0.130	0.081	0.041	0.011
	75	G	0.280	0.140	0.080	0.036	0.012	0.003
	75	W	0.280	0.170	0.120	0.077	0.040	0.011
Trees but no appreciable low brush- 13 feet fall height	25	G	0.420	0.190	0.100	0.041	0.013	0.003
	25	W	0.420	0.230	0.140	0.087	0.042	0.011
	50	G	0.390	0.180	0.090	0.040	0.013	0.003
	50	W	0.390	0.210	0.140	0.085	0.042	0.011
	75	G	0.360	0.170	0.090	0.039	0.012	0.003
	75	W	0.360	0.200	0.130	0.083	0.041	0.011

* G- grass type cover, W- broadleaf type cover

FIG. 8.16 Cover factor values for various land uses including woodlands. Source: USDA–NRCS, 1974.

and snowmelt on frozen or thawing soils are the main sources of erosion, c_{ur} should equal 0.00398 acre-in./lb and c_{us} should equal 0.000832 acre-in./lb.

B_{ur} = Mass density of both live and dead roots located in the top inch of soil, expressed as pounds per acre per inch.

B_{us} = Mass density of surface residue found in the upper inch of soil, expressed as pounds per acre per inch.

c_{uf} = The effectiveness of incorporated residue on soil consolidation; the recommended value is 0.5.

The B_u values represent the incorporated residue and roots (biomass) near the surface that can help reduce erosion. For example, if a recent earthwork operation affects 100% of the surface to a soil depth of 6 inches, the B_u values will represent the subsurface residue and root densities to the 6-inch depth. When using RUSLE, the minimum depth value is 2 inches, which establishes the least depth limit for B_u values. Input values are allowed for root mass to the 4-inch depth; however, RUSLE provides for 80 percent of that value between the 4- and 8-inch depths.

To illustrate how B_u values are determined, assume initially 5000 lb/acre of surface residue and 750 lb/acre of root mass in the top 4 inches. This results in B_{ur} having a value of 187.5 lb/acre in. (750/4) throughout the top 4 inches and a biomass density of 150 lb/acre in. (750 × 80%/4) for the layer between depths of 4 and 8 inches. Because no surface residue has been buried, B_{us} is assumed to be zero. If an earthwork operation now disturbs 100 percent of the surface while leaving 85 percent of the surface residue on the surface, and the disturbance depth (which is also the incorporation depth) is 6 inches, two new depths are formed. The first depth extends from the surface to the 6-inch elevation, and the second extends from the 6- to the 8-inch level. The 6-inch top layer has a total root mass of 187.5 × 4 + 150 × 2, or 1050, pounds per acre, or B_{ur} = 1050/6 = 175 lb/acre in. The top layer also contains 5000 × 0.15 = 750 lb/acre of incorporated surface residue, leaving 5000 × 0.85 = 4250 lb/acre on the surface and yielding B_{us} = 750/6 = 125 lb/acre in. If this operation is immediately followed by another earthwork operation where 100 percent of the soil is disturbed 2 inches in depth and 75 percent of the residue is left on the surface, three soil layers are formed (0–2 inches, 2–6, and 6–8 inches) where new B_{ur} and B_{us} must be determined. RUSLE Version 1.06 (Galetovic et al., 1998) accounts for the biomass in each soil layer as various sequential earthwork operations are introduced.

B. *Canopy Cover* Canopy cover is any vegetation acting to shield soil from direct raindrop impact; however, this cover does not touch the ground. The canopy cover subfactor can be estimated using Figure 8.17, where the vegetation height is not the total height, but rather a height from which drops re-form on canopy and then drop to soil.

C. *Surface Cover* Surface cover can reduce soil erosion by reducing the surface area to raindrop impact and also reducing the ability of runoff to transport soil, thus causing deposition in areas that are ponded. Surface cover includes crop residue, rocks not buried, and other nonerodible material that is in direct contact with the soil. Surface cover is defined by

$$SC = \exp\left[-bS_p\left(\frac{0.24}{R_u}\right)^{0.08}\right] \tag{8.15}$$

where
  SC = The surface cover subfactor.
   b = An empirical coefficient with the following recommended values: (1) rill erosion from irrigation, snowmelt, or construction sites—use 0.050; (2) field with interill erosion—use 0.025; (3) typical cropland—use 0.035. (Note: In RUSLE Version 1.06 this term is estimated automatically as a function of, e.g., land use.)
  S_p = The percentage of land area covered by surface cover.
  R_u = The surface roughness as defined by Equation 8.17.

D. *Surface Roughness* Earthwork operations can leave both oriented and random surface roughness features. When surface patterns result, they are the result of oriented roughness where runoff can be redirected as well as providing some sediment trapping. Oriented roughness is addressed when evaluating the support practice factor, P.

Random roughness is the variation from a plane after oriented roughness is quantified. It has no pattern and results from clods and aggregates randomly oriented on

FIG. 8.17   Effect of canopy cover and canopy height on the canopy cover subfactor. Canopy cover fraction is the fraction of land covered by canopy. Source: Renard et al., 1997.

the soil surface. Figure 8.18 shows various random roughness values for various earth operations. Roughness left as a result of field operations is smoothed by the action of rainfall, where the random roughness value approaches a minimum value of 0.24 over time. The amount of degradation can be estimated as follows:

$$D_r = \exp\left[\tfrac{1}{2}(-0.14P_t) + \tfrac{1}{2}(-0.012EI_t)\right] \qquad (8.16)$$

where D_r is the dimensionless roughness decay coefficient, P_t is the total inches of rainfall since the last earthwork operation that disturbed the entire surface, and EI_t is the total EI since the last earthwork operation that disturbed the entire surface. The random surface roughness before a field operation, R_u, can be estimated using

$$R_u = 0.24 + [D_r(R_i - 0.24)] \qquad (8.17)$$

where R_u is the current random roughness, D_r is the dimensionless roughness decay coefficient from Equation 8.16, and R_i is the initial random roughness for the time period since the last earthwork operation. The surface roughness subfactor can then be defined as

$$SR = \exp\left[-0.66(R_u - 0.24)\right] \qquad (8.18)$$

E. *Soil Moisture*   Soil moisture subfactors only apply to lands in the Northwest Wheat and Range Region of the United States. In other areas, soil moisture is assumed to be unity. The reader is referred to *Agricultural Handbook No. 703* (Renard et al., 1997) for a detailed discussion on this topic.

Field operations	Random roughness (in)	Fraction of Residue left on surface (%)	Depth of incorporation (in)	Soil surface disturbed (%)
Chisel, sweeps	1.2	70	6	100
Chisel, straight point	1.5	60	6	100
Chisel, twisted shovels	1.9	45	6	100
Cultivator, field	0.7	75	3	100
Cultivator, row	0.7	80	2	85
Cultivator, ridge till	0.7	40	2	90
Disk, 1-way	1.2	30	4	100
Disk, heavy plowing	1.9	35	6	100
Disk, tandem	0.8	50	4	100
Drill, double disk	0.4	90	2	85
Drill, deep furrow	0.5	70	3	90
Drill, no-till	0.4	80	2	60
Drill, no-till into sod	0.3	90	2	20
Fertilizer applicator, anhydrous knife	0.6	80	2	15
Harrow, spike	0.4	80	2	100
Harrow, tine	0.4	85	2	100
Lister	0.8	20	4	100
Manure injector	1.5	50	6	40
Moldboard plow	1.9	5	8	100
Mulch treader	0.4	75	2	100
Planter, no-till	0.4	85	2	15
Planter, row	0.4	90	2	15
Rodweeder	0.4	90	2	100
Rotary hoe	0.4	85	2	100
Vee ripper	1.2	80	3	20

From AH-703 - Predicting Soil Erosion by Water: A Guide to Conservation Planning with the Revised Universal Soil Loss Equation

FIG. 8.18   Attributes of typical tillage implements including estimates of random roughness. Source: Renard et al., 1997.

8.2.5   Support Practice Factor (*P*)

The support practice factor (*P*) is the ratio of soil loss using a specific type of land-disturbing scheme as compared to up- and downhill cultivation. Support practice factors reflect modifying patterns of flow, changing grades, redirecting where runoff travels, reducing runoff, and curtailing the runoff rate. Contouring, terracing, subsurface drainage, and strip cropping are methods of providing this type of support. In addition, the use of vegetation strips, deep ripping, and diversions positioned on or near the contour can be used to collect and store moisture, thus reducing the amount of runoff. On construction sites BMP structures such as bales of hay, gravel filters, silt fences, berms, and bench terraces are also considered part of the support activities influencing P. RUSLE allows one to differentiate between *frequently disturbed* and *infrequently disturbed* land options. The frequently disturbed option considers the use of permanent barriers, strips of close-growing vegetation, and concave hillside profiles. The infrequently disturbed option provides for consideration of mechanical disturbances. Both options, however, allow contouring and terracing to be considered. The following support practices have been categorized as subfactors of P:

(a)   Contouring and terracing.

(b)   Sediment control barriers and structures. These practices can include cross-slope strip cropping, buffer strips, filter strips, straw-bale barriers, gravel filters, sand bags, large-scale roughness, silt fences, berms, rock check-dams, and sediment basins which have been earlier categorized as BMPs. These types of structures installed on the contour cause water to pond resulting in the deposition of sediment. Their effectiveness is directly related to the length and volume of the water that is ponded based on the gradient of the slope. Figure 8.19 shows the width of pond that can be used to compute P values pertaining to sediment control barriers. The values are given in terms of percent of hillslope length above the barrier. RUSLE uses the width of the barrier strip plus the width of the pond obtained from this figure. Some typical P values associated with sediment control barriers are shown in Figure 8.20.

(c)   Subsurface drainage structures.

The overall P factor is the product of the individual subfactors used. Each subfactor will now be further examined:

A. *Contouring* Contouring subfactors associated with contour tillage can be obtained from Figure 8.21. Storm severity, estimating soil loss when slope lengths exceed critical slopes, and off-grade contouring factors are included in *Agricultural Handbook No. 703* (Renard et al., 1997) when refinements are needed for the contouring subfactor.

Effective width of barrier as a percent of hillslope length				
Hillslope Gradient (%)	Close-growing grasses	Straw bales, Gravel, Filter barriers	Stiff-grass hedges	Silt fences and berms
<5	5	8	12	15
5-10	3	5	8	10
10-15	2	3	4	5

FIG. 8.19   Width of pond used to compute P values for sediment control barriers. Values are given as a percent of hillslope length above the barrier. The width used in RUSLE is the width of the barrier strip plus the width of the pond obtained from this figure. Source: Galetovic et al., 1998.

Structure Type

Gradient %	Shortgrass Strip	Gravel Bag	Stiff Grass Hedge	Silt Fence
<5	0.37	0.21	0.11	0.08
5-10	0.55	0.37	0.21	0.15
10-15	0.67	0.55	0.45	0.37

FIG. 8.20 Some typical P values for barriers constructed on a silt loam soil at Lexington, Kentucky. Source: Galetovic et al., 1998.

The effectiveness of contouring diminishes on long slopes and the limit, referred to as the critical slope length, can be estimated using the following relationship:

$$\lambda_c = \frac{20182 n_t^{1.5}}{s^{1.1667} Q} \tag{8.19}$$

where λ_c is the critical slope length, n_t is Manning's coefficient of roughness, found in Figure 8.22, s is the slope expressed as the sine of the slope angle, and Q is the runoff amount from the 10-year storm EI.

To use Figure 8.22, one must first identify the type of site cover management conditions that are used in RUSLE as shown in Figure 8.23. The effective P factor can be computed for distances up to λ_c, where

$$P_{\text{eff}} = \left[1 - x_c^{m+1}(1 - P) \right] \tag{8.20}$$

where P_{eff} is the effective P subfactor for the average soil loss having a slope length $< \lambda_c$, x_c is the distance along the slope length where contouring is assumed to be fully effective in feet, m is as defined in Figure 8.11, and P is as defined in Figure 8.21.

RUSLE Version 1.06 (Galetovic et al., 1998) can generate P values which account for the various conditions presented here.

B. *Terracing* Terraces are used to reduce the length of slope into shorter increments (already reflected in the LS term) and at the same time help trap sediment eroded from areas between terraces. Terracing is often used in conjunction with contouring. The terrace P subfactor reflects both the benefit of deposition and the amount of sediment deposited being quantified into two categories when evaluating construction (reclamation) projects. When considering the first category, a portion of any sediment deposition in the terrace results in a credit being given because soil displacement is minimized. This credit reduces as the spacing between terraces increases to a point where those spaced at 300 feet receive negligible credit. When considering the second category, sediment yield from uniform-grade terraces can be estimated by

$$P_y = \begin{cases} 0.1e^{2.4s} & s < 0.9\% \tag{8.21} \\ 1.0 & s \geq 0.9\% \tag{8.22} \end{cases}$$

where P_y is the sediment delivery factor, s is the terrace slope grade in percent, and $e = 2.71828$.

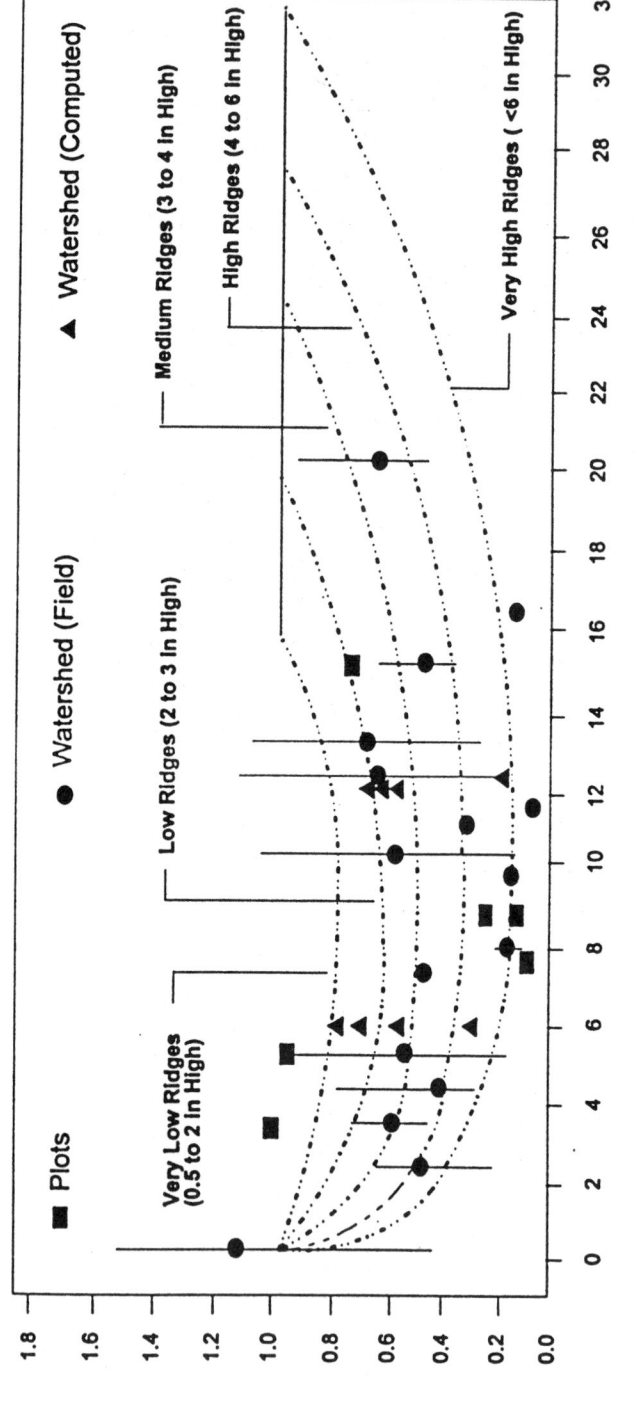

FIG. 8.21 Effects of contouring on soil loss. Source: Renard et al., 1997.

448

Cover-management condition[1]	Manning's n_t
C1	0.200
C2	0.110
C3	0.070
C4	0.040
C5	0.023
C6	0.014
C7	0.011

[1] Refer to Fig. 8.23 for a description of cover-management conditions.

FIG. 8.22 Values of Manning's n_t used in RUSLE cropland conditions (Renard et al., 1997).

The subfactor for terracing can be determined from the equation

$$P = 1 - B(1 - P_y) \qquad (8.23)$$

where P is the support practice factor for terracing, and B is the benefit assigned to deposition as indicated in Figure 8.24. (*Note:* The quantity $1 - P_y$ represents the amount of deposition that is comparable to the M factor in stripcropping computations.) Figure 8.25 shows terrace values that can be used for conservation planning where both the benefit of deposition and the amount of sediment deposited are considered. The loss of soil from areas positioned between terraces minus any soil deposited yields the net soil loss. Figure 8.26 shows values for estimating sediment yield from terraces.

C. *Cross-Slope Stripcropping, Buffer Strips, and Filter Strips* The practice of stripcropping relies on the use of strips of clean-tilled crops alternated with strips of closely growing vegetation such as grasses. Buffer strips consist of perennial vegetation oriented across the slope, and are comprised predominately with grass species that are left in place for extended periods or even permanently. Vegetated filter strips are positioned at the base of slopes in bands, while those strips located near streams or lakes are referred to as riparian filter strips. Because neither of these strip types trap sediment along the slope, they do not affect the P factor appreciably.

Because the P subfactor evaluates practice activities from the standpoint of soil erosion and conservation, the lowest P-subfactor values are achieved when stripcropping measures are employed. This is because most soil is trapped and deposited within a short distance (most often in the next downhill strip) from where it was mobilized. Buffer strips provide moderate P-subfactor values if the strip effectively reduces the transport capacity of the runoff as it leaves the strip; otherwise it is treated as giving no benefit. Base slope vegetative filter strips and riparian strips have the highest subfactors (1.0) because they are intended to reduce the amount of sediment reaching off-site rivers, streams, and lakes.

Categories of conditions	Description of condition
C1. Established meadow (very dense cover)	Grass is dense and runoff is very slow, about the slowest under any vegetative condition. Becomes condition 2 when mowed and baled.
C2. 1st yr meadow, hay (moderately dense cover)	Hay is a mixture of grass and legume just before cutting. Meadow is a good stand of grass that is nearing the end of 1st yr. Becomes condition 4 when mowed and baled.
C3. Heavy (dense) cover or very rough or both	Ground cover for this condition is about 75-95%. Roughness is like that left by a high-clearance moldboard plow on a heavy-textured soil. Roughness depressions appear 7 in or more deep. Vegetative hydraulic roughness like that from a good legume crop (such as lespedeza) that has not been mowed.
C4. Moderate cover or rough or both	Ground cover for this condition is about 40-65%. Roughness is like that left by a moldboard plow in a medium-textured soil. Depressions appear 4-6 in deep. Vegetative hydraulic roughness is similar to that produced by winter small grain at full maturity.
C5. Light cover or moderate roughness or both	Ground surface cover is 10-30%. Surface roughness is like that left by first pass of tandem disk over a medium-textured soil that has been moldboard plowed. This roughness could also be similar to that left after a chisel plow through a medium-textured soil at optimum moisture conditions for tillage. Roughness depressions appear 2-3 in deep. In terms of hydraulic roughness produced by vegetation, this condition is similar to that produced by spring small grain at about 3/4 maturity.
C6. No cover or minimal roughness or both	This condition closely resembles the condition typically found in row cropped fields after the field has been planted and exposed to a moderately intense rainfall. Ground cover is less than about 5%. Roughness is like that of a good seedbed for corn or soybeans. Surface is rougher than that of a finely pulverized seedbed for seeding vegetables.
C7. Clean-tilled, smooth, fallow	Surface is essentially bare, 5% or less of cover. Soil has not had a crop grown on it in the last 6 mo or more, so much of the residual effects of previous cropping has disappeared. Surface is smooth, similar to the surface that develops on a very finely pulverized seedbed exposed to several intense rainfalls. This condition is most likely found in fallow and vegetable fields.

FIG. 8.23 Description of cover management conditions used in RUSLE for estimating P-factor values. Source: Renard et al., 1997.

The P subfactor for strips can be determined using the equation

$$P_s = \frac{g_p - B}{g_p} \tag{8.24}$$

where P_s is the P subfactor for strips, g_p is the sediment load at the end of the slope that would occur if the strips caused no deposition, and B is the credit for deposition (see Figure 8.24). *Agricultural Handbook 703* (Renard et al., 1997) describes more fully four cases when evaluating each strip while sediment is routed from the top of the slope to the bottom:

1. The entire strip experiences net erosion.
2. The entire strip experiences net deposition.
3. The upper strip portion experiences net deposition, the lower net erosion.
4. The strip terminates runoff where no sediment or runoff leaves.

Terrace spacing (ft)	Benefit (B)
≤110	0.5
125	.6
160	.7
200	.8
260	.9
≥300	1.0

FIG. 8.24 Benefit assigned to deposition behind terraces. Source: Renard et al., 1997.

Horizontal terrace interval (ft)	Terrace P-factor values			
	Closed outlets[2]	Open outlets, with percent grade of[3]		
		0.1-0.3	0.4-0.7	>0.8
Less than 110	0.5	0.6	0.7	1.0
110-140	.6	.7	.8	1
140-180	.7	.8	.9	1
180-225	.8	.8	.9	1
225-300	.9	.9	1	1
More than 300	1	1	1	1

[1]Multiply these values by other P-subfactor values for contouring, stripcropping, or other support practices on interterrace interval to obtain composite P-factor value.

[2]Values for closed-outlet terraces also apply to terraces with underground outlets and to level terraces with open outlets.

[3]Channel grade is measured on the 300 ft of terrace closest to outlet or 1/3 of total length, whichever distance is less.

FIG. 8.25 Terrace P-factor values for conservation planning. Source: Renard et al., 1997.

RUSLE's frequently disturbed P-factor option allows one to take into consideration the effects of buffer strips where their coverage is greater than 5 percent of the slope length. When the effective widths for stiff-grass hedges are considered, Version 1.06 guidelines recommend using silt fence values.

Terrace grade	Sediment delivery subfactor
%	
Closed outlet	[2]0.05
0 (level)	.10
.1	.13
.2	.17
.4	.29
.6	.49
.8	.83
.9	1
[3]>1	1

[1]Includes terraces with underground outlet.
[2]From Foster and Highfill 1983. All other values from $P_y = 0.1e^{2.64g}$, where e = natural logarithm and g = terrace grade (%).
[3]Potential for net erosion in terrace channels, depending on flow hydraulics and soil erodibility in the channels. If net erosion occurs, P_y >1.

FIG. 8.26 Sediment delivery subfactor (P_y) for terraces. Source: Renard et al., 1997.

D. *Straw-Bale Barriers* Straw-bale barriers can be effective in trapping small amounts of sediment being transported by sheet and rill flows when kept in good working order. Initial trap efficiency may be as high as 95 percent; however, this initial value must be adjusted to account for some failure during operation. When evaluating this subfactor, use the RUSLE Version 1.06 permanent barriers, strips, and concave hillslope profile routine to estimate straw-bale barrier sediment delivery ratios.

E. *Gravel-Filter and Sand-Bag Barriers* Gravel-filter and sand-bag barriers offer temporary measures to trap sediments. Use the initial P value obtained for silt fences and adjust for the amount of expected failure. Figure 8.20 shows some typical P values for the gravel bag practice.

F. *Silt Fences* The effectiveness of a silt fence is similar to a straw-bale barrier where a properly installed and maintained fence can provide a low P value as indicated in Figure 8.20. When considering silt fence subfactors, the frequently disturbed RUSLE P option must be used where an initial P value based on contouring with zero furrow grade is determined while also selecting the vegetation strips or concave slope option. Next the P subfactor for a *permanent barriers, strips, and concave hillslope* profile is determined by using two strips where the width of the strip is selected from

Figure 8.19. Finally, choose the appropriate cover management condition for the land surface up-slope of the silt fence and a cover condition of 1 for the silt fence lower strip.

G. *Continuous Berms* Determining P values for berms (temporary perimeter or diversionary dikes) that are in good working order follows the same procedure as outlined above for silt fences.

H. *Rock Check Dams* Check dams made of rock, bushes, or brush can be used to reduce runoff velocities and provide some sediment trapping. For these practices P values can be estimated using the procedures applicable to sediment basins.

I. *Bench Terraces* As part of land development site improvements, bench terraces can be use to improve the aesthetics of roads and streets. These terraces can slope either outward toward the roadway surface or backward toward the hill slope. When considering outward-sloping bench terraces, the slope length is measured from the top of the slope. The P values are determined using the same procedure as outlined for barriers above. In addition, a weighted C value is needed and the irregular hillslope procedure is used to find the LS value. Backward-sloping bench terraces are treated where the interterrace interval is considered as an individual hillslope length when determining P.

J. *Sediment Basins* The effectiveness of sediment basins is estimated using the RUSLE terracing P-subfactor routine where the basin is treated as a closed-outlet terrace. Sediment delivery ratios for sediment basins are functions of the various particle sizes entering the basin. Figure 8.27 shows various sediment delivery ratios for well-designed sediment basins that are in good working order. As particle sizes decrease, the sediment delivery ratio increases because settling velocities are reduced. When selecting sediment particle sizes, one must be careful when considering a concave slope or a sediment control barrier because coarser particles may be deposited prior to reaching the basin.

Up-slope influences on the sediment delivery ratio for a sediment basin are depicted in Figure 8.28. Information contained in this figure can be used to adjust sediment delivery ratios computed by RUSLE for sediment basins or graded terrace channels.

K. *Subsurface Drainage* According to *Agricultural Handbook No. 703* (Renard et al., 1997), limited field data exist that quantify the effectiveness of subsurface drainage by reducing overland flow and erosion. Although some studies indicate an average value of 0.6 might be appropriate, it is stated in the handbook that more research is needed.

Because RUSLE Version 1.06 (Galetovic et al., 1998) allows one to consider sediment yields as part of the P-factor routines, the effect of any deposition resulting from the presence of concave slopes must be addressed (subject to conditions presented in Section 8.2.3 when the location of deposition is not evident). When the amount of sediment exceeds the transport carrying capacity of the runoff on a concave slope, deposition can take place. This frequently occurs at the base of a slope where the grade flattens. The lower end grade controls the amount of sediment being deposited and thus the amount of sediment still mobilized. RUSLE's P-factor frequently disturbed option provides an opportunity to define slopes using up to ten segments. Because the lower portion of the slope profile defines the amount of deposition, care must be taken to segment the concave slope where the lower portion is defined using at least four segment intervals. The degree of concavity, which is defined as the ratio of an individual

Soil texture	Sediment delivery ratio
Sand	0.01
Loamy sand	0.02
Sandy loam	0.03
Loam	0.05
Silt loam	0.06
Silt	0.07
Sandy clay loam	0.06
Clay loam	0.08
Silty clay loam	0.09
Sandy clay	0.10
Silty clay	0.12
Clay	0.14

FIG. 8.27 Sediment delivery ratios for sediment basins that are well designed, constructed, and maintained with full sediment storage capacity. Source: Galetovic et al., 1998.

Soil texture on upslope area producing sediment	Sediment delivery ratio of concave hillslope or barrier	Sediment delivery ratio for sediment basin	Sediment delivery ratio of second sediment basin in series
Silt loam	0.10	0.47	0.84
	0.50	0.11	0.75
High clay	0.10	0.90	0.90
	0.50	0.33	0.90
High sand	0.10	0.29	0.86
	0.50	0.06	0.84

FIG. 8.28 Effect of concave hillslope segments, sediment control barriers, and basin sequences on the effectiveness of sediment basins. Source: Galetovic et al., 1998.

segment's gradient to the entire slope's average gradient (e.g., uniform slope's degree of concavity ratio equals unity), also influences the sediment delivery ratio—as the degree of concavity increases, the sediment delivery ratio decreases. Cover management activities can also influence sediment delivery ratios associated with concave slopes. RUSLE provides the user an opportunity to take the effects of cover management into account when evaluating concave slopes.

In summary, RUSLE Version 1.06 allows the user to account for sediment yield with the *P*-factor subroutine, where the contouring subfactor is first determined, and then the effects of BMP devices, such as diversions or sediment basins, are estimated using the terrace function. Version 1.06 guidelines warn that if a diversion or terrace is located below an erosion control structure, such as a stiff-grass hedge or downslope of a concave hillslope element, the terrace subfactor should not be used. When barriers, strips, and concave hill slopes are involved, their individual effects are combined into a single P subfactor. The sediment delivery ratio shown on the RUSLE worksheet is the product of the contour P subfactor and the sediment delivery ratio (SDR) for sediment trapping efficiencies of the BMPs utilized, where the trapping efficiency equals 1 minus the SDR.

8.2.6 Erosion from Concentrated Flows

Erosion on construction sites that is not attributable to the sheet or rill erosion accounted for in RUSLE can be the result of ephemeral gully erosion, permanent gully erosion, and stream channel erosion.

Ephemeral gully erosion forms where runoff concentrates and produces features that are temporary in nature and are obscured during construction operations such as fine grading. The amount of erosion attributed to this, if left uncorrected, can be significant. This type of erosion produces cross sections that are usually wide in relation to the depth, while head cuts are not readily visible.

Ephemeral erosion can be effectively reduced by controlling runoff. The rate of runoff must be kept within the critical shear value of the soil; otherwise erosion will occur. Ephemeral gully erosion usually can be prevented by establishing temporary grassed waterways in areas susceptible to this type of activity (Laflen et al., 1986). Should ephemeral gully erosion continue unchecked, permanent gullies will form; therefore, effective management of this type of erosion is paramount.

Should permanent gully erosion occur, well-defined drainage walls with narrow cross sections are formed. Permanent gullies usually have steep side slopes that progress to the head of the gully, resulting in a prominent head cut. Permanent gully channels advance upstream if uncorrected.

Although there are methods available for predicting erosion caused by concentrated flows (including stream channel erosion), contract specifications should provide for preventive measures to inhibit the formation of ephemeral gullies (Watson et al., 1986). If they form, construction specifications should provide for rectification activities. With stringent field control, erosion from concentrated flows should be minimal.

8.3 CONSTRUCTION SITES

Since RUSLE yields anticipated average soil losses on an annual basis, EI values may be required to obtain R for construction durations of less than a year. Erosion calculations should also address the various subsoil horizon K factors that will be encountered during construction excavation. If chemical additives are to be applied to the soil to make it less erodible, the K values can be adjusted to reflect the improved performance. By analyzing various applications of the RUSLE to a construction site, the production of erosion can be studied by the design engineer for determining alternatives that minimize erosion potential and include BMPs contained in Section 8.1.

8.3.1   Upslope Contributions to Watershed Sediment Yield

The gross erosion within a watershed is the sum of all erosion that has taken place, including sheet, rill, gully, streambed, and streambank erosion. Not all eroded material calculated by RUSLE leaves the site, as eroded soils often are redeposited after only a short travel distance. To ascertain the amount of sediment delivered to any point within the system, the sediment delivery ratio must be determined using the P-factor routine contained in RUSLE Version 1.06. This information is necessary when calculating the amount of sediment being delivered to a pond, lake, or basin.

8.3.2   Determination of Sediment Production for Ponds, Lakes, and Basins

RUSLE Version 1.06 can be used to estimate the production of sediment from a completely denuded (not recommended in practice but used to illustrate maximum sediment production) construction site. To illustrate an example using the site depicted in Figure 4.58, the following parameters apply:

> Construction site area denuded (assuming for this example that site was completely cleared and grubbed: 5.13 acres minus 140 × 140-foot detention pond site) = 4.68 acres.
>
> Exposed soil horizon (see Figure 8.4 for reference) = Cahaba.
>
> Slope length = 340 feet.
>
> Duration of exposure = 12 months.
>
> Slope = 0.5 percent.
>
> Assumed R value = 400.

First the following factors must be determined:

$$K = 0.28 \text{ for Cahaba series from Figure 8.4}$$
$$LS = 0.10 \text{ from Figure 8.10}b \text{ (estimate high rill-to-interill erosion}$$
$$\text{because site has been cleared and grubbed)}$$
$$C, P = 1$$

Using Equation 8.1, the following calculation is made:

$$A = 400 \times 0.28 \times 0.10 = 11.20 \text{ tons per acre per year}$$

This gives the total annual runoff-related erosion estimated for this site as 52.4 tons per year. To convert tons per acre to cubic yards, use the factors contained in Figure 8.29. Since the Cahaba series falls into the loam category, an estimated conversion factor of 0.87 is used to determine the number of cubic yards lost. The total annual volume equates to 0.87 × 52.4, or 45.6, cubic yards.

If the area is only to be exposed between October 15 and May 1 of the following year, then the proportional amount of erosion is determined as a percentage of the annual energy times the intensity expected between these dates, which can be obtained by

Factors To Convert Eroded Materials From Weight To Volume	
Texture	Factor
Sands, Loamy Sands, Sand Loam	0.70
Sand Clay Loam, Silt Loams, Loams, and Silty Clay	0.87
Clay Loams, Sandy Clays, and Silty Clays	1.02

FIG. 8.29 Factors to convert tons per acre to cubic yards per acre of eroded materials. Source: USDA–NCRS, 1974.

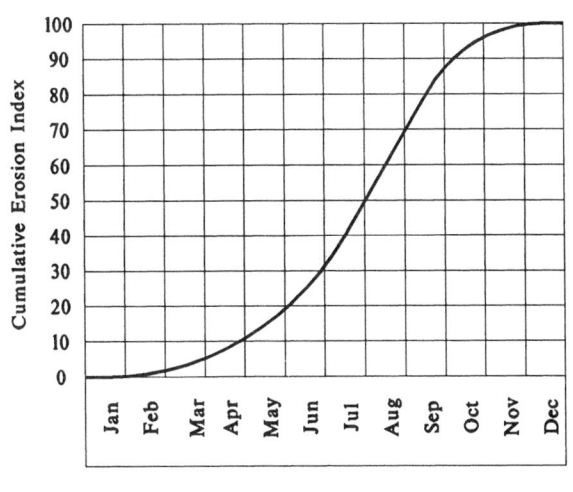

FIG. 8.30 EI graph for geographic area no. 117 of Figure 8.7.

using a graph of the cumulative erosion index. For this example geographic area 117 was chosen in Figure 8.7, and the resulting EI graph is Figure 8.30:

EI October 15	92
EI May 1	11
Total EI	$(100 - 92) + 11 = 19$

$$\text{Erosion}_{15 \text{ Oct}-1 \text{ May}} = 0.19 \times 45.6 = 8.7 \text{ cubic yards}$$

In this example, since C and P are equal to unity and P subfactors are not being used to estimate sediment delivery ratios, the amount of sediment delivered to the low elevation of the site can be estimated using sediment delivery ratio curves similar to those shown in Figure 8.31. This example's contributory area is 4.68 acres; therefore, the delivery ratio is 0.27 if the site is assumed to be located in coastal flatland. The corresponding amount of sediment delivered to the low area is $0.27 \times 45.6 = 12.3$ cubic yards per year.

Since the buoyant unit weight of the saturated sediment is less than the moist unit weight, the sediment basin must have adequate volume to accommodate this changed

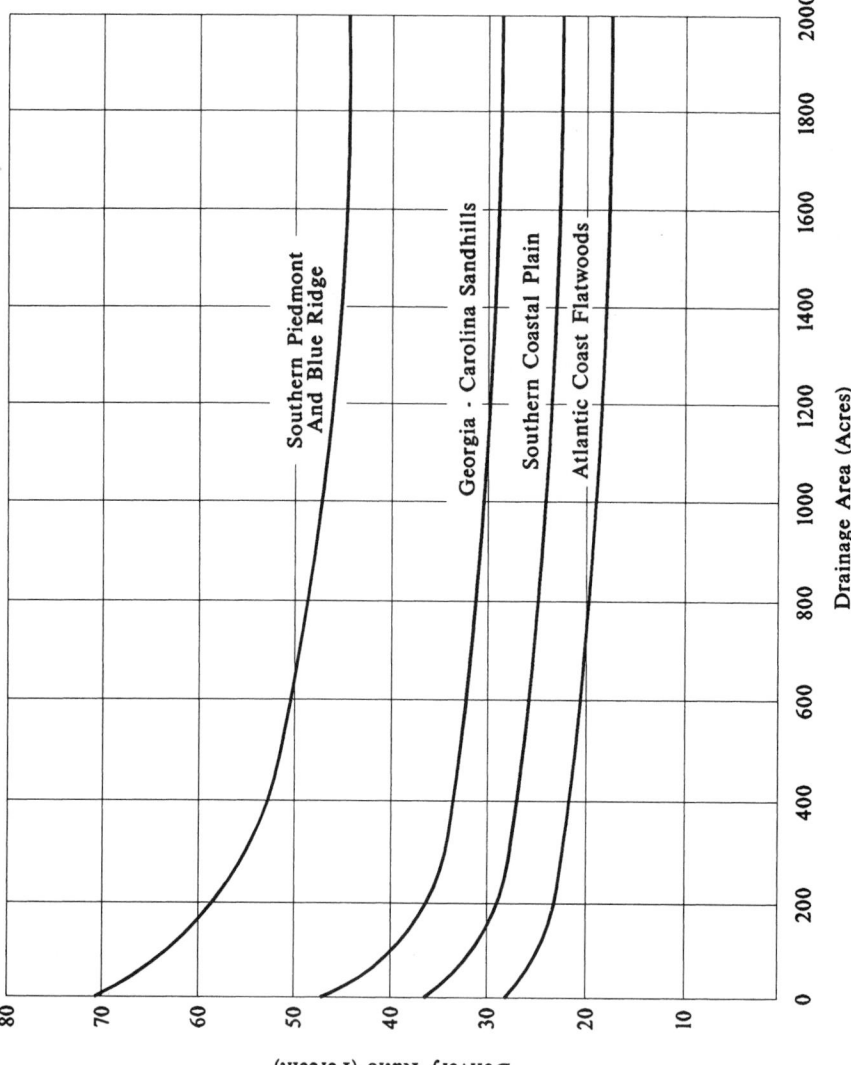

FIG. 8.31 Sediment delivery ratio curves. Source: Groce, 1981.

458

condition. Sediment exposed to alternating drying and wetting action within the basin will have a unit weight that approximates its moist condition. Figure 8.32 illustrates this relationship.

The engineer can use RUSLE Version 1.06 to evaluate various erosion control practices to estimate soil erosion production rates for the example given at the beginning of this section. Depending on the number and types of BMPs utilized, the net amount of sediment being delivered to the low point of the construction site will be reduced from the estimate made when no cover or practice measures were considered.

8.3.3 Sediment Basin Trap Efficiency

According to Ward et al. (1977), in 1904 Hazen developed what has become a classical approach for determining trap efficiencies of reservoirs when he studied the settling of soil particles under various hydraulic conditions. Hazen determined that trap efficiencies should be based on detention times, the velocities that particles fall, and their size

FIG. 8.32 Relation of watershed surface soil texture to surface soil and sediment. Source: Groce, 1981.

and other conditions. Since then various other methods for estimating trap efficiencies of reservoirs have evolved, such as the empirical method proposed by Brune (1953). Brune's method, which appears to provide little correlation to semidry reservoirs, utilizes a capacity-inflow relationship based on annual inflow, which may render the use of Brune's method questionable for small sediment basins, as shown in Figure 8.33.

Brune's procedure, based on data collected from 44 reservoirs, has nevertheless been used to generally predict trap efficiencies of pooled basins. To illustrate an application of the use of these curves, the capacity-inflow ratio must first be determined.

If, for example, the previously designed detention basin illustrated in Chapter 7 is to be used as a temporary sediment basin during construction, a perforated standpipe must be installed to act as the outlet. This outlet could be positioned close to the side of the detention pond, away from pipe 3. The 100-year storm analyzed in Chapter 7 for detention volume determination was found to require a working volume of 30,898 ft^3 (or 0.709 acre-foot).

The total annual rainfall inflow can be estimated by utilizing an average amount of precipitation for the development area, in this case assumed to be 48 inches. Applying a coefficient of runoff of 0.56, the inflow contributed from the 4.68-acre portion of the site is 4.68 × 48 inches × 0.56 coefficient of runoff/12 in./ft, or 10.48 acre-feet of estimated annual inflow. The capacity inflow ratio would be 0.709 acre-foot of storage/10.48 acre-feet of annual inflow, or 0.068. Figure 8.33 graphically indicates a corresponding trap efficiency of approximately 82 percent. [*Note:* For a more complete discussion on trap efficiency using Brune's curves, the reader is referred elsewhere (Renard, Heinemann, and Williams, 1979.)]

If the pond must, by local regulations, maintain 75 percent trap efficiency according to Brune's curves, the minimum volume required to produce that efficiency can be determined by obtaining the corresponding capacity-inflow required for 75 percent. By referring to Figure 8.33, one can see that a capacity-inflow rate of 0.047 corresponds to approximately 75 percent trap efficiency. The minimum storage volume would be

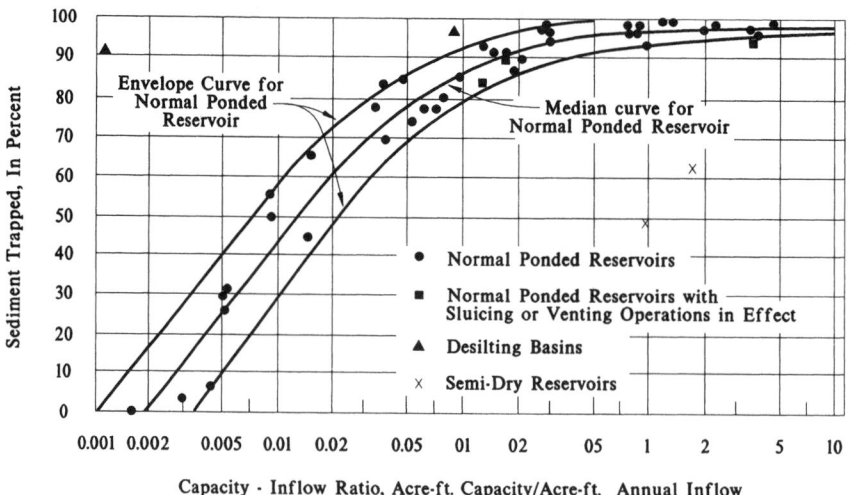

FIG. 8.33 Trap efficiency as related to capacity-inflow ratio, type of reservoir, and method of operation. Source: Gunnar M. Brune, *Transactions of the American Geophysical Union.* Volume 34(3), 1953. Copyright © 1953 by the American Geophysical Union.

0.047 × 10.48 acre-feet of annual inflow, or 0.49 acre-foot. The volume of sediment that can be allowed to accumulate is 0.709 acre-foot − 0.49 acre-foot, or 0.219 acre-foot (9540 ft³). Based on the geometry of the sediment basin, the elevation that allows 9540 ft³ of storage in its bottom can be calculated, and this value can be used to set a field stake with its top at this elevation. Using the top of the stake as a visual indicator, the basin would require cleaning when the stake is just covered.

Another empirical method to estimate trap efficiency is the use of curves produced by Churchill (1948). This method is event oriented and relates the percentage of sediment passing through a reservoir to the sediment index of the facility as defined by the ratio of retention period to mean velocity.

Figure 8.34 can be used to estimate the trap efficiency of the previous example while considering a two-year storm return period suggested by Walesh (1989). The mean outflow rate through the riser pipe's elevated entrance can be arbitrarily established at 1.5 ft³/s using the orifice equation (7.38). The corresponding detention time, or retention period, would be (0.709 acre-foot × 43,650 ft²/acre) (1.5 ft³/s), or approximately 20,590 seconds. The average cross-sectional area of the basin can be estimated using the volume of the basin divided by its length (using the water surface)—(0.709 acre-foot × 43,560 ft²/acre)/78.5 feet of assumed length when the pond is at capacity, or 393 ft² of cross-sectional area. In this particular situation, a baffle would probably have to be installed to prevent short-circuit flow patterns. The mean velocity of travel would be the flow rate/average cross-sectional area. In this example the mean velocity would be (1.5 ft³/s)/393 ft², or 0.004 ft/s. The corresponding sediment index (detention time/mean velocity) is 20,590 seconds/(0.004 ft/s), or 5.1×10^6 s²/ft. The corresponding estimated percent of sediment passing through the basin is obtained graphically from Figure 8.34 as 20 percent, resulting in an estimated trap efficiency of 80 percent. As can be seen, neither the Brune nor the Churchill method considers sediment characteristics.

Settling is usually categorized into four different types. Type I settling involves discrete particles that settle in low-concentration solutions where particles tend to

FIG. 8.34 Churchill's trap efficiency curves for reservoirs. Source: Churchill, 1948.

fall independently as defined by Stokes's law. Type II, or flocculation, settling occurs when dilute solutions of particles flocculate to form larger settleable particles. Type III, or zone, settling occurs where particles are so concentrated that forces between the particles hinder settlement activity, resulting in uniform settling within the zone. Type IV, or compression, settling refers to settling in which the particles are so close together that interparticle bridging has forced a stable structure, requiring compression for further settling activities.

Settling basins installed in land developments and on construction sites generally utilize Type I settling; therefore, suspended particles requiring flocculation to promote settling will pass through the basin where high turbidity can be expected. A Type I settling basin that conveys continuous flows is divided into four zones: the inlet zone, the settling zone, the bottom zone, and the outlet zone, as shown in Figure 8.35. Settlement paths of discrete particles in a horizontal flow basin are the vector sum of their settling velocities V_{settling} and the displacement velocity V of the basin. If particle settling velocities are less than V_{crit}, they will not all be removed because V_{crit} is the minimum velocity that allows a minimum-size particle to fall a vertical distance H equal to the basin depth during the time of detention.

The following example will serve to illustrate the design of a basin utilizing this basic concept. A rectangular settling basin is proposed to be placed where it would receive the relatively controlled discharge of the normally configured detention basin, as shown in Figure 8.36b. The maximum two-year return period flow rate obtained from Figure 7.53 of 3 ft^3/s will be utilized, along with a proposed sediment basin surface area of 25×75 feet. The overflow rate is W/A, or 3 ft^3/min/(25×75). This gives a velocity, V_{crit}, of 0.0016 ft/s. If a temperature of 68°F and a Reynolds number of less than 0.5 are assumed, the corresponding sphere diameter that is capable of being trapped is (Barfield et al., 1987)

$$d = \left[\frac{V_{\text{settling}}}{2.81}\right]^{0.5}, \quad \text{or 0.024 mm} \tag{8.25}$$

FIG. 8.35 Typical type I settling zones in a rectangular basin.

(a)

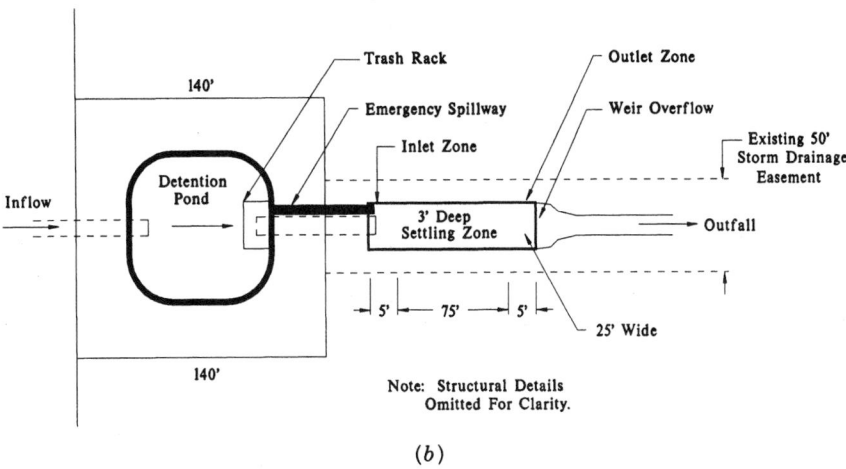

(b)

FIG. 8.36 Examples of sedimentation facilities. (a) Detention basin (showing final design elevations) being used temporarily as a sediment basin during the construction period. (b) Alternative approach using a 3-foot-deep sediment basin after the detention facility.

where V_{settling} is in feet per second and in this calculation is assumed be V_{crit} and d is in millimeters.

The fractional size of particles smaller than 0.024 mm (found using Equation 8.25) must be obtained from soil analysis representative of the site. Figure 4.74 will be used to illustrate the procedure. Only that portion of the semilog curve from 2 mm in size and smaller will be considered. If 100 percent of the material is finer than 2 mm, the percent smaller than 0.024 mm is found to be 10. Figure 8.37 shows only the 2-mm and smaller portion of the grain size distribution curve. The corresponding trap efficiency is the percentage of material greater than 0.024 mm plus proportional amounts of soils under 0.024 mm, based on the ratio of their settling velocity as compared to the critical. The incremental diameters shown in Figure 8.37 correspond to the average soil particle diameter of each 2 percent increment used in this example. Trap efficiency calculations are included as part of Figure 8.37.

If a 3-foot deep basin is used, the detention time would be the basin's volume divided by the flow rate, which is $(3 \times 27 \times 75)/(3 \text{ ft}^3/\text{s})$, or 1875 seconds. The corresponding velocity of plug flow would be the flow rate divided by the cross-sectional area, which is $(3 \text{ ft}^3/\text{s})/(3 \text{ feet} \times 25 \text{ feet})$, or 0.04 ft/s. To verify plug-flow conditions, the Reynolds number can be calculated using the flow-through velocity of 0.04 ft/s and the equation

$$\text{Re} = (4R_h V)/\text{kinematic viscosity of water} \qquad (8.26)$$

where R_h is the hydraulic radius, A/P, in feet and V is the mean (flow-through) velocity in feet per second. Kinematic viscosity for water is assumed for this example as $1.059 \text{ ft}^2/\text{s}$:

$$\text{Re} = (4)\,[(3 \times 25)/(3 + 3 + 25)]\,(0.04)/1.059$$

or 0.37, which indicates plug-flow conditions. When a 25- (or even 100-) year storm's controlled discharge (5 ft³/s) passes through this basin (Figure 7.53), the corresponding Reynolds number would be

$$\text{Re} = (4)\,[(3 \times 25)/(3 + 3 + 25)]\,(0.066)/1.059$$

Mean Dia, mm	% Inc.	$V_{settling}$, ft./sec. (Equation 8-5)	$(V_{settling} / V_{crit}) \times 0.02$
0.0023	2	1.5×10^{-5}	0.0002
0.005	2	7.0×10^{-5}	0.0009
0.008	2	1.8×10^{-4}	0.0022
0.011	2	3.4×10^{-4}	0.0043
0.018	2	9.1×10^{-4}	0.0114

Total % less than 0.024 mm = 10%   0.0190

Trap Efficiency = $(1 - 0.10) + (0.0002 + 0.0009 + 0.0022 + 0.0043 + 0.0114) = 0.919$
or 91.9 % Efficient

FIG. 8.37   Example of trap efficiency determination for Type I settling basin.

or 0.60, which still approximates the Reynolds number of 0.5 that was assumed in Equation 8.25.

If this basin were to be used during the course of a year while construction was underway, the annual estimated sediment delivery rate based on RUSLE (Section 8.3.2) would be 14.15 tons (52.4 tons per year × 0.27 delivery ratio). This equates to approximately 410 submerged cubic feet of sediment using a factor of 69 lb/ft^3 obtained from Figure 8.38 [(14.15 tons × 2000 lb/ton)/(69 lb/ft^3)]. This would amount to 410 ft^3/(25 feet × 75 feet), or 0.22 foot of sediment sludge accumulation per year in the sediment basin. Since the sediment basin is to be positioned after the detention facility in this case, the bottom of the preceding detention facility should be undercut by approximately 0.16 foot [(410 ft^3/(50 feet × 50 feet)] to allow for some additional sediment accrual during the year.

8.3.4 Summary

The placement and design of sediment basins are varied and must be tailored to individual site requirements. What has been included here are only examples of possible applications. More complete presentations on this important topic are included elsewhere (Barfield et al., 1987; Malcom and New, 1975; Walesh, 1989; Wanielista and Yousef, 1993).

8.4. EROSION CONTROL USING SOIL BIOENGINEERING

Land developments containing wetlands or streams may require some sort of bank stabilization to reduce or prevent erosion. Although retaining walls, bulkheads, revetments, riprap, and other constructed improvements might be considered, the use of vegetation as an alternative is increasing. Bioengineering relies on the placement of live cuttings and stems along the embankment slope where they help reinforce the soil to help reduce movement and also serve to remove soil moisture.

Porous revetments can be used in conjunction with the live cuttings to provide the slope with some initial protection while growth occurs. Depending on the application, vegetation consisting of tight dense grass provides excellent protection against rainfall and wind erosion, while deep-rooted vegetation can be better used to help improve slope stability, especially along a stream bank subject to moving water and scour.

Volume and Weight Relationships For Sediments In Selected Areas				
Land Resources Area	Aerated Sediment		Submerged Sediment	
	Volume (1)	Weight (2)	Volume (3)	Weight (4)
Blue Ridge	82	1786	55	1198
Southern Piedmont	82	1786	55	1198
Carolina-Georgia Sandhills	92	2004	80	1742
Southern Coastal Plain and Atlantic Coast Flatlands	88	1917	69	1503
Note: Columns (1) and (3) are given in pounds per cubic foot. Columns (2) and (4) are given as tons per acre foot.				

FIG. 8.38 Volume-weight relationships for sediment. Source: Groce, 1981.

8.5 CONSIDERATIONS FOR PERMANENT PONDS OR LAKES

If the detention pond is to be incorporated with a permanent standing pool of water (lake), then the detention basin should include dead storage for long-term sedimentation accumulation, possibly for as long as 20 years. RUSLE Version 1.06 can be used to estimate soil erosion over time by considering different land cover characteristics as the land development's landscaping vegetation establishes itself and the soil undergoes consolidation.

Consideration should be given to preventing solvents, degreasers, pesticides, fungicides, and herbicides that might be classified as hazardous wastes under the Resource Conservation and Recovery Act (RCRA) of 1976 from entering the storm drainage system through covenants. The best way to reduce the potential of sediment contamination is to have in place good housekeeping procedures where, if pollutants are known to be present in the runoff, they are cleaned up immediately before they settle to the bottom of the lake. When the sediment reaches the level where it must be removed, it must be checked to determine if hazardous wastes are present and, if so, handled and disposed following hazardous waste regulatory guidelines.

A permanent pond or lake has multiple benefits that can be an asset to a development. Most facilities of this nature improve the landscape quality of the site, as shown in Figure 8.39, and fire protection services can be augmented by installing dry hydrants into the pond for pumpers to have a source of replenishment water similar to that shown in Figure 8.40. In addition, recreational benefits can be obtained in the form of fishing, boating, and swimming. Wildfire usage and irrigation utilize permanent ponds to the benefit of all.

FIG. 8.39 Example of a permanent lake incorporated as part of a neotraditional development. Courtesy of R. H. Stout.

FIG. 8.40 Dry hydrant construction. Reprinted with permission from NFPA 1142, *Standard on Water Supplies for Suburban and Rural Fire Fighting*, p. B-5.3. Copyright © 1999 by the National Fire Protection Association, Quincy, MA, 02269. This reprinted material is not the complete and official position of the National Fire Protection Association on the referenced subject, which is represented only by the standard in its entirety.

REFERENCES

Barfield, B. J., R. C. Warner, and C. T. Haan. *Applied Hydrology and Sedimentology for Disturbed Areas*. Stillwater: Oklahoma Technical Press, 1987, pp. 400–407.

Brown, L. C., and G. R. Foster. "Storm Erosivity Using Idealized Intensity Distributions." *Transactions of the American Society of Agricultural Engineers*, Vol. 30, pp. 379–386, 1987.

Brune, G. M. "Trap Efficiency of Reservoirs." *Transactions of the American Geophysical Union*. Vol. 34, No. 3, pp. 407–418, 1953.

Churchill, M. A. "Discussion of 'Analysis and Use of Reservoir Sedimentation Data' by L.C. Gottschalk." In *Proceedings Federal Inter-Agency Sedimentation Conference*. Springfield, VA: U.S. Department of Commerce—National Technical Information Service (NTIS), 1948, pp. 131–140.

Galetovic, J. R., T. J. Toy, G. R. Foster, K. G. Renard, G. A. Weesies, S. A. Schroeder, W. F. Kuenstler, G. W. Wendt, and R. Warner. *Guidelines for the Use of the Revised Universal Soil Loss Equation (RUSLE) Version 1.06 on Mined Lands, Construction Sites, and Reclaimed Lands*. Denver: Office of Technology Transfer—Western Regional Coordinating Center of the Office of Surface Mining, 1998.

Groce, W. G. *Engineering 3 (Sedimentation)*. USDA-SCS-SC Technical Note. Columbia SC: 1981.

Laflen, J. M., D. A. Watson, and T. G. Franti. "Ephemeral Gully Erosion." In *Proceedings of the Fourth Inter-Agency Sedimentation Conference*, Vol. I, Section 3. Springfield, VA: U.S. Department of Commerce—National Technical Information Service (NTIS), 1986, pp. 29–37.

Laws, O. J., and D. A. Parsons. "The Relation of Raindrop-Size to Intensity." *Trans. AGU* 24: 452–460. 1943.

Malcom, H. R., and V. New. *Approaches for Stormwater Management in Urban Areas*, Part 4. Raleigh: North Carolina State University, 1975.

Renard, K. G., G. R. Foster, G. A. Weesies, D. K. McCool, and D. C. Yoder. *USDA Agricultural Research Service, Agricultural Handbook Number 703—Predicting Soil Erosion by Water: A Guide to Conservation Planning with the Revised Universal Soil Loss Equation (RUSLE)*. Washington: U.S. Government Printing Office, 1997.

Renard, K. G., H. G. Heinemann, and J. R. Williams. "Comment on 'Measurement of Sediment Control Impacts on Agriculture' by James C. Wade and Earl O. Heady." *AGU Water Resources Research*, Vol. 15, No. 5, pp. 1278–1280, 1979.

U.S. Department of Agriculture–National Resources Conservation Services (USDA–NRCS). "Erosion and Sediment Control in Developing Areas." An unpublished guide. Columbia, SC: NRCS, 1977.

U.S. Department of Agriculture–National Resources Conservation Services (USDA–NRCS). *Predicting Soil Losses Using the Universal Soil Loss Equation—South Carolina*. Columbia, SC: 1974.

Walesh, S. G. *Urban Surface Water Management*. New York: Wiley, 1989, pp. 297–313.

Wanielista, M. P., and Y. A. Yousef. *Stormwater Management*. New York: Wiley, 1993.

Ward, A. J., C. T. Haan, and B. J. Barfield. *Simulation of the Sedimentology of Sediment Detention Basins*. Research Report No. 103. Lexington: University of Kentucky, 1977, pp. 8–15.

Watson, D. A., J. M. Laflen, and T. G. Franti. "Estimating Ephemeral Gully Erosion." Paper No. 86-2020. St. Joseph, MI: American Society of Agricultural Engineers, 1986 summer meeting.

Wischmeier, W. H., and D. D. Smith. *Predicting Rainfall Erosion Losses—Agriculture Handbook # 537*. Washington: U.S. Government Printing Office, 1978.

SUGGESTED READINGS

National Fire Protection Association. *NFPA 1142—Standard on Water Supplies for Suburban and Rural Fire Fighting.* Quincy, Mass.: NFPA, 1999, p. B-5.3.

Renard, K. G., G. R. Foster, G. A. Weesies, and J. P. Porter. "RUSLE—Revised Universal Soil Loss Equation." *Journal of Soil and Water Conservation,* Vol. 46, pp. 30–33, 1991.

U.S. Department of Agriculture–National Resources Conservation Services (USDA–NRCS). *Ponds—Planning, Design, and Construction.* Washington: U.S. Government Printing Office, 1982.

U.S. Environmental Protection Agency (EPA). *Protecting Natural Wetlands—A Guide to Stormwater Best Management Practices.* Washington: EPA, 1996.

Water Environment Federation and the American Society of Civil Engineers. *Urban Runoff Quality Management. WEF Manual of Practice No. 23 and ASCE Manuals and Reports on Engineering Practice No. 87.* Reston: ASCE, 1998.

9 Site Grading and Appurtenance Design

Soil is the material most used to support and connect improvements made in a land development. It gives buildings and roads elevation to minimize the effects of flooding and provides continuity between points of unequal elevation. Soil is used to support imposed loads, and its surface can be shaped to form channels that control water movement. Naturally occurring granular soil usually seeks a surface slope that approaches its natural angle of repose, while erosion and weathering activities tend to smooth and round all exposed soil surfaces. Design professionals include rigorous requirements in their construction drawings to alter the shape of the soil's surface from its natural state to a reshaped (or graded) surface. This allows the various improvements made within a development to function properly and be pleasing to the eye. The developer's goals and constraints (Chapter 2) are considered by the design professional while evaluating site grading options. A site grading plan is used for this purpose. The plan is dependent upon the transportation, storm drainage, and sanitary sewer system designs.

9.1 IMPROVED LAND DEVELOPMENTS

Improved land developments consisting of detached single-family dwellings usually provide for the developer to arrange and pay for the grading of rights-of-way and drainage facilities, while requiring the builders of the dwelling units to undertake the grading of their respective lots. Because independent parties may provide lot grading services later, an overall plan is necessary to indicate how each lot will be graded. This plan should define general methods to grade lots with the storm drainage flowing to the street, to the lot rear, or in both directions. A lot grading detail sheet can be included in the construction drawings (similar to the one included in Chapter 16) keyed to a plan view drawing showing each lot (Fig. 16.3).

9.1.1 Lot Drainage Types

The U.S. Department of Housing and Urban Development has categorized three general types of lot surface runoff drainage plans (U.S. HUD, 1973). In instances where the rear-lot elevations are higher than the front, the lot should be graded to allow drainage toward the front of the lot and should incorporate provisions for diverting water around the dwelling unit. This general plan could be labeled type "A" on the detail sheet.

   If the lot is subject to high elevations at its center with low elevations at the street and rear-lot line, another type of grading plan must be used. A rear-lot line drainage easement can be incorporated to receive accumulated stormwater that is directed toward the lot rear. This general grading plan could be labeled type "B" on the detail sheet.

Should a lot be oriented where the topography is falling toward the rear of the lot, drainage must then generally flow in this direction. By including a rear-lot line drainage easement, stormwater can be intercepted and redirected before it has an opportunity to travel on abutting parcels. This grading scheme could be labeled type "C."

9.1.2   Lot Drainage Plans

To effectively identify and provide general guidance for the grading of each lot, a development plan is prepared that shows each lot along with the proposed type of drainage plan, signified by labeling with A, B, or C. Including this information in the construction plans (and later sharing it with individual lot builders) ensures the coordination of building lot improvements that complement and function in unison with the entire road and storm drainage systems.

9.1.3   Initial Grading Improvements

In a single-family development, grading improvements provided by the developer are a function of the road and storm drainage systems. The design professional would attempt to balance earthwork cut-and-fill volumes, as indicated in Section 9.3, when excavated material is suitable for construction use. Exceptions would be cases where proposed gravity ties to existing sanitary sewers or storm drains would necessitate the use of borrow material to provide adequate elevation for fall. The amount of borrow would be a function of cost. If the cost of borrow material is significant, pumping facilities to compensate for nongravity conditions would have to be considered. A specific site grading plan for this type of development normally is prepared only to show the erosion control plan, as pertinent grading information is already included on the road and storm drainage plans. The disposition of possible excess cut materials would be determined by the engineer in accordance with provisions included in the construction specifications. Any borrow material requirements would be similarly handled.

9.2   TOTAL DEVELOPMENT ACTIVITIES

A formal site grading plan is usually prepared when total development activities are undertaken to convert unimproved land into a finished product that includes all site improvements and structures.

9.2.1   Site Grading Plan Preparation

To prepare a site grading plan, the design professional must have the road and storm drainage plan well defined. In addition, the sitting of proposed structures must have been decided to properly allocate space for final contour lines. These are necessary to define topographically the final surface configuration.

The site grading plan should provide a minimum of soil cutting and filling operations consistent with good engineering practice. Advantage should be taken of natural site features when considering the plan and gravity flow systems. Most large sites cannot be clean cut and stripped because of tree and landscape ordinances, erosion control plans, and preservation of natural site features.

Many developments include concrete curb and gutter type road cross sections that require placement in a "cut" environment to allow proper drainage. The design professional can estimate the volume of any excess soil expected when constructing the road and storm drainage systems. If this excess soil is suitable for construction use, the designer can determine where to place it on site. This may require adjusting the final grades and contours on the grading plan to accommodate earthwork balancing.

9.2.2   Small Sites Associated with Total Development Activities

Some small sites associated with total development activities call for clearing the entire site. Landscaping provisions normally provide a means to reestablish natural growth in designated areas to enhance the finished product. In such cases the amount of earthwork associated with cut-and-fill operations is usually balanced. However, requirements for tying into an existing gravity sanitary sewer main or a storm drainage outfall may necessitate the inclusion of borrow material to elevate the site to provide for gravity flow to occur. Otherwise pumping facilities, which are expensive to construct and operate, may be required. Small development sites associated with total development activities, such as one containing several units of multifamily attached row or townhouses, may not require separate plan and profile sheets for the road and storm drainage systems. Information normally contained on those drawings could be incorporated on the site grading plan.

9.3   EARTHWORK CALCULATIONS

The amount of soil that a contractor moves and shapes directly affects the cost of construction. The goal is to minimize the amount of earthwork consistent with good engineering practice. To achieve this goal, an estimate of the amount of soil excavation (cutting) and embankment (filling) must be made. If the volume of cut material equals the volume of fill material (after soil expansion or shrinkage is considered), a balanced condition is achieved. Hence, there will be no soil surplus and no requirements for possibly expensive off-site borrow.

9.3.1   Earthwork Volume Determination Using Borrow Pit Method

One procedure to determine earthwork volume is by utilizing the borrow pit method. This method requires the use of an imaginary square grid system where depths of cut are required for each grid intersection point. These cut values can be obtained by subtracting the desired elevation of the bottom of the borrow pit after excavation from each individual existing grid intersection point spot elevation. Figure 9.1 illustrates a typical layout where each intersection point is designated by a row letter and column number. Once the cut values are known, the following methodology can be applied. Using Figure 9.2, where a 4 × 4-inch block with an irregular top surface is shown, the block is seated on its base. The vertical distance from each top corner to the base position represents the "depth" of cut. The volume of the block is the base

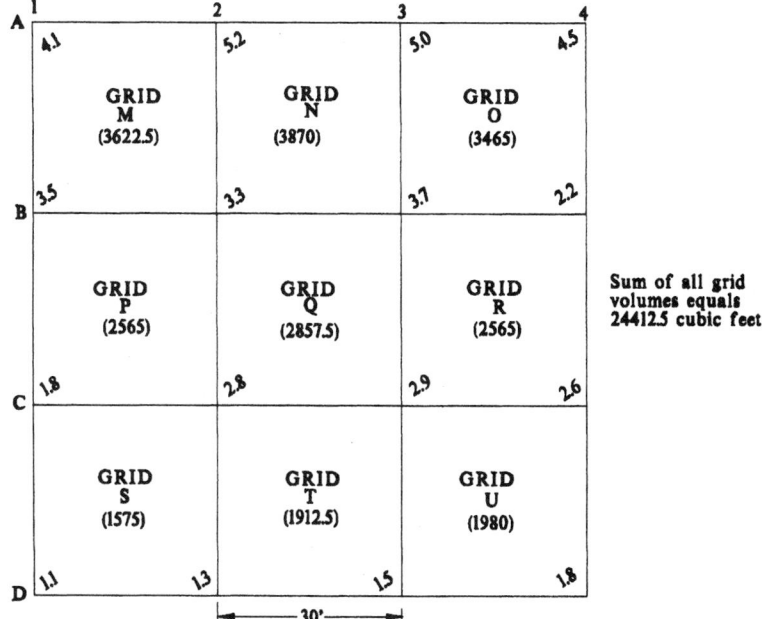

Short-cut: 1/4(30x30)[(4.1+4.5+1.1+1.8) +2(5.2+5.0+3.5+2.2+1.8+2.6+1.3+1.5)+4(3.3+3.7+2.8+2.9)]= 24,412.5 cubic feet.

FIG. 9.1 Example borrow pit grid showing (*a*) row letters and column numbers for point identification, (*b*) cut values for each point forming the grid, (*c*) volume of each grid block shown in parentheses, and (*d*) short-cut approach.

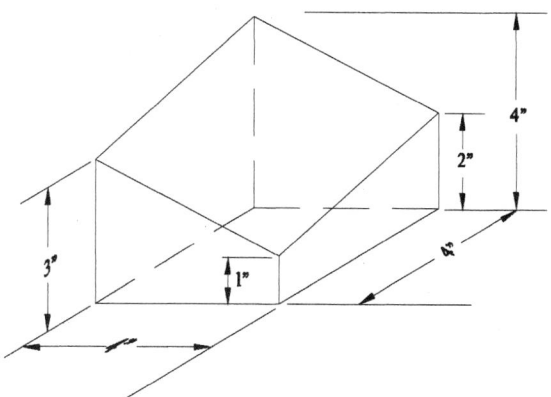

FIG. 9.2 Example of a 4 × 4-inch block having an irregular top to demonstrate the borrow pit method relationships for volume determination.

area, 4 × 4 inches, times the height of the block; however, in this case the average height must be used, which is $\frac{1}{4}(4+3+1+2)$, or 2.5, inches. The volume of the block is therefore 40 in^3.

Using the same procedure in conjunction with Figure 9.1 for each grid square, the corresponding volume for that grid block is shown in parentheses. The total volume of excavation is the sum of all the grid volumes. A more efficient way to determine the volume makes use of the fact that some of the cut values are only used once (A4), others are used only twice (A3), and the remaining are used four times. The total cut volume equals one-fourth of the base area multiplied by the sum of all the terms used once plus 2 times the sum of those used twice plus 4 times the sum of the remaining spot elevations.

9.3.2   Earthwork Volume Determination Using Contour Lines

Since a contour line represents points of equal elevation, the volume of soil encompassed between sequential contour lines can be determined using the average end-area method as expressed by the following formula:

$$\text{Volume} = (D_{1-2})(A_1 + A_2)/2 \tag{9.1}$$

where D_{1-2} is the vertical distance between end areas 1 and 2 in feet and A is the appropriate contributory end area, in square feet, that establishes the volume that is computed.

Figure 9.3 illustrates an application of Equation 9.1 to determine the volume of soil between elevations 10 and 15 feet.

9.3.3   Earthwork Volume Determination Using Cross Sections

Instead of calculating volumes based on horizontal reference surfaces, which is the case when using contour lines, vertical reference plans showing cross sections can be utilized. The procedure, however, is the same except that the distance, D, is the horizontal distance between end sections (cross sections), and A is the area of each cross section.

The cross-sectional area can be determined by plotting each section to scale and graphically determining the enclosed area using a planimeter. Usually the resulting area determined by the planimeter is in square inches and must be converted to square feet using a factor based on the scale with which the cross sections are plotted. For example, if the horizontal scale is 1 inch = 50 feet and the vertical scale is 1 inch = 10 feet, each square inch in the cross section represents 50 feet × 10 feet, or 500 ft². The equivalent end area in square feet is determined by multiplying the planimeter reading in square inches by the 500 equivalent square feet in this case.

Another method to determine the area of a cross section is to use section elevations as northings and horizontal reference distances as eastings while calculating the enclosed area using the coordinate method found in Figure 4.54. This procedure is illustrated in Figure 9.4.

9.3.4   Transition and Taper End Areas

The designer must only include in the calculations end areas that contribute to the desired volume. Often there are cross sections that require the use of different end areas depending on whether the volume is being calculated behind or ahead of the section. For example, Figure 9.5 illustrates both tapered and transition sections where an access road joins the end of a borrow pit. The amount of soil removed from the

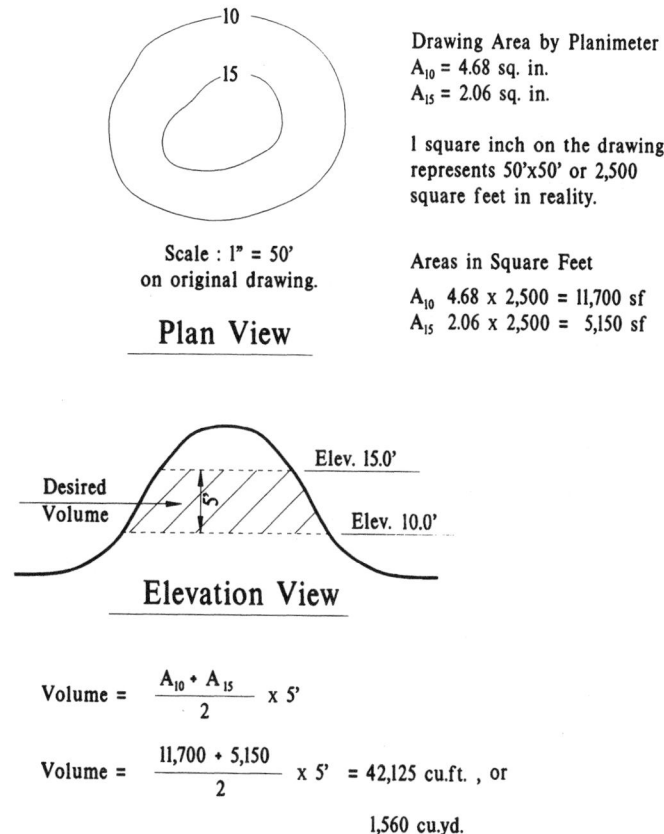

FIG. 9.3 Example of earthwork volume calculations using contour lines.

access road (a tapered section) is calculated using a beginning end area of zero cut at station $0 + 00$. At station $2 + 00$, the end area is 15 feet $\times$ 10 feet, or 150 ft^2. Using Equation 9.1, the corresponding volume of excavation is

$$\text{Volume}_{0-200} = (200 \text{ feet})(0 + 150)/2$$

or 15,000 ft^3 (555.55 yd^3). If the volume between stations $2 + 00$ and $3 + 00$ were determined, the corresponding end area for the transition section at station $2 + 00$ would be 10 feet $\times$ 300 feet, or 3000 ft^2, instead of the previously used 150 ft^2. Transition sections require special attention to ensure that the appropriate end area is used.

9.3.5 Earthwork Summary Sheet

The earthwork summary sheet shown in Figure 9.6 depicts calculations necessary to determine earthwork volumes in cubic yards and their balances. These calculations are prerequisites for constructing a mass diagram. End areas are shown at stations, while volumes are shown between stations. In these calculations, cut is designated as

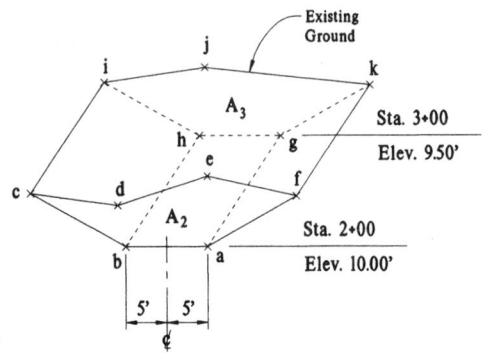

End Area for Sta. 2+00				
Point	N	E	+	−
a	10	+5		
b	10	-5	-50	+50
c	14	-13	-130	-70
d	12	-2	-28	-156
e	13	+6	+72	-26
f	12.5	+10	+130	+75
a	10	+5	+62.5	+100
Summation			+56.5	-27

End Area for Sta. 3+00				
Point	N	E	+	−
g	9.5	+5		
h	9.5	-5	-47.5	+47.5
i	15	-15	-142.5	-75
j	16	-4	-60	-240
k	14	+13	+208	-56
g	9.5	+5	+70	+123.5
Summation			+28	-200

Double Area = 83.5

A_2 = 41.75 square feet

Double Area = 228

A_3 = 114 square feet

$$\text{Volume} = \left[\frac{A_2 + A_3}{2}\right](100') = \left[\frac{41.75 + 114}{2}\right](100') = 7787.5 \text{ cubic feet}$$

FIG. 9.4 Example of earthwork calculations using cross sections.

a positive volume, while fill is denoted as a negative value. Shrinkage is the volume reduction that soil will undergo when compacted in the field. In this example shrinkage is assumed to be 10 percent. The mass ordinates shown are the cumulative amount of all earthwork from the beginning of the project up to the point under investigation.

9.3.6   Earthwork Mass Diagram

Earthwork mass diagrams are often used on large earth-moving projects such as highways, railroads, and airports. Occasionally they are used in conjunction with land developments to provide a means to identify the amount of excavation and embankment, the location of balance points, economical haul directions, and borrow placement.

There are numerous ways to estimate overhaul distances and volumes; however, the method of moments is illustrated in the mass diagram shown in Figure 9.7. Most land development contracts call for earthwork to be performed on a lump-sum basis where costs associated with transporting soil long distances (overhaul costs—costs for hauling soil distances in excess of the freehaul distance) are included in the project's base cost. To know what the estimated amount of earthwork overhaul can be, however, is useful to the design professional when estimating the cost of the project. A reasonable distance

FIG. 9.5 Transition and taper end sections associated with earthwork.

for an earthwork contractor to haul soil without incurring undue costs associated with hauling is called the freehaul distance. Often $\frac{1}{2}$ mile is used for the freehaul distance to estimate the best distribution of soil placement, although the freehaul distance can be defined by stations or miles.

In a mass diagram, a balance point is a station where the summations of volumes of excavation and embankment (including shrinkage) are equal. Any line that is constructed parallel to the mass diagram's base indicates balance points wherever the construction line intersects the diagram. For example, all earthwork conducted between stations $0 + 00$ and $44 + 00$ balances in Figure 9.7.

If balance lines are constructed parallel to the diagram's base using the freehaul distance to scale, all earthwork located between the balance points is to be accomplished with no overhaul being involved. In Figure 9.7 this occurs between stations $12 + 25$ and $38 + 65$. Overhaul, therefore, can be expressed as the number of cubic yards of soil moved outside of the freehaul distance and is expressed in this case as cubic yard half-miles. The required overhaul distance is found by using the mass diagram to obtain the distance between mass centers for excavation and embankment.

To determine the center of mass by moments, each volume within the overhaul area (on each side of the freehaul zone) is multiplied by the distance from a reference station (zero for one, 3865 feet for the other) to the midpoint between sequential volumes. Figure 9.7 shows the necessary calculations to determine the centers of mass by moments. The overhaul is calculated by multiplying the overhaul volume by the distance between mass centers while excluding the freehaul distance. The overhaul is therefore

$$\text{OH} = \frac{(\text{OH volume})(\text{center of mass right} - \text{center of mass left} - \text{freehaul})}{\text{freehaul}}$$

$$= \frac{1770(4065 - 813 - 2640)}{2640}$$

or 410 cubic yard half-miles. This value can aid the design engineer in estimating earthwork costs. Also indicated on the corresponding profile in Figure 9.7 are directions of haul and borrow placement.

	End Areas		Between	Volumes, Cubic Yards					Volume	Mass Ordinates	
Station	Cut,S.F.	Fill,S.F.	Stations	Cut	Fill	Shrinkage	Unusable	Embankment	Balance	Station	Ordinant
0	0	0	0-5	231.48	0.00	0.00	NONE	0.00	231.48	0	0.00
5	25	0	5-10	1064.81	138.89	13.89	NONE	152.78	912.04	5	231.48
10	90	15	10-15	2175.93	648.15	64.81	NONE	712.96	1462.96	10	1143.52
15	145	55	15-20	3148.15	1666.67	166.67	NONE	1833.33	1314.81	15	2606.48
20	195	125	20-25	3750.00	2129.63	212.96	NONE	2342.59	1407.41	20	3921.30
25	210	105	25-30	3453.70	2824.07	282.41	NONE	3106.48	347.22	25	5328.70
30	163	200	30-35	2324.07	4166.67	416.67	NONE	4583.33	-2259.26	30	5675.93
35	88	250	35-40	953.70	3175.93	317.59	NONE	3493.52	-2539.81	35	3416.67
40	15	93	40-45	138.89	1092.59	109.26	NONE	1201.85	-1062.96	40	876.85
45	0	25	45-50	0.00	231.48	23.15	NONE	254.63	-254.63	45	-186.11
50	0	0								50	-440.74
Summation				17240.74	16074.07	1607.41		17681.48			

FIG. 9.6 Example of an earthwork summary sheet.

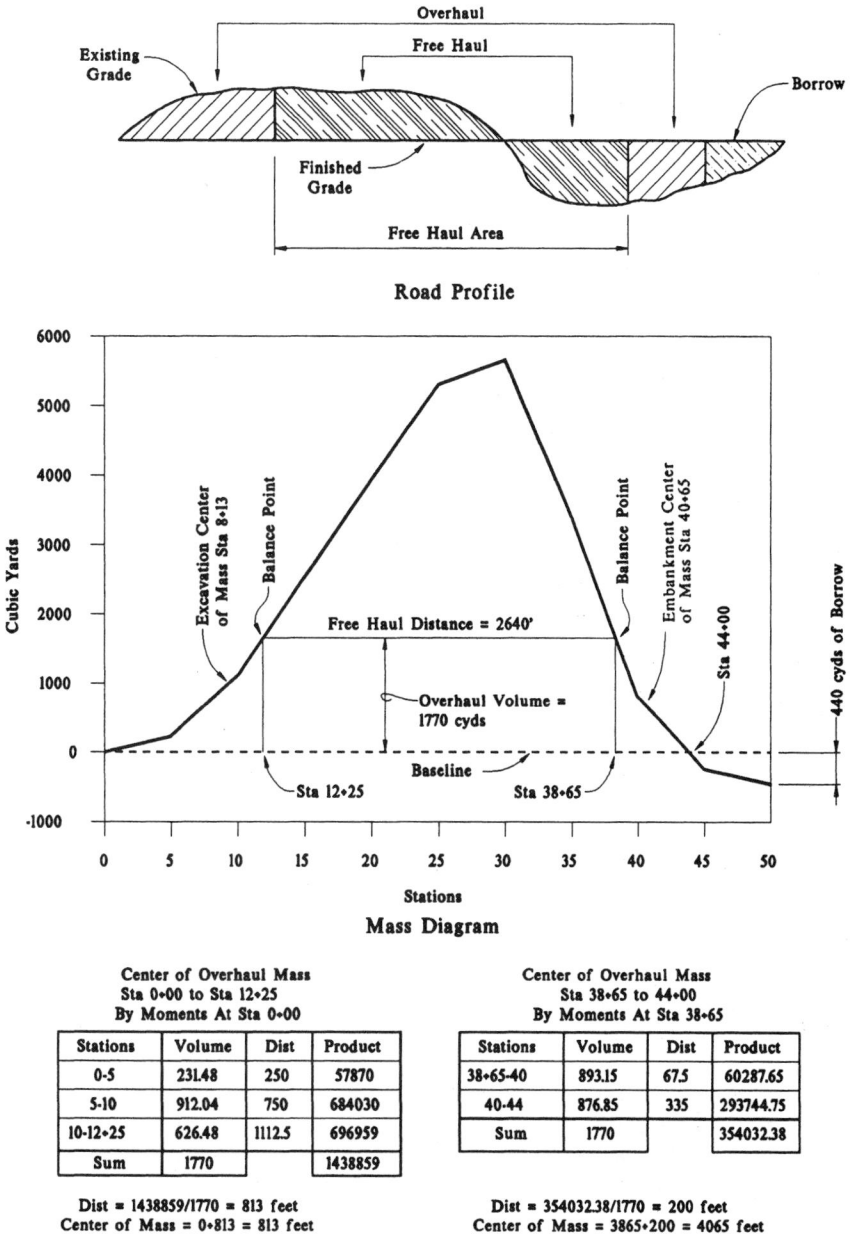

FIG. 9.7 Example of an earthwork mass diagram.

9.4 ELEMENTS OF THE GRADING PLAN

The site grading plan is an excellent place to show sedimentation and erosion control improvements along with temporary drainage improvements. This plan also includes existing contour lines and their proposed realignment, spot elevations, finished floor elevations for proposed buildings, elevations for recreational playing courts, parking lots, drainage channels, lakes, and ponds. Often a site's topography requires additional improvements to minimize design slope gradients and permit user access. Appurtenances such as stairs, ramps, and retaining walls can be used for such purposes.

The plan should also include considerations for laterally supporting adjoining property, especially if permanent excavations could affect this absolute right of the abutting landowner. Lateral support required for construction, however, is the responsibility of the contractor. Other elements of the grading plan could include delineation of areas requiring various amounts of compaction effort that would augment compaction requirements set forth in the construction specifications. The plan also could include provisions for stabilizing slopes and easement areas to protect slopes that would have to be shown on the final recorded plat.

Locations where poor soil conditions are expected to be found (as a result of geotechnical explorations and testing) can be delineated on the grading plan, along with provisions to improve or remove this undesirable soil. If the area contains expansive soils, then the grading plan could possibly accomplish the following:

(a) Indicate replacement of the upper soil layer along with provisions for backfilling with nonexpansive low-permeability soil. Provisions for the backfill to be compacted to a specified amount would be included in the construction specifications, as would requirements for shaping the compacted soil's surface to conform with the grading plan.

(b) Indicate the placement and mixing of nonexpansive soils with the existing expansive soils to reduce the effects of expansion. Provisions for the soil mixture to be compacted to a specified amount would be included in the construction specifications, along with requirements for shaping the compacted soil's surface to conform with the grading plan.

9.4.1 Finished Floor Elevations

The establishment of finished floor elevations is important to ensure that the proposed structure will not flood, is elevated enough to allow gravity flow sanitary sewer lines to drain into the installed main, and is high enough to keep soil from being transported into the structure.

Where the terrain is relatively flat, it is good practice to establish the finished floor elevation of a ground slab at least a foot above the elevation of the frontage road unless the expected flood elevation for the site requires the elevation to be set higher. If the site slopes appreciably, the finished floor elevation is a function of the driveway gradient. This driveway gradient should not usually exceed 10 percent. However, a slightly higher gradient can be used if the drive is upgrade from the street.

Slab on grade structures also require a minimum of 3 feet from the finished floor elevation to the invert of the sanitary sewer service line for the fittings and cleanout to have adequate space for installation. The invert elevation at the cleanout also must be high enough to allow the gravity service to slope to the main at a gradient equal to or greater than the minimum allowable slope for service lines according to local authority.

If the grading plan requires substantial earthwork operations to shape the site, extra care must be used if expansive soils are present to eliminate the possibility of differential heave under building foundations. This is especially true where a hilly site is regraded. Proposed building foundations located partly on fill-and-cut areas could experience this problem, according to Jones and Jones (1987). The grading plan should call for excavating problem areas to a depth greater than the one desired (called undercutting) so that nonexpansive backfill can be placed and compacted. Other methods of controlling the detrimental effects of expansive soils can include provisions for controlling the soil moisture or using chemicals. The reader is referred elsewhere for a more complete presentation on expansive soils (Nelson and Miller, 1992).

9.4.2 Spot Elevations

Spot elevations can represent either existing or proposed elevations on a grading plan. A legend is required to distinguish one elevation type from the other. Spot elevations are useful to indicate elevations where other means are not suitable. For example, if the design professional wishes to signify that the finished floor slab of a structure is 8 inches above the surrounding ground, spot elevations are used, as shown in Figure 9.8.

9.4.3 Slope Gradients

Maximum grades for finished surfaces within a development are based on functional use. Maximum gradients for curb cuts and ramps to accommodate handicap access are contained in Section 2.5.9. The gradients for streets are included in Chapter 6.

9.4.4 An Example of a Grading Plan

To illustrate an example of a site grading plan, the project shown in Figure 4.64 will be used. Existing contours are shown as fine lines, while contours requiring alterations (new contours) are shown thicker. By applying the principles in Section 4.12 concerning contour lines, the grading plan shown in Figure 9.9 is constructed.

Development sites can include topography that has great differences in elevation. To provide flexibility in accommodating these elevation changes within a development, appurtenances are often required, including stairs, ramps, handrails, and retaining walls (see Section 9.7 for retaining walls).

9.5 STAIRS, RAMPS, AND HANDRAILS

Steps and ramps provide pedestrians access between levels of unequal elevation. Steps provide incremental vertical movement while ramps provide uniform movements. Both steps and ramps must be designed to provide user comfort and safety.

9.5.1 Steps

Steps can be constructed of wood, reinforced concrete, brick, metal, or stone. The proportions are usually based on local regulatory code requirements. Steps must be placed so that their support is obtained below the frost line. Handrails are included to aid pedestrians; they provide security and balance if the gradient is not relatively flat.

FIG. 9.8   Spot elevations signifying finished grades.

9.5.2   Ramps

Ramps are usually made of materials similar to those for steps but allow access to either pedestrians or wheeled vehicles. As in the case of steps, ramps usually include handrails for user protection and restraint.

9.5.3   Handrails

Handrails provide positive restraint to pedestrians to help prevent accidental falls. They also give stability and guidance to the user. Handrails are usually constructed of metal; however, brick, stone, reinforced concrete, and wooden rails (walls) are also

FIG. 9.9 Typical site grading plan.

utilized. As with stairs, handrails must comply with local regulatory codes. In addition to being placed along steps and ramps, handrails are also used along the top edge of retaining walls if pedestrian exposure is contemplated.

9.6 SLOPE STABILITY

A soil slope may be stable under certain conditions but highly unstable if conditions change. Slope failure is usually expressed by sliding and occurs when the internal resistance of the soil to movement is less than the forces attempting to induce movement.

9.6.1 Slope Stability Analysis

Typically, if a 1-foot-thick section of embankment is analyzed, as shown in Figure 9.10a, the length of failure zone, L, is the corresponding length of arc. The moment attempting to induce sliding along the arc is the product of the weight of soil (W_t) and its lever arm (a). If the strength of the soil is expressed by its shear strength, S_s, in tons per square foot, the total shearing force is $S_s \times L \times 1$ foot. The total resisting moment is $S_s \times L \times 1 \times R$, as shown in Figure 9.10a, where R is the moment arm. The factor of safety against sliding is then the resisting moment divided by the overturning moment.

An increase in soil moisture from rain or the addition of overburden pressures at the

(a)

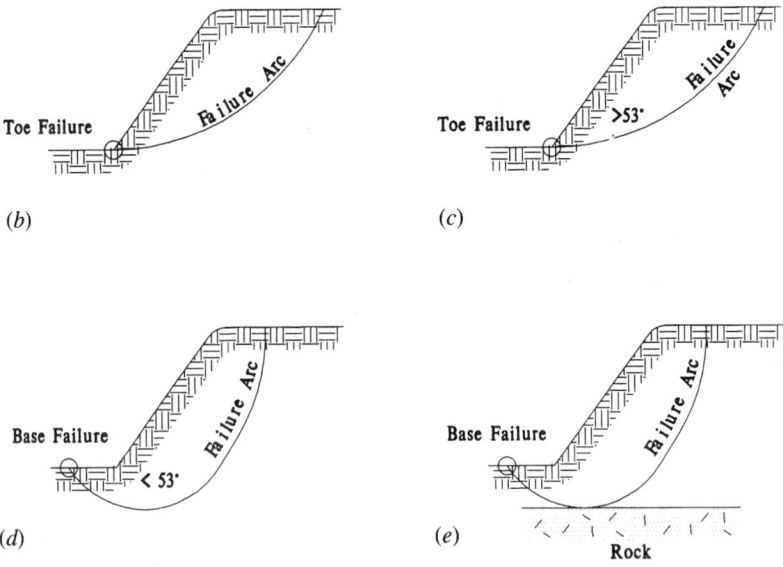

(b)   (c)

(d)   (e)

FIG. 9.10   Slope stability: (a) general relationships for slope stability analysis; (b) typical slope failure for soil containing both cohesive and noncohesive materials; (c) typical slope failure for cohesive soil with a step slope; (d) typical slope failure for cohesive soil with a mild slope; (e) typical slope failure with underlying rock. Source: Karl Terzaghi and Ralph B. Peck. *Soil Mechanics in Engineering Practice.* Copyright © 1966 by John Wiley & Sons, Inc. Reprinted by permission of John Wiley & Sons, Inc.

top of the slope can increase the moment attempting to induce soil movement. This will result in a decrease in the factor of safety. Likewise, a decrease in the soil's shear strength can result in possible slope failure. More specific reasons for slope failure can include

(a) an increase in W_t without a corresponding increase in S_s,
(b) additional fill or structures being placed at the top of the slope,
(c) increasing moisture in the soil,
(d) hydrodynamic forces such as piping attributed to running water or sudden drawdown (especially in earthen dams), and
(e) slope failure attributed to frost action, chemical changes, and foundation failure.

9.6.2 Typical Slope Failures

A stable slope comprised of cohesionless materials such as sand and gravel would take the form of the soil's natural angle of repose (or less) based on field conditions. If the soil is moist, it will stand on a slope greater than the natural angle of repose under dry conditions. However, this may not always be safe.

If the soil is cohesive, slope stability is difficult to determine and the analysis may be undertaken on a trial-and-error basis. Generally, when the soil obtains its strength from both cohesion and shear (both clay and sand or gravel are present), slope failures usually occur at the toe of the slope, as shown in Figure 9.10*b*. If the soil is entirely comprised of cohesive material, where it is essentially in a quick condition, two failures normally occur, as shown in Figure 9.10*c* and *d*:

(a) If the angle of slope is greater than 53°, the slope face is considered steep, and the failure usually occurs at the toe of slope.
(b) If the angle of slope is less than 53°, the failure usually occurs at the base of the toe.

Underlying soil conditions such as a rock layers can also cause failure at the base of the toe, as shown in Figure 9.10*e*.

Various methods are available to analyze slope stability. Interested readers are referred to Lambe and Whitman (1969) and Terzaghi and Peck (1966).

9.7 RETAINING WALLS

Structures to hold soil or other materials into position are called retaining walls. They can be constructed of wood, concrete, stone, masonry block, or steel. Figure 9.11 depicts various types of retaining structures.

9.7.1 Wall Pressures

The earth pressure exerted by a soil on a wall is not a constant value but depends on the yield of the wall. There are an infinite number of values, as depicted in Figure 9.12; however, one of three general conditions will occur:

FIG. 9.11 Examples of retaining structures. Source: Whitney Clark Huntington. *Earth Pressures and Retaining Walls.* John Wiley & Sons, Inc., 1957.

FIG. 9.12 Earth pressure relationships.

(a) Earth pressure at rest occurs when there is no wall movement.

(b) Active earth pressure occurs when the wall moves outward.

(c) Passive earth pressure occurs when the wall moves inward.

In the case of no wall movement either toward or away from the soil, the coefficient of at rest earth pressure is expressed as:

K_0 = coefficient of earth pressure at rest or horizontal pressure/vertical pressure

This value can range from 0.4 for loose sands to 0.5 for dense sands (Terzaghi and Peck, 1966). The horizontal pressure, as illustrated in Figure 9.12, is calculated at any given depth by

$$p_0 = K_0 wh \tag{9.2}$$

where K_0 is the coefficient of earth pressure at rest, w is the unit weight of soil in pounds per cubic foot, and h is any depth specified in feet.

If the pressure at the bottom of the wall is $K_0 wH$, where H is the total depth, the total horizontal force exerted on the wall's face is

$$E_0 = (K_0 wH)(H/2) \text{ per foot of wall length} \tag{9.3}$$

Here, E_0 acts $H/3$ from the wall's bottom.

Should the wall be allowed to yield slightly outward, the soil will follow the wall until on the verge of failure. At this point the soil's shear is maximum, and the pressure exerted on the wall has diminished since the soil is better able to hold itself. The resulting wall pressure is active earth pressure, which is commonly used in retaining wall analysis. In this case:

K_a = coefficient of active earth pressure or horizontal pressure/vertical pressure

The horizontal pressure at any depth is calculated by

$$p_a = K_a wh \tag{9.4}$$

where K is the coefficient of active earth pressure, w is the unit weight of soil in pounds per cubic foot, and h is any depth specified in feet. The total horizontal force exerted on the wall's face using H as the total depth is

$$E_a = (K_a w H)(H/2) \text{ per foot of wall} \qquad (9.5)$$

Should the wall be forced into the soil so that the soil is on the verge of failure, the soil resists, resulting in passive earth pressure. In the passive case,

$$K_p = \text{coefficient of passive earth pressure}$$
$$\text{or horizontal pressure/vertical pressure}$$

The horizontal pressure at any depth is

$$P_p = K_p w h \qquad (9.6)$$

where K_p is the coefficient of passive earth pressure, w is the unit weight of soil in pounds per cubic foot, and h is any depth specified in feet. The total horizontal force on the wall using H as the total depth is

$$E_p = (K_p w H)(H/2) \text{ per foot of wall} \qquad (9.7)$$

Two methods are commonly used to predict earth pressures: Coulomb's and Rankine's theories. In both theories assumptions are made that the wall is free to move and that water contained in the voids of the soil does not exert any appreciable seepage pressure (Terzaghi and Peck, 1966).

9.7.2 Coulomb's Theory

According to Huntington (1957), C. A. Coulomb published, in France during the year 1776, a theory for predicting earth pressures against a wall where an assumption is made that the wall's face is rough and that friction exists between the wall and backfill (see Figure 9.13).

The active earth pressure exerted against the wall's surface results from the tendency of the soil wedge to slide along *A–B*. The soil weight, *W*, is held in equilibrium by

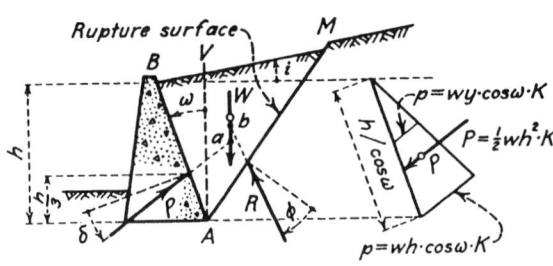

FIG. 9.13 Coulomb's active earth pressure relationships. Source: Whitney Clark Huntington. *Earth Pressures and Retaining Walls.* John Wiley & Sons, Inc., 1957.

a force on the plane of the retaining wall (P) and a force on the rupture surface of the soil (R). The angle of inclination of the rupture surface is assumed to be the plane that makes P a maximum value. To exert this pressure, the wedge must slide as the wall deflects away from the wedge, which further allows the soil to slide along the wall A–B, as illustrated in Figure 9.13. As a result, Coulomb assumed that the wedge has a failure surface A–M and a pressure surface A–B. In practice, the failure surface is curved rather than straight; however, the effect is negligible for active earth pressure calculations. To estimate the active earth pressure using Coulomb's theory, the following relationships apply:

$$P_a = \tfrac{1}{2}wh^2 \frac{\cos^2(\Phi - \omega)}{\cos^2 \omega \cos(\delta + \omega) \left[1 + \sqrt{\dfrac{\sin(\delta + \Phi)\,\sin(\Phi - i)}{\cos(\delta + \omega)\,\cos(\omega - i)}}\right]^2} \qquad (9.8)$$

where ω is the wall's angle referenced from the vertical, Φ is the soil's internal angle of friction, δ is the angle of friction between the soil and pressure surface, i is the backfill slope reference to the horizon, w is the unit weight of soil, and h is the vertical height equivalent of the pressure surface.

Although Coulomb did not include the analysis of passive earth pressures, his theory can be applied, with the resulting frictional forces acting in opposite directions. The equation for passive earth pressure would then become

$$P_p = \tfrac{1}{2}wh^2 \frac{\cos^2(\Phi + \omega)}{\cos^2 \omega \cos(\omega - \delta) \left[1 - \sqrt{\dfrac{\sin(\Phi + \delta)\,\sin(\Phi + i)}{\cos(\omega - \delta)\,\cos(\omega - i)}}\right]^2} \qquad (9.9)$$

9.7.3 Rankine's Theory

According to Huntington (1957) in 1857 W. J. M. Rankine developed a theory to estimate earth pressures pertaining to cohesionless soil deposits with a plane ground surface unlimited in lateral extent and depth. Rankine's theory is based upon the soil backfill having a horizontal surface and being supported by a frictionless vertical wall. Active earth pressure occurs as thin layers of soil parallel to the backfill's surface expand laterally. This expansion results in shear planes developing similar to those shown in Figure 9.14.

The coefficient of active earth pressure for Rankine's dry granular soil is defined by

$$K_a = \tan^2(45° - \Phi/2) \qquad (9.10)$$

and the passive coefficient by

$$K_p = \tan^2(45° + \Phi/2) \qquad (9.11)$$

where Φ is the soil's internal angle of friction. For a more complete treatment of active and passive earth pressures, the reader is referred elsewhere (Lambe and Whitman, 1969; Perloff and Baron, 1976; Terzaghi and Peck, 1966).

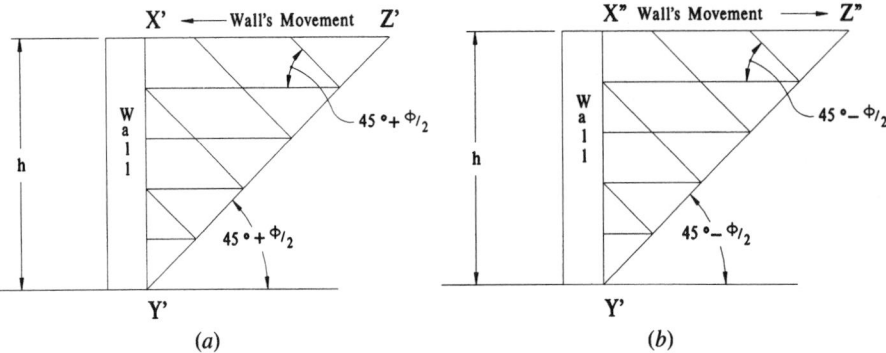

FIG. 9.14   Rankine's failure planes behind a frictionless wall: (*a*) active condition; (*b*) passive condition. Source: Karl Terzaghi and Ralph B. Peck. *Soil Mechanics in Engineering Practice.* Copyright © 1966 by John Wiley & Sons, Inc. Reprinted with permission of John Wiley & Sons, Inc.

9.7.4   Uncertainty of Cohesive Soil Backfills

Should cohesive soils be placed behind a retaining wall, active earth pressures would decrease while passive pressures would increase over those generated by cohesionless materials of similar unit weights. The presence of moisture could cause these soils to swell, creating very large pressures. Also, if adequate drainage is not maintained behind the wall, accumulated hydrostatic pressures can be significant in the analysis. It is therefore necessary that porous backfill be provided along with proper drainage design to eliminate these adverse field conditions.

9.7.5   Other Considerations

Most walls tilt outward when backfill is being installed. When walls are isolated, a small amount of wall tilting is usually acceptable. When the wall is long or consists of various height segments, wall joists may be necessary to allow differential movements. Extra care should be exercised to anticipate tilting if other rigid structures are to be connected to the retaining wall.

9.7.6   Stability Analysis Procedures

In analyzing a retaining wall several conditions must be satisfied, including safety against (a) overturning, (b) foundation failure, and (c) sliding.

The normal procedure for checking the stability of a retaining wall begins by choosing tentative wall dimensions and then estimating the earth pressures acting on the wall section developed.

The actual factor of safety against overturning is determined by comparing the total moment of forces resisting overturning about the toe of the wall with the moment of forces tending to cause overturning about that point.

The determination of safety against foundation failure involves three steps: (1) determining if the resultant of the footing contact pressure falls within the middle third of the footing width (but preferably within the middle sixth segment); (2) determining the actual factor of safety against bearing capacity failure by comparing the ultimate bearing capacity of the soil with the actual maximum footing contact pressure (or ensuring that the actual maximum footing contact pressure is less than the allowable soil bearing capacity); and (3) determining if anticipated total or differential settlements obtained using the principles of settlement analysis exceed criteria limits for settlement or tilt.

The actual factor of safety against sliding is determined by comparing the total horizontal forces resisting sliding with those tending to cause sliding.

Comparison of actual factors of safety determined above with values that are appropriate to the existing conditions are then made. Typical, acceptable factors are published in a number of current design texts. Should these calculations indicate an inadequate design, modifications to the initial wall's dimensions must be made and the process repeated until the design is found to be satisfactory. The structural design of the wall can then be undertaken using other guidance (Ferguson, 1979). Once the design is completed, it should be checked for slope stability, as indicated in Section 9.6.

9.7.7 An Example of a Retaining Wall Stability Analysis

Using Figure 9.15 as a guide, the footing thickness, t, can usually range from $H/8$ to $H/12$, with 12 inches customarily being the minimum. If $H/12$ is used for estimation purposes, the corresponding thickness is 1.42 feet. For ease of initial calculations 18 inches will be used. The corresponding stem height would be 15.5 feet (17.0 feet $-$ 1.5 feet). If the preliminary stem cap is assumed to be 12 inches, the stem base dimension can be estimated as 15.88 inches (12 inches $+$ 15.5 feet $\times$ 0.25 inch per foot). For convenience 16 inches will be preliminarily used. The footing width, W, should be as

FIG. 9.15 Example of cantilever retaining wall stability analysis.

small as possible for economic reasons; however, it must be large enough to provide stability and can range from $0.4H$ to $0.65H$, depending on the soil and surcharge conditions. To determine W using $0.65H$, calculate 0.65×17 feet, or 11.05 feet; therefore use 11 feet. The preliminary wall dimensions are shown in Figure 9.16.

The effective soil height (A–B) can now be determined by $H = 17$ feet $+ 5.67 \times \tan 18°$, or 18.84 feet. The active earth pressure can now be calculated using Equation 9.8, assuming that the wall friction (δ) has the same angle as the backfill surface slope (i):

$$P_a = \left[\frac{120 \times 18.84^2}{2} \right] \left[\frac{\cos^2(30 - 0)}{\cos^2(0) \cos(18 + 0) \left[1 + \sqrt{\dfrac{\sin(18 + 30) \sin(30 - 18)}{\cos(18 + 0) \cos(0 - 18)}} \right]^2} \right]$$

or 8408.2 pounds per foot of wall:

$$P_v = P_a \sin 18° = 2598.3 \text{ pounds}$$

$$P_h = P_a \cos 18° = 7996.6 \text{ pounds}$$

These forces act $H/3$ above the base (18.84 feet/3), or 6.28 feet.

Additional forces such as surcharge loads on the backfill surface, if present, would now be analyzed. Since none is included with this example, the active earth pressure forces would appear as shown in Figure 9.17.

FIG. 9.16 Preliminary wall dimensions.

FIG. 9.17 Active earth pressure forces acting on the wall.

The stem's base dimension now needs to be checked for preliminary structural adequacy. Interested readers are referred elsewhere for this analysis (Ferguson, 1979).

The wall's stability is determined taking moments about point A where all moments are counterclockwise:

Area	Vertical force (pounds)	Lever arm (feet)	Moment (foot-pounds)
1	$0.5 \times 5.67 \times 1.84 \times 120$	1.87	1,171
2	$5.67 \times 15.5 \times 120$	2.84	29,951
3	$1 \times 15.5 \times 150$	6.17	14,345
4	$(4/12) \times 0.5 \times 15.5 \times 150$	6.78	2,627
5	$11 \times 1.5 \times 150$	5.5	13,613
6	$2 \times 4 \times 120$	9.0	8,640
P_v	2598.3	0	0
P_h	7996.6	6.28	50,219
		Sum	120,566

To determine the location of the resultant using Figure 9.17, the following calculations are made:

Summation of all vertical forces = 19,918 pounds
Summation of moments at A = 120,566 foot-pounds
x = 120, 566 foot-pounds/19,918 pounds

or 6.05 feet from A is where the resultant acts.

The resultant's eccentricity is calculated by 6.05 feet − 5.5 feet, or 0.55 feet, and is within the middle third of the base (11 feet/3 = 3.67 feet situated with 1.83 feet on each side of the centerline). Therefore, the resultant is located in the middle third, but more importantly within the middle sixth. The distribution of soil pressure under the footing can be determined using

$$q = \left[\frac{\text{summation vertical forces}}{BW}\right]\left[1 \pm \frac{6e}{W}\right] \tag{9.12}$$

where B is equal to 1 foot of wall (Huntington, 1957);

$$q_{max} = \left[\frac{19,918}{1 \times 11}\right]\left[1 + \frac{6(0.55)}{11}\right]$$

or 2354 lb/ft²;

$$q_{min} = \left[\frac{19,918}{1 \times 11}\right]\left[1 - \frac{6(0.55)}{11}\right]$$

or 1268 lb/ft². Since $q_{allowable}$ was given as 3000 lb/ft², the supporting soil appears that it will not be overstressed.

The corresponding soil pressures at the base are shown in Figure 9.18a.

The wall must be checked for sliding, as shown in Figure 9.18b. Since the amount of embedment ($h = 3.5$ feet) is nominal and vegetation or erosion could affect the soil's strength, sliding should first be checked using the soil's friction with the sum of vertical forces constituting the normal load:

$$\text{Friction} = 19,918 \times 0.55 \text{ coefficient of friction}$$

$$= 10,955 \text{ pounds per foot of wall}$$

The horizontal component of the active earth pressure is 7996.6 pounds; therefore the factor of safety against sliding is

$$\text{FS} = \frac{19,918}{7996.6} = 1.37$$

Because the safety factor is less than the minimum recommendation of 1.5, a keyway needs to be incorporated to increase lateral sliding resistance, as illustrated in Figure 9.18c.

If it is desired to have a safety factor of 1.5 with soil removed to the bottom of the footing's toe, the keyway depth can be computed by

$$\text{FS} = \frac{\text{resisting force}}{\text{sliding force}}$$

$$1.5 = \frac{\text{resisting force}}{8996.6} \text{ or } 11,955 \text{ pounds}$$

The keyway must produce 1040 pounds (11,995 − 10,955).

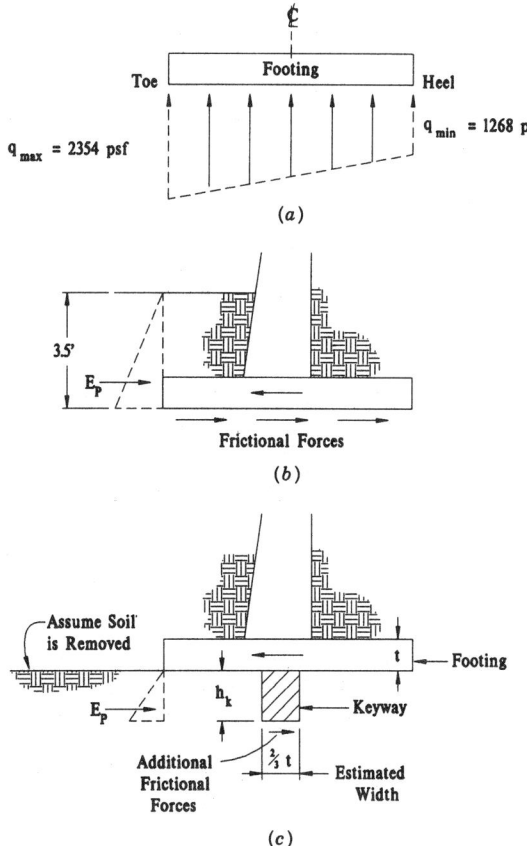

FIG. 9.18 Footing analysis for an example of a cantilever wall: (a) footing soil pressures; (b) passive earth pressure for sliding analysis; (c) keyway design.

Although there are a number of ways to design the key's depth, a conservative approach would be to use Rankine's formula with K_p being defined by Equation 9.11:

$$E_p = \frac{wH^2}{2}\left[\tan^2(45 + \Phi/2)\right]$$

and the additional frictional force produced by the reinforced concrete keyway's weight (150 lb/ft^3) using two-thirds of the footing thickness for its width is

$$1040 = \frac{120 \times H^2}{2}\tan^2 60° + (0.66)(1.5)(H)(150)(0.55)$$

$$= 60\,H^2\tan^2(60°) + 81.7H$$

$$180H^2 + 81.7H - 1040 = 0$$

$$H = \frac{-81.7 + 869}{360} = h_k$$

or $h_k = 2.2$ feet deep. The width is rounded to 1.5 feet. The keyway is usually located where the stem's reinforcement can extend into it. That requires the key's front face to be about 6 inches in front of the interior wall's face, as shown in Figure 9.19.

If the passive earth pressure produced by the full 3.5 feet of soil is considered, the factor of safety against sliding would be

$$E_p = \tfrac{1}{2}wH^2\left[\tan^2(45 + \Phi/2)\right]$$

or 5951 pounds, where H is comprised of the 3.5 feet of soil above the footing toe and 2.25 feet of soil in front of the keyway (5.75 feet total).

The resisting force is $(0.55)(19{,}918 + 150 \times 1.5 \times 2.25) + 5951$, or 17,185 pounds. The factor of safety is

$$\text{FS} = \frac{17{,}185}{7997}, \quad \text{or } 2.14$$

which is adequate. The final resultant force is determined by using:

Summation of moments previously calculated	120,566 foot-pounds
$1.5 \times 150 \times 2.25 \times 5.42$ keyway	2,744 foot-pounds
Total of all moments at A	123,310 foot-pounds

$$\text{summation vertical forces} = 19{,}918 + 1.5 \times 150 \times 2.25 = 20{,}424 \text{ pounds}$$

The resultant acts at distance x, or

$$\frac{123{,}310}{20{,}424} = 6.04 \text{ feet } - 5.5 \text{ feet or } 0.54 \text{ feet}$$

FIG. 9.19   Final cantilever wall dimensions.

$$q_{max} = \left[\frac{20,424}{1 \times 11}\right]\left[1 + \frac{6(0.54)}{11}\right]$$

or 2404 lb/ft^2;

$$q_{min} = \left[\frac{20,424}{1 \times 11}\right]\left[1 - \frac{6(0.54)}{11}\right]$$

or 1310 lb/ft^2.

Information about the size and placement of structural reinforcing steel following code requirements is available elsewhere (Ferguson, 1979).

The designed wall should now be checked for slope stability as outlined in Section 9.6.1.

REFERENCES

Ferguson, P. M. *Reinforced Concrete Fundamentals*, 4th ed. New York: Wiley, 1979, pp. 242–278.

Huntington, W. C. *Earth Pressures and Retaining Walls*. New York: Wiley, 1957, pp. 5–8, 51, 59–60, 70, 132, 408.

Jones, D. E., Jr., and K. A. Jones. "Treating Expansive Soils." *Civil Engineering*, pp. 62–65, August 1987.

Lambe, T. W., and R. V. Whitman. *Soil Mechanics*. New York: Wiley, 1969, pp. 122–185, 352–373.

Nelson, J. D., and D. J. Miller. *Expansive Soils—Problems and Practice in Foundation and Pavement Engineering*. New York: Wiley, 1992.

Perloff, W. H., and W. Baron. *Soil Mechanics—Principles and Applications*. New York: Wiley, 1976, pp. 100–149.

Terzaghi, K., and R. B. Peck. *Soil Mechanics in Engineering Practice*. New York: Wiley, 1966, pp. 140–141, 185, 188. Copyright ©1966 John Wiley & Sons, Inc. Reprinted by permission of John Wiley & Sons, Inc.

U.S. Department of Housing and Urban Development (HUD). *Land Planning Principles for Home Mortgage Insurance*. Washington: HUD, 1973, pp. 6-1 to 6-38.

SUGGESTED READINGS

Bergado, D. T., L. R. Anderson, N. Miura, and A. S. Balasubramaniam. *Soft Ground Improvement in Lowland and Other Environments*. Reston: American Society of Civil Engineers, 1996.

Carlson, R. F. (ed.). *Cold Regions Engineering—Proceedings of the Eighth International Conference on Cold Regions Engineering*. Reston: American Society of Civil Engineers, 1996.

Cernica, J. N. *Geotechnical Engineering: Soil Mechanics*. New York: Wiley, 1995.

Das, B. M. *Principles of Geotechnical Engineering*, 3rd ed. Boston: PWS-Kent, 1994.

Johnson, E. G., A. Phukan, and W. H. Haas (eds.). *Embankment Design and Construction in Cold Regions—ASCE Technical Council on Cold Regions*. Reston: American Society of Civil Engineers, 1988.

Jones, C. J. F. P. *Earth Reinforcement and Soil Structures*. London: Telford, 1996.

Liu, C., and J. B. Evett. *Soils and Foundations*, 4th ed. Englewood Cliffs, NJ: Prentice-Hall, 1998.

Sowers, G. F. *Soil Mechanics and Foundations: Geotechnical Engineering*, 4th ed. New York: Macmillan, 1979.

10 Sanitary Sewer System Design

Design of the sanitary sewer system usually takes place after the transportation and storm drainage systems are initially formulated, unless some unusual circumstances exist. For example, if an existing gravity sanitary sewer system's elevation at a point of proposed connection is high in relation to the ground elevation of the development site, the design of the sanitary system might be undertaken first. The cost of obtaining and placing borrow material to elevate the site (Section 9.2.2) must be evaluated in order to achieve the use of gravity flow when a situation such as this occurs. The designer must also determine how the fill will affect proposed street alignment, storm drainage configuration, and building siting.

The availability of new technology, however, allows various forms of alternative wastewater systems to be considered when dealing with difficult sites where gravity flow is not easily achieved. The U.S. Environmental Protection Agency (EPA)'s Small Community Outreach and Education (SCORE) Program promotes lower utility rates, conservation (of energy, water, materials, and land), and a reduction in urban sprawl. Some of these technologies are included in Section 10.3 and are further defined in a U.S. EPA bulletin (1992).

The design of a sanitary sewer can begin with the preparation of an engineer's report and the determination of wastewater flows, followed by the design of either individual disposal systems or a centralized wastewater collection and treatment system.

10.1 ENGINEER'S REPORT

Most regulatory review agencies require the design engineer to prepare a written report for a new sanitary sewer system prior to granting permits for the project's construction and operation. This report usually contains general project information (including phasing plans if an incremental approach is to be utilized), information pertaining to the collection system, a description of the anticipated waste along with its treatability and possible treatment processes, and a description of the treated wastewater outfall and impact on the receiving stream (or soil if land application is contemplated) along with emergency provisions. The report can also address how the proposed system would integrate with the Area Wide 208 Water Quality Management Plan (Section 2.5) as provided in the Clean Water Act. In addition, Section 401 of the Clean Water Act requires that each new point source discharge obtain a National Pollution Discharge Elimination System (NPDES) permit.

10.1.1 General Information

The project's owner is identified in the report to notify the regulatory review agency of the party responsible for the development. A detailed description of the proposed

498

facility, including collection, transport, and treatment, is included; all proposed unit operations and processes intended to be employed for treatment are described. Usually a flow diagram is provided to clarify the overall proposed sequencing of treatment units.

10.1.2 Description of Waste

The report can include the type of anticipated waste and all potential sources of waste generation. The amount of wastewater can be quantified in terms of average daily flow, minimum flow, and peak flow. Where proposed flows are anticipated, representative characteristics are usually based on comparable wastewater measurements. If some existing flows will be diverted into the proposed system, measurements of these flows should be obtained. Land developments usually include only domestic wastes; however, commercial and industrial wastes could be present. Combined wastes are described in terms of their proportions based on volumes and strength [biochemical oxygen demand (BOD) and chemical oxygen demand (COD) or total organic carbon (TOC)]. If raw or pretreated industrial wastewater is expected, a four-digit Standard Industrial code number must be provided, along with information on the process that generates the waste, its relative volume, and discharge frequency.

10.1.3 Characteristics of the Waste

Certain characteristics of the proposed untreated wastewater should be evaluated. (Section 10.9.1 includes typical domestic wastewater characteristics.) The wastewater's strength is described, including the BOD (five days, 20°C) and COD. Color can be included along with anticipated variations and treatability. The raw wastewater's pH is also of importance along with any anticipated fluctuations, as well as a description of the total and phenolphthalein alkalinity.

Should any heavy metals or noxious or toxic/hazardous compounds be anticipated, they must be identified and their concentrations quantified. Pretreatment of wastewater containing these types of materials is usually required.

Other materials resistant to biological degradation normally are identified along with their anticipated concentrations. Should the wastewater contain surfactants, nitrogen, or phosphorus, these should be identified and quantified.

10.1.4 Treatability of Waste

A description of the treatability of the anticipated wastewater is included where proposed pilot (field) or bench (laboratory) studies are described. Any applicable manufacturer's literature can be included along with a description of the anticipated quality and characteristics of the treated effluent.

10.1.5 Treatment Facility Location and Effluent Discharge

A description of the location of the proposed wastewater treatment plant is included along with the latitude and longitude of the proposed treated effluent discharge (including references to nearby landmarks). A location map may be provided for clarification. The discharge point could be some distance away from the treatment plant's site, in which case the effluent must be transported from the treatment plant in a long gravity outfall or pumped in a forcemain. Alternative effluent discharges can include irrigation or other land application practices.

10.1.6 Physical Characteristics of the Proposed Site

Should the project's scope include ground storage basins or provisions for spray irrigation, seepage, or composting, various prevailing site conditions must be investigated. These conditions can include identification of soils, percolation test results, drainage characteristics of the surrounding area, distance to the nearest existing wastewater treatment plant, distance to the nearest inhabited structure, distance to the nearest property line, and the location of all wells close to the water supply. Determination of the seasonal groundwater table depths also is important.

10.1.7 Development Configuration

The engineer's report usually includes a preliminary plan showing the layout of the proposed development. This plan can be a copy of the preliminary development plan that is described in Chapter 5.

10.1.8 Receiving Waters

A description of the receiving waters subjected to discharge of treated effluent usually is included, along with the minimum seven-day low flow that occurs on the average of once every 10 years (7Q10). Chemical and biological characteristics of the receiving waters should be included along with the results of any field sample surveys.

The engineer must identify all downstream areas that can be influenced by the treated effluent, including downstream water intakes, swimming areas, impoundments, and shellfish areas. It is important to identify any other discharges, both up- and downstream that might affect the receiving water's quality in combination with the proposed treated effluent.

10.1.9 Impact of Discharge on Receiving Waters

A description can be included of the quantitative and qualitative effects of the proposed discharge on the receiving stream based on local regulatory requirements as well as how the proposed discharge conforms to the Area Wide 208 Water Quality Management Plan. This assessment can consist of the amount of dissolved oxygen, dissolved solids, total suspended solids, and other parameters.

10.1.10 Emergency Provisions

The design engineer can describe provisions for standby power and routine maintenance when major system components are taken off-line. Should system failure or shutdown occur, a description of the anticipated consequences must include the expected quality of the effluent discharged during downtime and its effect on the receiving waters. The potential for health hazards, nuisance, and hazardous conditions to operators or the public also must be addressed.

10.1.11   Facility Consolidation

Other wastewater treatment facilities might exist close to the proposed new facility. In cases such as this, the Area Wide 208 Water Quality Management Plan may encourage consolidation of facilities; for example, the capacity of the existing plant might be upgraded to accommodate the additional flows. This could provide better service to end users and better environmental conditions. Alternatives such as this should be evaluated.

An analysis of the existing and ultimate plant capacities must be made to determine if an existing treatment facility is capable of accommodating the anticipated wastewater from the proposed development. Use of an existing facility can entail the installation of a wastewater pumping station to transport the waste either directly to the plant or to a sewer line that leads to the plant and that has adequate carrying capacity. Agreement with the existing facility's owner to accept and treat the proposed waste must be obtained by the developer. Evidence of this acceptance normally is included in the report.

10.2   WASTEWATER FLOWS

Wastewater flows can be comprised of domestic sewage (along with industrial wastes and processing water if applicable), inflow, infiltration, and possibly cooling water. In a new system, the engineer usually provides means to minimize the effects of inflow, infiltration, and cooling water. According to a Phase I Residential Water Use Study conducted in the Baltimore area (Geyer et al., 1963), approximately 89 percent of the water supplied for nonsprinkling domestic water use becomes contributory flow into sanitary sewers. Design of the wastewater system must account for anticipated peak-flow conditions.

10.2.1   Variations in Domestic Wastewater Flows

In the same Baltimore study, wastewater flows were found to be at a minimum between the hours of 2 A.M. and 6 A.M., while a morning peak occurred just before noon, and an evening peak between 6 P.M. and 10 P.M., as illustrated in Figure 10.1.

According to Imhoff (Novotny et al., 1989), typical per-capita sewage flows can range from 60 to 120 gallons per capita per day, with design peak flows between 1.5 and 10 times the average daily flow rate. Fair et al. (1966) suggest the ratios contained in Figure 10.2 can be used to estimate maximum and minimum flows, while other references suggest alternative values (Metcalf and Eddy and Tchobanoglous, 1981).

10.2.2   Flow Determination

Domestic wastewater flows can be estimated using fixture units when one fixture unit is equal to 7.5 gal/min. Figure 10.3 contains values for various fixtures.

Another method for estimating wastewater flows is the use of established unit loads similar to those contained in Figure 10.4.

The number of units of each establishment category gives the estimated averge daily

FIG. 10.1 Mean hourly domestic sewage flows. Source: Geyer et al., October 1963. Reproduced with permission.

flow, which can be adjusted by factors similar to those contained in Figure 10.2 to obtain maximum or minimum design flows.

If land developments similar to the one being designed exist near the proposed site, water usage records and/or field gaging surveys can be conducted to obtain data that can be used as a basis for design. This method is especially valuable if local jurisdictional requirements include the use of flow-reducing and water-saving devices. Other references contain fuller descriptions of this procedure (ASCE and WPCF, 1982; Metcalf and Eddy and Tchobanoglous, 1981).

10.2.3 Flows from Other Sources

Wastewater flows can originate from several sources, including commercial, industrial and process water, inflow, infiltration, and cooling water. Quantities of industrial,

Description Of Flow	Relationship To Average Daily Flow
Maximum Daily Flow	2 x Average Daily Flow
Maximum Hourly Flow	3 x Average Daily Flow
Minimum Daily Flow	0.67 x Average Daily Flow
Minimum Hourly Flow	0.33 x Average Daily Flow

FIG. 10.2 Residential sewage flows as ratios to the average. Source: Fair et al., 1996. Copyright © 1966 by John Wiley & Sons, Inc. Reprinted by permission of John Wiley & Sons, Inc.

process, and cooling water are dependent upon source activities, while some typical commercial flows are included in Figure 10.4. Inflow and infiltration often are difficult to determine. Inflow can occur from stormwater entering manhole covers or from downspouts and sump pumps, while infiltration can result from high groundwater entering the system through unsound joints in manholes or pipes. Older methods and materials used in the construction of sanitary sewer mains could produce pipe joints that are less tight than those customarily expected today. According to the ASCE and WPCF (1982), infiltration rates as high as 60,000 gallons per day per mile have been recorded. ASCE and WPCF (1982) also state that it is common to allow 30,000 gallons per day per mile infiltration for sanitary sewers, laterals, and house connections without regard to size on sewers not exceeding 24 inches in diameter. Improved construction materials and joint fittings have greatly reduced this potential for infiltration.

10.3 INDIVIDUAL ON-SITE WASTEWATER DISPOSAL SYSTEMS

In areas where public wastewater treatment or collection facilities are not available, individual waste disposal sytems can restore the wastewater prior to its return to groundwater. Adequate land area and soil depth are needed for conventional leaching (or absorption) systems. The site's groundwater must be deep enough to allow several feet of unsaturated flow below the absorption trenches and must have adequate separation from bedrock. Granular-type soils are most suitable for leaching system applications.

Safe disposal of domestic wastewater is necessary to protect the health of the public. An individual system such as this should not

(a) contaminate any drinking water supply;

(b) allow insects, rodents, or other carriers to create any health hazard;

(c) be accessible to children;

(d) pollute or contaminate any surface waters, including bathing, shellfish, or water supplies; and

(e) become a nuisance as a result of foul odor or unsightly appearances.

All components of an individual sewage disposal system, including septic tanks, absorption fields, and distribution boxes, should be at least 50 feet away from all ditches, wells, and bodies of water. No part of the individual disposal system should be closer than 10 feet to any building or parking area.

Suitable soil underlying leach fields can treat organic and inorganic materials and pathogens in the wastewater by filtration. The soil also acts as an exchanger, absorber,

Fixture or Group	Fixture Unit Value
Bathroom group consisting of a lavatory, bathtub or shower stall, and a water closet (direct flush, valve)	8
Bathroom group consisting of a lavatory, bathtub or shower stall, and a water closet (flush tank)...............................	6
Bathtub with $1\frac{1}{2}$ trap...	2
Bathtub with 2″ trap	3
Bidet with $1\frac{1}{2}$ trap...	3
Combination sink and wash tray with $1\frac{1}{2}$ trap........................	3
Combination sink and wash tray with food waste grinder unit (separate $1\frac{1}{2}$ trap for each unit)	4
Dental unit or cuspidor...	1
Dental lavatory	1
Drinking fountain	$\frac{1}{2}$
Dishwasher, domestic type	2
Floor drain	1
Kitchen sink, domestic type	2
Kitchen sink, domestic type with food waste grinder unit.........	3
Lavatory with $1\frac{1}{2}$ waste plug outlet....................................	2
Lavatory with $1\frac{1}{4}$ or $1\frac{3}{8}$ waste plug outlet	1
Lavatory (barber shop, beauty parlor, or surgeon's)	2
Lavatory, multiple type (wash fountain or wash sink), per each equivalent lavatory unit	2
Laundry tray (1 or 2 compartments)	2
Shower stall	2
Showers (group) per head...	3
Sink (surgeon's)	3
Sink (flushing rim type, direct flush valve)	8
Sink (service type with floor outlet trap standard)	3
Sink (service type with P trap)...	2
Sink (pot, scullery, or similar type)	4
Urinal (1 flush valve)...	8
Urinal ($\frac{3}{4}$ flush valve)	4
Urinal (flush tank)...	4
Water closet (direct flush valve)	8
Water closet (flush tank)	4
Swimming pools, per each 1000-gal capacity	1
Unlisted fixture, $1\frac{1}{4}$ or less fixture drain or trap size...............	1
Unlisted fixture, $1\frac{1}{2}$ fixture drain or trap size........................	2
Unlisted fixture, 2 fixture drain or trap size........................	3
Unlisted fixture, $2\frac{1}{2}$ fixture drain or trap size........................	4
Unlisted fixture, 3 fixture drain or trap size........................	5
Unlisted fixture, 4 fixture drain or trap size........................	6

Note: Values for continuous or intermittent flow. For a continuous or intermittent flow into a drain, as from a pump, ejector, air-conditioning equipment or similar equipment, a fixture value of 2 shall be assigned for each gpm of flow at rated capacity. The total discharge flow in gpm for any single fixture, divided by 7.5, provides the fixture unit value for that particular fixture.

FIG. 10.3 Sanitary drainage fixture units per fixture. Source: *New York State Uniform Fire Prevention and Building Code,* Table I-903, Division of Housing and Community Renewal, New York, NY. January 1984, pp. 276–277.

Activity	Gallons Per Day Per Person	Lbs. 5-Day BOD Per Day Per Person
Apartments - 3 Bedroom 4 Persons Each	100	0.17
- 2 Bedroom 3 Persons Each	100	0.17
- 1 Bedroom 2 Persons Each	100	0.17
- With Garbage Disposal Units	100	0.23
Churches - Per Seat	3	0.02
Clinics - Per Staff	15	0.03
- Per Patient	5	0.02
Country Club - Each Member	50	0.10
Food Service Operations		
- Ordinary Restaurant (Not 24 Hours)(Per Seat)	70	0.20
- 24-Hour (Per Seat)	100	0.30
- Curb Service (Drive In) (Per Car Space)	100	0.20
- Vending Machine Restaurant	70	0.12
Hotels - Per Bedroom (No Restaurant)	100	0.17
Laundries - Self Service (Per Machine)	400	0.68
Mobile Homes - 3 Persons Each	100	0.17
Offices - Per Person (no Restaurant)	25	0.05
Picnic Parks - Average Attendance	10	0.06
Residences - 4 Persons Each	100	0.17
- With Garbage Disposal Units	100	0.23
Rest Homes - Per Bed (No Laundry)	100	0.17
- Per Bed (With Laundry)	150	0.20
Schools - Per Person (No Showers, Gym, Cafe.)	10	0.04
- Per Person With Cafe. Only	15	0.05
- Per Person (With Showers, Gym, Cafe.)	20	0.06
Service Stations- Each Car Served	10	0.06
- Each Car Washed	75	0.03
- First Bay (per Day)	1000	2.00
- Each Additional Bay (Per Day)	500	1.00
Shopping Centers - Per 1,000 sq. ft. Floor Space (No Restaurant)	200	0.40
Swimming Pools - Per Person (With Sanitary Facilities and Showers)	10	0.04

FIG. 10.4 Typical unit contributory wastewater loadings. Source: South Carolina Department of Health and Environmental Control, 1972.

and a medium where chemical and biochemical processes can take place. These processes produce treated water that is acceptable for discharge into the groundwater under normal conditions of proper system operation. A thorough soils investigation must be conducted to study the soil characteristics in the proposed leaching area. This study can include the identification of the soil's texture, structure, and color as well as determining seasonal changes in the level of soil saturation. An observation well can also be installed for measuring seasonal groundwater fluctuations.

Individual wastewater disposal systems usually fail if the septic tank's sludge and septage are not pumped out periodically. Failure can also occur if the absorption field is subjected to more liquid than it is capable of assimilating or if the soil voids become biologically clogged. Systems subject to these failure conditions can exhibit undesirable attributes. The soil above the absorption field can remain soft and saturated and even hold standing black-colored water that appears oily. Offensive odors usually accompany conditions such as these. A well-designed system that is properly maintained will not have these symptoms. Other references are available for a more complete presentation on this important topic (Kaplan, 1987; Metcalf and Eddy et al., 1991; Salvato, 1992).

10.3.1 Septic Tanks

Individual wastewater disposal systems have usually been comprised of traditional septic tanks and absorption fields, similar to that illustrated in Figure 10.5.

The septic tank is a watertight container that receives wastewater from a dwelling unit. This wastewater contains (in addition to liquids and grease) solids such as grit, lint, food, and other materials, and is further defined in Section 10.9. Solids are separated from liquids in the tank. Floating material, including grease, is trapped along with accumulated liquids in the tank. Floating material, including grease, is trapped along with accumulated scum. Organic materials are digested and stored while clarified liquid is discharged to the absorption field for final treatment in the soil. Septic tank volumes can be 900 gallons for a one- or two-bedroom dwelling unit or 1000 gallons for a three-bedroom dwelling. This provides enough volume to accumulate sludge and scum. The tank can be inspected after the first year of operation to ensure that it is operating properly and then every five years when it can be pumped out if needed. Most septic tanks are constructed of precast concrete, fiberglass, or polyethylene. They must be watertight and must have sufficient weight or be anchored to prevent floating as a result of any buoyancy forces attributed to high groundwater.

The user must limit grease, disposable diapers, garments, and other sanitary products from being introduced into the waste system. Solvents, paints, chemicals, and sand can promote operational problems, as can garbage disposal wastes. When septage removal is required, the septage must be properly handled and disposed of, using approved regulatory procedures.

10.3.2 Conventional Gravel Absorption Field

Although requirements can vary from locale to locale, a conventional absorption field consists of a trench between 12 and 36 inches wide filled with at least 14 inches of clean coarse stone aggregate and a perforated 4-inch distribution pipe covered with a geotextile fabric. The trench, which is at least 24 inches deep, is covered by at least 9 inches of soil. The absorption field length is a function of the soil's ability to absorb and transport the wastewater away from the site. This can be determined by conducting a series of percolation tests on site and by analyzing the soil to determine the seasonal groundwater elevations. The amount of wastewater anticipated from the dwelling also affects the length of the absorption field. Local regulations usually provide guidance on the minimum leach field length based on percolation test results, the soil type, and the anticipated quantity of waste. Seasonal groundwater table elevation fluctuations also must be considered in the elevation and geometry of the absorption field. The maximum height of groundwater from the ground's surface should be at least 3 feet for proper treatment to take place.

FIG. 10.5 Individual wastewater disposal system, conventional method.

507

The slope of the trench bottom is made level. Parallel trenches should be separated by at least 5 feet to allow proper soil absorption. Usually 6 inches of coarse aggregate is placed under the distribution pipe and about 5 inches is allowed to surround the distribution pipe. Three inches of coarse aggregate can be placed on top of the distribution pipe and covered with geotextile fabric to prevent fine soil particles from clogging the coarse aggregate voids.

A conventional gravel absorption field is placed where gravity flow conditions can be obtained from the effluent of the septic tank throughout the perforated distribution lines in the absorption field. This method of effluent delivery to the absorption field will result in localized ponding in the absorption trench because of nonuniform liquid applications to the field. As a result, the absorption field develops a clogging layer on the soil interface. Clogging resulting from this phenomenon usually begins close to the absorption field inflow and moves progressively throughout the field. The clogging layer ensures unsaturated flow in the soil and helps reduce pathogens. One method used to reduce the clogging layer is to include a dosing tank (or pump) where the perforated distribution pipe is charged several times a day. This allows the distributed liquid to percolate into the soil and air to enter the absorption surface, thus reducing soil clogging.

A modification to the conventional system can be made where various absorption field trenches can be alternately used. This will prolong the useful life of the system and provide flexibility should a segment of the absorption field fail.

10.3.3   Serial Distribution Absorption Trenches

This system, which can be used on sloping sites, is depicted in Figure 10.6. A serial distribution system is comprised of a series of absorption fields that are filled with septic tank effluent beginning with the highest one. Once the trench completely fills, the effluent is allowed to overflow through one drop box to the next, which allows the effluent to flood all soil surfaces. This method allows the upper trenches to be used while not overusing the lower trenches. Upper trenches can be bypassed by capping to allow periods of rest.

FIG. 10.6   Septic tank with serial distribution. Source: U.S. EPA, 1992.

10.3.4   Leaching Chambers

Septic tank effluent is conveyed to bottomless concrete or arched plastic chambers similar to those shown in Figure 10.7. While stored in the chambers, the effluent floods the soil surface and seeps vertically through the chamber bottom.

10.3.5   Shallow-Trench Low-Pressure Pipe Distribution

This method utilizes small-diameter pipe that is placed closer to the ground surface than a conventional system. The septic tank effluent is pumped under pressure using a septic tank effluent pump (STEP) through small orifices to soak the entire trench network area. This method can be used on sites where elevated groundwater or shallow soils exist. Figure 10.8 illustrates a typical installation.

The STEP is located in a tank that has sufficient volume to allow a pumping duration of enough time for the pump to attain proper operating temperatures and allow lubrication to mobilize throughout the pump's moving parts. The number of times the pump is energized can range from three to five times per day, depending on the soil type, thus providing several absorption field doses that allow proper percolation and unsaturated flow in the soil to take place. Typically, the pump delivers at least 15 gal/min at the design head.

STEP installations require pump controls for "pump on," "pump off," and a high water alarm. The high water alarm is set just above the "pump on" level to alert the dwelling occupant if problems occur. Storage above the alarm provides time to correct a problem with the STEP. Because this system is installed underground, a waterproof electrical system is required.

FIG. 10.7   Septic tank with leaching chambers. Source: U.S. EPA, 1992.

FIG. 10.8   Shallow-trench low-pressure pipe distribution. Source: U.S. EPA, 1992.

STEP discharge piping is designed to discharge into a pressurized distribution device that splits the flow onto the absorption field or a pressurized distribution network. The pressured distribution device discharges to the gravity flow absorption field. The pressurized system's distribution piping contains pipe that has openings placed at intervals to uniformly distribute the liquid through the absorption field. This is accomplished by using small-diameter distribution piping and orifices that allow the distribution network to be charged with liquid prior to discharging into the absorption field. This provides uniform distribution, dosing features, and controlled absorption field loading, which promote aeration and absorption of the effluent.

When soil limitations prohibit the installation of a conventional gravel absorption field due to low percolation capability, high groundwater, or high bedrock, an artificially elevated absorption field can be considered. Aggregate in the raised bed must meet texture and uniformity limits while construction of the system requires specially trained and often regulatory certified contractors. This approach requires pumping the septic tank's effluent using a STEP from a low elevation to the elevated absorption field's higher elevation, as shown in Figure 10.9.

10.3.6   Evaporation and Adsorption Bed

On difficult sites located in areas where the climate consistently promotes water evaporation at a rate that exceeds the rainfall rate, the system shown in Figure 10.10 can be considered. In this system, septic tank effluent flows into gravel trenches or chambers that are placed on a mound of sandy soil. Some of the effluent may seep into the soil, while trees and vegetation that grow close to the absorption field pull liquid from the sand and transpire the water into the air.

10.3.7   Constructed Wetlands

This method can be used on difficult sites where septic tank effluent passes through a bed of rocks. Reeds can be planted on top of the rocks, as shown in Figure 10.11. The liquid evaporates and drains into a soil absorption system or is discharged after disinfection. This system can be used where additional treatment is required or where soils will not support absorption.

10.4   COLLECTION METHODS FOR OFF-SITE WASTEWATER TREATMENT

A centralized wastewater collection system can be considered in areas where individual wastewater disposal systems cannot be used due to limitations in topography, soil, groundwater table levels, space and setback requirements for drain field locations, or large numbers of dwelling units. If an existing centralized collection and treatment system is close to the site being developed or if the project under design is of sufficient size to support a centralized wastewater treatment facility, a centralized pressurized flow collection system or a vacuum collection system can be considered where a gravity system is not feasible. A pressurized collection system is comprised of individual septic tanks and STEP units or vacuum interfaces that discharge into a common pressure sanitary sewer main for treatment off-site. The pressure main terminates at a wastewater treatment plant or another collection system.

Cleanouts

Toe of Slope

Top of Slope

Top of Slope

Toe of Slope

Pressure Distribution Mound Plan View

Small Diameter Pressure Distribution Lines

From STEP

36"

Clean Out

Rock

Rock

24"

Natural Ground Level

Typical Section Of Filter Interior

Low Pressure Distribution Pipe

Granular Soil

Low Pressure Distribution System

Natural Soil

Pumped Effluent

1 - 1/2" Pressure Line

Check Valve

Outside Alarm with Red Flashing Light for Pump Malfunction

Ground Surface

Underground Electrical Conduit

Clear Well With STEP Pump

4" Gravity Flow Line From Septic Tank

Septic Tank

Dwelling

4" Gravity Flow Line

FIG. 10.9 Individual wastewater disposal system for high groundwater conditions.

511

FIG. 10.10 Evaporation and adsorption bed. Source: U.S. EPA, 1992.

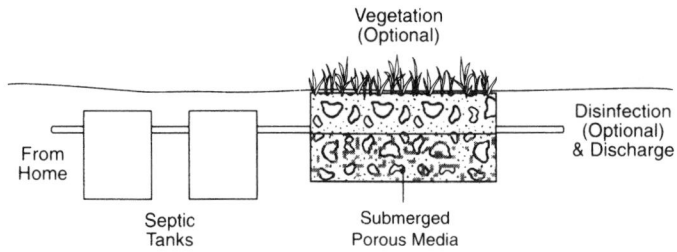

FIG. 10.11 Constructed wetlands wastewater disposal. Source: U.S. EPA, 1992.

An alternate system is comprised of a submersible grinder pump that pulverizes the fresh sewage and pumps the slurry into a common pressure main. Figure 10.12 illustrates a pressure flow collection system.

The design procedures for both STEP and grinder pumps are similar. The pressure main is the same in each case where small-diameter plastic pipe is buried below the frost line at least 36 inches deep. In either case, the design engineer usually selects one size pump for the project. This helps minimize confusion and provides compatibility between units for maintenance and repair work.

The pressure main should be designed where it is not a looped system, and cleanouts should be placed within at least every 1000 feet of pressure main. Where the pressure line is placed at a high point, an air release valve should be installed to vent trapped air and gas to the atmosphere. A vacuum breaker valve placed at the discharge side of the pump may also be required to prevent siphoning. Siphoning is undesirable because the liquid in various pump sumps can be removed when it occurs. This may result in solids being forced into the pump's intake when it is not in operation, resulting in clogging. Also, the removal of the liquid could induce a vapor lock within the pump's impeller, resulting in a loss of prime when the pump is later energized, causing pump failure. Size of the pressure main can be based on several criteria. One criterion could be to assume that each pump is capable of discharging at least 15 gal/min into the common pressure collection line on a random basis that is governed by individual dwelling demands. The common pressure collection line is then designed to maintain self-scouring velocities using contributory flow rates that equal approximately 90 percent (see Section 10.2) of the design peak hourly potable water demands, independent of lawn sprinkling and

Outside Alarm with Red Flashing Light for Pump Malfunction

Road R/W

¢ Road

Connection Box

1 - 1/2" Pressure Line

Sanitary Pressure Sewer

P_L

Check Valve

1 - 1/2" Pressure Line

Anti-Siphon Device When Vacuum Breaker is Needed

Ground Surface

Underground Electrical Conduit

Sanitary Pressure Sewer House Connection Details

Clear Well With STEP Pump

Scum Layer

Sludge

4" Gravity Flow Line From Septic Tank

Septic Tank

Dwelling

4" Gravity Flow Line

Elevated or "Dry" Gravel Filled Trench for Gas Scrubbing. Ground Water Must be Kept Below Trench. Same as French Drain, See Figure 7.61 for Detail.

Ground Surface

Air Release Valve

Forcemain

Air Release Valve Details

Sanitary Pressure Sewer Collection Line

45-deg Pipe Saddle

Gate Valve

Ground Surface

From STEP

Check Valve

Connection Box Details

FIG. 10.12 Centralized pressure flow wastewater collection system.

513

fire flow requirements (Chapter 11). This pressure collection system would therefore be sized to receive approximately 90 percent of the design peak hourly flows (minus sprinkling), whereas the potable water distribution system is used to deliver design peak hourly flows.

10.4.1   STEP System

Effluent pumped by the STEP system is extremely corrosive above the liquid level where hydrogen sulfide can exist to form sulfuric acid. This requires all valves, piping, and electrical controls to be made of materials not susceptible to degradation under these adverse conditions. Plastic is normally used.

Since the STEP is immersed in the liquid, standard cast iron parts can be used satisfactorily in the pump; however, the impeller should be made of bronze or plastic to prevent seizure. STEP system effluent has offensive odors that must be addressed where the delivered liquid is discharged, perhaps by the installation of a sewer gas scrubber. STEP effluent also gives off hydrogen sulfide and methane gas, both deadly in small quantities. This requires any work performed on a STEP system to incorporate forced clean air ventilation for worker safety.

Grease does not usually affect the operation of a STEP system because it is normally removed in the septic tank. A problem could result, however, if the septic tank is not periodically pumped of septage, resulting in the tank "burping" solids and/or grease into the STEP and pressure sewer. The discharge of solids may also affect the treatment systems, especially if a large soil absorption system is used.

10.4.2   Grinder Pump System

Grinder pumps are placed in a watertight chamber where the fresh wastewater is received from the dwelling. This wastewater can include grease that could possibly clog the pump and pressure lines. Velocities of flow greater than 2 ft/s must be maintained in the pressurized discharge lines during peak flows to help cleanse the pressure line.

One advantage of the grinder pump is that no sludge is formed because all wastes are evacuated; therefore, there is no need to periodically pump out the grinder pump's basin. Repair of grinder pumps can, however, be more costly than that for STEP pumps and can be required more often, especially if a high content of grit is present in the wastewater.

10.4.3   Vacuum Sewer Collection Lines

This system utilizes a storage well where wastewater accumulates until enough is present to activate a valve that allows air and sewage to enter the collection line and be transported toward the vacuum station. Normally one valve box serves a number of homes. (This is not the case with STEP or grinder pumps, which require individual power supplies.) A central vacuum pump in the vacuum station (shown in Figure 10.13) maintains a reduced pressure in collection lines. Sewage pumps are used to transport wastewater from the vacuum station to the treatment facility. Because both the vacuum pumps and sewage pumps exist in the vacuum station, electrical power requirements are centrally located. This system can offer certain economic advantages over traditional gravity collection systems.

FIG. 10.13 Vacuum sewer collection system. Source: U.S. EPA, 1992.

10.5 CENTRALIZED GRAVITY FLOW WASTEWATER COLLECTION SYSTEMS

For reasons similar to those cited in Section 10.4, to establish a need for centralized wastewater collection, a traditional approach can be to use a gravity flow collection system if the site's topography is adequate and it is economically feasible.

10.5.1 General Considerations for Gravity Flow Collection Systems

Sanitary sewer collection systems can be designed as gravity flow systems that operate under open-channel flow conditions. Factors affecting the design of a gravity flow collection system include horizontal control, vertical control, and hydraulic control.

Most regulations require the gravity sewage system to be placed some distance away from the potable water distribution system to protect the distributed water from contamination. Often local regulations provide for the sanitary sewer main to be located along one shoulder of the road while the potable water is placed along the other. For example, if a local road's geometric cross section provides for a 10-foot-wide strip between the back of curb to the right-of-way line, the sanitary sewer main could be placed along the center of this area, as shown in Figure 10.14.

10.5.2 Horizontal Control for Gravity Sewer Collection Systems

The horizontal alignment of the proposed system is established by placing manholes wherever the sewer main changes horizontal direction (and/or slope or pipe size). The road's centerline alignment can be used to aid calculations. Local regulations usually limit the maximum spacing between manholes to 300–400 feet on small sewers; therefore, on long straight runs of sewer main, manholes must be placed at prescribed intervals along the line.

Where sanitary sewer mains are located along street curves, chord bearings and distances between manhole centers can be determined that are referenced to the street's centerline alignment. Particular care must be exercised to ensure that the proposed gravity line segments do not cross over the edge of the right-of-way onto proposed building sites.

Figure 10.15 illustrates the establishment of horizontal control data for a sanitary sewer collection system located along a road. Key road stations such as PCs and PTs are excellent places to locate manholes because calculations and field staking during construction are simplified.

(Note differences between this gravity system and a pressure system shown in Figure 10.12)

FIG. 10.14 Example of the placement of a gravity sanitary sewer main.

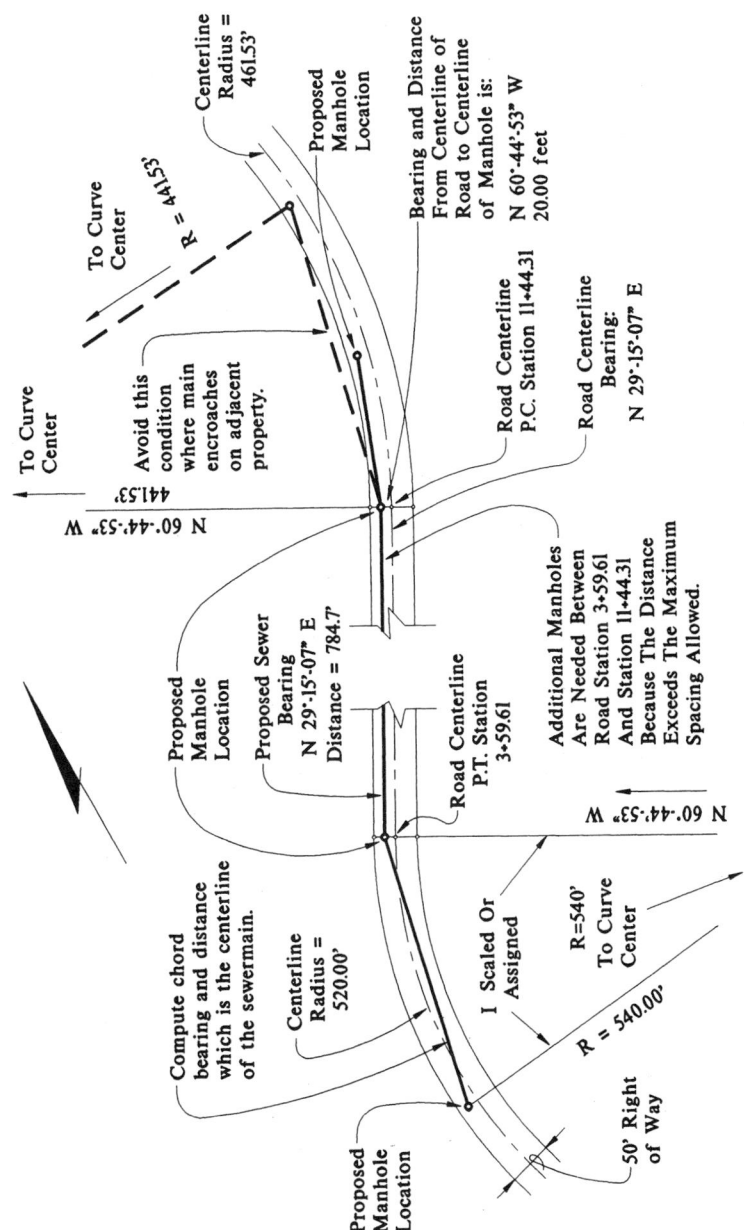

FIG. 10.15   Example of horizontal control relationships for gravity sewer main layout.

517

Gravity flow sanitary sewer mains need not extend the full length of all roads if smaller diameter lines can be used to provide service to each building site. The designer should keep in mind the slope of the topography in order to best locate the sewer mains for gravity flow. The site topography may require the engineer to place a portion of the proposed sewer main along various interior property lines of proposed building lots to take advantage of minimizing installation cuts. Mains placed in areas similar to this (off the road right-of-way) require a permanent easement for line protection and maintenance. Figure 10.16 illustrates several such conditions.

When completed, the horizontal alignment for the proposed sanitary sewer's centerline should be checked to ensure that the bearings and distances from centerline of manhole to centerline of manhole produce a closed traverse when tied to appropriate street horizontal control data. Additional horizontal control requirements for the wastewater system may exist if the point of discharge (connection) is into an existing sanitary sewer manhole several hundred feet away from the site along an existing road or through an adjoining landowner's property. In either case permission must be obtained for installation of the proposed new main prior to construction. Most highway departments require an encroachment permit to be submitted and approved that describes and shows the proposed system's features and its relationship

FIG. 10.16   Example of a sanitary sewer collection plan.

to the road. Where the connection must be made through another landowner's property, an easement must be obtained from that property owner. This easement usually entails a temporary construction easement where blanket ingress-egress is allowed during construction to provide working space for the contractor's workers and equipment, while a permanent easement of established width (possibly 20 feet with 10 feet along each side of the installed line) is provided. Final procurement of this easement is usually done by the developer with the technical assistance of the engineer.

10.5.3 Vertical Control for Gravity Sewer Collection Systems

Once horizontal control is established, the engineer can start design of the system's vertical control. According to Metcalf and Eddy and Tchobanoglous[*] (1981), sanitary sewer mains are usually designed for a minimum self-cleansing (scouring) velocity of at least 2 ft/s when flow is one-half or full depth, to prevent solids from settling. Some regulatory agencies recommend an 8-inch-diameter gravity sewer as a minimum size between manholes on all main lines. The minimum allowed slope for an 8-inch line may be regulated at 0.4 percent to provide this minimum self-scouring velocity. The maximum slope may be regulated to a slope that would minimize pipe erosion attributed to solids transport. Usually a maximum velocity of 8–10 ft/s will provide safety, according to Metcalf and Eddy and Tchobanoglous (1981). Another regulatory requirement might include provisions for at least 36 inches of cover to be placed on top of the main. If the soil cover is less than 36 inches, ductile iron must be utilized to protect against pipe collapse attributable to live traffic loads.

Vertical control is established by beginning with the invert elevation of the proposed point of connection. Each sewer manhole's invert elevation is calculated in sequence using an acceptable pipe slope and the horizontal distance between successive manholes. Figure 10.17*a* illustrates this process.

Should some economy of construction be available, the engineer can utilize a drop manhole to reduce the depth of cut on some lines, as illustrated in Figure 10.17*b*.

When a smaller diameter upstream sewer main joins a larger diameter downstream main at a manhole, the top inside pipe crowns are usually set at the same elevation to promote efficient hydraulic flows. In some instances where the site's topography is flat and the design engineer is attempting to conserve available fall, the pipe invert elevations can be matched if the partial-flow characteristics of the larger downstream main can provide adequate capacity. The full-flow discharge of the smaller upstream pipe must be conveyed by the larger line while adverse hydraulic flow conditions are prevented.

10.5.4 Hydraulic Considerations for Gravity Sewer Collection Systems

The hydraulic flow requirements must be examined once the system's horizontal and vertical control are established. Normally in a small land development the anticipated wastewater flows are much less than the full-flow capacity of a minimum-size 8-inch

---

[*] Source: Metcalf and Eddy, Inc. and George Tchobanoglous. *Wastewater Engineering: Collection and Pumping of Wastewater.* Copyright © 1981 McGraw-Hill Companies. Reprinted by Permission of McGraw-Hill Companies.

New Manhole No. 2

New Manhole No. 1

Existing Manhole Top Elev. = 100.00

New Manhole

Curb and Gutter
I.E. = 96.53

Edge of
Right of Way

Ground
Profile

Pavement

New
Manhole
No. 3

298.00'

198.65'

285.63'

300'

S = 0.4%

S = 0.4%

S = 0.4%

Existing
Gravity
Sewer
Main

I.E. = 96.13

I.E. = 95.34

I.E. = 94.20

I.E. = 93.00

Step 3

Compute the Invert
elevation of Manhole
No. 3 by 95.34 +
0.004x198.65= 96.13

Step 2

Compute the Invert
elevation of Manhole
No. 2 by :
94.20 +0.004x285.63=
95.34

Step 1

Begin computing invert
elevations at the
connection point by:
93.00+0.004x300=94.20

(*a*)

Ground Surface

Minimum cover must be maintained;
however, the drop allows this line
to be buried less deep than the
other lines using this manhole.

Drop
Manhole

Clean
Out

Additional
contributory
wastewater
flows

Upstream collection
lines require this
invert elevation for
gravity flow.

(*b*)

FIG. 10.17   Sanitary sewer gravity collection system, vertical control: (*a*) sample layout of a gravity sanitary sewer showing vertical control calculations; (*b*) drop manhole application.

sewer main. In large developments, however, flows can increase dramatically, and line sizing is critical. Each main must be sized for peak flows, and some regulatory agencies require a minimum peaking factor of 250 percent of the average daily flow to be used.

A certain quantity of water should be included along with domestic flows to account for groundwater infiltration. According to the ASCE and WPCF (1982), it is common practice to design for the peak anticipated rate of wastewater flow plus 30,000 gallons per day infiltration per mile of small to medium sewers and house connections. This allowance represents average conditions and can be revised by the engineer based on the physical characteristics of the area and the type of pipe joints used.

Figure 10.18 is used to illustrate the determination of peak flows in a proposed gravity system. For this example, it is assumed that inflow will not be a factor because proposed manhole covers will be constructed where stormwater cannot enter, and system users will not be allowed to place stormwater into the sanitary sewer system. The volume per capita per day for this example is based on Figure 10.4 where an amount of 100 gallons is obtained. Infiltration is assumed to be 200 gallons per inch of pipe diameter per mile of pipe per day based on pipe manufacturer's guidelines.

Collection line flows are calculated beginning at the remotest upstream point in the system. Contributory flows are accumulated while progressing downstream in the system to the point of connection. For example, the contributory flows for each line segment from manhole 1 to manhole 4 are determined as follows:

Flow from manhole 1 to manhole 2:

50 single-family dwellings $\times$ 100 gallons per person per day $\times$ 4 persons per dwelling $= 20,000$ gallons per day

$Q_{ave} = 13.88$ gal/min

$Q_{peak} = 2.5 \times 13.88 = 34.7$ gal/min

The length between manhole 1 and manhole 2 is 295.14 linear feet. Infiltration for an 8-inch-diameter line equals $(200 \times 8 \times 295.14)/(24 \times 60 \times 5280)$, or 0.062 gal/min. The total peak flow from manhole 1 to manhole 2 equals

$$Q_{tpeak} = 34.76 \text{ gal/min}$$

Flow from manhole 2 to manhole 3:

75 single-family dwellings $\times$ 100 gallons per person per day $\times$ 4 persons per dwelling $= 30,000$ gallons per day

$Q_{ave} = 20.83$ gal/min

$Q_{peak} = 2.5 \times 20.83 = 52.08$ gal/min

The length between manhole 2 and manhole 3 is 300 linear feet. Infiltration for this segment of 8-inch line equals $(200 \times 8 \times 300)/(24 \times 60 \times 5280) = 0.063$ gal/min. The total peak flow from manhole 2 to manhole 3 equals

$$Q_{tpeak} = 34.76 + 0.06 + 52.08 = 86.9 \text{ gal/min}$$

Flow from manhole 3 to manhole 4:

120 units apartments—3 bedrooms $\times$ 100 $\times$ 4 $= 48,000$ gallons per day

50 units apartments—1 bedroom $\times$ 100 $\times$ 2 $= 10,000$ gallons per day

$Q_{ave} = 40.3$ gal/min

$Q_{peak} = 2.5 \times 40.3 = 100.75$ gal/min

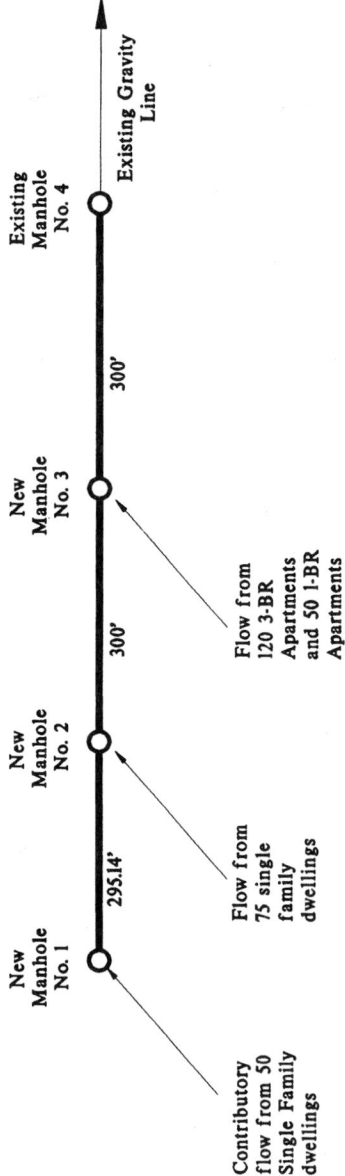

FIG. 10.18 Example of contributory flows into a sanitary sewer gravity collection system.

The length from manhole 3 to manhole 4 is 300 linear feet. Infiltration for this segment of 8-inch line equals $(200 \times 8 \times 300)/(24 \times 60 \times 5280)$, or 0.063 gal/min. The total peak flow from manhole 3 to manhole 4 equals

$$Q_{t\text{peak}} = 0.063 + 100.75 + 86.9 = 187.71 \text{ gal/min}$$

The full-flow rate under open-channel conditions for an 8-inch-diameter sewer main, using Equation 7.18 (Manning's $n = 0.011$) and a 0.4 percent slope, is

$$\begin{aligned}
Q &= \frac{1.486 A R^{0.67} S^{0.5}}{n} \\
&= \frac{1.486(0.349)(0.167)^{0.67}(0.004)^{0.5}}{0.011} \\
&= 0.89 \text{ ft}^3/\text{s (or 403 gal/min)}
\end{aligned}$$

It is evident that the flow capacity of an 8-inch-diameter main on a 0.4 percent slope is more than adequate to accommodate each of the above $Q_{r\text{peak}}$ flows. Should additional contributory flows be encountered further downstream, the analysis process would continue until the capacity of the 8-inch main was exceeded. Then the next available size main would need to be analyzed. If the regulatory agency having jurisdiction over the development requires gravity sewers less than 15 inches in diameter to be designed to carry peak flows when the pipe is half full, the designer must select a pipe size that accommodates this requirement.

10.5.5 Evaluation of the Preliminary Wastewater Sewer Collection System Design

After the required pipe sizes have been determined, the system must be studied relative to other previously designed improvements. This includes the road and storm drainage systems. First, the elevation of each storm drainage pipe (top outside of pipe and bottom outside of pipe) must be checked against the sanitary sewer main's top and bottom elevations where they cross on the plans. If there is an elevation conflict, either the storm drainage system must be raised or lowered, and/or the sanitary sewer system must be raised or lowered. In either case, the designer must repeat all design steps from the point of alteration to the point of solving the conflict in elevations. Usually if there is less than 1 foot of separation between the bottom of one pipe and the crown of a second, crossing pipe, a 20-foot section of cast iron or ductile-iron pipe is installed for the sewer main to protect the sewer main from being crushed.

The sanitary sewer should also be checked to ensure that at least 36 inches of cover is included over the top of all pipes. Otherwise ductile-iron pipe must be used, unless conditions warrant additional structural analysis of the pipe. On occasion ductile-iron pipe can be used to cross low topographical areas by placing the pipe on piers or pile bents for support at each pipe joint. Figure 10.19 illustrates this procedure.

An inverted siphon can be used in other areas such as ditch or river crossings (ASCE and WPCF, 1969).

A study of the relationships of the sanitary sewer gravity main elevations and those of proposed dwelling units should be undertaken to determine whether there is a potential

FIG. 10.19   Example of a ductile-iron pipe elevated gravity line.

for wastewater to back up into the dwellings if a stoppage occurs in the main line. Should the potential exist, check valves should be included on service lines leading to the vulnerable dwellings.

The design engineer should ensure that all building sites can be serviced by gravity flow. When a development site is comprised of sloping topography, buildings on the low side of a street (and the sanitary sewer main) may not have sufficient elevation to allow gravity flow to the sewer main located along the fronting street. This situation requires special provisions where the design engineer should investigate providing gravity access to the rear of the lots by using an easement through abutting property. If this cannot be accomplished, a STEP system can be considered for these problem lots.

This gravity collection system design assumes there is an existing gravity tie-in close to the site that has adequate hydraulic capacity. Many times this is not the case, and a sanitary sewer pumping station must be installed.

10.5.6   Extreme Slope Sewer Design

When sewers are designed for areas where slopes exceed 20 percent, drop manholes and changing horizontal alignments can be used to help moderate pipe flow velocities; however, the pipes and manholes must be anchored where the cost of installation is high when compared to sewers built in less steep terrain. An alternative design approach, using what is called a gravity forcemain (GFM) allows the designer to route the sewer along the shortest route where the gravity flow system operates under pressurized conditions (Orsatti, 1996). The design relies on the use of typical forcemain components and, in addition, an upstream screening device to prevent the line from becoming obstructed and the use of an energy-dissipating device (e.g., vortex chamber) prior to discharging into a conventional gravity sewer. The GFM pipe must be pressure rated and have restrained joints to resist thrust forces and anchoring devices. As flows vary throughout the day, there will be periods when solids will be deposited during low flow. Peak flows, however, should mobilize any settled wastes and flush the system. When designing a GFM, care must be taken to ensure that the pipeline profile is placed in such a position where the contractor can safely operate equipment, and still avoid extensive earthwork operations along the steep slope.

10.6   WASTEWATER PRESSURE TRANSPORT SYSTEM

A wastewater pressure transport system includes a centralized pumping facility and pressurized forcemain. This type of system relies on a collection system, which gathers the wastewater at a central location, and a destination point, where the wastewater can be discharged. This destination can be a wastewater treatment plant, another collection system, or another pumping facility. Most pumping facilities are nonclog centrifugal pumps of the submersible, flooded suction, or self-priming types, although small stations can utilize ejectors. A pumping station has a wetwell that is used for temporary wastewater storage to supply the pump during periods of pumping. The pumps can be located in a drywell next to the wetwell, above the wetwell, or in the wetwell if they are submersible.

10.6.1 Centrifugal Pump Characteristics

A centrifugal pump is used to add energy to a fluid flow system and is usually rated at a certain flow rate (capacity) at a particular pressure head. The pump's rotating impeller (or propeller) forces water to the pump's casing by centrifugal action. A centrifugal pump provides smooth transitional flows when system pressure levels change. When system pressures increase, the pump's capacity and power decrease. Likewise, when the system pressure decreases, the pump's capacity and power increase. At start-up, the required shaft torque initially is small.

A pump's rating can be governed by the pump's configuration or the motor providing energy to the pump. The water horsepower (HP_{water}) of a pump is represented by

$$HP_{water} = \frac{GPM \times TDH}{3960} \qquad (10.1)$$

where GPM is the flow rate in gallons per minute, TDH is the total dynamic head (when pumping at GPM) in feet, and 3960 is a constant.

A pump's efficiency is defined as the ratio of energy output to input. The energy output is the water horsepower while the energy input is the brake horsepower (HP_{brake}) applied to the pump. A pump's efficiency is therefore defined as

$$Eff = \frac{HP_{water} \times 100}{HP_{brake}} \qquad (10.2)$$

Pump capacity curves provided by the manufacturer show the relationships between the rate of pump discharge and head, efficiency, and brake horsepower. An additional curve showing the net positive suction head (NPSH) required indicates the head necessary to cause water to flow into the impeller's eye. The NPSH required varies based on the pump's speed and capacity.

A particular pump at a given speed has a unique set of performance values. When a pump is operated at maximum efficiency, the corresponding head and discharge provide the best efficiency point (BEP).

Some pumps operate at constant rotational speeds when in service while others can be operated at variable speeds. When shaft rotational speeds vary, the relationship between the pump's capacity, head, and brake horsepower also varies according to the affinity laws and specific speed relationships contained in ASCE's (1975) *Pipeline Design for Water and Wastewater.*[*] The affinity laws are

$$\frac{Q_1}{Q_2} = \frac{N_1}{N_2} \tag{10.3}$$

$$\frac{H_1}{H_2} = \left[\frac{N_1}{N_2}\right]^2 \tag{10.4}$$

$$\frac{HP_{brake1}}{HP_{brake2}} = \left[\frac{N_1}{N_2}\right]^3 \tag{10.5}$$

where Q is the pump's discharge capacity in gallons per minute, N is the pump's speed in revolutions per minute, H is the pump's head in feet, and HP_{brake} is the pump's brake horsepower when pumping Q gallons per minute.

The specific speed of a centrifugal pump is the relationship of a pump's discharge, head, and rotational speed when operating at its BEP. Specific speed can be used to predict the performance of one pump based on other single-stage pumps that are similar but of a different size:

$$\text{Specific speed} = \frac{RPM(Q)^{0.5}}{(H)^{0.75}} \tag{10.6}$$

where Q is the pump's discharge at BEP in gallons per minute, H is the pump's head in feet at the BEP, and RPM is the pump's rotational speed at the BEP.

A centrifugal pump can experience cavitation if the pressure at the edge of the impeller is reduced to the water's vapor pressure when vapor bubbles form. This reduces the pump's capacity and at the same time causes a pressure rise. The bubbles collapse and the cycle repeats rapidly with noise and vibrations resulting. Cavitation should be avoided because it can cause excessive impeller wear and premature pump failure.

10.6.2   General Considerations for Pumping Station Design

To design a sanitary sewer pumping station, several things must be known:

(a)   the elevation of the ground where the pump station is to be placed;

(b)   the invert elevation of the lowest influent line going into the wetwell;

(c)   the peak inflow going into the wetwell;

(d)   the distance the wastewater is to be pumped;

(e)   the high and low points in the discharge line (forcemain), which are usually shown in the form of a profile; and

(f)   criteria for pump control settings.

An example of the design of a suction lift (or self-priming) pumping station (pumps are located above the wetwell) follows, using as a general guide information contained in the DAVCO (1974) book:

---

*Reproduced by permission of the American Society of Civil Engineers.

A. *Determine the Required Rate of Pumping* If the average daily wastewater flow from a development is estimated at 200 gal/min and a peaking factor of 250 percent is to be used to etimate peak flows, the peak flow can be calculated as

$$Q_{max} = 2.5 \times Q_{ave} = 2.5 \times 200$$
$$= 500 \text{ gal/min}$$

Most regulatory agencies require small pumping stations containing constant-speed centrifugal pumps to maintain 100 percent standby pumping capability. To attain this, two pumps must be installed with each pump being capable of independently delivering 500 gal/min to the discharge point.

B. *Compute the Storage Requirements for the Wetwell* The wetwell should be sized for ultimate peak capacity (if economy dictates) even though the initial peak flows may be small. This can be accomplished by setting the pump control levels to accommodate the smaller flows initially. Normally, the wetwell should be pumped out every 10–15 minutes to prevent septic conditions; however, motors are not normally started more than three times per hour to minimize maintenance problems. Since this pumping station contains two pumps (providing 100 percent standby capacity in case of failure), a total of six cycles per hour are available to produce the maximum cycle rates. The time for one pump cycle is therefore

$$T = \frac{V}{Q - S} + \frac{V}{S} \tag{10.7}$$

where T is the time for one pump cycle in minutes, V is the effective volume of the wetwell in gallons, Q is the pumping rate in gallons per minute, and S is the flow into the wetwell in gallons per minute.

If Q is assigned a value of 250 percent of the average daily flow, or $2.5S$, and T equals 12 minutes to account for the time to cycle, the effective wetwell volume (volume of wetwell storage available for pumping) must be $7.2 \times S$. Therefore, the working volume to meet the criteria would be $7.2 \times Q_{ave}$. For this example, the volume would be 200 gal/min $\times$ 7.2, or 1440 gallons of working storage in the wetwell.

It is common to use reinforced concrete pipe culverts or precast manhole sections to construct circular wetwells similar to the one illustrated in this example. Figure 10.20 shows various volumetric relationships for several possible wetwell diameters.

For this example a 6-foot-diameter wetwell is selected. Using Figure 10.20 as a guide, each foot of vertical storage gives 212 gallons; therefore, the total required working depth is 1440 gallons of working volume/212 gallons per foot of depth, or 6.8 feet.

C. *Determine the Forcemain Diameter* The forcemain should be at least 4 inches in diameter and maintain a self-cleansing velocity greater than 2 ft/s. The maximum velocity should be less than 10 ft/s to minimize waterhammer problems. Using the continuity equation 7.18 with a flow rate of 500 gal/min, a maximum cross-sectional area of 0.557 ft^3 is required to provide this minimum velocity. If an 8-inch forcemain is chosen, the corresponding flow velocity at 500 gal/min is 3.19 ft/s, which is acceptable. This is determined as follows:

FIG. 10.20 Volumetric relationships for various wetwell diameters. Source: DAVCO—Division of Davis Water and Waste Industries. Inc., 1828 Metcalf Ave., Thomasville, Ga. 31792. Reproduced with permission.

(a)   An 8-inch pipe's cross-sectional area is determined:

$$A = 3.1415 \times (8/(2 \times 12))^2 = 0.349 \text{ ft}^2$$

(b)   The flow velocity using a rearranged form of Equation 7.18 is calculated:

$$V = Q/A$$
$$= (0.002228 \times 500 \text{ gal/min})/0.349 \text{ ft}^2$$
$$= 3.19 \text{ ft/s}$$

(The factor 0.002228 converts gallons per minute into equivalent cubic feet per second.)

Often the Hazen-Williams formula is used for pressure flow analysis, where

$$V = 1.318CR^{0.63}S^{0.54} \tag{10.8}$$

where V is the velocity of flow in feet per second, C is the Hazen-Williams coefficient of roughness, R is the hydraulic radius in feet equal to the cross-sectional area divided by the wetted perimeter, and S is the slope in feet per foot.

In cases where a pipe is circular and flowing full, Equation 10.8 can be rearranged in terms of the rate of flow, as illustrated in the equation

$$Q = 0.006756CD^{2.63}H^{0.54} \tag{10.9}$$

where Q is the discharge in gallons per minute, C is the Hazen-Williams coefficient of roughness, D is the inside diameter of the pipe in inches, and H is the head loss in feet per 1000 feet of pipe.

Another form of the basic Hazen-Williams equation used to solve directly for pipe frictional head loss is shown in the equation

$$H_{\text{loss per 1000}} = \left[\frac{Q}{(0.006756)(C)(D^{2.63})} \right]^{1.85} \tag{10.10}$$

where Q is the flow in gallons per minute, C is the Hazen-Williams coefficient of roughness, and D is the pipe diameter in inches.

Using Equation 10.10, the corresponding friction loss per 1000 feet of 8-inch pipe, using a Hazen-Williams C value of 140 for PVC, at the 500-gal/min rate, is 4.4 feet (or 0.0044 feet per foot of pipe).

D. *Determine the Pump Suction Line Diameter* The suction line should be sized for a velocity from 4 to 8 ft/s using an average of 6 ft/s. At 500 gal/min a 6-inch-diameter suction line yields 5.67 ft/s velocity according to Equation 7.18. This velocity falls within the acceptable range. The corresponding frictional head loss (assuming ductile-iron piping with a Hazen-Williams C value of 120) using Equation 10.10 is 23.75 feet per 1000 feet of suction line (or 0.02375 feet per foot of pipe).

E. *Compute the Total Dynamic Head (TDH) for the System Illustrated in Figure 10.21* The TDH is comprised of the dynamic suction lift and the discharge head. To calculate the liquid lift heights, the elevations must now be established for pump controls. In most duplex pumping stations similar to the one included in this example, an electronic alternator is provided in the motor control panel to alternate pumps through each subsequent cycle of operation. If the left-hand pump is labeled pump 1 and the right pump 2, the alternator would initially place pump 1 on line as the lead pump to pump down the wetwell when the inflow raises the liquid level to a predetermined normal maximum height designated as the "lead pump on" elevation in the wetwell. Upon the removal of accumulated wastewater in the wetwell to a predetermined low level called "pumps off" using pump 1 (pump 2 on standby), the alternator would reverse pump roles. Pump 2 would then become the lead pump for dewatering for the subsequent cycle (pump 1 would be placed on standby) once the inflow raises the liquid level in the wetwell up to the "lead pump on" elevation. By alternating the pumps, they both receive equal wear and are both kept operational by continuous exercise. If one pump was constantly designated as the lead pump, the standby pump would probably freeze from nonuse and would not be capable of coming on line when required for backup.

The standby pump, called the lag pump, is available if the lead pump should malfunction and trip out. The standby pump also can be energized if the wetwell inflow exceeds the lead pump's capacity and liquid continues to rise in the wetwell. Both constant-speed centrifugal pumps can be used simultaneously in this case to pump the wetwell down.

FIG. 10.21 Site conditions for a sample sanitary sewer pumping station design.

If local regulations require 6 inches of vertical distance between the influent invert elevation into the wetwell and the "lag pump on" elevation and 6 inches of vertical distance between the "lag pump on" and the "lead pump on" elevation, then the wetwell elevations can be established. The influent invert elevation for this example is calculated using the invert elevation of the closest manhole away from the pumping station while providing acceptable slope (0.4 percent) to the connection line. For this example the manhole's invert elevation is 54.65 feet, as shown in Figure 10.21. The influent invert elevation is calculated as 54.65—0.004 feet per foot × 271 feet, or 53.57 feet. Figure 10.22 shows the completed pump control elevations in the wetwell.

(a) Compute

　　(1) the dynamic suction head comprised of the suction lift distance from the low water level in the wetwell to the eye of the impeller that is 2 feet above the ground elevation and

FIG. 10.22 Example of wetwell elevation settings.

(2)   the frictional head loss in the suction line based on Step D as 0.0238 feet of head loss per foot of suction pipe (extending 2 feet below the pump's off elevation) at 500 gal/min flow.

$$\text{Static suction lift} = (64.05 - 45.77) + 2 = 20.28 \text{ feet}$$

$$\text{Friction head loss} = (0.0238)(2 + 64.05 - 43.77) = 0.53 \text{ foot}$$

Minor head loss determination requires knowledge of the velocity head in the pipe, determined using

$$\text{Velocity head} = V^2/2G \tag{10.11}$$

where V is the flow velocity in feet per second and $G = 32.2 \text{ ft/s}^2$.

Figure 10.23 shows various minor head loss coefficient values that can be multiplied by the velocity head to determine the minor head loss value. Minor losses assumed in this example in the suction line piping are as follows:

Minor Losses	Coefficient from Figure 10.23
2, each 90° ells	1.4
1, each gate valve	0.2
Bell entrance	0.23
Total	1.83

$$h_1 = K(V^2/64.4) = 1.83 \times (5.67^2/64.4) = 0.91 \text{ feet}$$

Total dynamic suction head then equals the sum of the component head losses, or 21.72 feet.

(b)   Compute the discharge head. The static discharge head is the difference between the elevation of the discharge point (also the high point in this case) and the elevation of the eye of the impeller. The frictional head loss attributed to the forcemain is found by utilizing the head loss per foot of forcemain found previously in Step C as 0.0044 feet of head loss per foot of forcemain when conveying 500 gal/min. The static discharge head equals $72.80 - 66.05$, or 6.75 feet, while the forcemain friction loss equals 1885 feet of forcemain × 0.0044 feet per foot, or 8.30 feet. Minor head losses assumed for this example are given below:

Minor Head Losses	Coefficient from Figure 10.23
Discharge into manhole	1.0
6, each 90° ells	4.2
1, each check valve	2.5
1, each gate valve	0.2
2, each 45° ells	1.0
Total	8.9

$$h_1 = K(V^2/64.4) = 8.9 \times (3.19^2/64.4) = 1.41 \text{ feet}$$

Fitting or Condition	Minor Headloss Coefficient, K
Gate Valve	
3/4 closed	24.00
1/2 closed	5.60
1/4 closed	1.20
Full Open	0.20
Angle Valve Open	2.50
Globe Valve Open	10.00
Swing Check Valve	0.6-2.5
Elbows	
90-deg standard	0.70
90-deg long radius	0.60
45-deg standard	0.50
Tee Flow Through Run	1.80
Sudden Contraction*	
d/D = 1/4	0.40
d/D = 1/2	0.30
d/D = 3/4	0.20
Sudden Enlargement*	
d/D = 1/4	0.90
d/D = 1/2	0.60
d/D = 3/4	0.20
Flow From Storage Tank	
Pipe Projecting Into Tank	0.78
Pipe Flush With Tank Wall	0.50
Slightly Rounded Entrance	0.23
Well Rounded Entrance	0.04
Flow Into Storage Tank	
All Conditions	1.00

* For Contractions and Enlargements, coefficients
are applied to the conditions in the smaller pipe
diameter.

FIG. 10.23   Minor head loss factors. Source: American Society of Civil Engineers, *Pipeline Design for Water and Wastewater,* 1975. Reproduced by permission of the American Society of Civil Engineers.

The total discharge head loss equals 6.75 + 8.30 + 1.41, or 16.46 feet.

The TDH equals the total dynamic suction head plus the total dynamic head: 21.72 + 16.46, or 38.18 feet.

A pump capable of pumping 500 gal/min at 38.18 feet TDH is required. According to Figure 10.24, the Burks Pumps (Pique, Ohio) performance curve, a model 4Cφ4D (Impellor Diameter 8D), 10 horsepower at 1650 rpm, will be sufficient with an efficiency of 55 percent.

F. *Compute the Required Brake Horsepower* The brake horsepower required to pump under design conditions using a pump of known efficiency can be determined using the equation

$$HP_{brake} = \frac{GPM \times TDH}{3960 \times Eff} \tag{10.12}$$

FIG. 10.24 Example of a self-priming pump performance curve. Courtesy of Burks Pumps, Inc., 420 Third Street, Piqua, OH.

534

where GPM is the flow rate in gallons per minute, TDH is the total dynamic head of the system in feet when delivering the required flow rate, 3960 is a constant, and Eff is the pump efficiency expressed in decimal form. For this example the required motor size is

$$HP_{brake} = \frac{GPM \times TDH}{3960 \times Eff} = \frac{500 \times 38.18}{3960 \times 0.55} = 8.76 \ HP$$

Ten horsepower is therefore adequate. For final motor selection there may be a requirement for the motor to be "nonoverloading" at every point along the performance curve. If this is a requirement, the engineer should check to see that the motor size is greater than the maximum horsepower required to drive the pump at all rates encompassed by the pump rating curve.

G. *Check the Net Positive Suction Head (NPSH)* The NPSH is the head required to lift the liquid into the eye of the pump's impeller so that the pump will remain primed during operation and help to eliminate cavitation. This relationship is shown in Figure 10.25:

$$Available \ air \ pressure = +33.83 \ feet$$
$$Dynamic \ suction \ lift = -21.72 \ feet$$
$$Vapor \ pressure \ at \ 74°F = -1.00 \ feet$$
$$Factor \ of \ safety = -2.00 \ feet$$
$$Total \ NPSH \ available = +9.11 \ feet$$

The selected Burks pump, a model 4Cφ4D (Impeller Diameter 8D), requires 13 feet of NPSH. There is not enough NPSH available to allow proper operation of this model pump; therefore, another pump must be chosen.

The curve for an alternative pump, Burks 6Cφ5D (not shown here) is checked and found to be capable of delivering 500 gal/min at 38.18 feet TDH using 1000 rpm. The efficiency is 55 percent; therefore, the 10 horsepower is adequate. The NPSH required for this particular model is approximately 4 feet, which is much less than the available 9.11 feet; therefore, this pump can be utilized.

H. *Compute the Buoyant Forces* The buoyant forces affecting the pump station must be computed to ensure that it will not float out of the ground as a result of upward pressure from high groundwater.

Upward forces attributed to groundwater, using the worst case situation, occur when the soil is saturated to the ground's surface. The total upward forces would equal the weight of displaced water, which is the wetwell's volume time the unit weight of water. For this example, the upward force is equal to the product of the π, the wetwell's radius squared, the wetwell height, and the unit weight of water. Assuming the 6-foot-diameter wetwell has a 7-inch-thick wall and bottom, the outside diameter of the wetwell is 7.17 feet. Numerically the upward force is equal to $3.14 \times 3.58^2 \times (64.05 - 43.27 \ feet) \times 62.4 \ lb/ft^3$, or approximately 52,210 pounds.

The downward forces contributed by the wetwell wall weight and bottom, using for the unit weight of concrete 150 lb/ft^3 foot, are as follows:

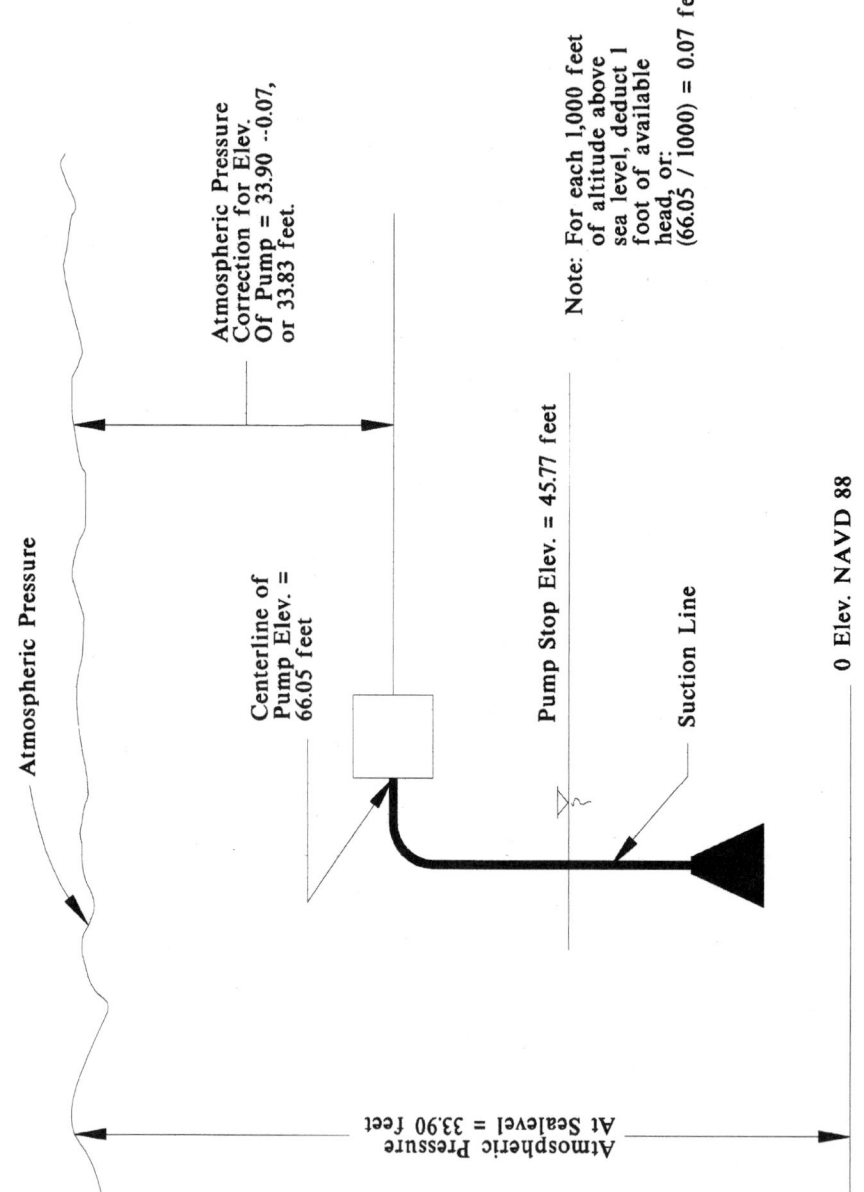

FIG. 10.25 Example of calculations for net positive suction head (NPSH).

Atmospheric Pressure

Atmospheric Pressure
Correction for Elev.
Of Pump = 33.90 --0.07,
or 33.83 feet.

Centerline of
Pump Elev. =
66.05 feet

Pump Stop Elev. = 45.77 feet

Suction Line

Note: For each 1,000 feet
of altitude above
sea level, deduct 1
foot of available
head, or:
(66.05 / 1000) = 0.07 feet.

0 Elev. NAVD 88

Atmospheric Pressure
At Sealevel = 33.90 feet

(a) The downward force contributed by the wetwell's wall is approximately equal to the product of PI, the wetwell's interior diameter, the wall thickness, the wetwell height, and the unit weight of concrete. Numerically the wall's downward force is 3.14×6 feet $\times 0.58$ foot $\times 20.78$ feet $\times 150$ lb/ft^3, or 34,077 pounds.

(b) The downward force contributed by the wetwell's bottom (assuming it to be 8 inches thick) is $3.14 \times 3.58^2 \times [8$ inches/(12 in./ft)$] \times 150$ lb/ft^3, or approximately 4026 pounds. The total downward force estimated for the wetwell and bottom is 34,077 pounds plus 4026 pounds, or 38,103 pounds. To ensure that this wetwell will not float, at least an additional 14,107 pounds must be provided by the pump station's floor slab, walls, and roof.

Several general considerations about the design of pump stations follow:

(a) The station should be sized for the ultimate "built out flow" expected during the life of the station if economically feasible. The wetwell may be oversized initially; however, pump control elevations can be adjusted to accommodate the actual flows encountered. These elevations are sensed by either mercury float switches or a pressure sensing system operated by air bubbles.

(b) The shutoff head (the head with the pump having its discharge closed off) of a pump should exceed the static lift from the low water level in the wetwell to the high point in the forcemain plus all upward gradient heads in the forcemain. On system start-up the pump must overcome all of these accumulated heads to charge the forcemain. Figure 10.26 illustrates this procedure.

(c) A system head curve must be made to check the performance of the selected pumps under various conditions. For example, to determine the pumping rate with both pumps on in parallel operation, a system curve first must be constructed. The static head needs to be plotted for zero flow. This can be based on the maximum static lift in the system. In this example it is equal to 72.80 feet − 45.77 feet, or 27.03 feet. The TDH values must now be computed for various pump flows based on this example's system. A summary of these calculations results follows, based on the low water level in the wetwell for static lift determination:

Q, gal/min	TDH, feet
0	27.03
100	27.56
200	29.00
300	31.27
400	34.33
500	38.18
600	42.79
700	48.17
800	54.28
900	61.15
1000	68.75

A similar curve using the high water level in the wetwell for static head determination can also be prepared but is not included here. Both curves provide a

FIG. 10.26 Example of minimum shutoff head determination.

range of values within which the pumps can be expected to operate, depending on where the wetwell liquid is located.

The pump performance curve for the chosen pump can now be superimposed on the graph. The intersection of the two curves yields the rated capacity of one pump and the system as shown in Figure 10.27. If both pumps are energized due to inflows greater than the capacity of one pump, the combined pumping capacity can be determined by using the system curve. For pumps in parallel, as illustrated by the example in Figure 10.28, the flows are added while the corresponding heads remain the same. Figure 10.27 illustrates that the pumping rate does not double since it increases from 500 gal/min to approximately 610 gal/min. If the pumps were able to be placed in series operation, as illustrated in Figure 10.29 (not usually the case in a normal duplex pumping station configuration), then the combined pumping curve would be obtained by adding the heads while the flow rates remain the same. In this case the pumping rate would increase from 500 gal/min to approximately 900 gal/min.

(d)   When flooded suction pumps are used, the low water level of the wetwell should be kept above the eye of the pump's impeller to provide positive internal pressure. This helps eliminate vapor locks and prevent cavitation.

(e)   If the station's interior piping is comprised of various pipe diameters, then any horizontal eccentric fittings must be oriented with the flat side up (eccentric down) to eliminate pockets that can trap vapor.

(f)   Discharge piping should be arranged to eliminate solids from being deposited into adjoining pipe segments where blockage could result. Figure 10.30 illustrates a condition that should be avoided.

FIG. 10.27   Example of pump and system curves.

FIG. 10.28 Elevation schematic of a flooded suction wastewater pump station using a parallel configuration. Source: ITT A-C Pump. *Pick A Pump*, 1981. Pump reproduced by permission of ITT A-C Pump, Cincinnati, OH.

(g) Surge control devices should be provided where long forcemains exist or on forcemains that have unusual flows or geometry. One method is to provide an air release/vacuum breaker valve at all high points in the forcemain and a surge relief valve (valve that vents excessive internal pressure from the forcemain) at the pumping station. Another method is to provide pump control valves that regulate the rate of valve closure. Other methods are also available. Tullis (1989) provides a more detailed discussion on hydraulic transients.

(h) To vent accumulated gas, air release valves must be installed along the forcemain where high points occur. Vacuum breaker valves may also be required at high points to relieve negative pressures in the forcemain that could collapse the pipe walls or create leaks around joint seals.

In addition, forcemains require thrust blocking at bends and junctions similar to that required on potable water distribution mains (Section 11.8.2) in order to maintain water-tight integrity of the forcemain.

(i) Proper interior ventilation in all pumping station drywells requires a complete change of air at least every 2 minutes. The wetwell must be vented to the atmosphere. An activated carbon air filter can be installed on the vent line to reduce complaints of foul odor from residents located close to the pumping station.

(j) Dehumidifier devices are required to protect equipment from moisture.

(k) Sump pumps are required to dewater pumping stations with drywells.

(l) Adequate access for equipment maintenance must be considered in the design process. Some pumping stations require the removal of roofs or hatches to allow access to installed equipment. A crane may also be required for equipment handling. Considerations for equipment repair and removal are paramount during the design process, and provisions for ease of access are important. Figure 10.28 shows a pumping station where the installed pumps can be easily removed by a truck boom hoist by simply opening an exterior roll-up door and removing a cast-iron gate in the ground-level floor slab of the station.

Due to long-term pump station operating expenses associated with labor and power, a cost-effective design is imperative. One method to analyze a pump station design is to perform a cost analysis on several combinations of wetwell and forcemain diameters. Each variation yields different system head losses, thus requiring different horsepower motors. The cost per gallon of pumping in dollars can be obtained using the following equation:

$$\text{Cost per gallon of pumping} = \frac{(\text{power cost})(\text{TDH})}{(317{,}460)(E_p)(E_m)} \qquad (10.13)$$

where power cost is the cost in dollars for electricity per kilowatt hour, TDH is the total

FIG. 10.29 Two flooded suction wastewater pumps in series. Source: ITT A-C Pump. *Pick A Pump*, 1981. Pump reproduced by permission of ITT A-C Pump, Cincinnati, OH.

FIG. 10.30 Elevation schematic of a parallel flooded suction pump station discharge piping system that should be avoided.

dynamic head in feet, E_p is the pump efficiency, E_m is the motor efficiency, and 317,460 is a constant. A graph can be constructed to show the various combined costs attributed to the alternatives; from this graph a desirable choice of system relationships can be made. In the example provided an 8-inch forcemain and a 6-foot-diameter wetwell yield the most economical solution for the analysis period used. The calculations utilizing a 6- and 10-inch forcemain are not included. Figure 10.31 illustrates this process.

Although the 4-foot-diameter wetwell curve is included in Figure 10.31, the NPSH available at the site (0.38 foot) would negate its use.

Local regulations may set forth other requirements; however, the general design process for a pumping station has been presented.

As design calculations are usually subject to outside review, it is imperative that the engineer conduct the analysis in a clear and logical manner so the reviewing party can follow the calculations without difficulty.

10.7 PUMPING STATION DISCHARGE

Discharge from a proposed pressure forcemain can be placed into another pumping station's wetwell, another forcemain, a gravity line, or directly into a wastewater treatment plant. The receiving facility should be capable of handling the proposed forcemain discharge along with other system flows.

10.7.1 Discharge into Another Pumping Station

If the pumping station under design is to discharge into another pumping station's wetwell, several factors must be investigated including:

(a) The determination of whether sufficient working storage capacity is available in the receiving pumping station's wetwell to accommodate the proposed additional flows.

Self Priming Pumping Station			
Job:			
Date:			
I. Site Conditions			
A. Finished Floor Elevation of Pumping Station, ft. =			64.05
B. I.E. Influent line, ft. =			53.57
C. Distance I.E. Influent to H.W. Alarm, ft. =			0
D. Distance High Water Alarm to 2nd Pump On =			0.5
E. Dist. 2nd Pump to 1st Pump On =			0.5
F. Distance Pumps Off to Low Water Alarm=			0.5
G. Average Daily Flow, GPM =			200
H. 'C' Value for Suction Line =			120
I. 'C' Value for Forcemain =			140
J. Hi Point Elev, in F.M., ft. =			72.8
K. Elevation of Discharge Point, ft., =			72.8
L. Length of Forcemain, ft. =			1885
M. Cost of Electricity, kwh =			0.06
N. Height Pump Above F.F., ft. =			2

II. Line Sizing

(Peak flow, gpm=)   500

A. Compute Suction Line Sizes

Pipe Dia	Velocity	Pump Station Cost Schedule		
(in.)	feet per second	Diameter	4	6	8

Pipe Dia (in.)	Velocity feet per second	Diameter	4	6	8
4"	12.77	Depth			
6"	5.67	8	$10,000.00	$14,000.00	$16,000.00
8"	3.19	12	$20,000.00	$26,000.00	$30,000.00
10"	2.04	16	$28,000.00	$32,000.00	$40,000.00
Use This, In. = 6		20	$32,000.00	$38,000.00	$46,000.00
		24	$37,000.00	$42,000.00	$50,000.00
B. Compute Forcemain Sizes		28	$44,000.00	$48,000.00	$55,000.00

Pipe Dia (in.)	Velocity, ft/ (FPS)	Cost
4"	12.77	$10.00
6"	5.67	$15.00
8"	3.19	$20.00
10"	2.04	$25.00
12"	1.42	$45.00
14"	1.04	$55.00
16"	0.80	$60.00

Use This Size Forcemain
For These Calculations=
8

III. Calculations For Wetwell Sizing

Wetwell Diameters, feet.	4	6	8
Working Volumes, Gallons	1440.00	1440.00	1440.00
Working Depths, feet.	15.32	6.79	3.83

Summary of Wetwell Elevations			
F.F. Elevation, feet	64.05	64.05	64.05
I.E. Influent Line, feet	53.57	53.57	53.57
H.W. Alarm, feet	53.57	53.57	53.57
2nd Pump On Elevation, feet	53.07	53.07	53.07
1st Pump On Elevation, feet	52.57	52.57	52.57
Pumps Off Elevation, feet	37.25	45.78	48.74
Low Water Alarm Elevation, feet	36.75	45.28	48.24
I.E. Suction Bell, feet	34.75	43.28	46.24
I.E. Bottom Wetwell, feet	34.25	42.78	45.74
Total Depth, feet	29.80	21.27	18.31

IV. Compute Total Dynamic Head				
A. Dynamic Suction Losses				
1 Static Suction Lift, feet		28.80	20.27	17.31
2. Friction HL,Suct. Pipe, feet		0.74	0.54	0.47
3. Hl in Fittings.	No.Fit.			
a. No. 90 Deg. Ells	2	0.70	0.70	0.70
b. No. 45 Deg Ells	0	0.00	0.00	0.00

FIG. 10.31   Wastewater pump station design alternatives.

c. No. Gate Valves	1	0.10	0.10	0.10
d. No. Check Valves	0	0.00	0.00	0.00
e. No. Plug Valves	0	0.00	0.00	0.00
f. Inlet Losses	1	0.11	0.11	0.11
Total Dynamic Suction Headloss		30.45	21.72	18.69
B. Discharge Piping Losses, feet				
1 Static Discharge Lift, feet.		6.75	6.75	6.75
2. Friction HL in Forcemain, feet.		8.30	8.30	8.30
3. HI in Fittings.	No. Fit.			
a. No. 90 Deg. Ells	6	0.66	0.66	0.66
b. No. 45 Deg Ells	2	0.16	0.16	0.16
c. No. Gate Valves	1	0.03	0.03	0.03
d. No. Check Valves	1	0.40	0.40	0.40
e. No. Plug Valves	0	0.00	0.00	0.00
f. Discharge Headloss-Subm.	1	0.16	0.16	0.16
Total Discharge Piping Losses, feet		16.46	16.46	16.46
Total Dynamic Head (TDH), feet		46.91	38.18	35.15
V. Compute NPSH Available At Site				
Wetwell Diameter, feet		4	6	8
a. Available Pressure		33.90	33.90	33.90
b. Dynamic Suct. Lift, feet		30.45	21.72	18.69
C. Vapor Press 74 degF, feet		1.00	1.00	1.00
d. Factor of Safety, feet		2.00	2.00	2.00
e. Loss to Site Elevation, feet		0.06	0.06	0.06
NPSH Available At Site, feet		0.38	9.11	12.15
IV. Horsepower Requirements				
a. Pump Efficiency		0.55	0.55	0.55
b. Motor Efficiency		0.95	0.95	0.95
Brake Horsepower Needed		10.77	8.77	8.07
VII. Cost Analysis Using 25 Year Analysis Period				
a. Assumed Interest Rate =	0.08			
b. Annual % Increase in Cost =	0			
c. Annual Pumping Cost		$1,783.76	$1,451.84	$1,336.51

Year	PW Factor		Cost 4' Dia.	Cost 6' Dia	Cost 8' Dia
1	0.92593		$1,651.63	$1,344.30	$1,237.51
2	0.85734		$1,529.29	$1,244.72	$1,145.85
3	0.79383		$1,416.01	$1,152.52	$1,060.97
4	0.73503		$1,311.12	$1,067.15	$982.38
5	0.68058		$1,214.00	$988.10	$909.61
6	0.63017		$1,124.07	$914.91	$842.23
7	0.58349		$1,040.81	$847.14	$779.84
8	0.54027		$963.71	$784.38	$722.08
9	0.50025		$892.33	$726.28	$668.59
10	0.46319		$826.23	$672.48	$619.06
11	0.42888		$765.03	$622.67	$573.21
12	0.39711		$708.36	$576.55	$530.75
13	0.36770		$655.89	$533.84	$491.43
14	0.34046		$607.30	$494.30	$455.03
15	0.31524		$562.32	$457.68	$421.32
16	0.29189		$520.66	$423.78	$390.12
17	0.27027		$482.10	$392.39	$361.22
18	0.25025		$446.38	$363.32	$334.46
19	0.23171		$413.32	$336.41	$309.69
20	0.21455		$382.70	$311.49	$286.75
21	0.19866		$354.35	$288.42	$265.51
22	0.18394		$328.11	$267.05	$245.84
23	0.17032		$303.80	$247.27	$227.63
24	0.15770		$281.30	$228.95	$210.77
25	0.14602		$260.46	$211.99	$195.15
d. Totals			$19,041.26	$15,498.09	$14,266.98
e. Forcemain Cost			$37,700.00	$37,700.00	$37,700.00
f. P.S. Const. Cost			$44,000.00	$42,000.00	$46,000.00
g. TOTAL P.W. COSTS			$100,741.26	$95,198.09	$97,966.98

FIG. 10.31   (continued)

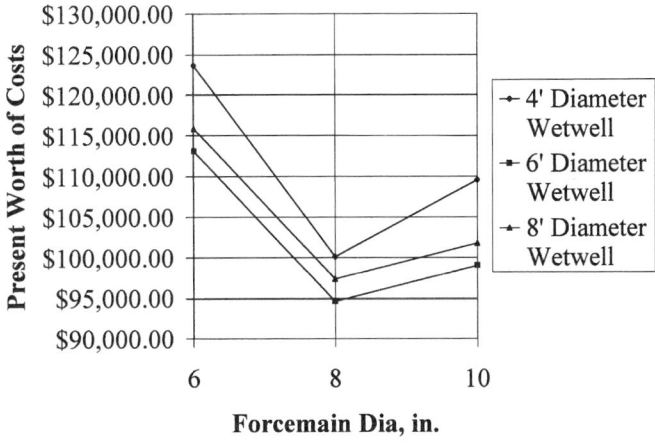

FIG. 10.31 (continued)

(b) The determination of whether the installed pumps in the existing pumping station are capable of handling the additional flows. Alternative impeller sizes, motor sizes, or speeds may be considered; otherwise, new pumps and piping may be required. Should new motors be required, the electrical entrance, motor control panel, starters, and distribution wiring must be checked for adequacy.

(c) The capacity of the receiving pumping station's forcemain must be checked to determine its adequacy. Should the existing station's forcemain be undersize, a forcemain upgrade to a larger diameter might be required.

10.7.2 Discharge into Another Forcemain

If the pumping station under design is to discharge into an existing forcemain of sufficient capacity, the results will be a pressure transport system comprised of multiple pumping stations. System head losses attributed to flows from various operating conditions necessitate that the designer analyze the forcemain under all possible conditions. These conditions include flows resulting from the various on-line combinations of pumping stations included in the network. For example, Figure 10.32 shows an existing forcemain and pumping station that are designated by points *A*, *B*, and *C*. A proposed new pumping station at *D* is to be tied into the existing forcemain at point *B*. Segment *A–B* will carry flows from either pumping station *C* or *D*, or both. Segment *B–C* carries only flows from pumping station *C*, while segment *B–D* carries only flows from pumping station *D*.

Because the existing system *A–B–C* is in place and currently handles only the flows from pumping station *C*, the design engineer should be concerned with the reduced pumping rate that will occur when pumping station *D* is also pumping, the frictional head loss for forcemain segment *A–B* is based on the combined flows of pumping stations *C* and *D*. The frictional head loss for forcemain segment *B–C* is based only on the flow from pumping station *C*. The increased frictional head loss attributed to the increased flow in segment *A–B* will alter the system curve and cause the pump performance curve for the pumps in station *C* to intersect at a lower pumping rate.

Plan View

FIG. 10.32   Multiple wastewater pump stations discharging into a common forcemain.

Likewise, the engineer must determine the TDH for pumping station D when pumping station C is on-line as well as when it is off-line. The frictional head loss component for the forcemain segment A–B when pumping station C is on-line at the same time as the pumps in station D is attributed to the sum of the flows from the two stations. If pump station C is not on-line, the flow through segment A–B is attributed to the flow from pump station B. For forcemain segment B–D, the frictional head loss is attributed only to the flow from pumping station D. Data obtained from these head loss analyses can be used to prepare a series of system curves reflecting various operational characteristics attributed to one or both stations being on-line. From these data, both pumping stations can be analyzed to ensure that the proper size pumps, motors, and impellers are provided.

High system head losses resulting from combined flows may prohibit multiple pumping stations from being connected to a common forcemain; however, an alternative would be to upgrade portions of the forcemain to larger pipe diameters, thus reducing frictional head losses in the system.

Another problem associated with multiple pumping stations in a common forcemain is that each pump station's performance must be analyzed where increased heads caused by the proposed new flow will alter previous operating points on their system curve. As a result, increased impeller diameter, motor size, and/or motor speeds may be required to overcome the new operating pressures. Analysis of waterhammer effects can be complicated, especialy when a multiple-station system on a common forcemain is involved.

10.7.3   Discharge into a Gravity Main

Forcemain discharges often are made into gravity sewer mains that flow to a wastewater treatment plant. The velocity of discharge from the forcemain can be significantly greater than the flow velocity that exists in the gravity sewer at the point of discharge. An expansion chamber can be used to reduce the discharge velocity from the forcemain and also allow for gas to escape from the system. Figure 10.33 illustrates a typical expansion chamber configuration.

The downstream capacity of the gravity lines must be checked to ensure that the proposed flow can be properly accommodated all the way to the treatment plant.

FIG. 10.33   Forcemain discharge expansion chamber.

10.7.4   Discharge into a Wastewater Treatment Plant

Forcemain discharges directly into a wastewater treatment plant can shock the treatment process if care is not exercised. Equalization facilities can be used to temporarily store the wastewater, so that it can be pumped into the plant at a lower than peak rate of delivery (over a sustained period of time). This helps provide more uniform hydraulic loadings to the plant.

10.8   SPECIAL FORCEMAIN CONSIDERATIONS

Sanitary sewer forcemains are not always located in level terrain or in areas where their discharge points are positioned at elevations equal to or above the elevation of the pumps. This makes system head calculations less simple. To further complicate matters, the installation of combination air release/vacuum breaker valves (valves that allow air into the forcemain when negative internal pressures occur and accumulated air out of the forcemain when positive internal pressures occur) at high points can alter the hydraulic performance characteristics of the forcemain, making it possible for some segments to promote gravity flow conditions at times. This is especially true if the forcemain is sized for large ultimate flows while smaller initial flows produce only minimum self-scouring velocities. The engineer must carefully determine the forcemain's system head at the desired pumping rate (or rates) because the flow might produce a transition from nonpressurized (open-channel) to pressurized flow at a forcemain high point. Changes in flow conditions will result in the system's static and friction heads being altered. This requires the use of good engineering judgment when analyzing a specific forcemain configuration.

The system head of the forcemain shown in Figure 10.34 can be estimated when the rate of flow is Q_x by using the following procedure. The forcemain's diameter and

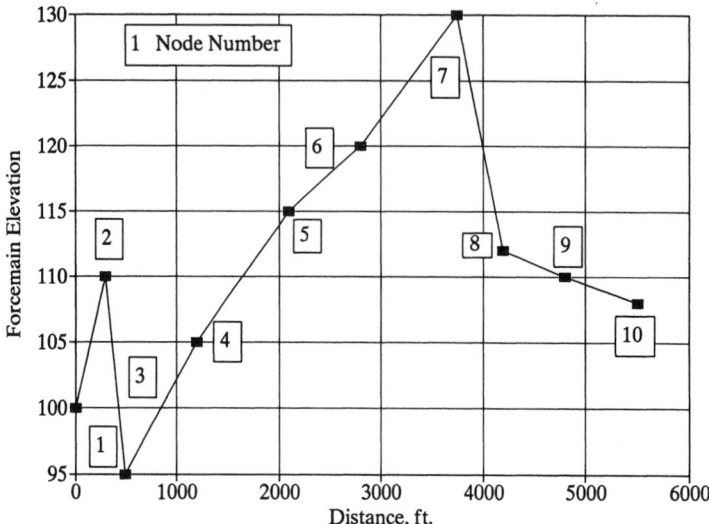

FIG. 10.34   Example of an uphill/downhill forcemain profile.

roughness coefficient are assumed to be uniform between nodes and air release/vacuum breaker valves are assumed to be provided at each high point. Node 1 is the beginning of the forcemain.

Step A: Separately calculate the friction head (attributed to Q_x) using Equation 10.10 and minor losses between node 1 and every downstream node including the discharge node.

Step B: For each friction head obtained in step A, calculate the static head by subtracting the elevation of node 1 from each downstream node's elevation.

Step C: Calculate the total dynamic head that is required to deliver Q_x from node 1 to each node in the forcemain by adding appropriate static and friction heads together.

Step D: Select the largest total dynamic head obtained from step C. This represents the head that the pump must deliver to the forcemain to provide a flow of Q_x.

To obtain enough values to construct the forcemain's system head curve, steps A–D must be repeated for each flow (Q_x) used to make the curve. Figure 10.35 illustrates the use of steps A–D when Q_x is 500 gal/min. The required forcemain TDH is 46.52 feet and is shown plotted in Figure 10.35*b* where the flow is 500 gal/min. Calculations for the other flows between 0 and 1000 gal/min are not included but follow the same procedure.

To obtain the total system head curve, the forcemain's system head curve must be adjusted to account for the level of the wastewater in the wetwell and all head losses in the pumping station between the wetwell and node 1.

Should the forcemain have a net negative slope (go downhill), a vacuum breaker valve can be installed at the beginning of the forcemain.

Flow (Qx)=		500	Gallons per Minute			Static	Friction	Cumulative	
Node	Elevation	Station	Segment	Diameter	Hazen-Williams	Head	Head	Frict. Head	TDH
	(feet)	(feet)	Nodes	(inches)	Coefficient	(feet)	(feet)	(feet)	(feet)
1	100	0	1-2	8	140	10	1.32	1.32	11.32
2	110	300	2-3	8	140	-5	0.88	2.20	-2.80
3	95	500	3-4	8	140	5	3.08	5.28	10.28
4	105	1200	4-5	8	140	15	3.96	9.25	24.25
5	115	2100	5-6	8	140	20	3.08	12.33	32.33
6	120	2800	6-7	8	140	30	4.18	16.52	46.52
7	130	3750	7-8	8	140	12	1.98	18.50	30.50
8	112	4200	8-9	8	140	10	2.64	21.14	31.14
9	110	4800	9-10	8	140	8	3.08	24.22	32.22
10	108	5500							

(a)

(b)

FIG. 10.35   (a) Example of TDH calculations for an uphill/downhill forcemain when flow is 500 gal/min. (b) Completed system curve.

10.9   CENTRALIZED WASTEWATER TREATMENT PLANT CONSIDERATIONS

A centralized wastewater treatment plant may be required to service a proposed land development if there are sufficient dwelling units to warrant a facility. This decision is usually a function of regulatory review agency actions and how the proposed facility interfaces with the area's 208 Water Quality Management Plan. If a facility is needed, its function would be to treat wastewater to a predetermined level of purity where the resulting treated water will not adversely affect the environment. Often field-erected packaged wastewater treatment plants can be used to provide the necessary treatment for the development.

10.9.1   Wastewater Composition

Wastewater includes human waste products and household waste products such as wash water, detergents, and food preparation wastes (especially if garbage disposal units are present). Water attributed to inflow and infiltration can also be included. Typical characteristics of wastewater are shown in Figure 10.36.

Wastewater has various types of solids present, including organic and inorganic solids. Total solids include all solids present in the wastewater, of which approximately 50 percent are organic. Organic solids usually originate from animals or plants although synthetic organic compounds exist. Organic solids contain carbon, hydrogen, and oxygen along with possibly nitrogen, phosphorus, and/or sulfur. Organic solids can decay or decompose under the activity of bacteria or other living organisms and are capable of being burned.

Inorganic solids are solids not subject to decay including sand, silt, and mineral salts contained in the water supply that produce water hardness. Usually inorganic solids are not combustible.

Suspended solids are solids that can be seen in suspension in wastewater. (Suspended solids can be defined as the solids retained by a filter mat in a Gooch crucible.) This type of solids can be removed physically or mechanically with sedimentation or filtration. Suspended particles can be composed of approximately 30 percent inorganic materials

Contaminants	Unit	Concentration		
		Weak	Medium	Strong
Solid, total (TS)	mg/L	350	720	1200
Dissolved, total(TDS)	mg/L	250	500	850
Fixed	mg/L	145	300	525
Volatile	mg/L	105	200	325
Suspended solids (SS)	mg/L	100	220	350
Fixed	mg/L	20	55	75
Volatile	mg/L	80	165	275
Settleable solids	mL/L	5	10	20
Biochemical oxygen demand, mg/L:				
5-day, 20-degC(BOD , 20-degC)	mg/L	110	220	400
Total organic carbon (TOC)	mg/L	80	160	290
Chemical oxygen demand (COD)	mg/L	250	500	1000
Nitrogen (total as N)	mg/L	20	40	85
Organic	mg/L	8	15	35
Free Ammonia	mg/L	12	25	50
Nitrites	mg/L	0	0	0
Nitrates	mg/L	0	0	0
Phosphorus (total as P)	mg/L	4	8	15
Organic	mg/L	1	3	5
Inorganic	mg/L	3	5	10
Chlorides	mg/L	30	50	100
Sulfate	mg/L	20	30	50
Alkalinity (as CaCO)	mg/L	50	100	200
Grease	mg/L	50	100	150
Total coliform	no/100 mL	10 - 10	10 - 10	10 - 10
Volatile organic compounds (VOCs)	ug/L	<100	100-400	>400

Source: Tchobanglous et al, 1991.

FIG. 10.36 Typical composition of untreated domestic wastewater. Source: Metcalf and Eddy, Inc. et al. *Wastewater Engineering: Treatment, Disposal and Reuse,* 3rd ed. Copyright © 1991 by McGraw-Hill Companies. Reprinted by permission of McGraw-Hill Companies.

such as grit and sand. Some of the suspended solids can be settled, if of sufficient size and weight, in an Imhoff cone during a 1-hour period. The amount of settled material is usually quantified as millimeters of solids per liter of sewage or parts per million by weight. Solids remaining in suspension after the 1-hour period are classified as colloidal solids. Colloidal solids can be highly organic and subject to rapid decay.

Dissolved solids can be defined as those solids that are not retained by a filter mat in a Gooch crucible, although some of the very fine solids passing through the filter mat are not dissolved (about 10 percent).

Wastewater can include various levels of dissolved gases including oxygen, carbon dioxide, nitrogen, hydrogen sulfide, and possibly some inorganic sulfur compounds. In addition, aerobic (needing oxygen) bacteria and anaerobic (no oxygen) bacteria can be present, along with facultative bacteria that can live in either environment. Water temperature can greatly affect the rate of bacteria growth and reproduction, which is of importance when treating wastewater. Parasitic or saprophytic bacteria can also be present, along with various other organisms including viruses.

10.9.2   Methods to Treat Wastewater

Wastewater can be treated using many methods including the following:

(a)   The removal of suspended inorganic material can be accomplished using sedimentation or clarification, flocculation, and filtration. To remove dissolved solids, reverse osmosis or electrodialysis can be used.

(b)   Organic materials can be treated using activated sludge methods, including extended aeration, contact stabilization, or other associative methods. Aerobic treatment can include trickling filters, lagoons, land treatment, and other methods. Anaerobic methods include mixed reactors or packed-bed reactors.

During the treatment of wastewater, gravity can be used to reduce the amount of suspended particles present. Biological and chemical changes allow colloidal particle to flocculate and settle.

During aerobic decomposition oxygen is combined with organic materials contained in the wastewater to produce carbon dioxide, water, nitrates, sulfates, phosphates, and other substances. Anaerobic decomposition includes the removal of oxygen from complex compounds to form simpler ones. The process continues in steps until complex compounds are broken down into stable substances of organic or inorganic materials.

The activated sludge process is an aerobic biological process that incorporates an aeration tank and a solid-liquid separator where the concentrated sludge can be recycled back to the aeration tank. This process is dependent upon the production of flocculent microorganisms that can be removed by gravity sedimentation. There are several variations of the activated sludge process, including extended aeration, contact stabilization, step aeration, and high-rate processes. Figure 10.37 contains various general loading and operational parameters that are used in various activated sludge processes. Figure 10.38 illustrates alternative activated sludge process flow diagrams that can be considered as a basis for wastewater treatment plant design.

Filters can be included to remove suspended particles as part of the treatment process. Nitrogen removal from the wastewater is also important. Most nitrogen present in the waste is in the form of ammonia. Nitrite nitrogen can be a product of oxidation

Process	BOD Loading $\left(\dfrac{\text{lb BOD/day}}{1000 \text{ cu ft}}\right)^a$	F/M Ratio $\left(\dfrac{\text{lb BOD/day}}{\text{lb MLSS}}\right)^b$	Sludge Age (days)	Aeration Period (hr)	Return Sludge Rates (percent)	BOD Removal Efficiency (percent)
Conventional	30–40	0.2–0.5	5–15	6.0–7.5	20–40	80–90
Step aeration	30–50	0.2–0.5	5–15	5.0–7.0	30–50	80–90
Contact stabilization	30–50	0.2–0.5	5–15	6.0–9.0	50–100	75–90
High rate	80 up	0.5–1.0	3–10	2.5–3.5	50–100	70–85
High-purity oxygen	120 up	0.6–1.5	3–10	1.0–3.0	30–50	80–90
Extended aeration	10–20	0.05–0.2	20 up	20–30	50–100	85–95

^{a} 1.0 lb/1000 cu ft day = 16.0 g, m^3·d
^{b} 1.0 lb/day·lb = 1.0 g, d·g

FIG. 10.37 General loading and operational parameters for activated sludge processes (F/M = food to microorganism ratio; MLSS = mixed liquor suspended solids). Source: Mark J. Hammer. *Water and Wastewater Technology,* 2nd. Copyright © 1986, p. 413. Reprinted by permission of Prentice-Hall, Englewood Cliffs, NJ.

of ammonia, while nitrate nitrogen is produced by the decomposition of nitrogenous materials. In wastewater "total Kjeldahl nitrogen" refers to ammonia and organic nitrogen whereas "total nitrogen" refers to ammonia, organic, nitrite, and nitrate nitrogen. Removal systems, such as a biological nitrification/denitrification system, can be used where ammonia nitrogen and nitrite nitrogen are converted into nitrates during the nitrification step. The denitrification step occurs where nitrates are converted into nitrogen gas.

Traditionally chlorine has been the method of choice for disinfecting wastewater because of its low relative cost and proven ability to kill pathogens and also provide odor control. Increased safety concerns relative to carcinogenic by-product formation, inadvertent toxicity to aquatic life, treatment plant personnel safety, ineffectiveness to control certain pathogens (*Giardia lamblia* and *Cryptosporidium*), and community safety have forced the environmental community to seek alternative ways to disinfect wastewater. Ultraviolet radiation and ozonation are the two primary treatment alternatives. Ultraviolet disinfection has higher capital but lower operating costs than chlorine systems. Because ultraviolet light physically disinfects, rather than chemically, no toxic compounds are formed. Unfortunately *G. lamblia* and *Cryptosporidium* are not controlled by ultraviolet radiation. Ozonation, however, is effective against these organisms, but it also has some drawbacks. It has a relatively high installation cost, is fairly expensive to operate, can increase BOD demands by transforming complex organic material into simple biodegradable forms, and even can help create carcinogenic type by-products.

Depending on the strength and composition of the waste, varying degrees or methods of treatment may be required to produce effluent that conforms to the waste load allocation set forth in issued permits. Other references on wastewater treatment plant design are available for the designer to consult. (Novotny et al., 1989, Metcalf and Eddy, et al., 1991).

FIG. 10.38 Flow diagrams for common activated sludge processes; (*a*) conventional plug-flow activated sludge plant; (*b*) step aeration plant; (*c*) contact stabilization activated sludge system; (*d*) complete mix plant. Source: Novotny et al., *Karl Imhoff's Handbook of Urban Drainage and Wastewater Disposal.* Copyright © 1989 by John Wiley & Sons, Inc. Reprinted by permission of John Wiley & Sons, Inc.

10.9.3   Waste Production from Treatment Processes

Sludge is produced as a by-product of the wastewater treatment process and must be handled properly. Some wastewater treatment plants use sludge drying beds to dewater the sludge prior to final disposal. Other facilities rely on dewatering equipment such as centrifuge and incineration for disposal. Regulatory agencies may require minimum set back distances from treatment plant facilities (especially chlorination storage and disinfection areas and sludge drying facilities) to the closest dwelling unit. Spatial requirements such as these place additional demands on the design engineer to allocate adequate land for treatment facilities.

10.10   WASTEWATER SYSTEM APPURTENANCES

Sanitary sewage collection systems are comprised of piping, fittings, laterals (or services), manholes and drops, and pipe bedding. Each of these components provides an important feature to the system.

10.10.1   Piping and Fittings

Most sanitary sewage piping and fittings utilized in land developments include small-diameter circular conduits. Typical diameters, however, can range from 4 to 144 inches, depending on the material of construction. Pipes used in sanitary sewage systems must be strong, durable, abrasion resistant, impervious, and resistant to corrosive action. Pipe joints must be watertight to prevent infiltration of groundwater, as well as exfiltration of wastewater. Sewage pipes can be either gravity flow nonpressurized pipes or pressurized pipes similar to those used in forcemains.

Gravity flow pipes most commonly are made of vitrified clay, polyvinyl chloride (PVC), polyethylene (PE), precast concrete, and ductile iron. Most pipe joints are of the slip type, where the barrel of one joint "slips" into a gasket contained in the bell of the adjoining pipe. Ductile-iron pipe can be made with mechanical joints where the gasket is mechanically held into place, flange joints where pipe sections are bolted together using end flanges, and restrained jointed pipe (river crossing pipe) where each joint provides positive restraint of the joined pipes, thus reducing the need for thrust blocking (Section 11.8.2). On occasion smooth-walled steel pipe, corrugated steel pipe, or bituminous fiber pipes have been utilized.

Pipes made of materials susceptible to corrosion must be protected with coatings or linings. This corrosion occurs when released hydrogen sulfide gas (H_2S) in the wastewater condenses on the interior pipe's crown, allowing sulfuric acid to form. This acid can corrode certain pipe materials. Epoxy, plastic, or bituminous coatings help to protect susceptible materials.

10.10.2   Manholes and Drops

Sanitary sewage manholes allow system access for maintenance during the life of the system. The minimum manhole size used on most small-diameter sanitary sewer mains is 4 feet inside; however, if acute angles of intersection of the mains joining the manhole exist, a larger diameter manhole can be used to allow more space inside to build an adequate invert to promote efficient hydraulic flow.

Manholes are often constructed of precast concrete manhole sections, which give superior water tightness over manholes constructed from masonry units.

10.10.3 Pipe Bedding

Proper installation of sanitary sewer mains requires that pipes be bedded properly in order to promote adequate performance. Bedding requirements are usually included in the construction specification (Chapter 13).

A properly bedded pipe will maintain its watertight integrity. Improperly bedded pipes can deflect at the joints, causing the joint seal to be lost. Also, properly bedded pipes can handle intended design live and dead loads without experiencing problems. Pipe strength should be investigated using methods similar to those included in Section 7.9.

REFERENCES

American Society of Civil Engineers (ASCE). *Pipeline Design for Water and Wastewater.* New York: ASCE, 1975, pp. 21, 33, 43. Referenced material is reproduced by permission of the American Society of Civil Engineers.

American Society of Civil Engineers (ASCE) and Water Pollution Control Federation (WFCF). *Gravity Sanitary Sewer Design and Construction.* ASCE Manual No. 60, MOP-FD-5. New York: ASCE, 1982, pp. 36–37.

DAVCO-Davis Water and Waste Industries. *DAVCO Equipment-Engineering: Pumping Station Design.* 1974. GA: Davco, 1974.

Fair, G. M., J. C. Geyer, and D. A. Okun. *Water and Wastewater Engineering,* Vol. 1. New York: Wiley, 1966, pp. 5–24.

Geyer, J. C., J. B. Wolff, F. P. Linaweaver, Jr., A. J. Duncan. "Report on Phase One, Residential Water Use Project. Final and Summary Report on Phase I of the Residential Water Use Research Project, January 1, 1959 to January 1, 1963." An unpublished report. Baltimore: Johns Hopkins University and Federal Housing Administration, October 1963, pp. 6–7, 6–8.

Kaplan, O. B. *Septic Systems Handbook.* Chelsea, MI: Lewis, 1987.

Metcalf and Eddy and G. Tchobanoglous. *Wastewater Engineering: Collection and Pumping of Wastewater.* New York: McGraw-Hill, 1981, pp. 109, 111.

Metcalf and Eddy, G. Tchobanoglous, and F. L. Burton. *Wastewater Engineering: Treatment, Disposal, and Reuse, 3rd Metcalf and Eddy, Inc.* New York: McGraw-Hill, 1991, p. 109.

Novotny, V., K. Imhoff, M. Olthof, and P. Krenkel (Eds.). *Karl Imhoff's Handbook of Urban Drainage and Wastewater Disposal.* New York: Wiley, 1989. pp. 21, 162–168. Copyright © 1989 by John Wiley & Sons, Inc. Reprinted by permission of John Wiley & Sons, Inc.

Orsatti, R. "Sewer Design on the Slant." In *Water Environment and Technology, March 1996.* Alexandria, VA: Water Environment Federation (WEF), 1996, pp. 39–43.

Salvato, J. A. *Environmental Engineering and Sanitation,* 4th ed. New York: Wiley, 1992, p. 508.

South Carolina Department of Health and Environmental Control (SCDHEC). *Guidelines for Unit Contributory Loadings to Wastewater Treatment Facilities.* Columbia: Water Pollution Control Division, SCDHEC, 1972.

Tullis, J. P. *Hydraulics of Pipelines.* New York: Wiley, 1989.

United States Environmental Protection Agency (EPA). *Small Wastewater Systems—Alternative Systems for Small Communities and Rural Areas.* 830/F-92/001. May 1992.

SUGGESTED READINGS

American Society of Civil Engineers (ASCE) and Water Pollution Control Federation (WPCF). *Design and Construction of Sanitary and Storm Sewers.* ASCE Manual No. 37, MOP-9. New York: ASCE, 1970. Reference material is reproduced by permission of the American Society of Civil Engineers.

American Society of Civil Engineers (ASCE) and Water Pollution Control Federation (WPCF). *Design of Municipal Wastewater Treatment Plans,* 4th ed. ASCE Manual No. 76, MOP-8. New York: ASCE, 1998.

Burks Pumps, Inc. Pump Curve-4C04D (Imp. Dia. 8D) 420 Third St., Piqua, Ohio 45356-0603.

Campbell, C. S., and M. H. Ogden. *Constructed Wetlands in the Sustainable Landscape.* New York: Wiley, 1999.

Environmental One Corporation. *Design Module No. 7—Low Pressure Sewer Systems WWBLDMO7—National Small Flows Clearing House.* Morgantown: West Virginia University, 1985.

Hammer, Mark J. *Water and Wastewater Technology,* 2nd ed. Englewood Cliffs, N.J.: Prentice Hall, 1986, p. 413.

ITT A-C Pump. *Pick-A-Pump,* 1981, p. 334. 1150 Tennessee Ave., Cincinnati, Ohio 45229.

J-M Manufacturing Company. "Green-Tite PVC Gravity Sewer Pipe ©—Installation Guide." An unpublished manual. Stockton, CA: J-M Manufacturing Company, 1987.

Otis, R. J. *Design Module No. 4—Onsite Wastewater Disposal Evapotranspiration and Evapotranspiration Absorption Systems WWBLDO4—National Small Flows Clearing House.* Morgantown: West Virginia University, 1990.

Otis, R. J. *Design Module No. 3—Distribution Networks WWBLDMO3—National Small Flows Clearing House.* Morgantown: West Virginia University, 1985a.

Otis, R. J. *Design Module No. 1—Artificially Drained Systems WWBLDMO1—National Small Flows Clearing House.* Morgantown: West Virginia University, 1985b.

Otis, R. J. *Design Module No. 18—Septic Tanks WWBLDM18—National Small Flows Clearing House.* Morgantown: West Virginia University, 1985c.

Reed, S. C. *Design Module No. 74—Subsurface Flow Constructed Wetlands for Wastewater Treatment: A Technology Assessment WWDKDM74—National Small Flows Clearing House.* Morgantown: West Virginia University, 1993.

Simmons, J. D., and J. O. Newman. *Design Module No. 13—Small Diameter Sewers WWBLDM13 —National Small Flows Clearing House.* Morgantown: West Virginia University, 1985.

South Carolina Department of Health and Environmental Control (SCDHEC). "Experimental Standard Elevated Infiltration System." An unpublished memorandum. Columbia: SCDHEC, 1987.

Submersible Waste Water Pump Association. *Submersible Sewage Pumping Systems Handbook.* Chelsea, MI: Lewis, 1985.

Tindall Concrete Products. "Underground Structures." An unpublished report. Spartanburg: Tindall Concrete Products, 1979.

Water Environment Federation (WEF). *Design of Wastewater and Stormwater Pumping Stations— MOP FD-4.* Alexandria, VA: WEF, 1993.

Water Environment Federation (WEF). *Wastewater Disinfection—MOP FD-10.* Alexandria, VA: WEF, 1996.

Water Environment Federation (WEF) and American Society of Civil Engineers (ASCE). *Design of Municipal Wastewater Treatment Plants. WEF Manual of Practice No. 8, 4th and ASCE Manuals and Reports on Engineering Practice No. 76.* Reston: ASCE, 1998.

11 Potable Water System Design

The last system to be designed in a land development is usually potable water because water mains are pressurized and placed where conflicts with other designed systems are minimal.

To properly design this system, the engineer must determine anticipated demands of flows and pressures. The source of supply along with storage and fire suppression requirements are of importance as well as the configuration and sizing of the distribution system. This chapter centers on the design of small domestic systems normally found in residential land developments.

11.1 ENGINEER'S REPORT

The design engineer may be required as part of the regulatory review process to prepare a written report for a new water system, or for the extension of an existing system, prior to obtaining a permit for construction and operation. This report usually contains general project information (including phasing plans if an incremental approach is to be utilized), surface water and groundwater source information, a discussion of the proposed treatment process, and a description of the proposed distribution system.

11.1.1 General Information

The name and address of the proposed system's owner and the name and address of the design engineer normally are included in the engineer's report. A general description of the area being served along with proposed land uses, terrain, location, and possibly consumer socioeconomic characteristics are included. The estimated number and type of potential customers are also described, and documentation of preliminary approvals from local regulatory agencies for the proposed development is provided. Should an existing public water supply be present, the distance to the project site is usually reported.

11.1.2 Surface Water Sources

The engineer's report may indicate the latitude and longitude of proposed raw water intake structures along with the name of the proposed water source and its classification. If a potential contributory watershed is involved, then its limits are usually described along with the expected low-flow stream characteristics. The engineer should include the names of all contributory discharges that may affect the quality of the water source for possibly as far as 10 miles upstream, indicating classifications of each type of discharge as agricultural, municipal, industrial, and so on.

557

This report could include a description of the proposed intake and pumping facilities (pumping rates and the proposed pumping distance) as well as alternate sources of water that might be available.

11.1.3   Groundwater Sources

Should groundwater sources be proposed, the engineer's report could indicate the latitude and longitude of each proposed well site and the type of proposed well along with its diameter and depth. The report could describe the proposed positioning of each well on the well site along with the location of the 100-year flood boundary, if applicable. In addition, potential groundwater contamination sources would be delineated along with directions and distances to each potential source.

11.1.4   Water Treatment Plant

If a water treatment plant is proposed, the engineer's report may include the maximum anticipated plant capacity at project completion along with an estimated completion date and interim plant phasing requirements. The location of the proposed plant and its proximity to the 100-year floodplain are of importance to most regulatory authorities. Also, the proposed treatment process is usually described along with proposed methods to handle waste by-products produced when treating water. Should the name of the operator of the proposed plant be known, it may be included along with the operator's qualifications.

11.1.5   Water Distribution System

The engineer's report for a residential land development includes anticipated average daily demands and peak requirements for domestic water use. Fire suppression flows using rating criteria formulated by the Insurance Services Office, Inc. (ISO) are usually considered along with types and capacities of proposed water storage facilities.

11.2   WATER SYSTEM DEMANDS

Rates of flow and volume demands in water systems vary based on geographical area, seasonal change, and user needs. Domestic, commercial, industrial, and fire protection requirements contribute to the required total system demands that must be met. According to Sweitzer and Flentje (1972), one month (during the annual seasonal cycle when the highest water usage occurs) can be designated as the maximum (water usage) month for that year. The maximum day for that year will likely be one of the days in the designated maximum month. A daily usage cycle that includes peak-flow periods of about 4 hours duration and even extreme peaks of an hour or less can be superimposed upon the seasonal cycle. Typical peak flow ratios are given in Figure 11.1, where the range of values reflects possible differences in system users and sizes. In smaller built-up areas, water users tend to have similar habits, as indicated in Section 11.2.1, while water use patterns in larger areas tend to be more diversified, so that peak-flow demands are reduced.

Typical System Peak Flow Characteristics As Ratios Of The Average Daily Flow	
Average Daily Flow =	1.0
Average Day During The Maximum Month =	1.3-1.8
Average Rate On The Maximum Day =	1.6-2.2
Average Rate During The Maximum Four Hour Period =	2.0-3.0
Maximum Rate Sustained For A Period Of One Hour =	2.5-6.0

FIG. 11.1 Typical water system peak flows. Source: Sweitzer and Flentje, 1972. Copyright © 1972 by American Concrete Pressure Pipe Association. Reproduced with permission.

According to the American Water Works Association's (AWWA's) *Manual M 31* (1998),[*] the demands or rates of water use placed on water systems are usually classified into three general categories, as follows:

(a) *Average Daily Demand*   The average amount of water used each day during a one-year period for the entire system.

(b) *Maximum Daily Demand*   Sometimes called the maximum daily consumption rate (MDC), this rate is the average rate of consumption on the maximum day. The maximum day is the 24-hour period during which the highest consumption total is recorded in the latest three-year period (ISO, 1998).

(c) *Maximum Hourly Demand*   The maximum amount of water used in a single hour, of any day, in a three-year period, normally expressed as gallons per day by multiplying the hourly peak by 24.

Generally, the flows used for distribution system design purposes in land developments are either the peak hourly flow or the needed fire flow plus the maximum daily demand, whichever is greater. A water system could be considered adequate by Insurance Services Office (ISO) if it can deliver the required fire flows to all points in the distribution system with the consumption at the maximum daily rate (average rate on maximum day of a normal year).

11.2.1 Domestic Water Use

Domestic water use includes water supplied to support residential activities such as flushing toilets, cleansing, drinking, cooking, and lawn/garden sprinkling. The average daily demand per capita ranges from approximately 50 to 150 gallons for metered systems and up to 250 gallons if unmetered. Domestic water rate of flow demands vary throughout the day and from day to day. According to a study by Wolff and Loos (1956) of the Baltimore area, two peak water use periods exist: one between 7 A.M. and 1 P.M. and a second between 5 P.M. and 9 P.M. Peak hourly flow requirements during these periods are reported to extend as much as 1500 percent above the average daily demand in residential areas with lots from $\frac{1}{2}$ to 3 acres in new and old neighborhoods

[*]Excerpted from *Distribution System Requirements for Fire Protection* AWWA, 1998. Copyright © 1998, American Water Works Association.

and from 500 to 600 percent in older neighborhoods comprised of well-settled small lots. New neighborhoods studied, consisting of lots ranging from $\frac{1}{4}$ to $\frac{1}{2}$ acre, are found to reach hourly peaks as high as 900 percent. Lawn sprinkling is identified as a major factor that contributes to a high percentage of the peak flows. Many water utilities today attempt to limit or prohibit consumers from irrigating their lawns during certain periods, hoping to reduce demands on the system.

In a first phase study of residential water use in the Baltimore metropolitan area between 1959 and 1962, four topics were investigated: residential water use, peak hourly design criteria, commercial water use, and coincident water and sanitary sewage flows (Geyer et al., 1963). Typical daily water use patterns obtained from this study are shown in Figure 11.2.

This initial study was later expanded into a second phase to develop improved methods for the design of water distribution systems that would be applicable to most locations in the United States (Linaweaver et al., 1966). This included 41 residential areas ranging from 44 to 410 dwelling units in various climatic regions in the United States, as indicated in Figure 11.3.

The study results provide evidence that residential water demands depend upon

(a)   the number of customers and their economic level,
(b)   the average irrigable lawn area,
(c)   the rate of transpiration,

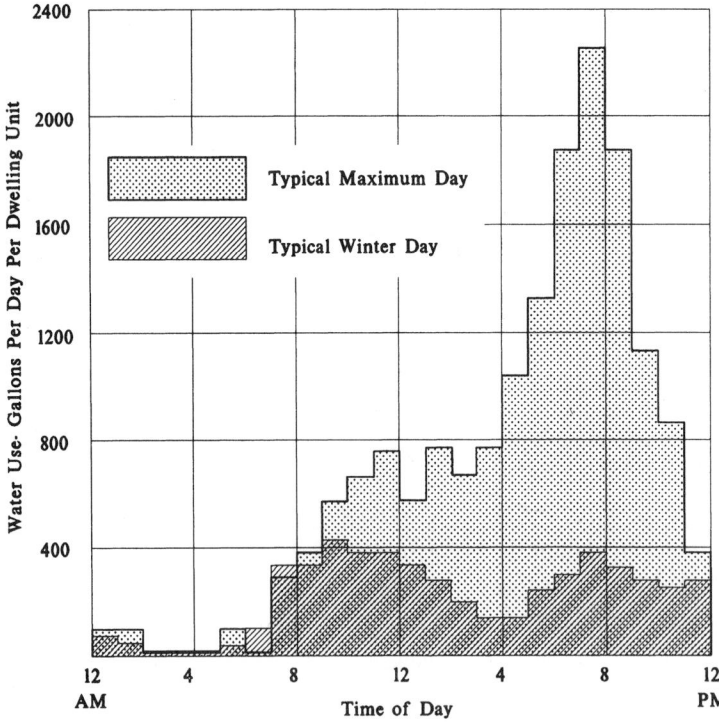

FIG. 11.2   Typical daily water use patterns: maximum day and winter day. Source: Geyer et al., 1963. Reproduced with permission.

FIG. 11.3 Study areas for Phase II residential water use. Source: Linaweaver et al., 1966. Reproduced with permission.

(d)   the amount of precipitation effective in reducing sprinkling, and

(e)   whether customers were metered or unmetered.

Figure 11.4 shows residential water use activities schematically.

It was further determined that water systems serving residential areas with gross densities of about seven homes per acre or less should be capable of supplying lawn sprinkling water approximately equal to evapotranspiration from lawns covering the entire area served.

Traditionally water system demand estimates have been performed by making population estimates and multiplying those estimates by an average per-capita daily use to estimate the total average use. Peaking factors based on entire service areas have then been used to estimate peak demands. According to the phase II study, "the peak to average ratios selected have been generally too low" (Linaweaver et al., 1966, p. 3). In 1965 the Federal Housing Administration adopted as part of the Minimum Design Standards a policy whereby:

(a)   If system operating data are available, the annual average demand should be ascertained on the basis of records or records of existing systems of similar nature in the area.

(b)   In the absence of reliable experience records, an average demand of 100 gallons per capita per day should be used along with four persons being allocated to each dwelling unit. This provides an average demand of 400 gallons per day per dwelling unit.

(c)   A maximum daily demand of 200 percent of the average demand was recommended.

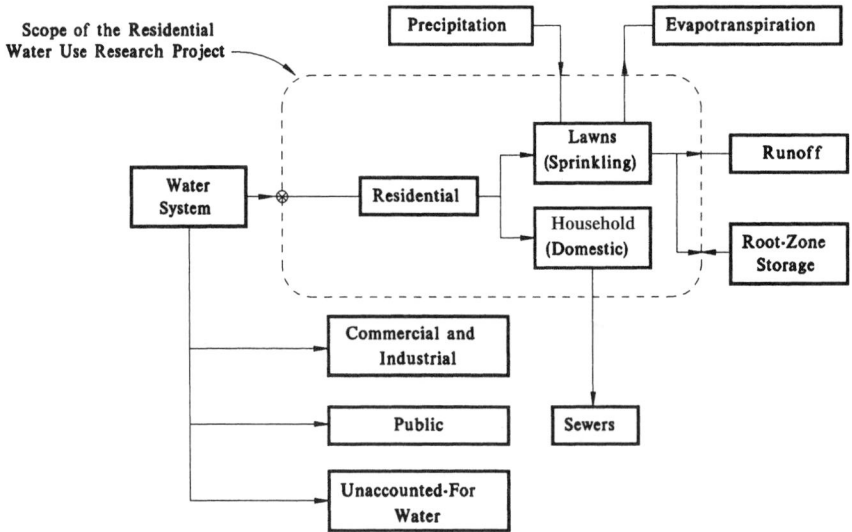

FIG. 11.4 Schematic diagram of residential water use. Source: Linaweaver et al., 1966. Reproduced with permission.

(d) A maximum hourly demand of 500 percent of the average demand was suggested, except in areas where "extensive" lawn irrigation is commonly practiced. Then a rate of 700 percent or more should be used.

According to Linaweaver et al. (1966, pp. 3-4), "These criteria are an improvement, but there are situations when the use could lead to under design and in other situations to over design of the water distribution system." The results of the phase II study are reported to provide a more equitable design approach.

A relationship for the design maximum daily demand and the design peak hourly demand are included in the Phase II study. The basic equation for predicting demands in metered and sewered residential areas are

$$\overline{Q} = \overline{Q_d} + 0.6ca\,\overline{L_s}\,(\overline{E_{\text{pot}}} - \overline{P_{\text{eff}}}) \quad \text{with } \overline{Q} \geq \overline{Q_d} \tag{11.1}$$

where

$\overline{Q}$ = Expected average demand for any period expressed in gallons per day.

$\overline{Q_d}$ = Expected average household use that equals the amount that is essentially returned to the wastewater system in gallons per day for all periods of a day or longer. This can be estimated as a function of the average market value of dwelling.

0.6 = Coefficient to adjust for the difference between actual and potential evapotranspiration from lawns.

c = Coefficient to adjust units, 2.72×10^4 gallons per acre-inch of water.

a = Number of dwelling units contributing to the demand.

$\overline{L_s}$ = Average irrigable area in acres per dwelling unit found by $L_s = 0.803W^{-1.26}$, where W is the gross density in dwelling units per acre.

$\overline{E_{\text{pot}}}$ = Estimated average potential evapotranspiration for the period of demand in inches of water per day, determined from climatological data. (In the case of the phase II study, the overall mean value for the study areas was found to be 0.28 inch of water.)

$\overline{P_{\text{eff}}}$ = Amount of natural precipitation effective in satisfying evapotranspiration for the period and thereby reducing the requirements for lawn sprinkling in inches of water per day.

During periods of high demands, precipitation becomes negligible and the equation takes the form

$$\overline{Q} = \overline{Q_d} + 0.6ca\overline{L_s}\,\overline{E_{\text{pot}}} \tag{11.2}$$

The domestic use in gallons per day per dwelling unit, Q_a/a, can be estimated from $Q_a/a = 157 + 3.46V$, where V is the average market value in $1000 per dwelling unit. Housing market values were correlated to the 1965 index.

Design curves that relate a 95 percent confidence interval were developed from Equation 11.1 and other data to show the relationship between numbers of dwelling units to peak system demands. Expected and design values for maximum daily and peak hourly demands are shown in Figures 11.5 and 11.6.

Peak-flow requirements have also been expressed in terms of fixture units demands, in which a fixture unit represents a flow 7.5 gal/min. For design using this criterion, the reader is referred to other sources (Konen and Chan, 1979).

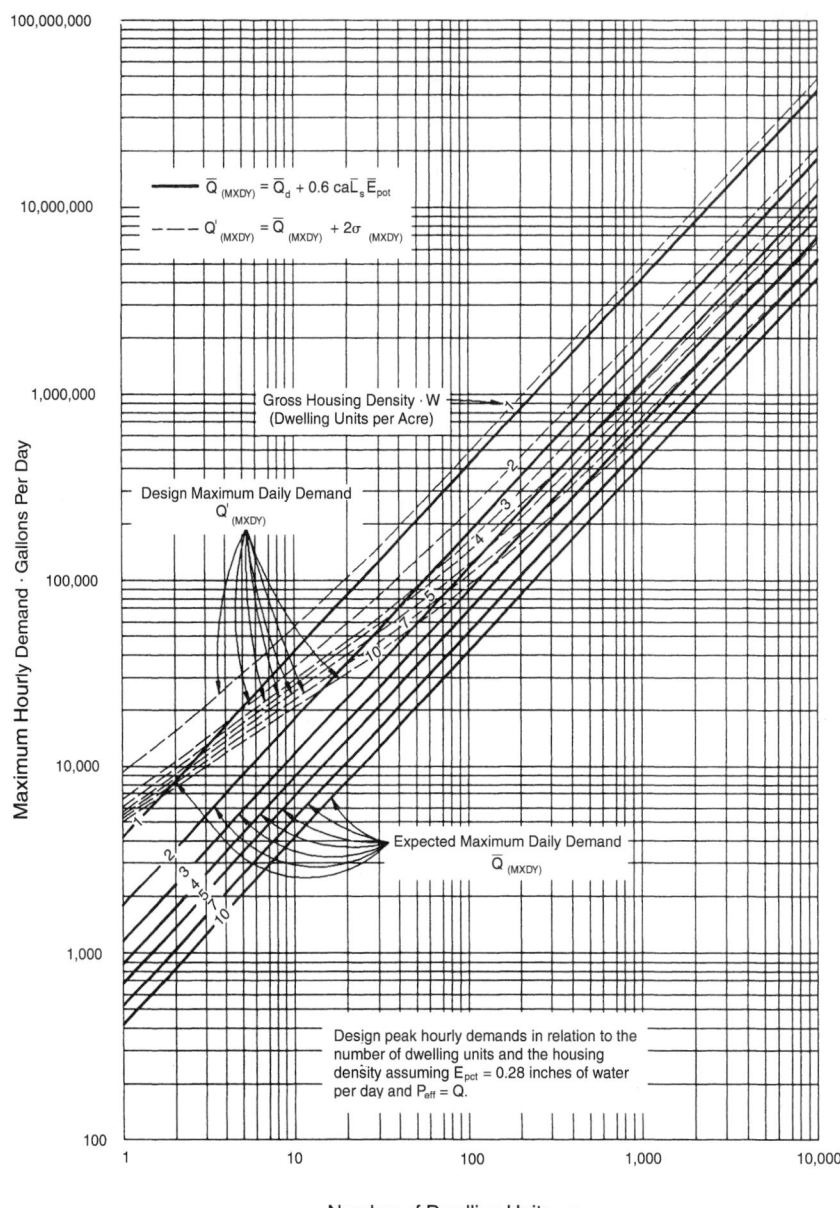

FIG. 11.5 Maximum daily demands. Source: Linaweaver et al., 1966. Reproduced with permission.

11.2.2   Commercial and Industrial Consumption

Although water demands for commercial-type residential rental units, such as apartments, can be included in domestic water estimates, other commercial establishments, such as offices, shopping facilities, and service stations, must be quantified. Some typical values are included in Figure 11.7.

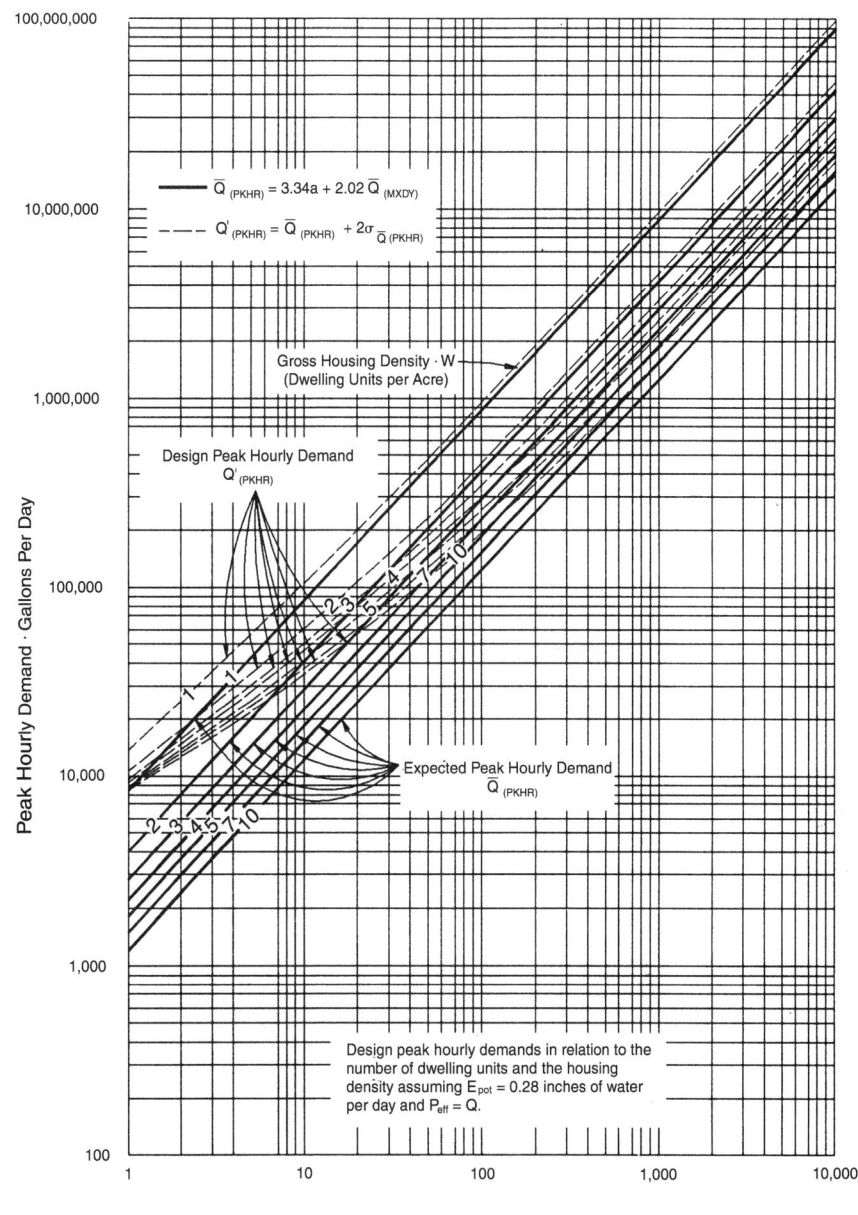

FIG. 11.6   Design peak hourly demands. Source: Linaweaver et al., 1966. Reproduced with permission.

Industrial water use varies based on makeup, process, and heating/cooling water needs. These demands vary greatly and can exceed other system demands; therefore, an accurate accounting of demand rates is essential.

Facility	Maximum Estimated Demands
Churches	0.4 Gallons per minute per member
Country Clubs	0.6 Gallons per minute per member
Beauty Shops	3.0 Gallons per minute per chair
Drug Store (No Fountain)	5.0 Gallons per minute
Launderette	8.0 Gallons per minute per unit
Motel	4.0 Gallons per minute per unit
Office Building	0.5 Gallons per minute per 100 s.f., or
	2.0 Gallons per minute per employee
Service Station	10.0 Gallons per minute per washrack

FIG. 11.7   Nonresidential peak water demands. Source: Ameen, 1971. Reproduced with permission.

11.2.3   Fire Protection Requirements

Today most potential buyers rely on the availability of fire protection when considering buying a dwelling in a development. Adequate water system design and performance are necessary to have fire fighting flows available of at least 500 gal/min with a minimum residual system pressure of not less than 20 lb/in.^{2} (AWWA, 1998).[*]

Insurance Services Office, Inc. (ISO) has prepared a schedule to rate public fire suppression facilities for insurance purposes on a basis from 1 to 10, with 1 being the best. This schedule, *Fire Suppression Rating Schedule—1998*, includes the evaluation of fire flow need, alarm handling, fire department equipment, personnel, training, and water supply. Although the schedule is "a fire insurance rating tool, and is not intended to analyze all aspects of a comprehensive public fire protection program" (ISO, 1998, p. 1), it does provide broad guidelines that generally can be incorporated into the water system design process where the installed system will enhance fire suppression capabilities.

According to the ISO, the needed fire flow (NFF) for selected locations throughout the protected area in gallons per minute is the rate of flow required to suppress a specific fire under certain conditions. For one- and two-family dwellings without wood shingles and not exceeding two stories in height, the NFF is given in Figure 11.8 along with the minimum durations.

For other types of residential units, or structures not fitting the above category, the NFF must be calculated from parameters addressing construction class, effective building area, occupancy, exposure, and communication factors associated with a building or within a fire district. In any case the NFF will not be less than 500 gal/min.

A public water system in a residential development that provides fire protection should be able to

(a)   supply the distribution system the NFF simultaneously with the maximum daily demand,

(b)   deliver throughout the distribution system the required NFF with a residual pressure of at least 20 pounds per square inch gauge (psig), and

---

[*] Excerpted from *Distribution System Requirements for Fire Protection* (AWWA Manual M31), by permission. Copyright ©1998, American Water Works Association.

Needed Fire Flows For 1 And 2 Family Dwelling Not Exceeding 2 Stories In Height	
Distance Between Buildings	Needed Fire Flow
Over 100 feet	500 Gallons per minute
31 feet to 100 feet	750 Gallons per minute
11 feet to 30 feet	1,000 Gallons per minute
10 feet or less	1,500 Gallons per minute
Needed Fire Flow Duration	
Needed Fire Flow	Duration In Hours
NFF ≤ 2,500 Gallons per minute	2 Hours
3,000 GPM ≥ Needed Fire Flow ≤ 3,500 GPM	3 Hours

FIG. 11.8   Needed fire flows (NFFs) for one- and two-family dwellings. Source: Insurance Services Office, 1998. *Fire Suppression Rating Schedule.* Copyright © 1998. Insurance Services Office, Inc. Reprinted with permission.

(c)   include three-way fire hydrants with standard fittings that are no more than 300 feet (closer in high-value districts) away from any dwelling measured by hose length.

Fire suppression capabilities of a water system are site specific and therefore require an in-depth analysis by ISO. Questions pertaining to system adequacy can *only be resolved on a case-by-case basis by the ISO, and the results of this analysis are governed solely by the ISO's independent application and interpretation of current ISO-produced documents.*

If the development includes total development activities, the design professional should incorporate adequate fire protection measures for all buildings. Various considerations of this nature are discussed by the National Fire Protection Association (NFPA, 1998a).

11.3   WAYS TO PROVIDE WATER TO A DEVELOPMENT

Potable water for the development should be obtained from an existing reliable municipal or public source, if available; otherwise, a well system must be installed or a suitable surface water source must be found. Wells or surface water sources also require treatment and storage facilities. Provisions must be made for the operation, maintenance, and control of these systems. Requirements are set forth in the Safe Drinking Water Act (PL 93–523) as amended, the National Primary Drinking Water Regulations (40 CFR 141), and the National Secondary Drinking Water Regulations (40 CFR 143) as amended. These requirements make the option of providing an extension to an existing municipal system very attractive.

Effective October 1999 all states must develop a program addressing what the EPA calls "capacity development" or lose 20 percent of their capitalization grant (annually) from the revolving loan fund. Specifically, any developer planning on constructing a development water system rather than connecting to an existing system must first prove that it is not feasible to make the connection. If successful in that endeavor, then the developer must submit a business plan for operating the proposed system which demonstrates technical, managerial, and financial ability to be in the water supply business (SCDHEC, 1999).

Should the developer rely either on wells or surface water for supply, provisions must be made to obtain the right of use for that source in accordance with regulatory permit requirements and to secure legal water rights for the source's use. Reliance on these types of supplies requires the absence of pollutant sources such as sewage lines, septic tanks and drain fields, sewer mains, abandoned unsealed wells, wastewater treatment and storage works, feedlots for animals, chemical handling and storage facilities, petroleum storage facilities (including those used for herbicides, pesticides, and fertilizers), waste disposal sites (including hazardous and radioactive wastes), and mining operations.

11.3.1 Extension to an Existing Water System

If the project is located close enough for the developer to economically provide an extension to an existing system, permission must be obtained from that system's owner. Because this extension becomes part of the existing system once installed and accepted, all proposed design methods, materials, and provisions must be approved. Often the existing water utility assumes the responsibility of providing an adequate supply of water and storage volume, although impact fees may be assessed for such purposes.

A field hydrant flow test must be conducted to inventory the existing system's capability. If the hydrant test is to check fire suppression flows, the test should be conducted during periods of ordinary system demands according to the National Fire Protection Association (NFPA, 1995b).[*] The hydrant flow test requires a minimum of two hydrants located close to the point of connection for the new extension. The location should be free from traffic, pedestrians, and other interferences. One hydrant is designated as the residual hydrant where normal pressure is initially observed with all other hydrants being closed (and later pressure is measured at the residual hydrant when the other designated hydrants are opened for flow). The residual hydrant is generally located between the hydrant to be flowed and the large mains that act as major sources of supply. The number of flowing hydrants utilized depends on the strength of the distribution system. The flow from one or several hydrants must produce a pressure drop of at least 25 percent in the residual hydrant for usable test results, or the observed rate of flow must be equal to the NFF (NFPA, 1995b).[*]

To generally describe the test procedure using Figure 11.9a as the example, initially the system's static pressure is observed at the residual hydrant with nothing flowing through hydrant 1. Next, hydrant 1 is opened for maximum flow. The rate of flow at hydrant 1 is determined using a Pitot tube to measure the pressure of the discharge stream, as illustrated in Figure 11.9b. According to the NFPA (1995b),[*] the corresponding flow can then be calculated using the Pitot reading by

$$Q = 29.83 C d^2 P^{0.5} \tag{11.3}$$

where Q is the discharge from the flowing hydrant in gallons per minute, C is the coefficient of discharge for various outlet configurations shown in Figure 11.9c, d is

---

[*] Reprinted with permission from NFPA 291, *Recommended Practice for Fire Flow Testing and Marking of Hydrants*. Copyright © 1995, National Fire Protection Association, Quincy, Mass. 02269. This reprinted material is not the complete official position of the National Fire Protection Association on the reference subject, which is represented only by the standard in its entirety.

FIG. 11.9 Fire hydrant flow testing details: (*a*) two-hydrant flow test layout; (*b*) Pitot tube position; (*c*) three general types of hydrant outlets and their coefficients of discharge. Reprinted with permission from NFPA 291, *Recommended Practice for Fire Flow Testing and Marking of Hydrants.* Copyright © 1995, National Fire Protection Association, Quincy, Mass. 02269. This reprinted material is not the complete and official position of the National Fire Protection Association on the referenced subject, which is represented only by the standard in its entirety. This figure includes copyrighted material of Insurance Services Office, Inc., with its permission. Copyright, Insurance Services Office,.Inc., 1963.

the diameter of the flowing outlet in inches, and *P* is the measured gauge pressure in pounds per square inch gauge from the Pitot.

The calculated flow is recorded at the time of day that the reading was obtained. Usually the Pitot reading is taken off the flowing hydrant's standard 2.5-inch outlet rather than a pumper outlet because better flow characteristics are obtained. While this peak flow continues, the system's dynamic (or residual) pressure is observed at the residual hydrant and recorded. With this field information the corresponding residual pressure at any flow can be computed using the equation (NFPA, 1995b)[*]

$$Q_r = Q_f \frac{h_r^{0.54}}{h_f^{0.54}} \tag{11.4}$$

where Q_r is the flow available at the desired residual pressure, Q_f is the maximum recorded flow at the flowing hydrant during the field test, h_r is static pressure at no flow minus the desired residual pressure, and h_f is the static pressure at no flow minus the residual pressure at maximum test flow.

Equation 11.4 is valuable for estimating available residual pressures at the connection point of an existing system when various design demands required by a potential development are delivered through the system.

Often a graph similar to Figure 11.10 is used instead of Equation 11.4. The static pressure is initially plotted where the flow equals zero. The maximum measured flow and corresponding observed residual pressure are then plotted. Residual pressures for any other flow can be obtained along the line that connects the plotted data.

Later the design professional can compute the head loss in the proposed water main extension from the proposed point of connection to the critical point (usually

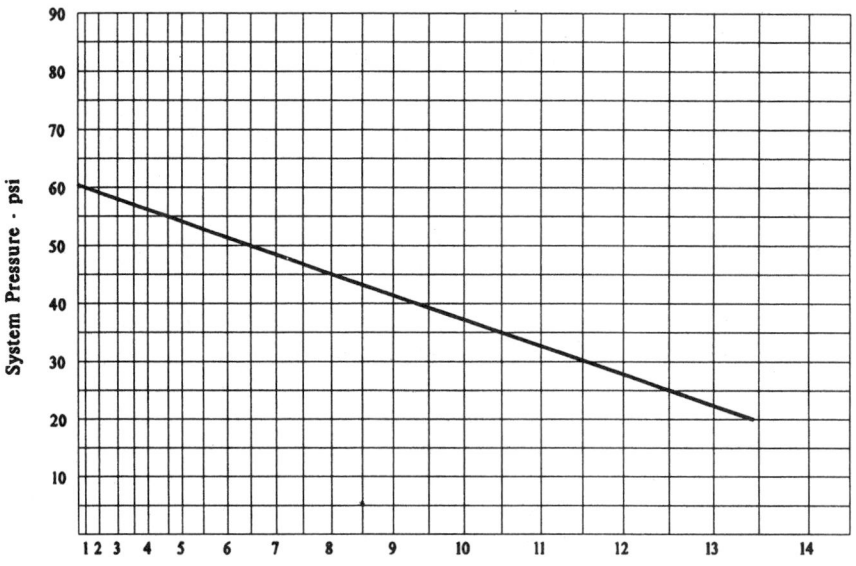

**Flow In Hundreds of Gallons Per Minute
At Flowing Hydrant**

FIG. 11.10 Pressure-flow graph for water supply analysis.

considered the location of the hydrant in the development that is most remote from the point of connection to the existing water system). If the difference between this calculated head loss in the proposed water main extension and the available system pressure when delivering the design flow is greater than the minimum residual pressure allowed (usually 20 lb/in.^{2}), the proposed system would be considered adequate. If this difference produces a residual pressure that is less than the allowable, the proposed system must be redesigned.

When conducting a flow test in the field, the following precautions must be followed:

(a)   Hydrants must be slowly flushed prior to installing any gauges or Pitot tubes to prevent rust or particles within the hydrant from clogging the testing device.

(b)   Hydrants must be opened and closed slowly to prevent waterhammer.

(c)   Hydrant valve stems must be completely closed in order to open the hydrant's weep hole for interior hydrant drainage when tests are completed.

As an example, data collected from a field hydrant flow test are shown below under conditions similar to Figure 11.9*a*:

> Observed static pressure   61 psig
> Observed maximum flow   1256 gal/min corrected for outlet discharge loss *C*
> Observed dynamic pressure   25 psig

Consider an example of a proposed residential development of approximately 100 dwelling units with a gross density of three dwelling units per acre; it is to be served by an extension to the system tested above. The average daily demand based on 24 hours (assuming 150 gallons per person per day) would be (150 gallons per person) × (3.4 persons per unit) × 100 units = 51,000 gallons flow per day. The corresponding maximum daily demand, using Figure 11.5, would be 170,000 gallons per day or 118 gal/min, while the peak-hour demand would be estimated at 360,000 gallons per day (or 250 gal/min) from Figure 11.6.

If the NFF is assumed to be 750 gal/min in this example, an acceptable design flow to demonstrate system adequacy would be to provide at least 750 gal/min + 118 gal/min, or 868 gal/min (which is greater than the peak-hour demand), at the most remote hydrant when testing after construction but before dwelling occupancy. The estimated residual pressure at the point of connection for that design flow can now be calculated using the flow test data and Equation 11.4, where

$$868 = 1256\frac{(61 - \text{residual})^{0.54}}{(61 - 25)^{0.54}}$$

Estimated residual pressure $= 42.8$, or 43 psig

Using a minimum allowable system pressure of 20 lb/in.^{2}, the sum of all static (elevation differences) and frictional head losses associated with the proposed extension and distribution network from the connection point to the most remote hydrant must not exceed 23 lb/in.^{2}. Otherwise the proposed system would be inadequate and would need to be redesigned by either increasing pipe sizes, adding a booster pump, or including an elevated tank to increase pressure.

11.3.2 Water Wells

A water well for use as a groundwater supply consists of the well itself and the pump, fittings, and piping. The well is a hole in an aquifer that is bored, dug, punched, or drilled and that can be lined with a casing. Most deep wells are drilled and are the type most commonly associated with land development water supplies. During the drilling operation, a driller's log is maintained where all subsurface formations encountered are recorded, with the data including coloring, mineralogy, rock types, grain size distribution, and geophysical and mechanical log data. Openings through perforations or screens allow water to enter the well while providing lateral support to prevent the walls from collapsing. Water that enters is removed by pumping (unless it flows naturally from artesian pressure), and replacement water is then allowed to enter. Provided the rate of withdrawal does not exceed the rate of recharge, the well operates within what is classically known as its safe yield. As water enters the well, the surrounding soil resists the flow, resulting in head loss. This causes the water's surface to form a cone of depression, known as drawdown. Figure 11.11 illustrates this phenomenon.

According to Hammer and MacKichan (1981), Darcy's law is generally used to define the velocity of water flowing through a granular soil by

$$v = Ki \tag{11.5}$$

where v is the velocity of moving water under laminar flow conditions in feet per second, K is a coefficient having the same units as the velocity, and i is the hydraulic slope gradient in feet per foot.

More commonly, the equation takes the form

$$Q = AKi \tag{11.6}$$

where, in addition to the previously defined parameters, Q is the flow rate in cubic feet per second and A is the cross-sectional flow area in square feet.

Factors affecting the coefficient of permeability include soil parameters, water viscosity, and varying subsurface conditions. According to Fair et al. (1966), the U.S. Geological Survey has chosen as its "standard coefficient of permeability" the flow of water in gallons per day at 60°F through a cross-sectional area of 1 ft^2 under a unit gradient. If the coefficient of permeability is obtained at field temperatures other than 60°F, the standard coefficient of permeability can be determined with the following relationship (Cedergren, 1989):

$$K_{\text{at any temp.}} = [K_{60F}] \left[\frac{\mu_{60}}{\mu_{\text{at any temp.}}} \right] \tag{11.7}$$

where K_{60F} is the standard coefficient of permeability, $K_{\text{at any temp.}}$ is the coefficient of permeability obtained under field conditions, μ_{60} is the dynamic viscosity of water at 60°F, and μ_{field} is the dynamic viscosity of water at field temperature.

According to Hammer and MacKichan (1981), for ideal unconfined flows (Figure 11.1a) the observed coefficient of permeability can be obtained after steady-state pumping conditions have been attained by

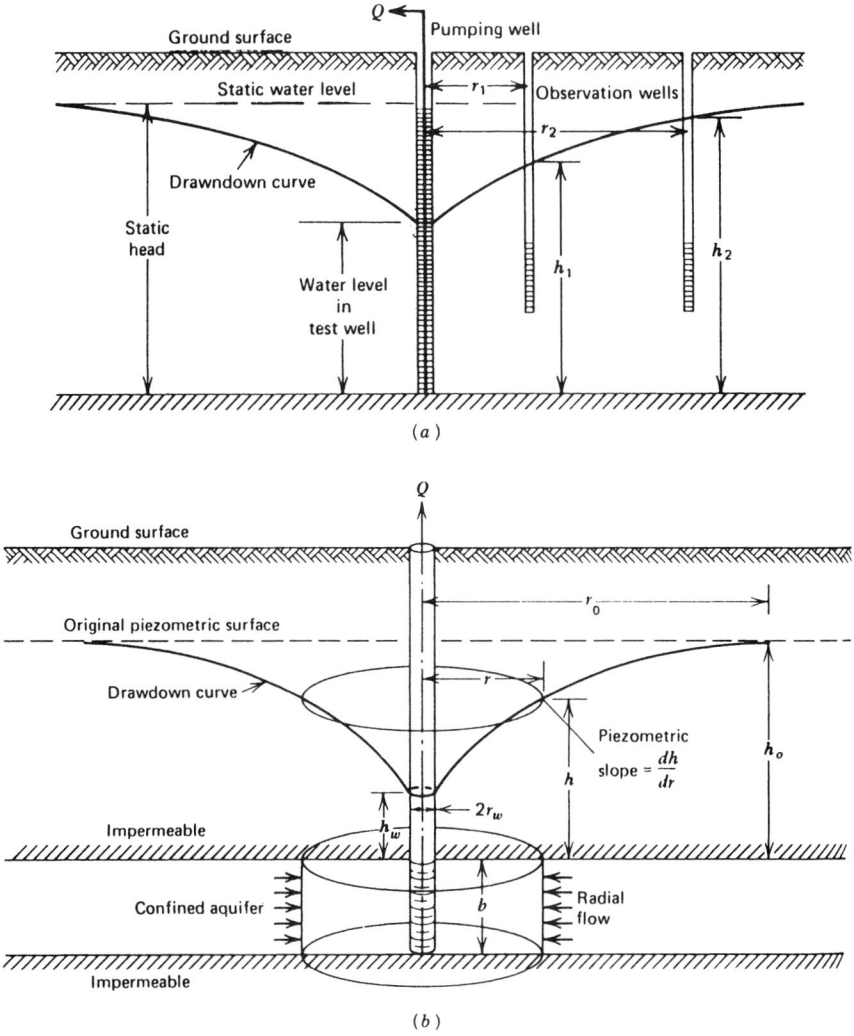

FIG. 11.11   Aquifer pumping: (*a*) flow of water toward a well during a pumping test in an unconfined coarse-grained aquifer; (*b*) steady radial flow to a well completely penetrating an ideal confined aquifer. Source: Mark J. Hammer and K. A. MacKichan. *Hydrology and Quality of Water Resources*. Copyright © 1981 by John Wiley & Sons, Inc.

$$K = \frac{Q \log_e(r_2/r_1)}{3.14(h_2^2 - h_1^2)} \qquad (11.8)$$

where K is the coefficient of permeability in feet per second, Q is Well's discharge in cubic feet per second, and r and h are the observed well data measured in feet, as indicated in Figure 11.11*a*. For confined aquifer conditions (Figure 11.11*b*), the observed coefficient of permeability after steady-state pumping conditions are attained is

$$K = \frac{Q \log_e(r_2/r_1)}{2(3.14)(b)(h_2 - h_1)} \qquad (11.9)$$

where Q is Well's discharge in cubic feet per second, K is the coefficient of permeability in feet per second, b is the aquifer's thickness in feet, and r and h are the observed well data measured in feet as defined in Figure 11.11a.

Other factors must be considered if the orientation of the original hydraulic grade line is not horizontal. A more complete presentation on this topic can be found elsewhere (Cedergren, 1989; Fair et al., 1966; Hammer and MacKichan, 1981).

Due to a number of conditions that can cause well yields to vary (including encountering non-steady-state flows), the design professional should consult other reference sources, as well as local well drillers and regulatory agencies, to determine additional design information including well sizes. Generally the larger the diameter of the well, the more it can produce. Several well sizes relative to possible production rates are depicted in Figure 11.12 as proposed by Ameen (1974).

Well screens that allow water to enter are positioned at depths (extending through the water bearing strata) where good water is available. Information necessary to position screens comes from soil samples obtained during drilling and the logged information on the well. Precautions must be taken to protect these screens. They can clog if improperly sized or if the inflow velocity exeeds a maximum safe velocity of approximately 1–2 in./min. An experienced well driller is needed to properly develop the well's production capability. A typical deep well with a vertical turbine pump installed is shown in Figure 11.13.

If several wells are to be installed, their locations should be dispersed to minimize the compounding effects of overlapping cones of depression resulting from drawdown. Wells also should be located where the possibility of surface water flooding, with its potential contamination, is remote and the intrusion of salt water is unlikely. Small-diameter test wells are often drilled to provide initial subsurface information that can be used to design the permanent production well. The production well can be constructed by either reaming and developing the initial test well or drilling a new hole. If a new hole is made, the test well must be properly sealed prior to abandonment to prevent groundwater contamination. Once the production well has been developed, it can be tested by pumping to determine its yield and specific capacity. Upon completion, it can be disinfected.

To determine the pump size that will deliver the desired yield, field drawdown and permeability data must be utilized. For example, if a well is to withdraw 350 gal/min from an unconfined aquifer, an estimated 12-inch-diameter well can be utilized, as shown in Figure 11.12. Once the test well or production well is installed and developed and steady-state conditions exist, a field pumping test can be undertaken. The following sample drawdown data are obtained from such a test:

Desired Daily Yield Of Well	Recommended Diameter Of Well
Flows up to 50,000 gallons per day	6-inch
50,000 to 100,000 gallons per day	8-inch
100,000 to 250,000 gallons per day	10-inch
250,000 to 500,000 gallons per day	12-inch
500,000 and above	Gravel packed wells

FIG. 11.12 Typical well sizes for various yields. Source: Ameen, 1971. Reproduced with permission.

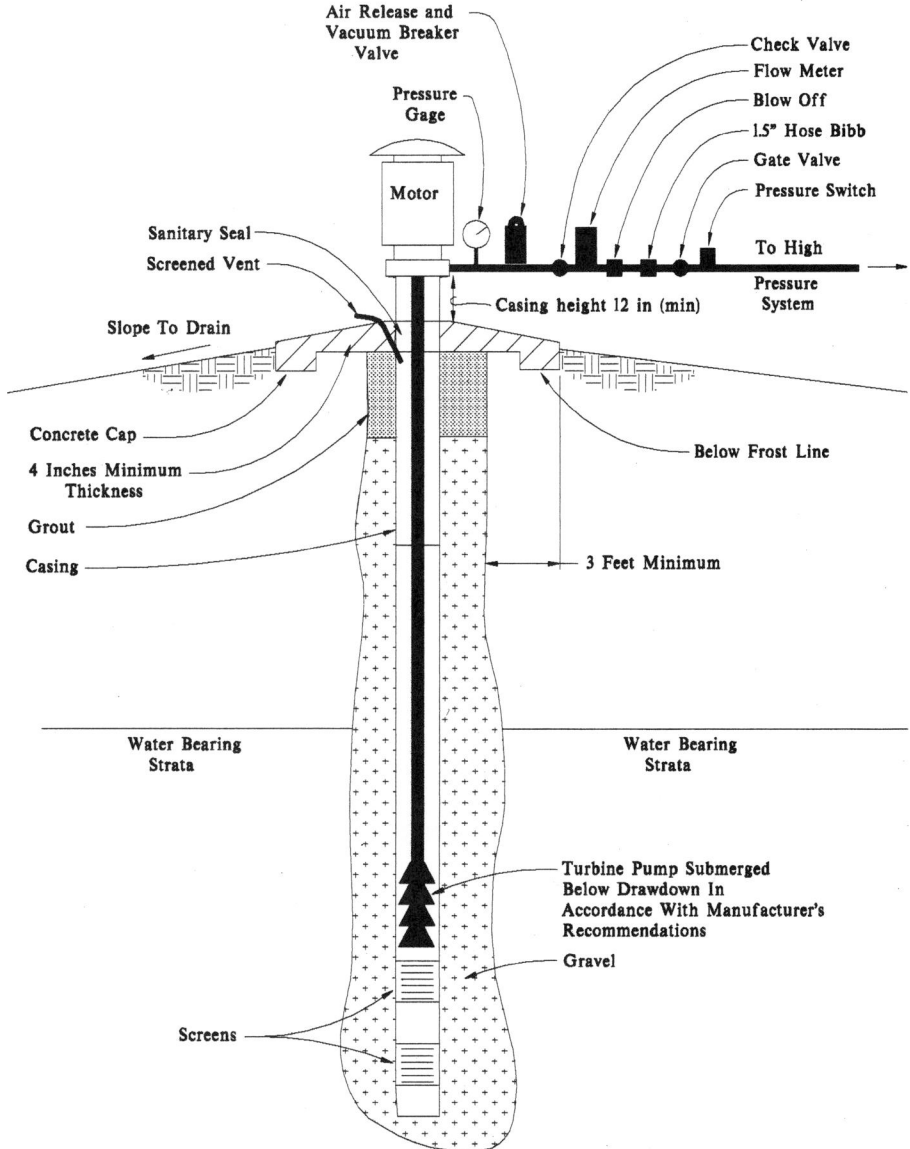

FIG. 11.13 Example of deep well details using a vertical turbine pump.

1. Steady-state pumping rate: 500 gal/min, or 1.114 ft³/s (0.002228 × 500 gal/min).
2. $r_1 = 15$ feet; $h_1 = 415$ feet.
3. $r_2 = 60$ feet; $h_2 = 440$ feet.
4. Depth from ground surface to bottom of aquifer: 450 feet.
5. Ground surface elevation: 50.0 feet (NGVD 29).

(a) Determine field permeability using Equation 11.8:

$$K = \frac{1.114 \text{ ft}^3/\text{s} \times \log_e(60/15)}{3.14(440^2 - 415^2)} = 0.000023 \text{ ft/s}$$

(b) Determine the amount of drawdown (at 1 foot from the well) when the required 350-gal/min yield (or 0.78 ft³/s) is withdrawn, using Equation 11.8:

$$0.000023 = \frac{0.78 \text{ ft}^3/\text{s} \times \log_e(60/1)}{3.14(440^2 - h_1^2)}$$

$$h_1 = 386.5 \text{ feet}$$

(c) Determine the elevation of drawdown when the yield is 350 gal/min:

$$\begin{array}{rl}
\text{Ground elevation:} & +50 \text{ feet (NGVD 29)} \\
\text{Depth to aquifer bottom:} & \underline{-450 \text{ feet}} \\
\text{Elevation of aquifer bottom:} & -400 \text{ (NGVD 29)} \\
\text{Height to drawdown at 350 gal/min:} & \underline{+386.5 \text{ feet}} \\
\text{Elevation of drawdown:} & -13.5 \text{ (NGVD 29)}
\end{array}$$

(d) Determine the TDH for the well pump. This example assumes the well pump will be submerged at least 35 feet and that the pump will discharge directly into a ground storage tank at the well site where the maximum height of water in the tank is estimated to be 24 feet.

The Hazen-Williams formula (Equation 10.10) is utilized to calculate the various frictional head losses (per 1000 feet of pipe) as shown below. The 6-inch pipe column used to convey water to the ground surface is 98.5 feet with a Hazen-Williams C of 100, where

$$H_{1\ 1000} = \left[\frac{350}{(0.006756)(100)(6^{2.63})} \right]^{1.85} = 17.2 \text{ feet}/1000$$

A summary of headlosses are:

$$\begin{array}{rl}
\text{Static system elevation head:} & \\
50 \text{ NGVD } 29 + 24 \text{ foot tank} - (-13.5) = & 87.5 \text{ feet} \\
\text{Friction in 6-inch pipe column—98.5 feet long:} & 1.7 \text{ foot} \\
\text{Minor losses (assumed for this example):} & \underline{0.5 \text{ feet}} \\
\text{Total:} & 89.7 \text{ feet} \\
\text{Use:} & 90.00 \text{ feet}
\end{array}$$

(e) Select possible vertical turbine pump sizes that can fit in the 12-inch well. Since various stages can be incorporated into the final pump configuration where each stage acts in series, the TDH can be incremented into various head segments that correspond to integer stage possibilities where the most efficient pump combination can be selected. To illustrate this point, the required 90 feet of head determined previously can be incremented into the following segments:

Number of Possible Stages	Corresponding Head/Stage
2	90/2 = 45 feet
3	90/3 = 30 feet
4	90/4 = 22.5 feet

The design professional can utilize various pump curves to select possible combinations of stages for various model pumps to obtain the most efficient system. Figure 11.14 illustrates one such typical set of curves where four stages, each having impellers of 5.469 inches and contributing 22.5 feet of head, can be satisfactorily used to pump the required 350 gal/min. The NPSH available must be at least 7 feet and the brake horsepower per stage is approximately 2.4 horsepower according to this figure.

Other factors relating to specific manufacturers' data must be evaluated including motor horsepower, thrust conditions, total efficiency based on possible corrections relating to the number of stages utilized, shaft sizing and stretch potential, and discharge size selection.

11.3.3 Surface Water Sources

Land developers might choose not to rely on surface water supply sources because substantial contributory drainage basins are required. If the development, however, does utilize this type of source, either a suitable surface water supply must be available with sufficient flow to meet system demands during dry periods, such as a river or lake, or an impoundment (or reservoir) must be constructed that has a storage capacity capable of accommodating withdrawal demands during dry periods based on sufficient inflows from streams, rainfall, and possibly melting ice and snow. The required storage volume of a reservoir is usually determined using a mass curve that shows the cumulative inflow; it must allow for loss of volume attributed to silting and loss of water due to evaporation and percolation. It must also feature protection from flooding. The design professional must determine whether any wells exist within the area proposed for inundation to ensure that they are sealed properly to prevent groundwater contamination.

According to Fair et al. (1971), the safe yield of a river or stream is its lowest dry weather flow, whereas with full development of storage, the safe yield approaches the mean annual flow. The attainable yield, which is of interest to the design professional when analyzing sources of supply, is affected by evaporation, bank storage, seepage, and silting. The reader is referred elsewhere for a more complete treatment of this subject (Fair et al., 1971).

11.4 TREATMENT CONSIDERATIONS

Potable water offered to the public for consumption in land developments must comply with the Safe Drinking Water Act of 1974 as amended (along with other federal and state regulations) and must be noncorrosive. Should the water meet all the regulatory quality standards, which is possible with water produced from some wells, then no

FIG. 11.14 Examples of vertical turbine pump performance curves. Courtesy of Floway Pumps, Inc., Fresno, Calif. 93707.

treatment may be required. Raw surface water, however, can contain suspended and dissolved solids, dissolved gases, and organic compounds that make the water unusable without some type of treatment to include the addition of a disinfectant such as chlorine, chlorine dioxide, chloramines, ultraviolet light, or ozone. The production of disinfection by-products (DBPs) must be minimized during the disinfection process, especially when organic precursors are present.

11.4.1 Suspended Solids

Organic or inorganic suspended solids are usually found in surface waters. They can be removed by physical or mechanical means using filtration or sedimentation. Organic suspended solids normally are products of plant or animal life but may include synthetic organic compounds. These solids can decay or decompose or be burned. Primarily, organic solids consist of oxygen, hydrogen, and carbon in possible combination with sulfur, phosphorus, and nitrogen.

Inorganic suspended solids do not decay as they are inert and include mineral substances that produce hardness and mineral levels in the raw water.

Suspended solids may be of sufficient size and weight to allow settling with time; however, smaller suspended solids (usually less than 1 μm) are considered colloidal and normally will remain in suspension unless combined into flocs.

11.4.2 Dissolved Solids

Dissolved solids are present more often in groundwater than in surface water as a result of subsurface minerals being dissolved in the underground water. These solids are removed using chemical processes, such as oxidation and precipitation, or physical processes, such as reverse osmosis.

Iron and manganese often are found in groundwater and anaerobic surface waters as Fe^{2+} and Mn^{2+}. Upon exposure to oxygen, they are transformed into more stable forms (Fe^{3+}, Mn^{4+}). These ions yield unpleasant tastes and odors in water and can produce stains in laundry and porcelain plumbing fixtures.

Water hardness is caused primarily by the presence of calcium and magnesium cations in the water. Hard water produces plumbing scale and requires large quantities of soap for cleansing. These ions are associated with bicarbonate or sulfate anions; however, they may also be associated with chlorides, nitrates, and silicates. Carbonate hardness refers to the calcium and magnesium compounds such as calcium or magnesium sulfates and chlorides. High fluoride concentrations are usually associated with groundwater supplies in several states. Two methods to lower fluoride concentrations are activated alumina absorption and reverse osmosis.

11.4.3 Dissolved Gases

Dissolved gases can be present in raw water supplies. For example, nitrogen and oxygen are both present in surface waters. Also, carbon dioxide can be present in groundwater and will begin to ionize into carbonic acid if the water pH is above 4.5. Carbon dioxide, which can be present in groundwater, usually lowers the pH of the water, resulting in distribution system corrosion.

Another dissolved gas, hydrogen sulfide, produces a rotten egg smell that often is offensive to consumers. If the pH is low, air stripping may remove the odor; however, chemical oxidation may be required if the pH is high. Ammonia also may be found dissolved in some surface waters. This results in unpleasant odors and tastes and can be stripped by aeration.

11.4.4   Trace Organic Compounds

Trace organic compounds can be found in both surface water and groundwater. These compounds are usually man made and can adversely affect consumer health. Volatile organic chemicals (VOCs) are essentially solvents, such as benzene and chloroform, that may have been used for industrial purposes and have escaped into the ground from underground storage tanks, leached from disposal sites, and been spilled. Other trace organic compounds include pesticides and herbicides that have been transported into surface waters by stormwater runoff or into groundwater aquifers by percolation. Other contaminants may exist in various forms and from various sources, all of which may require removal.

11.4.5   Microbiological Contamination

Microbial contamination may be encountered in surface waters and some groundwater sources; however, filtration and disinfection can eliminate most of these types of contamination.

11.4.6   Water Treatment Processes

Water treatment processes include aeration, chemical addition, mixing, coagulation, flocculation, sedimentation, filtration, adsorption, ion exchange, membrane separation, disinfection, and waste by-product handling. Most water treatment facilities utilize several of these processes to produce finished water. A typical treatment process is shown in Figure 11.15.

Fluorides, sodium, iron, manganese, magnesium, and calcium commonly are removed by the lime-soda ash process, reverse osmosis, or ion exchange. Chemical coagulation is also used to form flocs of suspended colloidal particles that, once removed from the system, reduce the color, odor, taste, and turbidity of the water. Alum is probably the most common coagulant used, although many others are available. Once the flocs are formed, the water normally is clarified and filtered. The general effectiveness of water treatment processes for contaminant removal is shown in Figure 11.16.

Although membrane processes have been used for many years to desalt water, new membranes are now capable of providing additional treatment capabilities. In addition to reverse osmosis, which has often been used to remove salts and natural/synthetic organic matter/chemicals, microfiltration and ultrafiltration are very low pressure systems that can effectively remove particles and reduce the levels of microorganisms such as *Giardia* and *Cryptosporidium* to levels below detection. Nanofilters are used to effectively remove calcium and magnesium ions.

Recent outbreaks of illness caused by *Cryptosporidium* have triggered the installation of ozone treatment units in various treatment facilities. Ozone effectively disinfects and

FIG. 11.15   Typical water treatment process.

Primary contaminants

Contaminant categories	Aeration and stripping	Coagulation, sedimentation or DAF* filtration	Precoat filtration	Lime softening	Chemical oxidation and disinfection	Membrane processes			Ion exchange		Absorption		
						Nanofiltration	Reverse osmosis	Electrodialysis/ED reversal	Anion	Cation	Granular activated carbon	Powdered activated carbon	Activated alumina
Inorganics													
Antimony		XO‡		XO			X†	X	X				X
Arsenic (+3)		X					X	X	X				X
Arsenic (+5)				X			X	X					
Barium		X		X			X	X		X			
Beryllium		X		X			X	X		X			
Cadmium		X		X			X	X		X			
Chromium (+3)							X	X					
Chromium (+6)							X	X	X				
Cyanide					X		X	X					
Fluoride				X			X	X					X
Lead§				X			X	X					
Mercury (inorganic)				X			X	X		X			
Nickel							X	X	X				
Nitrate							X	X	X				X
Nitrite							X	X	X				X
Selenium (+4)		X					X	X	X				X

Synthetic organics
Pesticides/Herbicides
Dissolved organic carbon
Radionuclides
Radium (226 + 228)
Uranium

Secondary contaminants and constituents causing aesthetic problems

Hardness
Iron
Manganese
Total dissolved solids
Chloride
Sulfate
Zinc
Color
Taste and odor

* DAF, dissolved air flotation.

† X, appropriate process for this contaminant.

‡ XO, appropriate when oxidation used in conjunction with this process.

§ Lead is generally a product of corrosion and is controlled by corrosion control treatment rather than removed by water treatment processes.

Fig. 11.16 General effectiveness of water treatment processes for removal of soluble contaminants. Particulate contaminants can be removed by the filtration processes shown and also with slow sand filtration, microfiltration, and ultrafiltration. Gaseous contaminates can be treated by aeration or air stripping. Source: *AWWA Water Quality and Treatment*, 5th ed. Copyright © 1999 by McGraw-Hill Companies. Reprinted by permission of McGraw-Hill Companies.

583

oxidizes public drinking water while preventing the formation of chlorinated disinfection by-products such as trihalomethanes (THMs) and haloacetic acids (HAAs). It can also destroy bacteria, inactivate viruses and protozoans, while improving filter performance if used prior to filtration. Ozone treatment is the only disinfectant technology today that kills *Cryptosporidium.*

Lack of space prohibits a more complete presentation of information pertaining to the purification of water for consumption as indicted in Figure 11.16. Other references on water treatment are available for the reader to consult (AWWA, 1999; Salvato, 1992; Weber 1972).

11.5   TREATED WATER STORAGE FACILITIES

Treated water storage facilities hold water for emergency reserve, fire fighting, equalization for peak demands, and providing idle time for supply and treatment works. These facilities can be elevated tanks, standpipes, ground-level high-pressure tanks, or low-pressure ground storage tanks. The most advantageous is the elevated tank because it is a ready source of water volume and pressure.

11.5.1   Elevated Water Tanks

Elevated water tanks are placed where the ground elevation is the highest in the surrounding service area and, if possible, near the center of areas of high demands or opposite the source of supply to help equalize pressures. Tanks can provide equalization storage ranging from approximately 15 to 35 percent of the maximum daily demand when constant pumping is utilized.

The engineer should obtain local information that reflects maximum daily demands and water use patterns for design purposes. If local information is not available, reasonable estimates should be made. In this example, Figure 11.5 is used to establish the maximum daily demand, and Figure 11.2 is used to allocate the demand throughout the maximum day.

To illustrate a method for sizing an elevated tank, consider an upscale development of 1000 persons with each person having an average demand of 150 gallons of water per day and a NFF requirement of 1000 gal/min for 2 hours. The required tank capacity can be determined as follows:

(a) Required fire storage reserve:

$$1000 \text{ gal/min } \times 2 \text{ hours } \times 60 \text{ min/h } = 120{,}000 \text{ gallons}$$

(b) Required one-day emergency reserve:

$$150 \text{ gallons per person per day } \times 1000 \text{ persons } = 150{,}000 \text{ gallons}$$

For this example two 75,000-gallon ground storage tanks will be provided (each 23.5 feet in diameter × 24 feet tall), with one located at each of the two proposed well sites. This allows the one-day emergency reserve to be turned over on a regular basis to keep it fresh.

(c) Required equalization volume using a graphical approach:

   (1)   For this example it is assumed that two wells will be at opposite sides of the development to provide two sources of water for 100 percent backup capability. If each well pumps 12 hours, providing effectively 24 hours of pumping, the required maximum daily demand will be provided as the pumping rate. The maximum daily demand can be determined using local data; however, if local data are lacking, the demand can be determined from Figure 11.5 unless the parameters used to construct Figure 11.5 are not germane. By entering Figure 11.5 with 300 dwelling units at a density of three dwelling units per acre (1000 persons/approximately 3.4 people per dwelling yields approximately 300 dwellings), about 500,000 gallons per day for the maximum daily demand is conservatively estimated. The required 24-hour pumping rate is calculated by

$$\frac{500{,}000 \text{ gallons per day}}{24 \text{ hours } \times 60 \text{ min/h}} = 347 \text{ gal/min}$$

which can be rounded to 350 gal/min.

   (2)   To obtain mass ordinates for constructing the diagram, a water use hydrograph based on local data must be obtained. For this example percentages of water use have been extrapolated from Figure 11.2 for the maximum day. Figure 11.17 contains the calculations.

   (3)   The calculated ordinates can now be used to construct the mass curve shown in Figure 11.18. Using the slope of the corresponding 24-hour pumping supply curve to construct tangents at the full-tank and depleted conditions, an equalization storage is determined graphically to be approximately 172,000 gallons.

(d) The total storage requirement is then:

Fire reserve:	120,000 gallons
Emergency reserve:	150,000 gallons
Equalization storage:	<u>172,000 gallons</u>
Total:	442,000 gallons

The elevated tank should include fire reserve and equalization storage of 292,000 gallons, with the one-day emergency reserve being placed in ground storage tanks in this example.

If the contribution of the well is included during fire flow, the required tank's capacity can be reduced by 350 gal/min × 2 hours × 60 min/h, or 42,000 gallons. The net elevated tank size would then be 292,000 − 42,000 = 250,000 gallons. Therefore, a standard 250,000-gallon tank would be used in conjunction with two 75,000-gallon ground storage tanks.

The design professional must determine the nominal water system operating pressure in order to establish the elevation of the tank's bottom, E_{tank}. Using for this example 40 lb/in.^{2} as the normal system operating pressure and an estimated elevation of the highest fixture in the proposed system of 70 feet NGVD, the tank's bottom must have an elevation of (40 psig design working pressure) × (2.31 feet of water per pound per square inch) + 70 feet = 162.4 or 162 feet NGVD 29. If the ground elevation at

Maximum Day Demands				
Time Interval-%	% of Total	Volume, gal	Time of Day	Cumulative Volume, gal.
12-1 A.M.	0.61	3050	Midnight	0
1-2	0.61	3050	1	3050
2-3	0.12	600	2	6100
3-4	0.12	600	3	6700
4-5	0.12	600	4	7300
5-6	0.61	3050	5	7900
6-7	0.12	600	6	10950
7-8	1.70	8500	7	11550
8-9	2.31	11550	8	20050
9-10	3.52	17600	9	31600
10-11	3.95	19750	10	49200
11-12	4.55	22750	11	68950
12-1 P.M.	3.46	17300	12 Noon	91700
1-2	4.55	22750	1	109000
2-3	4.01	20050	2	131750
3-4	4.63	23150	3	151800
4-5	6.25	31250	4	174950
5-6	8.07	40350	5	206200
6-7	11.41	57050	6	246550
7-8	13.66	68300	7	303600
8-9	11.41	57050	8	371900
9-10	6.80	34000	9	428950
10-11	5.10	25500	10	462950
11-12 P.M.	2.31	11550	11	488450
		500000	Midnight	500000

FIG. 11.17 Sample calculations for mass ordinates using maximum day demands.

the tank site is 50 feet NGVD 29, the height of the tank will be 112 feet. Figure 11.19 depicts the general features of an elevated tank.

The high-pressure service pumps delivering finished water into the system from the two 75,000-gallon ground storage tanks are sized in this example to deliver 350 gal/min at a head equal to the system head from the pump to the elevated tank at high water (194 feet).

The proposed supply and storage system is shown in Figure 11.20

11.5.2 Water Storage Standpipes

A standpipe for water storage is only a cylindrical water reservoir that has a height greater than its diameter. A standpipe is advantageous if the site's terrain includes an area where it can be located above the surrounding service area to provide adequate system working pressures.

The use of a standpipe can be considered instead of the elevated tank illustrated in Section 11.5.1 if the proposed tank site's elevation is 130 feet NGVD 29. A tank 32 feet, 10 inches in diameter × 64 feet tall, containing 400,000 gallons of storage (sum of all storage requirements equaling 442,000 gallons minus pump contributions during fire flow of 42,000 gallons), could be used, as shown in Figure 11.21.

If the highest plumbing elevation in the service area is the same 70 feet NGVD 29, the standpipe will approximately maintain the minimum system operating pressure of

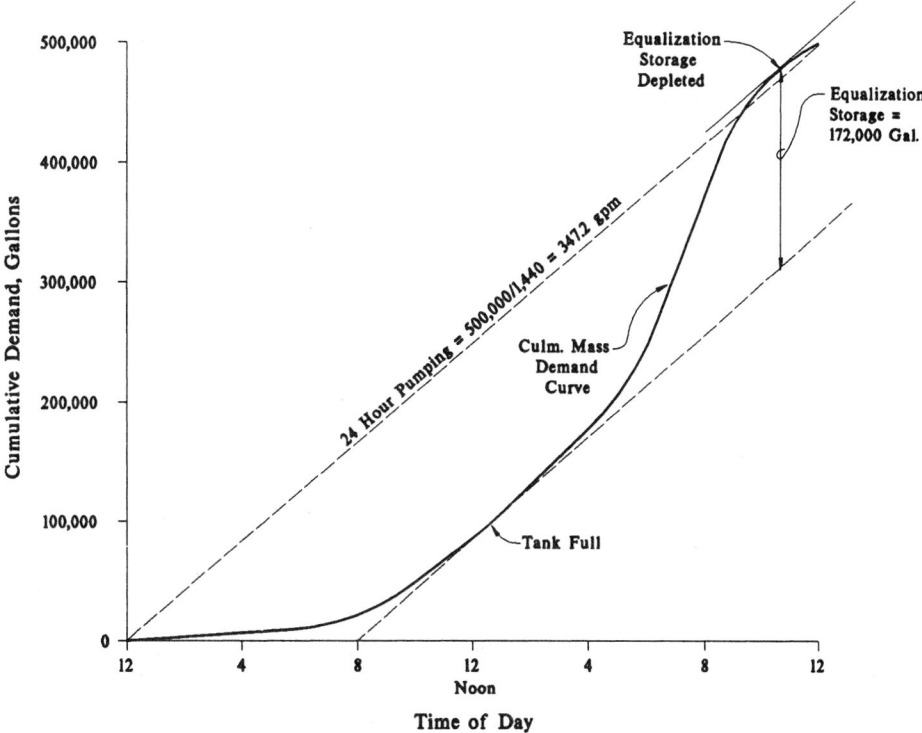

FIG. 11.18 Example of a mass diagram for equalization storage based on maximum day demands.

40 psig using (165.1 low equalization level − 70.00 feet of highest fixtures)/2.31 feet of water per pound per square inch). A high-capacity pump would have to be installed to increase the system pressure for flows below the equalization volume if the pressure of 40 lb/ft^2 is to be maintained for the 1000-gal/min fire flow.

11.5.3 High-Pressure Service Tanks

For small land developments without fire protection, high-pressure service tanks at ground level can be incorporated into the system to meet demands. Various approaches are used to design high-pressure service tanks. Some tanks (hydropneumatic) are designed to "float" on the system (Rothrock, 1974) and therefore add little storage capacity to the system. The value of this type of "floating" tank is that it allows the intermittent use of constant-speed electric pump motors and simple controls to respond to various system demands. A typical "floating" tank is shown in Figure 11.22.

This system requires the selected pump to be capable of delivering approximately 125 percent of the peak system demand at the high operating pressure. Use of this type of system requires high-capacity supply wells or the incorporation of low-pressure storage facilities to provide adequate standby reserve for pumping demands.

Some regulatory agencies require high-pressure service tanks to have adequate storage to accommodate peak system demands for 20–30 minutes when incorporated into small domestic water systems. According to Ameen (1974), each tank should be

250,000 Gal.
Elevated Tank

Elev. 194.0 NGVD 29
High Water

Elevation 162.0 NGVD 29
Low Water

Head
Range

32'

112' High

Highest Fixture
Elevation
70.0 NGVD 29

$\frac{162' - 70'}{2.31} \approx$ 40 psi
Working
Pressure

Ground Elevation 50.0 NGVD 29

FIG. 11.19   Typical 250,000-gallon elevated storage tank. Courtesy of CBI Na-Con (Chicago Bridge and Iron Company), Norcross, GA.

FIG. 11.20 Example of a water supply system feeder configuration. Tank courtesy of CBI Na-Con (Chicago Bridge and Iron Company), Norcross, GA.

FIG. 11.21   Typical 400,000-gallon potable water standpipe. Tank courtesy of CBI Na-Con (Chicago Bridge and Iron Company), Norcross, GA.

capable of storing the maximum flow demand for a 20-minute duration while allowing a reduction in storage volume for any contributions made by other sources (flows from low-pressure storage facilities) during a 20-minute period. This volume based on flow demands occupies approximately 25 percent of the total storage volume since adequate pressurized air must be included. In any case this storage should not be less than 20 times the pump's yield, with the pump's capacity being between 30 and 40 percent of the maximum flow demand (Ameen, 1974).

Consideration should also be given to pump cycling to prevent excessive wear of the pump. The tank should be sized so that the pump cycles a maximum of six times per hour (SCDHEC, State Primary Drinking Water Regulations, R.61-58, 1998).

It is important to prevent air from escaping the high-pressure storage tank into the distribution system at low water level and to minimize vortex action. An adequate dead water storage level must be maintained and can be determined using Figure 11.23.

A typical high-pressure storage facility is shown in Figure 11.24.

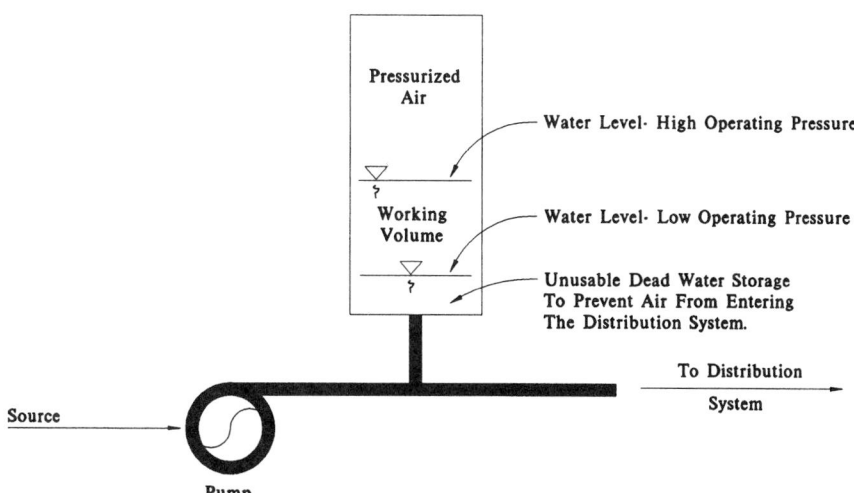

FIG. 11.22 Floating hydropneumatic high-pressure tank. Source: Rothrock, 1974. Copyright © 1974 by Donald F. Rothrock.

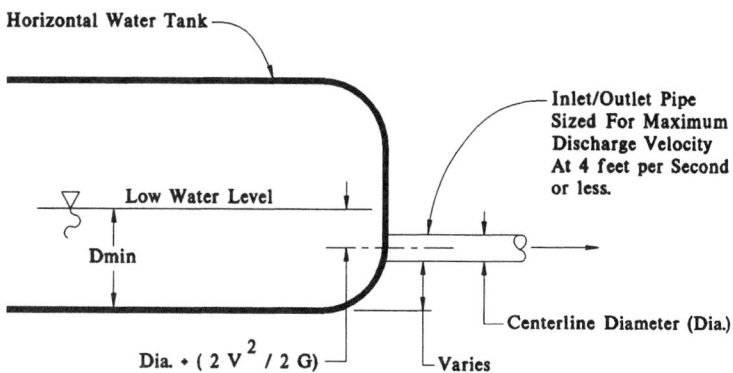

FIG. 11.23 Dead-storage determination for a high-pressure water tank. Source: Rothrock, 1974. Copyright © 1974 by Donald F. Rothrock.

11.5.4 Low-Pressure Storage

When the rate of water being supplied to the system is low, or when a large system demand is imposed for short periods of time, such as that generated by lawn sprinkling or schools, a low-pressure storage facility (at atmospheric pressure) can be provided close to the supply point. This facility can be designed to accommodate some equalization flows along with emergency reserve capacity to allow for maintenance downtime. This use has been illustrated in Section 11.5.1 in conjunction with elevated tanks.

Low-pressure storage facilities also can be used in conjunction with high-pressure storage facilities that are equipped with high service pumps capable of pumping at 30–40 percent of the peak demand to deliver the reserve directly into a high service storage

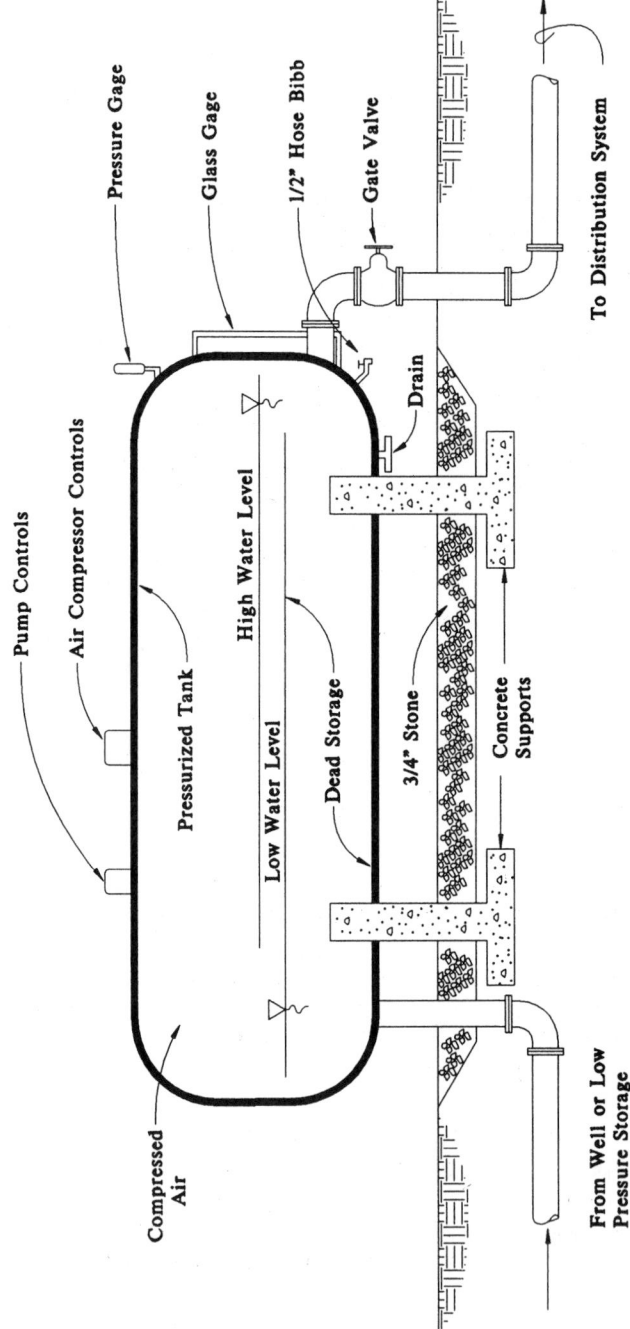

FIG. 11.24 Typical high-pressure storage water tank.

Pressure Gage

Glass Gage

1/2" Hose Bibb

Gate Valve

To Distribution System

Drain

Concrete Supports

3/4" Stone

From Well or Low Pressure Storage

Dead Storage

Low Water Level

High Water Level

Pressurized Tank

Air Compressor Controls

Pump Controls

Compressed Air

tank (Ameen, 1974). These pumps are usually chosen with run cycle times that extend from 15 minutes to 1 hour.

11.6   WATER DISTRIBUTION SYSTEMS

Water distribution systems consist of water lines and fittings, valves, service lines, meters, thrust blocks or restraining devices, and fire hydrants. The most desirable system provides adequate water service to all areas while minimizing unnecessary pipe length consistent with the incorporation of adequate redundant loops to reinforce the system. The most desirable network is a grid with a loop feeder to allow flexible system operation. Should a line rupture occur in a looped system, the whole system does not have to be secured since isolation valves and looped connections allow the system to continue operation outside the isolated problem area. Reinforced (or looped) systems also provide flow in more than one direction, resulting in pressures being more uniformly maintained. Branched systems do not exhibit these qualities and should be avoided if possible. Figure 11.25 illustrates these systems.

Gate valves are usually placed between the water main and fire hydrants and at water main intersections. The American Water Works Association (AWWA) recommends placing four isolation gate valves at crossings (where a four-way intersection of water

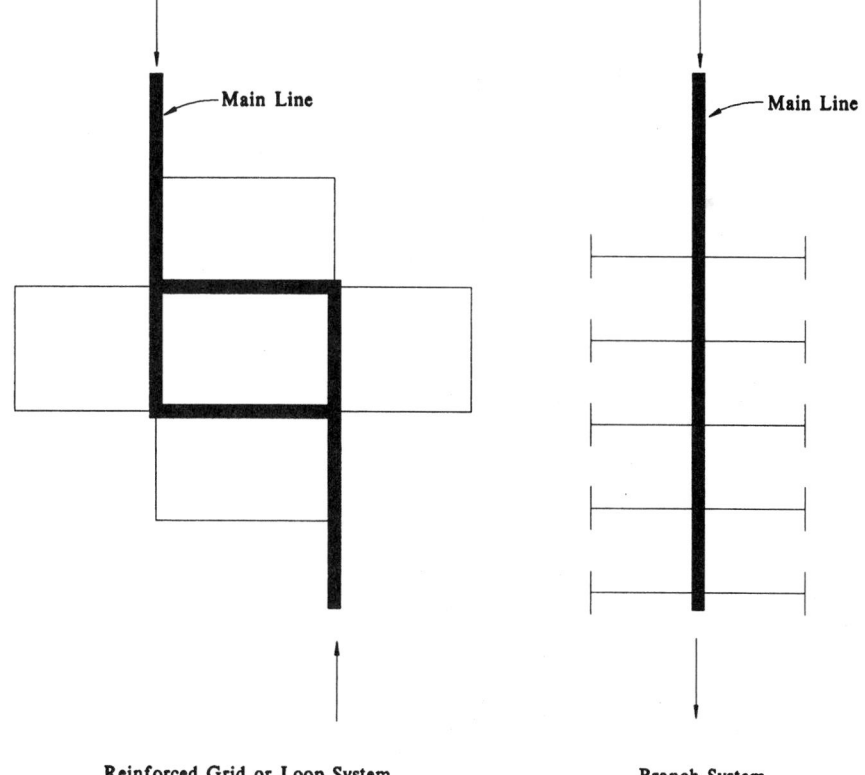

Reinforced Grid or Loop System   Branch System

FIG. 11.25   Typical water distribution system configurations.

mains occurs) and three at tee-type intersections. Other valves can be included along water mains, usually not more than 800 feet apart on long branches and 500 feet apart along mains in high-value districts (AWWA, M-31, 1998).*

The installed distribution system must be adequately protected from freezing during cold weather by using burial depths greater than the frost penetration depths experienced in the area. Figure 11.26 indicates the minimum recommended pipe burial depths where flows are low such as in a private service main. Municipal water mains, however, can be buried at depths that are approximately 6 inches less than those shown.

Other concerns associated with water main locations include

(a) that water mains have at least 36 inches of cover,

(b) that potable water mains are horizontally separated from sanitary sewer mains to avoid possible contamination of the water system in case of line rupture, and

(c) that proper vertical separation is maintained where water mains cross sanitary sewers.

These concerns are addressed in part 3.04 of the Project Manual's Section 02510, located in Chapter 13 (Section 13.3).

11.6.1 Water Distribution System Design Criteria

The AWWA recommends that water distribution systems with fire suppression capability include the following (AWWA 1998).*

Minimum line size in the network	6 inches
Smallest branching pipe for dead ends	8 inches
Largest spacing of 6-inch grid	600 feet
Smallest pipes in high-value district	8 inches
Smallest pipes on principal streets in central district	12 inches
Largest spacing of supply mains or feeders	3000 feet

To properly design a water distribution system, the design professional must have

(a) the peak system demands,

(b) fire flow requirements if applicable,

(c) the nominal water system operating pressure,

(d) the flow capacity of various pipe sizes, and

(e) a scale drawing (preliminary development plan) of the proposed development.

Information about peak system demands should be used from local sources if available. If local data are not available, the peak hourly demands shown in Figure 11.6 can be utilized, unless the criteria used to construct the figure are not applicable. Regardless of the values obtained from Figure 11.6, Ameen (1974) recommends 15 gal/min for one dwelling without reference to density; thus design demands should be at least:

---

*Excerpted from *Distribution System Requirements for Fire Protection* ©1998, American Water Works Association (AWWA Manual M-31), by permission.

Note:

Where frost penetration is a factor the depth of cover shown averages 6 inches greater than that usually provided by the municipal water works. Greater depth is needed because of the absence of flow in yard mains.

Recommended
Depth of Cover
in feet
Above Top of
Underground Yard Mains

FIG. 11.26 Water main burial depths. Source: Reprinted with permission from NFPA 24, *Standard for the Installation of Private Fire Service Mains and Their Appurtenances.* Copyright © 1995, National Fire Protection Association, Quincy, Mass. 02269. This reprinted material is not the complete and official position of the National Fire Protection Association on the reference subject, which is represented only by the standard in its entirety.

Number of Dwelling Units	Design Peak Hour Demand (gallons per minute)
1	15
2	16
3	17
4	18
5	19
6	20

A table containing the peak hourly demands is shown in Figure 11.27 for developments having a gross density of three dwelling units per acre and containing from 1 to 300 dwellings. The design peak-hour values for units numbering 4 and under reflect minimum values given above. Tables can be constructed for other dwelling unit densities. These tabular peak hourly flows coupled with the additional NFF provide a basis for designing water mains in a small residential system.

Before designing the system, the engineer must also determine the nominal water system operating pressure, as illustrated in Figure 11.19. For the following example, a 40 lb/in.^{2} working pressure is utilized.

Another table can be constructed using a spreadsheet for various pipe flows that can also aid in the design process. Figure 11.28 utilizes Equation 10.9 with a Hazen-Williams C of 130 as a design parameter, along with a working pressure of 40 lb/in.^{2} A minimum residual pressure of 20 psi is also used. Actual interior pipe diameters for various pipe lengths can be used to calculate flow delivery rates in gallons per minute when a corresponding 20 lb/in.^{2} pressure drop (40 lb/in.^{2} − 20 lb/in.^{2}) is experienced across each tabulated pipe length. Other tables can be constructed to represent different design parameters.

11.6.2 Water Distribution Design

Having the prerequisite information available, the designer can correlate peak hourly demands generated by each building site to the actual project layout. This is accomplished by placing a transparent paper overlay on top of the preliminary plan where the water mains are sketched, along with service lines to each building site. Often dual water services are used in single-family detached developments because they are more economical to install, operate, and maintain than single services even though the larger 1-inch service line is provided instead of the normal $\frac{3}{4}$-inch-diameter single service size. Figure 11.29 shows typical dual services for water and sewer.

To illustrate the design of a small residential development using single service taps (without fire protection), a method similar to that proposed by Ameen (1974) is utilized. This example uses the lot configuration shown in Figure 11.30a. The designer must sum the various cumulative maximum hourly demands using information contained in Figure 11.27 for each lot, beginning at the most remote point from the point of supply. Summation continues to the point of supply, with Figure 11.30b illustrating this process. If a nominal water system pressure of 40 lb/in.^{2} is chosen, information in Figure 11.28 can be utilized to match a pipe that can deliver the required maximum demand to all points along the line. The process begins with the smallest pipe diameter being initiated at the point of supply. By comparing demands with available capacities at specific distances from the supply

Unit	G.P.M.	Unit	G.P.M.	Unit	G.P.M.	Unit	G.P.M.	Unit	G.P.M.	Unit	G.P.M.
1	15	51	132	101	242	151	351	201	458	251	565
2	16	52	134	102	245	152	353	202	460	252	567
3	17	53	136	103	247	153	355	203	462	253	569
4	18	54	138	104	249	154	357	204	465	254	571
5	20	55	141	105	251	155	359	205	467	255	573
6	23	56	143	106	253	156	362	206	469	256	576
7	26	57	145	107	256	157	364	207	471	257	578
8	29	58	147	108	258	158	366	208	473	258	580
9	31	59	150	109	260	159	368	209	475	259	582
10	34	60	152	110	262	160	370	210	477	260	584
11	37	61	154	111	264	161	372	211	480	261	586
12	39	62	156	112	266	162	375	212	482	262	588
13	42	63	159	113	269	163	377	213	484	263	591
14	45	64	161	114	271	164	379	214	486	264	593
15	47	65	163	115	273	165	381	215	488	265	595
16	50	66	165	116	275	166	383	216	490	266	597
17	52	67	168	117	277	167	385	217	492	267	599
18	55	68	170	118	279	168	387	218	495	268	601
19	57	69	172	119	282	169	390	219	497	269	603
20	59	70	174	120	284	170	392	220	499	270	605
21	62	71	176	121	286	171	394	221	501	271	608
22	64	72	179	122	288	172	396	222	503	272	610
23	67	73	181	123	290	173	398	223	505	273	612
24	69	74	183	124	292	174	400	224	507	274	614
25	72	75	185	125	295	175	402	225	509	275	616
26	74	76	188	126	297	176	405	226	512	276	618
27	76	77	190	127	299	177	407	227	514	277	620
28	79	78	192	128	301	178	409	228	516	278	622
29	81	79	194	129	303	179	411	229	518	279	625
30	83	80	196	130	305	180	413	230	520	280	627
31	86	81	199	131	308	181	415	231	522	281	629
32	88	82	201	132	310	182	417	232	524	282	631
33	90	83	203	133	312	183	420	233	527	283	633
34	93	84	205	134	314	184	422	234	529	284	635
35	95	85	207	135	316	185	424	235	531	285	637
36	97	86	210	136	318	186	426	236	533	286	639
37	100	87	212	137	321	187	428	237	535	287	642
38	102	88	214	138	323	188	430	238	537	288	644
39	104	89	216	139	325	189	432	239	539	289	646
40	107	90	218	140	327	190	435	240	541	290	648
41	109	91	221	141	329	191	437	241	544	291	650
42	111	92	223	142	331	192	439	242	546	292	652
43	113	93	225	143	334	193	441	243	548	293	654
44	116	94	227	144	336	194	443	244	550	294	656
45	118	95	229	145	338	195	445	245	552	295	659
46	120	96	232	146	340	196	447	246	554	296	661
47	123	97	234	147	342	197	450	247	556	297	663
48	125	98	236	148	344	198	452	248	559	298	665
49	127	99	238	149	347	199	454	249	561	299	667
50	129	100	240	150	349	200	456	250	563	300	669

FIG. 11.27 Peak hourly demands in gallons per minute—three dwelling units per acre density.

Hazen Williams C	130						
Design Pres., psig=	40						
Residual Pres, psig	20						
Pipe Size,inches=	2	3	4	6	8	10	12
Length, ft.							
100	149	434	925	2686	5724	10293	16626
200	103	298	636	1847	3936	7079	11435
300	83	240	511	1484	3162	5687	9186
400	71	205	437	1270	2707	4869	7865
500	63	182	388	1126	2400	4316	6972
600	57	165	351	1021	2175	3911	6318
700	52	152	323	939	2001	3599	5813
800	49	141	301	874	1862	3349	5409
900	46	132	282	820	1747	3142	5076
1000	43	125	267	775	1651	2969	4795
1100	41	119	253	736	1568	2820	4554
1200	39	113	242	702	1496	2690	4345
1300	37	109	231	672	1433	2576	4162
1400	36	104	222	646	1376	2475	3998
1500	35	101	214	622	1326	2385	3852
1600	33	97	207	601	1281	2303	3720
1700	32	94	200	582	1239	2229	3600
1800	31	91	194	564	1202	2161	3491
1900	30	88	189	548	1167	2099	3390
2000	30	86	183˙	533	1135	2042	3298
2100	29	84	179	519	1106	1989	3212
2200	28	82	174	506	1078	1939	3132
2300	27	80	170	494	1053	1893	3058
2400	27	78	166	483	1029	1850	2989
2500	26	76	163	472	1006	1810	2923
2600	26	75	159	462	985	1772	2862
2800	25	72	153	444	947	1702	2750
3000	24	69	147	428	912	1640	2649
3200	23	67	142	413	881	1584	2559
3400	22	65	138	400	852	1533	2476
3600	22	63	134	388	827	1486	2401
3800	21	61	130	377	803	1444	2332
4000	20	59	126	366	781	1404	2268
4200	20	58	123	357	761	1368	2209
4400	19	56	120	348	742	1334	2154
4600	19	55	117	340	724	1302	2103
4800	18	54	114	332	708	1273	2055
5000	18	52	112	325	692	1245	2011
5500	17	50	106	309	657	1182	1910
6000	16	48	101	294	627	1128	1822
6500	16	46	97	282	601	1080	1745
7000	15	44	93	271	577	1038	1677
7500	15	42	90	261	556	1000	1615
8000	14	41	87	252	537	966	1560
8500	14	39	84	244	520	935	1510
9000	13	38	81	236	504	906	1464
9500	13	37	79	230	489	880	1422
10000	12	36	77	223	476	856	1383

FIG. 11.28 Pipe capacities in gallons per minute for various lengths for a 40-lb/in.^{2} system operating pressure.

FIG. 11.29 Typical double-utility services in a single-family residential development.

599

point, the beginning point of the pipe is systematically moved away from the supply point until all pipe capacities are capable of meeting the demands at each location. The next larger pipe diameter that is available is utilized to repeat the process for servicing the distance between the point of supply and the point where the initial pipe was found to be satisfactory. This larger pipe must also be capable of delivering a capacity equal to or exceeding the capacity of the previous pipe, as illustrated by the 3-inch main supplying 298 gal/min at a point 200 feet from the supply point, which exceeds the 149-gal/min capacity of the 2-inch main at 300 feet. (Refer to Figures 11.28 and 11.30.) This process continues until the entire line is designed.

The design of a more typical residential water system is illustrated in Figure 11.31, where fire protection is included. Because this development will be serviced by extending an existing water system, local authorities are requiring that the completed system provide flows equal to the NFF plus the maximum daily demand at the remotest hydrant with a residual pressure of at least 20 lb/in.[2]. The engineer may choose to design the proposed system using the peak hourly demands plus the NFF. The performance of the designed system is then analyzed by checking the static and frictional head losses with a flow equal to the NFF plus the maximum daily demand at the critical hydrant using Hardy Cross (1936) methods, equivalent pipes, or digital computer analysis.

The design peak hourly demands and NFF at various points in the system are determined as follows:

(a)   The design peak hourly demand at point B is the peak hourly demand for 6 lots, which is 23 gal/min.

(b)   The design peak hourly demand and NFF at point C is the peak hourly demand for 16 lots plus 750 gal/min or 50 + 750, or 800 gal/min. It should be noted that this total demand can be supplied to point C with flows coming through either point D or E, or both. Therefore, line $C-D-F$ must be sized to handle this flow as well as $C-E-F$.

(c)   At point D the peak hourly demand and NFF for 58 lots is 147 gal/min plus 750 gal/min, or 897 gal/min.

(d)   At point E the peak hourly demand and NFF for 58 lots is 147 gal/min plus the NFF, or 897 gal/min.

(e)   At point F the peak hourly demand for 100 lots (peak hourly demand of 240 gal/min) and the NFF is 990 gal/min.

Figure 11.32 contains a method to design the various water main segments.

11.6.3   Distribution System Analysis

For the aggregate 100 lots, the critical flow (Section 11.2.3) of 868 gal/min would have to be analyzed (750 gal/min NFF plus 118 gal/min maximum daily demand from Figure 11.5). If the available pressure at point G in Figure 11.31 is 43 psig when delivering 868 gal/min (using data from the example in Section 11.3.1), the amount of allowable system loss from point G to B would be (43 psig $-$ 20 lb/in.[2]),

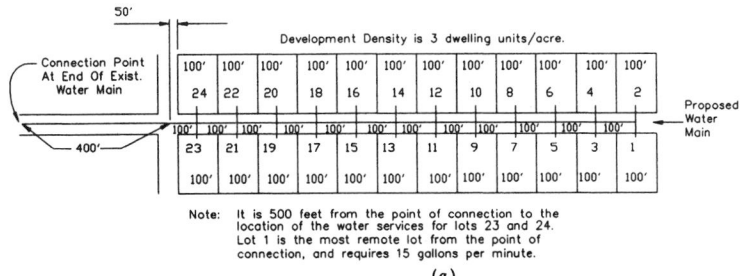

Development Density is 3 dwelling units/acre.

Connection Point At End Of Exist. Water Main

100'	100'	100'	100'	100'	100'	100'	100'	100'	100'		100'	100'
24	22	20	18	16	14	12	10	8	6		4	2

Proposed Water Main

100'	100'	100'	100'	100'	100'	100'	100'	100'	100'	100'	100'
23	21	19	17	15	13	11	9	7	5	3	1

| 100' | 100' | 100' | 100' | 100' | 100' | 100' | 100' | 100' | 100' | 100' | 100' |

Note: It is 500 feet from the point of connection to the location of the water services for lots 23 and 24. Lot 1 is the most remote lot from the point of connection, and requires 15 gallons per minute.

(*a*)

100	200	300	400	500	600	700	800	900	1000	1100	1200	1300	1400	1500	1600	Dist. From Connection, ft.
69	69	69	69	69	64	59	55	50	45	39	34	29	23	18	16	Cumulative Peak Hour Demand From Figure 11.27

(*b*)

100	200	300	400	500	600	700	800	900	1000	1100	1200	1300	1400	1500	1600	Dist. From Connection, ft.
149	103	83	71	63	57	52	49	46	43	41	39	37	36	35	33	2 inch line capacity from Figure 11.28.

✳

The capacity of the 2 inch watermain starting 500 feet from the connection is inadequate.

(*c*)

100	200	300	400	500	600	700	800	900	1000	1100	1200	1300	1400	1500	1600	Dist. From Connection, ft.
	149	103	83	71	63	57	52	49	46	43	41	39	37	36	35	2 inch line capacity from Figure 11.28.

✳

✳The capacity of a 2 inch watermain 600 feet from the connection point is inadequate because it can only deliver 63 gallons per minute when 64 are required.

(*d*)

100	200	300	400	500	600	700	800	900	1000	1100	1200	1300	1400	1500	1600	Dist. From Connection, ft.
	149	103	83	71	63	57	52	49	46	43	41	39	37	36		2 inch line capacity from Figure 11.28.

The capacity of a 2 inch watermain is satisfactory starting 200 feet from the connection point.

(*e*)

100	200		
434	298	------------> A 2 inch main is adequate. 3 inch watermain capacity.	

The 3 inch watermain is satisfactory from the connection point for a distance of 200 feet.

Point of Connection

1400 ' of 2" line

24	22	20	18	16	14	12	10	8	6	4	2

Install blow off at the end of the line

| 23 | 21 | 19 | 17 | 15 | 13 | 11 | 9 | 7 | 5 | 3 | 1 |

200 feet of 3" line

Note: Existing water main must be capable of supplying 69 gallons per minute with a residual system pressure of 40 pounds per square inch.

(*f*)

FIG. 11.30 Example of the design procedure for a simple water distribution system: (*a*) Simple development layout for water system design. (*b*) Cumulative system demands determined in gallons per minute. (*c*) Using a working pressure of 40 lb/in.^{2} to design the water mains, the capacity of a 2-inch water main, starting at the connection point, is shown. (*d*) The 2-inch water main will be analyzed using its beginning 100 feet from the connection point. (*e*) The 2-inch water main will be analyzed using its beginning 200 feet from the connection point. (*f*) A 3-inch line will be evaluated from the connection point to the beginning of the 2-inch lines. (*g*) The final system design would be as shown.

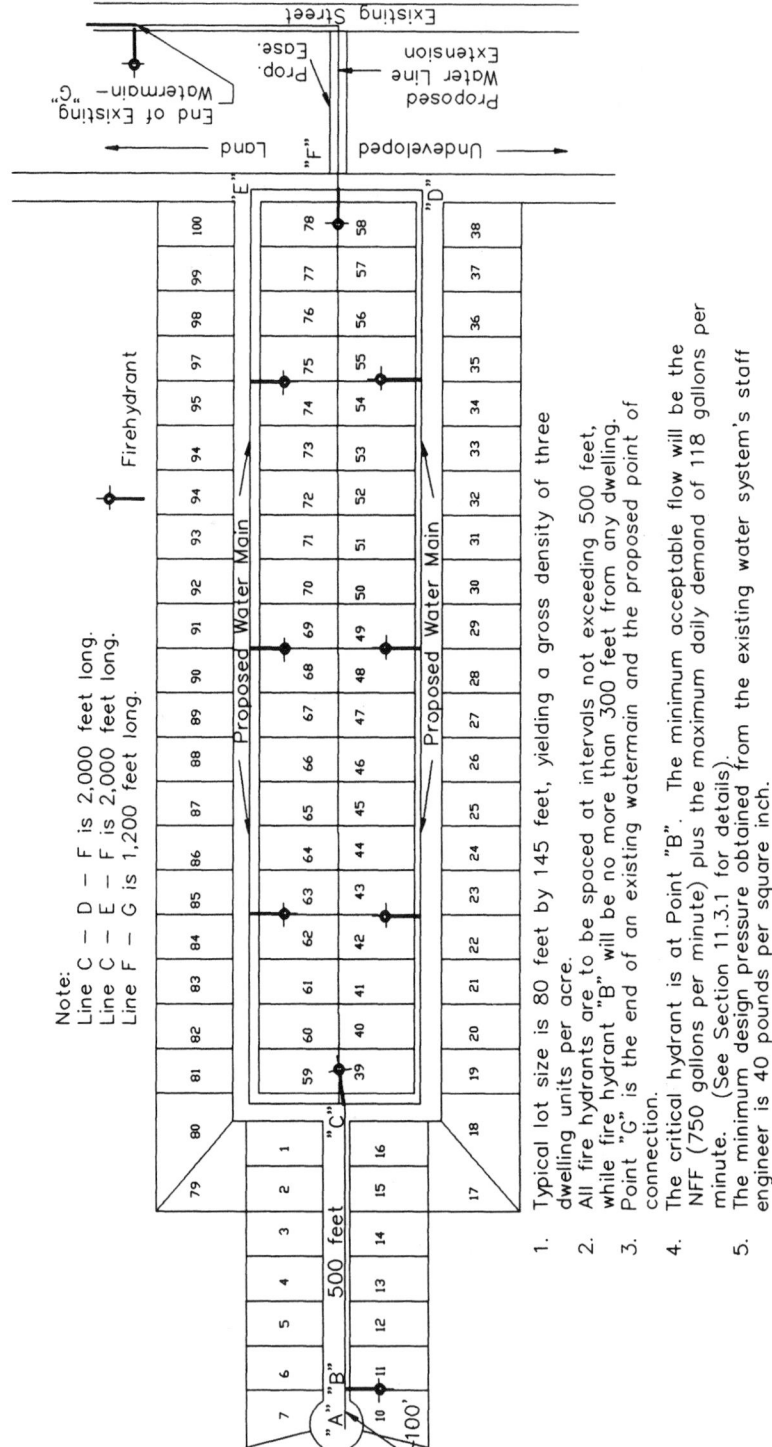

Note:
Line C – D – F is 2,000 feet long.
Line C – E – F is 2,000 feet long.
Line F – G is 1,200 feet long.

1. Typical lot size is 80 feet by 145 feet, yielding a gross density of three dwelling units per acre.
2. All fire hydrants are to be spaced at intervals not exceeding 500 feet, while fire hydrant "B", will be no more than 300 feet from any dwelling.
3. Point "G" is the end of an existing watermain and the proposed point of connection.
4. The critical hydrant is at Point "B". The minimum acceptable flow will be the NFF (750 gallons per minute) plus the maximum daily demand of 118 gallons per minute. (See Section 11.3.1 for details).
5. The minimum design pressure obtained from the existing water system's staff engineer is 40 pounds per square inch.

FIG. 11.31 Example of a potable water system layout with fire protection.

602

FIG. 11.32 Design of a typical water distribution system.

or 23 lb/in.^{2}, assuming that the ground elevation was relatively flat. Although analysis methods proposed by Hardy Cross (1936) or computer software can be utilized to analyze this system, the equivalent pipe approach is presented here. All pipes from B to G will be converted to an equivalent 8-inch single pipe using the following steps:

(a) Segment F–G is already in terms of an 8-inch pipe so the actual 1200 linear feet will be used for this segment.

(b) Segment $C–F$ is comprised of two parallel 8-inch mains of 2000 feet each, which must be replaced with a single 8-inch main. For parallel line replacement, a head loss is assumed between the terminal ends of the parallel pipes. In this example, 5 feet will be utilized. This amounts to 5 feet/2000 feet of 8-inch line, which is 2.5 feet/1000 feet for both pipes. Using Equation 10.9, each pipe can deliver:

$$Q = 0.006756 \times 130 \times 8^{2.63} \times 2.5^{0.54}$$

or 341.7 gal/min at the assumed head loss of 5 feet. The corresponding total flow of both pipes is 683.4 gal/min.

A single 8-inch pipe will have a corresponding head loss per 1000 feet of pipe for the 683.4-gal/min flow of

$$H_{1\,1000} = \left[\frac{683.4}{(0.006756)(130)(8^{2.63})} \right]^{1.85}$$

or 9 feet/1000 feet of 8-inch pipe, which yields an equivalent pipe of $(5/9) \times 1000$, or 555 feet.

(c) For the replacement of the 6-inch main from B to C (which constitutes series replacement), an arbitrary flow of 200 gal/min is assumed through $B–C$. For a 6-inch line with a Hazen-Williams C of 130, the corresponding head loss per 1000 feet of pipe is

$$H_{1\,1000} = \left[\frac{200}{(0.006756)(130)(6^{2.63})} \right]^{1.85}$$

or 3.76 feet per 1000 feet, which gives a total head loss for $B–C$ of $(3.76/1000) \times 500$ feet = 1.88 feet. For the equivalent 8-inch pipe at 200 gal/min, the head loss per 1000 would be

$$H_{1\,1000} = \left[\frac{200}{(0.006756)(130)(8^{2.63})} \right]^{1.85}$$

or 0.93 foot per 1000 feet of pipe. Therefore, the equivalent 8-inch line is $(1.88/0.93) \times 1000$, or 2022 feet.

(d) The equivalent 8-inch replacement pipe is equal to $1200 + 555 + 2022$, or 3777 feet.

(e) For 868 gal/min flowing through an 8-inch main with a Hazen-Williams C of 130, the head loss per 1000 feet of pipe is

$$H_{1\,1000} = \left[\frac{868}{(0.006756)(130)(8^{2.63})} \right]^{1.85}$$

or 14 feet per 1000 feet of pipe. The total head loss from G to B would be 14×3.777, or 52.88 feet of head (22.9 lb/in.^{2}).

(f) The calculated residual pressure at B while 868 gal/min is flowing would be $43.0 − 22.9$, or 20.1 lb/in.^{2}, which is acceptable since the assumption for this example was that the terrain was relatively flat. If the resulting residual pressure was below 20 lb/in.^{2}, various pipe sizes would have to be increased until an acceptable residual pressure was attained.

11.7   COMPUTER MODELING OF WATER SYSTEMS

Computer software that is made for modeling a water system is often used by civil engineers today when addressing water system needs. These software packages can be comprised of a graphical component where the user can construct the system on-screen where piping, tanks, pumps, valves, and other system components are shown. Hydraulic algorithms included as part of the software can then be used to solve network problems where hydraulic grades, system pressures, and flows can be computed throughout the system. Some options can also be provided where the user can conduct advanced water quality studies as well as automated fire flow analysis. Analysis results are often graphically displayed on graphs or maps or in report format.

One such software package, PIPE2000®, is offered by the Civil Engineering Software Center at the University of Kentucky (KYPipe/FlowNet Systems, Lexington, KY, www.kypipe.com). Figure 11.33 illustrates the graphical user interface for PIPE2000® applied to a hydraulic model for a typical residential development. EPANET is also a software program which can be used for hydraulic calculations where the focus of this program, however, is to address water quality analysis. Another widely used hydraulic modeling software package is Haestad Method's WaterCAD®, formerly known as Cybernet (www.haestad.com).

11.8   WATER SYSTEM COMPONENTS

Potable water distribution systems are comprised of pipes, fittings, thrust blocks, restraining joints, valves, blow-offs, laterals, meters, and back-flow preventors. Each plays an important role when incorporated into a water system.

11.8.1   Piping and Fittings

Most new residential water distribution systems rely on thermoplastic or ductile-iron pipe and fittings, although reinforced concrete pressure pipe is used in larger size mains. Asbestos cement pipe and cast-iron pipe have been widely used in the past and may be the type of pipe encountered when making extensions to existing water systems. Asbestos cement pipe is often present in areas where corrosive soil exists. Most smaller diameter pipes come in 18- to 20-foot lengths and are made to conform to various industrial standards.

Pipes must have adequate tensile and flexural strength to withstand internal and external loads resulting from backfill and other imposed loads. The interior must be smooth and noncorrosive to provide long service life and should provide satisfactory characteristics for the placement of field traps when installing service lines. Ductile-iron pipe and cast-iron pipe may exhibit increased interior wall friction as a result of

FIG. 11.33 Illustration of water system modeling software graphical user interface. Source: KYPIPE, Lexington, KY, 1999. Reproduced by permission.

tuberculation from aggressive water. Cement linings and coatings are effective in combating this problem. Pipe wall thicknesses are selected using the internal system pressure (including surge), external loadings, bedding conditions (or support conditions when the pipe is on piers or held by pipe hangers), corrosive thickness allowances, and factors of safety.

11.8.2   Thrust Blocking and Restraining Joints

Water systems consisting of slip joint piping and fittings require restraining systems called thrust blocks to transfer forces produced by moving water within the conduit to the soil. Failure to provide adequate bearing on the surrounding soil may result in ruptured lines caused by the separation of pipe segments. Thrust blocking is used to provide the necessary bearing. According to King et al. (1948), Figure 11.34 shows in plan view typical forces resulting at a bend having a deflection angle at Θ and a diameter that decreases from AB to CD. If p_1, A_1, and V_1 represent the pressure, area, and mean velocity at AB, then p_2, A_2, and V_2 can be used to represent similar variables at CD. The fluid is accelerated from V_1 to V_2 when flowing from AB to CD and the force producing this acceleration is the result of all forces acting on the mass $ABCD$. The symbols P_X and P_Y represent the X and Y components of the forces exerted by the closed channel upon the fluid and can be further defined by

$$P_X = A_1 p_1 - A_2 p_2 \cos \Theta + \frac{Qw}{g}(V_1 - V_2 \cos \Theta)$$

$$P_Y = A_2 p_2 \sin \Theta + \frac{Qw}{g} V_2 \sin \Theta$$

$$\text{Total force} = P = (P_X^2 + P_Y^2)^{0.5}$$

where p is the pressure in pounds per square inch; A is the cross-sectional area in square inches; Q is the flow rate in cubic feet per second; V is the velocity of flow in feet per second; w is the unit weight of water, 62.4 lb/ft^3; and g is acceleration due to gravity, 32.2 ft/s^2.

To illustrate the design of a thrust block for a 45° bend on a 12-inch water main

FIG. 11.34   Thrust block analysis. Source: King et al., 1948. Copyright © 1948 by John Wiley & Sons, Inc. Reprinted by permission of John Wiley & Sons, Inc.

where the operating pressure and surge pressure are 100 lb/in. and the flow rate is 9 ft³/s, the following example is given:

(a) The corresponding velocity for 9 ft³/s flow is calculated using Equation 7.18 rearranged as

$$V = Q/A = 9/\left[(3.14 \times 6^2)/(144 \text{ in.}^2/\text{ft}^2)\right]$$

or 11.46 ft/s.

(b) $P_X = (3.14 \times 36)(100) - (3.14 \times 36)(100)(0.707)$

$$+ \frac{9 \times 62.4}{32.2}(11.49 - 11.49 \times 0.707)$$

$$= 3371 \text{ pounds}$$

(c) $P_Y = (3.14 \times 36)(100)(0.707) + \dfrac{9 \times 62.4}{32.2}(11.49)(0.707)$

$$= 8134 \text{ pounds}$$

(d) $P_t = (3371^2 + 8134^2)^{0.5}$

$$= 8805 \text{ pounds}$$

(e) If a factor of safety of 1.25 is applied, the design load would be approximately 11,000 pounds.

(f) The bearing area required by the thrust block would be 11,000/(1000 lb/ft²) allowable soil pressure, or 11 ft² of bearing surface. If a 3 × 4-foot bearing area is used for ease of construction, the installed thrust block should look like Figure 11.35.

Should field conditions prohibit the thrust block from being placed on the outside of the pipe bend (this often occurs when the pipe is closed to a ditch bank), the thrust block must be located on the inside of the bend. The 45° fitting must then be secured to the thrust block by tie rods. If Portland cement concrete is used to anchor the fitting from the inside, the number of cubic yards required would be (11,000/150 lb/ft³ of concrete)/(27 ft³/yd³), or 2.7 yd³.

Another way to provide thrust restraint is to use either pipe restraining joints or tie rods that clamp and hold segments together.

11.8.3 Valves

Valves are used to control the direction or amount of flow in a water system, with globe, swing, slide, and rotating types being commonly available. Valves can be classified by use such as:

(a) *Shutoff Valves* Normally gate type valves or butterfly types are placed within the distribution system for repair and emergency purposes.

Undisturbed
Soil

Concrete 3' x 4'
Bearing Surface
Thrust Blocking

4'

3'

FIG. 11.35 Completed thrust block design.

(b) *Check Valves* These are used to control the direction of water flow in only one direction. The most common types of check valves are the swing check and lift check valves, which often incorporate a nonslamming feature.

(c) *Pressure-reducing Valves* These are used to automatically reduce inlet pressures to a controlled lower outlet pressure.

(d) *Altitude Control Valves* These valves are used to automatically control the flow into and out of elevated storage tanks, standpipes, or ground storage tanks. These valves normally are located in a pit close to the storage reservoir. Altitude control valves can be classified as either single acting or double acting.

A single-acting type closes on tank filling when water reaches the highest water level. Water must discharge through a separate discharge line. A double-acting valve closes on tank filling when the water reaches the high water level but opens for return flow when pressure on the valve inlet is less than that of the storage tank.

(e) *Air Release Valves* These are used to release trapped air within the water system. Pockets of air that restrict water flow accumulate at high points in water mains. Air release valves vent the accumulated air to the atmosphere and maintain use of the conduit's full-flow area.

(f) *Back-flow Preventers* These are antisiphon valves used to protect safe water systems from contamination as a result of cross connections made with systems containing contaminated or hazardous wastes or from siphoning action. They normally take the form of double check valves, air gap devices, vacuum breaker valves, or reduced-pressure back-flow preventers. Siphoning risk is minimized by maintaining normal system operating pressures within the distribution system; however, low pressures or ruptured water mains may cause siphoning action to attempt to take place. These valves will prohibit the action.

11.8.4   Blow-Offs

These are outlets that are used periodically to flush the distribution system of accumulated sediment, rust, scale, and other undesirable products. They usually consist of a manually operated gate valve and an orifice in a vault placed at the end of a dead-end line or at low points in the main distribution lines. Blow-offs should be sized to allow at least a 2 ft/s flushing velocity in the flushed pipe. Typically 6-inch blow-off valves and lines are used on all water mains over six inches in diameter. Four-inch valves and lines can be used on mains of 4 or 6 inches, and mains of less than 4 inches normally have blow-offs equal in size to the main's diameter.

Another problem that can occur in dead-end lines is associated with maintaining a minimum chlorine residual of no less than 0.2 mg/L (free chlorine) or 0.6 mg/L (combined chlorine). Dead-end lines may not have high enough demands to turn the water over where these minimum chlorine residuals can be maintained. One alternative is for the water company to implement a flushing program where field crews periodically flush the problem lines using manual labor. An innovative approach to solving this problem relies on the use of a Hydro-Guard automatic flushing device, as shown in Figure 11.36a, that operates on battery power that can be programmed for a 14-day 36-event flushing sequence. Figure 11.36b illustrates how daily flushing can help maintain chlorine residuals around the 2-mg/L level. In addition to helping maintain residual chlorine levels, this device is said to reduce bacteria presence, help control biofilm, and assist in the reduction of red water complaints.

11.8.5   Laterals and Meters

Laterals (service lines) are pipes that are used to convey water from the water main to the dwelling unit property line, with the minimum size for a single-family dwelling being $\frac{5}{8}$ inch, although $\frac{3}{4}$ inch is recommended. The lateral is connected to the main using a saddle or screwed fitting that is threaded through the wall, although other types of connections are available. Laterals usually terminate at a service meter box that is located on the edge of the road right-of-way. If the system is located in cold-weather areas, however, the meter is positioned inside the dwelling. The connection between the meter and the dwelling unit is installed and maintained by the dwelling unit owner. Meters are usually purchased and installed by the organization operating the water utility so that control over meter maintenance, spare parts, calibration, and the end user (meter locking for bill nonpayment) can be attained.

11.9   WATER SYSTEM PROTECTION

Water systems require protection from cold weather and corrosion. Cold-weather protection for piping utilizes burial depths discussed in Section 11.6; however, tanks and equipment must also be protected. Water heaters are often used to heat water in elevated tanks and standpipes, while space heaters can provide equipment protection from cold-weather damage.

A. Example of a typical HYDRO-GUARD field installation.

B. Example chlorine residual after automated daily line flushing.

FIG. 11.36 Example of an automatic flushing device. (*a*) Schematic of how Hydro-Guard™ is connected for warm-weather applications. Cold-weather versions are also available. (*b*) A graph showing chlorine residuals over an 18-day period where after the initial three-day flushing sequence, a daily maintenance flush duration of 4 minutes produced a chlorine residual of about 2 mg/L. Source: Reprinted by permission of Environmental Enhancement & Technologies USA, Inc. 715 Tenth Street South, Naples, FL, 1998.

Water systems also must be protected from corrosion attributed to the water itself, to the atmosphere, or to aggressive soil conditions. Lining water mains with cement or other coatings can protect the pipes from deterioration. Corrosion attributed to aggressive soils can be minimized by burying the mains in a protective barrier to prevent the soil from coming into contact with the mains. Tanks and equipment can be protected from corrosion by painting and by providing sacrificial metal sources to minimize corrosion. These precautions must be considered when corrosion is expected to pose a problem.

REFERENCES

Ameen, J. S. *Community Water System Source Book*. High Point: Technical Proceedings, 1974, pp. 19, 52–57, 62, 63, 70–75.

American Water Works Association (AWWA). *Distribution System Requirements for Fire Protection—Manual M 31*. Denver: AWWA, 1998, pp. 13, 18–21. Copyright © 1998 by American Water Works Association. Referenced material reprinted by permission of American Water Works Association.

American Water Works Association (AWWA). *Water Quality and Treatment*, 5th ed. New York: McGraw-Hill, 1999, pp. 3.4, 3.6–3.7.

Cedergren, H. R. *Seepage, Drainage, and Flow Nets*, 3rd ed. New York: Wiley, 1989, p. 41. Copyright © 1989 by John Wiley & Sons, Inc. Reprinted by permission of John Wiley & Sons, Inc.

Fair, G. M., J. C. Geyer, and D. A. Okun. *Water and Wastewater Engineering*, Vol. I. New York: Wiley, 1966, pp. 9–12. Copyright © 1966 by John Wiley & Sons, Inc. Reprinted by permission of John Wiley & Sons, Inc.

Fair, G. M., J. C. Geyer, and D. A. Okun. *Elements of Water Supply and Wastewater Disposal*, 2nd ed. New York: Wiley, 1971, p. 74. Copyright © 1971 by John Wiley & Sons, Inc. Reprinted by permission of John Wiley & Sons, Inc.

Geyer, J. C., J. B. Wolff, F. P. Linaweaver, Jr., and A. J. Duncan. Report on Phase One—Residential Water Use Project, "Final Summary Report on Phase I of the Residential Water Use Research Project, January 1, 1959 to January 1, 1963." An unpublished report. Baltimore: Johns Hopkins University and The Federal Housing Administration, 1963, pp. 1, 7.

Hammer, M. J., and K. A. MacKichan. *Hydrology and Quality of Water Resources*. New York: Wiley, 1981, pp. 45, 47, 59–60.

Hardy Cross. "Analysis of Flow in Networks of Conduits or Conductors." University of Illinois Bulletin, Vol. 286, 1936.

Insurance Services Office (ISO). *Fire Suppression Rating Schedule*. New York: ISO, 1998, pp. 1, 10, 32, 34–36.

King, H. W., C. O. Wisler, and J. G. Woodburn. *Hydraulics*, 5th ed. New York: Wiley, 1948, pp. 301–302. Copyright © 1948 by John Wiley & Sons, Inc. Reprinted by permission of John Wiley & Sons, Inc.

Konen, T. P., and W.-Y. W. Chan. *An Investigation of the Frequency and Duration of Use Parameters in Sizing Water Distribution Systems*. West Lane, CA: American Society of Plumbing Engineers, 1979.

Linaweaver, F. P., Jr., J. C. Geyer, and J. B. Wolff. "Report V on Phase II—Residential Water Use Research Project." An unpublished report. Baltimore: Johns Hopkins University and Federal Housing Administration, 1966.

National Fire Protection Association (NFPA). *NFPA 291—Recommended Practice for Fire Flow Testing and Marking of Hydrants.* Quincy, MA: NFPA, 1995b, Sections 2–2, 2–3, 2–6, 2–7, 2–10.1.

National Fire Protection Association (NFPA). *NFPA 1141—Standard for Fire Protection in Planned Building Groups.* Quincy, MA: NFPA, 1998a.

Rothrock, D. F. *Hydro-Pneumatic Tanks for Water Systems.* Phoenix: Red Coat Reports, 1974, pp. 1–4, 6.

Salvato, J. A. *Environmental Engineering and Sanitation,* 4th ed. New York: Wiley, 1992.

South Carolina Department of Health and Environmental Control (SCDHEC). *State Primary Drinking Water Regulations.* R.61–58. Columbia: SCDHEC, 1998.

South Carolina Department of Health and Environmental Control (SCDHEC). *Water System Viability Guidance—A Guide to Evaluating Viability for Proposed Public Water Systems—Bureau of Water.* Columbia: SCDHEC, 1999.

Sweitzer, R. J., and M. E. Flentje. *Basic Water Works Management.* Arlington: American Concrete Pipe Association, 1972, pp. 111–113.

Weber, W. J., Jr. *Physico Chemical Processes for Water Quality Control.* New York: Wiley, 1972.

Wolff, J. B., and J. F. Loos. "An Analysis of Public Water Demands." *Public Works,* September 1956, pp. 111–115.

SUGGESTED READINGS

American Water Works Association (AWWA) and the American Society of Civil Engineers (ASCE). *Water Treatment Plant Design,* 3rd ed. New York: McGraw-Hill, 1998.

Basil, A., and D. J. Wood. *PIPE2000®. Lexington, KY: KYPIPE/FlowNet Systems, 1999.* For more information on PIPE2000® see http://www.kypipe.com.

Benson, R.E. "Advanced Maintenance Flushing for Improving Water Quality in Remote Areas within Water Distribution Systems." In *Proceedings of Florida Section—American Water Works Association.* Orlando: Florida Section American Water Works Association, November 1998.

CBI Na-Con (Chicago Bridge and Iron Co.). P.O. Box 5650, Norcross, Ga.

Detay, M. *Water Wells Implementation, Maintenance, and Restoration.* New York: Wiley, 1997.

Environmental Enhancement & Technologies USA, Inc. *HYDRO-GUARD™.* Naples, Florida: Environmental Enhancement & Technologies USA, Inc., 1998.

Fowler, D. T., and D. H. Wegener. *Hydraulics of Pipelines.* Reston: American Society of Civil Engineers, 1994.

Mays, L. W. *Water Resources Handbook.* New York: McGraw-Hill, 1996.

National Fire Protection Association (NFPA). *NFPA 20—Standard for the Installation of Stationary Pumps for Fire Protection.* Quincy, MA: NFPA, 1999b.

National Fire Protection Association (NFPA). *NFPA 22—Standard for Water Tanks for Private Fire Protection.* Quincy, MA: NFPA, 1998b.

National Fire Protection Association (NFPA). *NFPA 24—Standard for the Installation of Private Fire Service Mains and Their Appurtenances.* Quincy, MA: NFPA, 1995a, A-8-1.1.

National Fire Protection Association (NFPA). *NFPA 1142—Standard on Water Supplies for Suburban and Rural Fire Fighting, NFPA 1142.* Quincy, MA: NFPA, 1999a, B-53(a).

Walski, T. M. *Analysis of Water Distribution Systems.* New York: Van Nostrand Reinhold Co, 1984.

12 Miscellaneous Design Considerations

Several independent topics related to land developments have yet to be presented. These include recreational playing courts, swimming pools, pipe jacking and boring, and fencing.

12.1 RECREATIONAL PLAYING COURTS

There are a number of recreational courts from which to choose for any facility plan. Tennis, basketball, volleyball, and shuffleboard courts are presented here.

12.1.1 Surface Composition and Orientation

Most paved courts are composed of reinforced Portland cement concrete, with some being constructed with hot-mix asphaltic concrete. All courts require stable subgrades with good drainage features. Courts constructed of Portland cement concrete are usually made with a minimum 4-inch-thick slab using a 3000-lb/in.^{2} concrete mix design along with temperature reinforcement being installed. The slab is usually formed over a polyethylene vapor barrier and includes construction joints to control expansion, contraction, and possible warping stresses depending on the total facility size. Expansion joints are placed at right angles across the court's center (net line) in most cases. Contraction joints are usually located outside of the playing area.

Each court included in this discussion is normally positioned with the long court dimension being parallel to a north-south line. This will minimize adverse player sight conditions resulting from the sun.

12.1.2 Tennis Courts

Typical tennis courts are illustrated in Figure 12.1*a*. The surface is sloped to drain from end to end or from side to side, since a uniform playing surface is required. The nets are supported so that the tops are 3.5 feet above the playing surface.

12.1.3 Basketball Courts

A typical college basketball court is depicted in Figure 12.1*b*. The backboards are comprised of 4 × 6-foot surfaces and are mounted 10 feet above the playing surface at each end, 4 feet from the end lines.

(a)

FIG. 12.1 Typical recreational playing courts: (a) tennis courts; (b) basketball court; (c) volleyball court; (d) shuffleboard court. Source: Harvey M. Rubenstein. *A Guide to Site and Environmental Planning,* 4th ed. Copyright © 1996 by John Wiley & Sons, Inc. Reprinted by permission of John Wiley & Sons, Inc.

12.1.4 Volleyball Courts

A typical volleyball court's dimensions are shown in Figure 12.1c. The center net is positioned where its top is 8.25 feet above the playing surface.

12.1.5 Shuffleboard

A typical shuffleboard court is illustrated in Figure 12.1d and requires a very smooth surface finish. The 1-inch striping used to mark the court must also provide a smooth riding surface.

12.2 SWIMMING POOL FACILITIES

Land developments often include swimming pools as part of the project's recreational facility plan. Such pools are normally considered "controlled public" because they

(b)

(c)

FIG. 12.1 (continued)

NOTE: WIDTH OF LINES A
      MAXIMUM OF 1½" AND
      A MINIMUM OF ¾".

(*d*)

FIG. 12.1 (continued)

are utilized by more than one family but are not open for general public use. These pools usually fall under regulatory controls that necessitate plan approval prior to construction (SCDHEC, 1995). During construction these pools are often subjected to regulatory inspections.

Controlled public pools are normally constructed of reinforced concrete, shotcrete (gunite), or similar materials and include National Sanitation Foundation approved materials. Pools exist in a variety of styles and shapes, with the most common being rectangular, ell, or tee shaped. They should be protected from dust, smoke, and soot. Decks and coping provisions, water circulation and treatment equipment, fencing, lighting, bath house and toilet facilities, chlorine room, water supply lines, and wastewater discharge lines must all be provided.

12.2.1   Pool Features

Pools are usually comprised of vertical walls, with a minimum of 6-inch-radius fillets being provided at all interior intersections. Wall tops are usually capped with coping tile that is suitable for gripping from within the pool, unless a perimeter guttering system is utilized. The pool's normal water level is usually delineated by installing 6-inch frost-free glazed tile under the coping tile. The interior is usually finished with a smooth white (or light-colored) plaster or nontoxic paint that is durable and water resistant. Depth markers are placed on the walls and deck at maximum and minimum points and at 2-foot-depth increments (with a maximum spacing being 15 feet apart). Typical water depths and spatial requirements are shown in Figure 12.2.

Other considerations include the positioning and sizing of auxiliary equipment, including

(a) ladders and steps,

(b) diving structures,

(c) surface skimmers or gutters,

(d) lighting,

(e) fittings,

(f) drains and inlets,

(g) pipe sizes and configuration,

(h) fill spout and valve, and

(i) surge chamber (if required).

The design professional usually includes in the plans the pool's volume, surface area, perimeter, deck area, turnover flow rate, and maximum user capacity.

12.2.2   Deck Area Features

The pool's apron, or deck area, is shown on the construction plans with dimensions. The deck is normally a minimum of 4 feet wide but often extends at least 10 feet. Its surface is made of nonslip material (often Portland cement concrete with a brush finish) and slopes away from the pool's coping using approximately $\frac{1}{2}$-inch fall per horizontal foot. Water depths are normally shown adjacent to corresponding wall marks. The position of area lighting fixtures and wiring is included. Piping for hose bibbs fitted with vacuum breaker valves, a foot shower, and a drinking fountain are also shown. Care should be exercised by the design professional to maintain clear air space for at least 10 feet horizontally from the edge of the pool to prevent adjacent balcony or rooftop access for diving; otherwise protective barriers need to be installed. Areas next to the deck should be paved or consist of grass to minimize potential pool water contamination.

12.2.3   Bath House Features

Local regulations can require separate facilities for each gender and facilities that are well drained and have adequate light and ventilation (SCDHEC, 1995). Unbreakable mirrors, soap dishes, tissue dispensers, a service sink, and hot water are included. To

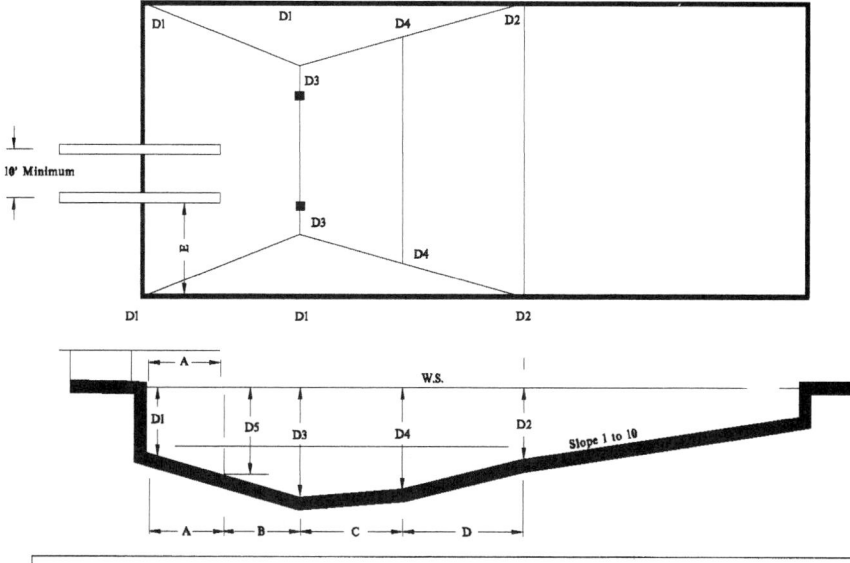

Depth - Minimum					
Stands & Boards Maximum Height to Water	D-1	D-2	D-3	D-4	D-5
3- Meter Board	6'-0"	4'-6"	12'-6"	12'-6"	12'-0"
1- Meter Board	6'-0"	4'-6"	10'-6"	10'-0"	10'-0"
Deck Level Board (Less than 26")	6'-0"	4'-6"	9'-0"	8'-6"	8'-6"
No Board	6'-0"	4'-6"	8'-6"	8'-0"	8'-0"
No Diving Pool	3'-0"	3'-0"	3'-0"	3'-0".	3'-0"

Length of Section - Minimum					
Stands & Boards Maximum Height to Water	A	B	C	D	E
3- Meter Board	5'-0"	6'-0"	9'-0"	23'-0"	13'-0"
1- Meter Board	5'-0"	6'-0"	9'-0"	17'-0"	11'-0"
Deck Level Board (Less than 26")	2'-6"	6'-0"	7'-6"	12'-0"	9'-0"
No Board	—	6'-0"	6'-0"	12'-0"	—
No Diving Pool	—	—	—	—	—

D-1 shall be no farther out than a maximum of 15" from pool wall. Slope of D shall not exceed 1' - 00" vertical to 3' - 00" horizontal.
The maximum values of A are 6' - 00" for 1-meter and 3-meter boards and 4'-00" for deck level boards. Clearance above the board
must exceed the entire length of sections B, C, & D. Depth D-5 is measured at midpoint of section B where a diving board is not provided
Where a diving board is provided D-5 shall be measured from the tip of the board. The minimum distance between the diving well wall
on the deep end and any opposite wall shall not be less than six (6) feet greater than the diving bowl dimensions (B, C, D). All diving
boards that are placed at a height above water between those listed shall be made to comply with the listing that is greater.

FIG. 12.2 Minimum pool safety setbacks and water depths. Source: South Carolina Department of Health and Environmental Control, 1995, as amended.

accommodate approximately 100 male swimmers (and every increment of 200 thereafter), one water closet, lavatory, and urinal are necessary. Female accommodations include two water closets and lavatories for each 100 users. Every shower installed in each facility can service approximately 50 swimmers. Privacy partitions are usually included in the female facilities.

12.2.4 Equipment House

The equipment house provides a safe location for all circulating and purification equipment necessary to efficiently operate the pool. The floor is usually constructed of concrete and sloped. This will allow positive drainage to protect both operators and equipment. Adequate ventilation is required to allow proper cooling of operating equipment when the door is secured. Adequate illumination is also desirable. All electrical circuits must be installed with grounds (and ground fault interrupters where required) to eliminate shock hazards. To promote safety, all circuits must comply with electrical code requirements throughout the entire facility area.

Diatomaceous earth, rapid sand, or cartridge filters are usually installed in conjunction with corrosion-resistant centrifugal pumps to cleanse and circulate the pool's water. Normally piping and valves are labeled to show flow directions and include pressure gauges, a flow meter, backwash sight glass, and connections for chemical feed dispensing. If PVC pipe and fittings are used, Schedule 40 or 80 classifications are satisfactory in most cases.

12.2.5   Chlorine Room Facility

There are several methods available to disinfect pool water. Chlorine in solid (calcium hypochlorite), liquid (sodium hypochlorite), or gas form is the most common disinfectant. Many states no longer allow the use of gas chlorine as a disinfectant; however, if chlorine gas is to be utilized, the dispensing and storage facility should be clearly marked as a chlorine facility with prominently displayed signs. This installation should be located no nearer than 200 feet to the closest dwelling, but located close to the main circulation pumps and filters to reduce the gas feed line distance. The room includes a corrosive-resistant exhaust fan located as close to the concrete floor as possible with its discharge to the out of doors. The fan should be able to provide at least one change of air per minute. Automated air intake louvers are needed to supply fresh air inside and are located as high as possible in the wall to allow circulation when the blower is energized. Switches to control interior lights and the exhaust fan are located outside the chlorine room for the operator's protection. A lockable door is provided that opens only to the outside. A viewing window is normally incorporated in this access door to allow observations of the interior without entry.

The installed dispensing equipment must be capable of feeding a minimum of 1 pound of free available chlorine per 10,000 gallons of pool volume per 24-hour period. A range of 1–5 parts per million must be maintained whenever the pool is in use. A typical installation of a calcium hypochlorite erosion feed chlorinator is shown in Figure 12.3

The pool treatment capacity of an erosion feed chlorinator can be determined as follows if the chlorinator output from the manufacturer is certified to have a maximum output value of 1720 grams of available chlorine per day. The equivalent number of

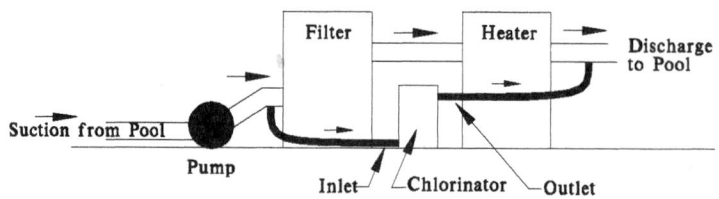

FIG. 12.3   Typical calcium hypochlorite erosion feed chlorinator installation.

pounds of free chlorine is given as (1720 g/day) (1 oz/28 g) (1 lb/16 oz) = 3.84 lb/day available chlorine. Let the variable X represent the maximum pool volume capable of being treated by this particular unit; then

$$X = \frac{3.84 \text{ lb/day chlorine available}}{1 \text{ pound available chlorine required/10,000 gallons}}$$

$$= 38{,}400 \text{ gallons}$$

To ensure an uninterrupted chlorine supply, if 150-pound chlorine cylinders are used, either an automatic changeover device capable of switching chlorine cylinders is required or a scale to determine when the cylinder is empty must be utilized.

A gas mask for emergencies is provided at a location outside the chlorine room. Also, a gas leak detector should be installed that has both audible and visual alarms to alert the operator of potential problems.

12.2.6   Chemical Dispensing Facilities

The pH of the pool's water must be maintained between 7.2 and 7.8 for good results. The addition of soda ash or muriatic acid may be necessary to keep the pH in balance. Installed corrosionless tanks with metered pumps can be utilized to inject the necessary doses directly into the filtered water's piping. This facility should be separated from the pump and filter area but could be part of the equipment house.

12.2.7   Facility Security

To provide controlled access use, the facility should be enclosed with a chain link type fence or other equivalent impenetrable barrier at least 4 feet high that is lockable. This fence may be required by law and will help to lower owner liability, since facilities such as pools are considered "attractive nuisances." Fencing also keeps pets and animals from entering the premises. Security lighting to illuminate the facility after hours is also helpful in reducing the chances of unauthorized use.

12.2.8   Miscellaneous Provisions

To increase swimmer safety, a lifeguard stand is usually provided for every 2000 ft^2 of pool surface area, along with life-saving and first-aid equipment. In addition, provisions for a telephone must be made where emergency phone numbers (including an ambulance) can be posted. The design must also include a place to conspicuously post and display the pool regulations.

12.2.9   An Example of Pool Design

For this example, it is assumed that the pool must be capable of serving a 50-family membership. Based on site constraints, this pool's shape will be in the form of an "L" with a diving well.

A. *Determine the Required Pool Surface Area* If 3.4 persons are estimated in each membership based on the development's targeted market, and maximum utilization is estimated for this example at 85 percent, the pool must accommodate (50 memberships) × (3.4 persons/membership) × 85% = 144.4, or 145 maximum users.

Allowing a maximum of three persons to occupy the 20-foot side diving well where a deck level board (less than 26 inches) is to be installed, the required surface area of the well is estimated by assuming that the board's tip will project horizontally past the coping by 3 feet. Providing an additional 12 feet of horizontal distance for safety, the resulting diving well's surface area is (3 feet for board projection + 12 feet for clearance) × (20 feet wide) = 300 ft^2.

If the percentage of users to occupy the nonswimming area is assumed to be 75 percent, and 10 ft^2 of surface area is allocated to each user, the corresponding nonswimming area is (145 maximum users) × (10 square feet per user) × 75% = 1087.5 ft^2.

Since 2 percent of the users are allocated to diving in this example and 75 percent to nonswimming, the swimming area must accommodate the remaining 23 percent. Allowing 25 ft^2 of surface area for each swimmer, the resulting area would be (145 maximum users) × (25 ft^2 per user) × 23% = 834 ft^2.

The pool's interior dimensions for wall clearances are determined by the following:

(a)   The diving well's width is 20 feet and the length is 15 feet to provide the required 300 ft^2 of area.

(b)   The swimming area that adjoins the diving well utilizes the same 20-foot width and would need a corresponding length of (834 ft^2 required)/(20 feet wide) = 41.7 feet.

(c)   The nonswimming area that adjoins the swimming area is assumed to have a width of 28 feet. The corresponding length is (1087.5 ft^2 required)/(28 feet wide) = 38.8 feet long.

If these values are rounded, an adjusted layout is obtained. Figure 12.4a shows these adjusted dimensions, which are used in this design. The configuration provides 2280 ft^2 of surface area with a corresponding perimeter of 236 linear feet.

B. *Determine the Corresponding Deck Area* If the deck area adjoining the pool is to be comprised of coping tile that is 1 foot wide and a concrete apron that is 11 feet wide (in addition to the 30 × 40-foot area contiguous to the pool), the resulting area is 4608 ft^2. The minimum deck area required, however, allowing a 33-ft^2 area for each swimmer, is 4785 ft^2. This necessitates an increase in the width of the apron adjoining the diving well to 13 feet, 1 inch, and of the apron contiguous to the shallow area to 11 feet, 1 inch. The final deck configuration is shown in Figure 12.4b; it contains 4790 ft^2.

C. *Establish the Pool Depths* The shallow depth of the pool in the nonswimming area usually ranges from 3 to 3.5 feet, with 3 feet being utilized in this example. A depth of 4.5 feet is normally used to separate swimming from nonswimming activities with the pool's bottom sloping between these two limits. If the diving board is to be located where its surface is approximately 20 inches above the deck, the swimming pool depths can be determined by using information similar to that shown in Figure 12.2.

Figure 12.5 shows the depths that are incorporated in this sample pool. They provide a corresponding volume of 84,300 gallons.

D. *Determine Pumping Rate* Using a turnover rate for the pool's water of once every 6 hours, the corresponding pumping rate is (84,300-gallon pool capacity)/[(6-hour turnover) × (60 min/h)] = 235 gal/min.

FIG. 12.4   Example of pool dimensions: (a) adjusted dimensions for pool's interior; (b) adjusted deck dimensions.

FIG. 12.5   Example of pool depths.

E. *Determine Filter Bed Area* By limiting the maximum flow rate through rapid sand filters to 15 gal/min per square foot of filter surface area, the corresponding minimum filter bed surface area for this example is (235 gal/min pumping rate) / (15 gal/min per square foot of filter area) = 16 ft^2, rounded.

F. *Determine Intake Skimmer Requirements* If the maximum flow rate through each skimmer is limited to 30 gal/min according to the manufacturer, the minimum number of required skimmers is (235 gal/min pumping rate) / (30 gal/min limit per skimmer) = 7.8, or 8 skimmers. Each skimmer includes a basket strainer, weir, and equalizer line with an intake that must be submerged more than a foot below the normal water surface level to be effective. These skimmers would be strategically located to allow complete pool surface coverage.

G. *Determine Pool Inflow Inlet Requirements* Since each wall inlet orifice is adjustable for both flow and direction, the spacing interval to be used will be not greater than 15 feet to provide complete circulation. A total of 17 inlet nozzles are provided.

H. *Determine Main Drain Requirements* Most states require that new public pools have at least two main drains that are interconnected. This ensures that if one drain becomes blocked, the recirculation pump can draw suction through the second interconnected drain. This minimizes the potential for swimmers to become entrapped on a drain that draws suction from a single source. If the maximum intake velocity is limited to 1.5 ft/s through each drain to minimize the effects of vortex when the desired pumping rate enters, the minimum clear way area of the drain grate can be calculated. First the flow rate must be converted from gallons per minute to cubic feet per second by (235 gal/min pumping rate) $\times$ (0.002228 ft^3/s gal/min) = 0.52 ft^3/s equivalent flow rate. The corresponding minimum clear way area for each grate is determined by (0.52 ft^3/s flow) / (1.5 ft/s limiting velocity) = 0.35 ft^2 of open grate area; therefore, each grate chosen from manufacturers' literature must have this area available.

Unless the pool's weight is sufficient to overcome buoyant forces resulting from the presence of groundwater, the pool could float. This can occur if the water level in the pool is sufficiently lower than that of the surrounding groundwater. Since most pools do not have enough weight to counter this buoyant force, care must be taken to keep the groundwater table low. This can be accomplished by installing a well point (or points) that can be pumped throughout construction. When construction is completed and the pool is filled, the weight of the pool and included water prevent flotation. The well point system can be left in place for dewatering that might be required in the future for pool maintenance.

As a precaution, a hydrostatic relief (check-type) valve is installed in the drain sump that allows groundwater to enter the pool until the levels are equalized if the pool's water surface drops below the surrounding groundwater level. Under normal operating conditions this valve remains closed.

I. *Determine Necessary Pump Size* Once the number and arrangement of intake and discharge outlets are known, the pipe sizes are designed. Normally the Hazen-Williams formula is used in conjunction with the following maximum head loss parameters:

(a)   Suction lines: 6 feet/100 feet of pipe.

(b)   Discharge line: 12 feet/100 feet of pipe.

The pump's TDH is calculated by including static, friction, filter media, and minor system head losses. This procedure is similar to that presented in Chapter 10. An efficient pumping unit can then be chosen from manufacturer's data.

It has proven beneficial to install a backup circulating pump and filter system that allows continuous operation when repairs are necessary. This keeps the membership satisfied throughout the swimming seasons for nominal additional capital expenditures. A typical piping schematic for this setup is shown in Figure 12.6.

J. *Determine Necessary Gas Chlorinator Size* An adequate chlorinator must be installed to maintain the proper chlorine residual. Chlorinators are adjustable to allow operator manipulation according to demands. If an average chlorine dose rate of 8 parts per million is used, the number of pounds of chlorine required per day is calculated by first determining the total number of pounds of water circulated each day. This is obtained by (235 gal/min) $\times$ (1440 min/day) $\times 10^{-6}$ = 0.338400 $\times 10^6$ gal/day. The equivalent number of million pounds of purified water each day is (0.3384 $\times 10^6$ gal/day) $\times$ (8.34 lb/gal of water) = 2.822 $\times$ 10^6 lb. Applying 8 parts per million, the daily chlorine demand is (2.822 $\times 10^6$ lb of water per day) $\times$ (8 parts per million chlorine) = 22.6 lb/day. A chlorinator capable of delivering from 0 to 100 lb/day would be satisfactory and would utilize one 150-pound cylinder in about a week.

K. *Determine Pool Lighting Requirements* Adequate pool lighting is obtained when each installed fixture is allowed to illuminate no more than 2000 ft^2 of the pool's subsurface

	Equipment In Use		Valve Status	
	Pump	Filter	Open	Closed
Normal Operation Valve 5 Stays Closed	1	1	1,3	2,4
	2	2	2,4	1,3
Emergency Operation Valve 5 Stays Open	1	2	1,3	2,4
	2	1	2,4	1,3

Note:
1. Pump No. 1 and Filter No. 1 are sized to accommodate design flows.
2. Pump No. 2 and Filter No. 2 are sized to accommodate design flows.
3. Each filter has a multiport valve that is placed in the "Closed" position when not in use.
4. Paint Pump No. 1, Filter No. 1, and Valves No. 1 and 3 GREEN to designate the green system.
5. Paint Pump No. 2, Filter No. 2, and Valves No. 2 and 4 RED to designate the red system.
6. Alternate between green and red systems each day under normal operation.

FIG. 12.6 Example of a pool piping schematic including backup pumps and filters.

626

area. In the case of this example, the minimum required number of underwater light fixtures is (2280 ft^2 of pool surface) / (2000 ft^2 effective area) = 1.14 fixtures; however, due to the configuration of this pool, 3 are provided. The amount of wattage required for each fixture based on 0.5 W/ft^2 of pool surface area is (2280 ft^2) $\times$ (0.5 W/ft^2) = 1140 watts total. Therefore, each fixture will be outfitted with a 400-watt bulb.

L. *Consider Miscellaneous Provisions*

(a) *Steps and ladders:* Usually two sets of steps and ladders are provided in each pool as a minimum, with additional ones being located approximately 75 feet apart. One set of steps in the shallow end and three sets of ladders are included in this example.

(b) *Fill spout:* The pool's fill spout is usually located in a safe place to prevent accidents and to make effective use of the area under the diving board. The pipe's outlet elevation is usually two pipe diameters or more above the pool's coping to prevent possible siphonage.

(c) *Surge tanks:* Water is displaced by each swimmer's body volume and by dynamic displacement from swimmer turbulence. Some pools require surge tanks to equalize this displaced water. This temporary excess of water volume is allowed to be stored until it can be circulated back into the pool without flooding. This example does not include a surge tank because the pool is relatively small and skimmer equalizer lines are provided to prevent loss of pump prime during peak pool usage.

(d) *Overflow protection:* The pool must be protected from flooding attributed to storm water or other sources. This requires that proper drainage be provided to eliminate this potential problem.

M. *Finalize Pool Layout* The final pool layout is shown in Figure 12.7, including the locations of lights, ladders and stairs, skimmers, inflow inlets, main drains, filling spout, and diving board.

N. *Determine Structural Design* The bottom of a swimming pool is usually made at least 6 inches thick when constructed of reinforced concrete or similar materials. The magnitude of earth pressures acting on the outside of the pool's buried surface and hydrostatic forces acting inside or out of the pool are used as criteria for structural design. The design must include adequate expansion joints that incorporate the use of waterstops to prohibit leaks. Structural design sources are available elsewhere that address this topic.

12.3 FENCING

Chain link–type fences are commonly used to protect development property, provide safety, and act as enclosures. Applications include protecting utility facilities (sanitary sewer pumping station or wastewater treatment plant), providing security around recreational areas such as swimming pools, and protecting hazardous areas such as deep stormwater detention facilities. Some fencing systems installed for security purposes include barbed wire strands along the fence top using special post caps. Figure 12.8 shows details of some typical installations.

FIG. 12.7   Final layout for a sample pool, with typical pool details: (*a*) layout details of sample pool; (*b*) stair details; (*c*) ladder details; (*d*) automatic hydrostatic relief valve mounted in main drain sump; (*e*) skimmer details; (*f*) diving board area details.

12.4   PIPE JACKING AND BORING

Today many regulatory authorities require underground pipelines to be installed under existing improvements by means other than using open trenching and backfilling. A commonly used procedure is the process of boring a pilot hole through the soil into which the pipe is forced under the improvement using a hydraulic jack.

12.4.1   The Installation Process

Initially a boring pit is excavated to a depth that allows the boring machine to be properly positioned. The machine is secured in place once it has been oriented to line and grade. The boring machine forces a rotating auger forward to prebore the casing's hole if a protective casing is to be installed; otherwise this is done to the carrier pipe's

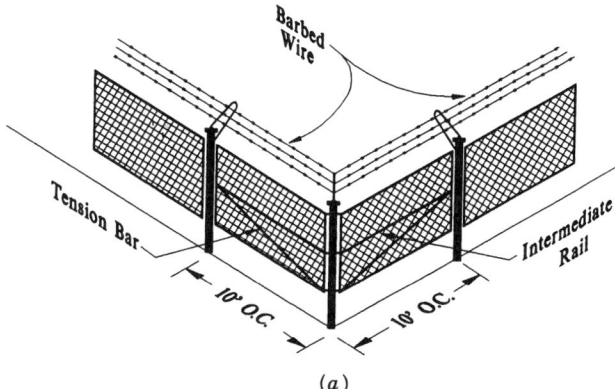

(a)

FIG. 12.8   Typical fencing details: (a) example of a security fence; (b) example of a single gate. Source: Harvey M. Rubenstein. *A Guide to Site and Environmental Planning*, 4th ed. Copyright © 1996 by John Wiley & Sons, Inc. Reprinted by permission of John Wiley & Sons, Inc.

hole. Periodically boring ceases so that sections of casing pipe can be welded to previous sections to form a continuous conduit and then be jacked forward. On completion, the casing (or uncased carrier pipe) extends to the desired location, at which point open trenching is continued.

12.4.2   Jacking and Boring Requirements

Depending on the situation, regulatory agencies have criteria for these activities. To illustrate an example using the American Railway Engineering Association's guidelines (AREA, 1991), a pressurized pipeline carrying nonflammable substances that crosses under a railroad may be required to horizontally intersect the track at an angle between 45° and 90°. Installation of the new pipe is usually not permitted through existing culverts or under railroad bridges if a reduction in flow capacity is realized or if any foundation has the potential of being weakened. Pressurized carrier pipe joints can be welded or of the mechanical type, with the pipe being installed so that it is not in tension. Casing pipe and joints are leak proof and capable of withstanding imposed railroad loads. Minimum 35,000-lb/in.² yield strength steel casing that is subjected to E-80 railroad loads must have a wall thickness of at least that shown in Figure 12.9.

If the casing does not have a protective coating and is not cathodically protected, these thicknesses are increased by at least 0.063 inch. The casing pipe's inside diameter must be at least 2 inches larger than the maximum outside diameter of the carrier pipe, including any joints or couplings, for carrier pipes less than 6 inches in diameter. For larger pipes a 4-inch minimum differential is required.

The top crown of most casings must be at least 5.5 feet below the main track rail and may be required to extend 2 feet beyond the toe of slope, 3 feet beyond ditches, otherwise to extend 25 feet from the track centerline. Figure 12.10*b* illustrates a typical pipe installation under a railroad.

For a more complete presentation of the requirements for pipeline installations within railroad rights-of-way, the reader is referred to Chapter 1, Parts 4 and 5 of the current AREA's (1991) *Manual for Railroad Engineering*.

FIG. 12.8 (continued)

NOTE: CHAIN LINK MESH FOR ALL HANDBALL COURT FENCE TO BE 1".

Nominal Diameter (in.)	Nominal Thickness (in.)
14 and Under	0.188
16	0.219
18	0.250
20 - 22	0.281
24	0.312
26	0.344
28	0.375
30	0.406
32	0.438
34 - 36	0.469
38	0.500
40	0.531
42	0.563

FIG. 12.9 Minimum wall thicknesses for casing pipe, E80 loading. Source: AREA, 1991. Copyright © 1991 by the American Railway Engineering Association. Reprinted by permission of the American Railway Engineering Association.

FIG. 12.10 Typical pipe jacking and boring details: (a) field setup for street bore; (b) typical pipeline installation under a railroad. Source: AREA, 1991. Copyright © 1991 by the American Railway Engineering Association. Reprinted by permission of the American Railway Engineering Association.

12.4.3   Potential Problems Encountered During Boring

Many factors contribute to problems associated with boring misalignment and progress, including groundwater interference, varying soil conditions, and obstacles (including rocks, roots, and buried debris). If the auger is unable to maintain line and grade, often the boring machine is located, with regulatory approval, at another position close by for another try. This minor relocation does not usually create operational problems if the line is pressurized. However, a gravity flow carrier pipe may require additional provisions to accommodate the pipe's grade.

To allow more flexibility during construction, for example, consideration should be given to the inclusion of a manhole at each end of an amply oversized casing pipe when gravity wastewater mains are being installed. A flange ductile-iron gravity carrier pipe can be inserted on intermediate supports and afforded room for position adjustments (to attain the desired grade) prior to being permanently secured in place.

On occasion poor boring conditions aggravated by groundwater and soft soil may cause the augered hole to collapse. Success under these conditions has been attained by using well points to dewater or by freezing the soil and then boring.

REFERENCES

American Railway Engineering Association (AREA). *AREA Manual for Railway Engineering, Chapter 1—Part 5 Pipelines.* Washington: AREA, 1991, pp. 1-5-7, 1-5-8, and 1-5-11. Referenced material is reprinted by permission of the American Railway Engineering Association.

South Carolina Department of Health and Environmental Control (SCDHEC). *Public Swimming Pools—Regulation 61-51.* Columbia: SCDHEC, Amended 1995.

SUGGESTED READINGS

American National Standards Institute (ANSI) and the National Spa and Pool Institute (NSPI). *1991 Standard for Public Swimming Pools ANSI/NSPI-1.* Alexandria: ANSI, 1991.

American Railway Engineering Association (AREA). *AREA Manual for Railway Engineering, Chapter 1—Part 4 Culverts.* Washington: AREA, 1989.

American Society of Civil Engineers (ASCE). *Pipeline Crossings—ASCE Manuals and Reports on Engineering Practice No. 89.* Reston: ASCE, 1996.

Castronovo, J. P., and J. A. Clark (Eds.). *Pipelines in the Constructed Environment.* Reston: American Society of Civil Engineers, 1998.

Rubenstein, H. M. *Site and Environmental Planning,* 4th ed. New York: Wiley, 1996, pp. 346, 349–351, 353.

13 Project Manual Including Specifications

There are several recognized methods of organizing and preparing construction documents and specifications that are used by civil engineers today. The Project Manual approach is presented here.

According to the Construction Specifications Institute's (CSI) *Manual of Practice, Construction Documents Fundamentals and Formats Module* (CSI, 1996), the term "Project Manual" originated with the American Institute of Architects (AIA) in 1964. The Project Manual includes bidding requirements, contract forms, conditions of the contract, and construction specifications. The Project Manual's construction specifications section is jointly categorized into 16 construction methods and materials divisions by the CSI in the United States and by Construction Specifications Canada (CSC) there, in an effort to foster better specification writing. The jointly prepared *Master-Format™* (CSI and CSC, 1995) contains a master listing of construction information that has been sequenced into groups comprised of introductory information, bidding requirements, contracting requirements, facilities and spaces, systems and assemblies, and construction products and activities divisions 1-16. Although this system evolved from architectural project applications, civil engineers have increased interest in the use of this format for land development projects since 1981 when CSI became a member of the Engineers Joint Contract Documents Committee (EJCDC). The American Society of Civil Engineers (ASCE) and the National Society of Professional Engineers (NSPE) are two of the EJCDC's members.

13.1 CONSTRUCTION AND CONTRACT DOCUMENTS

Construction and contract documents include all graphic and written documents that communicate the project's design intentions and construction administration activities. These documents, according to CSI's *Manual of Practice Construction Documents Fundamentals and Formats Module* (1996), include the following:

(a) Bidding requirements
  (1) Bid solicitation
  (2) Instructions to bidders
  (3) Information available to bidders
  (4) Bid forms and supplements
(b) Contract forms
  (1) Agreement
  (2) Performance bond

(3) Payment bond
(4) Certificates
(c) Conditions of the contract
    (1) General conditions
    (2) Supplementary conditions
(d) Specifications
(e) Drawings
    (1) Contract drawings
    (2) Resource drawings
(f) Addenda
(g) Modifications
    (1) Change orders
    (2) Field orders
    (3) Supplemental instructions

The relationships of these elements are illustrated in Figure 13.1.

The bidding requirements, contract forms, and conditions of the contract are not considered specifications and usually require the services of the developer's attorney and insurer when they are being prepared. Depending on the developer's approach to achieving project completions, various EJCDC standard forms can be incorporated as part of the contract documents. For example, Section 00700, General Conditions, could be comprised of EJCDC's Standard General Conditions of the Construction Contract (No. 1910-8). The EJCDC General Conditions, Instructions to Bidders, Bid Form, and Supplementary Conditions are written to correlate so that when one document is modified, the others must also be modified.

The appropriate location of subject matter within the Project Manual can be determined by using the EJCDC (and the American Institute of Architects) prepared document designated as the *Uniform Location of Subject Matter and Information in Construction Documents* (No. 1910-16). Utilization of this guideline will aid the specifier in coordinating the various elements that are included in the project manual.

The CSI/CSC (1996) format provides a standard set of rules for the arrangement, grammar, punctuation, and sequencing of material. The standard location for specific specification information is based primarily on 16 broad divisions, with each category being assigned a five-digit identification number. This system lends itself to computer adaptation through the use of this numbering system. The first two digits refer to the division within which the category is grouped. Each part is sequentially numbered beginning with the project title page labeled 00001. Preferred section numbers and titles are contained in CSI and CSC's periodically updated *MasterFormat*™ document, which specification writers can use to choose applicable entries when preparing specific Project Manuals.

MasterFormat™ contains classification numbers based on a four-level hierarchical system. Division titles are considered level 1 classifications. Level 2 is used for sections of broad scope that allow the most versatility in describing the work. Level 3 is used for sections where the scope of work is more limited and categorized as medium scope. Level 4 titles are applicable to work that is limited and very specific. Level 4 titles are numbered by the specifier using uncommitted numbers within level 3, thus allowing a flexible system of writing specifications. For example, under the level 2 broad scope category of Earthwork (02300), several level 3 medium scope categories are included, such as Grading (02310), Subgrade and Roadbed (02335), and Soil Stabilization

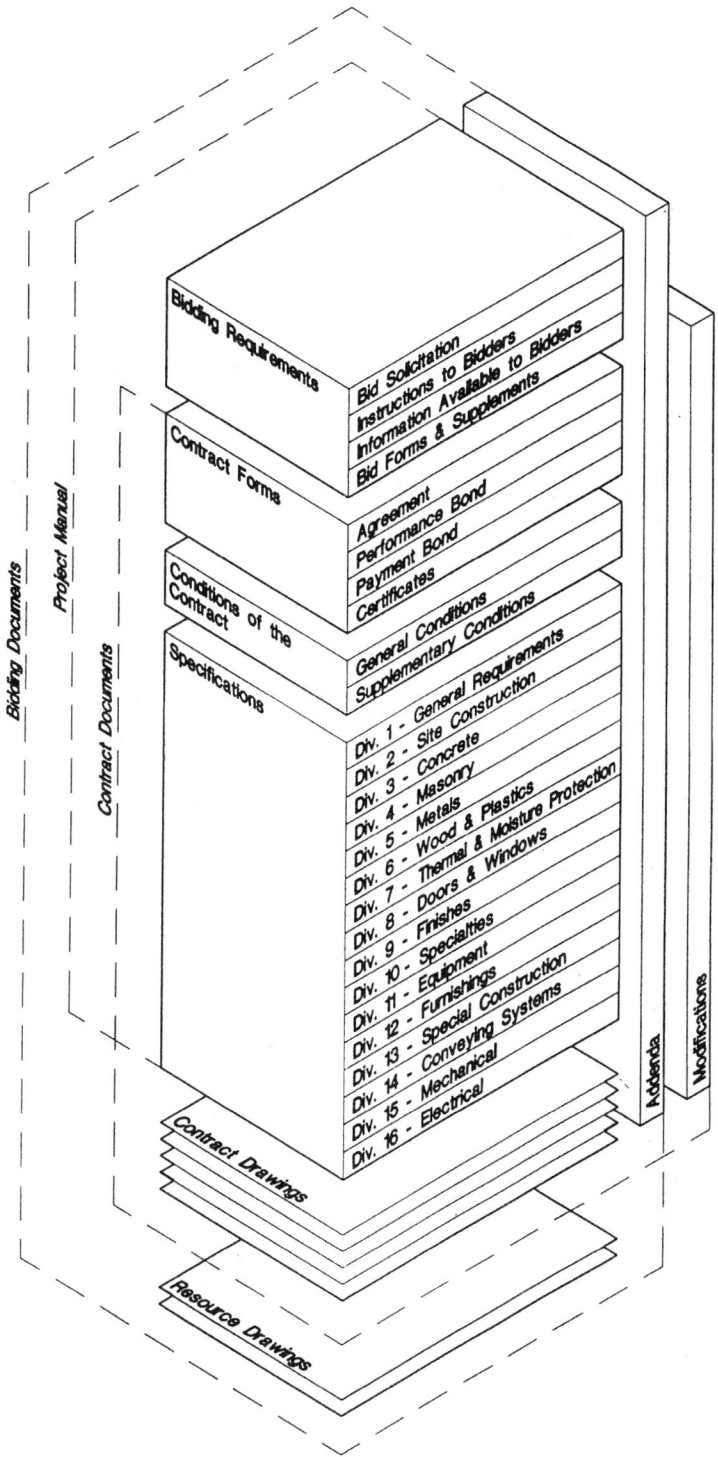

FIG. 13.1   Construction document relationships. Figure FF/CD-3 is from CSI's *Manual of Practice, Construction Documents Fundamentals and Formats Module*, is published by the Construction Specifications Institute (CSI), and is used with permission from CSI, 2001.

(02340). The specifier has the option of using only the level 2 category (02300) by incorporating all earthwork information together. However, if the specifier's approach is to address more categorically various aspects of earthwork, then several applicable level 3 categories can be utilized. The specifier also has the option of further refining the presentation of material by using a level 4 narrow scope approach, in which many applicable aspects of earthwork are narrowly defined while unused section numbers are assigned for their identification. The result is a document that is well organized and can be conveniently used to monitor construction activities and costs.

CSI /CSC also offers guidelines for specification section formats in its *SectionFormat*™ (CSI, 1997). In addition, guidelines are also available for arranging text on specification pages under the title *PageFormat*™ (CSI, 1992).

13.2   SPECIFICATIONS

Construction specifications are written instructions that provide information not found on the construction drawings. They convey to the contractor what final quality (performance) is expected in materials and workmanship used on the job. Specifications can be written using various formats, including:

(a) *Descriptive*   Describing methods and materials without the inclusion of proprietary names.
(b) *Performance*   Requiring specified end results that can be verified.
(c) *Reference Standard*   Requiring a product, procedure, or process that conforms with an established standard.
(d) *Proprietary*   Requiring a brand name or product.

Each of the 16 CSI/CSC divisions are subdivided into broad, medium, and narrow scope sections addressing specific topics; each of these sections are further categorized by CSI and CSC into a standard "Section Format" containing three parts:

Part 1: General
Part 2: Products
Part 3: Execution

Part 1 includes the administrative, procedural, and temporary requirements of the section. Procedures for measurement, payment, references, submittals, quality assurance, storage/handling, environmental precautions, and startup/warranties are just a few of the topics that can be addressed in this part. Part 1 also serves to amplify information contained in Division 1. Part 2 is used to describe technical provisions for all materials and equipment used in this section. Part 3 includes installation and standards relating to workmanship. If one or two of the parts are not germane, then the unnecessary part number(s) are stated along with the caption "Not Used" to indicate intentional exclusion. Figure 13.2 contains possible elements that can be included in the various parts.

The engineer who prepares the construction specifications must have a thorough knowledge of the work involved with the job, the kinds of materials and the methods

PART 1 GENERAL

SUMMARY
Section Includes
Products Supplied But Not Installed
    Under This Section
Products Installed But Not Supplied
    Under This Section
Related Sections
Allowances
Unit Prices
Measurement Procedures
Payment Procedures
Alternates

REFERENCES

DEFINITIONS

SYSTEM DESCRIPTION
Design Requirements,
    Performance Requirements

SUBMITTALS
Product Data
Shop Drawings
Samples
Quality Assurance/Control Submittals
    Design Data, Test Reports,
    Certificates,
    Manufacturers' Instructions,
    Manufacturers' Field Reports,
    Qualification Statements
Closeout Submittals

QUALITY ASSURANCE
Qualifications
Regulatory Requirements
Certifications
Field Samples
Mock-ups
Pre-installation Meetings

**DELIVERY, STORAGE,
AND HANDLING**
Packing, Shipping, Handling,
    and Unloading
Acceptance at Site
Storage and Protection
Waste Management and Disposal

PROJECT/SITE* CONDITIONS
Project/Site* Environmental
    Requirements
Existing Conditions

SEQUENCING

SCHEDULING

WARRANTY
Special Warranty

SYSTEM STARTUP

OWNER'S INSTRUCTIONS

COMMISSIONING

MAINTENANCE
Extra Materials
Maintenance Service

PART 2 PRODUCTS

MANUFACTURERS

EXISTING PRODUCTS

MATERIALS

MANUFACTURED UNITS

EQUIPMENT

COMPONENTS

ACCESSORIES

MIXES

FABRICATION
Shop Assembly
    Fabrication Tolerances

FINISHES
Shop Priming, Shop Finishing

SOURCE QUALITY CONTROL
Tests, Inspection
Verification of Performance

PART 3 EXECUTION

INSTALLERS

EXAMINATION
Site Verification of Conditions

PREPARATION
Protection
Surface Preparation

ERECTION

INSTALLATION

APPLICATION

CONSTRUCTION
Special Techniques
Interface with Other Work
Sequences of Operation
Site Tolerances

REPAIR/RESTORATION

RE-INSTALLATION

FIELD QUALITY CONTROL
Site Tests, Inspection
Manufacturers' Field Services

ADJUSTING

CLEANING

DEMONSTRATION

PROTECTION

SCHEDULES

* Project Conditions is the preferred term in the U.S.,
Site Conditions is the preferred term in Canada

FIG. 13.2   Construction Specification Institute's section format outline. Figure SF-1 is from *SectionFormat,* is published jointly by the Construction Specifications Institute (CSI) and Construction Specifications Canada (CSC), and is used with permission from CSI, 2001.

to be used in construction, and an ability to write clearly. The legal importance of specifications is that they usually govern and take precedence over drawings, unless the general conditions specifically state that the drawings govern. The specifier must select appropriate words to precisely impart the intended meaning to the reader. In addition, capitalization, punctuation, and grammar are important. Often the sentence structure takes the form of either the imperative or indicative mood. The imperative mood can be used when describing the installation of products or equipment where the verb is placed at the beginning of the sentence. The indicative mood, passive voice, requires the use of "shall"; however, CSI discourages its use. Examples of these methods are:

Imperative Mood   Compact soil in 6-inch lifts.
Indicative Mood   Soil shall be spread in 6-inch lifts with each lift being compacted.

An excellent presentation on the preparation of a project manual is contained in CSI's (1996) *Manual of Practice, Construction Documents Fundaments and Formats Module.* For those interested in a more in-depth explanation of the above documents and their use in the construction industry contact the Construction Specifications Institute, 99 Canal Center Plaza, Suite 300, Alexandria, VA 22314 (800-689-2900; 703-684-0300), www.csinet.org.

Development projects using public monies may be required to specify multiple source suppliers for material and products proposed for use in construction. Such provisions normally include at least two manufacturer suppliers and the phrase "or equal" when specifying products. "Or equal" normally refers to dimensional, mechanical, and material aspects of the desired product and can also include performance aspects such as operational speed, efficiency, and horsepower. CSI recommends the use of alternative phrases "as approved" or "approved substitute" and discourages use of "or equal" to reduce confusion because these phrases encourage the bidder to seek approval prior to bid opening for any substitute offered as part of a project bid.

Because materials, technology, and construction methods are changing rapidly, the specifier can obtain through the CSI computer-based CSI-formatted specification production and management systems. One system offered by Atlanta, Georgia based Building Systems Design (bsd@bsdsoftlink.com) is BSD SpecLink®. Another computer-based system is MASTERSPEC®, a product of the American Institute of Architects (AIA). MASTERSPEC® is published, supported, and licensed for the AIA by Architectural Computer Services (ARCOM) (arcom@arcomnet.com).

Should the specifier choose a computer-based system to prepare and manage specifications, a number of benefits can be gained. For example, BSD SpecLink® offers automated access to approximately 700 generic and proprietary sections and automatic quarterly database updates typically affecting anywhere from 25 to 65 percent of the database. In addition, the updates do not affect any changes (added or edited) by the user, since they are stored separately and merged with the master text when sections are viewed or printed.

13.3   AN EXAMPLE OF A PROJECT MANUAL

Each Project Manual must be individually prepared to meet the needs of a specific development. Examples of various typical sections are shown here to illustrate the general requirements of a typical privately funded improved land development. The following table of contents is applicable to a project that consists of the construction of roads, storm drainage facilities, potable water distribution facilities, and wastewater collection and flooded suction pumping systems similar to the one shown in Figure 10.28. CSI/CSC *MasterFormat*™ section designations used in this example follow:

Typical Project Manual

Table of Contents

Introductory Information

00001 Project Title Page
00005 Certifications Page
00007 Seals Page
00010 Table of Contents
00015 List of Drawings
00020 List of Schedules
00025 Project Consultants

Bidding Requirements

00101 Advertisement for Bids
00102 Bidder's Qualifications
00103 Invitation to Bid
00200 Instructions to Bidders
00320 Geotechnical Data
00330 Existing Conditions
00340 Environmental Assessment Information
00360 Permit Applications
00410 Bid Form
00431 Bid Security Form
00432 Estimated Quantities Form
00433 Proposed Subcontractor Form

Contracting Requirements

00500 Agreement
00611 Performance Bonds
00612 Payment Bonds
00621 Certificate of Insurance
00622 Affidavit of Prime Contractor[*]
00623 Release and Waiver of Claims[*]
00624 Contractor's General Warranty[*]

00700 General Conditions

00800 Supplementary Conditions

00900 Addenda and Modifications[*]

Specifications

Division 1—General Requirements

01110 Summary of Work

01140 Work Restrictions

01180 Project Utility Sources

01250 Contract Modification Procedures

01270 Unit Prices—Measurement & Payment

01310 Project Management and Coordination

01320 Construction Progress Documentation

01321 Survey and Layout Data

01331 Field Test Reports

01332 Shop Drawings, Product Data, and Samples

01333 Source Quality Control Reports

01410 Regulatory Requirements[*]

01431 Manufacturer's Qualifications

01432 Manufacturer's Field Services

01450 Quality Control

01500 Temporary Facilities and Controls[*]

01650 Product Delivery Requirements

01660 Product Storage and Handling Requirements

01721 Construction Layout

01722 Field Engineering

01740 Final Cleaning

01750 Starting and Adjusting

01760 Protecting Installed Construction

01770 Closeout Procedures

01781 Final Site Survey

01782 Operation and Maintenance Data[*]

01783 Product Warranties

01784 Project Record Documents

01785 Spare Parts

01810 System Performance Evaluation/Testing

01820 System Demonstration and Training

01830 System Operation and Maintenance

Division 2—Site Construction

02055 Soils

02060 Aggregate
02065 Cement and Concrete
02070 Geosynthetics
02081 Utility Materials—Fire Hydrants
02082 Utility Materials—Manholes
02083 Utility Materials—Meters and Boxes
02084 Utility Boxes
02085 Utility Materials—Valves and Boxes
02090 Pipe Joint Material
02100 Site Remediation
02230 Site Clearing*
02240 Dewatering*
02250 Shoring and Underpinning*
02260 Excavation Support and Protection
02310 Grading*
02315 Excavating and Fill*
02340 Soil Stabilization*
02370 Erosion and Sedimentation Control*
02445 Boring or Jacking Conduits
02510 Water Distribution*
02530 Sanitary Sewerage*
02600 Drainage and Containment
02711 Base Courses*
02741 Bituminous Concrete Pavement*
02750 Rigid Pavement
02770 Curbs and Gutters
02775 Sidewalks
02780 Unit Pavers
02800 Site Improvements and Amenities
02900 Planting

Division 3—Concrete

03300 Cast-in-Place Concrete
03400 Precast Concrete Specialties
03600 Grouts

Division 4—Masonry

04200 Masonry Units

Division 5—Metals

05510 Metal Stairs and Ladders

05520 Handrails and Railings
05560 Metal Castings

Division 6—Wood and Plastics

06110 Rough Carpentry—Wood Framing
06170 Wood Trusses

Division 7—Thermal and Moisture Protection

07100 Damproofing and Waterproofing
07300 Shingles, Roof Tiles, and Roof Coverings

Division 8—Doors and Windows

08100 Metal Doors and Frames
08330 Coiling Doors and Grilles

Division 9—Finishes

09900 Paints and Coatings

Division 10—Specialties

10200 Louvers and Vents

Division 11—Equipment

11310 Sewage and Sludge Pumps

Division 12—Furnishings (Omit)

Division 13—Special Construction

13150 Swimming Pools

Division 14—Conveying Systems (Omit)

Division 15—Mechanical

15490 Pool and Fountain Equipment
15810 Ventilator Ducts
15831 Power Ventilator

Division 16—Electrical

16050 Basic Electrical Materials and Methods
16100 Wiring Methods
16220 Motors and Generators
16440 Switch Boxes, Panelboards, and Control Centers
16500 Lighting

*Example sections follow.

Because of space limitations, an example of a complete Project Manual cannot be included; however, several illustrative sections (designated with asterisks above) have been included.

SECTION 00622

AFFIDAVIT OF PRIME CONTRACTOR

(PAYMENT STATUS—SUBCONTRACTORS AND MATERIAL SUPPLIERS)

Date: _____

We,   Name of Prime Contractor , do hereby certify that to the best of our knowledge and belief no claims or liens exist against any subcontractors or material suppliers who furnished labor and/or material for the construction of   Project's Name and Location . We shall hold harmless the Owner from any subsequent liens made of record. If any liens remain unsatisfied after final payment is made to the Contractor, the Owner will satisfy the lien, and all incurred costs including reasonable attorney's fees will be immediately due and paid to the Owner by the Contractor.

Sworn to and subscribed before me this _____ day of ____ 19 ___ .

_____
Signature

By: _____

Title: _____

Date: _____

Notarized By: _____

Date: _____

Seal:

Information contained in this section is subject to change and is intended for educational purposes only. CSI, CSC, BSD, and ARCOM did not prepare nor do they specifically endorse this material. The current format used in the above section is from CSI/CSC's *MasterFormat*™ (1995).

End of Section 00622

SECTION 00623

RELEASE AND WAIVER OF CLAIM

On_____, 19_____ there personally appeared before me the undersigned authority in and for said County of _____ and State of _____,   Authorized Person's Name , who, being duly sworn, states that all material bills, sales and other taxes, licenses, payrolls, state and federal unemployment insurance, and all other liabilities have been paid in full that were incurred in the performance of the contract for   Project's Name and Location , and waives any claims and releases the Owner from any rights or claims or debts due as a result of any material or supply liens.

Sworn to and subscribed before me this _____ day of _____19_____.

_____
Signature
By: _____
Title: _____
Date: _____

Notarized By: _____   Seal:
Date: _____

Information contained in this section is subject to change and is intended for educational purposes only. CSI, CSC, BSD, and ARCOM did not prepare nor do they specifically endorse this material. The current format used in the above section is from CSI/CSC's *MasterFormat*™ (1995).

End of Section 00623

SECTION 00624

CONTRACTOR'S GENERAL WARRANTY

PROJECT'S NAME

PROJECT'S LOCATION

The undersigned Contractor does hereby warrant in accordance with the provisions and terms of the Contract Documents, all materials and workmanship incorporated in _____Project's Name and Location_____, against any and all defects due to faulty materials, workmanship, or negligence for a period of 12 months for all streets and drainage facilities, and 12 months for all other improvements based on the effective date of this warranty. This Contractor further warrants that all work incorporated in this project will remain leakproof and watertight at all points for a period of 24 months from the effective date of this warranty.

This warranty shall be binding where defects occur due to normal use and does not cover wanton and malicious damages, damage caused by acts of God, or damage caused by casualties not controlled by the Contractor.

This warranty shall be in addition to all other warranties and guarantees set forth in the Contract Documents, and shall not act to constitute a waiver of additional protection of the Owner if afforded by consumer protection and product liability provisions of the law, and these stipulations shall not constitute waivers of any additional rights or remedies available to the Owner under the law.

_____
Signature
Name: _____
Title: _____

Date: _____

Subscribed and sworn before me this _____ day of ____19____ .

Notarized By: _____   Seal:

Date:_____

Information contained in this section is subject to change and is intended for educational purposes only. CSI, CSC, BSD, and ARCOM did not prepare nor do they specifically endorse this material. The current format used in the above section is from CSI/CSC's *MasterFormat*™ (1995).

End of Section 00624

SECTION 00900

ADDENDA

ADDENDUM NO. 1

PROJECT'S NAME

PROJECT SECTION OR PHASE

1. Modification to the contract documents pertaining to the Project cited above is as follows:
   A. *Drawings:* [Cite sheets with modifications or changes.]
   B. *Project Manual:* [Cite applicable Sections and paragraphs requiring modifications or changes.]
   C. *Other:* [State applicable information.]

Engineer's Signature
Engineer's Name

Date

Engineer's Seal

Information contained in this section is subject to change and is intended for educational purposes only. CSI, CSC, BSD, and ARCOM did not prepare nor do they specifically endorse this material. The current format used in the above section is from CSI/CSC's *MasterFormat*™ (1995).

End of Section 00900

SECTION 01410

REGULATORY REQUIREMENTS

Part 1—General

1.01 Summary:

    A. Section contains provisions for compliance with jurisdictional regulations affecting the Work.

    B. Related Sections include but are not necessarily limited to:

        1. General Conditions

        2. Supplementary Conditions.

        3. Sections in Division 1 of this Project Manual.

1.02 References:

    A. Regulatory requirements required by the Contract Documents include, but are not necessarily limited to, obtaining, maintaining, or complying with:

        1. Owner procured:

            a. Sedimentation and erosion control permits.

            b. Highway department encroachment permits.

            c. Potable water system construction permit.

            d. Permit to construct in navigable streams.

            e. Encroachment permits for public utilities.

            f. Permit to construct storm drainage system.

            g. Coastal zone construction permits.

            h. Permits to withdraw surface or subsurface waters.

            i. Permit to construct public swimming facilities.

        2. Contractor Procured:

            a. Current business license.

            b. Current general contractor's license for state in which the project is located.

            c. Building permits not included in this Project Manual for temporary offices and buildings.

            d. Burn permits.

            e. Fire, hazardous waste, safety, and OSHA standards.

            f. Permit for hauling waste and disposal.

            g. Permit for temporary utility services and construction coordination for Project utility connections.

            h. Permit to exceed load limits of certain bridges if they are on haul route or have very low load limits.

        3. Permits procured by the Owner and/or Contractor, if the Contractor undertakes temporary or independent activity.

            a. Filling permits.

            b. Permits to discharge water.

            c. Sign permits.

            d. Landscape and tree removal permits.

            e. Permits to restrict work in temporary construction easements.

1.03 Submittals:

    A. Comply with germane portions of Section 01332.

    B.   Provide required reports to regulatory agencies in accordance with provisions of issued permits.

1.04 Quality Assurance:

    A.   Designate in writing to the Owner a member of the Contractor's organization to monitor compliance with provisions of issued permits.

    B.   Conduct inspections of construction activities to ensure compliance.

1.05 Site Conditions:

    A.   Practice and be fully responsible for good housekeeping activities and procedures to prevent oil spills, hazardous waste contamination, unauthorized environmental pollution, and loss of safety.

    B.   Maintain drainage features while incorporating sedimentation and erosion control measures.

    C.   Maintain traffic using recognized traffic zone control measures.

1.06 Sequencing and Scheduling:

    A.   Provide ample notification to regulatory agencies for scheduled construction activities if a regulatory inspector is required for observation.

Part 2—Products (Not Used)

Part 3—Execution (Not Used)

Information contained in this section is subject to change and is intended for educational purposes only. CSI, CSC, BSD, and ARCOM did not prepare nor do they specifically endorse this material. The current format used in the above section is from CSI/CSC's *MasterFormat*™ (1995).

End of Section 01410

SECTION 01500

TEMPORARY FACILITIES AND CONTROLS

Part 1—General

1.01 Summary:

    A.   Section includes provisions for construction facilities and temporary controls necessary for and as part of the Work.

    B.   Related Sections include but are not necessarily limited to:

        1. General Conditions.

        2. Supplementary Conditions.

        3. Sections in Division 1 of this Project Manual.

        4. Section 02370.

1.02 Submittals:

    A.   Comply with provisions of Section 01332.

1.03 Quality Assurance:

    A.   Instruct personnel to maintain cleanliness throughout the construction area and prevent unauthorized pollution.

    B.   Conduct daily inspections to ensure that standards are being maintained.

1.04 Delivery, Storage, and Handling:
    A. Monitor and maintain construction facilities and temporary control in proper working order to promote safety while prosecuting the Work.

1.05 Site Conditions:
    A. Provide construction facilities and temporary controls necessary to conduct the Work, including but not necessarily limited to:
        1. Field office and sheds.
        2. Temporary storage areas.
        3. Temporary sanitary facilities.
        4. Temporary utilities.
        5. Temporary fencing, barriers, enclosures, and security devices.
        6. Temporary safety provisions.
        7. Temporary drainage facilities.
        8. Temporary signage for identification and traffic control.
        9. Temporary measures for dust control.
       10. Routine cleaning and waste removal.

1.06 Maintenance:
    A. Maintain all areas, improvements, work areas and access roads for clear passage and safe functional use.
    B. Locate field offices, temporary structures, and storage areas in accordance with the Engineer's directives.

Part 2—Products

2.01 Components:
    A. Provide temporary field offices and storage areas for use during construction.
    B. Provide temporary water supply to job site for use during construction.
        1. Arrange and pay for connection and fees.
        2. Install temporary piping and fittings.
        3. Pay for water used for construction.
        4. Remove temporary connection and piping at completion of job.
    C. Provide temporary electrical power to the job site during construction.
        1. Arrange and pay for connection and fees.
        2. Install temporary wiring and accessories.
        3. Pay for electricity used during construction.
        4. Remove temporary connection and wiring at completion of job.
    D. Provide temporary telephone service to the job site during construction.
        1. Arrange and pay for connection and fees.
        2. Install temporary wiring and accessories.
        3. Pay for telephone service used during construction.
        4. Remove temporary connection and wiring at completion of job.
    E. Provide and pay for temporary heat to the job site during construction.
    F. Provide warning signs, barricades, and temporary safety devices in compliance with regulatory requirements.
    G. Provide temporary fencing and security to prevent unauthorized entry by the public.

H. Provide signs to identify the project site and to control traffic. Signs must be acceptable to the Engineer.

I. Provide as part of the Contract Price all temporary drainage facilities necessary to prosecute the Work in a timely manner. The temporary drainage plan must conform to the overall drainage and sedimentation and erosion control plan for the site and be approved by the Engineer and any required regulatory authorities and provisions of Section 02370. This contractor prepared plan must meet all requirements of the NPDES General Permit for stormwater discharge pertaining to this project.

J. Provide hygroscopic materials for dust control that are acceptable to the Engineer.

Part 3—Execution

3.01 Erection, Installation, and Application:

A. Provide suitable on-site field offices and sheds to accommodate Contractor needs.
1. Facilities shall include adequate space for project meetings with tables, chairs, and utilities.
2. Provide and maintain adequate sanitary facilities for use by all personnel.

B. Maintain temporary facilities and control to provide safe working conditions and prevent unauthorized persons from entering the site.

C. Remove temporary controls and facilities when use is no longer required and grade area. Provide erosion control to the graded surface and clean as required in Section 01740.

D. Follow provisions of Section 02240 when providing temporary drainage by dewatering.

E. Provide temporary protection to existing improvement, including fences, poles, wires, pipes, property corners, and so on, that the Engineer deems necessary to preserve. If during the course of construction, the preserved items are injured, the Contractor shall restore them to original conditions immediately. When fences interfere with the Contractor's operation, permission must be requested to temporarily remove during construction and then immediately replace. No additional compensation shall be allowed to the Contractor for this work. Trees and shrubs so designated shall be saved and protected by the contractor from equipment, materials, and employee abuse.

F. Erect a temporary 6-foot high wire fence around unattended open trenches to seal off. Place bright plastic ribbon at random on the fence for warning and position hazard signs with blinking lights prominently around the perimeter for caution at conspicuous intervals.

G. Continuously maintain cleanliness and orderliness around storage areas during construction to allow safe passage, good drainage, and access.
1. pick up scrap, waste, debris, and spoil daily and place in temporary holding areas.

2. Straighten and restack storage piles at least once each week.
3. Remove debris, waste, spoil, and scrap material from the construction site and dispose of by approved means at least every other week, unless specifically directed in writing to do otherwise by the Engineer.
4. Responsibility for the disposition of all transported waste material shall be that of the Contractor only.
5. Inspect the construction site every seven days or within 24 hours after a rainfall of 0.5 inch or more in accordance with Section 02370.

H. Apply calcium chloride or other hygroscopic materials for dust control using application rates specified by the Engineer.

Information contained in this section is subject to change and is intended for educational purposes only. CSI, CSC, BSD, and ARCOM did not prepare nor do they specifically endorse this material. The current format used in the above section is from CSI/CSC's *MasterFormat*™ (1995).

End of Section 01500

SECTION 01782

OPERATION AND MAINTENANCE DATA

Part 1—General

1.01 Summary:
   A. Section includes provisions for providing information necessary to properly operate and maintain installed systems.
   B. Related Sections include but are not necessarily limited to:
      1. General Conditions.
      2. Supplementary Conditions.
      3. Sections contained in Division 1 of this Project Manual.
1.02 Submittals:
   A. Prepare and submit two copies of proposed operation and maintenance manuals for each system requiring operation and maintenance for review by the Engineer.
   B. Prepare and submit seven copies of each operation and maintenance manual, incorporating comments provided by the Engineer's review of the proposed manuals.
1.03 Quality Assurance:
   A. Utilize only knowledgeable personnel and manufacturer's recommendations to prepare operational and maintenance manuals.

Part 2—Products

2.01 Materials:
   A. Prepare operation and maintenance manuals using only $8\frac{1}{2} \times 11$-inch white bond paper with text being electronically or mechanically printed and bound on the left margin with front and back cover sheets.

B.   Label each manual to indicate the applicable system for which the manual was prepared, the Contractor's name, and a place for the Engineer's approval.

2.02 Components:
A.   Include in each manual:
1. Table of contents.
2. Complete operating instructions including assembly and disassembly procedures, lubrication procedures and lubricants, and required special tools and spare parts.
3. Means to identify and describe all components of each piece of equipment.
4. Parts listing of all components including supplier names and addresses.
5. Manufacturer's data sheets and drawings.
6. Troubleshooting guides for malfunctions.

Part 3—Execution

3.01 Preparation:
A.   Obtain manufacturer's recommended operation and maintenance procedures.

3.02 Application:
A.   Instruct designated operator(s) in the proper operation and maintenance of the system with appropriate manufacturer's representatives as provided for in Section 01432.

Information contained in this section is subject to change and is intended for educational purposes only. CSI, CSC, BSD, and ARCOM did not prepare nor do they specifically endorse this material. The current format used in the above section is from CSI/CSC's *MasterFormat*™ (1995).

End of Section 01782

SECTION 02230

SITE CLEARING

Part 1—General

1.01 Summary:
A.   Section includes procedures for clearing and grubbing areas shown on the Drawings and where required in these specifications as part of the Work.
B.   Related Sections include but are not necessarily limited to:
1. General Conditions.
2. Supplementary Conditions.
3. Sections contained in Division 1 of this Project Manual.

1.02 Quality Assurance:

    A. Detail sufficient numbers of qualified workmen that are experienced in executing the activities specified in this Section.

    B. Detail and maintain sufficient equipment capable of accomplishing the activities specified in this Section while producing sufficient progress on the Work.

    C. Operate all equipment with due caution.

    D. Instruct workmen in procedures specified in this Section.

1.03 Delivery, Storage, and Handling:

    A. Adhere to pertinent portions of Section 01500.

1.04 Maintenance:

    A. Restore material and property on which damage has been inflicted to predamaged conditions.

Part 2—Products

2.01 Materials:

    A. Provide incidental materials required for satisfactory completion of the activities specified in this Section.

Part 3—Execution

3.01 Examination:

    A. Examine areas where clearing and grubbing activities must be performed and verify that limits of clearing are field staked.

    B. Request underground utility location service to locate utilities within work areas.

    C. Determine prior to commencing clearing operations whether temporary drainage is required and incorporate in accordance with Section 01500 if needed.

    D. Review applicable provisions contained in Section 02370.

3.02 Preparation:

    A. Protect improvements from equipment, material, and employee abuse to include:

        1. Property corners, survey field markers, and monuments.

        2. Designated trees, shrubs, and grassed areas.

        3. Existing utilities, sedimentation and erosion control systems, property improvements, pavements, sidewalks, driveways, curbing, poles, wires, pipes, and fences, unless otherwise noted on the Drawings or specifically designated for removal by the Engineer. When fences interfere with construction operations, request permission from the Owner to temporarily remove during the time work is being conducted in the area and immediately replace upon completion at no additional cost to the Owner.

        4. All areas not within the construction site boundary subject to falling limbs, trees, and dust.

        5. Persons and personal property in or adjoining the work area.

    B. Provide for dust control if measures are warranted in accordance with Section 01500.

    C.   Ensure sedimentation and erosion control procedures provided for in Section 02370 are utilized.

3.03 Erection, Installation, Application:

    A.   Clear and grub the width of road rights-of-way, easements, drainage facilities, utility sites, and any other specified areas shown on the Drawings.

    B.   Restrict clearing and grubbing activities to within 5 feet in back of median curbs where divided entrance curb and gutter street sections are to be constructed. Protect trees and shrubs in addition to areas specifically noted on the Drawings or designated by regulatory authorities.

    C.   Burn trash, brush, roots, and unwanted trees in accordance with Contractor secured burn permit requirements or remove cleared materials off-site in accordance with Section 01410 unless otherwise directed by the Engineer.

    D.   Salable timber obtained from clearing operations shall become the property of the Contractor and shall be moved off-site while clearing operations are underway.

    E.   Remove all roots and vegetation 1 inch and larger from cleared areas.

    F.   Strip usable topsoil and stockpile in designated areas in accordance with Section 01500 for later use as provided for in Section 02900. Usable or waste material will either be placed on-site in accordance with written instructions from the Engineer or moved and be properly disposed of off-site by the Contractor at an approved location of the Contractor's choice in accordance with Section 01410.

Information contained in this section is subject to change and is intended for educational purposes only. CSI, CSC, BSD, and ARCOM did not prepare nor do they specifically endorse this material. The current format used in the above section is from CSI/CSC's *MasterFormat*™ (1995).

End of Section 02230

SECTION 02240

DEWATERING

Part 1—General

1.01 Summary:

    A.   Section includes provisions for the installation and operation of dewatering facilities required to safely complete the Work in an acceptable manner and on schedule as part of the Contract Price.

    B.   Related Sections include but are not necessarily limited to:

        1. General Conditions.

        2. Supplementary Conditions.

        3. Sections in Division 1 of this Project Manual.

1.02 Quality Assurance:

    A.   Detail sufficient numbers of qualified workmen that are experienced in executing the activities specified in this Section.

B. Detail and maintain sufficient equipment capable of undertaking the activities specified in this Section to produce sufficient progress on the Work.

C. Operate all equipment with due caution.

D. Instruct workmen in procedures specified in this Section.

Part 2—Products

2.01 Equipment:

A. Provide equipment necessary to install, maintain, and remove dewatering facilities.

Part 3—Execution

3.01 Examination:

A. Examine areas where dewatering activities must be performed and verify whether existing conditions are suitable to safely undertaking dewatering. Take any corrective action required prior to commencing dewatering.

B. Request underground utility service to locate all utilities within work area.

C. Determine whether temporary drainage is required and incorporate in accordance with Section 01500.

3.02 Preparation:

A. Protect improvements from equipment, material, and employee abuse including:

1. Property corners, survey field markers, and monuments.

2. Designated trees, shrubs, and grassed areas.

3. Existing utilities, sedimentation and erosion control systems, property improvements, pavements, sidewalks, driveways, curbing, poles, wires, pipes, and fences unless otherwise noted on the Drawings or specifically designated for removal by the Engineer. When fences interfere with construction operations, request permission from the owner to temporarily remove during the time work is being conducted in the area and immediately replace upon completion at no additional cost to the Owner.

4. All areas not within the construction site boundary subject to the discharge of pumped water.

5. Persons and personal property in or adjoining the work area.

3.03 Installation:

A. Provide well points, headers, piping, pumps, and other required equipment to safely remove unwanted water from the construction work area.

3.04 Protection:

A. Discharge pumped water into an approved storm drain, natural drainage area away from the construction area, or a specified area in accordance with Section 01410. Prevent dewatering discharges from being directed on construction surfaces such as road subgrade.

endorse this material. The current format used in the above section is from CSI/CSC's *MasterFormat*™ (1995).

End of Section 02240

SECTION 02250

SHORING AND UNDERPINNING

Part 1—General

1.01 Summary:
    A. Section includes procedures to provide shoring in all excavations not having sloped walls 5 feet or greater in depth to protect workers, material, real and personal property, and bystanders, unless poor soil conditions or overburden loads warrant shoring for excavations of less than 5 feet as part of the Contract Price.
    B. Related Sections include but are not necessarily limited to:
        1. General Conditions.
        2. Supplementary Conditions.
        3. Sections in Division 1 of this Project Manual.

1.02 System Description:
    A. Select an appropriate shoring and/or underpinning system to provide adequate support under all weather and work conditions.
    B. Provide securely held sheathing and bracing systems for trenches used for pipe laying that provide a clear bottom of the trench with a width that is at least 12 inches wider than the bell of the pipe unless otherwise permitted by the Engineer. In water-bearing soil, no portion of the sheathing below the level of 4 feet above the pipe shall be removed.

1.03 Submittals:
    A. Adhere to germane portions of Section 01332.
    B. Obtain regulatory approved shoring design with a Contractor-procured duly registered structural field engineer and submit the approved design to the Engineer for record purposes prior to installing the shoring system.
    C. Provide the Engineer with a report of an inspection conducted by the responsible Contractor procured field engineer under whose direction the installed shoring system has been designed. This report should certify the shoring installation once installed.

1.04 Quality Assurance:
    A. Provide workers who are experienced in the skills necessary to install and work around shoring.
    B. Detail and maintain sufficient equipment capable of undertaking the activities specified in this Section to produce sufficient progress on the Work.
    C. Operate all equipment with due caution.
    D. Instruct workmen in procedures specified in this Section.
    E. Employ an experienced licensed professional field engineer qualified to design and inspect a completed system.

F.   Require shoring plan to be coordinated with other proposed improvements and work space requirements.

1.05 Warranty:

A.   Warrant the safety, workability, adaptability, and utility of the installed shoring system in conjunction with the Contractor-procured field engineer while holding other parties harmless.

1.06 Maintenance:

A.   Monitor dewatering systems to ensure proper operation and provide maintenance to the installed shoring system when required.

Part 2—Products

2.01 Components:

A.   Provide design calculations with material and erection drawings for the proposed shoring system to include provisions necessary to provide lateral support to surrounding properties.

B.   Submit and obtain required approvals from regulatory agencies for the proposed shoring system. Pay regulatory fees at no additional cost to the Owner.

2.02 Materials:

A.   Provide materials required to install the shoring system as shown on Contractor-procured field engineer's approved plans.

2.03 Equipment:

A.   Provide equipment necessary to install, maintain, and remove the shoring system.

Part 3—Execution

3.01 Examination:

A.   Inspect the area where the shoring system is to be located and determine whether conditions are suitable for installation to begin.

B.   Inspect surrounding area to verify that conditions have not changed that would alter the shoring system design, such as newly placed overburden loads.

C.   Correct any deficiencies prior to commencing shoring installation.

3.02 Preparation:

A.   Request underground utility location service to locate underground utilities within the work area.

B.   Provide adequate access for soil handling and material.

C.   Provide adequate surface and subsurface drainage in accordance with Section 02240.

3.03 Installation:

A.   Install the proposed shoring system in accordance with the approved shoring plans.

Information contained in this section is subject to change and is intended for educational purposes only. CSI, CSC, BSD, and ARCOM did not prepare nor do they specifically endorse this material. The current format used in the above section is from CSI/CSC's *MasterFormat*™ (1995).

End of Section 02250

SECTION 02310

GRADING

Part 1—General

1.01 Summary:
- A. Section includes provisions for the conduct of excavation, backfill, compaction, and grading of earthwork required as part of the Work.
- B. Related Sections include but are not necessarily limited to:
  1. General Conditions.
  2. Supplementary Conditions.
  3. Sections in Division 1 of this Project Manual.
  4. Section 00320 for Geotechnical Data.

1.02 Submittals:
- A. Submit satisfactory results of field density tests conducted in accordance with Section 01331 and reviewed by the Geotechnical Consultant in accordance with Section 00320 certifying densities attained prior to requesting approval from the Engineer.
- B. Submit shoring plan when required in accordance with Section 02250.

1.03 Quality Assurance:
- A. Detail sufficient numbers of qualified workmen who are experienced in executing the activities specified in this Section.
- B. Detail and maintain equipment capable of undertaking the activities specified in this Section to produce sufficient progress on the Work.
- C. Check final graded surface for elevations and alignment. Use grades obtained from construction Drawings as finished grades unless directed otherwise. Account for pavement thickness requirements, foundation slab thicknesses, and other structural features to produce the final graded elevations.
- D. Limit extent of earthwork to authorized areas designated on the Drawings. Fill no wetlands unless authorized by permit. Allow no excavation under existing structure foundations, unless authorized in writing by the Engineer.
- E. Instruct workmen in procedures specified in this section and operate equipment with due caution.
- F. Ensure finished graded surfaces are free from roots, trash, and other foreign material.

1.04 Delivery, Storage, and Handling:
- A. Adhere to germane portions of Sections 01650 and 01660.
- B. Remove excess soil from the area being graded. Do not deposit or store soil closer than 5 feet from any drainage ditch, pond, or catchment.
- C. Utilize temporary storage areas provided in Section 01500.

1.05 Sequencing and Scheduling:
- A. Coordinate the requirements of this Section with Sections 02230 and 02370.

1.06 Maintenance:

A. Maintain completed areas by scarifying, reshaping, or recompacting to correct deficiencies attributed to construction-related activities or weather.

B. Maintain dust-free environment in accordance with Section 01500.

C. Maintain drainage.

Part 2—Products

2.01 Materials:

A. Classify soils into the following categories:
1. Embankment (fill) materials.
2. Unclassified (general) excavation materials.
3. Classified excavation materials, which are grouped into the following categories:
    a. Muck.
    b. Rock.
    c. Topsoil.

B. Embankment materials shall include:
1. Only soils that have a maximum 4-inch size and that are free from organic and deleterious material.
2. Only approved on-site soil or approved off-site borrow soil that is included as part of the Contract Price.

C. Suitable unclassified excavated materials shall be reused for embankments, with unsuitable material being handled as waste as part of the Contract Price.

D. Classified excavation materials include:
1. Extensive unusable material encountered during grading operations, which is classified as muck.
2. Soil extensively comprised of rocks larger than 6 inches in size; removal of this material is classified as rock excavation. No rock excavation is anticipated in the Work.
3. Topsoil, which shall be:
    a. Fertile organic soil, preferably of a loamy nature to support the planting of cover vegetation, and free from noxious weeds, debris, roots, litter, and other undesirable material.
    b. Provided normally from approved stockpiled on-site strippings resulting from Section 02230, unless additional approved off-site borrow is required to complete the Work.

Part 3—Execution

3.01 Examination:

A. Inspect areas to be graded and determine whether conditions are suitable to undertake the portions of work specified in this Section. Do not commence work under this Section until conditions are satisfactory.

B. Request underground utility service to locate all utilities within work area.

C. Verify line and grade stakes and report any obvious discrepancies to the Engineer for rectification.

3.02 Preparation:
  A. Identify, locate, and protect utilities located in areas to be graded.
  B. Protect traffic, pedestrians, and the public, including pavements, sidewalks, utilities, and real and personal property by barricading open cuts with positive restraints and marking as hazardous the work area with signs, high-visibility plastic ribbon, and blinking barricade lights for nighttime use.
  C. Provide adequate drainage to the work area as provided in Sections 01500 and 02600.
  D. Improve poor soil in areas shown on the Drawings or as directed by the Geotechnical Consultant using procedures contained in Section 02340.
3.03 Application:
  A. Excavate using these procedures:
    1. Conduct excavations to provide the required grades and elevations shown on the Drawings and required by these specifications.
      a. Shape cross sections in accordance with the drawings, including swales, ditches, slopes, and other areas shown.
      b. Grade swale sections beginning with zero cut at the point farthest from the discharge outlet and gradually taper the invert to the discharge point where the cut does not exceed $2\frac{1}{2}$ feet unless otherwise directed or shown on the Drawings.
    2. Excavate, transport, deposit, spread, and compact suitable general excavation material in embankment areas. Stockpile or remove from the construction sites excess material as designated by the Engineer.
    3. Handle unusable general excavation material as waste in accordance with Section 01410.
    4. Muck areas shown on plans. Mucked material will be excavated to limits established by the Geotechnical Consultant and handled as waste in accordance with Section 01410. For roadbeds, the roadway shall be removed for the entire width, ditch line to ditch line (or 2 feet in back of curb to back of curb), for a depth specified by the Geotechnical Consultant. Provide suitable backfill material and compact in accordance with these specifications to the desired grade.
    5. Construct side slopes no greater than a 1 : 1 gradient unless directed otherwise by the Geotechnical Consultant or in the Construction Documents. Provide and maintain shoring until excavation is backfilled in accordance with Section 02250 if sloping of embankment is not possible.
  B. Construct embankments using these procedures:
    1. Surface receiving fill shall be satisfactory to the Geotechnical Consultant and shall be free from frost, ice, mud, and extraneous material such as trash and debris.
    2. Install permanent bracing if required.
    3. Remove shoring as embankment progresses in accordance with the Contractor-procured field engineer's recommendations.

    4. Scarify existing surface and compact to minimum acceptable field density.

    5. Place loose fill in lifts not exceeding 8 inches thickness and aerate or moisten, if necessary, to attain optimum moisture conditions.

    6. Compact uniformly each layer to the required density and verify prior to proceeding with the next activity.

    7. Exercise care when placing and compacting fill next to structures.

C.   Compact soil using these procedures:

    1. Obtain and evaluate field densities in accordance with provisions contained within AASHTO T 191-93.[*]

    2. Maintain soil during compaction at optimum moisture content. Regulate by aeration and scarification or by adding water and mixing.

    3. Densify in accordance with the following requirements:

      a. Pavement subgrade areas, road shoulders, storm drainage detention pond sites and areas under sidewalks and structures— 95 percent Modified Proctor (AASHTO T 180-97) compaction to a depth of 8 inches of subgrade and for each layer of placed fill.

      b. All other areas—90 percent Modified Proctor (AASHTO T 180-97) compaction to a depth of 8 inches of subgrade and for each layer of placed fill.

D.   Grade using these procedures:

    1. Grade areas external to building lines to conform with typical cross sections using field control markers established by Owner-procured surveyor and elevations shown on the Drawings. Streets shall be fine graded to grade stakes set on the centerline and edge of right-of-way at intervals of not more than 100 feet on straight grades and not more than 50 feet on vertical curves. Create uniform surfaces as work progresses, ensuring that no detrimental dips or bumps exist that might affect positive drainage resulting in ponding or pavement imperfections.

    2. Grade areas adjacent to building lines to provide positive surface drainage and prevent ponding. Shape surface areas to provide a uniform transition throughout.

3.04 Field Quality Control:

A.   Coordinate and allow testing in accordance with Section 01331 for each compacted lift of fill material in embankments.

B.   Undertake at least three representative field density tests using AASHTO T 191-93 or other methods approved by the Engineer for each 2500 ft^2 of area to be paved or constructed upon.

C.   Recompact and regrade tested areas that indicate unsatisfactory densities in accordance with Section 01331.

D.   Check completed surface to verify acceptable density, shape, and grade.

[*] *Note:* Throughout example project manual, see References at end of chapter for complete reference.

3.05 Protection:
    A.   Protect all graded surfaces from erosion and maintain condition of shaped surfaces free from rutting, potholes, and depressions.
    B.   Repair immediately any utilities or other improvements damaged during grading operations at no cost to the Owner.
    C.   Report to the Engineer utilities found during grading operations that appear to require relocation. Engineer will investigate and coordinate in writing any required activities with the Contractor and utility owner.
    D.   Protect graded areas from pumped water discharges and stormwater action. Maintain moisture content of soil at optimum.
    E.   Protect air from excessive dust by controlling dust production in accordance with provisions contained in Section 01500.

Information contained in this section is subject to change and is intended for educational purposes only. CSI, CSC, BSD, and ARCOM did not prepare nor do they specifically endorse this material. The current format used in the above section is from CSI/CSC's *MasterFormat*™ (1995).

End of Section 02310

SECTION 02315

EXCAVATING AND FILL

Part 1—General

1.01 Summary:
    A.   Section includes procedures for trench and structural excavations, including requirements for pipe bedding and backfilling to accomplish the Work.
    B.   Related Sections include but are not necessarily limited to:
        1. General Conditions.
        2. Supplementary Conditions.
        3. Sections in Division 1 of this Project Manual.
        4. Section 00320 for Geotechnical Data.
        5. Section 02445 for Boring or Jacking Conduits where excavation is prohibited.

1.02 Submittals:
    A.   Prior to requesting approval from the Engineer submit satisfactory results of field density tests conducted in accordance with Section 01331 and reviewed by the Geotechnical Consultant in accordance with Section 00320 certifying densities attained.
    B.   Provide shoring submittal requirements for all nonsloped wall excavations greater than 5 feet deep and excavations under 5 feet where conditions warrant and for jacking and boring installations as specified in Section 02250.

1.03 Quality Assurance:

   A.   Detail sufficient numbers of qualified workmen who are experienced in
        executing the activities specified in this Section.
   B.   Detail and maintain equipment capable of undertaking the activities
        specified in this Section to produce sufficient progress on the Work and
        operate equipment with due caution.
   C.   Check final grades and alignment. Use grades obtained from the Draw-
        ings as finished grades unless directed otherwise.
   D.   Limit extent of earthwork to authorized areas designated on the Draw-
        ings. Fill no wetlands unless authorized by permit. Allow no excavation
        under existing structure foundations, unless authorized in writing by the
        Engineer.
   E.   Instruct workmen in procedures specified in this Section.
1.04 Delivery, Storage, and Handling:
   A.   Adhere to germane portions of Sections 01650 and 01660.
   B.   Remove excess soil from the area being graded. Do not deposit or store
        soil closer than 5 feet from any drainage ditch, pond, or catchment.
   C.   Utilize temporary storage areas provided in Section 01500.
1.05 Sequencing and Scheduling:
   A.   Coordinate the requirements of this Section with Sections 02230 and
        02310.
   B.   Excavations for trenching shall not extend more than 300 feet ahead of
        pipe laying or shall trenching be left open overnight unless permission
        is received from the Engineer and provisions are implemented in accor-
        dance with Section 01500.
   C.   Coordinate excavations with the requirements of Section 02250.
1.06 Maintenance:
   A.   Maintain completed areas by scarifying, reshaping, recompacting to cor-
        rect deficiencies attributed to construction-related activities or weather.
   B.   Maintain dust-free environment in accordance with Section 01500. Uti-
        lize power brooms and water flushing as necessary to initially clean pave-
        ments where excavations require pavement cutting and maintain a rela-
        tively dust free environment until resurfaced.
   C.   Maintain drainage and erosion control measures in accordance with
        Section 02370.
   D.   Correct immediately upon notification by the Engineer any settlement
        that occurs in trenches during the warranty period. Settlement occur-
        ring in asphalt pavement areas shall be repaired by resurfacing the entire
        width of the street with asphaltic concrete at the expense of the Contrac-
        tor.

Part 2—Products

2.01 Materials:
   A.   Use only new materials conforming to the requirements of these specifi-
        cations and approved by the Engineer. All materials proposed to be used
        may be tested and inspected at any time during their preparation and use.
        If, after trial, it is found that the source of supply has not been approved
        or does not furnish a uniform product or if the product becomes unac-

ceptable at any time, the Contractor shall furnish materials from another source. Approved materials that become unfit for use shall not be used in the Work and shall be removed from the job site. Materials indicated on the Drawings that are required in the Work but are not covered in detail in the specifications shall be of the best quality available and shall conform to the current specification of the American Society for Testing and Materials (ASTM) or the AASHTO, as approved by the Engineer.

   B. Backfill: Conform to the requirements of embankment materials under Section 02310.

   C. Pipe: Conform to the applicable requirements set forth in Sections 02510, 02530, and 02600.

   D. Stone bedding for PVC potable water or sanitary sewer pipe and special pipe bedding usage: Standard 789 stone.

Part 3—Execution

3.01 Examination:

   A. Inspect areas to be excavated to determine whether conditions are suitable for undertaking the portions of work specified in this Section. Do not commence work under this Section until conditions are satisfactory.

   B. Request underground utility service to locate all utilities within work area.

   C. Verify line and grade stakes and report obvious discrepancies immediately to the Engineer for rectification.

3.02 Preparation:

   A. Identify, locate, and protect utilities located in areas to be graded.

   B. Include adequate drainage to the work area as provided in Sections 01500 and 02240.

   C. Protect traffic, pedestrians, and the public including pavements, sidewalks, utilities, and real and personal property by barricading open cuts with positive restraints and adequately marking as hazardous the work area with signs, high-visibility plastic ribbon, and blinking barricade lights for nighttime use.

3.03 Application:

   A. Excavate using these procedures:

      1. Conduct excavation of materials as required to provide the grades and elevations shown on the Drawings and as required by these Specifications. Shape all cross sections in accordance with the drawings, including swales, ditches, slopes, and other areas as provided in Section 02310.

      2. Excavate the lower portion of the trench, including the area 2 feet above the pipe to a width that is not greater than the diameter of the bell of the pipe plus 12 inches to prevent undercutting the trench walls. If the trenches are excavated wider than specified or if they collapse, the Contractor shall be required to use special methods of bedding the pipe included in this Section at no additional cost to the Owner.

          Terminate mechanical excavations 4 inches above invert grade with the remaining depth being excavated manually, or after the mechanical excavation has been completed, return suitable fill material to the trench to achieve the proper grade. Mechanically tamp backfill material to ensure proper compaction.

3. Provide excavations for manholes, junction boxes, catch basins, and appurtenances so that at least 12–24 inches outside clearance is obtained on all sides.

4. Transport, deposit, spread, and compact excess suitable general excavation material in embankment areas elsewhere on-site once excavation is backfilled. Stockpile or remove from the construction site excess material as designated by the Engineer.

5. Handle unusable general excavation material as waste in accordance with Section 01410.

6. Muck areas shown on Drawings. Mucked material will be excavated to limits established by the Geotechnical Consultant and handled as waste in accordance with Section 01410. Provide suitable backfill material and compact in accordance with these specifications to the desired grade.

7. Construct side slopes no greater than a 1 : 1 gradient unless directed otherwise by the geotechnical consultant or directed differently in the Contract Documents. Provide and maintain shoring until excavation is backfilled in accordance with Section 02250 if sloping of embankment is not possible.

B. Install piping using these procedures:

1. Comply with applicable requirements set forth in Sections 02510, 02530, and 02600.

2. Ductile-iron pipe shall be installed in accordance with ANSI/AWWA C150/A21.50-96 and ANSI/AWWA C 105/A21.5-99.

3. PVC pressure pipe shall be installed in accordance with ANSI/AWWA C 900-97 and AWWA *Manual M23*.

4. Install PVC potable water and sanitary sewer mains using a "Class B" minimum 3-inch stone bedding depth between the pipe bottom and the trench bottom for pipe diameters up to 27 inches, a 4-inch stone bedding depth for pipe diameters up to 60 inches, and 6-inch stone bedding depth for pipe diameters 66 inches and greater.

Additional stone shall be added to the pipe springline elevation and compacted to at least 95 percent Proctor densities for areas where vehicles are expected to pass and 85 percent Proctor densities elsewhere.

5. Install Portland cement concrete storm drainage culverts (other than arch types) following the requirements for Type 1 standard installations for pipes buried under areas where vehicles are expected to pass as shown on the Drawings, elsewhere use Type 3 standard installations (*Note:* See Figure 7.60 for example.)

C. Install manholes and appurtenances once the soil is excavated where 6 inches of crushed stone bedding can be placed and compacted to allow the proper alignment and elevations to be established. Place a temporary 8-inch pipe into the stone bed and extend it to the ground surface prior to backfilling to allow temporary dewatering once the backfill is installed in accordance with Section 02240. Remove the temporary pipe at job completion.

D. Soil that is unstable in the opinion of the Geotechnical Consultant shall be removed from under pipe and structures as directed and replaced as specified for fill.

E. Install backfill using these procedures: Place suitable backfill that is approved by the Geotechnical Consultant over pipes immediately after the pipes have been laid and inspected by the Engineer. The initial backfill material shall be suitable for the type of bedding being provided as defined on detail sheets and tamped on both sides to prevent pipe displacement. Exercise care to ensure that compaction does not damage or deform the pipe. Backfill the remainder of the trench with general excavation material by placing it in 6-inch lifts and mechanically compacting it to 95 percent of its maximum density using the Modified Proctor (AASHTO T 180-97) method where pavements, sidewalks, structures, or heavy loads are expected. In all other areas place backfill in 12-inch lifts and mechanically compact to 90 percent of its maximum density using the Modified Proctor (AASHTO T 180-97) procedure. Provide in those areas not scheduled to be paved or covered by structures or sidewalks 4 inches of topsoil to sustain plant growth in accordance with Section 02900.

F. Mechanically compact soil using these procedures:
1. Obtain and evaluate field densities in accordance with provisions contained within AASHTO T 191-93.
2. Maintain soil at optimum moisture content during compaction. Regulate by aeration and scarification or by adding water and mixing.

G. Grade surface areas using these procedures:
1. Grade areas external to building lines using field control markers established by Owner-procured surveyor and elevations shown on the Drawings. Create uniform surfaces as work progresses, ensuring no detrimental dips or bumps exist that might affect positive drainage, resulting in ponding, or pavement imperfections.
2. Grade areas adjacent to building lines to provide positive surface drainage and prevent ponding. Shape all surface areas to provide a uniform transition throughout.

H. Pipe jacking and boring:
1. In areas where trenching is prohibited, the pipeline must be installed using procedures contained in Section 02445.

3.04 Field Quality Control:
A. Coordinate and allow field testing in accordance with Section 01331.
B. Undertake at least three representative field density tests using AASHTO T 191-93 or other methods approved by the Engineer for each 2500 ft^2 of areas to be paved or constructed upon.
C. Recompact and regrade tested areas indicating unsatisfactory densities in accordance with Section 01331.
D. Check completed surface to verify acceptable density, shape, and grade.

E. Check line and grade of all installed stormwater mains, potable water mains, and sanitary sewer mains. Clean interior of all conduits and appurtenances. Conduct integrity tests for leaks and correct any found prior to notifying the Engineer of readiness to conduct record tests.

3.05 Protection:

A. Protect graded surfaces from erosion in accordance with Section 02370 and maintain condition of shaped surfaces free from rutting, potholes, and depressions.

B. Repair immediately utilities or other improvements damaged during grading operations at no cost to the Owner.

C. Report to the Engineer utilities found during excavation operations that appear to require relocation. Engineer will investigate and coordinate in writing any required activities with the Contractor and utility owner.

D. Protect areas from pumped water discharges and stormwater action. Maintain moisture content of soil at optimum.

E. Protect air from excessive dust by controlling dust production in accordance with Section 01500.

F. Use extra care when trenching along the edge of an existing pavement. Resurface the entire width of the street if the pavement edge is damaged for the entire length of the damaged area plus 20 feet past each end in accordance with provisions of Section 02741.

Information contained in this section is subject to change and is intended for educational purposes only. CSI, CSC, BSD, and ARCOM did not prepare nor do they specifically endorse this material. The current format used in the above section is from CSI/CSC's *MasterFormat*™ (1995).

End of Section 02315

SECTION 02340

SOIL STABILIZATION

Part 1—General

1.01 Summary:

A. Section includes provisions for stabilizing a prepared subgrade to increase its strength by adding crushed stone, gravel, or slag using granular stabilization techniques in areas delineated on the Drawings or as directed by the Engineer as part of the Work.

B. Related Sections include but are not necessarily limited to:
   1. General Conditions.
   2. Supplementary Conditions.
   3. Sections in Division 1 of this Project Manual.

1.02 Submittals:

A. Comply with provisions of Sections 01332 and 01333.

B. Submit for review and approval to the Engineer within 30 working days of receipt of the Notice to Proceed a list of materials necessary for the work encompassed by this Section along with manufacturer's product data and specifications demonstrating compliance with these Specifications.

1.03 Quality Assurance:

A. Detail sufficient numbers of qualified workmen who are experienced in executing the activities specified in this Section.

B. Detail and maintain equipment capable of undertaking the activities specified in this Section to produce sufficient progress on the Work.

C. Check final graded surface for elevations and alignment. Use grades obtained from the Drawings as finished grades unless directed otherwise. Account for pavement thickness requirements, foundation slab thickness, and other structural features to produce the final graded elevations.

D. Limit extent of work to authorized areas designated on the Drawings. Fill no wetlands unless authorized by permit. Allow no excavation under existing structure foundations, unless authorized in writing by the Engineer.

E. Instruct workmen in procedures specified in this Section and operate equipment with due caution.

F. Ensure all finished graded surfaces are free from foreign material.

1.04 Sequencing and Scheduling:

A. Conduct stabilization procedures after the designated subgrade has been prepared to receive stabilized material and approved by the Engineer.

B. After stabilized material has been incorporated into the subgrade material, complete the preparation of the subgrade in accordance with Section 02310.

Part 2—Products

2.01 Materials:

A. Provide crushed stone composed of tough durable stone that is reasonably free from soft, thin, elongated, or laminated particles and that does not contain organic or deleterious material. Include only aggregates that have an abrasion loss of not more than 45 percent when subject to the Los Angeles Abrasion Test (AASHTO T 96-94) and have a gradation that conforms to:

Sieve Size	Percent Passing by Weight
2 inches	100
$1\frac{1}{2}$ inches	95–100
1 inch	70–100
$\frac{1}{2}$ inch	35–65
No. 4	10–40
No. 8	—

B. Provide gravel that is composed of hard durable particles of clean stone and that does not contain excessive amounts of thin or elongated pieces and organic or other deleterious materials. Follow other requirements specified for crushed stone under paragraph A above.

Part 3—Execution

3.01 Examination:
    A. Inspect prepared subgrade areas where stabilization is required, including soil that will be under roadway pavements, sidewalks, curbing, curb and guttering, base courses, and shoulders to ensure the requirements of this Section can be accomplished. Correct any deficiencies prior to commencing work under this Section.

3.02 Preparation:
    A. Protect all improvements of value from equipment, material, and employee abuse including:
        1. Property corners, survey field markers, and monuments.
        2. Designated trees, shrubs, and grassed areas.
        3. Existing utilities, sedimentation and erosion control systems, property improvements, pavements, sidewalks, driveways, curbing, poles, wires, pipes, and fences unless otherwise noted on the Drawings or specifically designated for removal by the Engineer. When fences interfere with construction operations, request permission from the owner to temporarily remove during the time work is being conducted in the area, and immediately replace upon completion at no additional cost to the Owner.
        4. All areas not within the construction site boundary subject to the discharge of pumped water.
        5. Persons and personal property in or adjoining the work area.

3.03 Application:
    A. Place aggregate upon the approved subgrade and spread to conform to the required shape and elevations. Mix the aggregate subbase material to the depths shown on the construction Drawings or directed by the Geotechnical Consultant.
    B. Adjust and maintain the moisture content of the blended mixture to optimum conditions to achieve proper compaction.
    C. Spread the mixture to a uniform thickness while at the optimum moisture content. Compact and consolidate the aggregates to provide bond with the subbase in accordance with subgrade requirements in Section 02310.
    D. The completed surface shall be compacted to required density and shall be smooth and free from loose material.

3.04 Field Quality Control:
    A. Follow provisions contained in Section 02310 for Field Quality Control.

3.05 Protection:
    A. Protect the completed stabilized surface by rolling and recompacting to preserve its integrity until the base is applied.

endorse this material. The current format used in the above section is from CSI/CSC's *MasterFormat™* (1995).

End of Section 02340

SECTION 02370

EROSION AND SEDIMENTATION CONTROL

Part 1—General

1.01 Summary:
  A. Section includes provisions for temporary erosion controls required throughout the Work to control erosion and water pollution by using temporary construction entrances, controlled discharge outlets, berms, sediment basins, sediment dams, silt fences, brush barriers, baled hay and straw, mulching, and temporary seeding.

  Provisions are also included for riprap and geosynthetics provided for in Section 02070 for permanent slope protection as part of the Work.
  B. Related Sections include but are not necessarily limited to:
      1. General Conditions.
      2. Supplementary Conditions.
      3. Sections in Division 1 of this Project Manual.

1.02 Submittals:
  A. Comply with provisions of Sections 01332 and 01333.
  B. At least 96 hours prior to breaking ground, submit to the Engineer all information required by the NPDES General Permit for stormwater pertaining to this project. This includes temporary drainage plans along with sedimentation and erosion control plans to the Engineer for approval. Include schedule of activities required by NPDES General Permit for stormwater, and keep schedule current.
  C. Submit for review and approval to the Engineer within 30 working days of receipt of the Notice to Proceed:
      1. A list of materials necessary for and the sequencing of activities to accomplish the work encompassed by this section along with manufacturer's product data and specifications demonstrating compliance with Section 02070 and these Specifications.
      2. Installation instructions from applicable manufacturers for products proposed to be included in the installation.
  D. Maintain on-site and keep current all required elements of the NPDES General Permit for stormwater pertaining to this project.

1.03 Quality Assurance:
  A. Detail qualified workmen that are experienced in executing the activities specified in this Section.
  B. Detail and maintain equipment capable of accomplishing the activities specified in this Section while producing sufficient progress on the Work.
  C. Operate all equipment with due caution.
  D. Instruct workmen in procedures specified in this Section.

1.04 Sequencing and Scheduling:
   A. Install temporary construction entrances as part of the job mobilization activities.
   B. Install temporary perimeter dikes where shown on erosion control plans and where needed to prevent erosion.
   C. Install silt fencing prior to clearing and grubbing to prevent sediment runoff.
   D. Construct sediment traps as soon as practical to control sediments.
   E. Install straw bales and temporary sediment traps as work progresses.
   F. Utilize detention ponds during construction as temporary sediment basins as shown on Drawings. Clean and place into operation as storm drainage detention facilities when the Work is nearing completion.

1.05 Maintenance:
   A. Routinely inspect the site and clean out sediment traps at least every seven days or within 24 hours after a rainfall of 0.5 inch or more.
   B. Maintain sediment traps, silt fencing, and dikes in good working order.
   C. Correct washouts and eroded areas by refurbishing and temporarily seeding in accordance with Section 02900.

Part 2—Products

2.01 Materials:
   A. Use only new materials conforming to the requirements of these specifications and approved by the Engineer. All materials proposed to be used may be tested and inspected at any time during their preparation and use. If, after trial, it is found that the source of supply has not been approved or does not furnish a uniform product or if the project become unacceptable at any time, the Contractor shall furnish materials from another source. Approved materials that become unfit for use shall not be used in the work and shall be removed from the job site.

   Materials indicated on the Drawings that are required in the Work but are not covered in detail in the Specifications shall be of the best quality available and shall conform to the current specification of the American Society for Testing and Materials (ASTM) or AASHTO, as approved by the Engineer.
   B. For sedimentation and erosion control use:
      1. Baled straw and hay that complies with the requirements contained in Section 02900.
      2. Silt fences manufactured of fabric specifically made for such purpose that complies with the requirements contained in Section 02070.
      3. Temporary CMP drainage pipes chosen by the Contractor that are adequate to perform under subject field conditions.
      4. Temporary seeding that complies with the requirements contained in Section 02900.
      5. Three-inch maximum-size coarse aggregate that is free from deleterious materials and does not contain more than 10 percent materials finer than the No. 4 sieve.
   C. Slope protection shall be provided by:

1. Using stones that weigh no more than 150 pounds with at least 20 percent weighing more than 60 pounds while no more than 20 percent weigh less than 20 pounds for hand-placed riprap.
2. Using geotextile fabric that is manufactured engineering fabric conforming to the requirements of Section 02070 for reducing soil erosion and that is approved by the Engineer.

Part 3—Execution

3.01 Examination:
    A. Examine all areas where slope protection and erosion control are required.
    B. Request underground utility service to locate all utilities within work areas.

3.02 Preparation:
    A. Protect all improvements of value from equipment, material, and employee abuse to include:
        1. Property corners, survey field markers, and monuments.
        2. Designated trees, shrubs, and grassed areas.
        3. Existing utilities, installed sedimentation and erosion control systems, property improvements, pavements, sidewalks, driveways, curbing, poles, wires, pipes, and fences unless otherwise noted on the drawings or specifically designated for removal by the Engineer. When fences interfere with construction operations, request permission from the Owner to temporarily remove during the time work is being conducted in the area. Immediately replace upon completion at no additional cost to the Owner.
        4. Persons and personal property in or adjoining the work area.
    B. Provide for dust control if measures are warranted in accordance with Section 01500.

3.03 Installation:
    A. Provide temporary construction entrances at all ingress/egress locations to reduce or eliminate transporting soil or mud by vehicles to roadways and public rights-of-way. Construct these entrances by installing a stone mat using three-inch maximum crushed stone. This mat shall be a minimum of 50 feet long by 12 wide and at least 6 inches thick. Drain wash water used to spray the underside of vehicles and tires when parked on the stone mat. Provide spent wash water drainage into a sediment basin or trap.
    B. Install temporary perimeter, diversionary, or interceptor dikes as shown on the Drawings and other places where required in the field. Their purpose is to divert sediment-laden storm runoff to an on-site sedimentation trapping facility or to direct stormwater away from exposed slopes.
    C. Provide temporary discharge controls that limit concentrations of stormwater flows by using level spreaders, gravel outlets, and temporary drop piping as shown on the Drawings.
    D. Provide sediment traps using silt fences, straw bales, or sediment basins as shown on the Drawings. Utilize detention pond facilities initially as

sediment basins by installing a temporary riser pipe as shown on the Drawings. Install a marker to indicate the maximum level of sediment accumulation to indicate when the facility requires cleaning. Remove and replace the riser piping when nearing the completion of the Work and approved by the Engineer. Remove accumulated residual sediment at that time and treat the residue as waste. Place the facility into a storm-water detention mode of operation.

E. Hand place riprap a minimum of 12 inches thick on designate slopes to comply with the Drawings. Begin placement 2 feet below the toe of slope while firmly imbedding each piece.

F. Install geotextile fabric for slope stability in accordance with Section 02070. Unless otherwise directed, install the fabric with the long dimension parallel to the toe of slope and lay where a smooth and wrinkle-free surface is obtained that is not under tension. Overlap subsequent fabric layers by at least 18 inches. Secure fabric using approved fasteners and cover within 30 days using materials approved by the Engineer.

3.04 Protection:

A. Repair damaged sediment control devices.

B. Ensure exposed soil surfaces are not eroding.

C. Inspect and maintain the temporary construction entrances' stone from accumulation of silt, mud, or other debris.

D. Prohibit vehicular traffic from crossing installed geotextile fabrics.

Information contained in this section is subject to change and is intended for educational purposes only. CSI, CSC, BSD, and ARCOM did not prepare nor do they specifically endorse this material. The current format used in the above section is from CSI/CSC's *MasterFormat*™ (1995).

End of Section 02370

SECTION 02510

WATER DISTRIBUTION

Part 1—General

1.01 Summary:

A. Section includes provisions for the potable distribution system shown on the Drawings.

B. Related Sections include but are not necessarily limited to:
   1. General Conditions.
   2. Supplementary Conditions.
   3. Sections in Division 1 of these Specifications.

1.02 References:

A. The following references are applicable in part or in total to the conduct of the work associated with this Section:
   1. Manufacturer's recommended installation instructions approved by the Engineer.
   2. ANSI/AWWA C 104/A21.4-95.

    3. ANSI/AWWA C111/A21.11-95.
    4. ANSI/AWWA C 150/A21.50-96.
    5. ANSI/AWWA C 151/A21.51-96.
    6. ANSI/AWWA C 200-97.
    7. ANSI/AWWA C 203-97.
    8. ANSI/AWWA C 205-95.
    9. ANSI/AWWA C 600-99.
    10. ANSI/AWWA C 900-97.
1.03 Submittals:
    A. Comply with provisions of Sections 01332 and 01333.
    B. Submit for review and approval to the Engineer within 30 working days
        of receipt of the Notice to Proceed:
        1. A list of materials necessary for the work encompassed by this
           Section and Sections 02081, 02083, 02085, and 02090.
        2. Names and addresses of the closest repair facilities and spare parts
           supply for products proposed to be included in the installation.
        3. Installation instructions from applicable manufacturers for prod-
           ucts proposed to be included in the installation.
1.04 Quality Assurance:
    A. Detail qualified workmen that are experienced in executing the activities
        specified in this Section.
    B. Detail and maintain equipment capable of undertaking the activities
        specified in this Section to produce sufficient progress on the Work.
    C. Instruct workmen in procedures specified in this Section.
    D. Ensure water mains and appurtenances are installed below potential frost
        penetration depths.
1.05 Delivery, Storage, and Handling:
    A. Adhere to germane portions of Sections 01650 and 01660.
    B. Store and handle materials and products in accordance with manufac-
        turer's recommended practices.

Part 2—Products

2.01 Materials:
    A. Use only new materials conforming to the requirements of these spec-
        ifications and approved by the Engineer. All materials proposed to be
        used may be tested and inspected at any time during their preparation
        and use. If, after trial, it is found that the source of supply has not been
        approved or does not furnish a uniform product or if the product be-
        comes unacceptable at any time, the Contractor shall furnish materials
        from another source. Approved materials that become unfit for use shall
        not be used in the Work.
            Materials indicated on the drawings that are required in the Work
        but are not covered in detail in the Specifications shall be of the best
        quality available and shall conform to the current specification of the
        American Society for Testing and Materials (ASTM) or the American
        Water Works Association (AWWA), as approved by the Engineer.
    B. Utilize only materials that are National Sanitation Foundation (NSF)

approved. Ductile-iron pressure pipe shall conform to ASTM A 377-95e1. The pipe shall be Class 150, cement lined, and have a bituminous coating. Mechanical or slip joints may be used. Ductile-iron pipe thickness shall be determined from plans in accordance with ANSI/AWWA C 150/A21.50-96. It shall comply also with the following requirements:

1. Use only pipe and fittings conforming to:
   a. Ductile-iron with cement mortar linings complying with ANSI/AWWA C 104/A21.4-95.
   b. Polyvinyl chloride (PVC) complying with ASTM D 1784-99a, SDR 27, and ANSI/AWWA C 900-97. PVC pipe used for water mains and fittings shall conform to ASTM D 2241-99 for standard dimension ratios 160 lb/in.^{2} SDR 26. The pipe shall be extruded from clean virgin approved Class 1254-A PVC compound conforming to ASTM D 1784-99a. Rubber rings shall conform to Section 02090.
   c. Slip joints shall be used on all pipes where thrust blocks can be installed. Ductile-iron pipes requiring restraints shall be of the mechanical joint type with tie rod restraints. Ductile-iron flange connections shall be used where shown on the Drawings.
2. Gate and check valves and boxes shall conform to the requirements of Section 02085. Use valves having a rated working pressure of 125 lb/in.^{2} Use adjustable valve boxes with extensions, if required. Include the label "WATER" cast into the cover.
3. Tapping sleeves shall be of the sleeve coupling type with the outlet being a flange connection using a standard 125 flange configuration.
4. Fire hydrants shall conform to the requirements of Section 02081.
5. Service lines, meters, and meter boxes shall conform to the Drawings and Section 02083.
6. Joint restraints, tie rods, and thrust blocks shall conform to the Drawings.
7. Back-flow prevention devices shall conform to ANSI/AWWA C 510-97 or ANSI/AWWA C 511-97 as indicated on the Drawings.

C. Coordinate, procure, and pay for potable water required to complete the system, including charging, flushing, and testing.

Part 3—Execution

3.01 Examination:
  A. Examine all areas where work under this Section is affected. Provide corrective actions to result in satisfactory working conditions prior to commencing work on the water distribution system.
  B. Ensure new right of way areas have been rough graded so that construction of the system takes place in soil that is at the approximate finished elevations.

3.02 Preparation:
  A. Request underground utility location service to mark all utilities.

B. Protect all improvements of value from equipment, material, and employee abuse to include:
  1. Property corners, survey field markers, and monuments.
  2. Designated trees, shrubs, and grassed areas.
  3. Existing utilities, sedimentation and erosion control systems, property improvements, pavements, sidewalks, driveways, curbing, poles, wires, pipes, and fences unless otherwise noted on the Drawings or specifically designated for removal by the Engineer. When fences interfere with construction operations, request permission from the Owner to temporarily remove during the time work is being conducted in the area. Immediately replace upon completion at no additional cost to the Owner.
  4. All areas not within the construction site boundary subject to falling limbs, trees, and dust.
C. Provide for dust control if measures are warranted in accordance with Section 01500.
D. Ensure sedimentation and erosion control procedures are utilized in accordance with Section 02370.
E. Transport pipe and fittings to trenched areas to protect them from damage. Carry all pipe lengths without dragging. Protect ends to ensure tight joint surfaces. Keep the interior clean and free from contamination.
F. Squarely cut pipe to lengths where installation can be easily assembled without placing pipe and fittings under strain. Remove burrs created during cutting process and prepare the pipe end in accordance with the approved manufacturer's recommended method of installation. Repair any damaged interior linings prior to installation.
G. Store rubber gaskets and pipe lubrication away from excessive heat and contamination such as dust and oil as provided in Section 02090.

3.03 Installation:
A. Pipes and fittings shall be installed in a prepared trench meeting the minimum bedding requirements as provided for in Section 02315.
  1. Backfilling the pipe immediately after bedding while leaving the joints exposed 1 foot on either side. Ensure that the backfill is compacted. The installed line should be pressure tested as soon as ready. Each portion of the line to be tested should be complete with thrust blocks and temporary end blocking, if necessary. After the line is satisfactorily pressure tested, complete backfilling the joints. Backfill thrust blocks and valve fitting after inspection by the Engineer.
  2. Temporarily plug the open pipe end when securing incomplete work to prevent debris and small animals from entering the pipe.
B. Valves and boxes shall be installed in accordance with Section 02085.
C. Fire hydrants shall be installed in accordance with the Drawings and Section 02081.
D. Thrust blocks or restraining joints shall be installed where water mains change directions (such as at ells and tees), change size, or terminate or where thrusts are expected to develop at valves and hydrants. Concrete used for thrust blocking shall conform to Section 03300.

1. Place thrust blocking in according with locations and dimensions shown on the Drawings.
2. Ensure that block facing rests on undisturbed native soil for good bearing.
3. When thrust blocking cannot be placed due to field limitations, locate thrust blocks with tie rods that anchor the joints or use tie rods in conjunction with restrained joints to link the pipe together.

E. Service lines, meters, and boxes shall be installed by:
1. Locating meter boxes to provide water service at each tap location shown on the Drawings.
2. Locating the box 3 feet from the property line within the road right-of-way. It shall be serviced by a $\frac{3}{4}$-inch line for a single service and a 1-inch line for a double service.
3. Installing a shutoff valve on supply side and linesetter in box. Meter will be set later by regulatory agency when dwelling service is requested.

F. Place air release valves at high points along the water mains as indicated on the Drawings.
G. Install blow-offs in accordance with the Drawings.
H. Install backflow prevention devices in accordance with the drawings and AWWA *Manual M14*.

3.04 Field Quality Control
A. Final inspection by regulatory agencies will be scheduled by the Engineer after the work is approved by the Engineer.
B. Test the integrity of all watertight joints by conducting a hydrostatic pressure test.
1. Pressure test all potable water lines in accordance with ANSI/ AWWA C 600-99,[*] Section 5.2. Conduct the test for at least 2 hours using a test pressure of at least 1.25 times the design pressure at the highest point along the test section, where the chosen test pressure does not exceed pipe or thrust restraint design pressures.
a. Calculate the allowable leakage per hour for ductile-iron pipe using

$$L = \frac{SD\sqrt{P}}{133,200}$$

where
L = testing allowance (makeup water) in gallons per hour
S = length of pipe tested in feet
D = nominal diameter of the pipe in inches.
P = average test pressure during the hydrostatic test in pounds per square inch (gauge).

b. Calculate the allowable leakage per hour for PVC pipe using the *Unibell Handbook for PVC Pipe*[*]

$$L = \frac{ND\sqrt{P}}{7400}$$

where

L = allowable leakage in gallons per hour
N = number of joints in length of line tested
D = nominal diameter of pipe in inches
P = average test pressure in pounds per square inch

2. Repair any leaks.

C. Test the delivery capability of the completed system by conducting hydrant flow test in conjunction with the Engineer and regulatory authorities.

D. Test the cleanliness of the installed system by conducting bacteria contamination tests. It is the responsibility of the Contractor to obtain two samples of potable water from the installed system and submit them to an approved sanitation laboratory for determination of contamination. Take the samples 24 hours apart. Obtaining satisfactory test results is the responsibility of the Contractor prior to scheduling the final water system inspection with the Engineer.

E. All work accomplished and materials used shall be subject to inspection by the Engineer and all jurisdictional regulatory authorities. All improper work disclosed by such inspection shall be reconstructed at the Contractor's expense. All rejected materials shall be removed from the site immediately.

F. Provide the necessary separation between sanitary sewers and potable water mains. Lay sewers at least 10 feet horizontally from any existing or proposed water main. Should local conditions prevent a lateral separation of 10 feet, a sewer may be laid closer than 10 feet to the water line if:

1. It is laid in a separate trench.
2. It is laid in the same trench with the water mains located at one side on a bench of undisturbed earth.
3. In either case the elevation of the crown of the sewer is at least 18 inches below the invert of the water main.

    Whenever sewers must cross under water mains, the sewer shall be laid at such an elevation that the top of the sewer is at least 18 inches below the bottom of the water main. When the sewer cannot be buried deeply enough to meet the above requirements, the water main shall be relocated to provide this separation or reconstructed with slip on or mechanical joint ductile-iron pipe for a distance of 10 feet on either side of the sewer. One full pipe length of water main should be centered over the sewer so that both joints will be as far from the sewer as possible.

---

[*]Reprinted from Unibell Handbook for PVC Pipe, 1991, by permission.

When it is impossible to obtain proper horizontal and vertical separation as stipulated above, the water main shall be constructed of slip on or mechanical joint ductile-iron pipe and the sewer constructed of mechanical joint ductile-iron pipe. Pressure test both lines to ensure watertightness.

3.05 Cleaning:

A. Do not charge any mains with water until a minimum of six working days after the last thrust block is installed.

B. Flush system slowly and discharge flushed water into an approved storm drainage system.

C. Disinfect system after satisfactory hydrostatic pressure and flow tests are conducted.

    1. Disinfect all water mains by sterilizing with a solution containing not less than 50 parts per million available chlorine.

    2. Sterilizing solution shall remain in the system for 24 hours. After the disinfection period, the residual chlorine shall not be less than 10 parts per million.

    3. After sterilization, the solution shall be flushed with clean water until the residual chlorine is not greater than 0.2 part per million.

3.06 Protection:

A. Protect all installed potable water system components by placing protective flagging around fire hydrants and keeping heavy equipment away from the area.

Information contained in this section is subject to change and is intended for educational purposes only. CSI, CSC, BSD, and ARCOM did not prepare nor do they specifically endorse this material. The current format used in the above section is from CSI/CSC's *MasterFormat*™ (1995).

End of Section 02510

SECTION 02530

SANITARY SEWERAGE

Part 1—General

1.01 Summary:

A. Section includes provisions for the installation of the sanitary sewerage system, including services, together with foundations, manholes, and appurtenances as shown on the Drawings or specified.

B. Related Sections include but are not necessarily limited to:

    1. General Conditions.

    2. Supplementary Conditions.

    3. Sections in Division 1 of this Project Manual along with Sections 02315 and 03400.

1.02 References:

A. Construction of sanitary sewers shall be in accordance with the American Society of Civil Engineers and the Water Pollution Control Federation *ASCE Manual No. 60/Manual of Practice FD-5.*

1.03 Submittals:
   A.   Comply with provisions of Section 01332.
   B.   Submit for review and approval to the Engineer within 30 working days of receipt of the Notice to Proceed:
      1. A list of materials necessary for the work encompassed by this Section along with manufacturer's product data and specifications demonstrating compliance with these specifications.
      2. Names and addresses of the closest repair facilities and spare parts supply for products proposed to be included in the installation.
1.04 Quality Assurance:
   A.   Detail qualified workmen that are experienced in executing the activities specified in this Section.
   B.   Detail and maintain equipment capable of undertaking the activities specified in this Section to produce sufficient progress on the Work.
   C.   Instruct workmen in procedures specified in this Section.
   D.   Operate equipment with due caution.
1.05 Delivery, Storage, and Handling:
   A.   Adhere to germane portions of Sections 01650 and 01660.
   B.   Store and handle materials and products in accordance with manufacturer's recommended practices.

Part 2—Products

2.01 Materials:
   A.   Use only new materials conforming to the requirements of these Specifications and approved by the Engineer. All materials proposed to be used may be tested and inspected at any time during their preparation and use. If, after trial, it is found that the source of supply has not been approved or does not furnish a uniform product or if the product becomes unacceptable at any time, the Contractor shall furnish materials from another source. Approved materials that become unfit for use shall not be used in the Work.
         Materials indicated on the Drawings that are required in the Work but are not covered in detail in the Specifications shall be of the best quality available and shall conform to the current specification of the American Society for Testing and Materials (ASTM), as approved by the Engineer.
         Install only sanitary sewer pipe and fittings where indicated on the Drawings that conform to the following:
      1. Use polyvinyl chloride (PVC) pressure sewer pipe that conforms to all requirements of ASTM D 1784-99a and ASTM D 2241-99. It shall be SDR 26. All PVC pipe shall be Class 1245-B (PVC 1120) suitable for 160 lb/in.^{2} working pressure and designed to withstand without failure a pressure of 400 lb/in.^{2} for 1000 hours. Joints shall be push-on type conforming to ASTM D 3139-98 as specified in Section 02090 using synthetic rubber ring gaskets and nontoxic water-soluble vegetable-type lubrication.
         Use PVC gravity sewer pipe and fittings made from clean virgin PVC compound conforming to ASTM resin specification ASTM

D 1784-99a. ASTM D 3034-98 shall be adhered to in all respects. The SDR shall be 35 in all diameters. Clean reworked material generated from the manufacturer's own pipe production may be used. Maximum allowable ordinate as measured from the concave side of the pipe shall not exceed $\frac{1}{16}$ inch per foot of length.

All pipe shall be suitable for use as a gravity conduit. Provisions must be made for contraction and expansion at each joint with a rubber ring. The bell shall consist of an integral wall section stiffened with two PVC retaining rings that securely lock the solid cross section rubber ring into position.

Use PVC plastic fittings and accessories that are manufactured and furnished by a pipe supplier and that have bell and spigot configurations identical to that of the pipe.

2. Use ductile-iron pipe that conforms to the latest requirements of the ANSI/AWWA C 150/A21.50-96 Type III mechanical joint. The slip joint, using rubber gaskets, may also be used. Flange piping used as shown on the Drawings must conform to ANSI/AWWA C 151/A21.51-96 and ANSI/AWWA C 110/A21.10-98. This pipe shall be Class 150 with a bituminous seal coat inside and out. Pressure pipe joints must conform to ANSI/AWWA C 111/A21.11-95. Ductile-iron gravity sewer pipe must conform to ASTM A 746-95.

B. Use precast concrete manholes conforming to Section 03400 and a minimum interior diameter of 48 inches except where larger diameters are shown on the Drawings. Provide a maximum chimney height of 12 inches to adjust the rim and cover top elevation.

C. Provide 3000 lb/in.^{2} concrete in accordance with Section 03300.

D. Provide mortar made of Portland cement, Type I, that conforms to ASTM C 150-99 and hydrated lime, Type "S," that conforms to ASTM C 207-91. Mortar sand shall conform to ASTM C 144-99, and the water shall be potable.

E. Use castings and fittings that conform to requirements set forth in Section 05560.

F. Use reinforcing steel that conforms to requirements set forth in Section 03300.

G. Use 789 stone for pipe bedding.

Part 3—Execution

3.01 Examination:

A. Examine all areas where work under this Section is affected. Provide corrective actions to result in satisfactory working conditions prior to commencing work on the sanitary sewerage system.

B. Ensure new right-of-way area has been rough graded so that construction of the system takes place in soil that is at the approximate finished elevations.

3.02 Preparation:

A. Request underground utility location service to mark all existing utilities.

B. Protect all improvements of value from equipment, material, and employee abuse to include:
    1. Property corners, survey field markers, and monuments.
    2. Designated trees, shrubs, and grassed areas.
    3. Existing utilities, sedimentation and erosion control systems, property improvements, pavements, sidewalks, driveways, curbing, poles, wires, pipes, and fences unless otherwise noted on the Drawings or specifically designated for removal by the Engineer. When fences interfere with construction operations, request permission from the Owner to temporarily remove during the time work is being conducted in the area. Immediately replace upon completion at no additional cost to the owner.
    4. All areas not within the construction site boundary subject to falling limbs, trees, and dust.
    5. Persons and personal property in or adjoining the work area.
C. Provide for dust control if measures are warranted in accordance with Section 01500.
D. Ensure sedimentation and erosion control procedures are utilized.
E. Transport pipe and fittings to excavated area to protect them from damage. Carry all pipe lengths without dragging. Protect ends to ensure tight joint surfaces. Keep the interior clean and free from contamination.
F. Squarely cut pipe to lengths where installation can be easily assembled without placing pipe and fittings under strain. Remove burrs created during cutting process and prepare the pipe end in accordance with the approved manufacturer's recommended method of installation. Repair any damaged interior linings prior to installation.
G. Store rubber gaskets and pipe lubrication away from excessive heat and contamination such as dust and oil in accordance with Section 02090.

3.03 Installation:
A. Lay pipes in trenches prepared in accordance with Section 02315. Pipe laying shall be done in a manner that will disturb as little as possible the pipe previously laid. Unless otherwise directed, the pipe shall be laid upgrade with the spigots pointing downgrade without any break in line or grade between manholes in accordance with the following procedures:
    1. Excavate trench in accordance with Section 02315.
    2. Before laying, the pipe shall be wiped clean of all soil and foreign matter, and the joints of all bells and spigots shall be clean and dry. The interior of the pipe shall be carefully freed from all soil and surplus material as the work proceeds.
    3. Lower pipe sections into trench to prevent ends from becoming soiled, and lay in prepared trench bottom to provide a minimum of Class B bedding with the bell end facing the direction of pipe laying.
    4. Bed pipe and backfill.
    5. Temporarily plug open pipe end when securing work to prevent debris from entering the pipe.

Measure sewer lines on the basis of the horizontal distance between manhole centers with no deduction for the space occupied by the manholes.

B.   Install ductile-iron pipe in accordance with ANSI/AWWA C 150/ A21.50-96 and thermoplastic sewer pipe in accordance with ASTM D 2321-89 and ASTM D 2774-94. Also use the approved pipe manufacturer's recommendations except as modified in these Specifications.

C.   Install manholes in accordance with Section 02315.

D.   Install manhole drops in accordance with the Drawings.

E.   Install service lines in accordance with the Drawings, with precautions being taken to provide and secure end caps for air testing capability.

F.   Provide piers in accordance with the Drawings on undisturbed soil and in accordance with Section 03300.

G.   Provide thrust blocks in accordance with Section 03300.

H.   Install forcemain gate valves in accordance with Section 02085.

3.04 Field Quality Control:

A.   Lamp installed gravity mains to ensure straight alignment from manhole to manhole.

B.   Verify grade of the installed lines to ensure compliance with the Drawings.

C.   Install gravity pipe so that the maximum diametric deflection of the pipe will not exceed 5 percent with verification by pulling a mandrel through completed segments as part of the final inspection.

D.   Conduct watertight integrity tests on gravity lines as determined by the Engineer using:
1. Infiltration.
2. Exfiltration.
3. Air test.

E.   Conduct watertight integrity tests by:
1. *Pressure Main Testing:* Make the pressure test after the forcemain has been laid and backfilled. Locate and repair any detected leaks at no expense to the Owner.

The purpose of the leakage test is to establish that the section of line to be tested—including all joints, fittings, and other appurtenances—will not leak or that the leakage is within tolerance. Leakage, if any, usually will be found involving joints at saddles, valves, transition joints, and adapters and not usually in the pipe joints.

Provide air release valves when setting up a section of line for testing. Air trapped in the line during testing will affect the results and can cause damage to the pipeline.

Test forcemains in accordance with ANSI/AWWA C 600-99,[*] Section 5.2. Conduct the test for at least two hours with the test pressure at least 1.25 times the design pressure at the highest point

*Excerpted from *AWWA Standard for Installation of Ductile-Iron Water Mains and Their Appurtenances* (ANSI/AWWA C600-99), by permission. Copyright © 1999, American Water Works Association.

along the test section, where the chosen test pressure does not exceed pipe or thrust restraint design pressures.

a. Calculate the allowable leakage per hour for ductile-iron pipe using

$$L = \frac{SD\sqrt{P}}{133{,}200}$$

where

L = testing allowance (makeup water) in gallons per hour
S = length of pipe tested in feet
D = nominal diameter of pipe in inches
P = average test pressure during hydrostatic test in pounds per square inch (gauge)

b. Calculate the allowable leakage per hour for PVC pipes using the *Unibell Handbook of PVC Pipe*[*] with:

$$L = \frac{ND\sqrt{P}}{7400}$$

where

L = allowable leakage in gallons per hour
N = number of joints in length of line tested
D = nominal diameter of pipe in inches
P = average test pressure in pounds per square inch

Check the forcemain for blockage and obstructions by flushing with water at a rate approved by the Engineer with the spent flush water being diverted into an approved drainage facility.

Repair any leaks and retest.

2. *Gravity Main Testing:* Test every section of pipe between manholes after it has been laid and backfilled using infiltration, exfiltration, or a low-pressure air test. When no groundwater exists at the time of the test, the pipe shall be subjected to an exfiltration test or an air test. The engineer shall designate the type of test to be conducted and the manner in which it shall be performed. When infiltration testing is permitted, the engineer shall give explicit instructions to be followed in carrying out the test. The maximum allowable amount of infiltration measured by testing shall be at a rate of not greater than 200 gal/in. of pipe diameter per mile per 24 hours. Should any test on any section of pipe line disclose an infiltration rate greater than that permitted, the contractor shall, at no expense to the Owner, locate and repair defective joints of pipe and retest until the infiltration is within specified allowance.

When an exfiltration test is required, the Engineer shall give explicit instructions to be followed in carrying out the test. The maximum allowable amount of exfiltration measured by the test

---

[*]Reprinted from Unibell Plastic Pipe Association (1991), by permission.

shall be at a rate not greater than 200 gal/in. of pipe diameter per mile per 24 hours. The average interval pressure of the system under test shall not be greater than five pounds per square inch (11.6 feet of head), and the maximum internal pressure in any part of the system under test shall not be greater than 10.8 lb/in.^{2} (25 feet of head).

Should any test on any section of pipe disclose an exfiltration rate greater than that permitted, the Contractor shall, at no expense to the Owner, locate and repair defective joints or pipes and retest until the exfiltration is within specified allowances.

When an air test is required, in lieu of the water exfiltration test, the Engineer shall give explicit instructions to be followed in carrying out the test using Section 306-1.4.4 of *"GREENBOOK"— Standard Specifications for Public Works Construction—2000 Edition* as a guide:[*]

Introduce air into the installed sewer main until 20 kPa (3.0 psi) gage pressure is attained, then reduce the inflow of air to maintain an internal pressure between 17 kPa and 24 kPa (2.5 and 3.5 psi) gage for at least 2 minutes. Do not allow the internal pressure to exceed 34 kPa (5 psi) gage at any time.

After the temperature of the air introduced into the pipeline has stabilized and the end plugs exhibit no leaks, the internal pressure can be allowed to drop to 17 kPa (2.5 psi). Begin timing the number of seconds required for the pressure to drop from 17 kPa (2.5 psi) to 10 kPa (1.5 psi) gage. If the observed lapsed time exceeds the allowable shown below, then the pipeline tested satisfactorily.

Main Line	Length m (feet)	100 mm (4 in) House Connections			150 mm (6 in) House Connections		
		30m (100 feet)	60m (200 feet)	90m (300 feet)	30m (100 feet)	60m (200 feet)	90m (300 feet)
Diameter,							
200 mm (8 in)	15 (50)	50	70	90	70	110	110
	30 (100)	90	100	100	110	120	110
	60 (200)	120	110	110	130	120	120
	90 (300)	130	120	110	130	120	120
	120 (400)	130	120	120	130	130	120
Diameter,							
250 mm (10 in)	15 (50)	70	90	100	90	120	120
	30 (100)	130	120	110	140	130	130
	60 (200)	150	140	130	150	140	140
	90 (300)	160	150	140	160	150	140
	120 (400)	160	150	150	160	150	150
Diameter,							
300 mm (12 in)	15 (50)	100	110	110	120	140	130
	30 (100)	170	150	140	170	150	140
	60 (200)	180	170	160	180	170	160
	90 (300)	190	180	170	190	180	170
	120 (400)	190	180	180	190	180	180

Should any test on any section of pipe disclose an air loss rate greater than that permitted, the Contractor shall, at his own ex-

[*] Reprinted from BNi ® Building News. (1999), by permission.

pense, locate and repair the defective joints of pipe and retest until the air loss is within the specified allowances as shown above.

F.   Before the sewer system is accepted, it shall be tested and cleaned to the satisfaction of the Engineer. If any obstruction is found, the Contractor will be required to clean the sewer by means of rods or other equipment. The pipe line shall be straight with a uniform grade between manholes. The bottoms of all manholes will be smooth to ensure free flow of sewage and shall be sloped to ensure drainage of any splashed material back into the line.

F.   Final inspection by regulatory agencies will be scheduled by the Engineer after the work is approved by the Engineer.

3.05 Cleaning:

A.   Inspect pipe interiors and structure inverts. Remove debris, trash, and unwanted material. Flush, scour, and drain the completed system to remove sediment.

3.06 Protection:

A.   Provide all precautions necessary to protect and maintain the buildings, fences, water lines, storm drains, power and telephone lines, and cable, and other structures. These precautions include the furnishing of suitable bridges and footways across intersected streets, replacing of pavement, sidewalks, curb and gutters, and cleaning away of all debris and surplus materials and any other work necessary to put in complete working order the specified sewers and appurtenances. Any damage to any utility lines, including overhead or underground, services, or structures of any nature (including any trees or shrubs so designated by the Engineer to remain protected), shall be repaired or replaced by the Contractor at no expense to the Owner.

B.   Prohibit unnecessary construction equipment from crossing installed pipe and structures to prevent displacement, cracking, or injury.

Information contained in this section is subject to change and is intended for educational purposes only. CSI, CSC, BSD, and ARCOM did not prepare nor do they specifically endorse this material. The current format used in the above section is from CSI/CSC's *MasterFormat*™ (1995).

End of Section 02530

SECTION 02600

DRAINAGE AND CONTAINMENT

Part 1—General

1.01 Summary:

A.   Section includes provisions for the installation of the storm drainage collection and detention systems as shown on the Drawings.

B.   Related Sections include but are not necessarily limited to:
   1. General Conditions.
   2. Supplementary Conditions.

3. Sections in Division 1 of this Project Manual
4. Section 02310 for ditch and swale construction.
5. Section 02310 for stormwater detention facility grading
6. State Highway Department's Standard Specifications for Highway Construction.

1.02 Submittals:
A. Comply with provisions of Sections 01332 and 01333.
B. Submit for review and approval to the Engineer within 30 working days of receipt of the Notice to Proceed the following:
   1. A list of materials necessary for the work encompassed by this Section along with manufacturer's product data and specifications demonstrating compliance with these Specifications.
   2. Installation instructions from applicable manufacturers for products proposed to be included in the installation.

1.03 Quality Assurance:
A. Detail qualified workmen who are experienced in executing the activities specified in this Section.
B. Detail and maintain equipment capable of accomplishing the activities specified in this Section while producing sufficient progress on the Work.
C. Instruct workmen in procedures specified in this Section.
D. Operate all equipment with due caution.

1.04 Delivery, Storage, and Handling:
A. Adhere to germane portions of Sections 01650 and 01660.
B. Store and handle materials and products in accordance with manufacturer's recommended practices.

1.05 Sequencing and Scheduling:
A. Coordinate activities under this Section with Sections 02230, 02310, 02315, and 02370.
B. Grade storm drainage detention facilities and temporarily use for sediment control by installing a temporary riser pipe as shown on the Drawings. Place detention facility into operation nearing completion of the Work as directed by the Engineer.
C. Coordinate the inspection of pipe joints by the Engineer and regulatory inspectors prior to covering with backfill.

Part 2—Products

2.01 Materials:
A. Use only new materials conforming to the requirements of these specifications and approved by the Engineer. All materials proposed to be used may be tested and inspected at any time during their preparation and use. If, after trial, it is found that the source of supply has not been approved, or does not furnish a uniform product or if the product becomes unacceptable at any time, the Contractor shall furnish materials from another source. Approved materials that become unfit for use shall not be used in the Work.

Materials indicated on the Drawings that are required in the Work but are not covered in detail in the Specifications shall be of the best quality available and shall conform to the current specification of the American Society for Testing and Materials (ASTM) or AASHTO, as approved by the Engineer.

B. Provide storm drainage piping as shown on the Drawings that conforms to the following:

 1. Reinforced concrete pipe shall conform to AASHTO M 170-98, Class II, Wall B, and all aspects of the State Highway Department's Standard Specifications for Highway Construction.

 2. Corrugated metal steel pipe shall be made of zinc-coated iron or steel sheets and conform to ASTM A929/A929M-97. All rivets shall conform to ASTM A 31-95, Grade A, coated with zinc. Bolts shall be zinc-coated, manufacturer's Grade C. All corrugations shall be 1 inch. All CMP culverts and connection bands shall be fully bituminous coated to a minimum thickness of 0.05 inch in accordance with AASHTO M 190M-95.

 3. Corrugated aluminum pipe shall conform to AASHTO M 196M-92 with thickness requirements shown on the Drawings and shall be bituminous coated.

C. Provide 3000 lb/in.^{2} concrete in accordance with Section 03300.

D. Provide mortar made of Portland cement, Type I, that conforms to ASTM C 150-99, and hydrated lime, Type "S," that conforms to ASTM C 207-91. Mortar sand shall conform to ASTM C 144-99, and the mix water shall be potable without contamination.

   Mortar, if not otherwise specified, shall be mixed in the proportions of one part Portland cement, one part dry hydrated lime or lime paste, and six parts sand by volume.

E. Castings and fittings shall conform to requirements set forth in Section 05560.

F. Reinforcing steel shall conform to requirements set forth in Section 03300.

G. Riprap shall conform to Section 02370.

H. French drains shall include:

 1. 789 crushed stone for drain filler.

 2. PVC underdrain pipe meeting AASHTO M 278-94 for Class PS 50 or bituminous-coated perforated CMP pipe meeting AASHTO M 36M-98, Type III.

 3. Geotechnical fabric suitable for underdrain applications approved by the Engineer must conform to applicable parts of Section 02070.

I. Stone bedding for pipe shall conform to 789 aggregate.

J. Precast concrete junction boxes, inlets, and manholes shall conform to provisions contained in Section 03400.

Part 3—Execution

3.01 Examination:

A. Examine all areas where work under this Section is affected. Provide corrective actions to result in satisfactory working conditions prior to commencing work.

    B. Ensure new right-of-way area has been rough graded so that construction of the system takes place in soil that is at the approximate finished elevations.

3.02 Preparation:

    A. Request underground utility service to locate utilities within the work area.

    B. Protect all improvements of value from equipment, material, and employee abuse to include:

        1. Property corners, survey field markers, and monuments.

        2. Designated trees, shrubs, and grassed areas.

        3. Existing utilities, sedimentation and erosion control systems, property improvement, pavements, sidewalks, driveways, curbing, poles, wires, pipes, and fences unless otherwise noted on the Drawings or specifically designated for removal by the Engineer. When fences interfere with construction operations, request permission from the owner to temporarily remove during the time work is being conducted in the area. Immediately replace upon completion at no additional cost to the Owner.

        4. Persons and personal property in or adjoining the work area.

    C. Provide for dust control if measures are warranted in accordance with Section 01500.

    D. Construct stormwater detention pond facilities in accordance with Section 02310.

    E. Ensure sedimentation and erosion control procedures are utilized.

    F. Transport pipe and fittings to excavated area to protect them from damage. Carry all pipe lengths without dragging. Protect ends to ensure tight joint surfaces. Keep the interior clean and free from contamination.

    G. Squarely cut pipe to lengths where installation can be easily assembled without placing pipe and fittings under strain. Remove burrs created during cutting process and prepare the pipe end in accordance with the approved manufacturer's recommended method of installation. Repair any damaged interior linings prior to installation.

    H. Store rubber gaskets and pipe lubrication away from excessive heat and contamination such as dust and oil in accordance with Section 02090.

3.03 Installation:

    A. Install pipes and fittings by:

        1. Excavating trenches and installing minimum bedding requirements in accordance with Section 02315.

        2. Lowering pipe sections into trenches to prevent the ends from becoming soiled and laying them in prepared trench bottoms with the bell end facing the direction of pipe laying.

        3. Bedding the pipe and backfilling temporarily while leaving joints exposed until approved by the regulatory inspectors and the Engineer.

        4. Temporarily plugging the open pipe end when securing incomplete work from debris from entering the pipe.

    B. Construct storm drainage structures in accordance with details shown on the drawings. Slope and taper inverts to provide smooth hydraulic flow characteristics.

C.   Install underdrains (French drains) $2\frac{1}{2}$ feet in back of the curb and properly connect to a permanent-type drainage outlet, such as a catch basin, junction box, or stormwater manhole. Make the invert of the drain at least 2 feet below the bottom of the curb as shown on the street profile. Provide suitable outlets for the pipe underdrains at designed elevations and complete the installation prior to placing the base course. Install pipe underdrains on both sides of the street where mucking operations have taken place and where the road cut is of such depth that the water table will temporarily be within 24 inches of the centerline subgrade during extended periods of rainfall.

Lay pipe underdrains on grade and in accordance with the State Highway Department's *Standard Specification* for highway construction. Do not cover until they have been inspected and approved by regulatory authorities and the Engineer.

D.   Place riprap as specified in Section 02370 to form a protective covering on slopes of embankments, at ends of bridges, around culvert inlets and outlets, on slopes and bottoms of ditches, and around foundations and other locations when necessary for the purpose of preventing scour, erosion, or slipping of embankments.

Install riprap at the open ends of all pipelines larger than 18 inches in diameter or span, at the open ends of all pipeline at cross ditches, at the outlet ends of pipe lines, at outlet ditches, lakes, and so on, and other locations necessary due to soil conditions.

Place riprap in open channels at points of intersection, at angle points where the angle of turn is 20 degrees or greater, and at other locations shown on the Drawings.

3.04 Field Quality Control:

A.   Final inspection by regulatory agencies will be scheduled by the Engineer after the system has been approved by the Engineer.

3.05 Cleaning:

A.   Inspect pipe interiors and structure inverts. Remove debris, trash, and unwanted material.

B.   Flush and scour the completed system to remove sediment.

3.06 Protection:

A.   Prohibit unnecessary construction equipment from crossing installed pipe and structures to prevent displacement, cracking, and injury.

Information contained in this section is subject to change and is intended for educational purposes only. CSI, CSC, BSD, and ARCOM did not prepare nor do they specifically endorse this material. The current format used in the above section is from CSI/CSC's *MasterFormat*™ (1995).

End of Section 02600

SECTION 02711

BASE COURSES

Part 1—General

1.01 Summary:
    A. Section includes provisions for macadam base courses composed of crushed stone filled and bound with screenings placed on a prepared subgrade to the specified shape, dimensions, and elevations as part of the Work.

       Included are provisions for an alternate sand asphalt base course of hot mix-asphalt (HMA) composed of fine aggregates and asphalt cement prepared in an approved plant and constructed on a prepared subgrade.

    B. Related Sections include but are not necessarily limited to:
       1. General Conditions.
       2. Supplementary Conditions.
       3. Sections in Division 1 of this Project Manual.

1.02 Submittals:
    A. Follow requirements set forth in Sections 01332 and 01333.
    B. Submit product data for materials and mix design to the engineer for approval 30 days prior to commencing paving operations.
    C. Include manufacturer's recommended asphalt cement mixing temperature for a viscosity of 170 centistokes ($\pm$20) and minimum compaction temperature for a viscosity of 280 centistokes ($\pm$30).

1.03 Quality Assurance:
    A. Detail sufficient numbers of qualified workmen who are experienced in executing the activities specified in this Section.
    B. Detail and maintain sufficient equipment capable of undertaking the activities specified in this Section to produce sufficient progress on the Work.
    C. Check final graded surface for elevations and alignment. Use grades obtained from construction Drawings as finished grades unless directed otherwise. Account for pavement thickness requirements, foundation slab thicknesses, and other structural features to produce the final graded elevations.
    D. Limit extent of work to authorized areas designated on the Drawings. Fill no wetlands unless authorized by permit. Allow no excavation under existing structure foundations unless authorized in writing by the Engineer.
    E. Instruct workmen in procedures specified in this Section and operate equipment with due caution.
    F. Ensure finished graded surfaces are free from foreign material.

1.04 Delivery, Storage, and Handling:
    A. Adhere to germane portions of Sections 01650 and 01660.
    B. Deliver sand asphalt base HMA at temperatures suitable for proper placement and compaction.
    C. Load delivery truck to prevent mix segregation during transport.

1.05 Sequencing:
   A. Coordinate activities of this Section with Sections 02310, 02315, and
       02741.

Part 2—Products

2.01 Materials:
   A. Use only new materials conforming to the requirements of these Specifi-
       cations and approved by the Engineer. All materials proposed to be used
       may be tested and inspected at any time during their preparation and use.
       If, after trial, it is found that the source of supply has not been approved
       or does not furnish a uniform product or if the product becomes unac-
       ceptable at any time, the Contractor shall furnish materials from another
       source. Approved materials that become unfit for use shall not be used in
       the Work. Materials indicated on the Drawings that are required in the
       Work but are not covered in detail in the Specifications shall be of the
       best quality available and shall conform to the current specification of
       the American Society for Testing and Materials (ASTM) or AASHTO,
       as approved by the Engineer.
   B. Macadam base materials:
       1. Use coarse aggregates retained on the No. 4 sieve that have hard
           durable particles of crushed stone and are reasonably free from
           thin or elongated pieces and organic or other deleterious materi-
           als. Include only aggregates that have an abrasion loss of not more
           than 45 percent when subjected to the Los Angeles Abrasion Test
           (AASHTO T 96-94).
       2. Provide choke stone consisting of fine aggregates passing the No. 4
           sieve produced by crushing operations. Sand shall not be permitted.
           Use fine aggregates that do not have a liquid limit greater than 25
           and a plasticity index greater than 6 when tested in accordance with
           AASHTO T 89-96 and AASHTO T 90-96 procedures.
       3. Use water that is potable and free from contaminants.
   C. Hot laid sand asphalt base materials shall include:
       1. Asphalt cement that is either AC-20 or AC-30, viscosity graded, and
           conforms to AASHTO M 226-80, Table 2.
       2. Rapid cure cutback asphalts that conform to AASHTO M 81-92 and
           medium cure cutback asphalts that conform to AASHTO M 82-75.
       3. Emulsified asphalts that conform to AASHTO M 140-88 or
           AASHTO M 208-96.
       4. Aggregates that are local sand or local sand containing crushed shell,
           blends of sand and stone, slag or limestone screenings, or other ma-
           terial approved by the Geotechnical Consultant. The approved ma-
           terial shall be nonplastic and shall provide sharp, hard, and durable
           grains that are free from deleterious substances while being reason-
           ably graded from coarse to fine.
2.02 Mixes:
   A. Hot laid sand asphalt base mix designs shall conform to the procedures

contained in AASHTO T 245-97 and shall meet the following requirements:

AGGREGATE GRADATION

Sieve Size	Percent Passing
No. 4	100
No. 8	75–100
No. 16	60–90
No. 30	45–75
No. 50	15–45
No. 100	5–15
No. 200	2–6

Asphalt content: 4.5 to 10 percent

MARSHALL DESIGN CRITERIA USING 50 BLOWS

Minimum stability	750
Flow, 0.01 inch	8–18
Percent of voids in total mix:	3–5

Part 3—Execution

3.01 Examination:

    A. Examine surface areas to ensure that no noticeable moisture exists and that they are properly prepared for HMA. Commence paving operations when deficiencies are corrected.

    B. Assess weather conditions for restrictions to paving operations. Sand asphalt base HMA courses can be placed anytime during the year except when the air temperature is below 40°F in the shade and falling or below 35° F in the shade and rising. Sand asphalt base HMA material shall not be placed on wet or frozen surfaces.

    C. Inspect delivery of HMA to ensure it is covered by tarpaulin and is at an acceptable temperature for proper placement and compaction.

3.02 Preparation:

    A. Macadam base courses shall be constructed on a subgrade approved by the Engineer.

    B. Sand asphalt base HMA courses shall be placed over a prepared subgrade that has been approved by the Engineer and primed with MC-30, MC-70, RC-30, or RC-70 cutback asphalt at a rate of 0.25–0.3 gal/yd^2. Once the prime has set, a tack coat of RS-1, RS-2, CRS-2, MS-1, MS-2, CRS-1, or CMS-2 emulsified asphalt shall be applied between 50°F and 160°F as directed by the Engineer depending on field conditions at a rate of 0.05–0.15 gal/yd^2. Allow emulsified asphalt to "break" prior to applying the sand asphalt base HMA.

3.03 Installation:

    A. Macadam base courses shall be constructed by:

        1. Spreading the coarse aggregate evenly to conform to the required shape, thickness, and elevation. Limit the maximum layer thickness to 8 inches. For base depths greater than 8 inches place in multiple layers of less than 8 inches each.

2. Adding water and mixing the materials without disturbing the prepared subbase. Respread for compaction.

3. Compacting the base material by rolling from the edges and proceeding toward the center, except on superelevated curves where rolling shall proceed from the lower to the upper side and shall continue until the aggregates are firmly keyed to each other.

4. Placing and spreading choke stone on the compacted base surface in sufficient quantities to fill the voids. Wet and broom the surface prior to rolling, and roll until the base is thoroughly compacted to 100 percent maximum density by AASHTO T 180-97 (Method D) and bonded for the full width and depth of the layer.

B. Sand asphalt base HMA courses shall be constructed by: Placing the sand asphalt base HMA on an approved primed surface using an approved paving machine to produce a compacted layer by rolling while the mixture is sufficiently hot. The minimum thickness after compaction shall be as shown on the Drawings.

3.04 Field Quality Control:

A. Check the finish of the macadam base course so that it does not vary more than $\frac{3}{8}$ inch from a straight edge 10 feet long placed parallel to the centerline, not more than $\frac{1}{2}$ inch perpendicular to the centerline.

B. Test macadam densities every 1000 feet of roadbed.

C. Use HMA plants to produce sand asphalt base HMA that conforms to AASHTO M 156-97. Heat asphalt cement to the manufacturer's recommended mixing temperature for plant mixing. Compact hot mix in the field at temperatures above the minimum recommended by the manufacturer. Provide haul trucks that have smooth metal beds and are covered. Provide a hole in the bed of each delivery truck for checking mix delivery temperatures. Use a paving machine that includes mechanical devices such as equalizing runners, level arms, and screeds. Place sand asphalt base in layers that do not exceed 3 inches in compacted thickness. Compact HMA using maximum roller speeds of 3 mph for static or vibratory rollers and of 5 mph for pneumatic rollers. Utilize three-wheeled steel wheel rollers of 10–12 tons in weight or 8–12 two-axle tandem rollers. Provide pneumatic rollers capable of varying tire pressures from 40 to 90 lb/in.^{2} gauge by adjusting ballast and inflation pressure.

D. Final inspection by regulatory agencies will be scheduled by the Engineer after the work is approved by the Engineer.

3.05 Protection:

A. Maintain completed macadam surface to keep a smooth and true grade and to prevent raveling.

B. Keep traffic off of primed and tack-coated surfaces.

C. Protect new HMA surfaces from traffic until sufficiently cooled, set, and cured to prevent rutting, scuffing, and pavement displacement.

End of Section 02711

SECTION 02741

BITUMINOUS CONCRETE PAVEMENT

Part 1—General

1.01 Summary:
  A. Section provides for installation of asphaltic concrete base, binder, or surface courses of HMA or intermediate course composed of mineral aggregates and asphalt cement, mixed in an approved plant and constructed on an approved subgrade or base course.

    Each pavement course shall be produced and constructed in accordance with the specified gradation (i.e., base, binder, or surface requirements).
  B. Related Sections include but are not necessarily limited to:
    1. General Conditions.
    2. Supplementary Conditions.
    3. Sections in Division 1 of this Project Manual.

1.02 Submittals:
  A. Follow requirements set forth in Sections 01332 and 01333.
  B. Submit product data for materials and mix design to the Engineer for approval 30 days prior to commencing paving operations.
  C. Include manufacturer's recommended asphalt cement mixing temperature for a viscosity of 170 centistokes ($\pm$20) and compaction temperature for a viscosity of 280 centistokes ($\pm$30).

1.03 Quality Assurance:
  A. Detail sufficient numbers of qualified workmen who are experienced in executing the activities specified in this Section.
  B. Detail and maintain sufficient equipment capable of undertaking the activities specified in this Section to produce sufficient progress on the Work.
  C. Instruct workmen in procedures specified in this Section.
  D. Operate equipment with due caution.
  E. Check the final surface for elevation, smoothness, and compaction. Use grades obtained from the Drawings as finished grades unless directed otherwise. Account for pavement thickness requirements, foundation slab thickness requirements, and other structural features to produce the desired elevations.
  F. Limit extent of work to authorized areas designated on the Drawings. Fill no wetlands unless authorized by permit.

1.04 Delivery, Storage, and Handling:
  A. Adhere to germane portions of Sections 01650 and 01660.
  B. Require HMA delivery trucks to be properly loaded to prevent mix segregation during transport.
  C. Use only an approved release agent to prevent the HMA from sticking to the truck body.
  D. Receive HMA at temperatures above the required compacting temperature to allow proper placement and compaction.

1.05 Sequencing and Scheduling:
  A. Coordinate activities of this Section with Sections 02711 and 02750.

Part 2—Products

2.01 Materials:

A. Use only materials conforming to the requirements of these Specifications and approved by the Engineer. All materials proposed to be used may be tested and inspected at any time during their preparation and use. If, after trial, it is found that the source of supply has not been approved or does not furnish a uniform product or if the product becomes unacceptable at any time, the Contractor shall furnish materials from another source. Approved materials that become unfit for use shall not be used in the Work.

Materials indicated on the Drawings that are required in the Work but are not covered in detail in the Specifications shall be of the best quality available and shall conform to the current specification of the American Society for Testing and Materials (ASTM) or AASHTO, as approved by the Engineer.

B. Use asphalt cements that are either AC-20 or AC-30 viscosity graded and conform to AASHTO M 226-80, Table 2.

C. Use rapid cure cutback asphalts that conform to AASHTO M 81-92, and medium cure cutback asphalts that conform to AASHTO M 82-75.

D. Use emulsified asphalts that conform to AASHTO M 140-88 or AASHTO M 208-96.

E. Use aggregates as follows:

1. Mineral filler that conforms to AASHTO M 17-95.

2. Fine aggregate that is uniformly graded from coarse to fine and is hard, sharp, angular, and free from clay and other deleterious substances.

3. Coarse aggregate that is clean, tough, angular, and free from excessive amounts of elongated particles or deleterious materials. Crushed stone and gravel must have an abrasion loss of not more than 45 percent when subjected to the Los Angeles Abrasion Test (AASHTO T 96-94).

2.02 Mixes:

A. Blended aggregate mixtures shall conform to the following specification and be approved by the Engineer:

Sieve Size	Type I	Type II	Binder	Base
1.5 inches	100	100	100	100
1 inch	100	100	100	85–100
$\frac{3}{4}$ inch	100	100	40–100	—
$\frac{1}{2}$ inch	97–100	97–100	72–90	60-80
$\frac{3}{8}$ inch	80–100	80–100	—	—
No. 4	58–75	53–73	42–62	40–55
No. 8	42–60	38–53	30–48	30–45
No. 30	19–40	20–34	—	—
No. 100	8–20	8–16	—	—
No. 200	3–8	3–8	—	—
Percent of Asphalt Cement	4.8–6.8	4.8–6.8	3.5–6.0	3.5–5.5
Minimum Stability	1400	1400	1000	—
Percent Air	3.5–6.0	3.5–6.0	—	—

B. The mix design shall include the percent of asphalt cement determined by the Marshall Method for (50 or) 75 blow criteria using procedures in AASHTO T 245-97 and AASHTO R 12-85.

2.03 Source Quality Control:

A. Provide to the Engineer in accordance with Section 01333 the HMA test results on materials processed and produced hot mix. These results will include stability, flow, air voids, voids in the mineral aggregate, and voids filled with asphalt.

Part 3—Execution

3.01 Examination:

A. Examine surface areas to ensure that no noticeable moisture exists and that they are properly prepared to receive HMA. Commence paving operations when deficiencies are corrected.

B. Assess weather conditions for restrictions to paving operations.

1. Bituminous plant mix surface courses generally shall not be placed between December 15 and March 1 (or other dates specified to conform with local regulations). Temperature and weather restrictions prohibit placement of surface courses when the air temperature is below 45°F in the shade and falling. Bituminous plant mix surface courses shall not be placed on a wet or frozen surface.

2. Bituminous plant mix base and binder courses can be placed anytime during the year except when the air temperature is below 40°F in the shade and falling. HMA base and binder courses shall not be placed on wet or frozen surfaces.

C. Inspect delivery of HMA to ensure it is covered by tarpaulin and is at an acceptable temperature to permit proper placement and compaction.

3.02 Preparation:

A. Place prime coat on previously untreated macadam base surfaces using MC-30 or RC-30 cut back asphalt at a rate of 0.25–0.3 gal/yd^2 and allow to cure. Under some circumstances, MC-70 or RC-70 may be used.

3.03 Application:

A. Apply a tack coat of RS-1, RS-2, CRS-2, MS-1, MS-2, CRS-1, or CMS-2 emulsified asphalt at a temperature between 50°F and 160°F as directed by the Engineer depending on field conditions and applied at a rate of 0.05–0.15 gal/yd^2. Permit emulsified asphalt to "break" prior to applying HMA.

B. Base, binder, and surface courses shall be laid hot using a standard asphalt paving machine to produce compacted layers shown on the Drawings. The compaction procedure shall consist of breakdown, intermediate, and finish rolling. Use a steel wheel roller and pneumatic tire roller to complete the compaction operation prior to reaching the allowable temperature limitations. Obtain required compaction along centerline joints and ensure smooth surface transitions between the various layers.

B. Apply striping neatly in accordance with striping plans contained in the Drawing using approved paint.

3.04 Field Quality Control

A. Use HMA plants to produce HMA that conforms to AASHTO M 156-97. Heat asphalt cement to the manufacturer's recommended mixing temperature for plant mixing. Compact in the field above the manufacturer's minimum compacting temperature. Provide haul trucks that have smooth metal beds and are covered. Provide a hole in each side of the bed of each delivery truck for checking mix delivery temperature. Use a paving machine that includes mechanical devices such as skis, traveling string lines, equalizing runners, level arms, and automatic screeds. Place HMA in layers that do not exceed 2 inches (compacted thickness) for surface courses, 3 inches (compacted thickness) for binder courses, and 4 inches (compacted thickness) for base courses. Compact HMA using maximum roller speeds of 3 mph for static or vibratory steel wheel rollers and 5 mph for pneumatic rollers. Utilize three-wheeled steel wheel rollers of 10–12 tons in weight or 8- to 12-ton two-axle tandem rollers. Provide pneumatic rollers capable of varying tire pressures from 40 to 90 lb/in.^{2} gauge by adjusting ballast and inflation pressure.

B. Coordinate presence of materials testing firm to verify truck delivery temperatures adequate for proper placement and compaction as provided in Section 01331.

C. Coordinate three random samples per 10,000 ft^2 of paved surface area for Marshall samples to verify air voids, stability, and other required test properties.

D. In parking lots, recreational courts, or curb and gutter street sections identify and correct low areas of $\frac{1}{8}$ inch per 10 feet that result in ponding when subjected to surface water.

E. Tack coat surfaces of previously placed asphalt pavement layers prior to placing subsequent layers unless subsequent layers are placed on the same day, in which case the tack coat may be omitted.

F. Final inspection by regulatory agencies will be scheduled by the Engineer after the work is approved by the Engineer.

3.05 Protection:

A. Protect tacked surfaces from traffic.

B. Protect new HMA surfaces from traffic until sufficiently cooled, set, and cured to prevent rutting, scuffing, and pavement displacement.

Information contained in this section is subject to change and is intended for educational purposes only. CSI, CSC, BSD, and ARCOM did not prepare nor do they specifically endorse this material. The current format used in the above section is from CSI/CSC's *MasterFormat*™ (1995).

End of Section 02741

REFERENCES

American National Standards Institute and the American Water Works Association. *ANSI/ AWWA C105/A21.5-99, American National Standard for Polyethylene Encasement for Ductile-Iron Pipe Systems.* Denver: AWWA, 1999.

American Water Works Association. *PVC Pipe-Design and Installation, AWWA Manual M23.* Denver: AWWA, 1980.

Construction Specifications Institute (CSI) and Construction Specifications Canada (CSC). *SectionFormat™*. Alexandria, VA: CSI, 1997.

Construction Specifications Institute (CSI) and Construction Specifications Canada (CSC). *MasterFormat™*. Alexandria, VA: CSI, 1995.

SUGGESTED READINGS

American Association of State Highway and Transportation Officials. *AASHTO M 17-95, Standard Specification for Mineral Filler for Bituminous Paving Mixtures*. Washington: AASHTO, 1995.

American Association of State Highway and Transportation Officials. *AASHTO M 36M-98, Standard Specification for Corrugated Steel Pipe, Metallic-Coated, for Sewers and Drains*. Washington: AASHTO, 1998.

American Association of State Highway and Transportation Officials. *AASHTO M 81-92(1996), Standard Specification for Cut-Back Asphalt (Rapid Cure Type)*. Washington: AASHTO, 1996.

American Association of State Highway and Transportation Officials. *AASHTO M 82-75(1996), Standard Specification for Cut Back Asphalt (Medium-Curing Type)*. Washington: AASHTO, 1996.

American Association of State Highway and Transportation Officials. *AASHTO M 140-88 (1994), Standard Specification for Emulsified Asphalt*. Washington: AASHTO, 1994.

American Association of State Highway and Transportation Officials. *AASHTO M 156-97, Requirements for Mixing Plants for Hot-Mixed, Hot-Laid Bituminous Paving Mixtures*. Washington: AASHTO, 1997.

American Association of State Highway and Transportation Officials. *AASHTO M 170-98, Standard Specification for Reinforced Concrete Culvert, Storm Drain, and Sewer Pipe*. Washington: AASHTO, 1998.

American Association of State Highway and Transportation Officials. *AASHTO M 190M-95, Standard Specification for Bituminous Coated Corrugated Metal Culvert Pipe and Pipe Arches*. Washington: AASHTO, 1995.

American Association of State Highway and Transportation Officials. *AASHTO M 196M-92(1995), Standard Specification for Corrugated Aluminum Pipe for Sewers and Drains*. Washington: AASHTO, 1995.

American Association of State Highway and Transportation Officials. *AASHTO M 208-96, Standard Specification for Cationic Emulsified Asphalt*. Washington: AASHTO, 1996.

American Association of State Highway and Transportation Officials. *AASHTO M 226-80 (1996), Standard Specification for Viscosity Graded Asphalt Cement*. Washington: AASHTO, 1996.

American Association of State Highway and Transportation Officials. *AASHTO M 278-94 (1995) Class PS 46 Polyvinyl Chloride (PVC) Pipe*. Washington: AASHTO: 1995.

American Association of State Highway and Transportation Officials. *AASHTO M 288-97, Standard Specification for Geotextile Specification for Highway Applications*. Washington: AASHTO, 1997.

American Association of State Highway and Transportation Officials. *AASHTO R 12-85(1993), Standard Recommended Practice for Bituminous Mixture Design Using the Marshall and Hveem Procedures*. Washington: AASHTO, 1993.

American Association of State Highway and Transportation Officials. *AASHTO T 89-96, Standard Specification for Determining the Liquid Limit of Soils*. Washington: AASHTO, 1996.

American Association of State Highway and Transportation Officials. *AASHTO T 90-96, Standard Specification for Determining the Plastic Limit and Plasticity Index of Soils*. Washington: AASHTO, 1996.

American Association of State Highway and Transportation Officials. *AASHTO T 96-94, Standard Specification for Resistance to Degradation of Small-Size Coarse Aggregate by Abrasion and Impact in the Los Angeles Machine*. Washington: AASHTO, 1994.

American Association of State Highway and Transportation Officials. *AASHTO T 180-97, Standard Specification for Moisture-Density Relations of Soils Using a 4.54 kg [10-lb] Rammer and a 457 mm [18-in.] Drop*. Washington: AASHTO, 1997.

American Association of State Highway and Transportation Officials. *AASHTO T 191-93 (1996), Standard Specification for Density of Soil In-Place by the Sand-Cone Method*. Washington: AASHTO, 1996.

American Association of State Highway and Transportation Officials. *AASHTO T 245-97, Standard Specification for Resistance to Plastic Flow of Bituminous Mixtures Using Marshall Apparatus*. Washington: AASHTO, 1997.

American National Standards Institute and the American Water Works Association. *ANSI/AWWA C104/A21.4-95, American National Standard for Cement-Mortar Lining for Ductile-Iron Pipe and Fittings for Water*. Denver: AWWA, 1995.

American National Standards Institute and the American Water Works Association. *ANSI/AWWA C110/A21.10-98, American National Standard for Ductile-Iron and Gray-Iron Fittings, 3 In. Through 48 In., for Water and Other Liquids*. Denver: AWWA, 1998.

American National Standards Institute and the American Water Works Association. *ANSI/AWWA C111/A21.11-95, American National Standard for Rubber-Gasket Joints for Ductile-Iron and Gray-Iron Pressure Pipe and Fittings*. Denver: AWWA, 1995.

American National Standards Institute and the American Water Works Association. *ANSI/AWWA C150/A21.50-96, American National Standard for the Thickness Design of Ductile-Iron Pipe*. Denver: AWWA, 1996.

American National Standards Institute and the American Water Works Association. *ANSI/AWWA C151/A21.51-96, American National Standard for Ductile-Iron Pipe, Centrifugally Cast in Metal Molds or Sand-Lined Molds, for Water or Other Liquids*. Denver: AWWA, 1996.

American National Standards Institute and the American Water Works Association. *ANSI/AWWA C200-97, Standard for Steel Water Pipe 6 In. and Larger*. Denver: AWWA, 1997.

American National Standards Institute and the American Water Works Association. *ANSI/AWWA C203-97, Standard for Coal-Tar Protective Coatings and Linings for Steel Water Pipes—Enamel and Tape-Hot Applied*. Denver: AWWA, 1997.

American National Standards Institute and the American Water Works Association. *ANSI/AWWA C205-95, Standard for Cement-Mortar Protective Lining and Coating for Steel Water Pipe—4 In. and Larger—Shop Applied*. Denver: AWWA, 1995.

American National Standards Institute and the American Water Works Association. *ANSI/AWWA C502-94, Standard for Dry-Barrel Fire Hydrants*. Denver: AWWA, 1994.

American National Standards Institute and the American Water Works Association. *ANSI/AWWA C510-97, Double Check Valve Backflow-Prevention Assembly*. Denver: AWWA, 1997.

American National Standards Institute and the American Water Works Association. *ANSI/AWWA C511-97, Reduced-Pressure Principle Backflow-Prevention Assembly*. Denver: AWWA, 1997.

American National Standards Institute and the American Water Works Association. *ANSI/AWWA C600-99, Standard for Installation of Ductile-Iron Water Mains and Their Appurtenances*. Denver: AWWA, 1999, pp. 21–22. Referenced material is reproduced by permission of the American Water Works Association.

American National Standards Institute and the American Water Works Association. *ANSI/ AWWA C900-97, Standard for Polyvinyl Chloride (PVC) Pressure Pipe, 4 In. Through 12 In., for Water.* Denver: AWWA, 1997.

American Society for Testing and Materials. *ASTM A 31-95, Specification for Steel Rivets and Bars for Rivets, Pressure Vessels.* Philadelphia: ASTM, 1995.

American Society for Testing and Materials. *ASTM A 377-95e1, Standard Index of Specifications for Ductile Iron Pressure Pipe.* Philadelphia: ASTM, 1995.

American Society for Testing and Materials. *ASTM A 929/A 929M-97, Standard Specification for Steel Sheet, Metallic-Coated by the Hot-Dip Process for Corrugated Steel Pipe.* Philadelphia: ASTM, 1997.

American Society for Testing and Materials. *ASTM A 674-95 Standard Practice for Polyethylene Encasement for Ductile Iron Pipe for Water or Other Liquids.* Philadelphia: ASTM, 1995.

American Society for Testing and Materials. *ASTM A 746-95, Standard Specification for Ductile-Iron Gravity Sewer Pipe.* Philadelphia: ASTM, 1995.

American Society for Testing and Materials. *ASTM C 144-99, Standard Specification for Aggregate for Masonry Mortar.* Philadelphia: ASTM, 1999.

American Society for Testing and Materials. *ASTM C 150-99, Standard Specification for Portland Cement.* Philadelphia: ASTM, 1999.

American Society for Testing and Materials. *ASTM C 207-91(1997), Standard Specification for Hydrated Lime for Masonry Purposes.* Philadelphia: ASTM, 1997.

American Society for Testing and Materials. *ASTM C 478-97, Standard Specification for Precast Reinforced Concrete Manhole Sections.* Philadelphia: ASTM, 1997.

American Society for Testing and Materials. *ASTM D 1784-99a, Standard Specification for Rigid Poly(vinyl Chloride) (PVC) Compounds and Chlorinated Poly(Vinyl Chloride) (CPVC) Compounds.* Philadelphia: ASTM, 1999.

American Society for Testing and Materials. *ASTM D 1869-95, Standard Specification for Rubber Rings for Asbestos-Cement Pipe.* Philadelphia: ASTM, 1995.

American Society for Testing and Materials. *ASTM D 2241-99, Standard Specification for Poly(Vinyl Chloride) (PVC) Pressure-Rated Pipe (SDR-Series).* Philadelphia: ASTM, 1999.

American Society for Testing and Materials. *ASTM D 2321-89, Standard Practice for Underground Installation of Thermoplastic Pipe for Sewers and Other Gravity-Flow Applications.* Philadelphia: ASTM, 1989.

American Society for Testing and Materials. *ASTM D 2774-94, Standard Practice for Underground Installation of Thermoplastic Pressure Piping.* Philadelphia: ASTM, 1994.

American Society for Testing and Materials. *ASTM D 3034-98, Standard Specification for Type PSM Poly(Vinyl Chloride)(PVC) Sewer Pipe and Fittings.* Philadelphia: ASTM, 1998.

American Society for Testing and Materials. *ASTM D 3139-98, Standard Specification for Joints for Plastic Pressure Pipes Using Flexible Elastometric Seals.* Philadelphia: ASTM, 1998.

American Society of Civil Engineers (ASCE) and The Water Pollution Control Federation. Gravity Sanitary Sewer Design and Construction: *ASCE Manual No. 60/Manual of Practice FD-5.* New York: ASCE, 1982, pp. 201–204. Referenced material is reproduced by permission of the American Society of Civil Engineers.

American Water Works Association. *AWWA C500-93, Metal Seated Gate Valves for Water Supply Service.* Denver: AWWA, 1993.

American Water Works Association. *Installation, Field Testing, and Maintenance of Fire Hydrants, AWWA Manual M 17.* Denver: AWWA, 1989.

American Water Works Association. *Recommended Practice for Backflow Prevention and Cross-Connection Control, AWWA Manual M 14.* Denver: AWWA, 1990.

ARCOM. *MASTERSPEC®*. Alexandria: ARCOM, 1999.

Building Systems Design. BSD *SpecLink®*. Atlanta: BSD, 1999.

Construction Specifications Institute (CSI). *Manual of Practice, Construction Documents Fundamentals and Formats Module*. Alexandria, VA: CSI, 1996, Parts I and II.

National Society of Professional Engineers, the American Consulting Engineers Council, the American Society of Civil Engineers (ASCE), and the Construction Specifications Institute. *Uniform Location of Subject Matter (EJCDC No. 1910-16)* New York: ASCE, 1995.

National Society of Professional Engineers, the American Consulting Engineers Council, the American Society of Civil Engineers (ASCE), and the Construction Specifications Institute. *Standard General Conditions of the Construction Contract (EJCDC No. 1910-8)* New York: ASCE, 1996a.

National Society of Professional Engineers, the American Consulting Engineers Council, the American Society of Civil Engineers (ASCE), and the Construction Specifications Institute. *Standard Form of Agreement between Owner and Contractor on the Basis of a Stipulated Price (EJCDC No. 1910-8-A-1)*. New York: ASCE, 1996b.

Public Works Standards. *"GREENBOOK"—Standard Specifications for Public Works Construction—2000 Edition*. Anaheim: **BNi** Building News, 1999, pp. 339–340. Referenced material is reprinted by permission of: **BNi**® Building News, (1-714-517-0970).

Unibell Plastic Pipe Association (UPPA). *Unibell Handbook of PVC Pipe*, 3rd ed . Dallas: UPPA, 1991, pp. 382–387. Copyright © 1991 by Unibell Plastic Pipe Association. Referenced material is reproduced by permission of Unibell Plastic Pipe Association.

United States Department of Agriculture—Soil Conservation Service. *Erosion and Sediment Control . . . in Developing Areas. Planing Guidelines/Design Aids*. Columbia: 1977.

14 Project Costs and Scheduling

Construction costs and schedules are factors vitally important to a successful development. They are necessary to effectively transform the existing site into a finished product. Plans and specifications reflect the amount of work the contractor must perform within an established period of time; this work is used as a basis for contractor compensation. Preliminary estimates formulated during the initial design phase do not contain details that are included in a final analysis because detailed drawings are not available when these estimates are made. The process of preparing preliminary and final estimates, however, is the same.

14.1 ESTABLISHING A DATABASE

The estimator should establish a database to reference project estimates. This database can draw from previous jobs (adjusted for economic trends), reference books (such as Mean's *Building Construction Cost Data*), other organizations' current bid information, and/or manufacturer's quotations. The availability of local equipment and manpower, along with prevailing industrial activities, also must be considered.

14.2 COST ANALYSIS

The cost analysis of a job should begin with the geographic location of the project. Prevailing wage and material rates vary with location. Transportation linked to geographic location can be a factor relative to the supply of materials, labor, and equipment. The availability of local labor, local regulations including union requirements, and local assessments will affect the cost of construction. The project's size, composition, and available time for construction can vary costs. Projected economic trends, new products, and new technology also could have an affect.

14.2.1 Background Information

The estimator should take an inventory of existing conditions at the job site. The project synopsis (Figure 5.4) contain pertinent information that can assist the estimator immensely. It is recommended that the estimator conduct a site visit so that familiarity with the project can be obtained firsthand. Topographic and site conditions, existing utilities, drainage features, accessibility, and the impact on adjoining property should be observed. The specifications need to be read and understood as to how they relate to the plans. The development design engineer who also provides an estimate would already be familiar with these conditions.

14.2.2   The Cost Estimating Process

A sound procedure for job estimating is to use felt-tipped pens of various colors for marking the plans and specifications. Their use can indicate each included item. Every trade usually is assigned a different color pen for clarity. The estimator should make sure that no information is omitted such as notes, addenda to specifications, and specialized reports. If pertinent factors are undefined, every effort should be made to quantify them using meaningful methods.

All estimating work should be neat and in logical order so that another person might easily review the estimate without guesswork. Forms are available for estimators to use; however, computer spreadsheets allow estimators more flexibility.

Normally the engineer's cost estimate is based on a unit price quantity that would include the contractor's cost of material, labor, overhead, profit, and other costs associated with construction. Any quotations solicited by the estimator should be examined for time limitations, escalation clauses, freight costs, taxes, and so on, prior to use. The divisions of cost entries vary, depending on whether a preliminary or final estimate is being made. Also entries vary based on the project's complexity, scope, and magnitude. Typical work units include square yards (SY), cubic yards (CY), linear feet (LF), and lump sum (LS). The estimate is prepared by determining the number of units in each category; these numbers are then multiplied by the corresponding unit prices and summed. An example of a cost estimate for an improved land development, prepared using a spreadsheet, is shown in Figure 14.1

There are several cost-estimating computer software programs available that can provide aid in estimating the costs of projects. One recent system that is also capable of modeling and managing design and construction costs is BSD *CostLink/CM™* (Building System Design, Atlanta, GA). This computer application allows the user to perform queries such as a listing of all project items costing more than a stipulated amount, items installed by a particular type crew, or items having zero quantities.

14.3   PROJECT SCHEDULING

To guide a project to completion in an orderly and timely manner, a work schedule must be formulated. Normally this schedule contains estimated costs and durations for various tasks associated with the job in their order of precedence. The contractor usually prepares the working project schedule based on developer-provided start and completion dates; however, design professionals often prepare a theoretical schedule to evaluate possible cost and time savings that can be reflected in the final construction documents prior to bidding.

In the early 1900s Henry L. Gantt and Frederick W. Taylor produced charts that graphically depicted the scheduling of similar activities. The critical path method (CPM), another method widely used today for project management, evolved in 1956 within the E.I. du Pont de Nemours Company for improved company management.

CPM is commonly used for managing larger land development projects. Activities are usually segregated by trades. The method includes the sequencing of events and time, which encompasses planning for manpower and equipment. These data are required for any project scheduling, and using a method such as CPM provides a logical and systematic approach to planning, scheduling, and monitoring the project.

I. ROADS AND STORM DRAINAGE	UNITS	UNIT PRICES	NO.UNITS	COSTS
a. Clearing and Grubbing	Acres	$4,000.00	1.30	$5,200.00
b. Curb Inlets	Each	$750.00	2.00	$1,500.00
c. Junction Boxes	Each	$600.00		$0.00
d. Ditch Excavation	L.F.	$4.00	600.00	$2,400.00
e. French Drains	L.F.	$5.50		$0.00
f. Riprap Outfalls	Each	$400.00	2.00	$800.00
g. Pavement	S.Y.	$11.50	870.00	$10,005.00
h. Curb and Gutter	L.F.	$6.25	1200.00	$7,500.00
i. Road Signs	Each	$100.00		$0.00
j. 15" RCP	L.F.	$35.00		$0.00
k. 18" RCP	L.F.	$37.50		$0.00
l. 24" RCP	L.F.	$40.50	175.00	$7,087.50
m. 36" RCP	L.F.	$46.25	30.00	$1,387.50
n. 48" RCP	L.F.	$55.00		$0.00
o. 18" x 29" CPM Arch	L.F.	$20.00		$0.00
p. 22" x 36" CPM Arch	L.F.	$25.00		$0.00
q. 27" x 43" CPM Arch	L.F.	$30.00		$0.00
r. Erosion Control	L.S.			$2,000.00
s. Earthwork	C.Y.	$5.00	1400	$7,000.00
t. Clean Up (10%)	L.S.			$4,488.00
SUBTOTAL				$49,368.00
II. POTABLE WATER SYSTEM				
a. 2 Inch Main	L.F.	$3.00		$0.00
b. 4 Inch Main	L.F.	$6.00		$0.00
c. 6 Inch Main	L.F.	$9.50	600.00	$5,700.00
d. 8 Inch Main	L.F.	$11.00		$0.00
e. 6" Valve and Box	Each	$475.00	2.00	$950.00
f. 8" Valve and Box	Each	$550.00		$0.00
g. 3-Way Hydrant	Each	$1,150.00	1.00	$1,150.00
h. Thrust Blocking	Each	$150.00	2.00	$300.00
i. Service Lines	L.F.	$4.00	300.00	$1,200.00
j. Meters	Each	$175.00	12.00	$2,100.00
k. Testing/Cleanup(10%)	L.S.			$1,140.00
SUBTOTAL				$12,540.00
III. SANITARY SEWER SYSTEM				
a. 8" Main 0'-5' Cut	L.F.	$10.00	300.00	$3,000.00
b. 8" Main 5'-7' Cut	L.F.	$12.00	160.00	$1,920.00
c. 8" Main 7'-9' Cut	L.F.	$13.00		$0.00
d. 8" Main 9'-11' Cut	L.F.	$15.00		$0.00
e. 8" Ductile (add on)	L.F.	$12.00		$0.00
f. 4' Manhole 0'-5' Cut	Each	$975.00	2.00	$1,950.00
g. 4' Manhole 5'-7' Cut	Each	$1,050.00		$0.00
h. 4' Manhole 7'-9' Cut	Each	$1,100.00		$0.00
i. 4' Manhole 9'-11' Cut	Each	$1,200.00		$0.00
j. Drop to Manholes	Each	$500.00		$0.00
k. 6" Service Lines	L.F.	$6.50	300.00	$1,950.00
l. 8"x8"x6" Wyes	Each	$50.00	6.00	$300.00
m. Testing/Cleanup(10%)	L.S.			$912.00
SUBTOTAL				$10,032.00
SUMMARY OF COSTS				
I. ROADS AND STORM DRAINAGE		$49,368.00		
II. POTABLE WATER SYSTEM		$12,540.00		
III. SANITARY SEWER SYSTEM		$10,032.00		
IV. MISCELLANEOUS		$3,500.00		
TOTAL CONSTRUCTION COSTS		$75,440.00		
ENGINEERING FEES (6%)		$4,526.40		
SURVEYING FEES (2%)		$1,508.80		
INSPECTION FEES (2%)		$1,508.80		
GRAND TOTAL		$82,984.00		

FIG. 14.1 Example of a cost estimate for a 5.13-acre development.

14.3.1 Scheduling Process

The estimator should go through this process at least twice. The first pass will in part be a learning process through which the estimator can establish a "feel" for the project. The second pass will allow the estimator to scrutinize each project element and allow work to be checked. The key is to be relaxed and not rush the job. This estimate should reflect what a qualified contractor would deem reasonable, although in practice the engineer does not normally control the contractor's work force or material procurement.

14.3.2 Time of Project Completion

Normally a developer prescribes a project completion time based on outside influences and constraints or on the design and/or professional construction manager's recommendations. Suppliers' performance, weather, labor, economic factors, and site conditions can affect the progress of a job. In most land development projects, using the customary approach for professional design services, as shown in Figure 3.1, the design professional formulates a theoretical construction schedule based on the developer's timetable and prevailing local conditions. As an example the designer might select construction production rates that best represent what a typical contractor could produce if awarded the job. If the resulting construction schedule indicates the developer's completion date can be met adequately, then the completion time remains reasonable and can be used as a basis for bidding and construction contract terms.

However, if the design professional's theoretical construction time exceeds the time hoped for by the developer, the engineer may have to reexamine the project schedule to determine whether adjustments can be made to shorten the project. If the time allowed for completion cannot be lengthened, an increase of activity intensity (e.g., requiring possibly longer than normal working hours and more crews) may result in higher construction costs being required to complete the project quickly.

For projects that involve professional services based on the customary approach that use either project management (Figure 3.2) or a professional construction manager (Figure 3.3), project milestone scheduling and job duration would be jointly established by the developer, the construction manager, and the design professional.

14.3.3 Critical Path Method

Principally, the CPM utilizes a logic diagram that shows the precedence of activities, with their durations and estimated costs. This logic diagram initially must be prepared manually; however, computer applications can be used for final drawings, calculations, and schedules. Once computerized, data may be utilized for project monitoring at the commencement of construction.

There are many considerations associated with CPM diagramming network logic. Most commonly in land developments, an activity-on-arrow network is utilized. Each project must be analyzed and dissected into incremental tasks or activities. An activity is shown in the network by an arrow that points in the direction that the work progresses relative to its duration. The scale of the arrow length is not necessarily related to the amount of time required to complete the activity. Normally in a network, the name or abbreviation of the activity is placed above the arrow shaft, and the estimated duration is placed in a box under the arrow, as shown in Figure 14.2.

FIG. 14.2 Example of an activity on an arrow.

An event is the instant in time when one or more activities start or end with no associated duration. Events are usually represented by circles at the beginning and end of each activity arrow. Various activities may share events, with several originating or terminating together. Each circle is called a node and each node contains either an "i" or "j" number. The "i" number represents the beginning of the activity, while the "j" is used to terminate an activity. An example of "i–j" numbering is shown in Figure 14.3.

Occasionally, activities require logical restraints to produce a workable and effective network. To produce this condition, a dummy activity is utilized that has no duration. The dummy activity is depicted by a dotted or dashed arrow. The complete CPM network consists of interconnected activities, nodes, and dummies.

14.3.4 Network Logic

Following are some logic conditions that may be encountered in CPM networking, as illustrated in Figure 14.4:

Condition I Activity E must be completed prior to commencing activity G.

Condition II Activity I must be completed prior to commencing activity J. Activity I must also be completed prior to commencing activity K.

Condition III Activity G must be completed prior to commencing activity I. Activity H must also be completed prior to commencing activity I.

Condition IV Activities A and B precede activities C and D.

Condition V Activity C must be completed prior to commencing activity E. Activity D must be completed prior to commencing activities E and F.

14.3.5 Diagramming Procedures

Network logic is assumed to flow in the direction of the arrow tip and can never flow against the arrow's direction. The logic also flows from left to right throughout the diagram, with each project beginning and ending at single nodes called the starting point and ending point, respectively. All activities that terminate at a node must be completed prior to the subsequent activities being started. To ensure clarity and accountability, each activity must have unique "i" and "j" numbers, which, if multiples of 10 are utilized, will allow the network to be modified or edited later. To denote activity progression, each "i" number is made smaller than the corresponding "j" value.

Restraints or dummy activities are utilized to prevent duplicate "i" and "j" numbers from occurring, as shown between activities I and L in Figure 14.4.

FIG. 14.3 Example of activity nodes.

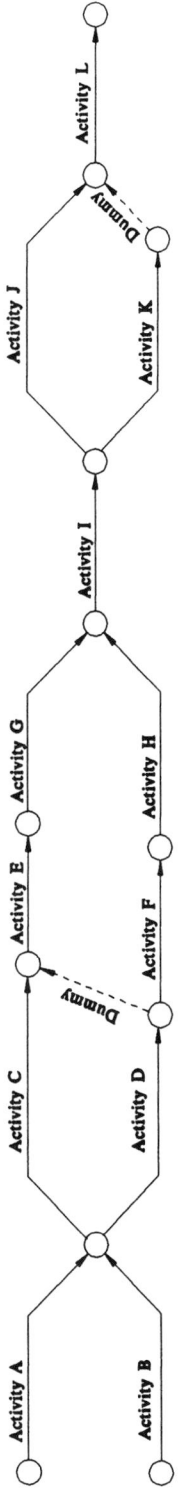

FIG. 14.4 Example of critical path network logic.

Dummies also can be utilized to denote logical restraints between nodes in parallel activity lines or as logic splitters to prevent unwanted flows from occurring, similar to that illustrated in Figure 14.5. Without the use of logic splitters activity A must be completed prior to starting activity B. Activities A and C must be completed prior to starting activity D. Also, activities A, C, and E must be completed prior to commencing activity F. With the use of logic splitters, activity A still must be completed prior to starting activity B, and activities A and C still need to be completed prior to starting activity D. With logic splitters activities C and E must be completed prior to commencing activity F while activity A does not affect the start of activity F.

14.3.6   Critical Path

The critical path of a network is the path (by time duration) through the diagram that produces the longest lapsed time. All activities that comprise this path are termed

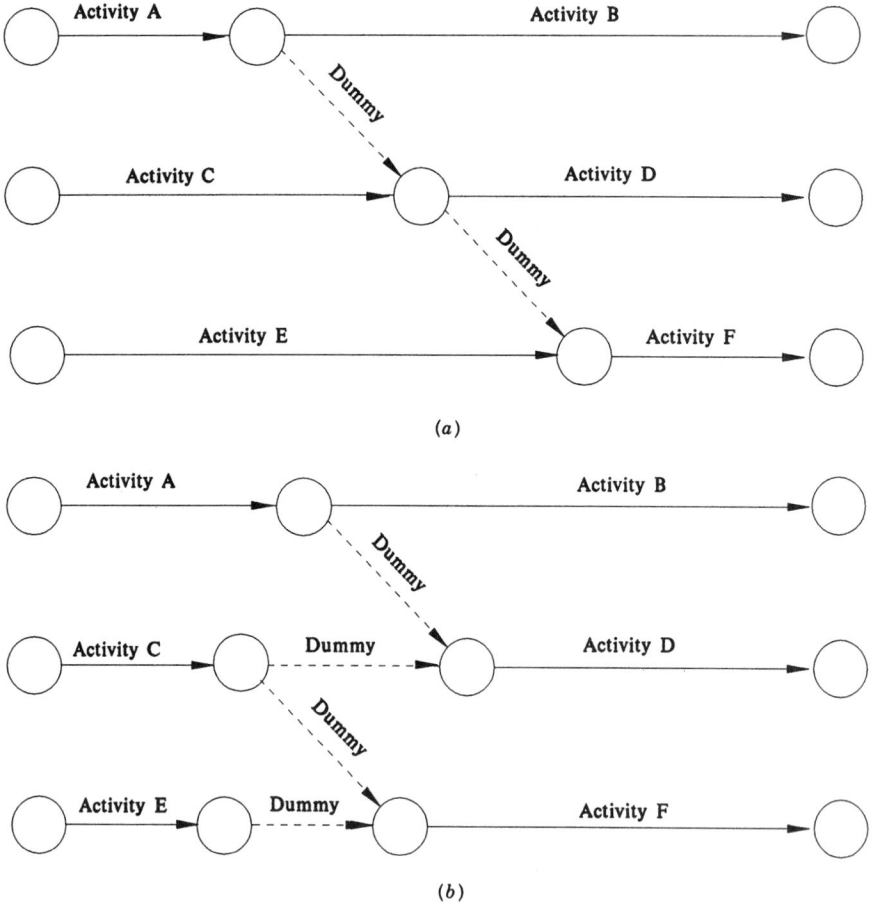

FIG. 14.5   Examples of critical path network logic splitting: (a) network without a logic splitter; (b) network with a logic splitter.

critical, which means that they must be completed on schedule as estimated or the job will lag. Other activities not considered critical may run over their estimated time schedule without resulting job delays. The amount of time an activity can slip without delaying the scheduled finish of the job is called float. Sometimes a series of activities may share float with other activities. If the float for an activity can be expended on that activity without delaying the earliest possible start of any other activity, that float is called free float. All the float available to an activity is called total float. Only that part of the total float that can be used without delaying the early start of any other activity is free. Free float is always some part of total float. Therefore, the critical path contains no float.

14.3.7   CPM Calculations

Given the simple five-activity network in Figure 14.6, various calculations will be performed to illustrate the critical path, early start, late start, early finish, late finish, and float. Information from the work is entered in tabular format to ease calculations. This is depicted in Figure 14.7*a*. Workdays are usually the time unit used; however, other units may be used as long as they are consistent throughout the diagram. When days are used, work will always start at the beginning of the first day and end at the end of the last day. Starts and finishes are instants in time (no duration) and therefore are associated with nodes rather than activity arrows.

From the network it can be seen that activity *A* is the first activity. Therefore it is assumed that it will commence on the first day of work at the beginning of the day and that the early start (ES) for that activity will be time 0. ES is the earliest time that the activity can be started.

Early finish (EF) is the earliest time that an activity can be completed. This is determined by the ES of the activity plus its duration. The EF of activity *A* is the end of day 2 ($ES + D = 0 + 2$).

The calculations are shown in Figure 14.7*b*. Note that the minimum length of time to complete the project is 10 days. Now the calculations must be continued in reverse order beginning with the EF of the last activity and using that as its late finish (LF). The LF is the latest time that an activity can be finished without delaying the completion of the project. If the LF of activity *E* is 10 days, then the late start (LS) of activity *E* is 9 ($LF - D$). LS is the latest time that an activity can be started without delaying the project. The calculations are continued in Figure 14.7*c*.

Now the float can be computed. Total float (TF) is the amount of time that the beginning or ending of an activity can be delayed without delaying the completion of the project ($TF = LF - EF$ or $LS - ES$). The TF of activity *A* is 0. Since there is no

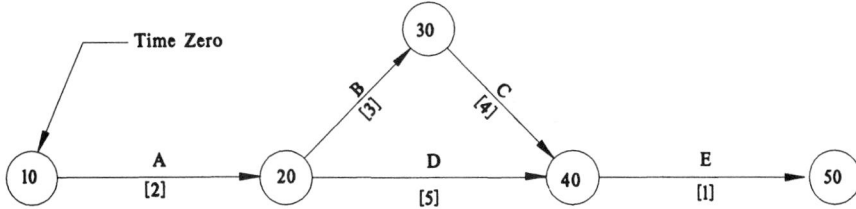

FIG. 14.6   Example of a critical path network diagram using activity on arrow methods.

A. CPM Calculation Data									
Activity	i	j	Duration	ES	LS	EF	LF	TF	FF
A	10	20	2						
B	20	30	3						
C	30	40	4						
D	20	40	5						
E	40	50	1						

B. Early Start and Early Finish Calculations									
Activity	i	j	Duration	ES	LS	EF	LF	TF	FF
A	10	20	2	0		2			
B	20	30	3	2		5			
C	30	40	4	5		9			
D	20	40	5	2		7			
E	40	50	1	9		10			

C. Late Start and Late Finish Calculations									
Activity	i	j	Duration	ES	LS	EF	LF	TF	FF
A	10	20	2	0	0	2	2		
B	20	30	3	2	2	5	5		
C	30	40	4	5	5	9	9		
D	20	40	5	2	4	7	9		
E	40	50	1	9	9	10	10		

D. Float Calculations									
Activity	i	j	Duration	ES	LS	EF	LF	TF	FF
A	10	20	2	0	0	2	2	0	0
B	20	30	3	2	2	5	5	0	0
C	30	40	4	5	5	9	9	0	0
D	20	40	5	2	4	7	9	2	2
E	40	50	1	9	9	10	10	0	0

FIG. 14.7   Example of critical path network calculations.

spare time, activity *A* becomes critical and is part of the critical path. Free float (FF) is the amount of time that the finish of an activity can be delayed without delaying the earliest starting time for following activities (FF = ES of next activity − EF of this activity). The calculations are completed in Figure 14.7*d*.

Upon completion of the schedules and charts, the network diagram should be re-drawn, plotting the activities to a time scale with actual calendar dates used. Weekends and holidays should be left in for this plot as they can represent a form of float for all activities. On a plot of this sort, start dates for activities with float can be adjusted to best use the resources available. This type of plot can be used to record project progress and is more effective for this purpose than a bar chart.

Economical computer software, such as CRITIC (Smith, 1985) can be utilized to accomplish the above calculations efficiently and can incorporate a calendar to provide meaningful output. A sample network is shown in Figure 14.8 for a residential subdivision that was analyzed using this program.

14.3.8   CPM Computer Application

To illustrate the use of a computer to assist in the management of a small improved land development, the construction costs developed in Figure 14.1 are utilized.

The activity network shown in Figure 14.8 represents in a general manner the sequencing of construction events necessary to complete improvements on a 5.13-acre

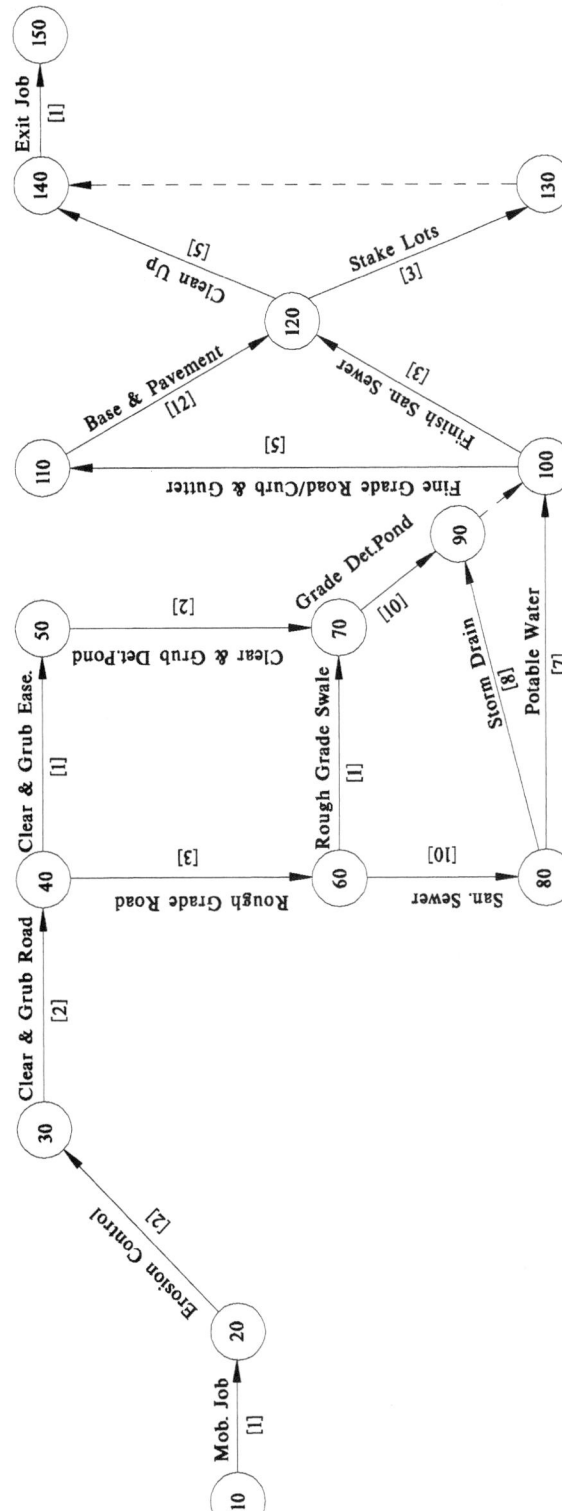

FIG. 14.8 Example of an activity on arrow network for the sample 5.13-acre development site.

development. The activities shown are broadly described for illustration purposes; however, further subdivision of these activities could be provided to include more detail. Most computer application programs require data similar to that shown in Figure 14.9. The included code listing are for the various accounts for administrative costs (Code 1 in this example), road and storm drainage construction costs (Code 2), water and sewerage construction costs (Code 3), and surveying costs (Code 4).

Figure 14.10 shows the results of the computer analysis in tabular form. This job consists of a total of 49 work days with work beginning the morning of May 10, 1999 and ending the afternoon of July 15, 1999. The total cost of the project is $75,440.00, and the work consists of 17 activities and 2 dummy activities. This particular program does not include in its construction calendar weekends and normal holidays.

Most programs provide bar charts similar to that shown in Figure 14.11. This enables one to see the progression of activities relative to the passing of time.

Work schedules for the various accounts also can be obtained, with Figure 14.12 depicting the activities of two accounts.

The job's progress relative to the accumulated percentage of cost can be shown in an S curve similar to the one in Figure 14.13. This curve allows projections of expenditures and a means of monitoring progress based on cash flow.

14.3.9 CPM Submission

Construction contracts may require the successful bidder to submit a work schedule upon award. This schedule, for example, could be required as part of the NPDES general permit for stormwater that pertains to the project. The engineer (construction manager) should ensure that this schedule reflects a reasonable approach and that the project's activities will be proportioned without inordinate time and expenses being

I	J	Activity	Duration	Cost	Code
10	20	Mobilize to Job	1	$1,000.00	1
20	30	Install Erosion Control	2	$2,000.00	2
30	40	Clear and Grub Road	2	$2,760.00	2
40	50	Clear and Grub Easement	1	$560.00	2
40	60	Rough Grade Road	3	$2,500.00	2
60	70	Rough Grade Swale	1	$2,400.00	2
50	70	Clear and Grub Det. Pond	2	$1,880.00	2
70	90	Grade Detention Pond	10	$4,500.00	2
60	80	Install Sanitary Sewer	10	$8,120.00	3
80	90	Install Storm Drainage	8	$10,775.00	2
80	100	Install Potable Water	7	$11,400.00	3
90	100	Dummy	0	$0.00	1
100	110	Fine Grade; Curb & Gutter	5	$7,500.00	2
110	120	Road Base and Pavement	12	$10,005.00	2
100	120	San. Sewer Invert & Covers	3	$1,000.00	3
120	130	Stake Lots	3	$1,500.00	4
120	140	Clean-Up and Inspections	5	$6,540.00	1
130	140	Dummy	0	$0.00	1
140	150	Exit Job Site	1	$1,000.00	1

FIG. 14.9 Example of critical path network data for a 5.13-acre development site. Source: Harrison S. Smith. *CRITIC,* 1985.

	5.13 Acre Development Project
Nodes				Early	Late	Early	Late	Total	Free	
I	J	Activities	Duration	Start	Start	Finish	Finish	Float	Float	
10	20	Mob Job	1	1 10-May-99	1	1	1 10-May-99	0	0	Critical
20	30	Erosion Control	2	2 11-May-99	2	3	3 12-May-99	0	0	Critical
30	40	C & G Road	2	4 13-May-99	4	5	5 14-May-99	0	0	Critical
40	50	C & G Easement	1	6 17-May-99	14	6	14 27-May-99	8	0	
40	60	Rough Grade Rd	3	6 17-May-99	6	8	8 19-May-99	0	0	Critical
50	70	C & G Detention Pond	2	7 18-May-99	15	8	16 31-May-99	8	1	
60	70	Rough Grade Swale	1	9 20-May-99	16	9	16 31-May-99	7	0	
60	80	San. Sewer Mains	10	9 20-May-99	9	18	18 2-Jun-99	0	0	Critical
70	90	Grade Det. Pond	10	10 21-May-99	17	19	26 14-Jun-99	7	7	
80	90	Storm Drain	8	19 3-Jun-99	19	26	26 14-Jun-99	0	0	Critical
80	100	Potable Water	7	19 3-Jun-99	20	25	26 14-Jun-99	1	1	
100	110	Fine Grade & Curb and Gutter	5	27 15-Jun-99	27	31	31 21-Jun-99	0	0	Critical
100	120	Fin. San Sewer	3	27 15-Jun-99	41	29	43 7-Jul-99	14	14	
110	120	Base & Pavement	12	32 22-Jun-99	32	43	43 7-Jul-99	0	0	Critical
120	130	Stake Lots	3	44 8-Jul-99	46	46	48 14-Jul-99	2	0	
120	140	Clean Up & Inspection	5	44 8-Jul-99	44	48	48 14-Jul-99	0	0	Critical
140	150	Exit Job	1	49 15-Jul-99	49	49	49 15-Jul-99	0	0	Critical

FIG. 14.10 Example of critical path network working schedule. Source: Harrison S. Smith. *CRITIC*, 1985.

allocated for certain activities. This schedule will probably vary from the theoretical schedule initially provided by the design professional as the contractor has actual control over specific labor, equipment, and materials. As an example, the contractor may plan on utilizing two construction crews to accomplish an activity that the design professional initially estimated would be accomplished with one. The contractor's construction schedule can be utilized to assist the developer and engineer in monitoring project progress. This is also helpful when construction management services are being provided and the contractor is obligated to attain various "milestone" events at predetermined times to maintain productivity. Should actual construction progress lag in the field, the engineer can confer and make recommendations to the contractor to ensure timely job completion. These recommendations are not to be construed as binding on the contractor because the engineer (construction manager) does not control the contractor's work force or equipment.

14.3.10 Progress Monitoring

Throughout the construction period the engineer (construction manager) is able to monitor the job's progress based on site observation and reports from the contractor.

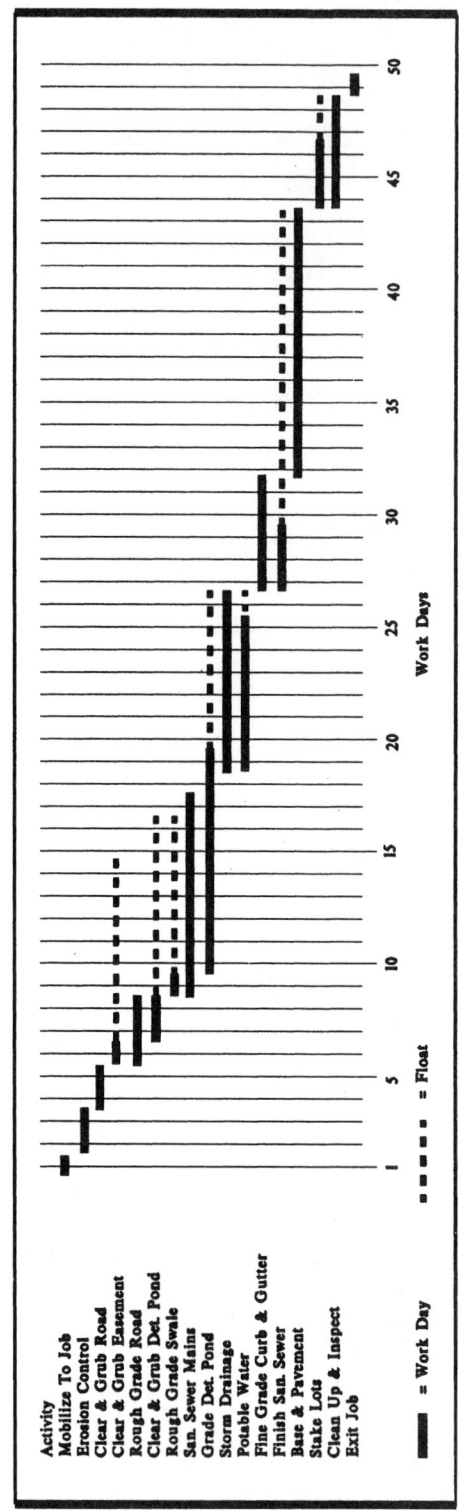

FIG. 14.11 Schedule bar chart for the sample project schedule. Source: Harrison S. Smith. *CRITIC*, 1985.

714

5.13 Acre Development For Account No. 1						
Activity	Scheduled Start	Actual Start	Scheduled Finish	Actual Finish	Estimated Cost	Costs to Date
Mob Job	Day 1 10-May-99		Day 1 10-May-99		$ 1,000.00	
Clean Up & Inspect	Day 44 8-Jul-99		Day 48 14-Jul-99		$ 6,540.00	
Exit Job	Day 49 15-Jul-99		Day 49 15-Jul-99		$ 1,000.00	

5.13 Acre Development For Account No. 2						
Activity	Scheduled Start	Actual Start	Scheduled Finish	Actual Finish	Estimated Cost	Costs to Date
Erosion Control	Day 2 11-May-99		Day 3 12-May-99		$ 2,000.00	
C & G Roads	Day 4 13-May-99		Day 5 14-May-99		$ 2,760.00	
C & G Ease.	Day 6 17-May-99		Day 6 17-May-99		$ 560.00	
R.Grade Roads	Day 6 17-May-99		Day 8 19-May-99		$ 2,500.00	
C & G Det. Pond	Day 7 18-May-99		Day 8 19-May-99		$ 1,880.00	
R. Grade Swale	Day 9 20-May-99		Day 9 20-May-99		$ 2,400.00	
Grade Det. Pond	Day 10 21-May-99		Day 19 3-Jun-99		$ 4,500.00	
Storm Drain	Day 19 3-Jun-99		Day 26 14-Jun-99		$ 10,775.00	
F. Grade C & Gutter	Day 27 15-Jun-99		Day 31 21-Jun-99		$ 7,500.00	
Base & Pavement	Day 32 22-Jun-99		Day 43 7-Jul-99		$ 10,005.00	

FIG. 14.12 Example of critical path network work schedule by account numbers. Source: Harrison S. Smith. *CRITIC*, 1985.

Should progress stall or lag behind the submitted contractor's schedule, alterations to the CPM network can be made by adding new activities or altering previously used ones to update the schedule. This is accomplished by changing all durations associated with completed activities to zero and modifying the durations of current activities to reflect actual remaining times until their estimated completion. These alterations originate from the contractor or are suggestions made by the engineer (construction manager) and adopted voluntarily by the contractor to reflect the contractor's desire to improve productivity. Reanalyzing the network will yield an updated schedule that might cause the critical path to change. This process can be repeated if the results are unacceptable and do not indicate punctual project completion.

14.3.11 Applying PERT to CPM

Another method to evaluate project scheduling is with the use of the performance evaluation and review technique (PERT), which was developed to manage the Polaris Fleet Ballistic Missile Program. For projects where activity durations are fairly well defined, the CPM method can be used. PERT, however, acknowledges the existence of

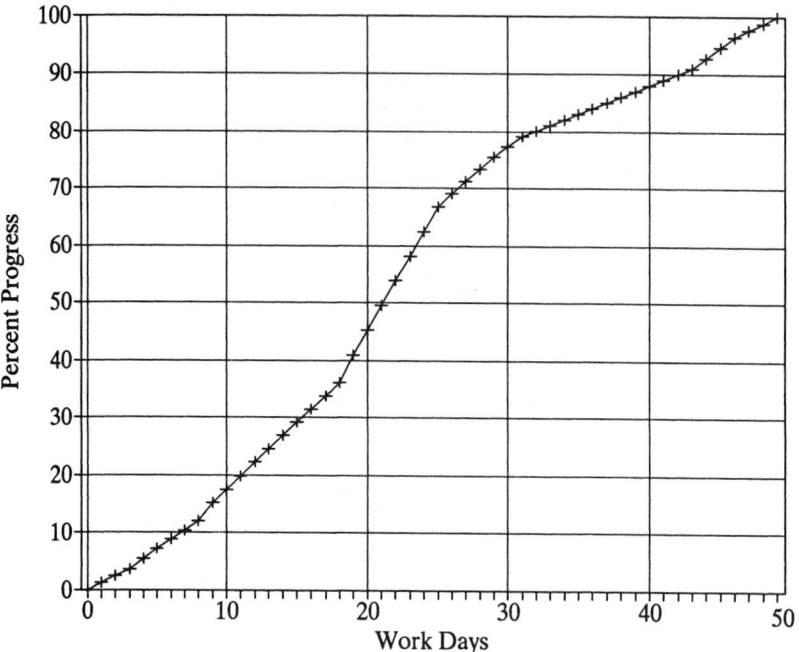

FIG. 14.13 The *S* curve for the sample critical path network. Source: Harrison S. Smith. *CRITIC,* 1985.

uncertainty associated with the completion times of project activities, and thus defines each duration using three different time estimates:

(a) optimistic time, or the minimum time to complete an activity;

(b) pessimistic time, or the maximum time to complete an activity; and

(c) most likely time, or the time most likely required to complete an activity, which is the same time used in CPM analysis.

The expected activity duration can be defined using

$$D_e = \tfrac{1}{6}(D_o + 4D_m + D_p) \tag{14.1}$$

where D_e is the expected duration, D_o is the optimistic duration, D_p is the pessimistic duration, and D_m is the most likely duration.

For example, using Figure 14.6, if activity D has the following characteristics, find the expected duration to complete the activity:

$$D_m = 5 \text{ days} \qquad D_o = 3 \text{ days} \qquad D_p = 8 \text{ days}$$

The expected duration is

$$D_e = \tfrac{1}{6}[3 + 4(5) + 8] = 5.166 \text{ days}$$

The associated variance for this activity can be determined using the following equation:

$$\sigma^2 = \left[\tfrac{1}{6}(D_p - D_o)^2\right] \qquad (14.2)$$

For this activity the variance is determined to be 0.6944.

If one wished to investigate the probability of completing the project on a specified date, the following equation can be used:

$$y = \frac{(T_{comp} - EF)}{\sqrt{S}} \qquad (14.3)$$

where y is a factor for determining the probability, T_{comp} is the time set for project completion, EF is the earliest expected completion time of the last network activity, and $S = \sum \sigma^2$ is the sum for all variances comprising the critical path.

To demonstrate this procedure, the CPM network shown in Figure 14.6 will be utilized along with the most likely durations for each activity. Because PERT requires optimistic and pessimistic durations, estimates must be made. Figure 14.14 shows each of these estimated values, the corresponding expected duration, and the associated variance for each activity.

The total variance along the critical path (A, B, C, and E) is 0.61 (0.25+0.11+0.25+0.00). If this project is set for completion (T_{comp}) in 12 days, what is the probability of completing it on time? (Note the early finish based on CPM analysis is 10 days.) Using Equation 14.3, the y factor can be determined as

$$y = \frac{12 - 10}{\sqrt{0.61}} = 2.56$$

Entering Figure 14.15, one can readily see that the probability of completing the job in 12 days is around 0.99, or 99 percent, when $y = 2,.56$.

Activity	Dm	Do	Dp	De	σ^2
A	2	1	4	2.166	0.25
B	3	2	4	3.00	0.11
C	4	3	6	4.166	0.25
D	5	3	8	5.166	0.6944
E	1	1	1	1.000	0.00

FIG. 14.14 Time estimates, expected times, and variances for each activity in the network shown in Figure 14.6 (D_m = most likely duration to complete the activity, D_o = optimistic duration to complete the activity, D_p = pessimistic duration to complete the actvity, σ^2 = associated variance).

y	Probability
-3	0.001
-2	0.02
-1	0.16
0	0.5
1	0.84
2	0.98
3	0.99

FIG. 14.15 Probability for standard normal distributions. (*Note:* Complete standard normal distribution tables are available in any standard reference book on statistical analysis.)

REFERENCES

Building Systems Design (BSD) *CostLink/CM™*. Atlanta, GA, BSD, 1999.
Smith, Harrison S. *Critic.* An unpublished public domain computer program. 1985.

SUGGESTED READINGS

American Society of Civil Engineers (ASCE). *Construction Cost Control.* ASCE Manual No. 65. New York: ASCE, 1985.
Fisk, E. R. *Construction Project Administration,* 3rd ed. New York: Wiley, 1988.
Horowitz, J. *Critical Path Scheduling—Management Control Through CPM and PERT.* New York: Ronald, 1967.
Kitchens, M. *Estimating and Project Management for Building Contractors.* Reston: American Society of Civil Engineers, 1996.

15 Development Guidelines

Development guidelines aid in project theme reinforcement by forcing landowners to coordinate individual preferences. These guidelines, which augment other local regulatory codes such as those required by zoning, are usually incorporated into project restrictions or protective covenants. They possibly include provisions for building setback distances, tree protection, and landscape regulations (Rubenstein, 1987).

15.1 PURPOSE OF GUIDELINES

Project guidelines, essential for successful improved land developments and beneficial to staged projects, are intended to protect a project's marketability and integrity during construction and postconstruction. Site improvements undertaken after the sale of improved property to others require control of quality, aesthetics, and conformity of use. In addition, acceptable improvements must be compatible with the development's installed infrastructure.

15.2 CONTROLLING ELEMENTS

Development guidelines are usually tied to the project's site through the development's master plan. For small projects, the final plan may suffice. Guidelines contained in covenants and restrictions usually include architectural, landscape, signage, lighting, and maintenance controls. Enforcement of these elements is usually accomplished through the actions of the homeowners' association. The developer may require the services of an architect, landscape architect, and attorney to assist in the preparation of guideline documents.

15.2.1 Homeowner's Association

The homeowner's association is an incorporated, nonprofit organization comprised of project property owners. Association rights and regulations are recorded in the public records office along with deeds of conveyance for common property from the developer to the association. This conveyance is completed prior to dwelling unit or parcel sales. The association is charged with the operation, maintenance, control, and ownership of all development common areas, recreational areas, streets, and easements not dedicated to the public. Property owners within the development are voice and vote members of this association with undivided interest in all association-controlled property. The developer usually maintains voice and vote based on the number of dwelling units remaining in the development. Each member is annually assessed a fee for operating revenue. Initial operational fees, however, are completely borne by the developer, while subsequent financial participation is reduced according to the developer's proportional dwelling unit holdings. Special assessments also are levied in case of emergency expenses or the need for capital improvement projects such as bottom and side replastering or pump and filter replacement associated with the association's swimming pool. Within the homeowner's association is usually a board that reviews all improvements that involve the development.

15.2.2 Board of Review

This board, initially appointed by the developer, ensures that new construction within the development meets established guidelines and criteria. As the project nears completion, the developer's involvement dwindles while the homeowners' involvement increases. This allows the homeowners an opportunity to appoint board members of their own choosing to carry out the organization's charge; the board should include an architect, landscape architect, or engineer to provide other board members with technical insight. Proposed plans and specifications associated with various development phases are reviewed for compliance with the project's approved master development plan and restrictions. Each site's submission is evaluated for architectural compatibility. Building plans that show the shape, size, color, and exterior treatments in addition to proposed fencing and walls are reviewed along with the proposed structure positioning on the parcel in relation to setback requirements. The proposed improvements also must be compatible with overall development transportation plans for vehicular, pedestrian, and bicycle travel. Parking areas must comply with development regulations.

The board of review ensures other types of site improvements are acceptable, including lighting for pedestrian, vehicular, and accent applications. Proposed grading and drainage plans must conform to the development's overall plan, as well as the utilization and protection of easements and facilities.

Site landscape plans pertaining to buildings, parking lots, and open space are reviewed for compatibility with drainage and maintenance master plans.

15.2.3 Project Restrictions or Protective Covenants

Restrictions should be initially prepared for the entire development and recorded prior to parcel or dwelling unit sales. Large phased developments that contain restriction variations that are not definable at the project's beginning should have each subsequent phase buffeted from previous phases to minimize conflicts. Each phase's restrictions are recorded prior to ownership transfers within that phase. Since protective covenants are recorded, they run with the land and are binding to all owners. Covenants, therefore, are an express agreement between the developer and purchaser with enforcement provisions. In addition to regulating site improvements, covenants are used to control noise and nuisances including erection of temporary shelters, tents, trailers, sheds, outbuildings, oil drilling, mining operations, kennels, and any harboring of farm animals.

Covenants can provide safety by regulating any alteration and excavation of slope control areas requiring stabilization and protection. In addition, sight control easements where corner lots must allow clear vision at street intersections for safe vehicular travel can be enforced through covenants. Finally, covenants can specify types of allowable business operations in commercial areas that are included in the development.

15.3 ADDITIONAL CONTROLS

The use of guidelines provides a means for activity coordination at the conclusion of improved land and total development activities; however, additional controls are sometimes helpful. For example, development guidelines provide guidance for the dwelling unit's builder as well as for the end user; however, Figure 15.1 illustrates a form of additional control that minimizes problems during building construction.

DWELLING UNIT BUILDER'S RESPONSIBILITY
1. Preservation of Improvements. The builder shall be responsible to the owner for preserving any and all land improvements made in the development. Employees of the builder shall be directed to use good housekeeping procedures to insure positive lot drainage and to minimize hazards.
2. Lot Improvements. After project contractors (e.g. road/sewer,etc) have completed their work, all remaining lot improvements shall be constructed in a self-sustained manner by the builder. No borrow material shall be obtained from any place on site unless such area is held in interest by the builder and removal does not alter the project plans. Waste material resulting from lot clearing shall be properly disposed of by the builder and not placed on adjoining lots or pushed into storm drainage facilities and subsequently covered with soil.
3. Service Taps. The tie-in area shall be maintained to allow the proper flow of stormwater when water and wastewater taps are exposed to make connections and visual inspections by the local regulatory authorities.
4. Delivery of Materials. It shall be the responsibility of the builder to insure that building material deliveries do not deface development improvements. For example, bricks shall be off-loaded from the delivery truck far enough back on the lot so as to not interfere with storm drainage facilities and ready-mix concrete trucks shall not deface curbing.
5. Responsibility of Maintenance. Each builder shall be responsible for maintaining all improvements which are on or contiguous to the building lot. All storm drains shall be kept open to run free, and all debris will be disposed of properly.
6. Conformance. Each builder shall conform to the requirements of all applicable project construction requirements which may include: a) Zoning   b) Building, electrical, and plumbing codes.   c) Sedimentation and erosion control plans.   d) Site grading plans.   e) Project restrictions and covenants.
Date: Builder's Signature   Owner's Signature

FIG. 15.1 Form of understanding between builder and developer.

REFERENCE

Rubenstein, H. M. *A Guide to Site and Environmental Planning*, 3rd ed. New York: Wiley, 1987, pp. 377–397.

SUGGESTED READING

U.S. Department of Housing and Urban Development (HUD). *Manual of Acceptable Practices*, 1973 ed., 4930.1 Vol. IV. Washington: HUD, 1973, Data sheets 40–41.

16 Completion of Design Activities

The final stage of the design process includes the assembly of all documents related to project construction and their submission to regulatory authorities for approval. Some of these documents are plans, specifications, and applications for permits to construct.

16.1 FINAL PLANS FORMAT

Engineering drawings are normally placed on standard size sheets, as shown in Figure 16.1.

Land development plans customarily utilize the standard D-size format because available drawing space is usually sufficient to adequately display information but not so large as to be cumbersome in the field. Record drawings are usually ink plotted on vellum or Mylar drafting media.

The presentation of lines normally follows acceptable characteristics portrayed in the alphabet of lines (French et al., 1970). In addition, emphasizing certain lines such as border and right-of-way edges by using various width pens (or line widths) produces a more pleasing display.

Engineering lettering is usually accomplished using CAD applications. Letter size conventions vary according to the importance of information being displayed, with the most general information being shown largest and the most specific information being shown smallest.

Depiction of information usually follows established graphic standards such as those contained in *Architectural Graphic Standards* (Hoke, 1994). Graphic presentations should embody a consistent style throughout the set of plans, including the use of

(a) a standard border format and

(b) a standard title block, placed toward the right side of each drawing, that includes the project's name and locations, date, scale, sheet number, revision table, engineer's signature and seal, engineer's address, and any other design professional's seal and signature.

16.1.1 Title Sheet

The cover sheet identifies the drawing set by the title, which is affixed using the largest size letters. The project's location is usually included along with a vicinity map that would aid a visitor in finding the site. The developer's name and address are also shown along with a listing of the internal drawings by sheet and topic. This cover sheet includes the engineer's name, address, seal, and signature.

Size	Architectural	Engineering	International	Metric
A	9" x 12"	8.5" x 11"	A4	210 x 297
B	12" x 18"	11" x 17"	A3	297 x 420
C	18" x 24"	17" x 22"	A2	420 x 594
D	24" x 36"	22" x 34"	A1	594 x 841
E	36" x 48"	34" x 44"	A0	841 x 1189

FIG. 16.1 Various standard drawing sheet sizes showing architectural, engineering, and international classifications.

Title sheets can have regulatory authority statements and project approval certifications affixed for endorsement and dating. Other information can include revision data, field survey book references, and general notations such as special conditions for emphasis to the contractor.

Other title page entries may be a summary sheet of the project's land use areas, including the number of various categories of building sites. Figure 16.2 shows the general layout of a typical title sheet.

16.1.2 Project Configuration Sheets

The first sheet in a set of construction plans is usually the approved preliminary plan upon which the set of drawings are based. This drawing should contain the data cited in Section 5.3.1.

Following the preliminary plan are drawings that contain lot and parcel dimensions similar to the final recorded development plat. This sheet is useful for orientation and field work.

16.1.3 Road and Storm Drainage Sheets

Design information required to construct the road and street system normally follows the lot layout drawings. These road drawings are normally prepared on standard plan and profile sheets. The plan view represents the street's horizontal control nomenclature, including centerline beginning and ending points; points of intersecting street centerlines (PIs) with tie equality stationing; centerline bearings; horizontal curve data; superelevation limits; stationing; right-of-way width data; lot line and property line intersections with the rights-of-way; easements; drainage facilities including catch basins, culverts, junction boxes, French drains, curb and guttering or ditching; benchmark locations and identification; horizontal scale; north arrow and meridian reference; and adjoining landowners if appropriate. A storm drainage culvert schedule may be included that summarizes in tabular form the pipe's identification, diameter, length, material, slope, and end invert elevations. If the job site is subject to flooding, the 100-year flood limits are often delineated. Where jobs are located close to existing facilities or in areas where the work must be confined, a line indicating the limits of construction (or clearing) can be provided in the plan view.

Each profile section includes road centerline stationing, existing ground profile, proposed road profile (being careful to specify either the subgrade or the finished grade), vertical curves, design elevations at each station (or half station) and at all

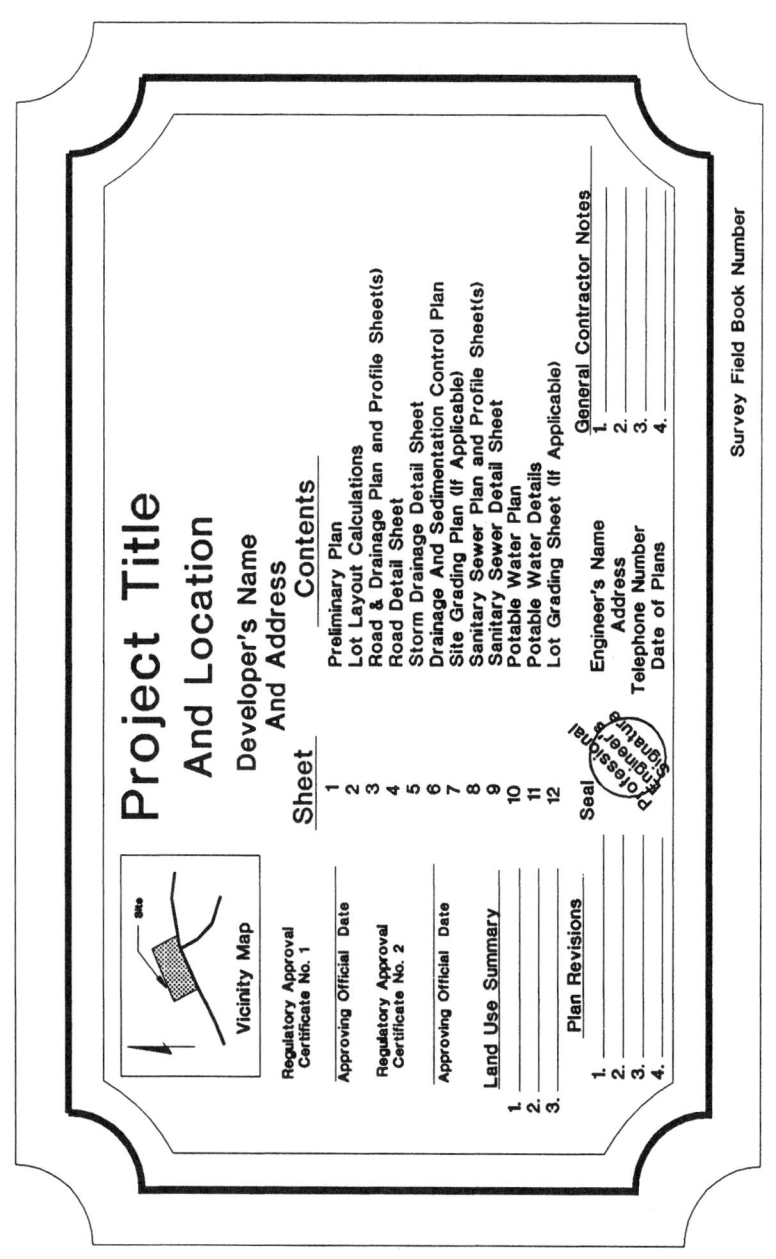

FIG. 16.2 Typical title sheet for a set of construction plans.

725

key horizontal control stations (curve tangent points and tie equalities), culverts, catch basins, French drains, and the horizontal and vertical scales.

Following the road plan and profile sheets are the road detail sheets. These sheets can include typical geometric cross sections, typical structural cross sections showing the pavement materials and thicknesses, cul-de-sac details, curb (and false curb) details, driveway and intersection details, signage details, and other standard details necessary for the complete construction of the road system.

Because a number of items included on road plan and profile sheets are associated with drainage, a storm drainage detail sheet is usually included. This sheet can include typical details for the construction of curb inlets, drop inlets, junction boxes, French drains, culvert headwalls, installation procedures for culverts (including bedding requirements), riprap placement, and typical channel cross sections.

16.1.4   Drainage and Sedimentation Control Plan

The drainage and sedimentation control drawings can be included in the set of plans after the road drawings since they are closely related. This plan typically shows the location of drainage easements, channels (including typical cross sections, slope, and material lining), culverts (including length, diameter, material, strength classification, invert elevations in and out, and slope), French drains, riprap placement, retention/detention facility locations and configurations, and outfall structure locations that can include stilling basin and splash pool locations.

Sedimentation and erosion control provisions can also be shown on this plan. These provisions can include locations where various control practices are to be employed during construction. For example, the location of a temporary construction entrance can be delineated along with how spray water will be provided (with a back-flow preventer) for washing a vehicle's underside. Silt fencing locations can also be shown so that the contractor can install the fencing before major clearing and grubbing operations begin. Temporary perimeter, interceptor, and diversionary dikes are included along with the location of temporary level spreaders for dispersing runoff. Other sediment control provisions can be shown, such as straw (or hay) bale barrier locations as well as sediment trap and basin locations. A sediment and erosion control detail sheet is helpful to show the contractor how to construct and install the various erosion control devices shown on the plans. If an NPDES general permit for stormwater is in effect, this drainage and sediment control plan must be kept current and on file at the construction site.

16.1.5   Site Grading Plan

If the project warrants a site grading plan, it is conveniently placed after the drainage and erosion control drawings because site grading closely relates to road and storm drainage facility construction. The site grading plan shows existing and proposed contours along with existing spot elevations and any finished floor elevations. Areas determined during the geotechnical investigation to contain poor soils, such as expansive clays, can be delineated on the site grading plan for removal or modification. Other on-site areas requiring protection from alteration, such as designated wetlands, can also be delineated on this plan to caution the contractor's field personnel.

16.1.6 Sanitary Sewer Sheets

The plan and profile sheets for the wastewater collection system number as many as needed to adequately display the project's system. The plan view shows necessary horizontal control data including bearings and distances between manhole centers, line diameters, and material used for manufacture. In addition, existing system structures are shown, along with information on new manholes, including stationing, rim and cover elevations (set above anticipated flood elevations), invert elevations, and drop heights. If the site is subject to flooding, the 100-year flood limits are often shown. Service lines, diameters, material, and tap locations also are shown in the plan view.

The profile includes the system's stationing orignating from the centerline of a manhole and progressing to subsequent manhole centers *while being completely independent of road stationing.* The profile shows the finished ground profile, the invert and crown line of the proposed main with the slope included along with any pipe crossings. Invert elevations for all manholes are included in addition to those associated with drops. A sanitary sewer detail sheet follows that shows approved construction techniques. Manhole details (including drops), pipe installation procedures (including pipe bedding requirements), piers, cleanouts, and service connections can be included.

Should a sanitary sewer pumping station and forcemain (or wastewater treatment plant) be part of the project's scope, these drawings are included after the collection system and could entail several sheets just for the pumping station construction.

16.1.7 Potable Water System Sheets

The potable water distribution system drawings usually follow those for the sanitary sewer because the water system is normally constructed after the deeper sanitary sewer system is installed. Water lines allow the contractor more flexibility than gravity sewer lines during installation. The water system is pressurized and does not have to be installed on a highly controlled grade where constant fall is required to provide efficient gravity flow conditions, such as are required in a gravity flow sanitary sewer collection system. The water system can be shown in a plan view where water main length, diameter, strength classification, and make are labeled along with valving, fire hydrants, thrust block locations, service lines, and tap locations. Blow-off and air release valve locations are also shown. If a profile of the water main is required, the finished surface profile must be shown in addition to pipe crossings, elevations, and water main stationing. The plans should also include any existing water works structures and facilities and any underground improvements that could affect the water system such as storm drains and sanitary sewers that are located close by. Where water mains cross streams, the normal high and low water elevations should be shown along with the 100-year flood elevation located on site. A potable water detail sheet provides information on installation and construction techniques required to properly install the water system. These details can include pipe installation procedures, valve and box details, fire hydrant details and their installation, thrust block positioning and sizing for various line sizes and working pressures, service and meter connections, and procedures for disinfecting the installed system, including the required pounds of chlorine per unit length of water main diameter. If potable water connections are to be made where possible contamination can occur, such as at a sanitary sewer pumping station, then details pertaining to back-flow preventers must be included. Drawings associated with any water source development or treatment would also be included in the set of project plans.

16.1.8   Lot Grading Sheet

Often, in an improved land development, individual parcel owners are responsible for lot clearing and landscaping. To provide guidance that coordinates with the development's storm drainage plans, a lot grading sheet can be prepared to show building site grading. This is accomplished by assigning each lot a particular generalized grading plan, either "A," "B," or "C," that is keyed to a lot drainage detail sheet similar to the one shown in Figure 16.3.

16.1.9   Document Reproduction and Assembly

Once the set of plans has been prepared and assembled, an initial set of blue or black line prints should be made for information verification by the engineer. This process includes the use of a felt-tipped marker or colored pencil to strike through each verified entry on every drawing sheet. If errors are detected, the engineer can note the corrections in red, and they can later be made on the original drawing. This check process is considered ordinary care to minimize errors and omissions. Once the plans have been checked and all necessary corrections made, corrected copies must be prepared for a final scan by the engineer.

16.2   COMPLETED PROJECT MANUAL

The project manual is normally bound into pamphlet or book form, having a cover sheet, table of contents, invitation to bid, bid information and proposal forms, conditions of the contract, insurance requirements, and construction specifications. Each segment requires checking for clarity, accuracy, and completeness. Any noted errors and omissions must be corrected prior to a final scan by the engineer.

16.3   PERMIT APPLICATIONS TO CONSTRUCT

Most regulatory agencies require the engineer to submit various forms, drawings, design calculations, engineering reports, and several sets of construction plans and specifications for review when seeking approval for project construction. Submittal requirements for these regulatory reviews must be obtained by the engineer so that when the applications are prepared by the engineer, they are complete. Often the developer must endorse the completed applications and simultaneously submit fees with these review documents when they are delivered to reviewing authorities.

16.4   FINAL DOCUMENT SCAN BY ENGINEER

After correcting any detected errors and adding any omitted information, the engineer should make a final scan of all project documents for a final check to ensure that

FIG. 16.3 Typical lot drainage detail sheet. Source: U.S. Department of Housing and Urban Development, 1973.

729

nothing is being overlooked. This last verification should include double-checking the accuracy of all information shown on the plans. In addition, the engineer should verify that ample room is available for the contractor to maneuver equipment and store materials on the job site while conducting the required work. If additional temporary construction easements are required, the engineer should communicate this need to the developer. The engineer also should verify that the plans and specifications stipulate who is to provide any temporary utility services necessary during the construction phase, and, if overlooked, to discuss the matter with the owner where appropriate action can be taken. Jobs that require temporary storage areas and stockpile areas that are provided for in the plans and specifications should be delineated on the drawings. Also, all specified material and equipment should be cross-checked for availability to minimize supply problems, unless specialty items are mandatory.

During this final review, the engineer should consider the types of construction required to complete the job and probable methods that might be employed by the contractor to accomplish the work. Minor modifications to the plans and specifications might allow the contractor additional flexibility that could result in reduced overall construction costs. An example of this might be the inclusion of sanitary sewer manholes on each end of an oversized jacked and bored casing placed under a roadbed. These manholes would allow the contractor added flexibility to compensate for boring inaccuracies in line and grade, which are critical for gravity sewer mains, resulting in reduced boring costs.

The engineer should reread all specifications to ensure clarity and continuity between what is shown on the plans and what is included in the specifications. This will help alleviate future confusion and grounds for possible construction claims.

16.5   DEVELOPER'S REVIEW

Prior to any final plans, project manuals (specifications), cost estimates, or applications for permits to construct being promulgated by the engineer, the developer should review and approve in writing all documents. This review will allow the engineer to make any minor last-minute alterations for the developer and allow the developer to consider the estimated construction costs based on completed drawings.

16.6   FINANCIAL INSTITUTION'S REVIEW

The developer usually forwards completed plans, project manuals (specifications), and estimate of construction costs to the financial institution to complete application requirements. Pending regulatory notification of the documents' approval for construction, the institution can complete loan processing.

16.7   REGULATORY REVIEWS

After the developer has approved all documents, the engineer is free to submit the required number of copies of the various documents to the jurisdictional reviewing authorities. The review process normally requires the engineer to periodically answer questions, provide additional information, make alterations, and confer with the reviewing authorities, along with the developer, to gain plan approval. Once approval is received, the developer can obtain a contractor for construction.

REFERENCES

French, T. E., C. J. Vierck, and R. J. Foster. *Graphic Science and Design*, 4th ed. New York: McGraw-Hill, 1970, p. 17.

Hoke, J. R. (Ed.). *Architectural Graphic Standards*, 9th ed. New York: Wiley, 1994.

U.S. Department of Housing and Urban Development (HUD). *Land Planning Principles for Home Mortgage Insurance*. Handbook, 4140.1. Washington: HUD, 1973, Section 6.

SUGGESTED READING

Earle, J. H. *Engineering Design*, 9th ed. Reading MA: Addison-Wesley, 1999.

PART IV
Construction-Related Activities

17 Contracts and Award

Once all construction documents have been approved, the developer has gained title to the development site, and all construction loan financing has been completed, the project enters the construction phase. During this phase a contractor is engaged to convert the plans and specifications into reality. On occasion the developer may not have received approval from all regulatory agencies or the financial lender but may wish to proceed with preparations to complete the job. If public funds are involved, the developer will be required to solicit competitive bids from several contractors. The competitive bid process is beneficial even if private funding is utilized, although a negotiated contract can be made.

For improved land development work, where several different types of construction are involved requiring different trades, *experience has shown that it is essential that only one prime contractor be awarded* (unless prohibited by law), who would then have the responsibility to coordinate and supervise the various subcontractors for the entire job. Cases where several individual prime contracts were let for work in a common area resulted in confusion, loss of efficiency, and loss of project control. Should several different development sections (that are separated from each other) be simultaneously developed, different prime contractors can be procured for each section; however, it is recommended that prime contracts not be awarded to any contractor who is simultaneously providing subcontract work in another section to minimize construction conflicts.

17.1 TYPES OF CONSTRUCTION CONTRACTS

Essential elements of an enforceable contract for engineering services presented in Chapter 3 also apply to contracts related to construction. Construction contracts can be awarded either as competitive bid or negotiated contracts. If public funds are involved, the impending work must be advertised to solicit competitive bids. Specified components within the project must include "or equal" provisions. The contractor is, in addition, usually obligated to provide minimum insurance amounts and include bid, performance, and payment bonds.

If the time in which the work is to be completed is not specified in the contract, it is understood that it will be performed within a "reasonable time." If "time is of the essence," it is clear that the completion date is a part of the consideration of the agreement, and the failure of a party to complete obligations within the time specified may result in contractual liability. The amount of damages may be determined by agreement; however, such damages often are difficult to determine, and usually an amount is stipulated to be a specific amount per day of delay as liquidated damages in lieu of determining actual damages. The injured party must be prepared to justify the amount stipulated. The contractor may be excused from liquidated damages if the contractor is entitled to a time extension or if the owner adds extra work to the contract. Bad weather conditions are assumed to be part of the anticipated working conditions and are not considered sufficient reason for waiving liquidated damages unless it can be demonstrated using climatological data that such weather was unseasonable and therefore an extraordinary problem.

Bonus payments of a specified amount per day for early completion of the contract and penalty assessments of an equal amount sometimes are included in the contract provisions as an incentive for the contractor to complete work in a timely manner.

17.1.1   Competitive Bid Contracts

Competitive bid contracts are usually based on a fixed price and can be classified as lump-sum or unit price contracts. The lump-sum contract provides that the contractor shall be paid a lump sum of money for all work covered by the plans, specifications, and issued permits; this lump-sum amount shall include all material and labor costs, overhead, profit, and so forth. The unit price contract provides for a breakdown of the number of units of each type of construction and a price for each unit.

Lump-sum contracts are used on projects requiring a variety of operations where it is impractical to break down the work into units. It is critical that finalized plans and specifications be comprehensive for this type of contract; they should show in complete detail the requirements of the work. If the plans are incomplete, change orders and extra work orders after the contract is signed are expensive and could lead to controversies. The contractor bidding on the job must allow ample funds to cover contingencies, which usually drives the bid price upward. Any performance risks, therefore, fall on the contractor.

When projects consist of large quantities of relatively few types of construction and the volume or limits of the work cannot be accurately determined in advance, the unit price contract may be advantageous. The unit price contract allows for minimum alterations in the scope of the work without the need of formal change orders.

On occasion, a combination of these two forms of contracts may be used, the lump-sum basis to account for a known base amount of work and the unit price basis for unknown amounts of work. For example, a lump-sum contract could be solicited for the construction of a development's clubhouse structure, while the foundation work may be based on unit prices as a result of the uncertainty of subsurface conditions. The EJCDC has prepared a set of guidelines (Smith 1987) that more fully addresses competitive bid activities.

17.1.2   Negotiated Contracts

A negotiated contract is awarded to a selected contractor who is chosen from a field of contractors. This method of award is not normally used when public funds are involved except under unusual conditions.

Negotiated contracts can include variations from cost plus a percentage of the cost to costs plus a fixed fee. One advantage of using a cost-plus contract is that construction can begin before plans are completed. Cost-plus contracts also allow the developer to make changes in the project at actual audited costs. The final project cost, however, is not guaranteed and a good deal of effort is required to audit the contractor.

Other cost-plus-type contract variations can include incentives with the contract, such as cost plus a fixed fee with profit sharing, a bonus clause, or a sliding-fee scale.

Cost plus a guaranteed maximum price (upset price contract) also allows certain contractual incentives. In this form of contract the contractor agrees to perform the required work with reimbursements being made on a basis of cost plus either a percent of the cost or a fixed fee up to an agreed-to threshold or guaranteed maximum. The contractor is required to complete all work even if costs exceed the agreed-to maximum.

A final type of negotiated contract would be a design-build contract. This type of arrangement provides for the prime contractor to design and construct the project under one contract. The construction portion of the agreement can be accomplished using either lump-sum, cost-plus, or a guaranteed maximum-price contract. A turn-key contract is a variation of the design-build contract because it usually includes provisions present in a design-build contract while possibly incorporating land acquisition, financing, and/or lease arrangement activities.

17.2 ENGINEER'S DUTIES

The engineer's contract for professional services requires the performance of various duties during the construction phase, which could include assisting the developer in securing, tabulating, and analyzing the bids, culminating in a recommendation for contractor engagement. In addition, the engineer can assist in the preparation of the formal contract documents.

17.3 JOB BIDDING AND/OR NEGOTIATION

The developer can solicit competitive bids for contract award by placing notice of the impending work in local newspapers, trade journals, and local trade offices. Also, the developer can notify contractors of the impending work and verbally invite their participation in the bid process. Should the project involve private funds, the developer may wish to negotiate or informally solicit proposals from one or several contractors who desire to perform the required work, from which one can be chosen.

17.3.1 Advertisement for Bids

The typical notice includes a description of the work along with its scope and location. Information concerning the type of bid (with formalities or the reserved right to reject any and all bids) and the bid opening date, place, and time are included along with the method that will be used to base the contract award. Requisite information pertaining to bonds, deposits, insurance, and payment schedules may be included. For further information, the notice lists, along with the owner, the engineer's firm, where copies of the applicable documents can be reviewed or obtained, and the amount of security deposit may be required.

All interested contractors who desire copies of the bid documents should be required to sign for a numbered document set when placing their document security deposit. This contractor roster, which includes the date when documents were issued, shows an accounting of what documents (and addenda) were issued and includes the contractor's address and telephone number in case subsequent addenda are issued to clarify questions or to modify the contract document's scope.

Advertising for bids is not in itself a contractual offer on the part of the developer; however, a bid response to such an advertisement is an offer that creates no legal rights for the bidder until it is accepted by the developer.

17.3.2   Prebid Conference

On occasion, a prebid conference is helpful to provide contractors with general information and an overview of the job requirements. Also, specific elements of the job can be emphasized, and other pertinent information can be issued. Care must be exercised to refrain from making comments contrary to material contained in the issued plans and specifications. An audio recording of the meeting from which a transcription can be prepared is invaluable. Copies should be made available to all potential bidders. The purpose of this conference is to promulgate information.

Should contractors encounter problems in interpreting the plans or specifications, the engineer can render an interpretation in the form of numbered addenda that can be provided to the party from which the issue arose. Copies are sent to all other potential bidders, and acknowledgment is documented, possibly by certified mail.

17.3.3   The Bid Package

The bid package issued to each bidder usually includes a copy of the invitation to bid, instructions for the bidder, a bid form to be used when preparing for submission, and all previously issued addenda. Included forms must be used by the bidder, otherwise the bid might be considered nonresponsive.

The instruction to bidders is a document that informs all bidders of the various aspects of the job. This provides uniform distribution of material and knowledge to all bidders from which they may prepare unbiased bids. The EJCDC has prepared a guide (NSPE et al., 1996c) that can be of value to the engineer when preparing these instructions.

Usually the instruction for bidders requests the bidder to provide evidence of construction experience along with a financial statement and equipment inventory or to have submitted prequalification data. The Association of General Contractors (AGC, undated) has available a standard form that can be used to obtain pertinent information about potential bidders.

The instructions to bidders can include procedures for the bidder to use when preparing and submitting the bid that reference the construction plans, specifications, and issued permits. If the contract is on a unit price basis, the estimated quantities of each unit of work are given; however, if a lump-sum contract relies on the total cost of the job for bid purposes, it is usually good practice for the engineer to provide a form that has the contractor account for the lump-sum bid. This information can be useful in determining whether the bid reflects consideration of all aspects of the job. Normally included would be a statement as to the accuracy of the information being provided relative to subsurface data, test borings, and so on. Information on bid formalities and provisions for the rejection of any bid or bids should be included. The information for bidders also includes the allowable length of time to complete the work.

To ensure that all bids will be prepared essentially in the same format, an appropriate bid form should be included. This will help in the bid analysis to detect any informalities. The bid form should include the price for which the contractor will perform the work. Bid surety requirements can be included. Two standard-form EJCDC documents for bid bonds are available (NSPE et al., 1996h, i). Bid bonds are not an expense to the developer. Bonding companies, hoping to later provide performance and payment bonds (if an award is made to their bid bonded contractor), usually provide the bid bond as a service to the bidding contractor.

Performance bonds, if required, ensure that the contractor will complete all terms of the contract in a satisfactory manner. Payment bonds are different in that they ensure that the contractor will pay all bills and will hold the developer harmless for any liens or financial claims relating to the job after the developer has satisfied all financial obligations with the contractor. An available EJCDC form (NSPE et al., 1996f) can be used as part of the contract documents when performance bonds are being stipulated along with another EJCDC form that relates to the payment of bonds (NSPE et al., 1996g), when required. Bid forms should list all addenda included with the plans and specifications. The contractor also should include a list of proposed subcontractors to be used on the job, if required. The form usually includes a certification that the site was examined and that the plans and specifications are understood by the bidder, in addition to certifying that no fraud or collusion was committed in the bid preparation. This form must be signed and witnessed prior to bid submission.

17.3.4 Bid Acceptance and Opening

On competitive bid jobs sealed bids are accepted by a designated receiver up to a stipulated deadline. For each bid that is received a written acknowledgment is provided the originator. At the designated time and place of bid opening, the developer, usually with the assistance of the engineer, opens all submitted sealed bids and reads aloud the bid contents. A record of all bid submittals is made that includes each participating contractor and the amount bid. Each bid will initially be examined for conformance with job specifications and checked to ensure that all addenda have been received and acknowledged on the required bid form. All bid bond information must be included. Also, any specification requirements for the bidder to provide potential subcontractors must be noted for subsequent review. If the bid complies with all criteria, it is accepted and tabulated. Once the tabulations are complete, the apparent low bidder is determined and announced, subject to later bid review and verification.

17.4 BID REVIEW AND ANALYSIS

The developer usually awards the contract to the lowest responsible bidder, while considering the contractor's experience in performing work similar to the solicited work. In addition, financial and organizational stability may be considered along with the firm's reputation, past performance, and safety record.

17.4.1 Verification of Bids

At a convenient time following the bid opening, the engineer reviews the submitted bid material and verifies all cost breakdowns and totals. If the contract is based on unit prices, care must be exercised by the engineer to ensure that the bid is balanced.

Occasionally contractors will raise the price on certain types of work and make corresponding reductions on other types, which produces an unbalanced bid.

All remaining information is double-checked, as a summary is prepared for the developer's use. Once the summary is completed, the engineer can make a considered determination of which contractor to recommend for the job to the developer, who will make the final determination.

17.5 CONTRACT AWARD

The developer usually awards the contract to the lowest responsible bidder on a competitive bid job. Circumstances may require the developer to abandon the project if the bid prices exceed a project's estimated construction cost, which is one reason for the developer to include in the bidding documents the right to reject any and all bids.

17.5.1 Postbid Activity

The developer notifies the successful bidder with either a letter of intent to award the contract and proceed with the project or a notice of award similar to one prepared by the EJCDC (NPSE et al., 1996e). The engineer is now free to return all unsuccessful bidders any bid bond that was specified except the bonds pertaining to the remaining two next lowest bidders and to notify all bidders of the successful bidder's name and bid price.

17.5.2 Preconstruction Conference

The engineer normally schedules a preconstruction conference with the successful bidder to discuss various housekeeping details associated with the job, such as payment requests, change orders, submittals, sedimentation and erosion control plan requirements, and shop drawing reviews. Topics could include a review of the project's scope and permit requirements along with the anticipated contract administration procedure that will be utilized. Temporary staging areas, storage areas, and construction easements are discussed. The engineer usually discusses the procedure to be used for field control establishment and the importance of marker protection. The contractor is usually requested to provide the engineer with an anticipated work and progress schedule that can be used to monitor the job and to satisfy regulatory requirements such as those associated with certain sedimentation and erosion control permits. Arrangements for a field office are discussed if specified as a requirement. During this conference the contractor is reminded of any contract administration submittals to be made prior to commencing work.

17.5.3 Review of Contractor's Submittals

The contractor must submit to the engineer for review all information specified in the bid documents. This process takes place between the notice of award being made and the time the actual work begins. Documents normally include performance and payment bonds, insurance certificates, the preliminary construction schedule, the name of the job superintendent, the field office telephone number, and the name of the designated job safety supervisor.

17.5.4   Contract Execution

As soon as reasonably possible, pending last-minute project approvals, a formal contract is delivered to the successful contractor, which must be executed along with the developer. Two typical forms of contracts prepared by the EJCDC that can be utilized are

(a)   *Standard Form of Agreement between Owner and the Contractor on the Basis of a Stipulated Price* (NSPE et al., 1996b) and

(b)   *Standard Form of Agreement between Owner and Contractor on the Basis of Cost Plus* (NSPE et al., 1996a).

The agreement, which must be signed and witnessed by the developer and contractor, establishes the rules of engagement between the two parties to finish the project. The contractor's bid bond is now returned upon execution of the contract. A "notice to proceed" can now be issued by the developer to the contractor similar to the one prepared by the EJCDC (NSPE et al., 1996e). This notice establishes the beginning time from which the allowable contract duration is reckoned (to achieve both substantial and final completion) and alerts the contractor to commence work. Once the agreement is signed, the remaining two next lowest bidders' bid bonds can be returned. Should the agreements not be signed by the successful bidder, then the next lowest responsible bidder can be awarded the job, while the owner utilizes the bid bond protection provided by the initial successful bidder.

Contractors sometimes place in the public record office a Notice of Project Commencement which must often be filed within a specified time period (15 days). This legal notice can help protect the contractor from mechanics liens filed by second- or third-tier subcontractors or suppliers with whom the prime contractor has no contact (S.C. §29-5-20). However, subcontractors and suppliers who are scheduled to work on the job have an opportunity to notify the prime contractor once the Notice of Project Commencement is on record so that they can ensure payment will be made for their involvement. Using a procedure such as this helps to ensure the satisfactory completion of the job.

REFERENCES

Associated General Contractors of America (AGC). *Contractors' Experience Questionnaire and Financial Statement—Standard Form No. 28.* Washington: AGC.

National Society of Professional Engineers (NSPE), the American Consulting Engineers Council, the American Society of Civil Engineers (ASCE), and the Construction Specifications Institute (CSI). *Standard Form of Agreement between Owner and Contractor on the Basis of Cost-Plus* (EJCDC No. 1910-8-A-2.) New York: ASCE, 1996a.

National Society of Professional Engineers (NSPE), the American Consulting Engineers Council, the American Society of Civil Engineers (ASCE), and the Construction Specifications

Institute (CSI). *Standard Form of Agreement between Owner and Contractor on the Basis of a Stipulated Price* (EJCDC No. 1910-8-A-1.) New York: ASCE, 1996b.

National Society of Professional Engineers (NSPE), the American Consulting Engineers Council, the American Society of Civil Engineers (ASCE), and the Construction Specifications Institute (CSI). *Guide to the Preparation of Instructions to Bidders* (EJCDC No. 1910-12.) New York: ASCE, 1996c.

National Society of Professional Engineers (NSPE), the American Consulting Engineers Council, the American Society of Civil Engineers (ASCE), and the Construction Specifications Institute (CSI). *Notice of Award* (EJCDC No. 1910-22.) New York: ASCE, 1996e.

National Society of Professional Engineers (NSPE), the American Consulting Engineers Council, the American Society of Civil Engineers (ASCE), and the Construction Specifications Institute (CSI). *Notice to Proceed* (EJCDC No. 1910-23.) New York: ASCE, 1996e.

National Society of Professional Engineers (NSPE), the American Consulting Engineers Council, the American Society of Civil Engineers (ASCE), the Construction Specifications Institute (CSI), the Surety Association of America, the Associated General Contractors of America (AGC), and the American Institute of Architects (AIA). *Construction Performance Bond* (EJCDC No. 1910-28-A.) New York: ASCE, 1996f.

National Society of Professional Engineers (NSPE), the American Consulting Engineers Council, the American Society of Civil Engineers (ASCE), the Construction Specifications Institute (CSI), the Surety Association of America, the Associated General Contractors of America (AGC), the American Institute of Architects (AIA), the American Subcontractors Association, and the Associated Specialty Contractors. *Construction Payment Bond* (EJCDC No. 1910-28-B.) New York: ASCE, 1996g.

National Society of Professional Engineers (NSPE), the American Consulting Engineers Council, the American Society of Civil Engineers (ASCE), and the Construction Specifications Institute (CSI). *Bid Bond—Penal Sum Form* (EJCDC No. 1910-28-C.) New York: ASCE, 1996h.

National Society of Professional Engineers (NSPE), the American Consulting Engineers Council, the American Society of Civil Engineers (ASCE), and the Construction Specifications Institute (CSI). *Bid Bond—Damages Form* (EJCDC No. 1910-28-D.) New York: ASCE, 1996i.

Smith, R. J. Prepared for the National Society of Professional Engineers (NSPE), the American Consulting Engineers Council, the American Society of Civil Engineers (ASCE), and the Construction Specifications Institute (CSI). *Recommended Competitive Bidding Procedures for Construction Projects* (EJCDC No. 1910-9-D.) New York: ASCE, 1987.

SUGGESTED READINGS

Abbett, R. W. *Engineering Contracts and Specifications*, 4th ed. New York: Wiley, 1965.

Levin, P. *Construction Contract Claims, Changes, and Dispute Resolution*, 2nd ed. Reston: American Society of Civil Engineers, 1998.

National Society of Professional Engineers (NSPE), the American Consulting Engineers Council, The American Society of Civil Engineers (ASCE), and the Construction Specifications Institute (CSI). *Guide to the Preparation of Supplementary Conditions* (EJCDC No. 1910–17.) New York: ASCE, 1996d.

O'Reilly, M. *Civil Engineering Construction Contracts*. London: Telford, 1996.

State of South Carolina. Code of Laws S.C. §29-5-20, as amended.

18 Layout, Observation, and Monitoring

The engineer's role during the construction phase may appear to conflict with interests of the developer or contractor. However, the engineer possesses professional knowledge that requires unbiased performance if the project is to be successful and safe for use. In addition, the developer should engage the engineer during construction to assist in the implementation of the plans that are inherently close to the engineer. This can help to protect the developer from inferior materials and workmanship. Having the engineer involved during the construction phase helps to maintain the construction schedule by resolving problems rapidly. Contract documents should therefore reserve the right for the engineer to make periodic job observations.

18.1 SCOPE OF ENGINEERING ACTIVITIES

The engineer's activities can fall under the categories of contract administration and construction administration during the construction phase. These activities are regulated by the terms of the engineering services contract contained in Chaper 3, depending on whether the customary approach or the customary approach with project management is used.

18.1.1 Customary Approach

After the construction contract award has been made, duties associated with its administration include the engineer's review of all job-related mill and equipment performance reports, documents, and submittals including construction, shop, and erection drawings. This is in addition to any reviews made by the designated safety supervisor responsible for site safety (Section 3.4.1). The *EJCDC Standard General Conditions of the Construction Contract* (NSPE et al., 1996a) requires the contractor to submit a schedule of values along with three preliminary schedules for progress, shop drawings, and sample submittals. If questions arise during construction, the engineer makes interpretations of the plans and specifications in a manner that is equitable for both the developer and the contractor to attempt to resolve potential claims. The engineer can also provide plan modifications to resolve conflicts caused by actual field conditions. These modifications could be reflected in change orders formulated by the engineer. The engineer makes periodic field visits to observe the progress of work for the purpose of processing payment requests and to view the work for general conformance with the approved plans, specifications, and issued permits. These intermittent visits do not place any responsibility on the engineer for the contractor's performance. It is the contractor's responsibility to maintain quality of materials, workmanship, and performance within the limits of the contract. According to Smith (1991), the *EJCDC Standard General Conditions of the Construction Contract* (NSPE et al., 1996a) stipulates that the contractor is responsible for the inspection of work. The engineer, however, warrants the design will work. The engineer also observes any specified field tests to ensure compliance with the construction documents. Upon completion of the final inspection, the engineer prepares record drawings for reference purposes and certifies that the constructed project conforms to the approved plans and specifications based on the engineer's knowledge of the project, including periodic observations and final inspection results.

18.1.2   Customary Approach with Project Management

Should the engineer be required to provide project management services in addition to design services, the engineer would in essence become the construction administrator. Construction administration includes all duties associated with contract administration and additional duties entailing the coordination, planning, scheduling, and closing out of the project. These duties are not undertaken to allow the engineer to dictate job activities or construction methods employed or to relieve the contractor of responsibility but are undertaken to ensure that certain milestone events are met. A resident project representative could be used to observe the contractor's prosecution of the work and to report any deviations from the approved construction plans, specifications, and construction permits. Although the designated safety supervisor might be contractually assigned to a party other than the design engineer, the engineer can possibly incur increased responsibility for job safety (over that associated with the customary approach) as a result of management-related issues.

The frequency and need for a resident project representative's services are governed by the project's size. For larger jobs a full-time representative may be justified; however, for smaller jobs, a part-time one might be sufficient.

18.1.3   Other Approaches

If a professional construction manager is engaged by the developer, the engineer would provide auxiliary services to support the project's management, as indicated in Chapter 3. If, however, the developer has engaged the engineer for a design-build or turn-key project, all factors would require the efforts of the engaged firm.

18.2   CONSTRUCTION LAYOUT

Site improvements must be staked on the ground to guide the contractor's work. Some construction documents require the contractor to provide construction stakeout as part of the work. Other contracts state that the owner will provide construction stakeout separately through an engineering firm that also provides surveying services or a land surveying firm. If the design professional has the capability and responsibility of providing field surveying services for the contractor to verify and use, then these services must be coordinated.

Construction stakeout using conventional means, such as a total station and data collector, requires access to three-dimensional coordinate values for those control points that are to be staked. These data, taken from the project's database or construction drawings, are stored in the data collector before starting the stakeout. In the field the station's location and orientation are set to the project's control, and the difference between a prism's shot location and corresponding stored coordinate location is used to set the control stakes. Total stations that operate robotically can be used for stakeout by a single person if the site is fairly free from obstructions.

Real time kinematic global positioning system (RTK GPS) surveying techniques (Section 4.10.3) can also be used if the site is fairly free from overhead obstructions. As with topographic surveying using RTK technology, the survey must be calibrated horizontally and vertically to established field controls. If the site was previously surveyed using RTK GPSs for gathering topographic data, then the calibration controls used previously can be reused, if preserved. As in the case with conventional surveying stakeout techniques, RTK GPSs require the surveyor to preload known three-dimensional coordinate values. After calibration, RTK GPS receiver positions are compared with corresponding coordinate values to guide setting construction control stakes. Periodic field checks to established control allow the surveyor a way to ensure quality control.

18.2.1   Clearing and Grubbing Control

The initial construction stakes set in the field for the contractor are flat stakes, usually 24 inches tall, to delineate the limits of clearing and grubbing (removal of roots and stumps) for all road rights-of-way, easements, and other areas requiring clearing. These stakes can also serve as limits for various sedimentation and erosion control practices. These stakes are marked with high-visibility material to aid the equipment operator's perception of areas to be cleared. Often these stakes are destroyed during the clearing and grubbing operation.

18.2.2   Rough Road Grading Stakes

Once the clearing and grubbing operation has been completed and all necessary topsoil has been stripped from the roadbed, the contractor requires control stakes to be set in the field for both line and grade. These stakes usually include centerline hubs (stakes set flush with the ground surface to locate a tack point of known position) and flat stakes (stakes similar to the ones used for clearing to guard and identify with inscribed information the adjacent hub). Both types are shown in Figure 18.1. Accurate staking of the centerline is essential because all horizontal and vertical control for other improvements relates to the road's centerline. Figure 18.2 illustrates a technique that can minimize field errors resulting from the use of short instrument backsights when establishing the proposed road's horizontal centerline control. Offset hubs are positioned normal to the centerline at each station or half station in addition to any intermediate control points such as PIs, PCs, and PTs. Cut-or-fill data are placed on the flat stake so that the contractor can shape the roadbed. If poor soil is encountered, the limits of mucking and backfill must be established to document the amount of classified excavation work required to provide an acceptable subgrade in that area. Slope stakes, if required, are set at this time.

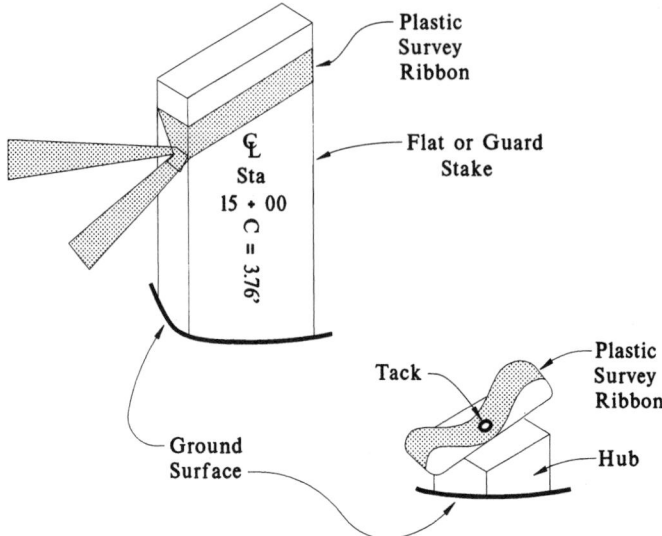

FIG. 18.1 Survey stakes for horizontal and vertical control in the field.

FIG. 18.2 Field procedure to correct for sighting error caused by using a short instrument backsight.

18.2.3 Sanitary Sewer Collection System

The gravity sanitary sewer collection system installation usually begins after the contractor has rough graded the roads. This is necessary because the sanitary system usually requires the deepest excavations. If the contractor is utilizing laser control equipment, hubs and tack points are usually set for each manhole's center with a cut distance marked on the accompanying flat stake to the manhole's invert. An offset stake or stakes may be required to preserve the manhole's centerline information, as the centerline stakes are removed when excavation begins. Tap locations and the terminus of service lines must also be marked in the field.

If the sanitary sewer collection system is a pressure type, it can be installed using the same general procedures that apply to potable water lines.

18.2.4 Storm Drainage Collection System

The storm drainage collection system is usually the second deepest system installed in a development. It normally is installed after the gravity sanitary sewer collection system is in place. The centerlines of all proposed catch basins, drop inlets, and junction boxes are staked in the field, with cut data being supplied on flat stakes to the inverts. If offset hubs are required, they too must be established.

Also, the engineer must establish control stakes along all culvert and open ditch locations for the contractor to use.

18.2.5 Potable Water System

One of the last systems to be installed is the potable water system, mainly because it is fairly shallow and has pressure flow that is not dependent on critical elevations other than at pipe crossings. Valve, hydrant, thrust block, blow-off, and service tap locations must be established in the field, along with the locations where pipe diameters change and terminate.

18.2.6 Fine Grading Road Stakes

Once the potable water system is installed, the engineer can reestablish all center-line control stakes for the roads and place offset blue-topped stakes along the proposed edge of the subgrade. These stakes are set at a known horizontal and vertical position and are called blue tops. Other necessary control stakes, including those for curb and gutter placement, can be established in the field to guide the contractor's activities.

18.2.7 Miscellaneous Field Control

Other field control markers can be set by the engineer to establish batter boards and finished floor elevations, if required. Batter boards are construction stakes that are commonly used in the field to locate building corners and walls. Figure 18.3 illustrates the use of these control devices.

18.3 ADMINISTRATIVE DUTIES

Certain activities are required of the engineer to minimize confusion, provide accurate accounting of events, and to ensure project success. These activities include administrative duties that are as important as other engineering duties associated with a development project.

FIG. 18.3   Examples of batter board locations for building layout.

18.3.1   Records Keeping

The engineer should maintain various types of records that are used for documentation and reference purposes, including a daily activity log. Telephone communication logs are important to record progressive activities. In addition, field observation records, delivery ticket stubs, and material test results normally performed by outside testing firms should be catalogued by the engineer. Each record should include various modifying information to render a complete and accurate portrayal of information. Other records kept for accurate administration of the contract include change order and extra work records. Additional records include payment requests, progress reports, and material deliveries.

18.3.2   Submittal Review

The engineer must review all working, shop, and erection drawings submitted by the contractor to ensure that they are in compliance with the approved plans and specifications. In addition, all "cut sheets" (manufacturers' specification sheets) and samples that are submitted must be reviewed for approval. Should the job require the contractor to submit concrete or asphalt mix designs, the engineer must carefully check these to ensure compliance with appropriate design criteria.

The specifications may require the contractor to submit for administrative purposes copies of all contractor procured engineering calculations and drawings pertaining to sheeting, bracing, and shoring systems that will be used on-site.

18.4   CONSTRUCTION CONFLICTS AND EXTRA WORK

Conflicts usually result in alterations and modifications being made to the construction drawings, which may translate into change orders. These change orders are formulated by the engineer through authority provided in the contract documents. The engineer must be vigilant to avoid any apparent or actual conflict of interest (or fraud) when exercising these duties. Figure 18.4 illustrates a typical change order document.

18.4.1   Differing Site Conditions

Contract plans and specifications usually include information on subsurface site conditions. Should this information prove not to be significantly representative of the actual field conditions, the contractor may be entitled to additional remuneration or time if unanticipated additional work is involved. On occasion, specifications may state that supplied subsurface information is for the developer's benefit and that the contractor is responsible for making whatever investigations are necessary to determine the extent of subsurface working conditions prior to bid.

According to Smith (1991, pp. 68–69), the revised EJCDC general conditions provide "that the owner (developer) should indemnify the contractor, its subcontractors, the engineer and its consultants for consequences of unanticipated hazardous conditions on the site. "These conditions can be attributed to the discovery of previously unknown asbestos, hazardous waste, petroleum, PCBs, or radioactive substances on site. If these types of materials are encountered, the contractor can stop work until the situation is rectified by the developer.

18.4.2   Plans and Specification Changes

The engineer may be required to alter the approved plans and specifications as a result of several reasons. If errors are detected in the plans or specifications during construction or, if necessary, information was omitted or is found to deviate substantially from that portrayed in the documents, changes may be required for amplification and correction. Other changes can result from jurisdictional regulatory review authority requirements or from the developer revising a portion of the project to address new needs. These changes may increase or decrease the scope of the project, resulting in additional compensation and/or time being given to the contractor.

18.4.3   Other Conditions

Other conditions affecting the contracted work might include a change in substantial or final completion times, resulting in the contractor being asked to accelerate work production. In addition, the scope can be altered because of damage being inflicted to completed work outside the contractor's control. Should the developer interfere with

CHANGE ORDER

No. _____

DATE OF ISSUANCE _____   EFFECTIVE DATE_____

OWNER _____
CONTRACTOR _____
Contract:_____
Project: _____
OWNER's Contract No. _____   ENGINEER's Contract No. _____
ENGINEER _____

You are directed to make the following changes in the Contract Documents:
Description:

Reason for Change Order:

Attachments: (List documents supporting change)

CHANGE IN CONTRACT PRICE:	CHANGE IN CONTRACT TIMES:
Original Contract Price $_____	Original Contract Times: Substantial Completion: _____ Ready for final payment: _____ (days or dates)
Net Increase (Decrease) from previous Change Orders No. ___ to ___: $_____	Net change from previous Change Orders No. ___ to No. ___: Substantial Completion: _____ Ready for final payment: _____ (days)
Contract Price prior to this Change Order: $_____	Contract Times prior to this Change Order: Substantial Completion: _____ Ready for final payment: _____ (days or dates)
Net increase (decrease) of this Change Order: $_____	Net increase (decrease) this Change Order: Substantial Completion: _____ Ready for final payment: _____ (days)
Contract Price with all approved Change Orders: $_____	Contract Times with all approved Change Orders: Substantial Completion: _____ Ready for final payment: _____ (days or dates)

RECOMMENDED: APPROVED: ACCEPTED:

By: _____   By: _____   By: _____
ENGINEER (Authorized Signature)   OWNER (Authorized Signature)   CONTRACTOR(Authorized Signature)

Date: _____   Date: _____   Date: _____

FIG. 18.4 Change order form. Source: Prepared by the Engineers Joint Contract Documents Committee and endorsed by The Associated General Contractors of America. NSPE, 1996b. American Consulting Engineers Council, American Society of Civil Engineers, The Construction Specifications Institute, Engineers' Joint Contract Documents Committee, (EJCDC No. 1910-8-B), 1996. Reproduced by permission of the American Society of Civil Engineers.

the contractor's prosecution of work, or suspend work, the outcome could result in contract changes.

18.5 WORK DIRECTIVES

A work directive is notice given to the contractor to concentrate efforts in a certain area due to emergency conditions. An example of a work directive is shown in Figure 18.5. This directive normally is used as a basis upon which a future change order is issued to account for the included work.

18.6 DELAYS

Time delays result in the passing of time when insufficient job progress is being made. Delays can result in the contractor either being granted an extension to complete the job or being suspended from work, depending on the cause of delay.

18.6.1 Owner-caused Delays

Owner-caused delays are compensable to the contractor. Such delays result from problems associated with construction plans and specifications that require revisions. In addition, the lack of available funds may result in the contractor not being paid on time, causing various delays. Other owner-caused delays can be attributed to inordinate time being required to respond to pertinent issues associated with the prosecution of the work, for example, not gaining access to the site, not issuing a timely notice to proceed, not certifying tests and evaluations in a timely manner, or not providing "owner-furnished" materials when required.

18.6.2 Contractor-caused Delays

Delays caused by the contractor can result from poor procurement of equipment and materials, manifested by inadequate lead times, or from financial problems. Delays resulting from poor scheduling and supervision or from the lack of subcontractor coordination can also cause the contractor problems. In addition, delays associated with contractor-procured permits for the job may result in problems.

18.6.3 Other Delays

Other delays that can keep the contractor from prosecuting project work in a timely manner range from unusual weather conditions (such as that associated with a hurricane) to labor strikes at the job site or supplier's location. Riots, wars, and natural disasters also cause unavoidable delays.

18.6.4 Disposition of Delays

Most delays caused by some event or someone other than the contractor result in the contractor seeking more compensation or an extension in time to complete the project.

WORK CHANGE DIRECTIVE

No. _____

DATE OF ISSUANCE _____   EFFECTIVE DATE _____

OWNER _____

CONTRACTOR _____

Contract: _____

Project: _____

OWNER's Contract No. _____   ENGINEER's Project No. _____

You are directed to proceed promptly with the following change(s):

Description:

Purpose of Work Change Directive:

Attachments: (List documents supporting change)

If OWNER or CONTRACTOR believe that the above change has affected Contract Price any Claim for a Change Order based thereon will involve one or more of the following methods as defined in the Contract Documents.

Method of determining change in
Contract Price:

☐ Unit Prices

☐ Lump Sum

☐ Cost of the Work_____

| Estimated increase (decrease) in Contract Price: $_____. If the change involves an increase, the estimated amount is not to be exceeded without further authorization. | Estimated increase (decrease) in Contract Times: Substantial Completion: _____ days; Ready for final payment: _____ days. |

RECOMMENDED: AUTHORIZED:

_____   _____
            ENGINEER   OWNER
By: _____   By: _____

FIG. 18.5 Work directive form. Source: Prepared by the Engineers Joint Contract Documents Committee and endorsed by the Associated General Contractors of America. NSPE, et al., Engineer's Joint Contract Documents Committee (EJCDC No. 1910-8-F), 1996d. Reproduced by permission of the American Society of Civil Engineers.

The engineer must use equity in resolving situations such as these by undertaking negotiations to issue appropriate change orders that reflect a mutual agreement between the developer and contractor. It is usually to the best interest of all parties involved to find agreement. Should negotiations fail, the contractor has the right to file a timely protest while performing the work in accordance with the contract documents. The unsettled issue can be arbitrated or settled in court as a result of litigation.

A contractor who relies on a CPM network analysis as a basis for filing a claim may find the analysis by itself provides insufficient proof. The contractor has the burden of proving that specific delays were the responsibility of the owner, that the delays resulted in the delay of the overall completion of the project, and that the owner-caused delays were not concurrent with delays triggered by the contractor, and the delays occurred along the critical path (Chapter 14).

18.7 CONSTRUCTION PROGRESS

Progress made when prosecuting the contractual work requires the engineer periodically to certify payment requests for completed work and material delivered. Requests for work payments are submitted only after work has been performed, using a form similar to that illustrated in Figure 18.6. The contractor's project schedule also is used by the engineer to monitor progress. The engineer reviews contractor-prepared revisions to this schedule to help ensure that the project will progress to completion in a timely manner.

18.7.1 Engineer's Observation

The contractor normally has responsibility for scheduling, prosecuting, and completing all work in accordance with the contractual documents unless a contract administrator is designated the coordinator. These documents usually call for the contractor to be responsible for job safety and to appoint a safety supervisor. The engineer who observes a job in progress has the obligation to cite anything that would contribute to loss of life, property, or limb regardless of responsibilities delineated within the contractual documents.

The observing engineer acts as the eyes and ears of the developer to monitor job progress and activities. While the contractor has the obligation to guarantee the quality of workmanship and installed materials, the developer has the responsibility of offering to the end user a product that is suitable for its intended use. This is part of the reason that the developer should secure a resident project representative and at the same time engage a reputable contractor for the job. The engineer has the obligation to report observed findings of defective or unacceptable work to the contractor so that corrective action can be undertaken as quickly as possible. Harboring critical information until the final inspection on the part of the engineer is unreasonable if the job is expected to be completed in a timely manner.

18.7.2 Regulatory Inspections

In most cases the installed improvements within a development are dedicated to the public for operation, maintenance, control, and ownership. Thus, the developer is

APPLICATION FOR PAYMENT NO. _____

To: _____ (OWNER)
From: _____(CONTRACTOR)
Contract: _____
Project: _____
OWNER's Contract No._____.   ENGINEER's Project No._____
For Work accomplished through the date of:_____.

1.	Original Contract Price:	$_____
2.	Net change by Change Orders and Written Amendments (+ or -):	$_____
3.	Current Contract Price (1 plus 2):	$_____
4.	Total completed and stored to date:	$_____
5.	Retainage (per Agreement):	
	_____% of completed Work: $_____	
	_____% of stored material: $_____	
	Total Retainage:	$_____
6.	Total completed and stored to date less retainage (4 minus 5):	$_____
7.	Less previous Application for Payments:	$_____
8.	**DUE THIS APPLICATION (6 MINUS 7):**	$_____

Accompanying Documentation:

CONTRACTOR'S Certification:

The undersigned CONTRACTOR certifies that (1) all previous progress payments received from OWNER on account of Work done under the Contract referred to above have been applied on account to discharge CONTRACTOR's legitimate obligations incurred in connection with Work covered by prior Applications for Payment numbered 1 through _____ inclusive; (2) title of all Work, materials and equipment incorporated in said Work or otherwise listed in or covered by this Application for Payment will pass to OWNER at time of payment free and clear of all Liens, security interests and encumbrances (except such as are covered by a Bond acceptable to OWNER indemnifying OWNER against any such Lien, security interest or encumbrance); and (3) all Work covered by this Application for Payment is in accordance with the Contract Documents and not defective.

Dated _____   _____
                                                              CONTRACTOR

                                        By:_____

State of _____
County of _____
Subscribed and sworn to before me this _____
day of _____, _____

_____
Notary Public
My Commission expires: _____

Payment of the above AMOUNT DUE THIS APPLICATION is recommended.

Dated _____   _____
                                                              ENGINEER

                                        By: _____

FIG. 18.6 Application for payment form. Source: Prepared by the Engineers Joint Contract Documents Committee and endorsed by the Associated General Contractors of America. NSPE, et al., Engineer's Joint Contract Documents Committee (EJCDC No. 1910-8-E), 1996c. Reproduced by permission of the American Society of Civil Engineers.

normally not the end user or long-term "owner." Many municipalities undertake some type of inspection and quality control program (possibly to include the developer paying for these inspection services) to ensure the finished project is acceptable. Should deficiencies be noted by the potential end user, the engineer and developer must rectify the unacceptable condition through the contractor such that the end product can be successfully dedicated (Baltz and Richardson, 1985).

REFERENCES

Baltz, L. J. III, and J. W. Richardson, Jr. "Inspection Costs Versus the Benefits Derived Avoiding Contract Disputes." In *Proceedings of a Symposium Sponsored by the Construction Division of the American Society of Civil Engineers in Conjunction with the ASCE Convention in Detroit, Michigan.* New York: American Society of Civil Engineers, 1985, pp. 92–99.

National Society of Professional Engineers (NSPE), the American Consulting Engineers Council, the American Society of Civil Engineers (ASCE), and the Construction Specifications Institute (CSI). *Standard General Conditions of the Construction Contract* (EJCDC No. 1910-8) New York: ASCE, 1996a.

National Society of Professional Engineers (NSPE), the American Consulting Engineers Council, the American Society of Civil Engineers (ASCE), and the Construction Specifications Institute (CSI). *Change Order* (EJCDC No.1910-8-B). New York: ASCE, 1996b.

National Society of Professional Engineers (NSPE), the American Consulting Engineers Council, the American Society of Civil Engineers (ASCE), and the Construction Specifications Institute (CSI). *Application for Payment.* (EJCDC No. 1910-8-E). New York: ASCE, 1996c.

National Society of Professional Engineers (NSPE), the American Consulting Engineers Council, the American Society of Civil Engineers (ASCE), and the Construction Specifications Institute (CSI). *Work Change Directive* (EJCDC No. 1910-8-F). New York: ASCE, 1996d.

Smith, R. J. "Construction Documents Revised." In *Civil Engineering.* New York: ASCE, 1991, pp. 68–69.

SUGGESTED READINGS

Davies, V. J., and K. Tomasin. *Construction Safety Handbook,* 2nd ed. London: Telford, 1996.

Heberle, D. *Construction Safety Manual.* New York: McGraw-Hill, 1998.

Hislop, R. D. *Construction Site Safety—A Guide for Managing Contractors.* New York: Lewis, 1999.

Levin, P. *Construction Contract Claims, Changes, and Dispute Resolution,* 2nd ed. Reston: American Society of Civil Engineers, 1998.

19 Completion and Start-Up

As the contractor nears completion of an improved land project, the engineer must check the various installed systems to see that they are operational. This requires the observation of tests and possibly making various field measurements to certify the job is installed in accordance with the approved construction documents. Usually these tests are coordinated with the overseeing regulatory agencies and ultimate system owner.

Total development activities include closeout procedures that are similar to those used for improved land projects and others associated with the additional improvements and structures.

19.1 SYSTEM TESTING AND START-UP

Utility systems controlled by health officials normally require testing for certification purposes. When system testing and evaluation are undertaken, most regulatory agencies send observers to ensure that the systems are in compliance prior to placing them into service.

19.1.1 Potable Water System

The water distribution system must meet various test criteria to be serviceable. First, the system must be capable of withstanding a predetermined internal pressure for a prescribed duration when a hydrostatic pressure test is conducted. Maintaining the system test pressure indicates that no leaks exist. Once the system has been determined to be tight, a flow test is conducted to ensure that proper fire fighting flows are available for protection. This test is similar to the hydrant flow test described in Chapter 11. Should this test's results prove unsatisfactory, the engineer should require the contractor to check to ensure that all new valves are in the open position. If this fails to produce satisfactory results, a recheck of the flow and pressure at the point of supply is necessary to determine if the current performance of the water system is approximately what it was when initially flowed while obtaining design field data. If the contractor tapped the existing system's water main, it is possible the contractor could have inadvertently used a drill bit appreciably smaller than the diameter of the new water line extension, resulting in a constriction. It is good practice to have the contractor wire the removed plug of pipe to the tapping sleeve when the connection is made to the existing line so that visual proof is available to show that the proper size hole is available to supply water to the new extension. If the hole is of proper size but the flows are still inadequate, the new system may have an obstruction such

756

as a rag, root, or hand tool. The contractor must reexcavate sequential segments of the line to systematically flush the system to attempt to find the problem and correct it.

A final test requirement prior to placing the water system into operation is to check the newly charged water main for bacteria-free water. Contamination of water mains resulting from open air exposure during construction can be eliminated by chlorination. Grab samples must be obtained and tested to ensure that a minimum chlorine residual exists and that there are no pathogenic organisms present. Should bacteria tests indicate contamination, the contractor must resterilize the system. If results are still unsatisfactory, the contractor must dismantle segments of the installed system to flush the mains, because small animals occasionally become trapped in the line during construction.

19.1.2   Sanitary Sewer Collection System

The watertight integrity of the completed sanitary sewer collection system must be checked by the contractor. This will ensure that no leaks exist in the system that would allow wastewater to leach into the surrounding area, especially at ditch crossing. It also prevents groundwater from entering the collection system. Any such groundwater would then have to be treated. Tests for the system's integrity include infiltration, exfiltration, or air pressure testing between manholes. These tests should indicate satisfactory joint and fitting tightness. If not, the contractor must reexcavate the line to locate and repair the leak. If a portion of the installed wastewater system consists of pressure forcemains, then a hydrostatic time test must be conducted similar to the one used on the potable water system.

Regulatory authorities often require the engineer to certify that the installed gravity system is in accordance with the approved plans and specifications. To certify alignment, the engineer must lamp (look through using a light beam) each line to ensure that a uniform and straight bore exists from manhole to manhole. Grade certification can be accomplished only if the engineer observes each invert elevation using a level and measures the horizontal distance between installed manholes. Should any section of line indicate less than minimum grade or any misalignment, the contractor must take corrective action. Other agency requirements may include a mandrel test, in which a steel sleeve is pulled through the installed sewer main to prove that the pipes have not deflected out of tolerance when certain pipe materials are used.

19.1.3   Sanitary Sewer Pumping Station

Should the project include a sanitary sewer pumping station, various component tests must be conducted to ensure proper operation. Initially, the electrical system must be checked for circuit continuity and resistance testing. Pump motors, controls, and appurtenances, such as sump pumps, ventilators, dehumidifiers, and so forth, must also be tested. Usually the equipment supplier is charged in the construction specifications with providing start-up supervision and inspection. Observational tests to ensure that the equipment is rotating properly without excessive vibration or overheating and without drawing excessive power are essential, as is a determination that the equipment is meeting the specified pumping rate and head.

19.2   FINAL INSPECTION

All installed improvements must be inspected by regulatory authorities that have indicated a willingness to accept the completed systems. Approval allows the developer to petition for dedication of the system and, at the same time, usually establishes the point in time from which the warranty duration is reckoned. As with system testing and start-up, regulatory agency representatives normally participate in the final inspection. At this inspection most authorities request copies of the prepared plats delineating all rights-of-way, easements, and other dedicated areas that will provide space for later operational and maintenance activities.

Regulatory authorities also want an accurate record of all improvements, called an as-built survey, for their permanent files. As-built surveys reflect any deviations or modifications to the approved construction drawings made during construction. Surveyors using conventional techniques (or GPS RTK technology) remeasure installed systems and then change the construction drawings to reflect actual conditions. Often a contractor will "red-line" a set of drawings to record data (such as the location of buried sewer tap wye fittings) that can serve as an as-built or be used to supply information that is combined with other as-built data. As-built surveys used to secure permanent mortgages must be performed to established standards where minimum quality control requirements must be met (Section 4.8.5). Authorities now often require electronic as-built or record drawing (see Section 19.4.3) files that can be added to a GIS database, if one exists.

19.2.1   Road and Storm Drainage Systems

Most road and storm drainage systems are inspected after final cleanup and after rights-of-way have been dressed with topsoil and seeded. The roadbed is inspected for uniformity and to ensure that low spots do not exist where water might puddle, especially if curb and gutter sections are used. Pavement core samples strategically obtained may be required to ascertain pavement thickness and base material depth. Storm drainage culverts, junction boxes, and catch basins are inspected for alignment and grade; these inspections are similar to those conducted for the gravity wastewater collection system. Inspectors also ensure that all features included on the plans are present, such as French drains, riprap, energy dissipators, and detention/retention facility components.

19.2.2   Potable Water System

Final inspection of the water system includes receiving copies of certified system test results and conducting a visual inspection to ensure that all valves are properly installed with their boxes, that fire hydrants are situated properly and anchored with adequate drain pit areas, and that thrust blocking is in place. This last item is accomplished by inspecting random locations or by requiring the contractor to keep the thrust blocks exposed until verified.

19.2.3 Sanitary Sewer Collection System

The sanitary sewer collection system must be flushed clean of sediment and debris and must be fully completed when the final inspection is conducted. All integrity tests results and certificates by the engineer that the installed mains are on line and grade are given to the local jurisdictional authorities. The entire system is visually inspected by lamping, and all manhole drops and inverts are inspected for smoothness and uniformity. In addition, position of each manhole's rim and cover, relative to the surrounding area, is inspected along with the manhole joints to ensure watertightness that will prevent infiltration. Should the rim be located in an area where appreciable stormwater might enter, a watertight gasket may be required on that manhole.

19.3 APPROVAL OF IMPROVEMENTS

If all improvements are acceptable to the developer, engineer, and regulatory agencies, then certificates of approval are issued by the engineer and the regulatory agencies having jurisdictional control. This is usually followed by a permit from the regulatory agency to place into operation the new system. The developer can file a certificate of completion with the local public records office for liens notice. The engineer is free to accept and process the final progress payment request from the contractor and to notify the developer that the facility is ready for use. If the deadline date is overrun during a "time is of the essence" construction contract, liquidated damages can be assessed by the developer and deducted from the money due the contractor.

19.3.1 Discrepancy Corrections

If discrepancies are detected during testing or inspection, the contractor must take corrective action to eliminate them. Usually these discrepancies are entered on a list, called a punch list, that is used as a basis for a follow-up inspection. Should the engineer issue a certificate of substantial completion, as illustrated in Figure 19.1, it would represent that the inspection revealed minor problems that are shown on a punch list; however, all other work is approved as is. With the issuance of this certificate the contractor continues to maintain responsibility and control over the job until it is fully acceptable.

If the engineer issues a beneficial occupancy certificate instead, it allows the developer to occupy and use the incomplete facility. The contractor, however, maintains the obligation to complete the contracted work while the developer assumes the responsibility of operation and maintenance of the facility, including the procurement of insurance, heat, and utilities. Product warranties for improvements installed in the facility usually begin at this time, although the contractor's warranty normally does not begin until a certificate of completion is issued.

19.4 OPERATIONAL NEEDS

If the installed improvements include systems that require operational guidance, the contractor must ensure that such guidance has been provided prior to completing the job.

CERTIFICATE OF SUBSTANTIAL COMPLETION

DATE OF ISSUANCE _____

OWNER _____
CONTRACTOR _____
Contract: _____
Project: _____

OWNER's Contract No. _____   ENGINEER's Project No. _____

This Certificate of Substantial Completion applies to all Work under the Contract Documents or to the following specified parts thereof:

To_____
                                    OWNER

And To _____
                                    CONTRACTOR

The Work to which this Certificate applies has been inspected by authorized representatives of OWNER, CONTRACTOR and ENGINEER, and that Work is hereby declared to be substantially complete in accordance with the Contract Documents on

_____
DATE OF SUBSTANTIAL COMPLETION

A tentative list of items to be completed or corrected is attached hereto. This list may not be all-inclusive, and the failure to include an item in it does not alter the responsibility of CONTRACTOR to complete all the Work in accordance with the Contract Documents. The items in the tentative list shall be completed or corrected by CONTRACTOR within _____ days of the above date of Substantial Completion.

FIG. 19.1 Certificate of substantial completion. Source: Prepared by the Engineer's Joint Contract Documents Committee and endorsed by The Associated General Contractors of America. NSPE, et al., Engineer's Joint Contract Documents Committee (EJCDC No. 1910-8-D), 1996. Reproduced by permission of the American Society of Civil Engineers.

The responsibilities between OWNER and CONTRACTOR for security, operation, safety, maintenance, heat, utilities, insurance and warranties and guarantees shall be as follows:

OWNER:_____

_____

_____

CONTRACTOR:_____

_____

_____

The following documents are attached to and made a part of this Certificate:

[For items to be attached see definition of Substantial Completion as supplemented and other specifically noted conditions precedent to achieving Substantial Completion as required by Contract Documents.]

This certificate does not constitute an acceptance of Work not in accordance with the Contract Documents nor is it a release of CONTRACTOR's obligation to complete the Work in accordance with the Contract Documents.

Executed by ENGINEER on _____
                              Date

_____
                ENGINEER

By: _____
          (Authorized Signature)

CONTRACTOR accepts this Certificate of Substantial Completion on _____
                                                                        Date

_____
                CONTRACTOR

By: _____
          (Authorized Signature)

OWNER accepts this Certificate of Substantial Completion on _____
                                                                  Date

_____
                OWNER

By: _____
          (Authorized Signature)

FIG. 19.1   (continued)

19.4.1   Instruction Manuals

All instruction and operation/maintenance manuals must be supplied and reviewed with the end user to ensure that the facility will function properly. If the specifications provide for operator training, which is necessary when mechanical equipment is involved, then provisions for on- or off-site instruction must be undertaken (usually by the equipment supplier).

19.4.2   Delivery of Specified Materials

Often with the installation of mechanical equipment various special tools, spare parts, and lubricants are provided by the supplier as part of the construction specifications. These items should be inventoried and turned over to the end user for safe storage. Facility keys should be logged and turned over to the owner for security purposes.

19.4.3   Peripheral Document Review

The contractor must supply the developer with all product warranties, such as the roof bond, and certificates of any other tests or inspections that pertain to the facility. In addition, the contractor must supply the engineer with all recorded measurements and field notations from which the engineer can prepare copies of record drawings, if specified. Most jobs today require electronic format record drawings that are prepared by the engineer or surveyor for creating reproducibles from which multiple copies can be made for distribution to all involved parties. The electronic drawing file is usually submitted where its contents can be added to the local GIS database.

19.5   FINAL STAKE-OUT OF PARCELS

After all improvements have been installed and the project has been cleaned up and dressed, the area is prepared for the final property monuments to be installed by a registered land surveyor. These monuments must be installed to established standards of precision and must reflect accurately on the ground the property corners that are represented on the final development plat to be recorded.

19.6   RELEASE OF RETAINAGE

Prior to releasing all retained funds due the contractor, a check of the public records office must be made to ensure that no liens have been filed against the project. In addition, the contractor may be requested to submit a waiver of liens and a certificate of surety before a recommendation for final payment can be made by the engineer. Upon receiving all monies due from the developer, the contractor can issue a certificate of completed payment for all work performed. Completion of these details allows the developer to obtain an unencumbered project that is free for dedication or conveyance.

19.7   ENGINEER'S ADMINISTRATIVE CLOSEOUT

The engineer must make a final accounting to the developer once the job is completed. This usually is accomplished in a final report. This report includes a summary of notices, records, and inspections. It constitutes a record file for the project that can be updated throughout the warranty period and closed out when the warranty period has passed. For development projects that fall under the jurisdiction of the NPDES general permits for stormwater, a notice of termination (NOT) can be filed once the site has been finally stabilized.

19.8   WARRANTY PERIOD

The warranty period for all improvements is normally included within the job specifications, during which time performance bond coverage is usually reduced to cover contingencies. The engineer should periodically inspect the completed job during this period to ensure that all is in order. Should corrective action be required, resulting from trench settlement or soil erosion, for example, the contractor must be contacted immediately to take corrective action. At the end of the warranty period, the engineer should receive final acceptance notices from all involved agencies, and copies should be sent to the developer to close out the project.

REFERENCE

National Society of Professional Engineers (NSPE), the American Consulting Engineers Council, the American Society of Civil Engineers (ASCE), and the Construction Specifications Institute (CSI). *Certificate of Substantial Completion* (EJCDC No. 1910-8-D). New York: ASCE, 1996.

APPENDIX
Conversion Factors

The following tables are from: Elwyn E. Seelye. *Data Book for Engineers—Design Volume I.* Adapted from Tables of Conversion Factors for Engineers by Dorr Oliver, Inc., Milford, CT. Copyright © 1960 by John Wiley & Sons, Inc. Reprinted by permission of John Wiley & Sons, Inc. and Dorr Oliver Inc.

GENERAL CONVERSION FACTORS—I

TABLES OF CONVERSION FACTORS FOR ENGINEERS

Data are arranged alphabetically.

Unless designated otherwise, the English measures of capacity are those used in the United States, and the units of weight and mass are avoirdupois units.

The word gallon, used in any conversion factor, designates the U. S. gallon. To convert into the Imperial gallon, multiply the U. S. gallon by 0.83267. Likewise, the word ton designates a short ton, 2,000 pounds.

The figures 10^{-1}, 10^{-2}, 10^{-3}, etc. denote 0.1, 0.01, 0.001, etc. respectively.

The figures 10^{1}, 10^{2}, 10^{3}, etc. denote 10, 100, 1000, etc. respectively.

With respect to the properties of water, it freezes at 32°F., and is at its maximum density at 39.2°F. In the conversion factors given below using the properties of water, calculations are based on water at 39.2°F. in vacuo, weighing 62.427 pounds per cubic foot, or 8.345 pounds per U. S. gallon.

"Parts Per Million," designated as P.P.M., is always by weight and is simply a more convenient method of expressing concentration, either dissolved or undissolved material. Usually P.P.M. is used where percentage would be so small as to necessitate several ciphers after the decimal point, as one part per million is equal to 0.0001 per cent.

As used in the Sanitary field, P.P.M. represents the number of pounds of dry solids contained in one million pounds of water, including solids. In this field, one part per million may be expressed as 8.345 pounds of dry solids to one million U. S. gallons of water. In the Metric system, one part per million may be expressed as one gram of dry solids to one million grams of water, or one milligram per liter.

In arriving at parts per million by means of pounds per million gallons or milligrams per liter, it may be mentioned that the density of the solution or suspension has been neglected and if this is appreciably different from unity, the results are slightly in error.

Multiply	By	To Obtain
Acres	43,560	Square feet
"	4047	Square meters
"	1.562×10^{-3}	Square miles
"	4840	Square yards
Acre-feet	43,560	Cubic feet
"	325,851	Gallons
" "	1233.49	Cubic meters
Atmospheres	76.0	Cms. of mercury
"	29.92	Inches of mercury
"	33.90	Feet of water
"	10,333	Kgs./sq. meter
"	14.70	Lbs./sq. inch
"	1.058	Tons/sq. ft.
Barrels-oil	42	Gallons-oil
" —cement	376	Pounds-cement
Bags or sacks-cement	.94	Pounds-cement
Board-feet	144 sq. in. x 1 in.	Cubic inches
British Thermal Units	0.2520	Kilogram-calories
" "	777.5	Foot-lbs.
" "	3.927×10^{-4}	Horse-power-hrs.
" "	107.5	Kilogram-meters
" "	2.928×10^{-4}	Kilowatt-hrs.
B.T.U./min.	12.96	Foot-lbs./sec.
" / "	0.02356	Horse-power
" / "	0.01757	Kilowatts
" / "	17.57	Watts
Centares (Centiares)	1	Square meters
Centigrams	0.01	Grams
Centiliters	0.01	Liters
Centimeters	0.3937	Inches
"	0.01	Meters
"	10	Millimeters
Centimtrs. of mercury	0.01316	Atmospheres
" "	0.4461	Feet of water
" "	136.0	Kgs./sq. meter
" "	27.85	Lbs./sq. ft.
" "	0.1934	Lbs./sq. inch

Multiply	By	To Obtain
Centimeters/second	1.969	Feet/min.
" / "	0.03281	Feet/sec.
Centimeters/second	0.036	Kilometers/hr.
" / "	0.6	Meters/min.
" / "	0.02237	Miles/hr.
" / "	3.728×10^{-4}	Miles/min.
Chain (Gunters)	66	Feet
Cms./sec./sec.	0.03281	Feet/sec./sec.
Cubic centimeters	3.531×10^{-5}	Cubic feet
"	6.102×10^{-2}	Cubic inches
"	10^{-6}	Cubic meters
"	1.308×10^{-6}	Cubic yards
"	2.642×10^{-4}	Gallons
"	10^{-3}	Liters
"	2.113×10^{-3}	Pints (liq.)
"	1.057×10^{-3}	Quarts (liq.)
Cubic feet	2.832×10^{4}	Cubic cms.
"	1728	Cubic inches
"	0.02832	Cubic meters
"	0.03704	Cubic yards
"	7.48052	Gallons
"	28.32	Liters
"	59.84	Pints (liq.)
"	29.92	Quarts (liq.)
Cubic feet/minute	472.0	Cubic cms./sec.
" / "	0.1247	Gallons/sec.
" / "	0.4720	Liters/sec.
" / "	62.43	Pounds of water/min.
Cubic feet/second	0.646317	Million gals./day
" " / "	448.831	Gallons/min.
" " / "	13.8/sq. mi. drainage area Inches/year	
Cubic inches	16.39	Cubic centimeters
"	5.787×10^{-4}	Cubic feet
"	1.639×10^{-5}	Cubic meters
"	2.143×10^{-5}	Cubic yards
"	4.329×10^{-3}	Gallons
"	1.639×10^{-2}	Liters
"	0.03463	Pints (liq.)
"	0.01732	Quarts (liq.)
Cubic meters	10^{6}	Cubic centimeters
"	35.31	Cubic feet
"	61.023	Cubic inches
"	1.308	Cubic yards
"	264.2	Gallons
Cubic meters	10^{3}	Liters
"	2113	Pints (liq.)
"	1057	Quarts (liq.)
Cubic yards	7.646×10^{5}	Cubic centimeters
"	27	Cubic feet
"	46,656	Cubic inches
"	0.7646	Cubic meters
"	202.0	Gallons
"	764.6	Liters
"	1616	Pints (liq.)
"	807.9	Quarts (liq.)
Cubic yards/min.	0.45	Cubic feet/sec.
" " / "	3.367	Gallons/sec.
" " / "	12.74	Liters/sec.
Decigrams	0.1	Grams
Deciliters	0.1	Liters
Decimeters	0.1	Meters
Degrees (angle)	60	Minutes
"	0.01745	Radians
"	3600	Seconds
Degrees/sec.	0.01745	Radians/sec.
" / "	0.1667	Revolutions/min.
" / "	0.002778	Revolutions/sec.
Dekagrams	10	Grams
Dekaliters	10	Liters
Dekameters	10	Meters
Drams	27.34375	Grains
"	0.0625	Ounces
"	1.771845	Grams
Fathoms	6	Feet
Feet	30.48	Centimeters
"	12	Inches
"	0.3048	Meters
"	1/3	Yards

GENERAL CONVERSION FACTORS – 2

Multiply	By	To Obtain
Feet of water	0.02950 ...	Atmospheres
" " "	0.8826 ...	Inches of mercury
" " "	304.8 ...	Kgs./sq. meter
" " "	62.43 ...	Lbs./sq ft.
" " "	0.4335 ...	Lbs./sq. inch
Feet/min.	0.5080 ...	Centimeters/sec.
" / "	0.01667 ...	Feet/sec.
" / "	0.01829 ...	Kilometers/hr.
" / "	0.3048 ...	Meters/min.
" / "	0.01136 ...	Miles/hr.
Feet/sec.	30.48 ...	Centimeters/sec.
" / "	1.097 ...	Kilometers/hr.
" / "	0.5921 ...	Knots
" / "	18.29 ...	Meters/min.
" / "	0.6818 ...	Miles/hr.
" / "	0.01136 ...	Miles/min.
Feet/sec./sec.	30.48 ...	Cms./sec./sec.
" / "	0.3048 ...	Meters/sec./sec.
Foot-pounds	1.286x10^{-3} ...	British Thermal Units
" "	5.050x10^{-7} ...	Horse-power-hrs.
" "	3.241x10^{-4} ...	Kilogram-calories
" "	0.1383 ...	Kilogram-meters
" "	3.766x10^{-7} ...	Kilowatt-hrs.
Foot-pounds/min.	1.286x10^{-3} ...	B. T. Units/min.
" / "	0.01667 ...	Foot-pounds/sec.
" / "	3.030x10^{-5} ...	Horse-power
" / "	3.241x10^{-4} ...	Kg.-calories/min.
" / "	2.260x10^{-5} ...	Kilowatts
Foot-pounds/sec.	7.717x10^{-2} ...	B. T. Units/min.
" / "	1.818x10^{-3} ...	Horse-power
" / "	1.945x10^{-2} ...	Kg.-calories/min
" / "	1.350x10^{-3} ...	Kilowatts
Gallons	3785 ...	Cubic centimeters
"	0.1337 ...	Cubic feet
"	231 ...	Cubic inches
"	3.785x10^{-3} ...	Cubic meters
"	4.951x10^{-3} ...	Cubic yards
"	3.785 ...	Liters
"	8	Pints (liq.)
"	4	Quarts (liq.)
Gallons, Imperial.	1.20095 ...	U.S. gallons
" U.S.	0.83267 ...	Imperial gallons
Gallons water.	8.3453 ...	Pounds of water
Gallons/min.	2.228x10^{-3} ...	Cubic feet/sec.
" / "	0.06308 ...	Liters/sec.
" / "	8.0208 ...	Cu. ft./hr.
" / "	8.0208 ...	Overflow rate (ft./hr.) Area (sq. ft.)
Gallons water/min	6.0086 ...	Tons water/24 hrs.
Grains (troy)	L	Grains (avoir.)
"	0.06480 ...	Grams
"	0.04167 ...	Pennyweights (troy)
"	2.0833x10^{-3} ...	Ounces (troy)
Grains/U.S. gal.	17.118 ...	Parts/million
/U.S. gal.	142.86 ...	Lbs./million gal.
/Imp. gal.	14.286 ...	Parts/million
Grams.	980.71 ...	Dynes
"	15.43 ...	Grains
"	10^{-3} ...	Kilograms
"	10^{3} ...	Milligrams
"	0.03527 ...	Ounces
"	0.03215 ...	Ounces (troy)
"	2.205x10^{-3} ...	Pounds
Grams/cm.	5.600x10^{-3} ...	Pounds/inch
Grams/cu. cm.	62.43 ...	Pounds/cubic foot
" / "	0.03613 ...	Pounds/cubic inch
Grams/liter	58.417 ...	Grains/gal.
" / "	8.345 ...	Pounds/1000 gals.
" / "	0.062427 ...	Pounds/cubic foot
" / "	1000 ...	Parts/million
Hectares.	2.471 ...	Acres
"	1.076x10^{5} ...	Square feet
Hectograms.	100 ...	Grams
Hectoliters.	100 ...	Liters
Hectometers.	100 ...	Meters
Hectowatts.	100 ...	Watts

Multiply	By	To Obtain
Horse-power.	42.44 ...	B.T. Units/min.
"	33,000 ...	Foot-lbs./min.
"	550 ...	Foot-lbs./sec.
"	1.014 ...	Horse-power (metric)
"	10.70 ...	Kg.-calories/min.
"	0.7457 ...	Kilowatts
"	745.7 ...	Watts
Horse-power (boiler)	33,479 ...	B.T.U./hr.
"	9.803 ...	Kilowatts
Horse-power-hours	2547 ...	British Thermal Units
"	1.98x10^{6} ...	Foot-lbs.
"	641.7 ...	Kilogram-calories
"	2.737x10^{5} ...	Kilogram-meters
"	0.7457 ...	Kilowatt-hours
Inches.	2.540 ...	Centimeters
Inches of mercury.	0.03342 ...	Atmospheres
"	1.133 ...	Feet of water
"	345.3 ...	Kgs./sq. meter
"	70.73 ...	Lbs./sq. ft.
"	0.4912 ...	Lbs./sq. inch
Inches of water.	0.002458 ...	Atmospheres
"	0.07355 ...	Inches of mercury
"	25.40 ...	Kgs./sq. meter
"	0.5781 ...	Ounces/sq. inch
"	5.202 ...	Lbs./sq. foot
"	0.03613 ...	Lbs./sq. inch
" " " /year	$\frac{\text{sq. mi.}}{13.8}$ ...	Cu. ft. / sec.
Kilograms.	980.665 ...	Dynes
"	2.205 ...	Lbs.
"	1.102x10^{-3} ...	Tons (short)
"	10^{3} ...	Grams
Kilograms-calories	3.968 ...	British Thermal Units
"	3086 ...	Foot-pounds
"	1.558x10^{-3} ...	Horse-power-hours
"	1.162x10^{-3} ...	Kilowatt-hours
Kilogram-calories/min.	51.43 ...	Foot-pounds/sec.
" / "	0.09351 ...	Horse-power
" / "	0.06972 ...	Kilowatts
Kgs./meter.	0.6720 ...	Lbs./foot
Kgs./sq. meter.	9.678x10^{-5} ...	Atmospheres
" / "	3.281x10^{-3} ...	Feet of water
" / "	2.896x10^{-3} ...	Inches of mercury
" / "	0.2048 ...	Lbs./sq. foot
" / "	1.422x10^{-3} ...	Lbs./sq. inch
Kgs./sq. millimeter.	10^{6} ...	Kgs./sq. meter
Kiloliters.	10^{3} ...	Liters
Kilometers.	10^{5} ...	Centimeters
"	3281 ...	Feet
"	10^{3} ...	Meters
"	0.6214 ...	Miles
"	1094 ...	Yards
Kilometers/hr.	27.78 ...	Centimeters/sec.
" / "	54.68 ...	Feet/min.
" / "	0.9113 ...	Feet/sec.
" / "	0.5396 ...	Knots
" / "	16.67 ...	Meters/min.
" / "	0.6214 ...	Miles/hr.
Kms./hr./sec.	27.78 ...	Cms./sec./sec.
" / " / "	0.9113 ...	Ft./sec./sec.
" / " / "	0.2778 ...	Meters/sec./sec.
Kilowatts.	56.92 ...	B.T. Units/min.
"	4.425x10^{4} ...	Foot-lbs./min.
"	737.6 ...	Foot-lbs./sec.
"	1.341 ...	Horse-power
"	14.34 ...	Kg.-calories/min.
"	10^{3} ...	Watts
Kilowatt-hours.	3415 ...	British Thermal Units
"	2.655x10^{6} ...	Foot-lbs.
"	1.341 ...	Horse-power-hrs.
"	860.5 ...	Kilogram-calories
"	3.671x10^{5} ...	Kilogram-meters
Link (Gunters)	66 ...	Feet
Liters.	10^{3} ...	Cubic centimeters
"	0.03531 ...	Cubic feet
"	61.02 ...	Cubic inches
"	10^{-3} ...	Cubic meters
"	1.308x10^{-3} ...	Cubic yards
"	0.2642 ...	Gallons
"	2.113 ...	Pints (liq.)
"	1.057 ...	Quarts (liq.)

Multiply	By	To Obtain
Liters/min.	5.886x10^{-4} ...	Cubic ft.-sec.
" / "	4.403x10^{-3} ...	Gals.-/sec.
Lumber Width (in.) x Thickness (in.) $\frac{}{12}$		Length (ft). Board Feet
Meters.	100 ...	Centimeters
"	3.281 ...	Feet
"	39.37 ...	Inches
"	10^{-3} ...	Kilometers
"	10^{3} ...	Millimeters
"	1.094 ...	Yards
Meters/min.	1.667 ...	Centimeters/sec.
" / "	3.281 ...	Feet/min.
" / "	0.05468 ...	Feet/sec.
" / "	0.06 ...	Kilometers/hr.
" / "	0.03728 ...	Miles/hr.
Meters/sec.	196.8 ...	Feet/min.
" / "	3.281 ...	Feet/sec.
" / "	3.6 ...	Kilometers/hr.
" / "	0.06 ...	Kilometers/min.
" / "	2.237 ...	Miles/hr.
" / "	0.03728 ...	Miles/min.
Microns.	10^{-6} ...	Meters
Miles.	1.609x10^{5} ...	Centimeters
"	5280 ...	Feet
"	1.609 ...	Kilometers
"	1760 ...	Yards
Miles/hr.	44.70 ...	Centimeters/sec.
" / "	88 ...	Feet/min.
" / "	1.407 ...	Feet/sec.
" / "	1.609 ...	Kilometers/hr.
" / "	0.8684 ...	Knots
" / "	26.82 ...	Meters/min.
Miles/min.	2682 ...	Centimeters/sec.
" / "	88 ...	Feet/sec.
" / "	1.609 ...	Kilometers/min.
" / "	60 ...	Miles/hr.
Milliers.	10^{3} ...	Kilograms
Milligrams.	10^{-3} ...	Grams
Milliliters.	10^{-3} ...	Liters
Millimeters.	0.1 ...	Centimeters
"	0.03937 ...	Inches
Milligrams/liter.	1 ...	Parts/million
Million gals./day.	1.54723 ...	Cubic ft./sec.
Miner's inches.	1.5 ...	Cubic ft./min.
Minutes (angle).	2.909x10^{-4} ...	Radians
Ounces.	16 ...	Drams
"	437.5 ...	Grains
"	0.0625 ...	Pounds
"	28.349527 ...	Grams
"	0.9115 ...	Ounces (troy)
"	2.790x10^{-5} ...	Tons (long)
"	2.835x10^{-5} ...	Tons (metric)
Ounces, troy	480 ...	Grains
"	20 ...	Pennyweights (troy)
"	0.08333 ...	Pounds (troy)
"	31.103481 ...	Grams
"	1.09714 ...	Ounces, avoir.
Ounces (fluid)	1.805 ...	Cubic inches
"	0.02957 ...	Liters
Ounces/sq. inch	0.0625 ...	Lbs./sq. inch
Overflow rate (ft./hr.) $\frac{0.12468 x}{1}$		area (sq. ft.) Gals./min.
"	8.0208 ...	Sq. ft./gal./min.
Overflow rate (ft./hr.)		
Parts/million.	0.0584 ...	Grains/U.S. gal.
" / "	0.07016 ...	Grains/Imp.gal.
" / "	8.345 ...	Lbs./million gal.
Pennyweights (troy).	24 ...	Grains
"	1.55517 ...	Grams
"	0.05 ...	Ounces (troy)
"	4.1667x10^{-3} ...	Pounds (troy)
Pounds.	16 ...	Ounces
"	256 ...	Drams
"	7000 ...	Grains
"	0.0005 ...	Tons (short)
"	453.5924 ...	Grams
"	1.21528 ...	Pounds (troy)
"	14.5833 ...	Ounces (troy)

GENERAL CONVERSION FACTORS—3

Multiply	By	To Obtain
Pounds (troy)	5760	Grains
" "	240	Pennyweights (troy)
" "	12	Ounces (troy)
" "	373.24177	Grams
" "	0.822857	Pounds (avoir.)
" "	13.1657	Ounces (avoir.)
" "	3.6735x10⁻⁴	Tons (long)
" "	4.1143x10⁻⁴	Tons (short)
" "	3.7324x10⁻⁴	Tons (metric)
Pounds of water	0.01602	Cubic feet
" " "	27.68	Cubic inches
" " "	0.1198	Gallons
Pounds of water/min.	2.670x10⁻⁴	Cubic ft./sec.
Pounds/cubic foot	0.01602	Grams/cubic cm.
" / "	16.02	Kgs./cubic meter
" / "	5.787x10⁻⁴	Lbs./cubic inch
Pounds/cubic inch	27.68	Grams/cubic cm.
" / " "	2.768x10⁴	Kgs./cubic meter
" / " "	1728	Lbs./cubic foot
Pounds/foot	1.488	Kgs./meter
Pounds/inch	178.6	Grams/cm.
Pounds/sq. foot	0.01602	Feet of water
" / " "	4.883	Kgs./sq. meter
" / " "	6.945x10⁻³	Pounds/sq. inch
Pounds/sq. inch	0.06804	Atmospheres
" / " "	2.307	Feet of water
" / " "	2.036	Inches of mercury
" / " "	703.1	Kgs./sq. meter
Quadrants (angle)	90	Degrees
" "	5400	Minutes
" "	1.571	Radians
Quarts (dry)	67.20	Cubic inches
Quarts (liq.)	57.75	Cubic inches
Quintal, Argentine	101.28	Pounds
" Brazil	129.54	Pounds
" Castile, Peru	101.43	Pounds
" Chile	101.41	Pounds
" Mexico	101.47	Pounds
" Metric	220.46	Pounds
Quires	25	Sheets
Radians	57.30	Degrees
"	3438	Minutes
"	0.637	Quadrants
Radians/sec.	57.30	Degrees/sec.
" / "	0.1592	Revolutions/sec.
" / "	9.549	Revolutions/min.
Radians/sec./sec.	573.0	Revolutions/min./min.
" / " / "	0.1592	Revolutions/sec./sec.
Reams	500	Sheets
Revolutions	360	Degrees
"	4	Quadrants
"	6.283	Radians
Revolutions/min.	6	Degrees/sec.
" / "	0.1047	Radians/sec.
" / "	0.01667	Revolutions/sec.
Revolutions/min./min.	1.745x10⁻³	Rads./sec./sec.
" / " / "	2.778x10⁻⁴	Revs./sec./sec.
Revolutions/sec.	360	Degrees/sec.
" / "	6.283	Radians/sec.
" / "	60	Revolutions/min.
Revolutions/sec./sec.	6.283	Radians/sec./sec.
" / " / "	3600	Revolutions/min./min.
Rods	16.5	Feet
Seconds (angle)	4.848x10⁻⁶	Radians
Square centimeters	1.076x10⁻³	Square feet
" "	0.1550	Square inches
" "	10⁻⁴	Square meters
" "	100	Square millimeters
Square Chains (Gunters)	16	Square rods

Multiply	By	To Obtain
Square feet	2.296x10⁻⁵	Acres
" "	929.0	Square centimeters
" "	144	Square inches
" "	0.09290	Square meters
" "	3.587x10⁻⁸	Square miles
" "	1/9	Square yards
1 / Sq. ft./gal./min.	8.0208	Overflow rate (ft./hr.)
Square inches	6.452	Square centimeters
" "	6.944x10⁻³	Square feet
" "	645.2	Square millimeters
Square kilometers	247.1	Acres
" "	10.76x10⁶	Square feet
" "	10⁶	Square meters
" "	0.3861	Square miles
" "	1.196x10⁶	Square yards
Square meters	2.471x10⁻⁴	Acres
" "	10.76	Square feet
" "	3.861x10⁻⁷	Square miles
" "	1.196	Square yards
Square miles	640	Acres
" "	27.88x10⁶	Square feet
" "	2.590	Square kilometers
" "	3.098x10⁶	Square yards
Square millimeters	0.01	Square centimeters
" "	1.550x10⁻³	Square inches
Square yards	2.066x10⁻⁴	Acres
" "	9	Square feet
" "	0.8361	Square meters
" "	3.228x10⁻⁷	Square miles
Temp. (°C.) + 273.1	1	Abs. temp. (°C.)
" " + 17.78	1.8	Temp. (°F.)
(°F.) + 460	1	Abs. temp. (°F.)
" " − 32	5/9	Temp. (°C.)
Tons (long)	1016	Kilograms
" "	2240	Pounds
" "	1.12000	Tons (short)
Tons (metric)	10³	Kilograms
" "	2205	Pounds
Tons (short)	2000	Pounds
" "	32000	Ounces
" "	907.18486	Kilograms
Tons (short)	2430.56	Pounds (troy)
" "	0.89287	Tons (long)
" "	29166.66	Ounces (troy)
" "	0.90718	Tons (metric)
1 / Tons dry solids/24 hrs.	Area (sq. ft.)	Sq.ft./ton/24 hrs.
Tons of water/24 hrs.	83.333	Pounds water/hour
" " " / "	0.16643	Gallons/min.
" " " / "	1.3349	Cu. ft./hr.
Watts	0.05692	B. T. Units/min.
"	44.26	Foot-pounds/min.
"	0.7376	Foot-pounds/sec.
"	1.341x10⁻³	Horse-power
"	0.01434	Kg.-calories/min.
"	10⁻³	Kilowatts
Watt-hours	3.415	British Thermal Units
"	2655	Foot-pounds
"	1.341x10⁻³	Horse-power-hours
"	0.8605	Kilogram-calories
"	367.1	Kilogram-meters
"	10⁻³	Kilowatt-hours
Yards	91.44	Centimeters
"	3	Feet
"	36	Inches
"	0.9144	Meters

INDEX